Tributes

Volume 16

Knowing, Reasoning, and Acting

Essays in Honour of Hector J. Levesque

Tributes Series Editor
Dov Gabbay

dov.gabbay@kcl.ac.uk

Knowing, Reasoning, and Acting
Essays in Honour of Hector J. Levesque

edited by

Gerhard Lakemeyer

and

Sheila A. McIlraith

© Individual author and College Publications 2011. All rights reserved.

ISBN 978-1-84890-044-8

College Publications
Scientific Director: Dov Gabbay
Managing Director: Jane Spurr
Department of Computer Science
King's College London, Strand, London WC2R 2LS, UK

http://www.collegepublications.co.uk

Original cover design by Laraine Welch
Illustration by Kathryn Finter
Printed by Lightning Source, Milton Keynes, UK

Hector J. Levesque

Table of Contents

List of Contributors

Luigia Carlucci Aiello, Dipartimento di Informatica e Sistemistica "Antonio Ruberti," Sapienza Università di Roma, Italy

Jorge A. Baier, Departamento de Ciencia de la Computacion, Pontificia Universidad Catolica de Chile, Chile

Vaishak Belle, Department of Computer Science, RWTH Aachen University, Germany

Alexander Borgida, Department of Computer Science, Rutgers University, USA

Craig Boutilier, Department of Computer Science, University of Toronto, Canada

Ronald J. Brachman, Yahoo! Labs, USA

Michael Brenner, Department of Computer Science, Albert-Ludwigs-Universität Freiburg, Germany

Diego Calvanese, KRDB Research Centre, Free University of Bozen-Bolzano, Italy

Amit K. Chopra, Department of Information Engineering and Computer Science, University of Trento, Italy

Philip R. Cohen, Adapx Inc., USA

Rogatien "Gatemouth" Cumberbatch, Gordie Howe Department of Issues and Distinctions, Floral University, Floral, Saskatchewan S1L 1L1, Canada

Fabiano Dalpiaz, Department of Information Engineering and Computer Science, University of Trento, Italy

James Delgrande, School of Computing Science, Simon Fraser University, Canada

Norman Foo, Computer Science and Engineering, University of New South Wales, Australia

Christian Fritz, Intelligent Systems Lab, Palo Alto Research Center Inc., USA

Alfredo Gabaldon, Center for Artificial Intelligence (CENTRIA), Universidade Nova de Lisboa, Portugal

Hojjat Ghaderi, Department of Computer Science University of Toronto, Canada

Giuseppe De Giacomo, Dipartimento di Informatica e Sistemistica "Antonio Ruberti," Sapienza Università di Roma, Italy

Paolo Giorgini, Department of Information Engineering and Computer Science, University of Trento, Italy

Carla P. Gomes, Department of Computer Science, Cornell University, USA

Yilan Gu, Department of Computer Science, University of Toronto, Canada

Joseph Y. Halpern, Department of Computer Science, Cornell University, USA

Sydney J. Hurtubise, Department of Issues, Distinctions, Controversies and Puzzles, University of Artificial Intelligence, North Bay, Ontario, Canada

Gerhard Lakemeyer, Department of Computer Science, RWTH Aachen University, Germany

Maurizio Lenzerini, Dipartimento di Informatica e Sistemistica "Antonio Ruberti," Sapienza Università di Roma, Italy

Yves Lespérance, Department of Computer Science and Engineering, York University, Canada

Hector J. Levesque, Department of Computer Science, University of Toronto, Canada

Vladimir Lifschitz, University of Texas at Austin, USA

Fangzhen Lin, Department of Computer Science, Hong Kong University of Science and Technology, China

Yongmei Liu, Department of Computer Science, Sun Yat-sen University, China

Sheila A. McIlraith, Department of Computer Science, University of Toronto, Canada

David G. Mitchell, School of Computing Science, Simon Fraser University, Canada

John Mylopoulos, Department of Information Engineering and Computer Science University of Trento, Italy

Daniele Nardi, Dipartimento di Informatica e Sistemistica "Antonio Ruberti," Sapienza Università di Roma, Italy

Abhaya Nayak, Department of Computing, Division of Information and Communication Sciences, Macquarie Univeristy, Australia

Bernhard Nebel, Department of Computer Science Albert-Ludwigs-Universität Freiburg, Germany

Maurice Pagnucco, School of Computer Science and Engineering, The University of New South Wales, Australia

Rafael Pass, Department of Computer Science, Cornell University, USA

Peter F. Patel-Schneider, Bell Labs Research, USA

Pavlos Peppas, Department of Business Administration, University of Patras, Greece

Ronald P. A. Petrick, School of Informatics University of Edinburgh, Scotland, United Kingdom

Fiora Pirri, Dipartimento di Informatica e Sistemistica "Antonio Ruberti," Sapienza Università di Roma, Italy

Matia Pizzoli, Dipartimento di Informatica e Sistemistica "Antonio Ruberti," Sapienza Università di Roma, Italy

Gabriele Randelli, Dipartimento di Informatica e Sistemistica "Antonio Ruberti," Sapienza Università di Roma, Italy

Riccardo Rosati, Dipartimento di Informatica e Sistemistica "Antonio Ruberti," Sapienza Università di Roma, Italy

Ashish Sabharwal, IBM Thomas J. Watson Research Center, USA

Scott Sanner, NICTA and the Australian National University, Australia

Sebastian Sardina, School of Computer Science and IT, RMIT University, Australia

Carlo Matteo Scalzo, Dipartimento di Informatica e Sistemistica "Antonio Ruberti," Sapienza Università di Roma, Italy

Bart Selman, Department of Computer Science, Cornell University, USA

Steven Shapiro, School of Computer Science and Information Technology, RMIT University, Australia

Mikhail Soutchanski, Department of Computer Science, Ryerson University, Canada

Eugenia Ternovska, School of Computing Science, Simon Fraser University, Canada

Stavros Vassos, Department of Informatics and Telecommunications, National and Kapodistrian University of Athens, Greece

Fangkai Yang, University of Texas at Austin, USA

Preface

It is fair to say that few people have shaped the field of Knowledge Representation and Reasoning (KR&R) over the past decades more than Hector Levesque. If one were to choose three words to characterize Hector's areas of interest, then *knowing*, *reasoning*, and *acting* seem to cover a good deal of them. In all these aspects of KR&R Hector has had profound impact, even to the extent that his work spawned whole new research directions. The paper "The Tractability of Subsumption in Frame-Based Description Languages," which appeared at AAAI 1984 and was co-authored by Ron Brachman, is essentially the birthplace of modern description logics. Together with the paper "A Logic of Explicit and Implicit Belief," which appeared at the same conference, it also established the idea of investigating forms of tractable reasoning, either by limiting the expressiveness of the representation language, or by limiting the inference capabilities in a principled way. Soon after, a seminal paper on belief, intention, and commitment, co-authored by Phil Cohen, started a long tradition of investigating the logical foundations of belief-desire-intention (BDI) architectures. In the early nineties, two papers, co-authored by Bart Selman and David Mitchell, kicked off what is now known as the area of SAT-solvers, and a few years later, together with Ray Reiter and other colleagues at the University of Toronto, Hector was instrumental in developing the action language Golog, which lies at the heart of the Toronto brand of cognitive robotics, which itself has been an active area of research ever since.

Given his many achievements so early in his career, it is not surprising that Hector received the prestigious *Computers and Thought Award* in 1985. We will not list the many other honours and distinctions Hector has received over the years except to say that the two papers mentioned in the previous paragraph not only won best paper awards in 1984 but the second also received the *Classic Paper Award* in 2004 for most influential work of the past twenty years in AI, with the first paper receiving an honourable mention!

It is with great pleasure that we present this collection of papers by former students and close colleagues in honour of Hector Levesque. The collection commences with personal remembrances from Hector's good friend and long-time collaborator, Ron Brachman. This is followed by papers spanning almost three decades and covering many of the research areas that owe so much if not their existence to Hector's work. The chapters by Borgida and by Calvanese et al. fall into the area of description logics. Also included is a re-print of a paper by Ron Brachman and Hector on the trade-off between tractable reasoning and expressiveness, exemplified in the context of description logics. The chapters by Patel-Schneider, and by Sanner and Boutilier, also address the issue of tractable reasoning. The chapters by Cohen and Levesque, by Dalpiaz et al., by Foo et al., by Sardina and Lespérance, and by Ghaderi et al. are concerned with BDI-related issues. The chapter by Gomes et al. focuses on SAT and the chapter by Mitchell and Ternovska on expressiveness. Belief is dealt with in the chapters by Belle and Lakemeyer, by Halpern and Pass, by Delgrande, and by Shapiro. Then there is a large number of chapters that focus on action, planning, and Golog. These are by Brenner and Nebel, by Pirri and Pizzoli, by Lin, by Liu and Lakemeyer, by Vassos and Sardina, by Gu and Soutchanski, by De Giacomo and Pagnucco, by Baier et al., by Gabaldon, by Lifschitz and Yang, by Aiello et al., and by Petrick. Finally, there is the paper by Hurtubise and Cumberbatch, preceded by personal reflections by Cumberbatch, two eminent scholars from North Bay, Ontario. Their ideas probably tell us as much about

Hector as any of the papers he wrote, if not more.

We would like to thank Vaishak Belle, who took on the thankless task of transforming the many contributions of colleagues and collaborators into what you see here. Kathryn Finter, artist and friend, deserves special thanks for the illustration on the cover of this book, which is based on the logo of the International Conference on Principles of Knowledge Representation and Reasoning, which she also designed more than twenty years ago. In this regard, we also thank Mirek Truszczynski and KR Inc. for permitting us to alter and reproduce the KR logo. We gratefully acknowledge Jane Spurr from College Publications for being so helpful and responsive throughout this project. Finally, we thank the authors who so enthusiastically agreed to contribute to this book.

And thank you, Hector, for being an inspiration to us all throughout these years. Happy Birthday!

Aachen and Toronto, June 2011
Gerhard Lakemeyer and Sheila McIlraith

1

A Brief, Fuzzy Account of Some of My Adventures with Hector

RON BRACHMAN

ABSTRACT. Hector Levesque and I have collaborated on a wide variety of projects, from technical work to conferences to books, over a period of more than thirty years. I consider him my best friend and an inspiration for all of the technical work I've done since 1980. Nothing I could say could do justice to the value I place on my relationship with him and the wealth of professional and personal contributions he's made to my career. But in honor of his 60th birthday I wanted to try to set down some rough recollections about our early days together and our collaborations in the world of Knowledge Representation and Reasoning. One's memory fades after so many years – there are no doubt many inaccuracies here – but our friendship hasn't. In homage to Hector, I offer this loose and informal interpretation of some facts and events that were certainly important in my life, I hope important in Hector's, and perhaps even significant in the history of our field.

While the exact details are lost in the mists of time and my increasing senility, it's likely that the first time I met Hector Levesque was at the COLING conference in Ottawa in 1976 – somewhat shockingly to me, more than a third of a century ago. I was working on my Ph.D. thesis at the time, and I had my first conference paper accepted there [Brachman 1976] – an analysis of semantic network knowledge representation formalisms, which complemented Bill Woods's original seminal paper, "What's in a Link" [Woods 1975]. There were a number of key people in the field at that conference, since semantic networks were popular for supporting natural language applications. There was lots of discussion of network formalisms, and Hector's ongoing Master's thesis work on Procedural Semantic Networks [Levesque 1977] was looking like an important contribution in the area. I wish I could remember something specific about our first meeting, but alas, I can't. I went on to finish my thesis at Harvard a year later and started working full-time at Bolt Beranek & Newman (BBN), where I had been working part-time since 1972. It's quite possible that Hector and I stayed in touch over the next year, but it's too long ago to remember clearly. Hector and his advisor, John Mylopoulos, had a paper on PSN's at IJCAI-77 in Cambridge [Levesque and Mylopoulos 1977], so I am sure we got together then.

The 1978 Canadian AI conference, sponsored by the Canadian Society for Computational Studies of Intelligence (CSCSI), was held in Toronto. That provided a great opportunity for discussion with Hector and others working in John's group at U. of T. Based on my thesis work on the foundations of "structured inheritance networks" [Brachman 1977], our team at BBN was in the process of developing the KL-ONE knowledge representation system [Brachman and Schmolze 1985], and there was lots to talk about with other KR researchers. To the best of my recollection, I also met Peter Patel-Schneider at that conference and probably met Phil Cohen then as well (Phil and I may have met at COLING'76 –

neither of us can exactly recall). Phil joined us in our AI department at BBN soon thereafter (early 1979), after he finished his Ph.D. thesis at the University of Toronto.

Whenever and however our first meetings were, I would love to go back in time to re-visit those initial moments, since they eventually came to be amongst the most important of my professional career. Chances are there was the usual excitement of graduate students/researchers about related work and sympatico approaches, and no doubt I fell prey to Hector's quirky sense of humor (I guess I should say, "humour," in deference to Hector; we've always had a good laugh about where to add an extra "u" or when to put "re" at the end of words like "theatre"). I've always enjoyed people with somewhat strange, but very intelligent senses of humor, and I am sure that was a big factor in us hitting it off when we first met.

There was an interesting coincidence in my initial connection with Hector concerning his thesis advisor, John Mylopoulos. When I first encountered John when Hector introduced us, I was actually fairly stunned: in seeing John I only then realized that he had actually been one of my graduate student instructors in a EE course when I was an undergrad at Princeton! He had helped with labs and graded my papers. That was a really fun connection and as such things tend to do, created a nice bond amongst the three of us. (I remember being somewhat shocked at John's retirement party a couple years ago to realize that I may have been the person in the room who had known him professionally the longest, since we met when he was still a graduate student; I even dug up from my attic some old lab reports that were graded by John in the late 1960's.) In any case, the work that Hector and John did together on PSN's was very closely related in spirit and substance to my own work, and was a key element to the establishment of our working relationship back in the early 1980's.

Having clearly enjoyed each other's company, Hector and I kept in touch, and he joined me for the summer in Cambridge in 1979 as part of our AI group at BBN. Work on KL-ONE had been underway for a while, and Hector made many key contributions. But more importantly, perhaps, that summer was a real blast. We had a fantastic time working in the group under Bill Woods, and hanging out with our friends from BBN (especially Phil, David Israel, Jeff Gibbons, Brad Goodman, Rusty Bobrow, and Candy Sidner, as well as Brian Smith from MIT). It was really tough to have Hector go home at the end of the summer, but that experience really set the stage for our future collaborations. I recall both productive and hysterically funny conversations about the things that KL-ONE could and couldn't do, Rusty's personal take on the meaning of KL-ONE structure, "minus nexuses," and a host of other things that at this point are simply very old inside jokes — but still funny to a small number of us. I still vividly remember David Israel's comment (he was standing at the whiteboard in my office) that a system using KL-ONE at that point was incapable of actually having any beliefs — it could only have the components out of which beliefs could be constructed. That was kind of a shocking revelation for us at the time,[1] but subsequently led to one of Hector's most significant contributions to the history of what have become known as Description Logics, which he brought to the second KL-ONE workshop in 1981 (I'll get to that in a moment).

My experience at BBN was formative and wonderful, but as AI started to grow in commercial interests, it was difficult to resist joining the migration of many of my friends and colleagues to the West Coast. In the summer of 1981, I had a great opportunity to join

---

[1] If you subscribe to the view that knowledge is justified true belief, then David's observation meant that we weren't really doing knowledge representation at all with the early version of KL-ONE.

an exciting new lab within Schlumberger, housed at its Fairchild subsidiary in Palo Alto. Peter Hart, Marty Tenenbaum, Harry Barrow, and Dick Duda had all left SRI to form the Fairchild Laboratory for Artificial Intelligence Research (FLAIR). With a focus on VLSI design and some interest in natural language processing, FLAIR was a place where a solid knowledge representation foundation could make a big difference. And it was very exciting to get in on the ground floor of a brand new, high-profile lab.

As my wife, Gwen, and I moved out west, one of my personal goals was to find a way to work with Hector again. Those who know Hector well know that he's a bit of a "home body" — all things being equal, he's happier to stay near home than travel (he's always told me he doesn't "travel well"), so a move to the West Coast was a long shot. Also, my close friends and colleagues at BBN would have loved to have had Hector join them, so recruiting had to be gentle and fair to those colleagues. I remember fairly vividly several long phone calls with Hector from the upstairs office in our house in Arlington, Massachusetts, before I left for California. Somehow, miraculously, after the team at FLAIR made him a nice offer, I convinced him to come join me for a new adventure far from home. I couldn't know it at the time, but it was probably the most important thing I've ever done in my professional life.

We had a fantastic time together in California. We learned about VLSI design and even designed our own chip (we implemented a text sorting routine and also left our few-micron-sized signature in silicon in the design). We tested it with software tools and even had the chip manufactured. I probably still have it in a drawer somewhere. But more importantly, we started to look at ways to use structured network and logical representations to capture human-style knowledge about circuits and their functions. We borrowed an idea that we had worked out together at BBN (the original idea of a "JARGON" lexical language for KL-ONE came from Bill Woods) and started using formalized noun phrases as a way to convey the meaning of structured concepts lexically. Building on Hector's recently completed Ph.D. thesis work, we also looked at the kinds of competence different parts of a knowledge representation system should have, using a functional approach. Using some of the insights we developed at BBN on the difference between term-forming operators and the use of complex terms to describe individuals (thanks again to David Israel's observations about KL-ONE), Hector proposed the idea of having two distinct components of a KR system, one to capture terminology and one to use that terminology to make assertions. For short, he called those two components the "TBox" and "ABox," and presented those ideas at the Second KL-ONE Workshop, which was held in the Fall of 1981 [Schmolze and Brachman 1982]. Amazingly, that terminology is still in common use 30 years later. Since we had moved to the West Coast (and our community included Bill Mark, Tom Lipkis, and the CONSUL team from ISI in Southern California), and some of our colleagues Back East had developed some slightly divergent views, we had some fun times in contrasting the East and West Coast views of various aspects of KL-ONE.[2] [3]

The view we developed of a KR system having multiple kinds of competence was captured in our first paper together, which appeared at AAAI-82 in Pittsburgh [Brachman and

---

[2] I remember, among other things, lots of discussion about the "East Coast QUA link" and the "West Coast QUA link."

[3] One of the more exciting developments at the workshop was when I got so sick that I had to be taken to a local hospital in the middle of the night, and on the ride back we discovered a burning building (this was in rural New Hampshire). David Israel and Jim Schmolze, who were kind enough to drive me, called the police and then credited themselves with saving the town. Of course I deserve the credit for getting sick in the first place. I can't remember if Hector joined us on that ride.

Levesque 1982].

As we were developing our KR competence ideas, we were fortunate to begin spending considerable time with Richard Fikes and Danny Bobrow, of Xerox PARC (and we continued our many exciting conversations with Brian Smith, who had moved to PARC from MIT). Structured, well-founded knowledge representation was the focus of a small group of extremely strong researchers, and we had a lot in common with the creators of KRL [Bobrow and Winograd 1977], a system that provided inspiration to many KR researchers of the time (I remember first hearing of "descriptors" in KRL from Brian and Terry Winograd, and in the end, that thinking had a substantial influence on what soon emerged as Description Logics). The many conversations we had on follow-ons from KL-ONE with Danny and Richard led to many exciting discussions at the KL-ONE workshops (I particularly remember some fun debate at one in Santa Barbara), and even more importantly for us, to a significant collaboration with Richard, in which we worked to build a system that implemented our ideas of separate TBox and ABox competences in an integrated, hybrid knowledge representation system. We called that system "Krypton" [Brachman, Fikes, and Levesque 1983], which while having a nice association with Superman, was actually inspired by the fact that the atomic symbol for the element Krypton is "Kr."[4] The work with Richard (and we were assisted by Victoria Pigman Gilbert, who was a student at Stanford at the time) led to some important papers that we co-wrote, and an exciting collaboration with Mark Stickel of SRI, who had developed an approach to extending logical theorem-proving with "theories," allowing a conventional first-order theorem-prover to defer to a special-purpose component that implemented some domain theory (e.g., arithmetic, temporal reasoning) and integrate that component's results in an elegant way [Stickel 1983]. Mark's "theory resolution" was the perfect framework in which to have a separate TBox, based on a well-founded implementation of subsumption and inconsistency reasoning, influence the computation in an ABox intuitively structured around conventional first-order logical reasoning.

The exploration of a hybrid KR system integrating taxonomic, "semantic-net-style" reasoning with concepts (which eventually became nicely interpretable as descriptions) and straightforward theorem-proving-based logical reasoning about objects and relationships in the world, captured the essence of Hector's and my wonderfully complementary partnership. I had a good feel for taxonomic hierarchies and reasoning with "concepts," and a gut feel for what a solid semantic foundation would look like — this was a major contribution of my Ph.D. thesis — but Hector was the real star with respect to formal logics, formal semantics, and all of the critical math that made our intuitive arguments really convincing. I don't think there is any chance in the world that I could have developed the basic semantic account for Krypton (which eventually became the core semantic framework for a huge amount of work on Description Logics) by myself. I like to think that it was my practical inclination and interest in applications that in exchange helped ground some of Hector's exciting work in real problems and real uses and led us to feel compelled to implement systems based on our ideas (Richard nicely spanned the two worlds and was a major contributor as well). One of my favorite quotes is, "In theory, there is no difference between

---

[4]Much of the time we kept the ambiguity of "Krypton" going on purpose. Our next two systems (largely driven by Peter Patel-Schneider) continued both the noble gas idea (ARGON) and the Superman mythology (KANDOR). We were particularly pleased with the KANDOR name, since it was used for an intentionally small system. Kandor was a city from the planet Krypton that had been miniaturized and kept in a bottle by the supervillain, Brainiac.

theory and practice; in practice, there is."[5] I've always felt it was critical to take into account the inspiration of use (in the manner of *Pasteur's Quadrant* [Stokes 1997]) when doing research, and that the best work comes from the back-and-forth between theoretical work and practical instantiation and application. However, I was never strong enough on the theory side to do this alone, and I somehow managed to drag Hector down into the muck of reality from time to time. Arguably, our relationship has reflected the incredible value of the partnership of deep, fundamental theoretical thinking and engineering in the context of real-world needs. I have certainly loved every moment of it and have found it much more satisfying to work together than to work alone.

The work we did together at FLAIR (and subsequently SPAR – Schlumberger Palo Alto Research) – eventually led us to perhaps the crowning technical contribution of our careers (well, mine at least; Hector's won so many awards it would be hard to pick a single high point), which was described in a AAAI paper in 1984 [Brachman and Levesque 1984], and expanded in a journal article that subsequently appeared in *Computational Intelligence*, which I include here in honor of Hector [Brachman and Levesque 1987]. But before I get to that, I have to mention another set of critical developments in our lives while in Palo Alto. We were in our very early thirties and both family-oriented people, but that doesn't explain the strange coincidence in 1982 when both of our wives got pregnant within a few weeks of one another. We were surprised to learn of this development from each other, but going through the pregnancies on the same timeline was great for all four of us. We had our first children, both girls, within a period of three weeks in the Fall of 1982. With our Rebecca and Hector and Pat's Michelle having built-in playmates and us all going through similar experiences, it brought our families close together. And in another inexplicable turn (unless it was Phil's competitive spirit), Phil Cohen, who had joined us in California not too long after Hector did, and his wife Sharon Oviatt, soon were expecting, and had a baby girl only a few months later. As Phil recalls, we began to think of ourselves as the "Knowledge representation, Reasoning, and Fertility" group.[6]

Given the formal foundation of our work on Krypton and similar languages, and the zeitgeist in Computer Science at least somewhat inspired by Steve Cook's work (again, at U. of Toronto) on computational complexity, perhaps it was inevitable we would have explored the intersection between KR and computational complexity. But whatever the inspiration, one day Hector started looking at the computation of subsumption, the core relationship in Description Logic systems, and investigating the formal complexity of the inference of subsumption. We had created a simple lexical format for our description languages, and that made it easy to imagine different variants on the basic language with added or subtracted concept-forming operators. In order to do the proofs about complexity, Hector had simplified the language to a very basic set of operators, and for the simplest meaningful language, which we came to call "$\mathcal{FL}$," we found a way to compute subsumption that was clearly polynomial time. But Hector played around with variants, and with one seemingly simple addition (an operator we called "RESTR," for the creation of a role built by restricting the fillers of another role; e.g., (RESTR child female) represented the role of "daughter" so that (ALL (RESTR child female) doctor) would represent the concept of something whose daughters were all doctors), Hector found a mapping to a problem that

---

[5] Attributed variously to Jan L. A. van de Snepscheut and Yogi Berra, of all people.

[6] This really put the pressure on Peter Patel-Schneider, who joined our team in late 1983. But Peter and his wife, Sandy, delivered, so to speak, and not too much later had their first child. But they went off-script and had a boy. In any case, at that point, anyone not interested in having children was clearly scared off from joining the KR & R group.

was the complement of a known NP-hard problem.

As it turned out, this was a seminal discovery. We wrote up the results in a paper for AAAI-84 [Brachman and Levesque 1984], showing that a seemingly small tweak to a concept representation language could result in a drastic change in the complexity of computing subsumption — what we felt was some kind of "computational cliff." While I have always felt that the basic discovery was Hector's, since I loved to write in a way that conveyed complex technical results to general audiences, I did a lot of writing and editing on that paper. We listed our authorship in alphabetical order to convey an equal partnership (which we have done on pretty much all of our joint work), but I still think Hector should get the bulk of the credit for the work. Somehow I also let him talk me into giving the talk at the conference, which in the end I greatly enjoyed; Hector had to give his own talk for another paper he wrote. We were very excited that our paper was given a Best Paper Award, but perhaps most impressively, Hector's other paper was also given a Best Paper Award, one of many to follow for him. (We were very proud of the fact that twenty years later, our AAAI-84 paper was given an honorable mention as a "Classic Paper" in AI by AAAI.)

It was impossible to anticipate at that point the future influence of the tradeoff paper, but it's very clear in hindsight that, for better or worse, it had a transformative effect on the field. Soon a growing number of papers appeared analyzing the complexity of Description Logic-based languages, and the same kind of analysis started to infiltrate other parts of AI. Ultimately it seemed like an avalanche. Between the simplicity and clarity of the concept language, which allowed researchers to dream up and include a huge variety of new operators, and the precedent of showing the complexity reduction, researchers all over the world (especially in Europe) investigated a growing number of variants and ultimately mapped out an incredibly extensive space of concept languages. In doing a retrospective analysis of this work for a talk in 2010, we looked over many years of subsequent AI conference proceedings, and it was quite remarkable to see the obvious change in work in knowledge representation through the later 80's and into the 90's.

Whether or not the seeming preoccupation with worst-case complexity analysis that our work spawned was totally a good thing is hard to say. But there were many, many exciting and surprising results along the way (including, for example, that reasoning in KL-ONE was undecidable), and by and large I think the field was better off for the increase in introspection and attention to solid formal foundations. One thing that was definitely not questionable about subsequent work in the Description Logic community was the fact that many groups around the world built and explored practical implementations of various versions of DL languages, applied them to real problems, and pursued innovative reasoning and implementation techniques that eventually made even very expressive Description Logic languages useful in practice [Baader, et al. 2003]. We were greatly honored when, roughly 25 years after the work was published, the organizers of the 2010 KR conference (in Toronto — where else?) named our work one of the "Great Moments in Knowledge Representation" and asked us to speak at a joint session of KR and the AAMAS conferences. Again, somehow, I got roped into giving the talk, but again, it was a great experience. It allowed us to do a fairly comprehensive rational reconstruction of the intellectual history of the ideas leading up to our 1984 paper, and we were pretty pleased with the talk. We still hope to turn it into an article so we can share it more generally. But the best part was the opportunity this gave us to work together again in 2009 and 2010 on a substantial project. It was incredibly fun.

Other great things were happening around the same time as our work on the expressiveness/tractability tradeoff. Hector and I agreed to give a tutorial in Knowledge Representation at the 1983 IJCAI conference in Karlsruhe, and once the tutorial was over we had a great time driving from Karlsuhe to Vienna, especially getting a chance to visit "Mad Ludwig"'s castle in Bavaria and one of Hector's relatives in Salzburg. I wish I could remember everything we laughed about during that trip, but there sure were many things. When we came home, while we were working on the tradeoff paper, we thought it would be nice to collect the key readings that we used as the basis for our tutorial into a book. Bonnie Webber and Nils Nilsson had set a precedent working with Mike Morgan of Morgan Kaufmann when they published a *Readings in Artificial Intelligence* book, and Mike and Nils enthusiastically supported our compilation of our *Readings in Knowledge Representation* book, which was published in 1985 [Brachman and Levesque 1985]. Given that there was no textbook in the field at the time, the book appeared to be valuable to people and we were pleased (and shocked) to sell several thousand copies. It also began a multi-decade collaboration with Mike.

Sadly, Hector and Pat decided to head back to Canada in 1984. For sure there were good family reasons to head home, and the CIAR gave Hector a spectacular fellowship that was impossible to turn down. I vividly remember seeing them off at San Francisco International Airport[7], and feeling rather empty and terribly sad. Fortunately, with email, the telephone, and airplanes, we were able to keep well in touch. Hector's move back to Toronto was one of the factors in our own family's heading back to the East Coast the following year, when I got the opportunity to create an AI group at Bell Labs in New Jersey. Once we discovered the many short flights available from Newark and LaGuardia to Toronto we got to see each other regularly, with Hector spending time with us in New Jersey and me traveling frequently up to Ontario. Along the way, we both had second children (first Hector and Pat's Marc, and then, our Lauren) and our whole families got together from time to time. Hector and Pat came down to New Jersey so we could work together full-time at AT & T for a whole summer.

Two other big and important projects followed, even with us in different places.

Late in the 1980's I got the somewhat insane idea of trying to create a new conference specific to knowledge representation and reasoning, which was increasingly inadequately represented at the main AI conferences, despite the central, critical role of the subject matter. I twisted Hector's arm to help, and then we ganged up on Ray Reiter, somehow convincing him to be the General Chair. I remember recruiting our first program committee by buttonholing many prominent people at the AAAI conference in Seattle in 1987. A lot of hard work and a great collaboration with Ray ensued, and out of this effort the First International Conference on Principles of Knowledge Representation and Reasoning (KR'89) was born. It was held at the Royal York Hotel in May of 1989, and seemed to work out very well.[8] We did it again in 1991 and 1992 in Cambridge, and then found a way to create a sustainable governance structure (eventually incorporating); somehow, miraculously, the conference is still going strong, every other year, to this day. KR-2010, at which we presented our "Great Moments" talk, was the twelfth international conference

---

[7]The day of the Levesques' departure is so seared in my memory that every time I end up in the United terminal at SFO I can't help but think back to that event, and Gate 82 will always be associated in my brain with what felt like the end of an era.

[8]Hector, Ray and I had the (somewhat agonizing) pleasure of creating an edited book based on the conference [Brachman, Levesque, and Reiter 1991] — another one of those projects that, if we knew in advance how difficult it would turn out to be, we might have declined to take on.

in the series.

Sometime in the late 80's Hector started teaching a first-year graduate course on knowledge representation and reasoning. We collaborated on parts of it and I regularly flew up to Toronto to present some lectures in the course. Over the years, we developed a comprehensive set of slides that fit the 13-week course well, and once again, in another fit of insanity, I got a crazy idea: with such well-developed lectures, wouldn't it be easy and fast to turn our notes into a textbook? I am not sure how I convinced Hector to tackle this one with me, but we were always eager to find interesting projects to do together, and we definitely shared a vision of a textbook that captured key principles of many styles of representation and reasoning, independent of the vagaries of any specific contemporary systems. We wanted it to be like a core math or economics textbook, capturing the basics in a simple, accessible, and timeless way. Nothing like that existed at the time, so we were both excited at the challenge.

Had we known at the beginning what it would have taken to write the book, I am sure we would never have started the project. We relatively quickly produced an outline that we both loved, and agreed (at Hector's suggestion − I thought this was one of his more brilliant ideas) to spec out in advance the length of each chapter such that we would produce a book of a reasonable length and with the right relative weighting of the chapters and subject matter. Morgan Kaufmann was thrilled with the idea and, not realizing how long we would keep them hanging, signed us to a contract.

As it turned out, I hopelessly underestimated how much of my time my management job at Bell Labs and subsequently AT & T Labs would take. Hector was able to pace himself better, but it wasn't fair to have him earnestly working away while I put writing on hiatus at various times. So the work proceeded in fits and starts, and as a couple years went by, we had to face the fact that it might very well never get done. But somehow, each time one of us was about to give up, the other found a way to save the day with words of encouragement and a concrete plan for the next steps. This kind of back-and-forth over what turned out to be a full ten years (!!) to me was a microcosm of our successful partnership. We both liked working with each other so much that we didn't really in our hearts want to give up on a big project together. But each of us had realities (and Hector had students and a huge amount of other technical work he was producing) and every once in a while, things looked very bleak to one of us. But it rarely looked that way to both of us at the same time, and somehow we found a way to scale down our aspirations, set some deadlines (which were usually missed, but they certainly motivated us), and miraculously deliver a final manuscript in late 2003. Along the way we developed a great methodology for collaboration − basically, each of us would write until we ran out of steam on a particular chapter, and at that point we passed the material to the other, who could be energized because he wasn't starting from scratch. Gwen and Pat were convinced that the book was a figment of our imaginations, and had a good laugh about it many times over the decade it took. Fortunately, the material was fundamental and not based on any current idiosyncratic work, or the book would have been a miserable failure when it finally appeared.

Throughout the period from 1980 to 2004, when our book was finally published, and even beyond, we found many other projects to do together, which allowed us to continue to have a great time collaborating and even do some useful work. But looking back, the major, huge efforts, like the early tradeoff work, our Readings book, the KR conferences, and our textbook were definitely high points (we also collaborated on IJCAI for many years − as Secretary-Treasurer, I helped twist Hector's arm to accept the invitation to be the General

Chair of IJCAI-01 in Seattle, and he subsequently became the President of the Board of Trustees — those were fun years together also). Each of those activities exemplified to me the depth and quality of our relationship — they were all true partnerships, which would have been unimaginable to either of us without the other.

It's a cliché, I'm sure, to lament the quickly passing years, but it is truly shocking to realize that it's been more than thirty years that Hector and I have been working together. One of the things that is so striking to me in retrospect is the fact that, while we collaborated on such a wide variety of activities over so many years, we only worked together on a day-to-day basis in the same place for something like three years (Hector's summer at AT & T adds a little, I guess). Every time we get together, except for some more gray hair in one of our cases and a little less hair in the other, it's like we were never apart. Even while facing retirement in the not-too-distant future, we're still scheming about new projects we can do together, and there is nothing as energizing to me in my professional life as the contemplation of some new adventure with Hector, my lifelong friend and collaborator.

References

Baader, F., D. Calvanese, D. McGuinness, D. Nardi, and P. Patel-Schneider, eds. [2003]. *The Description Logic Handbook: Theory, Implementation, and Applications.* Cambridge, UK: Cambridge University Press.

Bobrow, D. G., and T. Winograd [1977]. "An Overview of KRL, A Knowledge Representation Language," *Cognitive Science* 1(1): 3–46.

Brachman, R. J. [1976]. "What's in a Concept," *Sixth International Conference on Computational Linguistics (COLING 76)*, Ottawa, Ontario, June, 1976.

Brachman, R. J. [1977]. "A Structural Paradigm for Representing Knowledge," Ph.D. Dissertation, Harvard University, Cambridge, Massachusetts, May, 1977. Also, BBN Report No. 3605, Bolt Beranek & Newman, May 1978.

Brachman, R. J., and H. J. Levesque [1982]. "Competence in Knowledge Representation," *Proc. AAAI-82*, Pittsburgh, PA: 189–192.

Brachman, R. J., and H. J. Levesque [1984]. "The Tractability of Subsumption in Frame-Based Description Languages," *Proc. AAAI-84*, Austin, TX: 34–37.

Brachman, R. J., and H. J. Levesque [1985]. *Readings in Knowledge Representation.* Los Altos, CA: Morgan Kaufmann Publishers, Inc.

Brachman, R. J., and H. J. Levesque [1987]. "Expressiveness and Tractability in Knowledge Representation," *Computational Intelligence*, Vol. 3, No. 2, May, 1987: 78–93.

Brachman, R. J., and H. J. Levesque [2004]. *Knowledge Representation and Reasoning.* San Francisco: Morgan Kaufmann.

Brachman, R. J., and J. G. Schmolze [1985]. "An Overview of the KL-ONE Knowledge Representation System," *Cognitive Science* 9(2): 171–216.

Brachman, R. J., R. E. Fikes, and H. J. Levesque [1983]. "Krypton: Integrating Terminology and Assertion," *Proc. AAAI-83*, Washington, DC, August, 1983: 31–35.

Brachman, R. J., H. J. Levesque, and R. Reiter [1991]. *Knowledge Representation.* Amsterdam: Elsevier/Cambridge: MIT Press.

Levesque, H. J. [1977]. *A Procedural Approach to Semantic Networks.* Technical Report No. 105, Department of Computer Science, University of Toronto, Toronto, Ontario.

Levesque, H. J., and J. Mylopoulos [1977]. "An Overview of a Procedural Approach to Semantic Networks." *Proc. IJCAI-77*, Cambridge, MA: 283.

Schmolze, J. G., and R. J. Brachman [1982]. *Proceedings of the 1981 KL-ONE Workshop.* FLAIR Technical Report No. 4, Fairchild Laboratory for Artificial Intelligence Research, Palo Alto, CA, May, 1982.

Stickel, M. E. [1983]. "Theory Resolution: Building-In Nonequational Theories," *Proc. AAAI-83*, Washington, DC: 391–397.

Stokes, D. E. [1997]. *Pasteur's Quadrant: Basic Science and Technological Innovation.* Washington, DC: Brookings Institute Press.

Woods, W. A. [1975]. "What's in a Link: Foundations for Semantic Networks," in *Representation and Understanding: Studies in Cognitive Science.* D. G. Bobrow and A. M. Collins (eds.), New York: Academic Press.

Expressiveness and tractability in knowledge representation and reasoning[1]

HECTOR J. LEVESQUE[2]

Department of Computer Science, University of Toronto, Toronto, Ont., Canada M5S 1A4

AND

RONALD J. BRACHMAN

AT&T Bell Laboratories, 600 Mountain Avenue, 3C-439, Murray Hill, NJ 07974, U.S.A.

Received November 3, 1986

Revision accepted April 8, 1987

A fundamental computational limit on automated reasoning and its effect on knowledge representation is examined. Basically, the problem is that it can be more difficult to reason correctly with one representational language than with another and, moreover, that this difficulty increases dramatically as the expressive power of the language increases. This leads to a tradeoff between the expressiveness of a representational language and its computational tractability. Here we show that this tradeoff can be seen to underlie the differences among a number of existing representational formalisms, in addition to motivating many of the current research issues in knowledge representation.

Key words: knowledge representation, description subsumption, complexity of reasoning, first-order logic, frames, semantic networks, databases.

Cet article étudie une limitation computationnelle fondamentale du raisonnement automatique et examine ses effets sur la représentation de connaissances. A la base le problème tient en ce qu'il peut être plus difficile de raisonner avec un langage de représentation qu'avec un autre et que cette difficulté augmente considérablement à mesure que croît le pouvoir expressif du langage. Ceci donne lieu à un compromis entre le pouvoir expressif d'un langage de représentation et sa tractibilité computationnelle. Nous montrons que ce compromis peut être vu comme l'une des causes fondamentales de la différence qui existe entre nombre de formalismes de représentation existants et peut motiver plusieurs recherches courantes en représentation de connaissances.

Mots clés : représentation de connaissances, complexité du raisonnement, logique du premier ordre, schémas, réseaux sémantiques, bases de données.

[Traduit par la revue]

Comput. Intell. 3, 78–93 (1987)

1. Introduction

This paper examines from a general point of view a basic computational limit on automated reasoning, and the effect that it has on knowledge representation (KR). The problem is essentially that it can be more difficult to reason correctly with one representational language than with another and, moreover, that this difficulty increases as the expressive power of the language increases. There is a tradeoff between the expressiveness of a representational language and its computational tractability. What we attempt to show is that this tradeoff underlies the differences among a number of representational formalisms (such as first-order logic, databases, semantic networks, and frames) and motivates many current research issues in KR (such as the role of analogues, syntactic encodings, and de-

[1] This is a revised and substantially augmented version of "A Fundamental Tradeoff in Knowledge Representation and Reasoning," by Hector J. Levesque, which appeared in the Proceedings of the Canadian Society for Computational Studies of Intelligence Conference, London, Ontario, May 1984. It includes portions of two other conference papers: "The Tractability of Subsumption in Frame-Based Description Languages," by Ronald J. Brachman and Hector J. Levesque, which appeared in the Proceedings of the American Association for Artificial Intelligence Conference, Austin, Texas, August 1984; and "What Makes a Knowledge Base Knowledgeable? A View of Databases from the Knowledge Level," by the same authors, which appeared in the Proceedings of the First International Workshop on Expert Database Systems, Kiawah Island, South Carolina, October 1984. Much of this paper appeared as a chapter in *Readings in Knowledge Representation* (Morgan Kaufmann Publishers Inc., 1985), edited by the authors.

[2] Fellow of the Canadian Institute for Advanced Research.

faults, as well as systems of limited inference and hybrid reasoning).

To deal with such a broad range of representational phenomena we must, of necessity, take a considerably simplified and incomplete view of KR. In particular, we focus on its computational and logical aspects, more or less ignoring its history and relevance in the areas of psychology, linguistics, and philosophy. The area of KR is still very disconnected today and the role of logic remains quite controversial, despite what this paper may suggest. We do believe, however, that the tradeoff discussed here is fundamental. As long as we are dealing with computational systems that reason automatically (without any special intervention or advice) and correctly (once we define what *that* means), we will be able to locate where they stand on the tradeoff: They will either be limited in what knowledge they can represent or unlimited in the reasoning effort they might require.

Our computational focus will not lead us to investigate specific algorithms and data structures for KR and reasoning, however. What we discuss is something much stronger, namely, whether or not algorithms of a certain kind can exist at all. The analysis here is at the *knowledge level* (Newell 1981) where we look at the content of what is represented (in terms of what it says about the world) and not the symbolic structures used to represent that knowledge. Indeed, we examine specific representation schemes in terms of what knowledge they can represent, rather than in terms of how they might actually represent it.

In the next section, we discuss what a KR system is for and what it could mean to reason correctly. Next, we investigate how a KR service might be realized using theorem proving in first-order logic and the problem this raises. Following this, we present various representational formalisms and examine the special kinds of reasoning they suggest. We concentrate in particular on frame-based description languages, examining in some detail a simple language and a variant. In the case of this pair of languages, the kind of tradeoff we are talking about is made concrete, with a dramatic result. Finally, we draw some tentative general conclusions from this analysis.

2. The role of knowledge representation

While it is generally agreed that KR plays an important role in (what have come to be called) knowledge-based systems, the exact nature of that role is often hard to define. In some cases, the KR subsystem does no more than manage a collection of data structures, providing, for example, suitable search facilities; in others, the KR subsystem is not really distinguished from the rest of the system at all and does just about everything: make decisions, prove theorems, solve problems, and so on. In this section, we discuss in very general terms the role of a KR subsystem within a knowledge-based system.[3]

2.1. The knowledge representation hypothesis

A good place to begin our discussion is with what Brian Smith has called the *knowledge representation hypothesis* (Smith 1982):

> Any mechanically embodied intelligent process will be comprised of structural ingredients that (a) we as external observers naturally take to represent a propositional account of the knowledge that the overall process exhibits, and (b) independent of such external semantical attribution, play a formal but causal and essential role in engendering the behaviour that manifests that knowledge.

This hypothesis seems to underlie much of the research in KR, if not most of the current work in Artificial Intelligence in general. In fact, we might think of *knowledge-based systems* as those that satisfy the hypothesis by design. Also, in some sense, it is only with respect to this hypothesis that KR research can be distinguished from any number of other areas involving symbolic structures such as database management, programming languages, and data structures.

Granting this hypothesis, there are two major properties that the structures in a knowledge-based system have to satisfy. First of all, it must be possible to interpret them as *propositions* representing the overall knowledge of the system. Otherwise, the representation would not necessarily be of *knowledge* at all, but of something quite different, like numbers or circuits. Implicit in this constraint is that the structures have to be expressions in a language that has a *truth theory*. We should be able to point to one of them and say what the world would have to be like for it to be true. The structures themselves need not *look* like sentences—there are no syntactic requirements on them at all, other than perhaps finiteness—but we have to be able to understand them that way.

A second requirement of the hypothesis is perhaps more obvious. The symbolic structures within a knowledge-based system must play a *causal role* in the behaviour of that system,

as opposed to, say, comments in a programming language. Moreover, the influence they have on the behaviour of the system should agree with our understanding of them as propositions representing knowledge. Not that the system has to be aware in any mysterious way of the interpretation of its structures and their connection to the world;[4] but for us to call it knowledge-based, *we* have to be able to understand its behaviour as if it believed these propositions, just as we understand the behaviour of a numerical program as if it appreciated the connection between bit patterns and abstract numerical quantities.

2.2. Knowledge bases

To make the above discussion a bit less abstract, we can consider a very simple task and consider what a system facing this task would have to be like for us to call it knowledge-based. The amount of knowledge the system will be dealing with will, of course, be very small.

Suppose we want a system in PROLOG that is able to print the colours of various items. One way to implement that system would be as follows:

```
printColour(snow) :- !, write("It's white.").
printColour(grass) :- !, write("It's green.").
printColour(sky) :- !, write("It's yellow.").
printColour(X) :- write("Beats me.").
```

A slightly different organization that leads to the same overall behaviour is

```
printColour(X) :-
   colour(X,Y), !, write("It's "), write(Y), write(".").
printColour(X) :- write("Beats me.").

colour(snow,white).
colour(grass,green).
colour(sky,yellow).
```

The second program is characterized by explicit structures representing the (minimal) knowledge[5] the system has about colours and is the kind of system that we are calling knowledge-based. In the first program, the association between the object (we understand as) referring to grass and the one referring to its colour is implicit in the structure of the program. In the second, we have an explicit *knowledge base* (or KB) that we can understand as propositions relating the items to their colours. Moreover, this interpretation is justified in that these structures determine what the system does when asked to print the colour of a particular item.

One thing to notice about the example is that it is not the use of a certain programming language or data-structuring facility that makes a system knowledge-based. The fact that PROLOG happens to be understandable as a subset of first-order logic is largely irrelevant. We could probably read the first program "declaratively" and get sentences representing some kind of knowledge out of it; but these would be very strange ones

---

[3] We should emphasize that we are concentrating on the kind of knowledge representation system that would be used as a component of a larger AI program. KR research also seems to encompass attempts at general models of cognitive behavior. Although some of our comments are applicable even to such models, we are generally ignoring that part of the field here.

[4] Indeed, part of what philosophers have called the *formality condition* is that computation at some level has to be uninterpreted symbol manipulation.

[5] Notice that typical of how the term "knowledge" is used in AI, there is no requirement of *truth*. A system may be mistaken about the colour of the sky but still be knowledge-based. "Belief" would perhaps be a more appropriate term, although we follow the standard AI usage in this paper.

dealing with writing strings and printing colours, not with the colours of objects.

2.3. The knowledge representation subsystem

In terms of its overall goals, a knowledge-based system is not directly interested in what specific structures might exist in its KB. Rather, it is concerned with what the application domain is like—for example, what the colour of grass is. How that knowledge is represented and made available to the overall system is a secondary concern and one that we take to be the responsibility of the KR subsystem. The role of a KR subsystem, then, is to manage a KB for a knowledge-based system and present a picture of the world based on what it has represented in the KB.[6]

If, for simplicity, we restrict our attention to the yes—no questions about the world that a system might be interested in, what is then involved is being able to determine what the KB says regarding the truth of certain sentences. It is not whether the sentence itself is present in the KB that counts, but whether its truth is *implicit* in the KB. Stated differently, what a knowledge representation system has to be able to determine, given a sentence α, is the answer to the following question:

Assuming the world is such that what is believed is true, is α also true?

We will let the notation KB $\models \alpha$ mean that α is implied (in this sense) by what is in the KB.

One thing to notice about this view of a KR system is that an understanding of the service it provides to a knowledge-based system depends only on the truth theory of the language of representation. Depending on the particular truth theory, determining if KB $\models \alpha$ might require not just simple retrieval capabilities, but also *inference* of some sort. This is not to say that the *only* service to be performed by a KR subsystem is question-answering. If we imagine the overall system existing over a period of time, then we will also want it to be able to augment the KB as it acquires new information about the world.[7] In other words, the responsibility of the KR system is to use appropriate symbolic structures to represent knowledge, and to use appropriate reasoning mechanisms both to answer questions and to assimilate new information, *in accordance with the truth theory of the underlying representation language*.

So our view of KR makes it depend only on the semantics of the representation language, unlike other possible accounts that might have it defined in terms of a set of formal symbol manipulation routines (e.g., a proof theory). This is in keeping with what we have called elsewhere a *functional* view of KR (see Levesque (1984b) and Brachman et al. (1983)), where the service performed by a KR system is defined separately from the techniques a system might use to realize that service.

3. The logical approach

To make a lot of the above more concrete, it is useful to look at an example of the kinds of knowledge that might be available in a given domain and how it might be represented in a KB. The language that will be used to represent knowledge is that of a standard first-order logic (FOL).[8]

3.1. Using first-order logic

The first and most prevalent type of knowledge to consider representing is what might be called simple *facts* about the world, such as

- Joe is married to Sue.
- Bill has a brother with no children.
- Henry's friends are Bill's cousins.

These might be complicated in any number of ways, for example, by including time parameters and certainty factors.

Simple observations such as these do not exhaust what might be known about the domain, however. We may also have knowledge about the *terminology* used in these observations, such as

- *Ancestor* is the transitive closure of parent.
- Brother is *sibling* restricted to males.
- *Favourite-cousin* is a special type of cousin.

These could be called definitions except for the fact that necessary and sufficient conditions might not always be available (as in the last example above). In this sense, they are much more like standard dictionary entries.

The above two sets of examples concentrate on what might be called *declarative* knowledge about the world. We might also have to deal with *procedural* knowledge that focuses not on the individuals and their interrelationships, but on *advice* for reasoning about these. For example, we might know that

- To find the father of someone, it is better to search for a parent and then check if he is male, than to check each male to see if he is a parent.
- To see if x is an ancestor of y, it is better to search up from y than down from x.

One way to think of this last type of knowledge is not necessarily as advice to a reasoner, but as declarative knowledge that deals implicitly with the combinatorics of the domain as a whole.

This is how the above knowledge might be represented in FOL:

1. The first thing to do is to "translate" the simple facts into sentences of FOL. This would lead to sentences like

$$\forall x \text{ Friend(henry, } x) \equiv \text{Cousin(bill, } x)$$

2. To deal with terminology in FOL, the easiest way is to "extensionalize" it, that is, to pretend that it is a simple observation about the domain. For example, the *brother* statement above would become[9]

$$\forall x \forall y \text{ Brother}(x, y) \equiv (\text{Sibling}(x, y) \wedge \text{Male}(y))$$

3. Typically, the procedural advice would not be represented explicitly at all in a FOL KB, but would show up in the *form* of (1) and (2) above. Another alternative would be to use

---

[6] As hinted earlier, this is not the only role that we could imagine for such a subsystem, but this approach is consonant with the majority of work in the field.

[7] It is this management of a KB over time that makes a KR subsystem much more than just the implementation of a static deductive calculus.

[8] The use of FOL per se is not an essential feature of the arguments to follow. Any language that allows us to express what we can in FOL would suffice.

[9] This is a little misleading since it will make the *brother* sentence appear to be no different in kind from the one about Henry's friends, though we surely do not want to say that Henry's friends are *defined* to be Bill's cousins.

extra-logical annotations like the kind used in PROLOG or those described in Moore (1982).

The end result of this process would be a first-order knowledge base: a collection of sentences in FOL representing what was known about the domain. A major advantage of FOL is that given a yes—no question also expressed in this language, we can give a very precise definition of KB $\models$ α (and thus, under what conditions the question should be answered *yes*, *no*, or *unknown*):

KB $\models$ α iff every interpretation satisfying all of the sentences in the KB also satisfies α.[10]

There is, moreover, another property of FOL that makes its use in KR even more simple and direct. If we assume that the KB is a finite set of sentences and let KB stand for their conjunction, it can be shown that

KB $\models$ α iff $\vdash$ (KB $\supset$ α)

In other words, the question as to whether or not the truth of α is implicit in the KB reduces to whether or not a certain sentence is a *theorem* of FOL. Thus, the question-answering operation becomes one of *theorem proving* in FOL.

3.2. The problem

The good news in reducing the KR service to theorem proving is that we now have a very clear, very specific notion of what the KR system should do. The bad news is that it is also clear that *this service cannot be provided*. The sad fact of the matter is that deciding whether or not a sentence of FOL is a theorem (i.e., the decision problem) is unsolvable. Moreover, even if we restrict the language practically to the point of triviality by eliminating the quantifiers, the decision problem, though now solvable, does not appear to be solvable in anywhere near reasonable time.[11] It is important to realize that this is not a property of particular algorithms that people have looked at but of the *problem* itself: there *cannot* be an algorithm that does the theorem proving correctly in a reasonable amount of time. This bodes poorly, to say the least, for a service that is supposed to be only a part of a larger knowledge-based system.

One aspect of these intractability results that should be mentioned, however, is that they deal with the *worst case* behaviour of algorithms. In practice, a given theorem-proving algorithm may work quite well. In other words, it might be the case that for a wide range of questions, the program behaves properly, even though it can be shown that there will always be short questions whose answers will not be returned for a very long time, if at all.

How serious is the problem, then? To a large extent this depends on the kind of question you would like to ask of a KR subsystem. The worst case prospect might be perfectly tolerable if you are interested in a mathematical application and the kind of question you ask is an open problem in mathematics. Provided progress is being made, you might be quite willing to stop and redirect the theorem prover after a few months if it seems to be thrashing. Never mind worst case behaviour; this might be the *only* case you are interested in.

But imagine, on the other hand, a robot that needs to know about its external world (such as whether or not it is raining outside or where its umbrella is) before it can act. If this robot has to call a KR system utility as a subroutine, the worst case prospect is much more serious. Bogging down on a logically difficult but low-level subgoal and being unable to continue without human intervention is clearly an unreasonable form of behaviour for something aspiring to intelligence.

Not that "on the average" the robot might not do alright. The trouble is that nobody seems to be able to characterize what an "average" case might be like.[12] As responsible computer scientists, we should not be providing a general inferential service if all that we can say about it is that by and large it will probably work satisfactorily.[13]

If the KR service is going to be used as a utility and is not available for introspection or control, then it had better be *dependable* both in terms of its correctness and the resources it consumes. Unfortunately, this seems to rule out a service based on full theorem proving (in full first-order logic).

3.3. Two pseudosolutions

There are at least two fairly obvious ways to minimize the intractability problem. The first is to push the computational barrier as far back as possible. Research in automatic theorem proving has concentrated on techniques for avoiding redundancies and speeding up certain operations in theorem provers. Significant progress has been achieved here, allowing open questions in mathematics to be answered (Winker 1982; Wos *et al.* 1984). Along similar lines, VLSI and parallel architectural support stands to improve the performance of theorem provers at least as much as it would any search program.

The second way to make theorem provers more usable is to relax our notion of correctness. A very simple way of doing this is to make a theorem-proving program always return an answer after a certain amount of time.[14] If it has been unable to prove either that a sentence or its negation is implicit in the KB, it could assume that it was independent of the KB and answer *unknown* (or maybe reassess the importance of the question and try again). This form of error (i.e., one introduced by an incomplete theorem prover) is not nearly as serious as returning a *yes* for a *no*, and is obviously preferable to an answer that never arrives. This is of course especially true if the program uses its resources wisely, in conjunction with the first suggestion above.

[10] The assumption here is that the semantics of FOL specify in the usual way what an interpretation is and under what conditions it will satisfy a sentence.

[11] Technically, the problem is now co-NP-complete, meaning that it is strongly believed to be computationally intractable.

[12] This seems to account more than anything for the fact that there are so few average case results regarding decidability.

[13] As we noted earlier, not all KR-related research is aimed at providing an inferential component for a larger, knowledge-based system. Work in a similar spirit sometimes has as its goal the realistic modeling of cognitive agents, imperfections and all. While the central concern of this paper is directed less at such research than at that KR work intent on providing a KR service, we believe that this issue of carefully and precisely characterizing the KR system holds equally well for both. To be informative, cognitive models must be correct and timely, no matter what they are modeling. There is a big difference between a precise and predictable model of (say) sloppy reasoning, and a sloppy model of perfect (or other) reasoning. If the model itself is not well understood or may not even be working properly, it is not going to inform us about anything. Thus, while the system being modeled (e.g., a human) may only "by and large work satisfactorily," the *model* must work reliably, predictably, and completely, or it will not do its job.

[14] The resource limitation here should obviously be a function of how important it might be to answer the question either quickly or correctly.

However, from the point of view of KR, both of these are only pseudosolutions. Clearly, the first alone does not help us guarantee anything about an inferential service. The second, on the other hand, might allow us to guarantee an answer within certain time bounds, but would make it very hard for us to tell how seriously to take that answer. If we think of the KR service as reasoning according to a certain logic, then the logic being followed is immensely complicated (compared to that of FOL) when resource limitations are present. Indeed, the whole notion of the KR system calculating what is implicit in the KB (which was our original goal) would have to be replaced by some other notion that went beyond the truth theory of the representation language to include the inferential power of a particular theorem-proving program. In a nutshell, we can guarantee getting an answer, but not necessarily the one we want.

One final observation about this intractability is that it is *not* a problem that is due to the formalization of knowledge in FOL. If we assume that the goal of our KR service is to calculate what is implicit in the KB, then as long as the truth theory of our representation language is upward-compatible with that of FOL, we will run into the same problem. In particular, using English (or any other natural or artificial language) as our representation language does not avoid the problem as long as we can express in it at least what FOL allows us to express.

4. Expressiveness and tractability

It appears that we have run into a serious difficulty in trying to develop a KR service that calculates what is implicit in a KB and yet does so in a reasonable amount of time. One option we have not yet considered, however, is to *limit* what can be in the KB so that its implications are more manageable computationally. Indeed, as we will demonstrate in this section, much of the research in KR can be construed as trading expressiveness in a representation language for a more tractable form of inference. Moreover, unlike the restricted dialects of FOL typical of those analyzed in the logic and computer science literatures (e.g., in terms of nestings of quantifiers), the languages considered here have at least proven themselves quite useful in practice, however contrived they may appear on the surface.

4.1. Incomplete knowledge

To see where this tradeoff between expressiveness and tractability originates, we have to look at the use of the expressive power of FOL in KR and how it differs from its use in mathematics.

In the study of mathematical foundations, the main use of FOL is in the formalization of infinite collections of entities. So, for example, we have first-order number and set theories that use quantifiers to range over these classes, and conditionals to state what properties these entities have. This is exactly how Frege intended his formalism to be used.

In KR, on the other hand, the domains being characterized are usually finite. The power of FOL is used not so much to deal with infinities, but to deal with *incomplete knowledge* (Moore 1982; Levesque 1982). Consider the kind of facts[15] that might be represented using FOL:

1. ¬Student(john).

This sentence says that John is not a student without saying what he is.

2. Parent(sue,bill) $\lor$ Parent(sue,george).

This sentence says that either Bill or George is a parent of Sue, but does not specify which.

3. $\exists x$ Cousin(bill, x) $\land$ Male(x).

This sentence says that Bill has at least one male cousin but does not say who that cousin is.

4. $\forall x$ Friend(george, x) $\supset$ $\exists y$ Child(x, y).

This sentence says that all of George's friends have children without saying who those friends or their children are or even if there are any.

The main feature of these examples is that FOL is not used to capture complex details about the domain, but to avoid having to represent details that may not be known. *The expressive power of FOL determines not so much what can be said, but what can be left unsaid.*

For a system that has to be able to acquire knowledge in a piecemeal fashion, there may be no alternative to using all of FOL. But if we can restrict the kind of the incompleteness that has to be dealt with, we can also avoid having to use the full expressiveness of FOL. This, in turn, might lead to a more manageable inference procedure.

The last pseudosolution to the tractability problem, then, is to restrict the logical form of the KB by controlling the incompleteness of the knowedge represented. This is still a pseudosolution, of course. Indeed, provably, there cannot be a *real* solution to the problem. But this one has the distinct advantage of allowing us to calculate exactly the picture of the world implied by the KB, precisely what a KR service was supposed to do. In what follows, we will show how restricting the logical form of a KB can lead to very specialized, tractable forms of inference.[16]

4.2. Database form

The most obvious type of restriction to the form of a KB is what might be called *database form*. The idea is to restrict a KB so that it can only contain the kinds of information that can be represented in a standard database. Consider, for example, a very simple database that talks about university courses. It might contain a relation (or record type or whatever) like

COURSE

ID	NAME	DEPT	ENROLL-MENT	INSTRUCTOR
csc248	Programming Languages	Computer Science	42	S. J. Hurtubise
mat100	History of Mathematics	Mathematics	137	R. Cumberbatch
csc373	Artificial Intelligence	Computer Science	853	T. Slothrop
	...			

---

[15] The use of FOL to capture *terminology* or laws is somewhat different. See Brachman and Levesque (1982) for details.

[16] As we have mentioned, there are other ways of dealing with the tradeoff, but the tactic of limiting the form of the representation seems to account for the vast majority of current practice. Indeed, the only style of representation proven so far to scale up to realistic sizes—database technology—falls under this account.

If we had to characterize in FOL the information that this relation contained, we could use a collection of function-free atomic sentences like[17]

COURSE(csc248)   DEPT(csc248,ComputerScience) ENROLLMENT(csc248,42) ...
COURSE(mat100) DEPT(mat100,Mathematics) ...
   ...

In other words, the tabular database format characterizes exactly the positive instances of the various predicates. But more to the point, since our list of FOL sentences never ends up with ones like

DEPT(mat100,Mathematics) $\bigvee$ DEPT(mat100,History),

the range of uncertainty that we are dealing with is quite limited.

There is, however, additional information contained in the database not captured in the simple FOL translation. To see this, consider, for instance, how we might try to determine the answer to the question:

How many courses are offered by the Computer Science Department?

The knowledge expressed by the above collection of FOL sentences is insufficient to answer this question; nothing about our set of atomic sentences implies that computer science has at least two courses (since csc373 and csc248 could be names of the same individual), and nothing implies that it has at most two courses (since there could be courses other than those mentioned in the list of sentences). On the other hand, from a database point of view, we could apparently successfully answer our question using our miniature database by phrasing it something like

Count c in COURSE where c.DEPT = ComputerScience;

this yields the definitive answer, "2". The crucial difference here, between failing to answer the question at all and answering it definitively, is that we have actually asked *two different questions*. The formal query addressed to the database must be understood as

How many tuples in the COURSE relation have Computer Science in their DEPT field?

This is a question *not* about the world being modelled at all, but about the *data* itself. In other words, the database retrieval version of the question asks about the structures in the database itself, and not about what these structures represent.[18]

To be able to reinterpret the database query as the intuitive question originally posed about courses and departments

(rather than as one about tuples and fields), we must account for additional information taking us beyond the stored data itself. In particular, we need FOL sentences of the form

$c_i \neq c_j$

for distinct constants c_i and c_j, stating that each constant represents a unique individual. In addition, for each predicate, we need a sentence similar in form to

$\forall x[\text{COURSE}(x) \supset x = \text{csc248} \bigvee \ldots \bigvee x = \text{mat100}]$

saying that the only instances of the predicate are the ones named explicitly.[19] If we now consider a KB consisting of all of the sentences in FOL we have listed so far, a KR system could, in fact, conclude that there were exactly two computer science courses, just like its database management counterpart. We have included in the imagined KB all of the information, both explicit and implicit, contained in the database.

One important property of a KB in this final form is that it is much easier to use than a general first-order KB. In particular, since the first part of the KB (the atomic sentences) does not use negation, disjunction, or existential quantifications, we know the exact instances of every predicate of interest in the language. There is no incompleteness in our knowledge at all. Because of this, *inference reduces to calculation*. To find out how many courses there are, all we have to do is to count how many appropriate tuples appear in the COURSE relation. We do not, for instance, have to reason by cases or by contradiction, as we would have to in the more general case. For example, if we also knew that either csc148 or csc149 or both were computer science courses but that no computer science course other than csc373 had an odd identification number, we could still determine that there were three courses, but not by simply counting. But a KB in database form does not allow us to express this kind of uncertainty and, because of this expressive limitation, the KR service is much more tractable. Specifically, we can represent what is known about the world using just these sets of tuples, exactly like a standard database system. From this perspective, a database is a knowledge base whose limited form permits a very special form of inference.

This limitation on the logical form of a KB has other interesting features. Essentially, what it amounts to is making sure that there is very close structural correspondence between the (explicit) KB and the domain of interest: For each entity in the domain, there is a unique representational object that stands for it; for each relationship that it participates in, there is a tuple in the KB that corresponds to it. In a very real sense, the KB is an *analogue* of the domain of interest, not so different from other analogues such as maps or physical models. The main advantage of having such an analogue is that it can be used directly to answer questions about the domain. That is, the calculations on the model itself can play the role of more general reasoning techniques much the way arithmetic can re-

---

[17] This is not the only way to characterize this information. For example, we could treat the field names as function symbols or use ID as an additional relation or function symbol. Also, for the sake of simplicity, we are ignoring here integrity constraints (saying, for example, that each course has a unique enrollment), which may contain quantificational and other logical operations, but typically are only used to verify the consistency of the database, not to infer new facts. None of these decisions affect the conclusions we will draw below.

[18] The hallmark, it would appear, of conventional database management is that its practitioners take their role to be providing users access to the data, rather than using the data to answer questions about the world. The difference between the two points of view is especially evident when the database is very incomplete (Levesque 1984a).

[19] This is one form of what has been called the *closed-world assumption* (Reiter 1978b).

place reasoning with Peano's axioms. The disadvantage of an analogue, however, should also be clear: Within a certain descriptive language, it does not allow anything to be left unsaid about the domain.[20] In this sense, an analogue representation can be viewed as a special case of a propositional one where the information it contains is relatively complete.

4.3. Logic-program form

The second restriction on the form of a KB we will consider is a generalization of the previous one that is found in programs written in PROLOG, PLANNER, many production systems, and related languages. A KB in logic-program form also has an explicit and an implicit part. The explicit KB in a PROLOG program is a collection of first-order sentences (called Horn sentences) of the form

$$\forall x_1 \ldots x_n [P_1 \wedge \ldots \wedge P_m \supset P_{m+1}]$$

where $m \geq 0$ and each P_i is atomic. In the case where $m = 0$ and the arguments to the predicates are all constants, the logic-program form coincides with the database form. Otherwise, because of the possible nesting of functions, the set of relevant terms (whose technical name is the *Herbrand universe*) is much larger and may be infinite.

As in the database case, if we were only interested in the universe of terms, the explicit KB would be sufficient. However, to understand the KB as being about the world, but in a way that is compatible with the answers provided by a PROLOG processor, we again have to include additional facts in an implicit KB. In this case, the implicit KB is normally infinite since it must contain a set of sentences of the form $(s \neq t)$, for any two distinct terms in the Herbrand universe. As in the database case, it must also contain a version of the closed-world assumption which is now a set containing the negation of every ground atomic sentence not implied by the Horn sentences in the explicit KB.

The net result of these restrictions is a KB that once again has complete knowledge of the world (within a given language), but this time, may require inference to answer questions.[21] The reasoning in this case, is the *execution* of the logic program. For example, given an explicit PROLOG KB consisting of

    parent(bill,mary).
    parent(bill,sam).
    mother(X,Y) :- parent(X,Y), female(Y).
    female(mary).

we know exactly who the mother of Bill is, but only after having executed the program.

In one sense, the logic-program form does not provide any computational advantage to a reasoning system since deter-

mining what is in the implicit KB is, in general, undecidable.[22] On the other hand, the form is much more manageable than in the general case since the necessary inference can be split very nicely into two components: a *retrieval* component that extracts (atomic) facts from a database by pattern-matching and a *search* component that tries to use the nonatomic Horn sentences to complete the inference. In actual systems like PROLOG and PLANNER, moreover, the search component is partially under user control, giving him the ability to incorporate some of the kinds of procedural knowledge (or combinatoric advice) referred to earlier. The only purely automatic inference is the retrieval component.

This suggests a different way of looking at the inferential service provided by a KR system (without even taking into account the logical form of the KB). Instead of automatically performing the full deduction necessary to answer questions, a KR system could manage a *limited form of inference* and leave to the rest of the knowledge-based system (or to the user) the responsibility of intelligently completing the inference. As suggested in Frisch and Allen (1982), the idea is to take the "muscle" out of the automatic component and leave the difficult part of reasoning as a problem that the overall system can (meta-)reason about and plan to solve (Genesereth 1983; Smith and Genesereth 1985).

While this may be a promising approach, especially for a KB of a fully general logical form, it does have its problems. First of all, it is far from clear what primitives should be available to a program to extend the reasoning performed by the KR subsystem. It is not as if it were a simple matter to generalize the meager PROLOG control facilities to handle a general theorem prover, for example.[23] The search space in this case seems to be much more complex.

Moreover, it is not clear what the KR service itself should be. If all a KR utility does is perform explicit retrieval over sentences in a KB, it would not be much help. For example, if asked about $(p \vee q)$, it would fail if it only had $(q \vee p)$ in the KB. What we really need is an automatic inferential service that lies somewhere between simple retrieval and full logical inference. But finding such a service that can be motivated *semantically* (the way logical deduction is) and defined independently of how any program actually operates is a nontrivial matter, though we have taken some steps towards this in Levesque (1984c) (and see Patel-Schneider (1985, 1986)).

4.4. Semantic-network form

Semantic networks and similar hierarchic representational frameworks have been in common use in AI for perhaps 20 years. The form of such representations has been apparently dictated by need and a somewhat natural fit to problems under consideration in the field. But, as we shall see next, semantic networks can also be viewed as making a trade of expressive power for a kind of computational tractability.

A first observation about a KB in what we will call "semantic-network form" is that it contains only unary and binary predicates. For example, instead of representing the fact that John's grade in cs100 was 85 by

---

[20] The same is true for the standard analogues. One of the things a map does not allow you to say, for example, is that a river passes through one of two widely separated towns, without specifying which. Similarly, a plastic model of a ship cannot tell us that the ship it represents does not have two smokestacks, without also telling us how many it does have. This is not to say that there is no *uncertainty* associated with an analogue, but that this uncertainty is due to the coarseness of the analogue (e.g., how carefully the map is drawn) rather than to its content.

[21] Notice that it is impossible to state in a KB of this form that $(p \vee q)$ is true without stating which, or that $\exists x \, P(x)$ is true without saying what that x is. However, see the comments below regarding the use of encodings.

[22] In other words, determining if a ground atomic sentence is implied by a collection of Horn sentences (containing function symbols) is undecidable. This is not true, however, if the Herbrand universe is finite, the case that arises almost exclusively in the type of production system used in expert systems. In fact, in the propositional case, it is not hard to prove that the implicit KB can be calculated in linear time.

[23] Though see Stickel (1984) for some ideas in this direction.

Grade(john, cs100, 85)

we would postulate the existence of objects called "grade-assignments" and represent the fact about John in terms of a particular grade-assignment g-a1 as

Grade-assignment(g-a1) $\land$ Student(g-a1,john)
$$\land \text{ Course}(g\text{-}a1,cs100) \land \text{Mark}(g\text{-}a1,85)$$

This part of a KB in semantic-network form is also in database form: a collection of function-free ground atoms, sentences stating the uniqueness of constants, and the closed-world assumption.

The main feature of a semantic net (and of the frame form below), however, is not how individuals are handled, but the treatment of the ("generic") predicates (the unary ones we will call *types*, the binary ones we will call *attributes*[24]). First of all, the types are organized into a taxonomy, which, for our purposes, can be represented by a set of sentences of the form[25]

$$\forall x [B(x) \supset A(x)]$$

Thus the basic skeleton of the taxonomy is provided by a universally quantified conditional, or "*is-a*" connection. For example, "Student IS-A Person" would amount to a statement that all Students are Persons, or

$$\forall x [\text{Student}(x) \supset \text{Person}(x)]$$

The second kind of sentence in the generic KB places a constraint on an attribute as it applies to instances of a type:

$$\forall x [B(x) \supset \exists y (R(x,y) \land V(y))]$$
or
$$\forall x [B(x) \supset R(x,c)]^{26}$$

This latter form corresponds to "value restriction" in KL-ONE and other languages. For example, equating Graduate with "Person with an Undergraduate Degree" would be the equivalent of

$$\forall x [\text{Graduate}(x) \supset \exists y (\text{Degree}(x,y)$$
$$\land \text{ UndergraduateDegree}(y))]$$

This completes the semantic-network form.

One property of a KB in this form is that it can be represented by a labelled directed graph (and displayed in the usual way). The nodes are either constants or types, and the edges are either labelled with an attribute or with the special label *is-a*.[27] The significance of this graphical representation is that it allows certain kinds of inference to be performed by simple graph-searching techniques. For example, to find out if a particular individual has a certain attribute, it is sufficient to search from

the constant representing that individual, up *is-a* links, for a node having an edge labelled with the attribute. By placing the attribute as high as possible in the taxonomy, all individuals below it can *inherit* the property. Computationally, any mechanism that speeds up this type of graph-searching can be used to improve the performance of inference in a KB of this form.

In addition, the graph representation suggests different kinds of inference that are based more directly on the structure of the KB than on its logical content. For example, we can ask how two nodes are related and answer by finding a path in the graph between them. Given, for instance, Clyde the elephant and Jimmy Carter, we could end up with an answer saying that Clyde is an elephant and that the favourite food of elephants is peanuts which is also the major product of a farm owned by Jimmy Carter. A typical method of producing this answer would be to perform a "spreading activation" search beginning at the nodes for Clyde and Jimmy. Obviously, this form of question would be very difficult to answer for a KB that was not in semantic-network form.[28]

For better or worse, the appeal of the graphical nature of semantic nets has led to forms of reasoning (such as default reasoning (Reiter 1978a)) that do not fall into standard logical categories and are not yet very well understood (Etherington and Reiter 1983).[29] This is a case of a representational notation taking on a life of its own and motivating a completely different style of use not necessarily grounded in a truth theory. It is unfortunately much easier to develop an algorithm that appears to reason over structures of a certain kind than to *justify* its reasoning by explaining what the structures are saying about the world.

This is not to say that defaults are not a crucial part of our knowledge about the world. Indeed, the ability to abandon a troublesome or unsuccessful line of reasoning in favour of a default answer intuitively seems to be a fundamental way of coping with incomplete knowledge in the presence of resource limitations. The problem is to make this intuition precise. Paradoxically, the best formal accounts we have of defaults (such as Reiter (1980)) would claim that reasoning with them is even *more difficult* than reasoning without them, so research remains to be done (but see Lifschitz (1985)).

One final observation concerns the elimination of higher arity predicates in semantic networks. It seems to be fairly commonplace to try to sidestep a certain generality of logical form by introducing special representational objects into the domain. In the example above, a special "grade-assignment" object took the place of a 3-place predicate. Another example is the use of encodings of sentences as a way of providing (what appears to be) a completely extensional version of modal logic (Moore 1980).[30] Not that exactly the same expressiveness is preserved in these cases; but what *is* preserved is still fairly

---

[24] We use "type" and "attribute" here for consistency. In some systems (like KL-ONE (Brachman and Schmolze 1985)) the former are called "concepts" and the latter "roles."

[25] See Brachman (1983) for a discussion of some of the subtleties involved here.

[26] There are other forms possible for this constraint. For example, we might want to say that *every* R rather than *some* R is a V. See also Hayes (1979). For the variant we have here, however, note that the KB is no longer in logic-program form.

[27] Note that the interpretation of an edge depends on whether its source and traget are constants or types. For example, from a constant c to a type B, *is-a* means $B(c)$, but from a type B to a type A, it is a taxonomic sentence (again, see Brachman (1983)).

[28] Quillian (1968) proposed a "semantic intersection" approach to answering questions in his original work on semantic nets. See also Collins and Loftus (1975) for follow-up work on the same topic.

[29] A simple example of a default would be to make *elephant* have the colour *grey* but to allow things below elephant (such as *albino-elephant*) to be linked to a different colour value. Determination of the colour of an individual would involve searching up for a value and stopping when the first one is found, allowing it to preempt any higher ones. See also Brachman (1985) and Touretzky (1986).

[30] Indeed, some modern semantic network formalisms (such as Shapiro (1979)) actually include all of FOL by encoding sentences as terms.

mysterious and deserves serious investigation, especially given its potential impact on the tractability of inference.

4.5. Frame-description form

The final form we will consider, the frame-description form, is mainly an elaboration of the semantic-network one. The emphasis, in this case, is on the structure of types themselves (usually called *frames*), particularly in terms of their attributes (called *slots*). Typically, the kind of detail involved with the specification of attributes includes

1. *values*, stating exactly what the attribute of an instance should be. Alternatively, the value may be just a *default*, in which case an individual inherits the value provided he does not override it.

2. *restrictions*, stating what constraints must be satisfied by attribute values. These can be *value* restrictions, specified by a type that attribute values should be instances of, or *number* restrictions, specified in terms of a minimum and a maximum number of attribute values.

3. *attached procedures*, providing procedural advice on how the attribute should be used. An *if-needed* procedure says how to calculate attribute values if none have been specified; an *if-added* procedure says what should be done when a new value is discovered.

Like semantic networks, frame languages tend to take liberties with logical form and the developers of these languages have been notoriously lax in characterizing their truth theories (Brachman 1985; Etherington and Reiter 1983; Hayes 1979). What *we* can do, however, is restrict ourselves to a non-controversial subset of a frame language that supports descriptions of the following form:

> (Student
> with a dept is computer-science and
> with ≥ 3 enrolled-course is a
> (Graduate-Course
> with a dept is a Engineering-Department))

This is intended to be a structured type that describes computer science students taking at least three graduate courses in departments within engineering. If this type had a name (say *A*), we could express the type in FOL by a "meaning postulate" of the form

$$\forall x\, A(x) \equiv [\text{Student}(x) \wedge \text{dept}(x, \text{computer-science}) \wedge$$
$$\exists y_1 y_2 y_3 (y_1 \neq y_2 \wedge y_1 \neq y_3 \wedge y_2 \neq y_3 \wedge$$
$$\text{enrolled-course}(x, y_1) \wedge \text{Graduate-Course}(y_1) \wedge$$
$$\exists z(\text{dept}(y_1, z) \wedge \text{Engineering-Department}(z)) \wedge$$
$$\text{enrolled-course}(x, y_2) \wedge \text{Graduate-Course}(y_2) \wedge$$
$$\exists z(\text{dept}(y_2, z) \wedge \text{Engineering-Department}(z)) \wedge$$
$$\text{enrolled-course}(x, y_3) \wedge \text{Graduate-Course}(y_3) \wedge$$
$$\exists z(\text{dept}(y_3, z) \wedge \text{Engineering-Department}(z)))]$$

Similarly, it should be clear how to state equally clumsily[31] in FOL that an individual is an instance of this type.

One interesting property of these structured types is that we do not have to state explicitly when one of them is below another in the taxonomy. The descriptions themselves implicitly define a taxonomy of *subsumption*, where type A sub-

sumes type B if, by virtue of the form of A and B, every instance of B must be an instance of A. For example, without any world knowledge, we can determine that the type *Person* subsumes

> (Person with every male friend is a Doctor)

which in turn subsumes

> (Person with every friend is a
> (Doctor with a specialty is surgery))

Similarly,

> (Person with ≥2 children)

subsumes

> (Person with ≥3 male children).

Also, we might say that two types are *disjoint* if no instance of one can be an instance of the other. An example of disjoint types is

> (Person with ≥3 young children)

and

> (Person with ≤2 children).

Analytic relationships like subsumption and disjointness are properties of structured types that are not available in a semantic net where all of the types are atomic.

There are very good reasons to be interested in these analytic relationships (Brachman and Levesque 1982). In KRYPTON (Brachman *et al.* 1983, 1985), a full first-order KB is used to represent facts about the world. However, subsumption and disjointness information is made available without having to enlarge the KB with a collection of meaning postulates representing the structure of the types, but rather via a separate "terminological component" based on a language in frame-description form. ·This is significant because, while subsumption and disjointness can be defined in terms of logical implication,[32] there are good special-purpose algorithms for calculating these relationships in KRYPTON's frame-description language.[33] Again, because the logical form is sufficiently constrained, the required inference can be much more tractable.

4.6. A detailed example of the tradeoff

As it turns out, frame-description languages and the subsumption inference provide a rich domain for studying the tradeoff between expressiveness and tractability. To illustrate this, we will consider in some detail a simple frame-description language, which we will call $\mathcal{FL}$.[34]

$\mathcal{FL}$ has the following grammar:

> ⟨*type*⟩ ::= ⟨*atom*⟩
> | (AND ⟨*type₁*⟩ . . . ⟨*typeₙ*⟩)
> | (ALL ⟨*attribute*⟩ ⟨*type*⟩)
> | (SOME ⟨*attribute*⟩)
>
> ⟨*attribute*⟩ ::= ⟨*atom*⟩
> | (RESTRICT ⟨*attribute*⟩ ⟨*type*⟩)

---

[31] What makes these sentences especially awkward in FOL is the number restrictions. For example, the sentence *"There are a hundred billion stars in the Milky Way Galaxy"* would be translated into an FOL sentence with about 10^{22} conjuncts.

[32] Specifically, type A subsumes type B iff the meaning postulates for A and B logically imply the sentence, $\forall x[B(x) \supset A(x)]$.

[33] In particular, see the next section. Also, see Stickel (1985) for details on speedups achieved in this fashion.

[34] Please note that the style of this section constitutes a significant departure from that of previous sections, presenting enough technical detail to make the point truly concrete.

19

Intuitively, we think of types in $\mathscr{FL}$ as representing (sets of) individuals, and attributes as representing relations between individuals.

While the linear syntax is a bit unorthodox, $\mathscr{FL}$ is actually a distillation of the operators in typical frame languages; in particular, it is the frame-description kernel derived from years of experience with languages like KL-ONE (Brachman and Schmolze 1985) and KRYPTON (Brachman and Levesque 1982; Brachman et al. 1983, 1985):

• Atoms are the names of primitive (undefined) types.
• AND constructions represent conjoined types, so, for example, (AND adult male person) would represent the concept of something that was at the same time an adult, a male, and a person (i.e., a man). In general, x is an (AND $t_1 t_2 \ldots t_n$) iff x is a t_1 and a t_2 and ... and a t_n. This allows us to put several properties (i.e., supertypes or attribute restrictions) together in the definition of a type.
• The ALL construct provides a type-restriction on the values of an attribute (x is an (ALL a t) iff each a of x is a t). Thus (ALL child doctor) corresponds to the concept of something all of whose children are doctors. It is a way to *restrict* the value of a slot at a frame (a "value restriction" in KL-ONE).
• The SOME operator guarantees that there will be at least one value for the attribute named (x is a (SOME a) iff x has at least one a). For instance, (AND person (SOME child)) would represent the concept of a parent. This is a way to *introduce* a slot at a frame.

Note that in the more common frame languages, the ALL and SOME are not broken out as separate operators, but instead, either every attribute restriction is considered to have *both* universal and existential import, or exclusively one or the other (or it may even be left unspecified).[35] Our language allows for arbitrary numbers of attribute values, and allows the SOME and ALL restrictions to be specified independently.

• Finally, the RESTRICT construct accounts for attributes constrained by the types of their values, e.g., (RESTRICT child male) for a child who is a male, that is, a son (in general, y is a (RESTRICT a t) of x iff y is an a of x and y is a t).

The $\mathscr{FL}$ language can be considered a simplified (though less readable) version of the frame-based language used in the previous section. So, for example, where we would previously have written a description like

(person **with every** male friend **is a**
    (doctor **with a** specialty))

the equivalent $\mathscr{FL}$ type is written as

(AND person (ALL (RESTRICT friend male)
    (AND doctor (SOME specialty)))))

To specify exactly what these constructs mean, we now briefly define a straightforward extensional semantics for $\mathscr{FL}$. As a result, we will provide a precise definition of subsumption. This will be done as follows: imagine that associated with each description is the set of individuals (individuals for

types, pairs of individuals for attributes) that it describes. Call that set the *extension* of the description. Notice that by virtue of the structure of descriptions, their extensions are not independent (for example, the extension of (AND $t_1 t_2$) should be the intersection of those of t_1 and t_2). In general, the structures of two descriptions can imply that the extension of one is always a superset of the extension of the other. In that case, we will say that the first *subsumes* the second (so, in the case just mentioned, t_1 would be said to subsume (AND $t_1 t_2$)).

More formally, let $\mathscr{D}$ be any set and $\mathscr{E}$ be any function from types to subsets of $\mathscr{D}$ and attributes to subsets of the Cartesian product, $\mathscr{D} \times \mathscr{D}$. So

$$\mathscr{E}[t] \subseteq \mathscr{D} \qquad \text{for any type } t$$

and

$$\mathscr{E}[a] \subseteq \mathscr{D} \times \mathscr{D} \qquad \text{for any attribute } a$$

We will say that $\mathscr{E}$ is an *extension function* over $\mathscr{D}$ if and only if

1. $\mathscr{E}[(\text{AND } t_1 \ldots t_n)] = \cap_i \mathscr{E}[t_i]$
2. $\mathscr{E}[(\text{ALL } a\ t)] = \{x \in \mathscr{D} | \forall y \text{ if } \langle x, y \rangle \in \mathscr{E}[a] \text{ then } y \in \mathscr{E}[t]\}$
3. $\mathscr{E}[(\text{SOME } a)] = \{x \in \mathscr{D} | \exists y [\langle x, y \rangle \in \mathscr{E}[a]]\}$
4. $\mathscr{E}[(\text{RESTRICT } a\ t)] = \{\langle x, y \rangle \in \mathscr{D} \times \mathscr{D} | \langle x, y \rangle \in \mathscr{E}[a] \text{ and } y \in \mathscr{E}[t]\}$

Finally, for any two types t_1 and t_2, we can say that t_1 *is subsumed by* t_2 if and only if for any set $\mathscr{D}$ and any extension function $\mathscr{E}$ over $\mathscr{D}$, $\mathscr{E}[t_1] \subseteq \mathscr{E}[t_2]$. That is, one type is subsumed by a second type when all instances of the first—in all extensions—are also instances of the second. From a semantic point of view, subsumption dictates a kind of necessary set inclusion.

Given a precise definition of subsumption, we can now consider algorithms for calculating subsumption between descriptions. Intuitively, this seems to present no real problems. To determine if s subsumes t, what we have to do is make sure that each component of s is "implied" by some component (or componen) of t. Moreover, the type of "implication" we need should be fairly simple since $\mathscr{FL}$ has neither a negation nor a disjunction operator.

Unfortunately, such intuitions can be nastily out of line. In particular, let us consider a slight variant of $\mathscr{FL}$—call it $\mathscr{FL}^-$. $\mathscr{FL}^-$ includes all of $\mathscr{FL}$ except for the RESTRICT operator. On the surface, the difference between $\mathscr{FL}^-$ and $\mathscr{FL}$ seems expressively minor.[36] But it turns out that it is computationally very significant. In particular, we have found an $O(n^2)$ algorithm for determining subsumption in $\mathscr{FL}^-$, but have proven that the same problem for $\mathscr{FL}$ is intractable. In the rest of this section, we sketch the form of our algorithm for $\mathscr{FL}^-$ and the proof that subsumption for $\mathscr{FL}$ is as hard as testing for propositional tautologies, and therefore most likely unsolvable in polynomial time. A more formal version of the algorithm and the full proofs can be found in the Appendix.

4.7. Subsumption algorithm for $\mathscr{FL}^-$

1. Flatten both arguments s and t by removing all nested

---

[35] See Hayes (1979) for some further discussion of the import of languages like KRL. As it turns out, the universal/existential distinction is most often moot, because most frame languages allow only single-valued slots. Thus the slot's meaning is reduced to a simple predication on a single-valued function (e.g., the slot/value pair age : integer means $integer(age(x))$).

[36] It is the case, however, that there are concepts that can be expressed in $\mathscr{FL}$ that cannot be expressed in $\mathscr{FL}^-$, such as the concept of a person with at least one son and at least one daughter: (AND (SOME (RESTRICT *child male*)) (SOME (RESTRICT *child female*))). In $\mathscr{FL}^-$ all attributes are primitive, so sons and daughters cannot play the same role (child) and yet be distinguished by their types.

AND operators. So, for example,

(AND x (AND y z) w)   becomes   (AND x y z w)

2. Collect all arguments to an ALL for a given attribute. For example,

(AND (ALL a (AND u v w)) x (ALL a (AND y z)))

becomes

(AND x (ALL a (AND u v w y z)))

3. Assuming s is now (AND $s_1 \ldots s_n$) and t is (AND $t_1 \ldots t_m$), then return T iff for each s_i,
(a) if s_i is an atom or a SOME, then one of the t_j is s_i.
(b) if s_i is (ALL a x), then one of the t_j is (ALL a y), where x subsumes y, calculated recursively.

This algorithm can be shown to compute subsumption correctly (see Appendix A.2.2, Lemma 10). For the purposes of this paper, the main property of the algorithm that we are interested in is that it can be shown to calculate subsumption for $\mathcal{FL}^-$ in $O(n^2)$ time (where n is the length of the longest argument, say). This can be shown roughly as follows (see Appendix A.2.1, Lemma 9 for details): Step 1 can be done in linear time. Step 2 might require a traversal of the expression for each of its elements, and Step 3 might require a traversal of t for each element of s, but both of these can be done in $O(n^2)$ time.

We now turn our attention to the subsumption problem for full $\mathcal{FL}$. The proof that subsumption of descriptions in $\mathcal{FL}$ is intractable is based on a correspondence between this problem and the problem of deciding whether a sentence of propositional logic is implied by another. Specifically, we define a mapping π (see Appendix A.1) from propositional sentences in conjunctive normal form to descriptions in $\mathcal{FL}$ that has the property that for any two sentences α and β, α logically implies β iff $\pi[\alpha]$ is subsumed by $\pi[\beta]$. π itself can be calculated quickly.

What this mapping provides is a way of answering questions of implication by first mapping the two sentences into descriptions in $\mathcal{FL}$ and then seeing if one is subsumed by the other. Moreover, because π can be calculated efficiently, any good algorithm for subsumption becomes a good one for implication.

The key observation here, however, is that there can be no good algorithm for implication. To see this, note that a sentence implies $(p \wedge \neg p)$ just in case it is not satisfiable. But determining the satisfiability of a sentence in this form is NP-complete (Cook 1971). Therefore, a special case of the implication problem (where the second argument is $(p \wedge \neg p)$) is the complement of an NP-complete one. The correspondence between implication and subsumption, then, leads to the observation that subsumption over $\mathcal{FL}$ is co-NP hard. In other words, since a good algorithm for subsumption would lead to a good one for implication, subsumption over descriptions in $\mathcal{FL}$ is intractable.[37]

5. Conclusions and morals

In this final section, we step back from the details of the specific representational formalisms we have examined and attempt to draw a few conclusions.

An important observation about these formalisms is that we

cannot really say that one is *better* than any other; they simply take different positions on the tradeoff between expressiveness and tractability. For example, full FOL is both more expressive and less appealing computationally than a language in semantic-net form. Nor is it reasonable to say that expressiveness is the primary issue and that the other is "merely" one of efficiency. In fact, we are not really talking about efficiency here at all; that, presumably, is an issue of algorithm and data structure, concerns of the Symbol Level (Newell 1981). The tractability concern we have here is much deeper and involves whether or not it makes sense to even think of the language as computationally based.

From the point of view of those doing research in KR, this has a very important consequence: We should continue to design and examine representation languages, *even when these languages can be viewed as special cases of FOL*. What really counts is for these special cases to be interesting both from the point of view of what they can represent, and from the point of view of the reasoning strategies they permit. All of the formalisms we have examined above satisfy these two requirements. To dismiss a language as *just* a subset of FOL is probably as misleading as dismissing the notion of a context-free grammar as just a special case of a context-sensitive one.

What truth in advertising does require, however, is that these special cases of FOL be identified as such. Apart from allowing a systematic comparison of representation languages (as positions on the tradeoff), this might also encourage us to consider systems that use more than one sublanguage and reasoning mechanism (as suggested for equality in Nelson and Oppen (1979)). The KRYPTON language (Brachman *et al.* 1983, 1985), for example, includes all of FOL *and* a frame-description language. To do the necessary reasoning, the system contains both a theorem prover and a description subsumption mechanism, even though the former could do the job of the latter[38] (but much less efficiently). The trick with these *hybrid systems* is to factor the reasoning task so that the specialists are able to cooperate and apply their optimized algorithms without interfering with each other.

These considerations for designers of representation languages apply in a similar way to those interested in populating a KB with a theory of some sort. A good first step might be to write down a set of first-order sentences characterizing the domain, but it is somewhat naive to stop there and claim that the account could be made computational after the fact by the inclusion of a theorem prover and a few well-chosen heuristics. What is really needed is the (much more difficult) analysis of the logical form of the theory, keeping the tradeoff clearly in mind. An excellent example of this is the representation of *time* described in Allen (1983). Allen is very careful to point out what kind of information about time cannot be represented in his system, as well as the computational advantage he gains from this limitation.

It should be noted here that we have addressed only one approach to dealing with the tradeoff. While the tactic of limiting the form of a representation[39] seems to account for almost all current practice in knowledge representation, other ways of

---

[37] As mentioned in Section. 3.2, the co-NP-complete problems are strongly believed to be unsolvable in polynomial time.

[38] This is true only to a certain extent. See Footnote 9, Brachman and Levesque (1982), and Brachman *et al.* (1983, 1985).

[39] Note that "restricting the form" does not confine us to only simple, obvious types of restrictions. Useful forms of limited languages may have no obvious syntactic relationship to standard logical languages.

avoiding undue complexity should be considered. Especially worthy of attention are weaker logics, having expressive languages but limited power to make inferences (see, for example, Patel-Schneider (1985, 1986) and Levesque (1986)). Another tack to take is to use assumptions as much as possible to produce a tractable, "vivid" knowledge base (Levesque 1986) (being subsequently prepared to undo the effects of assumptions that turn out to be unwarranted).

Finally, one should be aware that the issues addressed here are significant only when concerned with serious scaling up of representations. If there are only a small number of complex sentences (i.e., involving only a small amount of incompleteness), then the tradeoff is a manageable issue. Here we are looking toward representation systems capable of rivalling current database management systems in the number of items stored.

For the future, we still have a lot to learn about the tradeoff. It would be very helpful to accumulate a wide variety of data points involving tractable and intractable languages.[40] Especially significant are crossover points where small changes in a language change its computational character completely (such as that illustrated in Sect. 4.6). Moreover, we need to know more about what *people* find easy or hard to handle. There is no doubt that people can reason when necessary with radically incomplete knowledge (such as that expressible in full FOL) but apparently only by going into a special problem-solving or logic puzzle mode. In normal commonsense situations, when reading a geography book, for instance, the ability to handle disjunctions (say) seems to be quite limited. The question is what forms of incomplete knowledge *can* be handled readily, given that the geography book is not likely to contain any procedural advice on how to reason.

In summary, we feel that there are many interesting issues to pursue involving the tradeoff between expressiveness and tractability. Although there has always been a temptation in KR to set the sights either too low (and provide only a data-structuring facility with little or no inference) or too high (and provide a full theorem-proving facility), this paper argues for the rich world of representation that lies between these two extremes. We should not despair that no matter what we try to do we are faced with intractability, but rather move ahead with the investigation of ways to integrate limited forms of languages and reasoning, with the goal of forging a powerful system out of tractable parts.

Acknowledgements

Many of the ideas presented here originally arose in the context of the KRYPTON project (undertaken mainly at the AI Lab of Schlumberger Palo Alto Research). We are indebted to SPAR for making this research possible and to Richard Fikes, Peter Patel-Schneider, and Victoria Gilbert for major contributions. We would especially like to thank Peter, and also Jim des Rivières and David Etherington for providing very helpful comments on drafts of this paper, and S. J. Hurtubise for not.

ALLEN, J. 1983 Maintaining knowledge about temporal intervals. Communications of the ACM, **26**: 832–843.
BRACHMAN, R. J. 1983. What IS-A is and isn't: an analysis of tax-

onomic links in semantic networks. IEEE Computer, **16**(10): 30–36.
——— 1985. I Lied about the Trees. AI Magazine, **6**(3): 80–93.
BRACHMAN, R. J., and LEVESQUE, H. J. 1982. Competence in knowledge representation. Proceedings of the National Conference on Artificial Intelligence, Pittsburgh, PA, pp. 189–192.
BRACHMAN, R. J., and SCHMOLZE, J. G. 1985. An overview of the KL-ONE knowledge representation system. Cognitive Science, **9**(2): 171–216.
BRACHMAN, R. J., FIKES, R. E., and LEVESQUE, H. J. 1983. Krypton: a functional approach to knowledge representation. IEEE Computer, **16**(10): 67–73.
BRACHMAN, R. J., GILBERT, V. P., and LEVESQUE, H. J. 1985. An essential hybrid reasoning system: knowledge and symbol level accounts of KRYPTON. Proceedings of the International Joint Conference on Artificial Intelligence 1985, Los Angeles, CA, pp. 532–539.
COLLINS, A. M., and LOFTUS, E. F. 1975. A spreading-activation theory of semantic processing. Psychological Review, **82**: 407–428.
COOK, S. A. 1971. The complexity of theorem-proving procedures. Proceedings of the 3rd Annual ACM Symposium on Theory of Computing. Association for Computing Machinery, New York, NY, pp. 151–158.
ETHERINGTON, D., and REITER, R. 1983. On inheritance hierarchies with exceptions. Proceedings of the National Conference on Artificial Intelligence, Washington, DC, pp. 104–108.
FRISCH, A., and ALLEN, J. 1982. Knowledge representation and retrieval, for natural language processing. TR 104, Computer Science Department, University of Rochester, Rochester, NY.
GENESERETH, M. R. 1983. An overview of meta-level architecture. Proceedings of the National Conference on Artificial Intelligence, Washington, DC, pp. 119–123.
HAYES, P. J. 1979. The logic of frames. In Frame conceptions and text understanding. Edited by D. Metzing. Walter de Gruyter and Company, Berlin, West Germany, pp. 46–61.
LEVESQUE, H. J. 1982. A formal treatment of incomplete knowledge bases. Technical Report No. 3, Fairchild Laboratory for Artificial Intelligence Research, Palo Alto, CA.
——— 1984a. The logic of incomplete knowledge bases. In On conceptual modelling: perspectives from artificial intelligence, databases, and programming languages. Edited by M. L. Brodie, J. Mylopoulos, and J. Schmidt. Springer-Verlag, New York, NY, pp. 165–186.
——— 1984b. Foundations of a functional approach to knowledge representation. Artificial Intelligence, **23**: 155–212.
——— 1984c. A logic of implicit and explicit belief. Proceedings of the National Conference on Artificial Intelligence, Austin, TX, pp. 198–202.
——— 1986. Making believers out of computers. Artificial Intelligence, **30**: 81–108.
LIFSCHITZ, V. 1985. Computing circumscription. Proceedings of the International Joint Conference on Artificial Intelligence 1985, Los Angeles, CA, pp. 121–127.
MOORE, R. C. 1980. Reasoning about knowledge and action. Technical Note 191, SRI International, Menlo Park, CA.
——— 1982. The role of logic in knowledge representation and commonsense reasoning. Proceedings of the National Conference on Artificial Intelligence, Pittsburgh, PA, pp. 428–433.
NELSON, G., and OPPEN, D. C. 1979. Simplification by cooperating decision procedures. ACM Transactions on Programming Languages and Systems, **1**: 245–257.
NEWELL, A. 1981. The knowledge level. AI Magazine, **2**(2): 1–20.
PATEL-SCHNEIDER, P. F. 1985. A decidable first-order logic for knowledge representation. Proceedings of the International Joint Conference on Artificial Intelligence 1985, Los Angeles, CA, pp. 455–458.
——— 1986. A four-valued semantics for frame-based description languages. Proceedings of the National Conference on Artificial

---

[40] Some progress has been made on this already at the University of Toronto (Gullen, A., M.Sc. thesis, Department of Computer Science, University of Toronto, Toronto, Ontario, in preparation).

Intelligence, Philadelphia, PA, pp. 344–348.

QUILLIAN, M. R. 1968. Semantic memory. *In* Semantic Information Processing. *Edited by* M. Minsky. MIT Press, Cambridge, MA. pp. 227–270.

REITER, R. 1978a. On reasoning by default. Proceedings of Theoretical Issues in Natural Language Processing-2, University of Illinois at Urbana-Champaign, Urbana-Champaign, IL, pp. 210–218.

—— 1978b. On closed world data bases. *In* Logic and Data Bases. *Edited by* H. Gallaire and J. Minker. Plenum Press, New York, NY, pp. 55–76.

—— 1980. A logic for default reasoning. Artificial Intelligence, 13: 81–132.

SHAPIRO, S. C. 1979. The SNePS semantic network processing system. *In* Associative networks: representation and use of knowledge by computers. *Edited by* N. V. Findler. Academic Press, New York, NY, pp. 179–203.

SMITH, B. C. 1982. Reflection and semantics in a procedural language. Ph.D. thesis and Technical Report MIT/LCS/TR-272, MIT, Cambridge, MA.

SMITH, D. E., and GENESERETH, M. R. 1985. Ordering conjunctive queries. Artificial Intelligence, 26: 171–215.

STICKEL, M. E. 1984. A Prolog technology theorem prover. Proceedings of the 1984 Symposium on Logical Programming, Atlantic City, NJ, pp. 211–217.

—— 1985. Automated deduction by theory resolution. Proceedings of the International Joint Conference on Artificial Intelligence 1985, Los Angeles, CA, pp. 1181–1186.

TOURETZKY, D. 1986. The mathematics of inheritance systems. Morgan Kaufmann Publishers, Inc., Los Altos, CA.

WINKER, S. 1982. Generation and verification of finite models and counterexamples using an automated theorem prover answering two open questions. Journal of the ACM, 29: 273–284.

WOS, L., WINKER, S., SMITH, B., VEROFF, R., and HENSCHEN, L. 1984. A new use of an automated reasoning assistant: open questions in equivalential calculus and the study of infinite domains. Artificial Intelligence, 22: 303–356.

Appendix: proofs

In this appendix, we present the details of the proofs of the complexity of subsumption for the languages $\mathcal{FL}$ and $\mathcal{FL}^-$. First, we treat the intractability of $\mathcal{FL}$ by showing a direct relation between subsumption in the language and satisfiability in propositional logic. Because propositional satisfiability is difficult, we determine that there can be no good algorithm for subsumption in $\mathcal{FL}$. Subsequently, we show that there exists a sound and complete algorithm for computing subsumption in $\mathcal{FL}^-$ that operates in $O(n^2)$ time.

A.1. The intractability of subsumption for $\mathcal{FL}$

We prove the intractability of subsumption for $\mathcal{FL}$ by showing that for any two propositional formulas α and β, in conjunctive normal form (CNF), there are types in $\mathcal{FL}$, $\pi[\alpha]$, and $\pi[\beta]$ such that

$$\models (\alpha \supset \beta) \text{ iff } \pi[\alpha] \text{ is subsumed by } \pi[\beta]$$

where $\pi[\alpha]$ is roughly the same length as α. Thus, a good algorithm for subsumption in $\mathcal{FL}$ would imply a good algorith for CNF implication. However, CNF implication is difficult, and thus there can be no good algorithm for subsumption.

We being our proof with a lemma relating CNF implication to propositional satisfiability.

Lemma 1

CNF implication is co-NP-hard.

Proof

$\models (\alpha \supset [p \wedge \neg p])$ iff α is unsatisfiable, since otherwise, if

$v(\alpha) = T$ and $\models (\alpha \supset [p \wedge \neg p])$ then $v(p \wedge \neg p)$ would have to be T, which is impossible. But determining if α is satisfiable is NP-hard (Cook 1971). So, determining if $\models (\alpha \supset [p \wedge \neg p])$ is co-NP-hard. Since determining if $\models (\alpha \supset [p \wedge \neg p])$ is a special case of determining if $\models (\alpha \supset \beta)$, then the latter (CNF implication) is co-NP-hard. ∎

Next we define the mapping, π, which takes formulas of propositional logic in CNF into types of $\mathcal{FL}$. We assume throughout that clauses do not use the special propositional letters *SELF* or *BOTTOM*.

For any clause c, where $c = (p_1 \vee \ldots \vee p_n \vee \neg p_{n+1} \vee \ldots \vee \neg p_{n+k})$, define $\pi[c] =$

```
(AND (ALL (RESTRICT SELF p₁) BOTTOM)
          (ALL (RESTRICT SELF pₙ) BOTTOM)
          (SOME (RESTRICT SELF pₙ₊₁))
            ...
          (SOME (RESTRICT SELF pₙ₊ₖ)))
```

For any well-formed formula (wff) α in CNF, where $\alpha = (c_1 \wedge \ldots \wedge c_m)$, define $\pi[\alpha] =$

```
(AND (ALL (RESTRICT SELF (SOME (RESTRICT
                               SELF BOTTOM))) BOTTOM)
          (ALL (RESTRICT SELF π[c₁]) BOTTOM)
            ...
          (ALL (RESTRICT SELF π[cₘ]) BOTTOM))
```

Note that for simplicity we will treat single clauses as degenerate conjunctions (so that a wff that is a single clause will be mapped by π into a type that begins with (AND (ALL (RESTRICT *SELF* (SOME (RESTRICT *SELF BOTTOM*))) *BOTTOM*) ...).

Before giving an example of the mapping π, let us introduce a notational convention: Let (NEG p) stand for (ALL (RESTRICT *SELF* p) *BOTTOM*) and (POS p) stand for (SOME (RESTRICT *SELF* p)). So, if, for example, $\alpha = (p \vee q \vee \neg r) \wedge (s \vee \neg t) \wedge (u \vee v)$, then $\pi[a]$ is

```
(AND (NEG (POS BOTTOM))
          (NEG (AND (NEG p) (NEG q) (POS r)))
          (NEG (AND (NEG s) (POS t)))
          (NEG (AND (NEG u) (NEG v)))).
```

A.1.1. From subsumption to implication

Now, define the following extension function, $\mathcal{C}_0$, over a domain, $\mathcal{D}_0$:

$\mathcal{D}_0 = [\text{LETTERS} \rightarrow \{T,F\}]$ (that is, the domain is all functions that take propositional letters into T and F).

$\mathcal{C}_0[BOTTOM] = \{ \}$

$\mathcal{C}_0[SELF] = \{\langle v, v \rangle | v \in \mathcal{D}_0\}$.

$\mathcal{C}_0[p] = \{v | v(p) = T\}$.

Lemma 2

(a) $v(p) = T$ iff $v \in \mathcal{C}_0(\text{POS } p)$.

(b) $v(p) = F$ iff $v \in \mathcal{C}_0(\text{NEG } p)$.

Proof

(a) $v(p) = T$ iff $v \in \mathcal{C}_0[p]$ iff
$\langle v, v \rangle \in \mathcal{C}_0[(\text{RESTRICT } SELF \ p)]$ iff
$v \in \mathcal{C}_0[(\text{POS } p)]$.

(b) $v(p) = F$ iff $v \notin \mathcal{C}_0[p]$ iff
$\langle v, v \rangle \notin \mathcal{C}_0[(\text{RESTRICT } SELF \ p)]$ iff
$v \in \mathcal{C}_0[(\text{ALL } (\text{RESTRICT } SELF \ p) \ BOTTOM)]$
(since $\mathcal{C}_0[BOTTOM] = \{ \}$)
iff $v \in \mathcal{C}_0[(\text{NEG } p)]$. ∎

Lemma 3

For any clause c, and any valuation v, $v(c) = $ F iff $v \in \mathcal{E}_0[\pi[c]]$.

Proof

$v(p_1 \vee \ldots \vee p_n \vee \neg p_{n+1} \vee \ldots \vee \neg p_{n+k}) = $ F iff
$\forall i (1 \leq i \leq n) \; v(p_i) = $ F and $\forall i (1 \leq i \leq k) \; v(p_{n+i}) = $ T iff
$\forall i (1 \leq i \leq n) \; v \in \mathcal{E}_0[(\text{NEG } p_i)]$ and
$\forall i (1 \leq i \leq k) \; v \in \mathcal{E}_0[(\text{POS } p_{n+i})]$ by Lemma 2) iff
$v \in \mathcal{E}_0[(\text{AND (NEG } p_i) \ldots (\text{NEG } p_n) (\text{POS } p_{n+1})$
$\ldots (\text{POS } p_{n+k}))]$ iff
$v \in \mathcal{E}_0[\pi[p_1 \vee \ldots \vee p_n \vee \neg p_{n+1} \vee \ldots \vee \neg p_{n+k}]]$.   ∎

Lemma 4

$\mathcal{E}_0[(\text{NEG (POS } BOTTOM))] = \mathcal{D}_0$.

Proof

Let $v \in \mathcal{D}_0$. Then $v \notin \mathcal{E}_0[BOTTOM]$. So $\langle v, v \rangle \notin$
$\mathcal{E}_0[(\text{RESTRICT } SELF \; BOTTOM)]$, so $v \notin \mathcal{E}_0[(\text{POS } BOTTOM)]$, so $\langle v, v \rangle \notin \mathcal{E}_0[(\text{RESTRICT } SELF \; (\text{POS } BOTTOM))]$, so $v \in \mathcal{E}_0[(\text{NEG (POS } BOTTOM))]$.   ∎

Lemma 5

If $\pi[\alpha]$ is subsumed by $\pi[\beta]$ then $\models \alpha \supset \beta$.

Proof

Assume $\pi[\alpha]$ is subsumed by $\pi[\beta]$ and $v(\alpha) = $ T, where v is any valuation. So, for each clause c_i of α, $v(c_i) = $ T. Thus, by Lemma 3, $v \notin \mathcal{E}_0[\pi[c_i]]$. So, $\langle v, v \rangle \notin \mathcal{E}_0[(\text{RESTRICT } SELF \; \pi[c_i])]$, and so, $v \in \mathcal{E}_0[(\text{NEG } \pi[c_i])]$. Also, by Lemma 4, $v \in \mathcal{E}_0[(\text{NEG (POS } BOTTOM))]$. Thus, $v \in \mathcal{E}_0[(\text{AND (NEG (POS } BOTTOM)) (\text{NEG } \pi[c_1]) \ldots (\text{NEG } \pi[c_m]))]$; that is, $v \in \mathcal{E}_0[\pi[\alpha]]$. Since $\pi[\alpha]$ is subsumed by $\pi[\beta]$, $v \in \mathcal{E}_0[\pi[\beta]]$. So $v \in \mathcal{E}_0[(\text{NEG } \pi[d_i])]$, for each clause d_i of β. Thus, $v \notin \mathcal{E}_0[\pi[d_i]]$, and by Lemma 3, $v(d_i) = $ T. Thus $v(\beta) = $ T. So for any v, if $v(\alpha) = $ T then $v(\beta) = $ T, and so $\models (\alpha \supset \beta)$.   ∎

A.1.2. From implication to subsumption

Given an extension function $\mathcal{E}$ over $\mathcal{D}$ and an element $d \in \mathcal{D}$, define $v_d \in [\text{LETTERS} \rightarrow \{\text{T,F}\}]$ by

$v_d(p) = $ T iff $d \in \mathcal{E}[(\text{POS } p)]$.

Lemma 6

If $\mathcal{E}$ is an extension function over $\mathcal{D}$ and $d \in \mathcal{D}$ where $d \notin \mathcal{E}[(\text{POS } BOTTOM)]$, then for any clause c, $v_d(c) = $ F iff $d \in \mathcal{E}[\pi[c]]$.

Proof

Since $d \notin \mathcal{E}[(\text{POS } BOTTOM)]$, $\forall d^* \langle d, d^* \rangle \in \mathcal{E}[SELF] \Rightarrow d^* \notin \mathcal{E}(BOTTOM)$. So, $\forall d^* \langle d, d^* \rangle \in \mathcal{E}[SELF] \Rightarrow d^* \notin \mathcal{E}[p]$ iff $\forall d^* \langle d, d^* \rangle \in \mathcal{E}[SELF] \Rightarrow d^* \notin \mathcal{E}[p]$ or $d^* \in \mathcal{E}[BOTTOM]$. Thus, $d \notin \mathcal{E}[(\text{POS } p)]$ iff $d \in \mathcal{E}[(\text{NEG } p)]$. Now, by definition of v_d, $v_d(p_i) = $ T iff $d \in \mathcal{E}[(\text{POS } p)]$ and so $v_d(p_i) = $ F iff $d \in \mathcal{E}[(\text{NEG } p)]$. So $v_d(p_1 \vee \ldots \vee p_n \vee \neg p_{n+1} \vee \ldots \vee \neg p_{n+k}) = $ F iff $d \in \mathcal{E}[(\text{AND (NEG } p_1) \ldots (\text{NEG } p_n) (\text{POS } p_{n+1}) \ldots (\text{POS } p_{n+k}))]$; that is, $d \in \mathcal{E}[\pi[p_1 \vee \ldots \vee p_n \vee \neg p_{n+1} \vee \ldots \vee \neg p_{n+k}]]$.   ∎

Lemma 7

If $\models (\alpha \supset \beta)$ then $\pi[\alpha]$ is subsumed by $\pi[\beta]$.

Proof

Suppose $\models (\alpha \supset \beta)$ and $\mathcal{E}$ is any extension function over some $\mathcal{D}$. Suppose $x \in \mathcal{E}[\pi[\alpha]]$. Then

1. $x \in \mathcal{E}[(\text{NEG (POS } BOTTOM))]$, so if $\langle x, y \rangle \in \mathcal{E}[SELF]$ and $y \notin \mathcal{E}[BOTTOM]$, then $y \notin \mathcal{E}[(\text{POS } BOTTOM)]$.

2. $x \in \mathcal{E}[(\text{NEG } \pi[c_i])]$ for each c_i in α, so if $\langle x, y \rangle \in \mathcal{E}[SELF]$ and $y \notin \mathcal{E}[BOTTOM]$, then $y \notin \mathcal{E}[\pi[c_i]]$.

Let d_j be any clause of β and suppose $y \notin \mathcal{E}[BOTTOM]$ and $\langle x, y \rangle \in \mathcal{E}[SELF]$. By (1), $y \notin \mathcal{E}[(\text{POS } BOTTOM)]$; by (2), $y \notin \mathcal{E}[\pi[c_i]]$ for every c_i in α, and so by Lemma 6, $v_y(c_i) = $ T. Thus, $v_y(\alpha) = $ T. But $\models (\alpha \supset \beta)$, so $v_y(\beta) = $ T and so $v_y(d_j) = $ T. Then, by Lemma 6 again, $y \notin \mathcal{E}[\pi[d_j]]$. So, if $\langle x, y \rangle \in \mathcal{E}[SELF]$ and $y \in \mathcal{E}[\pi[d_j]]$, then $y \in \mathcal{E}[BOTTOM]$. Thus $x \in \mathcal{E}[(\text{NEG } \pi[d_j])]$ and overall, $x \in \mathcal{E}[\pi[\beta]]$. Since this applies to any $\mathcal{E}$ and any x, $\pi[\alpha]$ is subsumed by $\pi[\beta]$.   ∎

Theorem 1

Subsumption of $\mathcal{FL}$ is co-NP-hard.

Proof

Consider the special case of determining if $\pi[\alpha]$ is subsumed by $\pi[\beta]$. By Lemmas 5 and 7, this is true iff $\models (\alpha \supset \beta)$. But by Lemma 1, this problem is co-NP-hard. Since the size of the expressions are within a polynomial of each other, the first problem is co-NP-hard as well.   ∎

A.2. The tractability of subsumption for $\mathcal{FL}^-$

Our proof of the tractability of subsumption for $\mathcal{FL}^-$ will proceed as follows: first we provide an alternative, "flat" form for types of $\mathcal{FL}^-$, which makes the definition of a subsumption algorithm straightforward. We then show that the combination of translation of types into the flat form, coupled with the subsumption algorithm for flat types, yields an algorithm that operates in $O(n^2)$ time. Finally, we prove that the algorithm presented does indeed compute subsumption.

A.2.1. Complexity of the subsumption algorithm

Define a subset of $\mathcal{FL}^-$ as follows:

Definition

A type t of $\mathcal{FL}^-$ is a *flat type* iff it is of the form, (AND $t_1 \ldots t_n$), where each t_i is a *flat factor*. A type is a flat factor iff it is atomic, or of the form (SOME a), or of the form (ALL $a \; t$), where t is a flat type. In addition, we assume that $t_i \neq t_j$ for $i \neq j$, and that if $t_i = (\text{ALL } a \; u)$ and $t_j = (\text{ALL } b \; v)$ where $i \neq j$, then $a \neq b$.

Lemma 8

If t is a type, then there is an $O(n^2)$ algorithm that converts t to a flat type t' such that $\mathcal{E}[t] = \mathcal{E}[t']$ for any $\mathcal{E}$, and t' is not longer than t.

Proof

First replace (AND x (AND y) z) by (AND $x \; y \; z$), working from the inside out. This clearly does not change any extensions and can be done in linear time. Next, collect arguments to all ALL types, replacing

(AND $\ldots$ (ALL $a \ldots$) $\ldots$ (ALL $a \ldots$) $\ldots$)

everywhere by

(AND $\ldots$ (ALL a (AND $\ldots \ldots$)) $\ldots$),

which requires traversing the type at most once per factor. Moreover, this preserves extensions since

$\mathcal{E}[(\text{AND (ALL } a \; t) (\text{ALL } a \; u))]$
$= \{x \mid \forall y \; \langle x, y \rangle \in \mathcal{E}[a] \Rightarrow y \in \mathcal{E}[t], \text{ and } \forall y \; \langle x, y \rangle \in \mathcal{E}[a] \Rightarrow y \in \mathcal{E}[u]\}$
$= \{x \mid \forall y \; \langle x, y \rangle \in \mathcal{E}[a] \Rightarrow y \in \mathcal{E}[t] \text{ and } y \in \mathcal{E}[u]\}$
$= \mathcal{E}[(\text{AND (ALL } a \; (\text{AND } t \; u)))]$.   ∎

The algorithm for subsumption given flat types is as follows:

SUBS?[(AND $x_1, x_2, \ldots x_n$), (AND $y_1, y_2, \ldots y_m$)]:
do
  let $i \leftarrow 1$
  let covered $\leftarrow$ true
  while (($i \leq n$) $\wedge$ covered)
  do
    let $j \leftarrow 1$
    let found $\leftarrow$ false
    while (($j \leq m$) $\wedge \neg$found)
    do
      if $x_i \neq$ (ALL a t)
        then found $\leftarrow (x_i = y_j)$
        else found $\leftarrow ((y_j = $ (ALL a u))$\wedge$ SUBS?[t, u])
      $j \leftarrow j + 1$
    end
    covered $\leftarrow$ found
    $i \leftarrow i + 1$
  end
  return covered
end

By Lemma 8, we now need only consider subsumption for flat types (SUBS?). It should be clear from the body of the procedure SUBS? as defined above, that, for flat factors x_i and y_j,

SUBS?[(AND $x_1 \ldots x_n$), (AND $y_1 \ldots y_m$)] returns T iff
$\forall i\, 1 \leq i \leq n\, \exists j\, 1 \leq j \leq m$
  if $x_i \neq$ (ALL a t) then $x_i = y_j$
otherwise $y_j =$ (ALL a u) and SUBS?[t, u].

Lemma 9
  SUBS?[x, y] runs in $O(|x| \times |y|)$ time.

Proof
By induction on the depth of ALL operators in x:
1. If depth $= 0$, then for each x_i, we must scan all the y_j looking for equal factors, taking $|x| \times |y|$ steps.
2. Assume true for depth $\leq k$.
3. Suppose x has maximum depth $= k + 1$ and let x_i be a factor. If $x_i \neq$ (ALL a t) then as before, we must scan y in $|y|$ steps. Suppose there are l factors (ALL a_i t_i) in x. For each such factor, we must find a corresponding one in y, taking $|y|$ steps, and then call SUBS? recursively. By induction, this can be done in roughly $|t_i| \times |y|$ steps, so the total effort for the l factors is

$$\sum_{i=1}^{l} (|y| + |t_i||y|) = \sum_{i=1}^{l} (|t_i| + 1) |y|$$

But $\sum_{i=1}^{l} (|t_i| + 1)$ is the total length of these factors, so overall, the procedure is completed in $|x| \times |y|$ steps. ∎

A.2.2. Correctness of the subsumption algorithm
Now, we move on to the proof that this algorithm indeed calculates subsumption: first we must show that if SUBS?[x, y] is T then x indeed subsumes y (soundness); then we must show the converse (completeness). Before beginning, note that the first two steps of the algorithm do not change the extensions of x and y for any extension function, and so do not affect the correctness of the algorithm.

Informally, to see why the algorithm is sound, suppose that SUBS?[x, y] is T and consider one of the conjuncts of x—call it x_i. Either x_i is among the y_j or it is of the form (All a t). In the latter case, there is a (ALL a u) among the y_j, where SUBS?[t, u]. Then, by induction, any extension of u must be

a subset of t's and so any extension of y_j must be a subset of x_i's. So no matter what x_i is, the extension of y (which is the conjunction of all the y_j's) must be a subset of x_i. Since this is true for every x_i, the extension of y must also be a subset of the extension of x. So, whenever SUBS?[x, y] is T, x subsumes y.
More formally, we have the following lemma.

Lemma 10 (Algorithm soundness)
  If SUBS?[(AND $x_1 \ldots x_n$), (AND $y_1 \ldots y_m$)] = T, then (AND $x_1 \ldots x_n$) subsumes (AND $y_1 \ldots y_m$).

Proof
Suppose SUBS? returns T and let $\mathcal{E}$ be any extension function. We will show that $\forall i$, $1 \leq i \leq n$, $\exists j$, $1 \leq j \leq m$, such that $\mathcal{E}[y_j] \subseteq \mathcal{E}[x_i]$, by induction on the depth of ALL operators in the x_i factors.
1. Suppose x_i does not contain an ALL; then, since SUBS? returns T, $\exists j$ such that $y_j = x_i$; so $\mathcal{E}[y_j] = \mathcal{E}[x_i]$.
2. Assume true for depth $\leq k$.
3. If $x_i =$ (ALL a t) then, since SUBS? returns T, $\exists j$ such that $y_j =$ (ALL a u), where SUBS?[t, u] is T. By induction, t must subsume u, so $\mathcal{E}[u] \subseteq \mathcal{E}[t]$. But then $\mathcal{E}[($ALL a u)] $\subseteq \mathcal{E}[($ALL a t)]$.
Now suppose that, for some t, $t \in \mathcal{E}[($AND $y_1 \ldots y_m$)]$. Then $\forall j$, $1 \leq j \leq m$, $t \in \mathcal{E}[y_j]$. By the above, $\forall i$, $1 \leq i \leq n$, $\exists j$, $1 \leq j \leq m$ such that $\mathcal{E}[y_j] \subseteq \mathcal{E}[x_i]$, so $t \in \mathcal{E}[x_i]$. Thus, $t \in \mathcal{E}[($AND $x_1 \ldots x_n$)]$. Since this holds for any t and any $\mathcal{E}$, x subsumes y. ∎

Now, we turn to the completeness of the subsumption algorithm. Here we have to be able to show that anytime SUBS?[x, y] is F, there is an extension function that does not assign x to a superset of what it assigns y (i.e., in some possible situation, a y is not an x). Prior to the proof itself, we set up two lemmas. The formal completeness proof will hinge on our ability in all cases where SUBS? returns F to find a factor that is in the "lower" type but not in the "higher" type. The first lemma (11) allows us to construct an extension function over a domain with a distinguished object d_1 that will be in the extension of every type except for a few critical exceptions. This will be used as a counterexample to subsumption in the proof. The second lemma (12) is used in one of the case analyses in the proof.

Lemma 11
  Suppose $\mathcal{E}$ is an extension function over $\mathcal{D}$. Suppose that d_0, $d_1 \in \mathcal{D}$, that Q is a primitive type, S is an attribute, and C is a flat type. Furthermore, suppose $\mathcal{E}$ satisfies

1. $d_0 \in \mathcal{E}[p]$ for every primitive p;
2. $d_1 \in \mathcal{E}[p]$ for every primitive p except perhaps Q;
3. $\langle d_0, d \rangle \in \mathcal{E}[a]$ iff $d = d_0$, for every attribute a;
4. $\langle d_1, d \rangle \in \mathcal{E}[a]$ iff $d = d_0$, for every attribute a except perhaps S;
5. $\langle d_1, d \rangle \in \mathcal{E}[S]$ only if $d \in \mathcal{E}[C]$.

Then, for any flat factor t,
(A) $d_0 \in \mathcal{E}[t]$;
(B) if $t \neq Q$ and $t \neq$ (SOME S), and for any u such that u does not subsume C, $t \neq$ (ALL S u), then $d_1 \in \mathcal{E}[t]$.

Proof
(A) By induction on $|t|$:
1. If t is primitive, then true by (1) above.
2. If t is (SOME a), then $\langle d_0, d_0 \rangle \in \mathcal{E}[a]$ by (3), so $d_0 \in \mathcal{E}[($SOME a)]$.

3. If t is (ALL a u), then since $\forall d \langle d_0, d \rangle \in \mathscr{E}[a] \Rightarrow d = d_0$ by (3), we have that $\forall d \langle d_0, d \rangle \in \mathscr{E}[a] \Rightarrow d \in \mathscr{E}[u]$ by induction. So $d_0 \in \mathscr{E}[(\text{ALL } a \, u)]$.

(B) By cases on t:

1. If t is primitive, then if $t \neq Q$, $d_1 \in \mathscr{E}[t]$ by (2).
2. If t is (SOME a), then if $a \neq S$, $\langle d_1, d_0 \rangle \in \mathscr{E}[a]$ by (4), so $d_1 \in \mathscr{E}[(\text{SOME } a)]$.
3. If t is (ALL a u) where $a \neq S$, then by (4) $\forall d \langle d_1, d \rangle \in \mathscr{E}[a] \Rightarrow d = d_0$. So by part A, $\forall d \langle d_1, d \rangle \in \mathscr{E}[a] \Rightarrow d \in \mathscr{E}[u]$. So, $d_1 \in \mathscr{E}[(\text{ALL } a \, u)]$.
4. If t is (ALL S u), then by (5), $\forall d \langle d_1, d \rangle \in \mathscr{E}[S] \Rightarrow d \in \mathscr{E}[C]$. As long as u subsumes C, then $\forall d \langle d_1, d \rangle \in \mathscr{E}[S] \Rightarrow d \in \mathscr{E}[u]$. Thus $d_1 \in \mathscr{E}[(\text{ALL } S \, u)]$. ∎

Lemma 12

For any flat type t, there is an $\mathscr{E}$, a $\mathscr{D}$, and a $d \in \mathscr{D}$ such that $d \notin \mathscr{E}[t]$. (In other words, no type is "tautologous," i.e., a *summum genus*.)

Proof

Suppose $t = (\text{AND } p_1 \ldots p_l \, (\text{SOME } a_1) \ldots (\text{SOME } a_m) \, (\text{ALL } b_1 \, u_1) \ldots (\text{ALL } b_n \, u_n))$. If $l \neq 0$ let $\mathscr{D} = \{0\}$, $\mathscr{E}[p_1] = \emptyset$; then $0 \notin \mathscr{E}[t]$. If $m \neq 0$ let $\mathscr{E}[a_1] = \emptyset$, $\mathscr{D} = \{0\}$; then $0 \notin \mathscr{E}[t]$. Otherwise $n \neq 0$ and by induction $\exists \mathscr{E}^*, \mathscr{D}^*, d^*$ such that $d^* \notin \mathscr{E}^*[u_n]$. Let $\mathscr{D} = \mathscr{D}^* \cup \{0\}$ (assuming $0 \notin \mathscr{D}^*$), $\mathscr{E} = \mathscr{E}^*$ except $\mathscr{E}[b_n] = \mathscr{E}^*[b_n] \cup \{\langle 0, d^* \rangle\}$; since $d^* \neq 0$, $d^* \notin \mathscr{E}[u_n]$. Thus, $0 \notin \mathscr{E}[(\text{ALL } b_n \, u_n)]$, so $0 \notin \mathscr{E}[t]$. ∎

Lemma 13 (Algorithm completeness)

If SUBS?[(AND $x_1 \ldots x_n$), (AND $y_1 \ldots y_m$)] $= F$, then (AND $x_1 \ldots x_n$) does not subsume (AND $y_1 \ldots y_m$).

Proof

Since SUBS? returns F, there must be an x_i for which there is no corresponding y_j. Given this, we will show by induction on the depth of ALL operators in x_i, how to define an $\mathscr{E}$, a $\mathscr{D}$, and a $d_1 \in \mathscr{D}$ such that $d_1 \notin \mathscr{E}[x_i]$, but $\forall j$, $1 \leq j \leq m$, $d_1 \in \mathscr{E}[y_j]$. Thus, $\mathscr{E}[(\text{AND } y_1 \ldots y_m)] \nsubseteq \mathscr{E}[(\text{AND } x_1 \ldots x_n)]$, and so the former is not subsumed by the latter.

Case 1. Suppose x_i is a primitive, Q. Define $\mathscr{E}$ and $\mathscr{D}$ by
Let $\mathscr{D} = \{0, 1\}$.
Let $\mathscr{E}[t] = \begin{cases} \{0, 1\} & \text{if } t \neq Q \\ \{0\} & \text{otherwise.} \end{cases}$
Let $\mathscr{E}[a] = \{\langle 0, 0 \rangle, \langle 1, 0 \rangle\}$.
Let $d_1 = 1$. Clearly, $1 \notin \mathscr{E}[Q]$; thus, $1 \notin \mathscr{E}[x_i]$. Now consider any y_j. Let $d_0 = 0$, $C = (\text{AND } Q)$, and $S = $ any attribute. The conditions of Lemma 11 are thus satisfied, so unless y_j is one of the named exceptions to Lemma 11B, we know that $1 \in \mathscr{E}[y_j]$. Moreover, y_j cannot be Q since SUBS? returns F; $1 \in \mathscr{E}[(\text{SOME } S)]$ since $\langle 1, 0 \rangle \in \mathscr{E}[S]$; and $1 \in \mathscr{E}[(\text{ALL } S \, u)]$ since, by Lemma 11, $0 \in \mathscr{E}[u]$ for any u. Thus, no matter what y_j is, $1 \in \mathscr{E}[y_j]$.

Case 2. Suppose x_i is (SOME S). Define $\mathscr{E}$ and $\mathscr{D}$ by
Let $\mathscr{D} = \{0, 1\}$.
Let $\mathscr{E}[t] = \{0, 1\}$.
Let $\mathscr{E}[a] = \begin{cases} \{\langle 0, 0 \rangle, \langle 1, 0 \rangle\} & \text{if } a \neq S \\ \{\langle 0, 0 \rangle\} & \text{otherwise.} \end{cases}$
Again, let $d_1 = 1$, so that $1 \notin \mathscr{E}[x_i]$. Now consider any y_j. Let $d_0 = 0$, $C = $ any flat type, and $Q = $ any primitive. Again, the conditions of Lemma 11 are satisfied, so unless y_j is one of the named exceptions, we know that $1 \in \mathscr{E}[y_j]$. Moreover, $1 \in \mathscr{E}[Q]$, y_j cannot be (SOME S) since SUBS? returns F, and $1 \in \mathscr{E}[(\text{ALL } S \, u)]$ for any u. Thus, no matter what y_j is, $1 \in \mathscr{E}[y_j]$.

Case 3. Suppose $x_i = (\text{ALL } S \, u)$, but there is no (ALL S v) among the y_j. By Lemma 12, there are $\mathscr{E}^*, \mathscr{D}^*, d^*$ such that $d^* \notin \mathscr{E}^*[u]$. Define
$\mathscr{D} = \mathscr{D}^* \cup \{0, 1\}$ (assuming 0 and 1 do not appear in $\mathscr{D}^*$).
$\mathscr{E}[t] = \mathscr{E}^*[t] \cup \{0, 1\}$.
$\mathscr{E}[a] = \mathscr{E}^*[a] \cup \begin{cases} \{\langle 0, 0 \rangle, \langle 1, 0 \rangle\} & \text{if } a \neq S \\ \{\langle 0, 0 \rangle, \langle 1, d^* \rangle\} & \text{if } a = S. \end{cases}$
As before, let $d_1 = 1$. Since $d^* \notin \{0, 1\}$ and $d^* \notin \mathscr{E}^*[u]$, $d^* \notin \mathscr{E}[u]$. So, $1 \notin \mathscr{E}[x_i]$. Now consider any y_j. Let Q be any primitive type and C be any flat type. Again, the conditions of Lemma 11 are satisfied. Moreover, $1 \in \mathscr{E}[Q]$, $1 \in \mathscr{E}[(\text{SOME } S)]$, and $y_j \notin (\text{ALL } S \, v)$ for any v. So no matter what y_j is, $1 \in \mathscr{E}[y_j]$.

Case 4. Suppose $x_i = (\text{ALL } S \, u)$ and some $y_j = (\text{ALL } S \, v)$ but SUBS?[u, v] = F. By induction, there is a $\mathscr{D}^*$ and an $\mathscr{E}^*$ and a $d^* \in \mathscr{D}^*$ such that $d^* \in \mathscr{E}^*[v]$ but $d^* \notin \mathscr{E}^*[u]$.
Let $\mathscr{D} = \mathscr{D}^* \cup \{0, 1\}$ (assuming 0 and 1 do not appear in $\mathscr{D}^*$).
Let $\mathscr{E}[t] = \mathscr{E}^*[t] \cup \{0, 1\}$.
Let $\mathscr{E}[a] = \mathscr{E}^*[a] \cup \begin{cases} \{\langle 0, 0 \rangle, \langle 1, 0 \rangle\} & \text{if } a \neq S \\ \{\langle 0, 0 \rangle, \langle 1, d^* \rangle\} & \text{if } a = S. \end{cases}$
As before, let $d_1 = 1$. As in the previous case, $1 \notin \mathscr{E}[x_i]$. Now consider any y_j. Let Q be any primitive type and $C = v$. Again, the conditions of Lemma 11 are satisfied. Moreover, $1 \in \mathscr{E}[Q]$, $1 \in \mathscr{E}[(\text{SOME } S)]$, and for any z different from v, and thus for any z that does not subsume v, $y_j \neq (\text{ALL } S \, z)$. So no matter what y_j is, $1 \in \mathscr{E}[y_j]$. ∎

Theorem 2

There is an algorithm to calculate subsumption for $\mathcal{FL}^-$ in $O(n^2)$ time.

Proof

By Lemma 8, types can be flattened in $O(n^2)$ time. The procedure SUBS? works on flattened types in $O(n^2)$ time by Lemma 9, and by Lemmas 10 and 13, SUBS? returns T iff its first argument subsumes its second. ∎

2

Suppose You Have a Robot

LUIGIA CARLUCCI AIELLO, DANIELE NARDI, GABRIELE RANDELLI, AND CARLO MATTEO SCALZO

ABSTRACT. In this paper we address the use of knowledge representation and reasoning (KR&R) techniques in current research and implementation of robotic systems. We quickly overview the work on knowledge representation and reasoning, since the area became popular in the eighties, starting from the assumption that entitles the paper: "Suppose you have a robot". Then, we survey the relatively few recent approaches that try to improve the capabilities of robotic systems, by exploiting an explicit representation of knowledge. The availability of cheap and powerful robots makes it feasible to embody intelligence in physical agents, and the motivations for embedding KR&R components in robotic agents become more and more compelling. Unfortunately, there is still a gap in performance that knowledge representation and reasoning have not yet filled in. We argue that one possibility to improve this state of affairs is to build systems that, by exploiting several AI technologies, support human robot interaction to make robots more knowledgeable with the help of humans.

1 Introduction

The beginning of AI research is characterized by a strong emphasis on autonomous robots (consider the wheeled Stanford cart or the famous Speaky cart developed at SRI [Nilsson 1969]). Moreover, the role of knowledge representation and reasoning (KR&R) in the contribution of autonomous robots has been advocated by AI researchers since the early days of this discipline. A strong impulse is due to John McCarthy, starting with his early papers of the late fifties and early sixties. The incipit of the '69 paper [Hayes and McCarthy 1969] is still a guideline for the construction of intelligent agents nowadays.

A computer program capable of acting intelligently in the world must have a general representation of the world in terms of which its inputs are interpreted.

Then McCarthy goes on making precise the three main axes according to which a representation should be measured to assess its adequacy, namely metaphysical, epistemological and heuristic adequacy, which inspired all the subsequent developments of KR&R, culminating with the purely declarative characterization of the knowledge representation approach [Levesque 1984].

Throughout the years only the language for speaking about KR&R has evolved, and subtle issues of the representation have beed addressed, but the main principles informing the process of the choice of a representation formalism and an inference systems stay there as stated in the foundational papers. After a significant growth of the field, robots have only been used in classroom examples that, at the time, but also today, are far from being actually applicable on real life problems. The title of the paper "Suppose you have a robot[1]" is

---

[1] Not to be confused by non-English speakers with the "Suppose you have a rowboat" of the missionaries and cannibals puzzle.

usually the beginning of a problem description aiming at justifying complex representation and/or reasoning problems. For example, *"Suppose you have a robot that can move blocks on a table ... "*. The blocks world example and related research soon lost any connection with the real robot implementation, and, unless we take some care in selecting the blocks of the proper size, color, maybe put in the environment some camera to help with the perception, even today we are not able to build a system that satisfies the generality advocated in the example.

In the nineties, a renewed effort to ground AI research, and KR&R in particular, to physical robotic agents was boosted by the Cognitive Robotics Manifesto [Levesque and Reiter 1998], from the Toronto group. Unfortunately, the results of these efforts that are nowadays embedded in real robots are limited (while the theoretical work has grown steadily) and the attention of the AI research community towards robotic applications is not significant. On the other hand, research and technology in robotics is progressing at a quick pace: nowadays various types of robots are available, including humanoid robots, that can even play soccer [Kitano and Asada 1998], but are also expected to enter our homes as assistants and/or companions. This availability of robotic platforms has fostered the development of a number of alternative approaches that negate the need for an explicit model of the world, or rely on a different world modeling. Despite significant progress, new groundbreaking approaches are not yet clearly emerging. Morever, in a large class of applications, we undoubtedly would like to interact with robots using symbolic terms like *Pass the ball to your right*, or *Go to the kitchen*, as opposed to $goto\ [x, y, \theta]$. This suggests that at least knowledge representation should play a role in the design of such system (and why not reasoning, to figure out, for example the kind of objects that are likely to be found in a kitchen).

Likely because we are biased by our previous research in knowledge representation, we believe that robotic systems can not scale up in performance if they can not suitably handle common sense knowledge about the surrounding environment as suggested by John Mc Carthy long ago. Such a view is not fully in line with the current research trends, and only a limited number of research groups are pursuing it. The goal of the paper is to review the recent research aiming at making robots intelligent (or at least less stupid), by relying on an explicit representation of the world, trying to highlight promising achievements and opportunities for research developments. A key, recognized, obstacle to this endeavor is in limited capabilities in perception, that severely restrict the acquisition of knowledge about the environment. Consequently, we focus on recent research in Human Robot interaction (HRI), which attempts to rely upon an explicit representation of knowledge to support the interaction between human and robot. Finally, we sketch a framework, where the combined use of several AI components, as well as new technologies, can improve the interaction with people and make the robot more knowledgeable with the help of the human.

The paper is organized as follows. In Section 2, we address the general framework provided by Cognitive Robotics. In Section 3, we survey various implementations of KR&R on real robots. In Section 4 we first address KR&R in HRI, and then outline a framework to facilitate human robot interaction by high-level exchange of knowledge as a key step towards robots with commonsense.

2 Cognitive architectures

The question of how the basic structure of the autonomous agent/robot structure can be implemented has been subject to a long debate and is still under investigation. Agents, and specifically robots, usually present various kinds of sensing and acting devices. The flow of data from the sensors to the actuators is processed by several modules and the description of the interaction among these modules defines the *agent's architecture*.

The first, purely *deliberative*, architectures [Fikes and Nilsson 1971; Nilsson 1984] view the robot as an agent embedding a high-level representation of the environment and of the actions that it can perform. Perceptual data are interpreted for creating a model of the world, a planner generates the actions to be performed, and the execution module takes care of executing these plans. In practice, a sense-plan-act cycle is repeatedly executed. The problem is that building a high-level world model and generating a plan are time consuming activities and thus these systems have shown to be inadequate for agents embedded in dynamic worlds. In fact, building and maintaining a high-level symbolic representation of the environment as outlined above may lead to poor reactive capabilities in the agent, since all data must be processed by (usually computationally intensive) interpretation and decision-making procedures.

Reactive architectures focus on the basic functionalities of the robot, such as navigation or sensor interpretation, and propose a direct connection between stimuli and response. Brooks's *subsumption architecture* [Brooks 1986] is composed by levels of competence containing a class of *task-oriented behaviors*. Each level is in charge of accomplishing a specific task (such as obstacle avoidance, wandering, etc.) and the perceptual data are interpreted only for that specific task. Reactive architectures suitably address the dynamics of the environment and do not generally require building a world model. Thus reasoning is "compiled" into the structures of the executing program. Reactivity to the perception of the environment is achieved, but the lack of a world model makes the system inflexible, general aspects of perception (not related to a specific behavior) are missed, and the system has very limited capabilities in terms of projecting into the future the effects of actions and, consequently, the achievement of a goal.

The above considerations gave rise to a renewed effort to combine a representational view of the robot as an intelligent agent, with a suitably reactive behaviour. This endeavor led to *Cognitive Robotics* (see [Levesque and Lakemeyer 2008]). The name was first introduced by the research group at the University of Toronto [Lesperance, Levesque, Lin, Marcu, Reiter, and Scherl 1994], then put forward in the Cognitive Robotics Manifesto [Levesque and Reiter 1998].

Cognitive Robotics aims at designing and realizing actual agents (in particular mobile robots) that are able to accomplish complex tasks in real, and hence dynamic, unpredictable and incompletely known environments, without human assistance.

Arguably, the intended meaning of the new terminology was to re-design a knowledge centered approach to intelligent agents, in such a way as to address the physical embodiment into robots, and, consequently, the inaccuracies of perception, reactivity and all the limitations arising from a pure deliberative approach.

The label cognitive robotics is nowadays accepted as a general umbrella for the embedding of an intelligent agent into a real robot; therefore, it is used in a more general perspective, by looking at the perception/action cycle in a broader sense, including approaches without an explicit world model, or with a different view on world modeling. In the first case, the robot behavior is generated not by the robot controller alone, but it emerges by means of the interactions between the robot with its body and the environment, as in bio-inspired *evolutionary* systems (see e.g., [Nolfi and Floreano 2000]) or in *embodied intelligence* (see e.g., [Pfeifer and Scheier 1999]).

A popular approach to action representation in robots is based on decision making techniques, which maximize the utility of the actions selected by the robot, depending on the operational context (see for example [Russell and Norvig 2005]). This approach does not provide an explicit representation of the properties that characterize the dynamic system, and focusses on the action selection mechanism.

In this paper, we concentrate on the Toronto approach to Cognitive Robotics, where a robot can be controlled at a high/symbolic level, by providing it with a description of the world and by expressing the tasks to be performed in the form of goals to be achieved. In this view the characterizing feature of a cognitive robot is the presence of cognitive capabilities for reasoning about the information sensed from the environment and about the actions it can perform. Specifically, this is achieved through a system *architecture*, which includes a symbolic representation of knowledge and suitable model of the dynamics of the system, in terms of an *action* theory.

There are many features that are considered essential in the design of agent architectures, and each proposal describes a solution that provides for some of these features. Approaches to architectures that try to combine symbolic and reactive reasoning are called *Hybrid Architectures* (see for example in [Iocchi 1999]). We can roughly describe a layered hybrid architecture of an agent with two levels: the deliberative level, in which a high-level state of the agent is maintained and decisions on which actions are to be performed are taken, and the operation level, in which conditions on the world are verified and actions are actually executed.

At the core of the Cognitive Robotics approach is the representation of *dynamical* systems (action theory) developed in the Situation Calculus, as reformulated in [Reiter 2001]. A large stream of work has been generated by the central role given to the Situation Calculus in Cognitive Robotics. Several aspects of action representation including non-determinism, persistence and sensing [Scherl and Levesque 1993], concurrency [De Giacomo, Lesperance, and Levesque 1997] have been developed; Situation Calculus has been further extended with probabilistic representations, representations of time etc.

In addition, a number of theories of actions have been developed in order to represent the agent's knowledge: A-Languages (e.g., [Giunchiglia, Kartha, and Lifschitz 1997]), Dynamic Logics (e.g., [De Giacomo, Iocchi, Nardi, and Rosati 1999]), Fluent [Thielscher 2005] and Event Calculi [Shanahan 1997]. They are characterized by the expressive power, that is the ability of representing complex situations, by the deductive services allowed, and by the implementation of automatic reasoning procedures.

However, much of the work carried out on action theories has been disconnected from applications on real robots, with some notable exceptions, that we specifically address in the next section.

3 KR in Robotic Systems

In this section we present some significant attempts to embody an explicit representation of knowledge into a robotic system. The representation of knowledge addresses both the modeling of the actions of the robot and the modeling of the environment where the robot is operating. Obviously both models are needed, but the different emphasis given in the literature to the two aspects allows us to address them separately.

3.1 Representing dynamical systems

We start by addressing the few implementations of action theories that have been actually experimented on robotic platforms. As we recalled above, the Cognitive Robotics Manifesto places a theory of actions at the core of the system and the Situation Calculus as a basis for such theory. Moreover, Golog [Levesque, Reiter, Lespérance, Lin, and Scherl 1997] is proposed as a suitable language for expressing high level robot programs, that rely on an underlying action theory. High level programs expressed in Golog contain primitive actions and tests of predicates that are domain-dependent. An interpreter for such programs must reason about the preconditions and effects of the actions in the program, in order to find a legal terminating execution.

Golog provides limited support for writing reactive programs, which is a basic requirement for robotic applications: a robot must often react to events and exceptional conditions by suspending its current plan and selecting a new plan that is appropriate to the situation. Consequently, many extensions of the Golog system and its underlying theory.

ConGolog [De Giacomo, Lesperance, and Levesque 1997] is an extension of Golog that provides concurrent processes (possibly with different priorities) and interrupts. Concurrent processes are modeled as interleavings of the primitive actions involved. ConGolog allows the designer to specify concurrent processes with priorities: the process with a lower priority level may only execute when the high-priority process is completed or blocked. ConGolog can also specify interrupts: this allows the designer to write reactive programs, i.e., programs that will suspend whatever task they are doing to handle exogenous events as they arise.

ConGolog is used for the design of high-level reactive control modules in robotics applications [Lésperance, Tam, and Jenkin 2000]. The robot uses a hierarchical architecture to provide real time response as well as high-level planning. At the lowest level, a reactive control system performs time-critical tasks such as collision avoidance and straight line path execution. The middle layer contains modules for path planning, map building, and so on. The upper level is represented by the ConGolog-based control module that supports high-level plan execution to accomplish the robots tasks. This high level controller runs asynchronously with the rest of the architecture, so that other tasks can be attended to while the robot is navigating towards a destination. The main control loop handles exogenous events by using prioritized interrupts. The system has been tested in a mail delivery application and ported to a RWI B12 mobile robot. The experiments confirmed the system's ability to deal with navigation failures and to interrupt the current task when an urgent shipment order is made.

A variant of ConGolog, called ccGolog, is presented in [Grosskreutz and Lakemayer 2000]. The authors define a new extension of the Situation Calculus, which forms the basis for the definition of ccGolog.

Actions in the Situation Calculus cause discrete changes, i.e., there is no notion of time. However, robotics application must face processes like navigation, which causes the robots location and orientation to change continuously over time. To model continuous change and time, four new sorts are added to the Situation Calculus: *real* (ranging over real numbers), *time* (ranging over the reals), *t-function* (functions of time), and *t-form* (temporal formulas). In addition, a special function *val* is introduced in order to evaluate t-functions. The new framework models the time evolution using a special *waitFor* action.

A key feature of ccGolog, as defined on this extended version of the Situation Calculus, is the ability to have part of a program wait for an event (like the battery voltage dropping dangerously low) while other parts of the programs run in parallel. This mechanism allows very natural formulations of robot controllers.

The ccGolog framework has been implemented by using Prolog and a built-in constraint solver library, and tested in the mail delivery application domain. The experiments show that the robot is able to employ reactive behaviors as well as integrating them with a high-level control program.

A different extension of Golog, called Readylog, is presented in [Ferrein and Lakemayer 2008]. This Golog dialect has been developed to support the decision making of robots acting in dynamic real-time domains (like robotic soccer), and allows for decision-theoretic planning in a continuously changing world.

Readylog features non-standard constructs employing Markov Decision Process theories in the logical framework: the Readylog interpreter chooses the best action alternative based on the underlying utility theory.

The Readylog system has been implemented on the robotic platform of the Middle-size RoboCup Team AllemaniACs, and used in several robotic soccer competitions. The high-level control program included a decision-theoretic planning step: the reward function gives high rewards for situations in front of the opponent goal, and negative rewards for situations in front of the own goal. The agent has the choice to calculate the best action among kicking the ball, combining a dribbling with a goal shot, and a cooperative play in which the robot plays a pass to its teammate, which in turn plans to intercept the ball and try a goal shot. The authors also applied Readylog in the service robotics domain, i.e., a setting in which the robot operates as a tour-guide in a local bank.

A different approach to the representation of dynamic systems, based on the correspondence between Description Logics and Propositional Dynamic Logics, is given in [De Giacomo, Iocchi, Nardi, and Rosati 1999]. The authors use Epistemic Description Logics to specify static axioms, precondition axioms and effect axioms, thus obtaining a framework in which the dynamics of the system is specified in terms of what the robot knows of the world, instead of what is true in the world. The proposed approach has been implemented on a mobile robot based on a two-level architecture combining reactive control and planning. Using the framework, the robot is able to generate and execute a plan to navigate in an office environment. The approach has also been discussed and implemented in the RoboCup setting [Castelpietra, Guidotti, Iocchi, Nardi, and Rosati 2002], together with a layered hybrid architecture. The system has been successfully implemented in the ART and SPQR RoboCup teams (both in the Legged and in the Middle-size League), leading to good performances in a task (soccer playing) where the environment is very dynamic. The design of a system based on a formal high level specification has also contributed to speed up the software development, thus making easier to port the framework to different robotic platforms.

An approach that tries to combine a theory of actions with a more operational approach to the representation and execution monitoring of robot plans is the Petri Net Plans framework [Ziparo, Iocchi, Lima, Nardi, and Palamara 2010]. Petri Net Plans (PNP) are a behavior representation framework based on Petri Net semantics, but providing special constructs that are inspired by a theory of actions. In particular, they include constructs for modeling non-instantaneous actions, sensing, loops, concurrency, action failures, and action-synchronization in a multi-agent context. As Golog, PNPs do not follow a generative approach: they are a tool for the representation of plans, whose execution is specified through an operational semantics.

PNPs are also used to provide an implementation for robotic systems of the the joint commitment theory presented in [Levesque, Cohen, and Nunes 1990]. The theory analyzes the implications for a group of agents to jointly commit to a common goal, as well as the impacts of this decision on the individual commitments of the agents. The theory is expressed in a modal language which takes into account the agents believes and goals, as well as mutual believes and sequences of events.

The implementation of the Joint Commitment Theory is based on the multi-robot sync and interrupt operations to model the specifications. The multi-robot interrupt operator is used to consistently interrupt action execution among robots engaged in a cooperation, while the successful conclusion of individual actions is implemented through a hard-sync operator.

This approach has been implemented in the RoboCup robotic soccer environment, where explicit cooperation is required to execute complex tasks like a pass between two robots. If the condition for a pass holds, a commitment is established, and the robots need to agree on the allocation of the required tasks, i.e., passing and receiving the ball. If the passer robot loses the ball, the failure needs to be communicated to the receiving robot, and the execution of cooperative behaviors must be interrupted. A proper Petri Net Plan has been successfully defined to model this scenario, following the guidelines provided by the Joint Commitment Theory.

PNPs have been used in a number of implemented robotic systems for rescue applications, as well as for robotic soccer. The implementation of the Joint Commitment Theory has been tested on a system where two AIBO robots play passing the ball to each other [Palamara, Ziparo, Iocchi, Nardi, Lima, and Costelha 2008].

3.2 Representing the operational environment

Autonomous robots typically incorporate two different types of processes: high-level cognitive processes and processes related to sensors and actuators. The former perform abstract reasoning and planning, the latter observe the external world and execute actions in it. These two types of processes refer to the same physical objects in the environment, although in very different ways. Cognitive processes typically use symbols to denote objects, while lower-level processes use perceptual data (i.e., data acquired by the sensors). We now focus on the problem of building and using a knowledge base representing the knowledge about the operational environment; this aspect is addressed in ad-hoc ways in all the above discussed approaches.

Any robotic system that incorporates a symbolic component must address the problem of connecting symbols and sensor data that refer to the same physical objects in the world. This is known as *symbol grounding* [Harnad 1990] and is a fundamental problem regardless of the specific robotic architecture to be used: Solving it means providing these two types of processes a way to communicate and work properly. Moreover, some of the properties of physical objects may evolve over time, so the grounding process must take this temporal dimension into account, in order to cope with the flow of continuously changing input from its sensors.

Creating and maintaining the correspondence between symbols and sensor data that refer to the same physical objects has been referred to as the *anchoring* problem [Coradeschi and Saffiotti 2000]. Their formal framework for anchoring includes a symbol system and a perceptual system. The symbol system manipulates individual symbols denoting physical objects (like *cup22*) and associates each symbol with a set of properties (symbolic predicates, like *red*). The perceptual system generates percepts (i.e., collections of sensor data from the same physical object) and associates each percept with the values of a set of measurable attributes (like hue values in images).

The framework assumes the presence of a predicate grounding relation, which encodes the correspondence between predicate symbols and the observed attribute values. This relation can be quite complex in the general case, and the framework makes no assumption about its origin: for instance, it can be hand-coded by the designer, or learnt from examples.

The anchoring process can now be rephrased as the problem to use the grounding relation to connect individual symbols in the symbol system to percepts in the perceptual system. Such a connection is an anchor, and its time-dependent.

In order to obtain a complete anchoring process, the framework defines three abstract functionalities: Find, Track, and Reacquire. The Find functionality creates an anchor the first time that an object is perceived, the Track functionality continuously updates the anchor description while observing the object, and the Reacquire functionality updates the anchor when the robot needs to reacquire the object because it has not been observed for some time.

The authors implemented their framework and tested it in the robot navigation domain, by using a Nomad 2000 robot equipped with an array of sonar sensors and a simple STRIPS-like planner. The perceptual system extracts linear contours from sonar measurements, in order to detect walls and corridors, as well as some attributes like lenght and width. The planner specifies the symbolic description of the objects (like *corridor-1*) to be used for a task, and the anchoring system manages them by using the Find and Track functionalities. Experiments show that the robot is able to recover from an erroneous initial anchoring of a corridor. By updating the anchoring information with the Track functionality, the robot is able to handle spurious sonar readings produced by a peculiar configuration of obstacles, and to obtain a correct anchor, which is used to easily match subsequent percepts.

In [Chella, Coradeschi, Frixione, and Saffiotti 2004] the formal framework for the anchoring problem is recast by using conceptual spaces [Gärdenfors 2000]. A conceptual space has dimensions that are related with the quantities processed by the sensors as well as with the concepts managed at the symbol level, and can therefore be used as a middle layer to integrate symbolic and sensor-based information.

A conceptual space has dimensions (called qualities) which are related with the quantities managed by the robot sensors (e.g., possible dimensions are color coordinates or spatial coordinates). Points in a conceptual space (called *knoxels*) represent primitive elements: for instance, a knoxel can represent an object, which is characterized by its color and its position (i.e., specific values on the dimensions of the conceptual space). In this setting, concepts can be represented by regions in the conceptual space, i.e., as sets containing of all their instances. This representation provides the basis to link sensor data to symbolic predicates.

The authors tested this new framework on a mobile robot. The robotic platform included a conditional planner, a fuzzy behavior-based navigation planner, sonars for obstacle detection, and a vision system to perceive objects. The robot is given the task to go into a room in a building and look for victims. The robot must also look for dangers like bottles containing explosive gases. The experiments show how the robot is able to use anchoring to cross doors and handle gas bottles, by creating suitable anchors and updating them over time.

In [LeBlanc and Saffiotti 2008] the cooperative anchoring problem is introduced, in order to extend the anchoring framework to multi-robot settings. In cooperative settings, robots can receive information from other agents. This is particularly important when dealing with smart environments, where many devices (not just robots) can provide information. This richness of information adds fundamental challenges like representing, communicating, comparing and fusing information.

The authors define cooperative anchoring as the problem of creating and maintaining the correspondence between information referring to the same physical objects, in a distributed robot system. The computation framework for the cooperative anchoring problem relies on the concept of a global anchor space, i.e., a conceptual space into which information from individual anchor spaces can be mapped (notice that each robot manages its own anchor space).

The framework is discussed through an illustrative experiment in which one of the robots must fetch a specific parcel from the floor. This robot is equipped with a symbolic task planner and a vision system, while the other robot uses only the vision system. The environment also contains a RFID reader (to detect the parcels' RFID tags) and a fixed ceiling camera. The experiment shows how the robot is able to integrate information from all the agents in the environment, and accomplish its task by using cooperative anchoring.

Another approach to the problem of integrating symbolic knowledge and sensor data is described in [Tenorth and Beetz 2009]. The authors describe KnowRob, a knowledge processing system for autonomous personal robots. The system provides knowledge representation capabilities by using description logics, and supports the acquisition of grounded concepts through observation. The knowledge base uses description logics to represent a taxonomy of encyclopedic knowledge, i.e., a model of the classes of objects in the environments (like *Cup* and *Cupboard*), as well as general concepts like action and event. The knowledge base also contains information about specific instances (objects, actions, events). Objects are created by analyzing sensor data, thus creating instances in the most appropriate categories (like *Dishwasher*).

To interface the observation system and load observations into the knowledge base, a set of computable classes and properties is defined. Computables create either instances or relations between instances: this setup keeps the representation of the knowledge itself separated from technical issues related to the automatic creation of instances. Moreover, it allows us to define several computables for the same property (e.g., one that reads object information from the vision system, and another one using RFID tags).

The system is implemented on a mobile robot acting in a sensor-equipped kitchen environment, which includes cameras, magnetic sensors (to detect if a cupboard is open), laser range finders, and RFID tag readers (to identify objects). The robot is also equipped with perception modules to create 3D environment maps.

The anchoring problem and the use of description-logic knowledge base form the basis for the use of semantic maps in task planning. In [Galindo, Fernandez-Madrigal, Gonzalez, and Saffiotti 2008] a specific type of semantic map is defined, in order to integrate hierarchical spatial information and semantic knowledge.

A semantic map comprises a spatial box (or S-Box) and a terminological box (or T-Box). The S-Box contains factual knowledge about the state of the environment and the objects inside it, while the T-Box contains general semantic knowledge about the domain, in terms of general conceptions and relations. Notice that this structure is reminiscent of the structure of a description logics knowledge base, since the S-Box is the extension of the assertional component (A-Box). The S-Box indeed contains also the associations between individuals and sensor-level information, as well as information about the spatial structure of the environment (i.e., morphology of the space, connectivity among places, and so on).

The S-Box is actually structured as a three-level hierarchy. The lower level, called *appearance level*, contains sensor signatures perceived from the environment, as well as information about the robot position from where the sensor information is collected. The middle level, called *occupancy level*, represents the partitioning of the free space in the environment into bounded areas corresponding to rooms and corridors. Elements in these two levels can be connected by links to indicate the area where a sensor signature is perceived. They can also be linked to the T-Box, if perception processes are able to give a specification to classify a spatial entity, like rooms and objects. Notice that this classification can be seen as part of an anchoring process. The upper level in the S-Box, called *symbolic level*, maintains a symbolic representation of the space: individuals represent objects in the environment (as stored in the appearance level) and areas (as stored in the occupancy level), as well as relations representing the connection between different areas and the presence of an object in a given area.

The T-Box contains concepts and relations structured into a hierarchy, which provides an abstract description of the entities in the domain, thus giving meaning to the terms used in the S-Box (e.g., the individual *area-2*, representing a specific area in the S-Box, can be classified as an instance of the concept *Kitchen* in the T-Box). The T-Box can also be used to perform inference: for instance, if the T-Box contains axioms stating that *Kitchen* is a subconcept of *Room*, and that *Room* must have at least one door, then the robot can infer that *area-2* has a door, even if the door has not been perceived yet.

The whole framework is implemented and tested on a mobile robot operating in a home-like environment. The robot is an Activemedia PeopleBot robot equipped with a PTZ color camera and a laser range finder. In the experiments, the robot is able to build a semantic map of the environment. The robot then uses this knowledge to accomplish tasks like "go to the kitchen". Notice that nothing in the S-Box is classified as a kitchen: the robot is able to use the knowledge in T-Box to infer that one of the areas is a kitchen, thus obtaining all the information needed to accomplish the task.

In many service robotics applications, the robot is expected to accomplish an open-ended set of tasks: this is particularly true in the case of household robots, as they need to understand under-specified commands given by humans. The focus moves then on the construction of the *intensional* component of the knowledge base, the T-Box, which was engineered in ad hoc ways in the previous approaches, whose goal was mainly to populate the knowledge base with *extensional* knowledge.

[Kunze, Tenorth, and Beetz 2010] propose a system to address this issue. Their system converts commonsense knowledge from the large Open Mind Indoor Common Sense (OMICS) database from natural language into a Description Logic representation, thus making it available to automated reasoning processes. The OMICS project collects commonsense knowledge described in natural language from Internet users, covering areas like the objects found in different rooms, the correct action to take in a situation, or possible problems that may occur while performing a task.

The system first applies natural language processing techniques (like part-of-speech tagging and syntax parsing), then it resolves the words' meanings to cognitive synonyms (synsets) in WordNet and exploits mappings between these synsets and concepts in Open-Cyc. Based on OpenCyc's concept definitions, the system generates a formal representation in Description Logic which becomes part of the robot's knowledge base. By integrating the OMICS knowledge into an ontology, the robot can perform reasoning about similar objects and/or sub-classes of objects.

Another key aspect in the use of common sense knowledge, that is also addressed in this work, is related with the use of natural language in the human robot interaction. Indeed, a number of tools like speech processing systems or dialogue managers are available, and they can be exploited both to build an explicit representation of knowledge and to improve the interaction with users.

An approach related to natural language is discussed in [Gold, Doniec, Crick, and Scasellati 2009], where the TWIG system is introduced. The main concern here is to allow the robot to learn compositional meanings for new words that are grounded in its sensory capabilities.

TWIG is divided into two parts. The first part is represented by an extension finder that parses an utterance and matches it to a fact (an atomic sentence of first-order logic) observed in the environment: this part is related to the problem of finding an extension of the word, i.e., its meaning under particular circumstances. Notice that if a word is not understood but the sentence is close to matching a fact in the environment, the new word is inferred to refer to the object or relation in the environment that satisfies the meaning of the rest of the sentences. The second part of the TWIG system is the definition tree generator: it takes as input the new words that the robot has heard and all the atomic sentences in the environment description that mentions those words extensions, and generates decision trees representing their intensional meanings (the intension of a word is its general meaning, abstracted from any particular referent).

The TWIG system assumes that the robots environment is represented as a first-order logic knowledge base; however, the predicates are allowed to mention continuous values, for which the system later generates thresholds. This makes the system more useful in the real world, and puts less burden on the designer.

TWIG has been implemented on Nico, a humanoid non-mobile robot equipped with video cameras, duel channel microphones, and an ultrasound signaling indoor location system to sense the world. Nicos vision system is used to find faces and determine the directions they faced: for each face, a new symbol personL or personR is added to the knowledge base (depending on the person being detected on the left or on the right side of the visual scene). Much more information is added to the knowledge base, including the information about the direction a person is looking at, as well as the distance between each entity in the environment. The experimenters move objects to locations in the room, and then speak a simple utterance like "This is a ball". The TWIG system has been able to learn definitions for pronouns like "you" and "he": each pronoun is defined by a decision tree containing conditions on atomic formulas (facts) in the knowledge base (e.g., "you" is a person whom the speaker is looking at).

Finally we specifically look at how to exploit the knowledge represented in the system, in such a way that a measurable increase of the performance can be assessed.

In [Aiello, Cecchi, and Sartini 1986] an in-depth discussion of the combination of knowledge and meta-level knowledge is provided. The authors describe various paradigms to use meta-knowledge in the design of knowledge-based systems, and address the issue of self-descriptive systems. Such a system could employ knowledge about itself to perform self-evaluation and self-modification. The use self-evaluation and self-modification in robotics is addressed in [Calisi, Iocchi, Nardi, Scalzo, and Ziparo 2008], where the concept of Context-Based Architecture is introduced. The authors present an approach to the design of robotic systems (including a formal model of robotic architecture) that is based on the explicit representation of knowledge about context. The goal of the approach is to improve the system performance, by dynamically tailoring the functionalities of the robot to the specific features of the situation at hand. The architecture enables a new design methodology for robotic systems, in which the robot is equipped with a high-level feedback controller exploiting three broad classes of contextual knowledge: mission-related, environmental, and introspective. Knowledge is represented in the form of a Horn knowledge base: the system continuously acquires data from the robot's sensing modules, extracts symbolic knowledge from this data, updates the knowledge-base, and uses reasoning procedures to infer the controlled values to be passed to the robot's submodules (e.g., the planning module or the navigation module). This way, the knowledge base is always referring to the current situation, and the knowledge acquired can be used to control all the robot's subsystems. Notice that using a common representation for the robot's knowledge can improve the overall design, since in most robotic architectures symbolic knowledge is used only inside a specific subsystem (i.e., the anchoring system or the planning system).

4 KR and Human-Robot Interaction

In the previous section we introduced the problem of symbol grounding. In particular, we focused on the autonomous capability of robots to adopt explicit knowledge representation and ground symbols with perceived elements in the operational environment. However, robots are, at present, not enough skilled, and the grounding process can be affected by various types of errors that decrease the robot performance in this task. First, errors and noise may affect the perception itself; second, the information extracted from raw data could be incomplete, hence leading to ambiguities. Finally, even when grounding a single concept, inconsistencies may arise when adding the symbol to the knowledge base. Grounding also requires to cope with additional challenges, such as combining multiple perceptions coming from different sensors, or managing multi-robot sources. As a result, current autonomous grounding implementations handle a narrow set of aspects, which do not guarantee any system flexibility or robustness to new concepts, unexpected events, or re-planning in case of failure.

On the other hand, humans own an innate attitude in grounding symbols with physical elements in the environment. Even more, according to Normal, our own cognitive processes are somehow related to the way we perceive physical objects [Norman 1991]. Because of these considerations, the robotic community has recently investigated how to leverage these humans skills to support robotic systems in the grounding process. *Human-centered* symbol grounding is more advantageous with respect to autonomous robotic approaches, and contributes to acquire effective knowledge, which can be in turn exploited by robotic systems to enhance their performance and the overall mission status.

Human-centered symbol grounding can be applied to different robotic domains. For example, it may effectively contribute in *human augmented mapping*, a novel approach to semantic mapping where a human operator is moved in the middle of the acquisition and grounding process, cooperating with the robotic system. Another relevant topic is the human's capability to control a robot through natural commands. Humans tend to provide spatial information about places by using *spatial relations* directly within the environment. For example, commanding the robot to reach "the second door on the left side of the corridor" is a natural instruction for a user, yet it involves a significant grounding effort for a robotic system, such as grounding the concepts of door, corridor, as well as the spatial hint *left* with respect to the perceived objects.

However, moving humans in the middle of the grounding process poses several challenges with respect to robot-centered grounding. In fact, it is important to design natural processes for humans, in order to select elements of the operational environment. First, such a process should be continuous, that is, not isolated or limited in time. Second, the acquisition should be quick, to guarantee a high number of acquired elements without frustrating the user. Finally, it should leverage humans' innate skills and comfortable interaction means. Sitting in front of a PC meanwhile the robot is navigating somewhere within the environment is not the best acquisition means. Labeling objects through mouse and keyboard is a repetitive and not human friendly task. Thereby, traditional human-computer interaction paradigms are not effective for this activity, and this has been for a long time a major drawback for this human-centered grounding.

4.1 Innovative interaction means for human-centered symbol grounding

Recent technological improvements in the field of interaction means has renovated the interest in human-centered symbol grounding in a different light, and fostered the design of innovative human-robot interactions. Nowadays, knowledge representation formalisms can be coupled with human-robot interaction technologies, such as: *(i)* natural language, *(ii)* speech technologies and *(iii)* gesturing and pointing through tangible user interfaces. Concerning the adoption of natural language, in [Kruijff, Zender, Jensfelt, and Christensen 2006] the grounding process of a mobile robot is supported by asking humans for clarifications using natural language. However, the human role is limited only to solve ambiguous situations previously detected by autonomous robot modules, and not to provide new knowledge. Marciniak *et al.* design a natural language interface for a mobile robot, whose knowledge base contains spatial information describing static situations and actions, such as projective relations (e.g. *behind, in_front_of, left_of, right_of*) and relative distance (e.g. *far, close, in*) [Marciniak and Vetulani 2002]. Skubic *et al.* investigate the use of spatial relations to ease human-robot communications [Skubic, Perzanowski, Blisard, Schultz, Adams, Bugajska, and Brock 2004]. Such information is extracted by grid map, and adopted for human feedback and for human-robot commands. Loutfi *et al.* validate how a knowledge representation and reasoning system (KR&R) can improve anchoring and human-robot interaction-based tasks [Loutfi, Coradeschi, Daoutis, and Melchert 2008]. Moreover, this is one of the few attempts in robotics to reuse generic knowledge, since they adopt a general purpose upper level ontology, DOLCE (A Descriptive Ontology for Linguistic and Cognitive Engineering), instead of a content dependent one. This effort is significant, since it enhances the integration between robotic systems and other devices and fosters their deployment in everyday activities.

Addressing speech technologies, Theobalt *et al.* combine a sophisticated low-level robot navigation with a symbolic high-level spoken dialogue system [Theobalt, Bos, Chapman, Espinosa-Romero, Fraser, Hayes, Klein, Oka, and Reeve 2002]. Kruijff *et al.* present an ontology-based approach to multi-layered conceptual spatial mapping that provides a common ground for human-robot dialogue [Kruijff, Zender, Jensfelt, and Christensen 2007]. It is thus possible to establish references to spatial areas in a situated dialogue between a human and a robot about their environment.

Finally, we look at tangible interaction, in particular referring to gesturing and pointing metaphors. The robotic community has recently investigated the role of tangible user interfaces, which have been widely adopted in human-computer interaction, but are still emerging in robotics. Hasanuzzaman *et al.* define a frame-based knowledge model for person-centric gesture interpretation [Hasanuzzaman, Zhang, Ampornaramveth, Gotoda, Shirai, and Ueno 2007]. Their knowledge-based management system SPAK acquires knowledge from various software agents (e.g. gesture recognizer) and, through reasoning, determines the actions to be taken and submits the corresponding commands to the target robot control. Manifold TUIs are portable devices; for example, Nintendo Wiimote, Sony Move, or Microsoft XWand. This allows users to move within the environment, and to accomplish tasks while co-located with robots. Even more, nowadays there are many cheap commercial off-the-shelf (COTS) products, whose adoption fosters their use in common activities. Tangible interfaces exhibit different interaction metaphors, which share a common human attitude: gesturing. In particular, spot pointing is relevant to naturally refer to elements of the operational environment. Since spot pointing in 3D environments requires to point specific places or objects, this can be effectively accomplished using TUIs acting as *tangible pointing devices*. In fact, most of tangible interfaces are equipped with accelerometer sensors, and sometimes with gyroscopes and magnetometers too. Information coming from these sensors is used to detect the position and attitude of the tangible pointer, which is a preliminary step to point objects within the environment. Still, one problem arises: these sensors do not provide any feedback about the distance between the interface and the selected objects, hence they are not valuable for localizing the object position. This problem is relatively easy to solve in virtual environments, where metric distances between objects are known a priori. For example, a common technique is *ray casting*. A virtual ray originating at the user's device shoots out in the direction she is pointing. Typically, the first object to be hit by the ray is selected. Gallo *et al.* adopt this method to manipulate 3D objects reconstructed in a virtual environment from medical data [Gallo, De Pietro, and Marra 2008]. In real environments the problem is much more challenging. A first approach to the general problem is to couple tangible interfaces with distance sensors (e.g. sonars, laser rangefinders, and so on). However, this does not guarantee users' mobility, since these sensors are typically not designed as portable devices. Another approach is to equip environments with sensors to detect the position of selected spots. For example, Sko and Gardner adopt multiple IR bars to enhance the Wiimote capabilities as controller in a virtual reality theatre [Sko and Gardner 2009]. Yet, such a solution would require significant changes in the environment.

It is worth mentioning that the robotic community is moving towards multimodal interfaces, by combining, for example, vision, speech and tactile techniques, to improve the overall interaction process. Since each of them exhibits specific advantages and disadvantages, their combination enhances the overall system robustness. Needless to say, such integration fosters human-centered grounding as well. In fact, it has been proved by Messing and Campbell that combining utterances and gestures is an effective interaction mean for humans, especially when dealing with spatial relations [Messing and Campbell 1999]. For example, consider the following instruction "go over *there*". This would be useless without recognizing the user pointing to a specific place. Perzanowski *et al.* propose a multimodal interface where gestures disambiguates speech commands [Perzanowski, Schultz, Adams, Marsh, and Bugajska 2005].

4.2 A multimodal framework for HRI-based semantic knowledge acquisition

In the rest of this section we sketch out how the aforementioned human-robot interaction metaphors can be combined in a multimodal interface, and integrated with established AI approaches, to design innovative human-centered grounding methodologies. In particular, our proposal relies on a tight synergy between tangible user interfaces and knowledge representation; we refer to this aspect as *semantic-driven tangible interfaces*. Further details about such multimodal framework can be found in [Randelli 2011]. The leading principle we are considering is to take the best from each of the HRI technologies so far considered, with a twofold goal: achieving effective grounding, and designing a comfortable acquisition methodology for humans. We analyze those aspects where each interaction mean is more effective, in order to evince their effective integration. We start our investigation with an extensive example of our human-centered framework for grounding, reported in Table 1, and analyzed in the rest of this section. Our approach is to decompose human-centered grounding into two activities: pointing relevant spots of the operational environment, and grounding symbols with the acquired perceptions. Tackling the former aspect, we already introduced two classes of approaches: coupling tangible interfaces with distance sensors, or equipping environments with sensors that detect the position of selected spots. The former solution does not permit any user's mobility within the environment, while the latter requires significant changes in the environment. Consequently, we decouple the pointing process into a *selection* and a *detection* component, and demand the latter to the robotic platform. In particular, tangible interfaces are a valuable tool for spot selection, while spot detection has been demanded to vision-based systems, which allow for relatively easy object recognition. The human operator is equipped with a commercial-off-the-shelf small pointer, which is used to highlight a relevant spot with a green dot. A vision subsystem, composed of a stereo camera and a pan/tilt unit, is delegated to recognize the green dot, hence detecting the selected spot. Our framework for grounding supports the operator throughout the whole acquisition process. At the beginning, the user selects the desired robot behavior or task (step 1), and the grounding framework is initialized for the grounding process (step 2). The human operator can thus point at spots in the environment (steps 3 and 16), whose location can be determined by the vision system on the robot (steps 5 and 18). It is worth noting that the user activates the vision subsystem when she is ready (steps 4 and 17) and that every utterance for the robot is prefixed by the token "*Robot*", which is useful for the robot's automatic speech recognition (ASR) module. In this way, we boost the overall system robustness to utterance noises or to other people speaking. In addition to pointing elements in the environment, the tangible interface is useful to provide contextual information (step 6), which is mapped onto information for the pan/tilt unit, to narrow the search space for the green dot (step 7). For example, moving the arm outwards, the operator suggests to the vision component that the dot is farther, hence the system will react by tilting up the camera. Moving the arm towards the left implies that the spot is more on the left, and the pan will be activated accordingly. Our framework supports a tight integration with the knowledge representation and reasoning system. The robot dynamically activates smart behaviors in accordance to the user's actions. For example, once detected, the robot accesses relevant information about windows or doors to further interact with the user, in order to acquire further details that will boost its patrolling performance (steps 8-9 and 19-22). On the other hand, in case of failure, the robot asks the user for support (steps 13-15), in order to identify the type of problem and recover from the failure, by re-planning its task.

Once the relevant spots in the environment have been selected, a second activity is how to effectively ground symbols with the acquired perceptions. We have already mentioned that pure vision-based approaches suffer from several limitations under this point of view. Gestures are typically mapped onto restricted vocabularies, hence they are not expressive enough to solve grounding. On the other hand, conventional approaches, such as tagging, or

Time	Human		Robot	
	Comp.	**Action**	**Comp.**	**Action**
1	U	"Plan a surveillance task"		
2			S	"Select the elements to patrol"
3	T	[The user points a window]		
4	U	"This is a window"		
5			V	[The robot activates the vision-based detection algorithm]
6	T	[The user performs a left gesture with the TUI to suggest that the window is more on the left]		
7			V	[The pan/tilt unit rotates on the left, according to the user's hint]
8			S	"Window detected" [The robot retrieves knowledge about patrolling windows.] "Should I patrol the window in daylight?"
9	U	"No"		
10			S	"Select the next element"
11	U	"Follow me" [The user approaches the second element]		
12			S	"OK." [The robot activates the vision-based user tracking.]
13			S	[The robot gets stuck because of an undetected obstacle and asks for clarifications] "What did I bump into?"
14	U	"This is a chair"		
15			S	[The robot adds in its map the position of the obstacle and recovers from the failure] "OK"
16	T	[The user points a door]		
17	U	"This is a door"		
18			V	[The robot activates the vision-based detection algorithm]
19			S	"Door detected. [The robot retrieves knowledge about patrolling doors.] Will the door be open during the night?"
20	U	"No"		
21			S	"Is this an entry door?"
22	U	"Yes"		
21	...	...	...	...

Table 1. An example of our grounding framework with the involved time ordered activities. The *component* column (Comp.) reports the specific human interaction mean or robotic module adopted: *U* stands for *utterance*, *T* for *tangible interaction*, *V* for *vision subsystem*, and *S* for *robot speakers*.

using graphical interfaces, are less comfortable for humans, in particular if symbol grounding is regarded as a continuous process (not restricted to system set-up). Our framework adopts a speech recognition system that converts the operator's utterance into a symbolic representation, which is in turn grounded to the element previously detected by the vision subsystem (in the aforementioned example, the terms *window*, *door* and *chair* are expanded with unique IDs and grounded with the detected objects). Once grounded, the symbol is stored in the knowledge base, and will be exploited to trigger robot behaviors.

To summarize, in order to obtain a timely and effective knowledge transfer from the human to the robot, the most suitable means must be used: pointing, gesture and speech. Indeed, this flow of information is built upon the common knowledge of the domain of human and robot. Moreover, the interaction should be supported by the ability of the robot to use commonsense knowledge to focus the user on the knowledge that is needed for a successful task accomplishment.

Acknowledgments: Doing research in Knowledge Representation and Reasoning and in Cognitive Robotics has been and is challenging and rewarding for us. Hector has been through these years a constant source of inspiration and insight and we sincerely thank him, wishing, on his sixtieth birthday, all the best and many happy and productive years to come.

References

Aiello, L., C. Cecchi, and D. Sartini [1986]. Representation and use of metaknowledge. In *Proceedings of the IEEE*, pp. 1304 – 1321.

Brooks, R. A. [1986]. A robust layered control system for a mobile robot. *IEEE Journal of Robotics and Automation RA-2*(1), 14–23.

Calisi, D., L. Iocchi, D. Nardi, C. Scalzo, and V. A. Ziparo [2008, November). Context-based design of robotic systems. *Robotics and Autonomous Systems (RAS) - Special Issue on Semantic Knowledge in Robotics 56*(11), 992–1003.

Castelpietra, C., A. Guidotti, L. Iocchi, D. Nardi, and R. Rosati [2002]. Design and implementation of cognitive soccer robots. In *RoboCup 2001: Robot Soccer World Cup V*, pp. 312–318.

Chella, A., S. Coradeschi, M. Frixione, and A. Saffiotti [2004]. Perceptual anchoring via conceptual spaces. In *Proc. of the AAAI-04 Workshop on Anchoring Symbols to Sensor Data*, Menlo Park, CA. AAAI Press. Online at http://www.aass.oru.se/~asaffio/.

Coradeschi, S. and A. Saffiotti [2000]. Aaai symbols to sensor data: preliminary report. In *Proceedings of the 17th AAAI Conference (AAAI-00)*, Menlo Park, CA, pp. 129–135. AAAI Press. Online at http://www.aass.oru.se/~asaffio/.

De Giacomo, G., L. Iocchi, D. Nardi, and R. Rosati [1999]. A theory and implementation of cognitive mobile robots. *Journal of Logic and Computation 5*(9), 759–785.

De Giacomo, G., Y. Lesperance, and H. J. Levesque [1997]. Reasoning about concurrent execution, prioritized interrupts, and exogenous actions in the Situation Calculus. In *International Joint Conference on Artificial Intelligence*, Volume 15, pp. 1221–1226.

Ferrein, A. and G. Lakemayer [2008]. Logic-based robot control in highly dynamic domains. *Journal of Robotics and Autonomous Systems, Special Issue on Semantic Knowledge in Robotics 56*(11), 980 – 991.

Fikes, R. and N. Nilsson [1971]. STRIPS: A new approach to the application of theorem proving to problem solving. *Artificial Intelligence 2*, 189–208.

Galindo, C., J. Fernandez-Madrigal, J. Gonzalez, and A. Saffiotti [2008]. Robot task planning using semantic maps. *Robotics and Autonomous Systems 56*(11), 955–966. Online at http://www.aass.oru.se/~asaffio/.

Gallo, L., G. De Pietro, and I. Marra [2008]. 3D interaction with volumetric medical data: experiencing the Wiimote. In *Proceedings of the 1st international conference on Ambient media and systems*, pp. 1–6. ICST (Institute for Computer Sciences, Social-Informatics and Telecommunications Engineering).

Gärdenfors, P. [2000]. *Conceptual Spaces*. Cambridge, MA: MIT Press, Bradford Books.

Giunchiglia, E., G. N. Kartha, and V. Lifschitz [1997]. Representing action: Indeterminacy and ramifications. *Artificial Intelligence 95*(2), 409–438.

Gold, K., M. Doniec, C. Crick, and B. Scasellati [2009]. Robotic vocabulary building using extension inference and implicit contrast. *Artificial Intelligence* (173), 145–166.

Grosskreutz, H. and G. Lakemayer [2000]. ccGolog: Towards more realistic logic-based robot controllers. In *Proceedings of the 7th Conference on Artificial Intelligence (AAAI-00)*, Menlo Park, CA, pp. 476–482. AAAI Press.

Harnad, S. [1990]. The symbol grounding problem. *Physica D: Nonlinear Phenomena 42*(1-3), 335–346.

Hasanuzzaman, M., T. Zhang, V. Ampornaramveth, H. Gotoda, Y. Shirai, and H. Ueno [2007]. Adaptive visual gesture recognition for human-robot interaction using a knowledge-based software platform. *Robotics and Autonomous Systems 55*(8), 643–657.

Hayes, P. and J. McCarthy [1969]. Some philosophical problems from the standpoint of artificial intelligence. *Machine Intelligence 4*, 463–502.

Iocchi, L. [1999]. *Design and Development of Cognitive Robots*. Ph.D. thesis, Univ. "La Sapienza", Roma, Italy, On-line ftp.dis.uniroma1.it/pub/iocchi/.

Kitano, H. and M. Asada [1998]. RoboCup humanoid challenge: That's one small step for a robot, one giant leap for mankind. In *Intelligent Robots and Systems, 1998. Proceedings., 1998 IEEE/RSJ International Conference on*, Volume 1, pp. 419–424. IEEE.

Kruijff, G., H. Zender, P. Jensfelt, and H. Christensen [2006, March]. Clarification dialogues in human-augmented mapping. In *Proceedings of the 1st Annual Conference on Human-Robot Interaction (HRI'06)*, Salt Lake City, UT.

Kruijff, G., H. Zender, P. Jensfelt, and H. Christensen [2007]. Situated dialogue and spatial organization: What, where... and why. *International Journal of Advanced Robotic Systems 4*(2), 125–138.

Kunze, L., M. Tenorth, and M. Beetz [2010]. Putting people's common sense into knowledge bases of household robots. In *KI 2010: Advances in Artificial Intelligence, 33rd Annual German Conference on AI, Karlsruhe, Germany, September 21-24, 2010. Proceedings*, pp. 151–159. Springer.

LeBlanc, K. and A. Saffiotti [2008]. Cooperative anchoring in heterogeneous multi-robot systems. In *Proc. of the IEEE Int. Conf. on Robotics and Automation (ICRA)*, Pasadena, CA, pp. 3308–3314. Online at http://www.aass.oru.se/~asaffio/.

Lesperance, Y., H. Levesque, F. Lin, D. Marcu, R. Reiter, and R. Scherl [1994]. A logical approach to high-level robot programming. In *AAAI FAll Symposium on Control of the Physical World by Intelligent Systems*, pp. 79–85.

Lésperance, Y., K. Tam, and M. Jenkin [2000]. Reactivity in a logic-based robot programming framework. In *Intelligent Agents VI: Agent Theories, Architectures, and Languages*, pp. 173–187.

Levesque, H. [1984]. A logic of implicit and explicit belief. In *Proceedings of the National Conference on Artificial Intelligence (AAAI-84)*, pp. 198–202.

Levesque, H., P. Cohen, and J. Nunes [1990]. On acting together. In *Proceedings of the Eighth Conference on Artificial Intelligence (AAAI-90)*, Boston, MA, pp. 94–99.

Levesque, H. and G. Lakemeyer [2008]. Chapter 23 cognitive robotics. In V. L. Frank van Harmelen and B. Porter (Eds.), *Handbook of Knowledge Representation*, Volume 3 of *Foundations of Artificial Intelligence*, pp. 869 – 886. Elsevier.

Levesque, H. and R. Reiter [March 1998]. Beyondplanning. In *Working notes of the AAAI Spring Symposiumon Integrating Robotics Research*.

Levesque, H., R. Reiter, Y. Lespérance, F. Lin, and R. Scherl [1997]. Golog: A logic programming language for dynamic domains. *Journal of Logic Programming 31*, 59–84.

Loutfi, A., S. Coradeschi, M. Daoutis, and J. Melchert [2008]. Using knowledge representation for perceptual anchoring in a robotic system. *Int. Journal on Artificial Intelligence Tools 17*, 925–944.

Marciniak, J. and Z. Vetulani [2002]. Ontology of Spatial Concepts in a Natural Language Interface for a Mobile Robot. *Applied Intelligence 17*(3), 271–274.

Messing, L. and R. Campbell [1999]. *Gesture, speech, and sign*. Oxford University Press.

Nilsson, N. [1984]. Shakey the robot. Technical note 323.

Nilsson, N. J. [1969]. A mobile automaton: an application of Artificial Intelligence techniques. In *Proc. of Int. Joint Conf. on Artificial Intelligence (IJCAI)*, pp. 509–520.

Nolfi, S. and D. Floreano [2000]. *Evolutionary Robotics*. MIT Press.

Norman, D. [1991]. *Cognitive artifacts*, pp. 17–38. Cambridge Univ Pr.

Palamara, P., V. Ziparo, L. Iocchi, D. Nardi, P. Lima, and H. Costelha [2008, May]. A robotic soccer passing task using Petri Net Plans (demo paper). In P. M. Padgham and Parsons (Eds.), *Proc. of 7th Int. Conf. on Autonomous Agents and Multiagent Systems (AAMAS 2008)*, Estoril, Portugal, pp. 1711–1712. IFAAMAS Press.

Perzanowski, D., A. Schultz, W. Adams, E. Marsh, and M. Bugajska [2005]. Building a multimodal human-robot interface. *Intelligent Systems, IEEE 16*(1), 16–21.

Pfeifer, R. and C. Scheier [1999]. *Understanding intelligence*. The MIT Press.

Randelli, G. [2011]. *Improving Human-Robot Awareness through Semantic-driven Tangible Interaction.* Ph.D. thesis, "Sapienza" University of Rome, Department of Computer and System Sciences, Rome, Italy (forthcoming).

Reiter, R. [2001]. *Knowledge in action: Logical foundations for describing and implementing dynamical systems.* MIT Press.

Russell, S. J. and P. Norvig [2005]. *Artificial Intelligence: A Modern Approach.* Pearson Education. Third Edition.

Scherl, R. B. and H. J. Levesque [1993]. The frame problem and knowledge-producing actions. In *AAAI*, pp. 689–695.

Shanahan, M. [1997]. *Solving the frame problem: A mathematical investigation of the common sense law of inertia.* MIT Press.

Sko, T. and H. Gardner [2009]. The Wiimote with multiple sensor bars: creating an affordable, virtual reality controller. In *CHINZ '09: Proceedings of the 10th International Conference NZ Chapter of the ACM's Special Interest Group on Human-Computer Interaction*, New York, NY, USA, pp. 41–44. ACM.

Skubic, M., D. Perzanowski, S. Blisard, A. Schultz, W. Adams, M. Bugajska, and D. Brock [2004]. Spatial language for human-robot dialogs. *Systems, Man, and Cybernetics, Part C: Applications and Reviews, IEEE Transactions on 34*(2), 154–167.

Tenorth, M. and M. Beetz [2009]. KnowRob - Knowledge Processing for Autonomous Personal Robots. In *IEEE/RSJ International Conference on Intelligent RObots and Systems*, pp. 4261–4266.

Theobalt, C., J. Bos, T. Chapman, A. Espinosa-Romero, M. Fraser, G. Hayes, E. Klein, T. Oka, and R. Reeve [2002]. Talking to Godot: Dialogue with a mobile robot. In *Proceedings of IEEE/RSJ International Conference on Intelligent Robots and Systems (IROS 2002)*, pp. 1338–1343.

Thielscher, M. [2005]. *Reasoning Robots*, Volume 33 of *Applied Logic Series*. Springer.

Ziparo, V., L. Iocchi, P. Lima, D. Nardi, and P. Palamara [2010]. Petri Net Plans. *Autonomous Agents and Multi-Agent Systems*, 1–40. 10.1007/s10458-010-9146-1.

3

Golog-Style Search Control for Planning

Jorge A. Baier, Christian Fritz, and Sheila A. McIlraith

ABSTRACT. Domain control knowledge (DCK) has proven effective in improving the efficiency of plan generation by reducing the search space for a plan. *Procedural* DCK is a compelling type of DCK that supports a natural specification of the skeleton of a plan. Unfortunately, most state-of-the-art planners do not have the machinery necessary to exploit procedural DCK. To resolve this deficiency, we propose to compile procedural DCK directly into the Planning Domain Definition Language (PDDL), specifically PDDL2.1. PDDL is the de facto standard input language for state-of-the-art automated planning systems. Our compilation enables any PDDL2.1-compatible planner to exploit procedural DCK without the need for special-purpose computational machinery. The contributions of this paper are threefold. First, inspired by the logic programming language GOLOG, we propose a PDDL-based semantics for an Algol-like procedural language that can be used to specify procedural DCK as a program. Second, we provide a polynomial algorithm that translates an ADL planning instance and a DCK program, into an equivalent, program-free PDDL2.1 instance whose plans are only those that adhere to the program. Third, we argue that the resulting planning instance is well-suited to being solved by domain-independent heuristic planners. To this end, we propose three approaches to computing domain-independent heuristics for our translated instances, sometimes leveraging properties of our translation to guide search. In our experiments on familiar PDDL planning benchmarks we show that the proposed compilation of procedural DCK can significantly speed up the performance of a heuristic search planner. Our translators are implemented and available on the web.

Foreword (by Sheila McIlraith)

When Hector, Ray Reiter, and colleagues at the University of Toronto first introduced the GOLOG logic programming language, it was viewed as a means of specifying high level control for robots and software agents, as well as for industrial processes and discrete event simulations. Part of GOLOG's elegance was its situation calculus semantics which enabled GOLOG programs to reason about the complex dynamics of the world, as specified in situation calculus. From an outsider's perspective, GOLOG was inextricably tied to the situation calculus and to the simple interpreter that allowed it to reason over the situation calculus specification of dynamical systems.

Our motivation with this work was to bring the same basic GOLOG philosophy to bear on dynamical systems described in the less expressive Planning Domain Definition Language (PDDL) [McDermott 1998], using a GOLOG-style language to specify programs that either imparted procedural search control on the plan generation process or that specified what could be viewed as a complex, temporally extended plan objective. In contrast to GOLOG, the semantics of this GOLOG-style language was provided in PDDL. More importantly, rather than use a GOLOG interpreter to extract a plan, the mechanism by which we specified our PDDL semantics enabled us to *compile away* our GOLOG-like programs and to exploit highly effective state-of-the-art planning technology to synthesize plans that adhered to the constraints of the GOLOG-like programs.

Our initial work in this vein is reflected in the article that follows, a version of which originally appeared under the title "Exploiting Procedural Domain Control Knowledge in State-of-the-Art Planners" in the Proceedings of the Seventeenth International Conference on Automated Planning and Scheduling (ICAPS2007). A follow-on paper at KR2008 examined related issues, extending the expressivity of our GOLOG-like language to include a number of ConGolog constructs.

The article that follows was written for a planning audience and as such extols the virtues and shortcomings of the work from a plan generation perspective. Indeed, an important benefit of our work is that it provides a means for state-of-the-art planners to exploit procedural domain control knowledge (DCK) specified in a GOLOG-like language. This benefit is also shared by GOLOG researchers in so far as the work provides a computationally effective means of synthesizing a class of GOLOG program executions with respect to a restricted class of dynamical systems. However, the benefits of this work do not stop there. The technique used to specify the semantics, here in PDDL, in the KR2008 paper in the situation calculus, is interesting because it avoids the need for reification – something Hector particularly liked about the work. Further, from our understanding of the relationship between PDDL and situation calculus and between GOLOG and other DCK formalisms such as Hierarchical Task Networks (HTNs) (elaborated upon by Alfredo Gabaldon in this volume), this body of work enables not only GOLOG, but also HTN-like DCK, or even hybrid GOLOG-HTN DCK to be compiled away in a similar fashion and for plans and program executions to be synthesized using efficient automated planning technology.

1 Introduction

Domain control knowledge (DCK) imposes domain-specific constraints on the definition of a valid plan. As such, it can be used to impose restrictions on the course of action that achieves the goal. While DCK sometimes reflects a user's desire to achieve the goal a particular way, it is most often constructed to aid in plan generation by reducing the plan search space. Moreover, if well-crafted, DCK can eliminate those parts of the search space that necessitate backtracking. In such cases, DCK together with blind search can yield valid plans significantly faster than state-of-the-art planners that do not exploit DCK. Indeed most planners that exploit DCK, such as TLPLAN [Bacchus and Kabanza 1998] or TALPLANNER [Kvarnström and Doherty 2000], do little more than blind depth-first search with cycle checking in a DCK-pruned search space. Since most DCK reduces the search space but still requires a planner to backtrack to find a valid plan, it should prove beneficial to exploit better search techniques. In this paper we explore ways in which state-of-the-art planning techniques and existing state-of-the-art planners can be used in conjunction with DCK, with particular focus on *procedural* DCK.

As a simple example of DCK, consider the `trucks` domain of the 2006 International Planning Competition, where the goal is to deliver packages between certain locations using a limited capacity truck with restricted access. Once a package reaches its destination it must be delivered to the customer. We can write simple and natural procedural DCK that significantly improves the efficiency of plan generation for instance: *Repeat the following until all packages have been delivered: Unload everything from the truck, and, if there is any package in the current location whose destination is the current location, deliver it. After that, if any of the local packages have destinations elsewhere, load them on the truck while there is space. Drive to the destination of any of the loaded packages. If there are no packages loaded on the truck, but there remain packages at locations other than their destinations, drive to one of these locations.*

Procedural DCK, as used in HTN [Nau, Cao, Lotem, and Muñoz-Avila 1999] or GOLOG [Levesque, Reiter, Lespérance, Lin, and Scherl 1997], is action-centric. It is much like a programming language, and often times like a plan skeleton or template. It can (conditionally) constrain the order in which domain actions should appear in a plan. In order to exploit it for planning, we require a procedural DCK specification language. To this end, we propose a language based on GOLOG that includes typical programming languages constructs such as conditionals and iteration as well as nondeterministic choice of actions in places where control is not germane. We argue that these action-centric constructs provide a natural language for specifying DCK for planning. We contrast them with DCK specifications based on linear temporal logic (LTL) which are state-centric and though still of tremendous value, arguably provide a less natural way to specify DCK. We specify the syntax for our language as well as a PDDL-based semantics following Fox and Long [2003].

With a well-defined procedural DCK language in hand, we examine how to use state-of-the-art planning techniques together with DCK. Of course, most state-of-the-art planners are unable to exploit DCK. As such, we present an algorithm that translates a PDDL2.1-specified Action Description Language (ADL) [Pednault 1989] planning instance and associated procedural DCK into an equivalent, program-free PDDL2.1 instance whose plans provably adhere to the DCK. Any PDDL2.1-compliant planner can take such a planning instance as input to their planner, generating a plan that adheres to the DCK.

Since they were not designed for this purpose, existing state-of-the-art planners may not exploit techniques that optimally leverage the DCK embedded in the planning instance. As such, we investigate how state-of-the-art planning techniques, rather than planners, can be used in conjunction with our compiled DCK planning instances. In particular, we propose domain-independent search heuristics for planning with our newly-generated planning instances. We examine three different approaches to generating heuristics, and evaluate them on three domains of the 2006 International Planning Competition. Our results show that procedural DCK improves the performance of state-of-the-art planners, and that our heuristics are sometimes key to achieving good performance.

2 Background

In this section, we review the subset of PDDL2.1 that we use to define the semantics of our GOLOG-like language. PDDL is the de facto standard specification language for input to most state-of-the-art planners, providing a means of specifying planning domains (roughly, predicates and actions) and planning instances (roughly, objects, initial state, and goal specification).

2.1 A Subset of PDDL 2.1

In PDDL, a *planning instance* is a pair $I = (D, P)$, where D is a domain definition and P is a problem. To simplify notation, we assume that D and P are described in an ADL subset of PDDL. The difference between this ADL subset and PDDL 2.1 is that no concurrent or durative actions are allowed [Fox and Long 2003].

Following convention, domains are tuples of finite sets $(PF, Ops, Objs_D, T, \tau_D)$, where PF defines domain predicates and functions, Ops defines operators, $Objs_D$ contains domain objects, T is a set of types, and $\tau_D \subseteq Objs_D \times T$ is a type relation associating objects to types. An operator (or action schema) is also defined by a tuple $\langle O(\vec{x}), \vec{t}, Prec(\vec{x}), Eff(\vec{x})\rangle$, where $O(\vec{x})$ is the unique operator name and $\vec{x} = (x_1, \ldots, x_n)$ is a vector of variables. Furthermore, $\vec{t} = (t_1, \ldots, t_n)$ is a vector of types. Each variable x_i ranges over objects associated with type t_i. Moreover, $Prec(\vec{x})$ is a boolean formula with quantifiers (BFQ) that specifies the operator's preconditions. BFQs are defined inductively as follows. Atomic BFQs are either of the form $t_1 = t_2$ or $R(t_1, \ldots, t_n)$, where t_i ($i \in \{1, \ldots, n\}$) is a term (i.e. either a variable, a function literal, or an object), and R is a predicate symbol. If φ is a BFQ, then so is $Qx\text{-}t\,\varphi$, for a variable x, a type symbol t, and Q is either $\exists$ or $\forall$. BFQs are also formed by applying standard boolean operators over other BFQs. Finally $Eff(\vec{x})$ is a list of conditional effects, each of which can be in one of the following forms:

$$\forall y_1\text{-}t_1 \cdots \forall y_n\text{-}t_n.\, \varphi(\vec{x}, \vec{y}) \Rightarrow R(\vec{x}, \vec{y}), \tag{1}$$

$$\forall y_1\text{-}t_1 \cdots \forall y_n\text{-}t_n.\, \varphi(\vec{x}, \vec{y}) \Rightarrow \neg R(\vec{x}, \vec{y}), \tag{2}$$

$$\forall y_1\text{-}t_1 \cdots \forall y_n\text{-}t_n.\, \varphi(\vec{x}, \vec{y}) \Rightarrow f(\vec{x}, \vec{y}) = obj, \tag{3}$$

where φ is a BFQ whose only free variables are among $\vec{x}$ and $\vec{y}$, R is a predicate, f is a function, and obj is an object After performing a ground operator – or *action* – $O(\vec{c})$ in a certain state s, for all tuples of objects that may instantiate $\vec{y}$ such that $\varphi(\vec{c}, \vec{y})$ holds in s, effect (1) (resp. (2)) expresses that $R(\vec{c}, \vec{y})$ becomes true (resp. false), and effect (3) expresses that $f(\vec{c}, \vec{y})$ takes the value obj. As usual, states are represented as finite sets of atoms (ground formulae of the form $R(\vec{c})$ or of the form $f(\vec{c}) = obj$).

Planning problems are tuples $(Init, Goal, Objs_P, \tau_P)$, where $Init$ is the initial state, $Goal$ is a sentence with quantifiers for the goal, and $Objs_P$ and τ_P are defined analogously as for domains.

Semantics: Fox and Long [2003] gave a formal semantics for PDDL 2.1. In particular, they define when a sentence is *true* in a state and what *state trace* is the result of performing a set of *timed actions*. A state trace intuitively corresponds to an execution trace, and the sets of timed actions are ultimately used to refer to plans. In the ADL subset of PDDL2.1, since there are no concurrent or durative actions, time does not play any role. Hence, state traces reduce to sequences of states and sets of timed actions reduce to sequences of actions.

Building on Fox and Long's semantics, we assume that $\models$ is defined such that $s \models \varphi$ holds when sentence φ is true in state s. Moreover, for a planning instance I, we assume there exists a relation $Succ$ such that $Succ(s, a, s')$ iff s' results from performing an executable action a in s. Finally, a sequence of actions $a_1 \cdots a_n$ is a plan for I if there exists a sequence of states $s_0 \cdots s_n$ such that $s_0 = Init$, $Succ(s_i, a_{i+1}, s_{i+1})$ for $i \in \{0, \ldots, n-1\}$, and $s_n \models Goal$.

3 A Language for Procedural Control

In contrast to state-centric languages, that often use LTL-like logical formulae to specify properties of the states traversed during plan execution, procedural DCK specification languages are predominantly action-centric, defining a plan template or skeleton that dictates *actions* to be used at various stages of the plan.

Procedural control is specified via *programs* rather than logical expressions. The specification language for these programs is based on GOLOG [Levesque, Reiter, Lespérance, Lin, and Scherl 1997] and thus incorporates desirable elements from imperative programming languages such as iteration and conditional constructs. However, to make the language more suitable to planning applications, it also incorporates nondeterministic constructs.

These elements are key to writing flexible control since they allow programs to contain missing or open program segments, which are filled in by a planner at the time of plan generation. Finally, our language also incorporates property testing, achieved through so-called *test actions*. These actions are not real actions, in the sense that they do not change the state of the world, rather they can be used to specify properties of the states traversed while executing the plan. By using test actions, our programs can also specify properties of executions similarly to state-centric specification languages.

The rest of this section describes the syntax and semantics of the procedural DCK specification language we propose to use. We conclude this section by formally defining what it means to plan under the control of such programs.

3.1 Syntax

Our procedural search control language is based on GOLOG. In contrast to GOLOG, our language supports specification of types for program variables, but does not support procedures.

Programs are constructed using the implicit language for actions and boolean formulae defined by a particular planning instance I. Additionally, a program may refer to variables drawn from a set of program variables V. This set V will contain variables that are used for nondeterministic choices of arguments. In what follows, we assume $\mathcal{O}$ denotes the set of operator names from Ops, fully instantiated with objects defined in I or elements of V.

The set of programs over a planning instance I and a set of program variables V can be defined by induction. In what follows, assume ϕ is a boolean formula with quantifiers on the language of I, possibly including terms in the set of program variables V. Atomic programs are defined as follows.

1. *nil*: the empty program.
2. o: a single operator instance, where $o \in \mathcal{O}$.
3. **any**: "any action".
4. ϕ?: a *test action*, that tests for the truth of formula ϕ.

If σ_1, σ_2 and σ are programs, so are the following:

1. $(\sigma_1; \sigma_2)$: a sequence of programs.
2. **if** ϕ **then** σ_1 **else** σ_2: a conditional sentence.
3. **while** ϕ **do** σ: a while-loop.
4. σ^*: nondeterministic iteration.
5. $(\sigma_1 | \sigma_2)$: nondeterministic choice between two programs.
6. $\pi(x\text{-}t)\,\sigma$: nondeterministic choice of variable $x \in V$ of type $t \in T$.

Before we formally define the semantics of the language, we show some examples that give a sense of the language's expressiveness and semantics.

- **while** $\neg clear(B)$ **do** $\pi(b\text{-}block)\,putOnTable(b)$: while B is not clear choose any b of type block and put it on the table.
- $\mathbf{any}^*; loaded(A, Truck)$?: Perform any sequence of actions until A is loaded in $Truck$. Plans under this control are such that $loaded(A, Truck)$ holds in the final state.
- $(load(C, P); fly(P, LA) \mid load(C, T); drive(T, LA))$: Either load C on the plane P or on the truck T, and perform the right action to move the vehicle to LA.

3.2 Semantics

The problem of planning for an instance I under the control of program σ corresponds to finding a plan for I that is also an execution of σ from the initial state. In the rest of this section we define what those legal executions are. Intuitively, we define a formal device to check whether a sequence of actions $a_1 \cdots a_n$ corresponds to the execution of a program σ. The device we use is a nondeterministic finite state automaton with ε-transitions (ε-NFA).

For the sake of readability, we remind the reader that ε-NFAs are like standard nondeterministic automata except that they can transition without reading any input symbol, through the so-called ε-transitions. ε-transitions are usually defined over a state of the automaton and a special symbol ε, denoting the empty symbol.

An ε-NFA $A_{\sigma, I}$ is defined for each program σ and each planning instance I. Its alphabet is the set of operator names, instantiated by objects of I. Its states are *program configurations* which have the form $[\sigma, s]$, where σ is a program and s is a planning state. Intuitively, as it reads a word of actions, it keeps track, within its state $[\sigma, s]$, of the part of the program that remains to be executed, σ, as well as the current planning state after performing the actions it has read already, s.

Formally, $A_{\sigma, I} = (Q, \mathcal{A}, Tr, q_0, F)$, where Q is the set of program configurations, the alphabet $\mathcal{A}$ is a set of domain actions, the transition function is $Tr : Q \times (\mathcal{A} \cup \{\varepsilon\}) \rightarrow 2^Q$, $q_0 = [\sigma, Init]$, and F is the set of final states.

Our definition of Tr closely follows the definition of $Trans$ and $Final$ from GOLOG's transition semantics [De Giacomo, Lespérance, and Levesque 2000].

The transition function Tr is defined as follows for atomic programs.

$$Tr([a, s], a) = \{[nil, s']\} \quad \text{iff } Succ(s, a, s'), \text{ for any } a \in \mathcal{A}, \tag{4}$$

$$Tr([\mathbf{any}, s], a) = \{[nil, s']\} \text{ iff } Succ(s, a, s'), \text{ for any } a \in \mathcal{A}, \tag{5}$$

$$Tr([\phi?, s], \varepsilon) = \{[nil, s]\} \quad \text{iff } s \models \phi. \tag{6}$$

Equations 4 and 5 dictate that actions in programs change the state according to the $Succ$ relation described in the previous section. Finally, Equation 6 defines transitions for $\phi?$ when ϕ is a sentence (i.e., a formula with no program variables). It expresses that a transition can only be carried out if the plan state so far satisfies ϕ.

Now we define Tr for non-atomic programs. In the definitions below, assume that $a \in \mathcal{A} \cup \{\varepsilon\}$, and that σ_1 and σ_2 are subprograms of σ, where occurring elements in V may have been instantiated by any object in the planning instance I.

$$Tr([(\sigma_1; \sigma_2), s], a) = \{[(\sigma_1'; \sigma_2), s'] \mid [\sigma_1', s'] \in Tr([\sigma_1, s], a)\} \text{ if } \sigma_1 \neq nil, \tag{7}$$

$$Tr([(nil; \sigma_2), s], \varepsilon) = \{[\sigma_2, s]\}, \tag{8}$$

$$Tr([\mathbf{if}\ \phi\ \mathbf{then}\ \sigma_1\ \mathbf{else}\ \sigma_2, s], \varepsilon) = \begin{cases} [\sigma_1, s] & \text{if } s \models \phi, \\ [\sigma_2, s] & \text{if } s \not\models \phi, \end{cases} \tag{9}$$

$$Tr([(\sigma_1 | \sigma_2), s], \varepsilon) = \{[\sigma_1, s], [\sigma_2, s]\} \tag{10}$$

$$Tr([\mathbf{while}\ \phi\ \mathbf{do}\ \sigma_1, s], \varepsilon) = \begin{cases} \{[nil, s]\} & \text{if } s \not\models \phi, \\ \{[\sigma_1; \mathbf{while}\ \phi\ \mathbf{do}\ \sigma_1, s]\} & \text{if } s \models \phi, \end{cases} \tag{11}$$

$$Tr([\sigma_1^*, s], \varepsilon) = \{[(\sigma_1; \sigma_1^*), s], [nil, s]\}, \tag{12}$$

$$Tr([\pi(x\text{-}t)\ \sigma_1, s], \varepsilon) = \{[\sigma_1 |_{x/o}, s] \mid (o, t) \in \tau_D \cup \tau_P\}. \tag{13}$$

where $\sigma_1|_{x/o}$ denotes the program resulting from replacing any occurrence of x in σ_1 by o. We now give some intuitions for the definitions. First, a transition on a sequence corresponds to transitioning on its first component first (Eq. 7), unless the first component is already the empty program, in which case we transition on the second component (Eq. 8). A transition on a conditional corresponds to a transition in the *then* or *else* part depending on the truth value of the condition (Eq. 9). A transition of the nondeterministic choice leads to the consideration of either of the programs (Eq. 10). A transition of a while-loop corresponds to the *nil* program if the condition is false, and corresponds to the body followed by the while-loop if the condition is true (Eq. 11). On the other hand, a transition of σ_1^* represents two alternatives: executing σ_1 at least once, or stopping the execution of σ_1^*, with the remaining program *nil* (Eq. 12). Finally, a transition of the nondeterministic choice corresponds to a transition of its body when the variable has been replaced by any object of the right type (Eq. 13).

To end the definition of $A_{\sigma,I}$, Q corresponds precisely to the program configurations $[\sigma', s]$ where σ' is either *nil* or a subprogram of σ such that program variables may have been replaced by objects in I, and s is any possible planning state. Moreover, Tr is assumed empty for elements of its domain not explicitly mentioned above. Finally, the set of accepting states is $F = \{[nil, s] \mid s$ is any state over $I\}$, i.e., those where no program remains in execution. We can now formally define an execution of a program.

DEFINITION 1 (Execution of a program). A sequence of actions $a_1 \cdots a_n$ is an execution of σ in I iff $a_1 \cdots a_n$ is accepted by $A_{\sigma,I}$.

We use the symbol $\vdash$ to represent a single computation of the automaton. We say that $q \vdash q'$ iff there exists an a such that $q' \in Tr(q, a)$. The symbol $\vdash^*$ represents the reflexive and transitive closure of $\vdash$. Finally, $q_0 \vdash^k q_k$ iff there are exist states $q_1, \ldots, q_{k-1}$ such that $q_0 \vdash q_1 \vdash q_2 \vdash \ldots \vdash q_{k-1} \vdash q_k$.

Before defining what we mean by planning in the presence of control, we prove a number of results that justify the correctness of our automata-based semantics. The detailed proofs can be found in Baier's Ph.D. thesis [Baier 2010]. The first result proves that the definition of the sequence is intuitively correct, i.e., the execution of $\sigma_1; \sigma_2$ corresponds to the execution of σ_1 followed by σ_2.

PROPOSITION 2. *Let σ_1 and σ_2 be programs. If*

$$[\sigma_1; \sigma_2, s] \vdash q_1 \vdash q_2 \vdash \ldots \vdash q_{k-1} \vdash q_k = [nil, s'],$$

then for some $i \in [1, k]$, $q_i = [\sigma_2, s']$ and $[\sigma_1, s] \vdash^* [nil, s']$.

Our second result establishes that the semantics for the execution of an **if - then - else** is intuitively correct.

PROPOSITION 3. *Let ϕ be a BFQ and let σ_1 and σ_2 be programs, then the following holds:*

$$[\textbf{if } \phi \textbf{ then } \sigma_1 \textbf{ else } \sigma_2, s] \vdash^* [nil, s']$$

iff

$$s \models \phi \text{ and } [\sigma_1, s] \vdash^* [nil, s'], \text{ or } s \not\models \phi \text{ and } [\sigma_2, s] \vdash^* [nil, s'].$$

The execution of a nondeterministic choice of programs has the intended meaning too, as shown by the following result.

PROPOSITION 4. *Let σ_1 and σ_2 be programs, then the following holds:*

$$[(\sigma_1|\sigma_2), s] \vdash^* [nil, s'] \quad iff \quad [\sigma_1, s] \vdash^* [nil, s'] \ or \ [\sigma_2, s] \vdash^* [nil, s'].$$

Now we prove that the execution of the while loop correspond to a repeated execution of the body of the loop.

PROPOSITION 5. *Let ϕ be a BFQ and σ be a program. If*

$$[\textbf{while } \phi \textbf{ do } \sigma, s] \vdash q_1 \vdash q_2 \vdash \ldots \vdash q_k \vdash [nil, s'],$$

then:

1. *for all $i \in [1, k]$, q_i is either of the form $q_i = [\sigma_r; \textbf{while } \phi \textbf{ do } \sigma, r_i]$, or of the form $q_i = [\textbf{while } \phi \textbf{ do } \sigma, r_i]$.*

2. *for all $i \in [1, k]$, if q_i is of the form $q_i = [\textbf{while } \phi \textbf{ do } \sigma, r_i]$ then $i < k$ iff $r_i \models \phi$. State q_k is of the form $q_k = [\textbf{while } \phi \textbf{ do } \sigma, r_k]$*

3. *Finally, let n be the number of states q_i ($i \in [1, k]$) of the form $q_i = [\textbf{while } \phi \textbf{ do } \sigma, r_i]$. Then, $[\sigma^n, s] \vdash^* [nil, s']$, where σ^n represents the sequence that repeats σ n times.*

In the GOLOG language [Levesque, Reiter, Lespérance, Lin, and Scherl 1997], the if - then - else construct is defined by macro expansion, in terms of test actions and non-deterministic choices. Below we prove that our semantics for the if - then - else and for the GOLOG macro expansion of such a construct are equivalent.

PROPOSITION 6. *Let ϕ be a BFQ and let σ_1 and σ_2 be programs, then the following holds:*

$$[\textbf{if } \phi \textbf{ then } \sigma_1 \textbf{ else } \sigma_2, s] \vdash [\sigma, s] \quad iff \quad [(\phi?; \sigma_1)|(\neg\phi?; \sigma_2), s] \vdash^3 [\sigma, s].$$

Now that we have justified the correctness of the semantics of the control language, we return to planning. We are now ready to define the notion of planning under procedural control.

DEFINITION 7 (Planning under procedural control). A sequence of actions $a_1 a_2 \cdots a_n$ is a *plan for instance I under the control of program σ* iff $a_1 a_2 \cdots a_n$ is a plan for I and is an execution of σ in I.

4 Compiling Control into the Action Theory

This section describes a translation function that, given a program σ in the DCK language defined above together with a PDDL2.1 domain specification D, outputs a new PDDL2.1 domain specification D_σ and problem specification P_σ. The two resulting specifications can then be combined with any problem P defined over D, creating a new planning instance that embeds the control given by σ, i.e. that is such that only action sequences that are executions of σ are possible. This enables any PDDL2.1-compliant planner to exploit search control specified by any program.

To account for the state of execution of program σ and to describe legal transitions in that program, we introduce a few bookkeeping predicates and a few additional actions. Figure 1 graphically illustrates the translation of an example program shown as a finite state

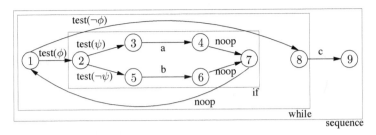

Figure 1: Automaton for **while** ϕ **do** (**if** ψ **then** a **else** b); c.

automaton. Intuitively, the operators we generate in the compilation define the transitions of this automaton. Their preconditions and effects condition on and change the automaton's state.

The translation is defined inductively by a function $C(\sigma, n, E)$ which takes as input a program σ, an integer n, and a list of program variables with types $E = [e_1\text{-}t_1, \ldots, e_k\text{-}t_k]$, and outputs a tuple (L, L', n') with L a list of domain-independent operator definitions, L' a list of domain-dependent operator definitions, and n' another integer. Intuitively, E contains the program variables whose scope includes (sub-)program σ. Moreover, L' contains restrictions on the applicability of operators defined in I, and L contains additional control operators needed to enforce the search control defined in σ. Integers n and n' abstractly denote the program state before and after execution of σ.

We use two auxiliary functions. $Cnoop(n_1, n_2)$ produces an operator definition that allows a transition from state n_1 to n_2. Similarly $Ctest(\phi, n_1, n_2, E)$ defines a similar transition, but conditioned on ϕ. They are defined as:[1]

$$Cnoop(n_1, n_2) = \langle noop\_n_1\_n_2(), [\,], state = s_{n_1}, [state = s_{n_2}]\rangle$$
$$Ctest(\phi, n_1, n_2, E) = \langle test\_n_1\_n_2(\vec{x}), \vec{t}, Prec(\vec{x}), Eff(\vec{x})\rangle \text{ with}$$
$$(\vec{e\text{-}t}, \vec{x}) = mentions(\phi, E), \ \vec{e\text{-}t} = e_1\text{-}t_1, \ldots, e_m\text{-}t_m,$$
$$Prec(\vec{x}) = \big(state = s_{n_1} \wedge \phi[e_i/x_i]_{i=1}^m \wedge \bigwedge_{i=1}^{m} bound(e_i) \to map(e_i, x_i)\big),$$
$$Eff(\vec{x}) = [state = s_{n_2}] \cdot [bound(e_i), map(e_i, x_i)]_{i=1}^m.$$

Function $mentions(\phi, E)$ returns a vector $\vec{e\text{-}t}$ of program variables and types that occur in ϕ, and a vector $\vec{x}$ of new variables of the same length. Bookkeeping predicates serve the following purposes: $state$ denotes the state of the automaton; $bound(e)$ expresses that the program variable e has been bound to an object of the domain; $map(e, o)$ states that this object is o. Thus, the implication $bound(e_i) \to map(e_i, x_i)$ forces parameter x_i to take the value to which e_i is bound, but has no effect if e_i is not bound.

Consider the inner box of Figure 1, depicting the compilation of the if statement. It is

---

[1] We use $A \cdot B$ to denote the concatenation of lists A and B.

defined as:

$$C(\text{if } \phi \text{ then } \sigma_1 \text{ else } \sigma_2, n, E) = (L_1 \cdot L_2 \cdot X, L_1' \cdot L_2', n_3) \text{ with}$$
$$(L_1, L_1', n_1) = C(\sigma_1, n+1, E),$$
$$(L_2, L_2', n_2) = C(\sigma_2, n_1+1, E),\ n_3 = n_2 + 1,$$
$$X = [\, Ctest(\phi, n, n+1, E),\ Ctest(\neg\phi, n, n_1+1, E),$$
$$Cnoop(n_1, n_3),\ Cnoop(n_2, n_3) \,]$$

and in the example we have $\phi = \psi, n = 2, n_1 = 4, n_2 = 6, n_3 = 7, \sigma_1 = a$, and $\sigma_2 = b$.
The inductive definitions for other programs σ are:

$$C(nil, n, E) = ([\,], [\,], n)$$
$$C(O(\vec{r}), n, E) = ([\,], [\langle O(\vec{x}), \vec{t}, Prec'(\vec{x}), Eff'(\vec{x})\rangle], n+1) \text{ with}$$
$$\langle O(\vec{x}), \vec{t}, Prec(\vec{x}), Eff(\vec{x})\rangle \in Ops,$$
$$\vec{r} = r_1, \ldots, r_m,$$
$$Prec'(\vec{x}) = (state = s_n \wedge \bigwedge_{i \text{ s.t. } r_i \in E} bound(r_i) \rightarrow map(r_i, x_i) \wedge \bigwedge_{i \text{ s.t. } r_i \notin E} x_i = r_i),$$
$$Eff'(\vec{x}) = [\, state = s_n \Rightarrow state = s_{n+1}] \cdot$$
$$[state = s_n \Rightarrow bound(r_i) \wedge map(r_i, x_i)]_{i \text{ s.t. } r_i \in E}$$
$$C(\phi?, n, E) = (\, [Ctest(\phi, n, n+1, E)],\ [\,],\ n+1)$$
$$C((\sigma_1; \sigma_2), n, E) = (L_1 \cdot L_2,\ L_1' \cdot L_2',\ n_2) \text{ with}$$
$$(L_1, L_1', n_1) = C(\sigma_1, n, E), (L_2, L_2', n_2) = C(\sigma_2, n_1, E)$$
$$C((\sigma_1 | \sigma_2), n, E) = (L_1 \cdot L_2 \cdot X, L_1' \cdot L_2', n_2 + 1) \text{ with}$$
$$(L_1, L_1', n_1) = C(\sigma_1, n+1, E),$$
$$(L_2, L_2', n_2) = C(\sigma_2, n_1+1, E),$$
$$X = [\, Cnoop(n, n+1),\ Cnoop(n, n_1+1),$$
$$Cnoop(n_1, n_2+1),\ Cnoop(n_2, n_2+1)\,]$$
$$C(\text{while } \phi \text{ do } \sigma, n, E) = (L \cdot X, L', n_1+1) \text{ with}$$
$$(L, L', n_1) = C(\sigma, n+1, E),\ X = [Ctest(\phi, n, n+1, E),$$
$$Ctest(\neg\phi, n, n_1+1, E), Cnoop(n_1, n)]$$
$$C(\sigma^*, n, E) = (L \cdot [Cnoop(n, n_2), Cnoop(n_1, n)], L', n_2) \text{ with}$$
$$(L, L', n_1) = C(\sigma, n, E),\ n_2 = n_1 + 1$$
$$C(\pi(x\text{-}t, \sigma), n, E) = (L \cdot X, L', n_1 + 1) \text{ with}$$
$$(L, L', n_1) = C(\sigma, n, E \cdot [x\text{-}t]),$$
$$X = [\langle free\_n_1(x), t, state = s_{n_1},$$
$$[state = s_{n_1+1}, \neg bound(x), \forall y. \neg map(x, y)]\rangle\,]$$

The atomic program **any** is handled by macro expansion to above defined constructs.

As mentioned above, given program σ, the return value $(L, L', n_{\text{final}})$ of $C(\sigma, 0, [\,])$ is such that L contains new operators for encoding transitions in the automaton, whereas L' contains restrictions on the applicability of the original operators of the domain. Now we are ready to integrate these new operators and restrictions with the original domain specification D to produce the new domain specification D_σ.

D_σ contains a constrained version of the operators $O(\vec{x})$ of the original domain D also mentioned in L'. Let $[\langle O(\vec{x}), \vec{t}, Prec_i(\vec{x}), Eff_i(\vec{x})\rangle]_{i=1}^n$ be the sublist of L' that contains additional conditions for operator $O(\vec{x})$. The operator replacing $O(\vec{x})$ in D_σ is defined as:

$$\langle O'(\vec{x}), \vec{t}, \; Prec(\vec{x}) \wedge \bigvee_{i=1}^n Prec_i(\vec{x}), \; Eff(\vec{x}) \cup \bigcup_{i=1}^n Eff_i(\vec{x})\rangle$$

Additionally, D_σ contains all operator definitions in L. Objects in D_σ are the same as those in D, plus a few new ones to represent the program variables and the automaton's states s_i ($0 \le i \le n_{\text{final}}$). Finally D_σ inherits all predicates in D plus $bound(x)$, $map(x, y)$, and function $state$.

The translation, up to this point, is problem-independent; the problem specification P_σ is defined as follows. Given any predefined problem P over D, P_σ is like P except that its initial state contains condition $state = s_0$, and its goal contains $state = s_{n_{\text{final}}}$. Those conditions ensure that the program must be executed to completion.

As is shown below, planning in the generated instance $I_\sigma = (D_\sigma, P_\sigma)$ is equivalent to planning for the original instance $I = (D, P)$ under the control of program σ, except that plans on I_σ contain actions that were not part of the original domain definition (*test*, *noop*, and *free*).

THEOREM 8 (Correctness). *Let Filter(α, D) denote the sequence that remains when removing from α any action not defined in D. If α is a plan for instance $I_\sigma = (D_\sigma, P_\sigma)$ then Filter(α, D) is a plan for $I = (D, P)$ under the control of σ. Conversely, if α is a plan for I under the control of σ, there exists a plan α' for I_σ, such that $\alpha = Filter(\alpha', D)$.*

Proof. Appears in [Baier, Fritz, and McIlraith 2007]. □

Now we turn our attention to analyzing the size of the output planning instance relative to the original instance and control program. Assume we define the size of a program as the number of programming constructs and actions it contains. Then we obtain the following result.

THEOREM 9 (Succinctness). *Let σ is a program of size m, and let k be the maximal nesting depth of $\pi(x\text{-}t)$ statements in σ, then $|I_\sigma|$ (the overall size of I_σ) is $O((k+p)m)$, where p is the size of the largest operator in I.*

Proof. Appears in [Baier, Fritz, and McIlraith 2007]. □

The encoding of programs in PDDL2.1 is, hence, in worst case $O(k)$ times bigger than the program itself. It is also easy to show that the translation is done in time linear in the size of the program, since, by definition, every occurrence of a program construct is only dealt with once.

5 Exploiting DCK in State-of-the-Art Heuristic Planners

Our objective in translating procedural DCK to PDDL2.1 was to enable *any* PDDL2.1-compliant state-of-the-art planner to seamlessly exploit our DCK. In this section, we investigate ways to best leverage our translated domains using domain-independent heuristic search planners.

There are several compelling reasons for wanting to apply domain-independent heuristic search to these problems. Procedural DCK can take many forms. Often, it will provide explicit actions for some parts of a sequential plan, but not for others. In such cases, it will contain unconstrained fragments (i.e., fragments with nondeterministic choices of actions) where the designer expects the planner to figure out the best choice of actions to realize a sub-task. In the absence of domain-specific guidance for these unconstrained fragments, it is natural to consider using a domain-independent heuristic to guide the search.

In many domains it is very hard to write deterministic procedural DCK, i.e. DCK that restricts the search space in such a way that solutions can be obtained very efficiently, even using blind search. An example of such a domain is one where plans involve solving an optimization sub-problem. In such cases, procedural DCK will contain open parts (fragments of nondeterministc choice within the DCK), where the designer expects the planner to figure out the best way of completing a sub-task. However, in the absence of domain-specific guidance for these open parts, it is natural to consider using a domain-independent heuristic to guide the search.

In other cases, it is the choice of action arguments, rather than the choice of actions that must be optimized. In particular, fragments of DCK may collectively impose global constraints on action argument choices that need to be enforced by the planner. As such, the planner needs to be *aware* of the procedural control in order to avoid backtracking. By way of illustration, consider a travel planning domain comprising two tasks "buy air ticket" followed by "book hotel". Each DCK fragment restricts the actions that can be used, but leaves the choice of arguments to the planner. Further suppose that budget is limited. We would like our planner to realize that actions used to complete the first task should save enough money to complete the second task. The ability to do such lookahead can be achieved via domain-independent heuristic search.

In the rest of the section we propose three ways in which one can leverage our translated domains using a domain-independent heuristic planner. These three techniques differ predominantly in the operands they consider in computing heuristics.

5.1 Direct Use of Translation (*Simple*)

As the name suggests, a simple way to provide heuristic guidance while enforcing program awareness is to use our translated domain directly with a domain-independent heuristic planner. In short, take the original domain instance I and control σ, and use the resulting instance I_σ with any heuristic planner. We call this the *Simple* heuristic.

Unfortunately, when exploiting a relaxed graph to compute heuristics, two issues arise. *bound* predicates are relaxed, whatever value is already assigned to a variable, will remain assigned to that variable. This can cause a problem with iterative control. For example, assume program $\sigma_L \stackrel{\text{def}}{=}$ **while** ϕ **do** $\pi(c\text{-}crate)\ unload(c, T)$, is intended for a domain where crates can be only unloaded sequentially from a truck. As soon as c is assigned a value, such a value will be considered in all possible iterations of the while loop, which is not what is intended by the program, and has the potential of returning misleading estimates.

The second issue is one of efficiency. Since fluent *state* is also relaxed, the benefits of the reduced branching factor induced by the programs is lost. This has an effect on the time required to compute the heuristic.

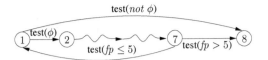

Figure 2: *H-ops* translation of **while** loops. While computing the heuristics, pseudo-fluent *fp* is increased each time no new effect is added into the relaxed state, and it is set to 0 otherwise. The loop can be exited if the last five (7-2) actions performed didn't add any new effect.

5.2 Modified Program Structure (*H-ops*)

The *H-ops* approach addresses the two issues potentially affecting the computation of the *Simple* heuristic. It is designed to be used with planners that use heuristics based on the relaxed planning graph. The input to the planner in this case is a pair $(I_\sigma, HOps)$, where $I_\sigma = (D_\sigma, P_\sigma)$ is the translated instance, and $HOps$ is an additional set of planning operators. The planner uses the operators in D_σ to generate successor states while searching. However, when computing the heuristic for a state s it uses the operators in $HOps$.

Additionally, function *state* and predicates *bound* and *map* are *not* relaxed. This means that when computing the relaxed graph we actually delete their instances from the relaxed states. As usual, *deletes* are processed before *adds*. The expansion of the graph is stopped if the goal or a fixed point is reached. Finally, a relaxed plan is extracted in the usual way, and its length is reported as the heuristic value. In the computation of the length, auxiliary actions such as tests and noops are ignored.

The "un-relaxing" of *state*, *bound* and *map* addresses the problem of reflecting the reduced branching factor provided by the control program while computing the heuristics. However, it introduces other problems. Returning to the σ_L program defined above, since *state* is now un-relaxed, the relaxed graph expansion cannot escape from the loop, because under the relaxed planning semantics, as soon as ϕ is true, it remains true forever. A similar issue occurs with the nondeterministic iteration. Furthermore, we want to avoid state duplication, i.e. having *state* equal to two different values at the same time in the same relaxed state. This could happen for example while reaching an **if** construct whose condition is both true and false at the same time (this can happen because p and *not-p* can both be true in a relaxed state).

This issue is addressed by the *HOps* operators. To avoid staying in the loop forever, the loop will be exited when actions in it are no longer adding effects. Figure 2 provides a graphical illustration. An important detail to note is that the loop is not entered when ϕ is not found true in the relaxed state. (The expression *not* ϕ should be understood as negation as failure.) Moreover, the pseudo-fluent fp is an internal variable of the planner that acts as a real fluent for the *HOps*. A similar approach is adopted for nodeterministic iterations, whose description we omit here.

Since loops are guaranteed to be exited, the computation of *H-ops* is guaranteed to finish because at some relaxed state the final state of the automaton will be reached. At this point, if the goal is not true, no operators will be possible and a fixed point will be produced immediately.

For **if**'s, if the condition is both true and false at the same time, the **then** part is processed first, followed by the **else** part. The objective of this is avoidance of state

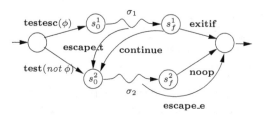

Figure 3: *H-ops* translation for **if - then - else**. Action $\mathbf{testesc}(\phi)$ is possible if condition ϕ is true. If condition $\neg\phi$ is also true in the relaxed state, the $\mathbf{testesc}(\phi)$ adds a fact *escape_active* that will enable the execution of **continue** and **escape_t** and **escape_e**. Actions **escape_t** and **escape_e** are possible only when no other actions are possible. This is checked using the pseudo-fluent fp described in Figure 2. Action **exitif** is only possible if *escape_active* is true. Both the **noop** and the **escape_e** actions delete the fact *escape_active*. Nested **if** constructs are handled using a parameterized version of the *escape_active* predicate.

duplication. However, this new interpretation of the **if** introduces a new problem. This problem occurs when, while performing the actions of one of the parts, no action is possible anymore. Intuitively, this could happen because the heuristics has chosen the wrong subprogram to execute actions from. Indeed, if there exists an execution of the program from state s that executes the "then" part of the **if**, it can happen that, during the computation of the heuristic for s, the "else" part forces some actions to occur that are not possible. Under normal circumstances, the non existence of any possible action produces a fixed point. Because the goal is not reached on such a fixed point, the heuristic regards the goal as unreachable, which could be a wrong estimation.

To solve this problem, *H-ops* considers new "escape" actions, that are executable only when no more actions are possible. Escapes can be performed only inside "then" or "else" bodies. After executing an escape, the simulation of the program's execution jumps to the "else" part if the escape occurs in the "then" part, or to the end of the **if**, if the escape occurs in the "else" part. Figure 3 provides a graphical depiction.

5.3 A Program-Unaware Approach (*Basic*)

Our program-unaware approach (*Basic*) completely ignores the program when computing heuristics. Here, the input to the planner is a pair (I_σ, Ops), where I_σ is the translated instance, and Ops are the *original* domain operators. The Ops operators are used exclusively to compute the heuristic. Hence, *Basic*'s output is not at all influenced by the control program. Note that although *Basic* is program unaware, it can sometimes provide good estimates, as we see in the following section. This is especially true when the DCK characterizes a solution that would be naturally found by the planner if no control were used. It is also relatively fast to compute.

6   Implementation and Experiments

Our implementation takes a PDDL planning instance and a DCK program and generates a new PDDL planning instance. It will also generate appropriate output for the *Basic* and *H-ops* heuristics, which require a different set of operators. Thus, the resulting PDDL instance may contain definitions for operators that are used only for heuristic computation using the `:h-action` keyword, whose syntax is analogous to the PDDL keyword `:action`.

Our planner is a modified version of TLPLAN, which does a best-first search using an FF-style heuristic. It is capable of reading the PDDL with extended operators.

We performed our experiments on the *trucks, storage* and *rovers* domains (30 instances each). We wrote DCK for these domains. For details of the GOLOG code used for these examples, see [Baier 2010]. We ran our three heuristic approaches (*Basic, H-ops,* and *Simple*) and cycle-free, depth-first search on the translated instance (*blind*). Additionally, we ran the original instance of the program (DCK-free) using the domain-independent heuristics provided by the planner (*original*). Table 1 shows various statistics on the performance of the approaches. Furthermore, Figure 4 shows times for the different heuristic approaches.

Not surprisingly, our data confirms that DCK helps to improve planner performance, solving more instances across all domains. In some domains (i.e., storage and rovers) blind depth-first cycle-free search is sufficient for solving most of the instances. However, quality of solutions (plan length) is poor compared to the heuristic approaches. In trucks, DCK is only effective in conjunction with heuristics; blind search can solve very few instances.

We observe that *H-ops* is the most informative (expands fewer nodes). This fact does not pay off in time in the experiments shown in the table. Nevertheless, it is easy to construct instances where the *H-ops* performs better than *Basic*. This happens when the DCK control restricts the space of valid plans (i.e., prunes out valid plans). We have experimented with various instances of the storage domain, where we restrict the plan to use only one hoist. In some of these cases *H-ops* outperforms *Basic* by orders of magnitude.

7 Related Work and Discussion

DCK can be used to constrain the set of valid plans and has proven an effective tool in reducing the time required to generate a plan. Nevertheless, many of the planners that exploit it use arguably less natural state-centric DCK specification languages, and their planners use blind search. In this paper we examined the problem of exploiting procedural DCK with state-of-the-art planners. Our goal was to specify rich DCK naturally in the form of a program template and to exploit state-of-the-art planning techniques to actively plan towards the achievement of this DCK. To this end we made three contributions: provision of a GOLOG-like procedural DCK language syntax and PDDL semantics; a polynomial-time algorithm to compile DCK and a planning instance into a PDDL2.1 planning instance that could be input to any PDDL2.1-compliant planner; and finally a set of techniques for exploiting domain-independent heuristic search with our translated DCK planning instances. Each contribution is of value in and of itself. The language can be used without the compilation, and the compiled PDDL2.1 instance can be input to any PDDL2.1-compliant state-of-the-art planner, not just the domain-independent heuristic search planner that we propose. Our experiments show that procedural DCK improves the performance of state-of-the-art planners, and that our heuristics are sometimes key to achieving good performance.

Much of the previous work on DCK in planning has exploited state-centric specification languages. In particular, TLPLAN [Bacchus and Kabanza 1998] and TALPLAN-NER [Kvarnström and Doherty 2000] employ declarative, state-centric, temporal languages based on LTL to specify DCK. Such languages define necessary properties of states over fragments of a valid plan. We argue that they could be less natural than our procedural specification language.

Though not described as DCK specification languages there are a number of languages

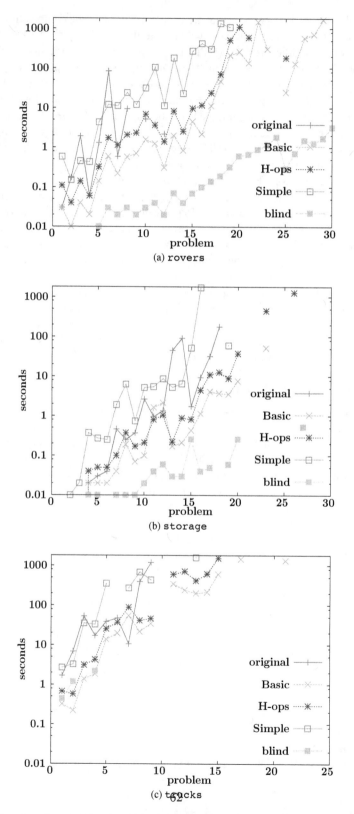

(a) rovers

(b) storage

(c) trucks

Figure 4: Running times of the three heuristics and the original instance: logarithmic scale.

		original	Simple	Basic	H-ops	blind
Trucks	#n	1	0.31	0.41	0.26	19.85
	#s	9	9	15	14	3
	ℓ_{min}	1	1	1	1	1
	ℓ_{avg}	1.1	1.03	1.02	1.04	1.04
	ℓ_{max}	1.2	1.2	1.07	1.2	1.07
Rovers	#n	1	0.74	1.06	1.06	1.62
	#s	10	19	28	22	30
	ℓ_{min}	1	1	1	1	1
	ℓ_{avg}	2.13	1.03	1.05	1.21	1.53
	ℓ_{max}	4.59	1.2	1.3	1.7	2.14
Storage	#n	1	1.2	1.13	0.76	1.45
	#s	18	18	20	21	20
	ℓ_{min}	1	1	1	1	1
	ℓ_{avg}	4.4	1.05	1.01	1.07	1.62
	ℓ_{max}	21.11	1.29	1.16	1.48	2.11

Table 1: Comparison between different approaches to planning (with DCK). #n is the average factor of expanded nodes to the number of nodes expanded by *original* (i.e., #n=0.26 means the approach expanded 0.26 times the number of nodes expanded by original). #s is the number of problems solved by each approach. ℓ_{avg} denotes the average ratio of the plan length to the shortest plan found by any of the approaches (i.e., ℓ_{avg}=1.50 means that on average, on each instance, plans where 50% longer than the shortest plan found for that instance). ℓ_{min} and ℓ_{max} are defined analogously.

from the agent programming and/or model-based programming communities that are related to procedural control. We have mentioned the GOLOG logic programming language. EAGLE is a somewhat related goal language designed to also express intentionality [dal Lago, Pistore, and Traverso 2002]. Further, languages such as the Reactive Model-Based Programming Language (RMPL) [Kim, Williams, and Abramson 2001] – a procedural language that combines ideas from constraint-based modeling with reactive programming constructs – also share expressive power and goals with procedural DCK. Finally, HTN specification languages such as those used in SHOP [Nau, Cao, Lotem, and Muñoz-Avila 1999] provide domain-dependent hierarchical task decompositions together with partial order constraints, not easily describable in our language.

A focus of our work was to exploit state-of-the-art planners and planning techniques with our procedural DCK. In contrast, well-known DCK-enabled planners such as TLPLAN and TALPLANNER use DCK to prune the search space at each step of the plan and then employ blind depth-first cycle-free search to try to reach the goal. Unfortunately, pruning is only possible for maintenance-style DCK and there is no way to plan towards achieving other types of DCK as there is with the heuristic search techniques proposed here.

Similarly, GOLOG interpreters, while exploiting procedural DCK, have traditionally employed blind search to instantiate nondeterministic fragments of a GOLOG program. Most recently, Claßen, Eyerich, Lakemeyer, and Nebel [2007] have proposed to integrate an incremental GOLOG interpreter with a state-of-the-art planner. Their motivation is similar to ours, but there is a subtle difference: they are interested in combining *agent programming* and efficient planning. The integration works by allowing a GOLOG program to make explicit calls to a state-of-the-art planner to achieve particular conditions identified by the user. The actual planning, however, is not controlled in any way. Also, since the GOLOG interpreter executes the returned plan immediately without further lookahead, backtracking does not extend over the boundary between GOLOG and the planner. As such, each fragment of nondeterminism within a program is treated independently, so that actions selected locally are not informed by the constraints of later fragments as they are with the approach that we propose. Their work, which focuses on the semantics of ADL in the situation calculus, is hence orthogonal to ours.

Finally, there is related work that compiles DCK into standard planning domains. Baier and McIlraith [2006], Cresswell and Coddington [2004], Edelkamp [2006], and Rintanen [2000], propose to compile different versions of LTL-based DCK into PDDL/ADL planning domains. The main drawback of these approaches is that translating full LTL into ADL/PDDL is worst-case exponential in the size of the control formula whereas our compilation produces an addition to the original PDDL instance that is linear in the size of the DCK program. Son, Baral, Nam, and McIlraith [2006] further show how HTN, LTL, and GOLOG-like DCK can be encoded into planning instances that can be solved using answer set solvers. Nevertheless, they do not provide translations that can be integrated with PDDL-compliant state-of-the-art planners, nor do they propose any heuristic approaches to planning with them.

8 Postlude

The focus of this research, as described here, has been to improve planner performance while also enabling standard planners to plan for a richer class of goals. In addition to this practical achievement, there are a number of interesting theoretical insights in this work. We explore these in more detail in [Fritz, Baier, and McIlraith 2008]. In this later work we go beyond the simplified GOLOG-like language discussed here, and consider full GOLOG with procedures, including its extension for concurrency, ConGolog [De Giacomo, Lespérance, and Levesque 2000].

The main technical contribution of this follow-on work is the definition of a compilation scheme that takes a ConGolog program and a basic action theory of the situation calculus as input and outputs a new basic action theory that represents the program in the context of the original theory. I.e., the resulting basic action theory describes the same tree of situations as the tree of situations induced by the program in *conjunction* with the original basic action theory. This compilation eliminates the need for a ConGolog interpreter. The resulting theory can be interpreted by any implementation of the situation calculus. Further, providing semantics to ConGolog programs in this way eliminates the need for reification in the specification of the semantics.

This *compiled semantics* is useful for a variety of purposes. First and foremost, as discussed in the previous sections, standard PDDL-compliant planners can plan using ConGolog programs to specify search control and/or temporally extended goals. While the target language in [Fritz, Baier, and McIlraith 2008] was that of basic action theories of the situation calculus, in certain cases it is possible to also compile to other representations, such as PDDL. This requires certain restrictions to keep the state space finite, but important features such as concurrency *can* be represented. Intuitively, in the context of the previously described compilation to PDDL, this is achieved by allowing the state machine to be in multiple states at the same time. The type of concurrency achieved is interleaving, just as it is in the original ConGolog.

Secondly, it is possible to represent a significant subset of HTNs [Ghallab, Nau, and Traverso 2004] in ConGolog, which given this work means that it is now possible to reason about such HTNs in the situation calculus as well. Given ConGolog's ability to represent and reason about recursive procedures, these HTNs may similarly make use of recursive method definitions.

Thirdly, since the semantics of programs in the compilation result is captured via fluents, action preconditions and action effects, it is possible to reason about program executions using regression. Loosely speaking, one can "regress programs", i.e., reduce all the constraints that are imposed by a program on the legal executions of that program into constraints about the initial state of the world, when presented with a sequence of actions. This is different from regressing a formula over a program. Rather, since after compilation all constraints that the program imposes on the evolution of the world (the tree of situations described by the underlying basic action theory) are now explicitly expressed as conditions over fluents, these program constraints can be regressed. Just as this was the original motivation for Reiter's use of regression, this use of regression allows applying regular theorem provers that only need to reason about the initial state. Hence, reasoning about the truth values of fluents amounts to mere database look-up. Practically this means that, when given a specific sequence of actions, one can test whether or not it constitutes a legal execution of the regressed program without temporal projection and without the need for program interpretation. This has significant implications with respect to execution monitoring of (Con)Golog programs, building on the unifying perspective of execution monitoring developed in Fritz's Ph.D. thesis [Fritz 2009].

Acknowledgements   We are grateful to Yves Lespérance, Hector Levesque and ICAPS anonymous reviewers for their feedback. This research was funded by the Natural Sciences and Engineering Research Council of Canada (NSERC) and by the Ontario Ministry of Research and Innovation (MRI).

References

Bacchus, F. and F. Kabanza [1998]. Planning for temporally extended goals. *Annals of Mathematics and Artificial Intelligence 22*(1-2), 5–27.

Baier, J. [2010]. *Effective search techniques for non-classical planning via reformulation*. Ph.D. thesis, University of Toronto, Toronto, Canada.

Baier, J., C. Fritz, and S. McIlraith [2007]. Exploiting procedural domain control knowledge in state-of-the-art planners (extended version). Technical Report CSRG-565, University of Toronto.

Baier, J. A. and S. A. McIlraith [2006]. Planning with first-order temporally extended goals using heuristic search. In *Proc. of the 21st National Conference on Artificial Intelligence (AAAI-06)*, Boston, USA, pp. 788–795.

Claßen, J., P. Eyerich, G. Lakemeyer, and B. Nebel [2007]. Towards an integration of Golog and planning. In *Proc. of the 20th Int'l Joint Conference on Artificial Intelligence (IJCAI-07)*, Hyderabad, India, pp. 1846–1851.

Cresswell, S. and A. M. Coddington [2004]. Compilation of LTL goal formulas into PDDL. In *Proc. of the 16th European Conference on Artificial Intelligence (ECAI-04)*, Valencia, Spain, pp. 985–986.

dal Lago, U., M. Pistore, and P. Traverso [2002]. Planning with a language for extended goals. In *Proc. of AAAI/IAAI*, Edmonton, Alberta, Canada, pp. 447–454.

De Giacomo, G., Y. Lespérance, and H. Levesque [2000]. ConGolog, a concurrent programming language based on the situation calculus. *Artificial Intelligence 121*(1-2), 109–169.

Edelkamp, S. [2006]. On the compilation of plan constraints and preferences. In *Proc. of the 16th Int'l Conference on Automated Planning and Scheduling (ICAPS-06)*, Lake District, UK, pp. 374–377.

Fox, M. and D. Long [2003]. PDDL2.1: An extension to PDDL for expressing temporal planning domains. *Journal of Artificial Intelligence Research 20*, 61–124.

Fritz, C. [2009]. *Monitoring the Generation and Execution of Optimal Plans*. Ph.D. thesis, University of Toronto, Toronto, Canada.

Fritz, C., J. A. Baier, and S. A. McIlraith [2008]. ConGolog, Sin Trans: Compiling ConGolog into basic action theories for planning and beyond. In *Proc. on the 11th Int'l Conference on Principles of Knowledge Representation and Reasoning (KR-08)*, Sydney, Australia, pp. 600–610.

Ghallab, M., D. Nau, and P. Traverso [2004]. *Automated Planning: Theory and Practice*. Morgan Kaufmann.

Kim, P., B. C. Williams, and M. Abramson [2001]. Executing reactive, model-based programs through graph-based temporal planning. In *Proc. of the 17th Int'l Joint Conference on Artificial Intelligence (IJCAI-01)*, Seattle, USA, pp. 487–493.

Kvarnström, J. and P. Doherty [2000]. TALPlanner: A temporal logic based forward chaining planner. *Annals of Mathematics and Artificial Intelligence 30*(1-4), 119–169.

Levesque, H., R. Reiter, Y. Lespérance, F. Lin, and R. B. Scherl [1997]. GOLOG: A logic programming language for dynamic domains. *Journal of Logic Programming 31*(1-3), 59–83.

McDermott, D. V. [1998]. PDDL — The Planning Domain Definition Language. Technical Report TR-98-003/DCS TR-1165, Yale Center for Computational Vision and Control.

Nau, D. S., Y. Cao, A. Lotem, and H. Muñoz-Avila [1999]. SHOP: Simple hierarchical ordered planner. In *Proc. of the 16th Int'l Joint Conference on Artificial Intelligence (IJCAI-99)*, Stockholm, Sweden, pp. 968–975.

Pednault, E. P. D. [1989]. ADL: Exploring the middle ground between STRIPS and the situation calculus. In *Proc. of the 1st Int'l Conference on Principles of Knowledge Representation and Reasoning (KR-89)*, Toronto, Canada, pp. 324–332.

Rintanen, J. [2000]. Incorporation of temporal logic control into plan operators. In *Proc. of the 14th European Conference on Artificial Intelligence (ECAI-00)*, Berlin, Germany, pp. 526–530. IOS Press.

Son, T. C., C. Baral, T. H. Nam, and S. A. McIlraith [2006]. Domain-dependent knowledge in answer set planning. *ACM Transactions on Computational Logic 7*(4), 613–657.

4

Multi-Agent Only-Knowing*

VAISHAK BELLE AND GERHARD LAKEMEYER

ABSTRACT. Levesque introduced the notion of only-knowing to precisely capture the beliefs of a knowledge base. He also showed how only-knowing can be used to formalize non-monotonic behavior within a monotonic logic. Despite its appeal, all attempts to extend only-knowing to the many-agent case have undesirable properties. A belief model by Halpern and Lakemeyer, for instance, appeals to proof-theoretic constructs in the semantics and needs to axiomatize validity as part of the logic. It is also not clear how to generalize their ideas to a first-order case. In this paper, we propose a new account of multi-agent only-knowing which, for the first time, has a natural possible-world semantics for a quantified language with equality. We then provide, for the propositional fragment, a sound and complete axiomatization that faithfully lifts Levesque's proof theory to the many agent case.

Prologue by Gerhard Lakemeyer

Only-knowing has been among Hector's many research interests literally for decades, with the essential ingredients of this concept already present in his 1981 Ph.D. thesis. I remember hand-written slides by Hector from the early eighties with a first version of the semantics of only-knowing. He published his first paper on this topic in 1987, which had a tremendous influence on my own work, including my dissertation under Hector's supervision. Ever since we started collaborating on the book "The Logic of Knowledge Bases," which really is about only-knowing, we have never quite stopped working on it until today.

Only-knowing first and foremost is about characterizing what a knowledge-based agent capable of introspection believes. Whereas other proposals such as autoepistemic logic needed to resort to meta-theoretic formalizations including fixed points, Hector has always cared more about finding the right truth conditions. After all, as he and Ron Brachman write in their book on knowledge representation and reasoning [Brachman and Levesque 2004], "the system should believe p if, according to the beliefs it has represented, the world it is imagining is one where p is true." In other words, knowledge is inconceivable without a notion of truth. In many ways, Hector's semantics of only-knowing is a beautiful example of how strongly he feels about that.

So it seems very fitting to include in this collection another paper on only-knowing. While the emphasis is on multi-agent settings, the spirit and techniques used here owe much if not everything to what Hector began more than thirty years ago. Hector, we hope you enjoy it!

---

* This chapter is a slightly revised version of a paper [Belle and Lakemeyer 2010] that appeared in *Principles of Knowledge Representation and Reasoning: Proceedings of the Twelfth International Conference*, Fangzhen Lin, Ulrike Sattler, Miroslaw Truszczynski (Eds.), 2010.

1 Introduction

When Levesque introduced his logic of only-knowing [Levesque 1990], one of the motivations was to capture precisely what a knowledge base knows in terms of valid sentences of the form

$$O\text{KB} \supset K\alpha,$$

which can be read as "if the KB is all that is known then α is also known." What is particularly interesting about the operator O is that it does not only allow to draw conclusion about what is known but also what is not. For example, $Op \supset \neg Kq$ and, by introspection, $Op \supset K\neg Kq$ both come out valid. Note that this is quite different from classical epistemic logic, that is, if we replace O by K, then neither of these sentences is valid.

Levesque showed that only-knowing can be given a very simple possible-world semantics: roughly, the truth conditions are defined wrt a set of worlds e, and a KB is only-known just in case it is true in all worlds in e and one cannot add more worlds without falsifying at least one of the formulas of the KB.

When the KB itself refers to the agent's beliefs,[1] only-knowing also exhibits a form of non-monotonic reasoning. For example, consider the classical Tweety example with the following KB:

$$\{\forall x.\ Bird(x) \land \neg K\neg Fly(x) \supset Fly(x), Bird(Tweety)\}.$$

Here the first sentence expresses the default that birds fly unless known otherwise. Then $O\text{KB} \supset KFly(Tweety)$ comes out valid, which is as it should be. The obvious similarity with Moore's autoepistemic logic (AEL) [Moore 1985] is not accidental. In fact, Levesque showed that only-knowing captures AEL precisely. Remarkably, this reconstruction of a major branch of non-monotonic reasoning happens within a *monotonic* logic.

While Levesque only considered a single agent, it seems natural to ask whether these ideas can be extended to the many agents. For example, in the case of non-monotonic reasoning, agents should believe that other agents are also able to draw non-monotonic conclusions: if Alice knows that Bob only-knows the birds-fly default and that Tweety is a bird, then Alice should believe that Bob believes that Tweety can fly.

It turns out that existing approaches to formalize multi-agent only-knowing are suprisingly complex compared to Levesque's simple possible-world account. Most extensions so far make use of arbitrary Kripke structures [Fagin, Halpern, Moses, and Vardi 1995] with explicit accessibility relations for each agent. Moreover, they are restricted to the propositional case and also have some undesirable properties, perhaps invoking some caution in their usage. For instance, in a canonical Kripke model approach by Lakemeyer [1993], certain types of epistemic states cannot be constructed. Note also that canonical models are not usable in a practical way — not only are there uncountably many worlds, but each world is characterized by an infinite set of formulas and so cannot be described easily. Besides, proof theoretic constructs appear in the semantics via maximally consistent sets. Halpern [1993] proposes another Kripke approach, and here the modalities do not seem to interact in an intuitive manner. Although an approach by Halpern and Lakemeyer [2001] does successfully model multi-agent only-knowing, it also makes use of canonical models. Moreover, it forces us to have the semantic notion of validity directly in the language. Precisely for this reason, that proposal is not natural, and it is matched with a proof theory that

---

[1]Throughout we will be concerned with agents who may have false beliefs. Nevertheless we will use the terms "knowledge" and "belief" interchangeably.

has a set of new axioms to deal with these new notions. It is also not clear how one can extend their ideas to the first-order case. Lastly, an approach by Waaler [2004] avoids such an axiomatization of validity, but the model theory also has problems [Waaler and Solhaug 2005]. Technical discussions on their semantics are deferred to later.

The goal of this chapter is to show that there is indeed a semantics for multi-agent only-knowing for a first-order language, which extends Levesque's approach in a natural way. For the propositional subset, there is also a sound and complete axiomatization that faithfully generalizes Levesque's proof theory.[2] Moreover, different from [Halpern and Lakemeyer 2001], we do not enrich the language any more than necessary (modal operators for each agent), and we do not make use of canonical Kripke models. We also obtain a first-order multi-agent generalization of AEL, defined using notions of classical logical entailment, as an analogue to Levesque's result. Finally, we review a comparison to [Halpern and Lakemeyer 2001].

Organization is as follows. We begin by reviewing Levesque's logic of only-knowing.[3] We then define a semantics for multi-agent only-knowing with so-called *k-structures*. Next, we compare the framework to earlier attempts while going over its properties. Following that, we provide a sound and complete axiomatization for the propositional fragment. In the last sections, we prove that k-structures and [Halpern and Lakemeyer 2001] agree on valid sentences and sketch the multi-agent (first-order) generalization of AEL. Then, we conclude and end. In the rest of the paper we abbreviate [Lakemeyer 1993] as L93, [Halpern 1993] as H93, and [Halpern and Lakemeyer 2001] as HL.

2 Levesque's logic of only-knowing

Levesque's logic[4] $\mathcal{ONL}$ is a first-order modal dialect with $=$ and a countably infinite set of standard names $\mathcal{N}$, which are like constants syntactically but will also serve as the universe of discourse in the semantics. The logical connectives are $\vee$, $\neg$, and $\forall$, together with modal operators K and N, which can be nested arbitrarily. Other connectives like $\vee$, $\supset$, $\equiv$, and $\exists$ are taken for their usual syntactic abbreviations. To keep matters simple, function symbols are not considered in this language. We call a predicate other than $=$, applied to first-order variables or standard names, an *atomic* formula or *atom*. A *ground atom* is an atom whose arguments are standard names only. We write α_n^x to mean that the variable x is substituted in α by standard name n. A formula without free variables is called a *sentence*. A formula not mentioning any modalities is called *objective*. A formula is called *subjective* if every atom is in the scope of a modality.

The semantics is defined wrt *worlds*, which are simply sets of ground atoms. The set of all possible worlds is denoted as $\mathcal{W}$. The standard names are thus *rigid designators*, and denote precisely the same entities in all worlds.

An *epistemic state* e is any set of possible worlds over $\mathcal{W}$. A pair (e, w), where w is a world and e an epistemic state, is called a model and the truth of sentences in $\mathcal{ONL}$ is defined as follows:

1. $e, w \models p$ iff $p \in w$ for a ground atom p,

---

[2]Levesque's proof theory for a quantified language is *incomplete*. It is also known that any complete axiomatization cannot be *recursive* [Halpern and Lakemeyer 1995; Levesque and Lakemeyer 2001].

[3]There are other notions of "all I know", such as [Halpern and Moses 1984] and [Ben-David and Gafni 1989], which will not be discussed here. See [Rosati 2000; Halpern and Lakemeyer 2001] for discussions.

[4]We name the logic following [Halpern and Lakemeyer 2001] for ease of comparisons later on. It is referred to as $\mathcal{OL}$ in [Halpern and Lakemeyer 1995; Levesque and Lakemeyer 2001].

2. $e, w \models (m = n)$ iff m and n are identical standard names,

3. $e, w \models \neg\alpha$ iff $e, w \not\models \alpha$,

4. $e, w \models \alpha \vee \beta$ iff $e, w \models \alpha$ or $e, w \models \beta$,

5. $e, w \models \forall x.\ \alpha$ iff $e, w \models \alpha_n^x$ for all standard names n,

6. $e, w \models K\alpha$ iff for all $w' \in e$, $e, w' \models \alpha$,

7. $e, w \models N\alpha$ iff for all $w' \notin e$, $e, w' \models \alpha$.

We read $K\alpha$ as "at least α is believed" since $K\alpha$ certainly does not preclude $K(\alpha \wedge \beta)$ from holding, in general. Of course, this is the classical knowledge operator, and (at least) α is believed iff it is true at all worlds considered possible. We read $N\neg\alpha$ as "at most α is believed" since if the agent knows more, he would not consider all worlds where α holds epistemically possible. Thus, (at most) α is believed to be false iff it is true at all worlds considered *impossible*. Now we say that an agent only-knows α, written $O\alpha$ and is a syntactic abbreviation for $K\alpha \wedge N\neg\alpha$, when worlds in e are precisely those where α is true. That is,

- $e, w \models O\alpha$ iff for all w', $w' \in e$ iff $e, w' \models \alpha$.[5]

Validity and satisfiability are defined as usual, that is, a formula α is *valid* ($\models \alpha$) iff for every set of worlds e and every $w \in \mathcal{W}$, $e, w \models \alpha$.[6]

Here we will not go into details about the properties of $\mathcal{ONL}$ but focus instead on three properties of the semantics which HL argue should also hold when generalizing the logic to the multi-agent case:

P1. Evaluating formulas of the form $N\alpha$ does not affect an agent's possibilities, *i.e.* the epistemic state e remains *fixed*. So if α is a complicated formula involving the nesting of K operators, then the corresponding subformulas are interpreted wrt the same e.

P2. A union of the epistemically possible worlds in e, wrt which K is evaluated, and the impossible worlds, wrt which N is evaluated, is *absolute* and *independent* of e, and is the set of all *conceivable* worlds $\mathcal{W}$. The intuition is that the exact complement of an agent's possibilities is used in evaluating N.

P3. For every set of conceivable worlds, there is always a model where *precisely* this set is the epistemic state.

Although these notions seem clear enough in the single agent case, generalizing them to the many agent case is non-trivial [Halpern and Lakemeyer 2001]. We return to analyze the properties after we consider our account of multi-agent only-knowing in the next section.

---

[5]Note that the only difference to the semantics for K is that an "if" becomes an "iff".

[6]Originally, Levesque only considered what he calls maximal sets of worlds when defining validity, but this was shown not to be necessary in [Levesque and Lakemeyer 2001].

3   Multi-Agent Only-Knowing

Let us begin by extending the language. Let $\mathcal{ONL}_n$ be a first-order modal language that enriches the non-modal subset of $\mathcal{ONL}$ with modal operators K_i and N_i for $i \in \{a, b\}$. For ease of exposition, we only have two agents a (Alice) and b (Bob). Extensions to more agents is straightforward. By analogy to the single agent case, we freely use O_i, such that $O_i\alpha$ abbreviates $K_i\alpha \wedge N_i\neg\alpha$, and is read as "all that i knows is α". The terms *objective* and *subjective* are now understood relative to an agent i. A formula is called *i-objective* if all epistemic operators which do not occur within the scope of another epistemic operator are of the form M_j for $i \neq j$, where M denotes K or N. A formula is called *i-subjective* if every atom is in the scope of an epistemic operator and all epistemic operators which do not occur within the scope of another epistemic operator are of the form M_i. Intuitively, *i*-subjective formulas refer to what is true about *i*'s beliefs about the world whereas *i*-objective formulas refer to what is true about the world from *i*'s perspective, which may include beliefs of agents other than *i*.

We will continue to call a formula *objective* if it does not contain any modal operators. A formula is *basic* if it does not mention any N_i for $i \in \{a, b\}$.

In the following it will be useful to refer to the degree of nestings of modal operators within a formula, where we lump together consecutive nestings of operators for the same agent.

DEFINITION 1.   The *i*-depth of a formula α, denoted $|\alpha|_i$, is defined inductively as:

1. $|\alpha|_i = 1$ for atoms,

2. $|\neg\alpha|_i = |\alpha|_i$,

3. $|\forall x.\, \alpha|_i = |\alpha|_i$,

4. $|\alpha \vee \beta|_i = max(|\alpha|_i, |\beta|_i)$,

5. $|M_i\alpha|_i = |\alpha|_i$,

6. $|M_j\alpha|_i = |\alpha|_j + 1$, for $j \neq i$.

A formula has a depth k if $max(a\text{-depth}, b\text{-depth}) = k$.

To illustrate the notion of depth, a formula of the form $K_aK_bK_ap \vee K_bq$ has a depth of 4, a a-depth of 3 and a b-depth of 4. K_bq is both b-subjective and a-objective.

3.1   The k-structures Approach

The main purpose of the semantics is to interpret *i*-subjective and *i*-objective formulas. Our idea will be to keep separate the worlds Alice believes from the worlds she considers Bob to believe, to some depth k.

DEFINITION 2.   A k-structure ($k \geq 1$), say e^k, for an agent is defined inductively:

 — $e^1 \subseteq \mathcal{W} \times \{\{\}\}$,

 — $e^k \subseteq \mathcal{W} \times \mathbb{E}^{k-1}$, where $\mathbb{E}^m$ is the set of all m-structures.

A e^1 for Alice, denoted as e_a^1, is intended to represent a set of worlds $\{(w, \{\}), \ldots\}$. A e^2 is of the form $\{(w, e_b^1), (w', e_b'^1), \ldots\}$, and it is to be read as "at w, a believes b considers

71

worlds from e_b^1 possible but at w', a believes b to consider worlds from e'^1_b possible". This conveys the idea that Alice has only partial information about Bob, and so at different worlds, her beliefs about what Bob knows differ.

We define a e^k for Alice, a e^j for Bob and a world $w \in \mathcal{W}$ as a (k, j)-model (e_a^k, e_b^j, w). Only sentences of a maximal a-depth of k, and a maximal b-depth of j are interpreted wrt a (k, j)-model. The complete semantic definition is:

1. $e_a^k, e_b^j, w \models p$ iff $p \in w$ for a ground atom p,

2. $e_a^k, e_b^j, w \models (m = n)$ iff m and n are identical standard names,

3. $e_a^k, e_b^j, w \models \neg\alpha$ iff $e_a^k, e_b^j, w \not\models \alpha$,

4. $e_a^k, e_b^j, w \models \alpha \vee \beta$ iff $e_a^k, e_b^j, w \models \alpha$ or $e_a^k, e_b^j, w \models \beta$,

5. $e_a^k, e_b^j, w \models \forall x.\ \alpha$ iff $e_a^k, e_b^j, w \models \alpha_n^x$ for all standard names n,

6. $e_a^k, e_b^j, w \models \mathbf{K}_a\alpha$ iff for all $(w', e_b^{k-1}) \in e_a^k,\ e_a^k, e_b^{k-1}, w' \models \alpha$,

7. $e_a^k, e_b^j, w \models \mathbf{N}_a\alpha$ iff for all $(w', e_b^{k-1}) \notin e_a^k,\ e_a^k, e_b^{k-1}, w' \models \alpha$.

Now, since $\mathbf{O}_a\alpha$ syntactically denotes $\mathbf{K}_a\alpha \wedge \mathbf{N}_a\neg\alpha$, it follows that

- $e_a^k, e_b^j, w \models \mathbf{O}_a\alpha$ iff for all worlds w', for all e^{k-1} for Bob, $(w', e_b^{k-1}) \in e_a^k$ iff $e_a^k, e_b^{k-1}, w' \models \alpha$.

The semantics for $\mathbf{K}_b\alpha$ and $\mathbf{N}_b\alpha$ are given analogously. A formula α (of a-depth of k and of b-depth of j) is *satisfiable* iff there is a (k, j)-model such that $e_a^k, e_b^j, w \models \alpha$. The formula is *valid* ($\models \alpha$) iff α is true at all (k, j)-models. Satisfiability is extended to a set of formulas Σ (of maximal a, b-depth of k, j) in the manner that there is a (k, j)-model e_a^k, e_b^j, w such that $e_a^k, e_b^j, w \models \alpha'$ for every $\alpha' \in \Sigma$. We write $\Sigma \models \alpha$ to mean that for every (k, j)-model (e_a^k, e_b^j, w), if $e_a^k, e_b^j, w \models \alpha'$ for all $\alpha' \in \Sigma$, then $e_a^k, e_b^j, w \models \alpha$.

Validity is not affected if models of a depth greater than that needed are used. This is to say, if α is true wrt all (k, j)-models, then α is true wrt all (k', j')-models for $k' \geq k, j' \geq j$. We obtain this result by constructing for every $e_a^{k'}$, a k-structure $e_a{\downarrow}_k^{k'}$, such that they agree on all formulas of max a-depth k. An analogous construction holds for every $e_b^{j'}$.

DEFINITION 3. Given $e_a^{k'}$, we define a k-structure $e_a{\downarrow}_k^{k'}$ for $k' \geq k \geq 1$:

- $e_a{\downarrow}_1^1 = e_a^1$,

- $e_a{\downarrow}_1^{k'} = \{(w, \{\}) \mid (w, e_b^{k'-1}) \in e_a^{k'}\}$, and

- $e_a{\downarrow}_k^{k'} = \{(w, e_b{\downarrow}_{k-1}^{k'-1}) \mid (w, e_b^{k'-1}) \in e_a^{k'}\}$.

LEMMA 4. *Let $k' \geq k, j' \geq j$. For all formulas α of maximal a, b-depth of k, j:*
$$e_a^{k'}, e_b^{j'}, w \models \alpha \quad iff \quad e_a{\downarrow}_k^{k'}, e_b{\downarrow}_j^{j'}, w \models \alpha.$$

Proof. By induction on the depth of formulas. It is immediate for atomic formulas, disjunctions and negations since we have the same world w. Assume that the result holds for formulas of a, b-depth 1. Let α such a formula, and suppose $e_a^{k'}, e_b^{j'}, w \models \boldsymbol{K}_a\alpha$, i.e. $\boldsymbol{K}_a\alpha$ has a, b-depth of $1, 2$. Then, for all $(w', e_b^{k'-1}) \in e_a^{k'}, e_a^{k'}, e_b^{k'-1}, w' \models \alpha$ iff (by hypothesis) $e_a\!\downarrow_1^{k'}, e_b\!\downarrow_1^{k'-1}, w' \models \alpha$ iff $e_a\!\downarrow_2^{k'}, \{\}, w \models \boldsymbol{K}_a\alpha$. Because α has depth 1, we also have $e_a\!\downarrow_1^{k'}, \{\}, w \models \boldsymbol{K}_a\alpha$. Lastly, since $\boldsymbol{K}_a\alpha$ is a-subjective, b's structure is irrelevant, and thus, $e_a\!\downarrow_1^{k'}, e_b\!\downarrow_2^{j'}, w \models \boldsymbol{K}_a\alpha$.

Conversely, suppose $e_a\!\downarrow_1^{k'}, e_b\!\downarrow_2^{j'}, w \models \boldsymbol{K}_a\alpha$. Then for all $w' \in e_a\!\downarrow_1^{k'}, e_a\!\downarrow_1^{k'}, \{\}, w' \models \alpha$ iff (since α has depth 1) for all $(w', e_b^{k'-1}) \in e_a^{k'}, e_a^{k'}, e_b^{k'-1}, w' \models \alpha$ iff $e_a^{k'}, \{\}, w \models \boldsymbol{K}_a\alpha$. Since b's structure is irrelevant, we have $e_a^{k'}, e_b^{j'}, w \models \boldsymbol{K}_a\alpha$. The proof is analogous for $\boldsymbol{K}_b\alpha$, $\boldsymbol{N}_a\alpha$ and $\boldsymbol{N}_b\alpha$. ∎

THEOREM 5. *For all formulas α of a, b-depth of k, j, if α is true at all (k, j)-models, then α is true at all (k', j')-models with $k' \geq k$ and $j' \geq j$.*

Proof. Suppose α is true at all (k, j)-models. Given any (k', j')-model, by assumption $e_a\!\downarrow_k^{k'}, e_b\!\downarrow_j^{j'}, w \models \alpha$ and by Lemma 4, $e_a^{k'}, e_b^{j'}, w \models \alpha$. ∎

Before moving on, let us briefly reflect on the fact that k-structures have finite depth. So suppose a only-knows KB of depth k. Using k-structures alone allows us to reason about what is believed up to depth k. Also, the logic correctly captures that a is completely ignorant about beliefs at depth greater than k. For example, let *true* (depth 1) be all that a knows. It is easy to see that both $\boldsymbol{O}_a(true) \supset \neg\boldsymbol{K}_a\neg\boldsymbol{K}_b p$ and $\boldsymbol{O}_a(true) \supset \neg\boldsymbol{K}_a\boldsymbol{K}_b p$ are valid sentences in the logic, that is, a knows neither that b knows p nor that b does not know p. So, although the KB has finite depth, we are able to ask queries α of any depth (in the sense of determining whether $\boldsymbol{O}_a\text{KB} \supset \boldsymbol{K}_a\alpha$ is valid).

There is, however, one aspect which previous approaches to multi-agent only-knowing can handle but we cannot: simultaneously satisfy an infinite set of sentences of unbounded depth. Indeed, k-structures cannot be used for this purpose simply because, for fixed k, the satisfaction relation is undefined for formulas beyond depth k. While this is certainly a restriction, we are willing to pay that price because in return we get, for the first time, a very simple possible-world style account of only-knowing for many agents.

3.2 Properties

Knowledge with k-structures satisfy $\boldsymbol{K45}_n$ properties, and the Barcan formula [Hughes and Cresswell 1972].

LEMMA 6. *If α is a formula, the following are valid wrt models of appropriate depth ($\boldsymbol{M}_i$ denotes $\boldsymbol{K}_i$ or $\boldsymbol{N}_i$):*

1. $\boldsymbol{M}_i\alpha \wedge \boldsymbol{M}_i(\alpha \supset \beta) \supset \boldsymbol{M}_i\beta$,

2. $\boldsymbol{M}_i\alpha \supset \boldsymbol{M}_i\boldsymbol{M}_i\alpha$,

3. $\neg\boldsymbol{M}_i\alpha \supset \boldsymbol{M}_i\neg\boldsymbol{M}_i\alpha$,

4. $\forall \boldsymbol{x}.\, \boldsymbol{M}_i\alpha \supset \boldsymbol{M}_i(\forall \boldsymbol{x}.\, \alpha)$.

Proof. We show item (3); the others are similar. Let M_i be K_a (N_a is analogous). Suppose $e_a^k, e_b^j, w \models \neg K_a \alpha$. There is some $(w', e_b^{k-1}) \in e_a^k$ such that $e_a^k, e_b^{k-1}, w' \models \neg \alpha$. Consider any $(w'', e_b^{\prime k-1}) \in e_a^k$. Then, $e_a^k, e_b^{\prime k-1}, w'' \models \neg K_a \alpha$. Thus, $e_a^k, e_b^j, w \models K_a \neg K_a \alpha$. ∎

We now return to the features **P1**, **P2** and **P3** of only-knowing discussed earlier and verify that the new semantics reasonably extends them to the multi-agent case. We also briefly discuss earlier attempts at capturing these features. H93 and L93 independently attempted to extend $\mathcal{ONL}$ to the many agent case.[7] There are some subtle differences in their approaches, but the main restriction is they only allow a propositional language. Henceforth, to make the comparison feasible, we shall also speak of the propositional subset of $\mathcal{ONL}_n$ with the understanding that the semantical framework is now defined for propositions (from an infinite set Φ) rather than ground atoms.[8]

In interpreting these features, we are mainly concerned with the notion of a *possibility*. In the single agent case, each world represents a possibility. Thus, from a logical viewpoint, a possibility is the set of objective formulas true at some world, or more simply, a truth assignment. Analogously, the set of epistemic possibilities is a set of truth assignments. However, in the many agent case, this notion is far from direct. In fact, both p and $K_b p$ are objective as far as a is concerned. Halpern and Lakemeyer correctly argue that the appropriate generalization of the notion of possibility in the many agent case are *i-objective formulas*. Intuitively, a possible state of affairs according to a includes the state of the world as well as what b is taken to believe.

As hinted in the introduction, earlier attempts are based on Kripke structures. Formally, a Kripke structure S is a tuple $(W, \pi, \mathcal{K}_a, \mathcal{K}_b)$, where W is a set of worlds, π associates worlds $w \in W$ with a truth assignment for the propositions in Φ, and $\mathcal{K}_i$ is a accessibility relation that determines which worlds are epistemically possible for i from a given world. For an introspective logic such as **K45**, the accessibility relations are Euclidean and transitive [Hughes and Cresswell 1972], where $\mathcal{K}_i$ is Euclidean if $(w, w') \in \mathcal{K}_i$ and $(w, w'') \in \mathcal{K}_i$ implies that $(w', w'') \in \mathcal{K}_i$, and $\mathcal{K}_i$ is transitive if $(w, w') \in \mathcal{K}_i$ and $(w', w'') \in \mathcal{K}_i$ implies that $(w, w'') \in \mathcal{K}_i$. A model for the logic is such a structure S and a world $w \in W$. By analogy to the single agent case, the notion of epistemic possibilities given a model (S, w) is defined as $\{obj_i(S, w') \mid w' \in \mathcal{K}_i(w)\}$, where $obj_i(S, w')$ is a set consisting of i-objective formulas true at (S, w'). (Note that, in the single agent case, i-objective formulas are simply objective formulas.) With this in hand, Halpern and Lakemeyer reexamine the features wrt their formalisms. As it turns out, **P3** does not hold in L93 and **P2** does not hold in H93. Consequently, these approaches show some peculiar properties, which we will look at shortly.

Note that while the definition of epistemic possibilities certainly seems intuitive, even for the propositional subset of $\mathcal{ONL}$, a Kripke world is a completely different entity from what Levesque supposes. Perhaps, one consequence is that the semantic proofs in earlier approaches are very involved. In contrast, we define worlds exactly as Levesque does. And, here is how we capture epistemic possibilities:

DEFINITION 7. Suppose $M = (e_a^k, e_b^j, w)$ is a (k, j)-model. Let

---

[7] For space reasons, we do not review all aspects of these approaches. Both HL and a proposal in [Waaler 2004; Waaler and Solhaug 2005] are motivated by the proof theory. Hence, discussions on these approaches are deferred to after we review the axiomatization.

[8] Somewhat surprisingly, when the set of propositions Φ is finite, then [Halpern and Lakemeyer 2001] show that Levesque's axiomatization is incomplete also for the propositional language. But they also show that a completeness result still holds, for the given semantics, with a few extra axioms.

74

1. $obj_a(M) = \{a\text{-objective } \phi \text{ of } b\text{-depth} \le j \mid M \models \phi\}$;

2. $obj_b(M) = \{b\text{-objective } \phi \text{ of } a\text{-depth} \le k \mid M \models \phi\}$;

3. $Obj_a(e_a^k) = \{obj_a(\{\}, e_b^{k-1}, w) \mid (w, e_b^{k-1}) \in e_a^k\}$;

4. $Obj_b(e_b^j) = \{obj_b(e_a^{j-1}, \{\}, w) \mid (w, e_a^{j-1}) \in e_b^j\}$.

All the a-objective formulas true at a model M, which are those describing what is true about the world and b's beliefs, are given by $obj_a(M)$. Note that these formulas do not really correspond to a's possibilities, which are determined strictly by e_a^k. Hence, we define $Obj_a(e_a^k)$, which gives us the set of all a-objectives formulas that a considers possible. We now argue that the intuition of Levesque's properties is maintained.[9]

P1. In the single agent case, this property ensured that an agent's epistemic possibilities are not affected on evaluating N. This is immediately the case here. Given a model, say (e_a^k, e_b^j, w), a's epistemic possibilities are determined by $Obj_a(e_a^k)$. To evaluate $N_a\alpha$, we consider all models (e_a^k, e_b^{k-1}, w') such that $(w', e_b^{k-1}) \notin e_a^k$. Here too, a's possibilities are fixed by $Obj_a(e_a^k)$ and so **P1** holds.

P2. In the single agent case, this property ensured that evaluating $K\alpha$ and $N\alpha$ is always wrt the set of all possibilities, and completely independent of e. As discussed, in the many agent case, possibilities mean i-objective formulas and analogously, if α is a possibility in a's view, say an a-objective formula of maximal b-depth of k, then we should interpret $K_a\alpha$ and $N_a\alpha$ wrt all a-objective possibilities of maximal depth k: the set of $(k + 1)$-structures. Clearly then, the result is independent of any e_a^{k+1}. The following lemma is a direct consequence of the definition of the semantics.

LEMMA 8. *Let α be a-objective of b-depth k. Then, the set of $(k + 1)$-structures that evaluate $K_a\alpha$ and $N_a\alpha$ is $\mathbb{E}^{k+1}$. (Analogously stated for b.)*

P3. The third property, by analogy, must allow us to characterize epistemic states from any set of i-objective formulas. Intuitively, given such a set of formulas, we must have a model where *precisely* this set represents the beliefs of an epistemic state. Earlier attempts at clarifying this property involved constructing a *set* of maximally $\mathbf{K45}_n$-consistent sets of *basic* i-objective formulas, and showing that there exist an epistemic state (in an appropriate Kripke structure) that corresponds to this set. But, defining possibilities via $\mathbf{K45}_n$ proof-theoretic machinery inevitably leads to some limitations, as we shall see. We instead proceed semantically, and go beyond basic formulas.

Let Σ be a satisfiable set of i-objective (not necessarily basic) formulas, say of maximal j-depth k, for $j \neq i$. Suppose γ is i-objective formula of maximal j-depth k. If $\Sigma \cup \{\gamma\}$ is satisfiable, then let $\Sigma_1 = \Sigma \cup \{\gamma\}$. Otherwise, let $\Sigma_1 = \Sigma$. By considering all i-objective formulas of maximal j-depth k, let us construct $\Sigma_2, \ldots$ *i.e.* where formulas are added iff the resultant set remains satisfiable. Denote the limit as Σ^* and this is what we shall call a

---

[9]It is interesting to note that such a formulation of Levesque's properties is not straightforward in the first-order case. That is, for the quantified language, it is known that there are epistemic states that can not be characterized using only objective formulas [Levesque and Lakemeyer 2001]. Thus, it is left open how one must correctly generalize the features of first-order $\mathcal{ONL}$.

maximally satisfiable i-objective set.[10] (Note: there may be many maximally satisfiable i-objective sets beginning from some Σ.) We show that given *any* set of maximally satisfiable i-objective sets, there is a model where precisely this set characterizes the epistemic state.

THEOREM 9. *Let S_i be a set of maximally satisfiable sets of i-objective formulas, and σ a satisfiable objective formula. Suppose S_a is of max b-depth k and S_b is of max a-depth j. Then there is a model $M^* = (e_a^{*k+1}, e_b^{*j+1}, w^*)$ such that $M^* \models \sigma$, $S_a = Obj_a(e_a^{*k+1})$ and $S_b = Obj_b(e_b^{*j+1})$.*

Proof. Consider S_a. Each $S' \in S_a$ is a maximally satisfiable a-objective set, and thus by definition, there is a (w', e_b^k) such that $\{\}, e_b^k, w' \models S'$. Define $e_a^{*k+1} = \{(w', e_b^k) \mid \{\}, e_b^k, w' \models S' \text{ and } S' \in S_a\}$. It is immediate to verify that $Obj_a(e_a^{*k+1}) = S_a$. Analogously, construct e_b^{*j+1} using S_b. Finally, let w^* be any world where σ holds. ∎

Thus, (arguably) the properties are captured. The question we want to now answer is this: How does the semantics compare to earlier approaches?

3.3 A Comparison to Lakemeyer [1993] and Halpern [1993]

For this part, we are concerned with formulas that come out valid, besides the usual $\mathbf{K45}_n$ properties, in L93, H93, and k-structures. In L93, where a semantics based on $\mathbf{K45}_n$-canonical models [Hughes and Cresswell 1972] is proposed, it is shown that the formula $\neg O_a \neg O_b p$ (denote as γ_1) for any proposition p is valid. Intuitively, it says that all that Alice knows is that Bob does not only know p, and as Lakemeyer argues, the validity of $\neg O_a \neg O_b p$ is unintuitive. After all, Bob could *honestly* tell Alice that he does not only know p.

The negation of this formula, on the other hand, is satisfiable in the Kripke structure approach of H93, called the i-set approach.[11] It is also satisfiable in the k-structure semantics. Interestingly, the i-set approach and k-structures agree on one more notion. The formula $K_a(false) \supset \neg N_a \neg O_b \neg O_a p$ (denote as γ_2) is valid in both, while $\neg \gamma_2$ is satisfiable wrt L93. (It turns out that the validity of γ_2 in our semantical framework is implicitly related to the satisfiability of $O_b \neg O_a p$, so this property is not unreasonable.)

However, we remark that the i-set approach and k-structures do not share too many similarities beyond those presented above. In fact, the i-set approach does not truly satisfy Levesque's second property; for instance, in the approach $N_a \neg O_b p \wedge K_a \neg O_b p$ (denote as γ_3) is satisfiable. Recall that, in this property, the union of models that evaluate $N_i \alpha$ and $K_i \alpha$ must lead to the full set of possibilities. So, the satisfiability of γ_3 leaves open the question as to why $O_b p$ is not considered since $\neg O_b p$ is true at all conceivable states! We show that, in contrast, γ_3 is not satisfiable in the k-structures approach.

THEOREM 10. *The following are properties of the semantics (in models of appropriate depth):*

1. $O_a \neg O_b p$, for any $p \in \Phi$, is satisfiable.

---

[10]A maximally satisfiable set is to be understood as a semantically characterized *complete* description of a possibility, analogous to a proof theoretically characterized notion of maximally consistent set of formulas. In the single agent case, both (objective) maximally satisfiable sets and (objective) maximally consistent sets are clearly definable, owing to the standard semantics and axiomatization of propositional logic. But it is not immediate what a i-objective maximally consistent set of formulas should look like, since we do not have a background proof theory. This is the reason for choosing a semantic alternative.

[11]In his original formulation, Halpern [1993] uses a different terminology. We build on discussions in HL.

2. $K_a(false) \supset \neg N_a \neg O_b \neg O_a p$ is valid.

3. $N_a \neg O_b p \wedge K_a \neg O_b p$ is not satisfiable.

Proof. *Item (1)*. Let $\mathcal{W}_p = \{w \mid w \models p\}$ and let E be all subsets of $\mathcal{W}$ except the set $\mathcal{W}_p$. It is easy to see that if $e_b^1 \in E$, then $\{\}, e_b^1, w \not\models O_b p$, for any world w. Now, define a e^2 for a that has all of $\mathcal{W} \times E$. Thus, $e_a^2, \{\}, w \models O_a \neg O_b p$.

Item (2). Suppose $e_a^k, \{\}, w \models K_a(false)$ for any $w \in \mathcal{W}$. Then, for all $(w', e_b^{k-1}) \in e_a^k$, $e_a^k, e_b^{k-1}, w' \models false$, and thus, $e_a^k = \{\}$. Suppose now $e_a^k, \{\}, w \models N_a \neg O_b \neg O_a p$. Then, wrt all of $(w', e_b^{k-1}) \notin e_a^k$ i.e. all of $\mathbb{E}^k$, $\neg O_b \neg O_a p$ must hold. That is, $\neg O_b \neg O_a p$ must be valid. From above, we know this is not the case (by switching the agent indices).

Item (3). Suppose $e_a^k, \{\}, w \models K_a \neg O_b p$, for any w. Then, for all $(w', e_b^{k-1}) \in e_a^k$, $e_a^k, e_b^{k-1}, w' \models \neg O_b p$. Since $O_b p$ is satisfiable, there is a e_b^{*k-1} such that $\{\}, e_b^{*k-1}, w^* \models O_b p$, and $(w^*, e_b^{*k-1}) \notin e_a^k$. Then, $e_a^k, \{\}, w \models \neg N_a \neg O_b p$. ∎

Thus, k-structures seem to satisfy our intuitions on the behavior of only-knowing. To understand why, notice that γ_1 and γ_3 involve the nesting of N_i operators. L93 makes an unavoidable technical commitment. In his formalism, an i-objective possibility is formally a maximally $K45_n$-consistent set of *basic i-objective* formulas. The restriction to basic formulas is an artifact of a semantics based on the canonical model. Unfortunately, there is more to agent i's possibility than just basic formulas. In contrast, Theorem 9 shows that we allow non-basic formulas and by using a strictly semantic notion, we avoid problems that arise from the proof-theoretic restrictions. In the case of H93, the problem seems to be that N_i and K_i do not interact naturally, and that the full complement of epistemic possibilities is not considered in interpreting N_i. In our case, however, since the semantics faithfully complies with the second property, γ_3 is not satisfiable.

4 Proof Theory

The natural question is if there are axioms that characterize the semantics. We show that the answer is affirmative. Let us begin with a review of $\mathcal{ONL}$'s proof theory.

Axioms:

    A1. All instances of propositional logic,

    A2. $K(\alpha \supset \beta) \supset (K\alpha \supset K\beta)$,

    A3. $N(\alpha \supset \beta) \supset (N\alpha \supset N\beta)$,

    A4. $\sigma \supset K\sigma \wedge N\sigma$ for any subjective σ,

    A5. $N\alpha \supset \neg K\alpha$ if $\neg\alpha$ is a propositionally consistent objective formula.

Inference Rules:

    MP. From α and $\alpha \supset \beta$ infer β.

    NEC. From α infer $K\alpha$ and $N\alpha$.[12]

---

[12]Strictly speaking, this is not the proof theory introduced in [Levesque 1990], where an axiom replaces the inference rule **NEC**. Here, we consider an equivalent formulation by HL.

Essentially, axioms **A2 – A4** tell us that K and N separately have all properties of a **K45** modal operator. In addition, the axiom **A4** tells us that K and N are mutually introspective, that is, $K\alpha \supset NK\alpha$ is valid. Perhaps, the most interesting axiom is **A5** which establishes the connection between K and N, and gives only-knowing the desired properties. In fact, on revisiting the features of $\mathcal{ONL}$, we observe that **P1** is required for **A4**, and **P2** is required for **A5** because K and N range over all possible world states.

The other detail worth noting is that the soundness of **A5** appeals to falsifiability in classical propositional logic. Precisely for this reason only that axiom is controversial. Intuitively, when generalizing Levesque's proof theory to the many agent case we have to extend the applicability of this axiom from consistent objective formulas to consistent i-objective formulas. Unfortunately, establishing the consistency of i-objective formulas is non-trivial and even circular, since to have a notion of consistency implies that we have already defined an axiom system.

To this end, L93 proposes to resolve the circularity by appealing to the existing logic **K45**$_n$. That is, his variant of a generalized **A5** is the axiom $N_i\alpha \supset \neg K_i\alpha$, where α is any formula that is falsifiable in (classical) **K45**$_n$. Of course, this means that he has to restrict his attention to basic formulas.

Axioms:

    A1$_n$. All instances of propositional logic,

    A2$_n$. $K_i(\alpha \supset \beta) \supset (K_i\alpha \supset K_i\beta)$,

    A3$_n$. $N_i(\alpha \supset \beta) \supset (N_i\alpha \supset N_i\beta)$,

    A4$_n$. $\sigma \supset K_i\sigma \wedge N_i\sigma$ for i-subjective σ,

    A5$_n$. $N_i\alpha \supset \neg K_i\alpha$ if $\neg\alpha$ is a **K45**$_n$-consistent i-objective basic formula.

Inference Rules:

    MP. From α and $\alpha \supset \beta$ infer β.

    NEC. From α infer $K_i\alpha$ and $N_i\alpha$.

Clearly L93 has appropriately generalized all of Levesque's axioms, except for **A5**. Let us refer to the above set of schemas as **AX**$_n$. As we shall see, one significant consequence of restricting to basic formulas in **A5**$_n$ is that **AX**$_n$ is not complete for formulas in $\mathcal{ONL}_n$. However, **AX**$_n$ is sound and complete for a sublanguage $\mathcal{ONL}_n^- \subsetneq \mathcal{ONL}_n$ wrt both the canonical model and the i-set approach, which is defined as follows:

DEFINITION 11. $\mathcal{ONL}_n^-$ consists of all formulas α in $\mathcal{ONL}_n$ such that no N_j may occur in the scope of a K_i or a N_i, for $i \neq j$.

For example, $N_a K_a \neg N_a p$ is in $\mathcal{ONL}_n^-$, but $N_a N_b p$ is not. We now show that **AX**$_n$ is sound and complete for formulas in $\mathcal{ONL}_n^-$ wrt our semantics as well.

4.1   **AX**$_n$ is sound and complete for $\mathcal{ONL}_n^-$

Let us begin by observing that the soundness of **A5**$_n$ appeals to satisfiability in **K45**$_n$. However, our semantics does not make use of Kripke structures. Therefore the soundness proof is not immediate.

Nevertheless, there is a well-known property in modal logic that every consistent formula is satisfiable in the canonical model [Fagin, Halpern, Moses, and Vardi 1995]. So we proceed as follows. We propose a construction called the *correspondence* (k, j)-*model* which appeals to the accessibility relations in a Kripke structure in the manner that every formula of maximal a, b-depth of k, j is satisfiable in the Kripke structure iff the correspondence model satisfies the formula. Then by constructing a correspondence model for the canonical Kripke structure, the satisfiability of $\mathbf{K45}_n$-consistent formulas is implied.

In what follows, we will need to disambiguate worlds as considered by Levesque on the one hand and Kripke worlds on the other. So we shall refer to the former as *propositional valuations*.

DEFINITION 12. The $\mathbf{K45}_n$ canonical model $M^c = (\mathcal{W}^c, \pi^c, \mathcal{K}_a^c, \mathcal{K}_b^c)$ is defined as:

- $\mathcal{W}^c = \{w \mid w$ is a maximally consistent set of basic formulas wrt $\mathbf{K45}_n\}$,

- for all $p \in \Phi$ and worlds $w \in \mathcal{W}^c$, $\pi^c(w)(p) = \mathit{true}$ iff $p \in w$,

- $(w, w') \in \mathcal{K}_i^c$ iff $w \backslash \mathbf{K}_i \subseteq w'$, where $w \backslash \mathbf{K}_i = \{\alpha \mid \mathbf{K}_i \alpha \in w\}$.

DEFINITION 13. Given M^c, define a set of propositional valuations $\mathcal{W}$ such that for each world $w \in \mathcal{W}^c$, there is a valuation $\|w\| \in \mathcal{W}$ such that $\|w\| = \{p \in \Phi \mid p \in w\}$.

DEFINITION 14. Given M^c and $w \in \mathcal{W}^c$, construct a (k, j)-model $(e_{\|w\|_a}^k, e_{\|w\|_b}^j, \|w\|)$ from valuations $\mathcal{W}$ inductively:

- $e_{\|w\|_a}^1 = \{(\|w'\|, \{\}) \mid w' \in \mathcal{K}_a^c(w)\}$,

- $e_{\|w\|_a}^k = \{(\|w'\|, e_{\|w'\|_b}^{k-1}) \mid w' \in \mathcal{K}_a^c(w)\}$

  where $e_{\|w'\|_b}^{k-1} = \{(\|w''\|, e_{\|w''\|_a}^{k-2}) \mid w'' \in \mathcal{K}_b^c(w')\}$.

Further, $e_{\|w\|_b}^j$ is constructed analogously. Let us refer to this model as the *correspondence* (k, j)-*model of* (M^c, w).

That is, a 1-structure for a is precisely the set of valuations corresponding to the worlds $\{w' \mid w' \in \mathcal{K}_a^c(w)\}$. Inductively, a k-structure is the set of all $(\|w'\|, e^{k-1})$, where $w' \in \mathcal{K}_a^c(w)$ as before with e^{k-1} constructed in an analogous fashion (to the appropriate depth) from all $w'' \in \mathcal{K}_a^c(w')$. By an induction on the depth of a *basic* formula α, we obtain a theorem that α of maximal a, b-depth k, j is satisfiable at (M^c, w) iff the correspondence (k, j)-model of (M^c, w) satisfies α.

THEOREM 15. *For all basic formulas* α *in* $\mathcal{ONL}_n^-$ *and of maximal* a, b-*depth of* k, j:

$$M^c, w \models \alpha \ \ \mathit{iff} \ \ e_{\|w\|_a}^k, e_{\|w\|_b}^j, \|w\| \models \alpha.$$

Proof. The proof is immediate for propositional formulas, disjunctions and negations. Assume that the result holds for formulas of a, b-depth 1. Now suppose $M^c, w \models \mathbf{K}_a \alpha$. Note that $\mathbf{K}_a \alpha$ has a, b-depth of $1, 2$. Then for all $w' \in \mathcal{K}_a^c(w)$, $M^c, w' \models \alpha$ iff by hypothesis $e_{\|w'\|_a}^1, e_{\|w'\|_b}^1, \|w'\| \models \alpha$ iff by definition $e_{\|w\|_a}^2, \{\}, \|w\| \models \mathbf{K}_a \alpha$. Since α has a depth of 1 and b's structure is irrelevant, $e_{\|w\|_a}^1, e_{\|w\|_b}^2, \|w\| \models \mathbf{K}_a \alpha$ as well.

Conversely, suppose $e_{\|w\|_a}^1, e_{\|w\|_b}^2, \|w\| \models \mathbf{K}_a \alpha$. Then for all $\|w'\| \in e_{\|w\|_a}^1$, clearly $e_{\|w\|_a}^1, \{\}, \|w'\| \models \alpha$ iff by hypothesis, for all $w' \in \mathcal{K}_a^c(w)$, $M^c, w' \models \alpha$ iff $M^c, w \models \mathbf{K}_a \alpha$. The proof is analogous for $\mathbf{K}_b \alpha$. ∎

LEMMA 16. *Every* $\mathbf{K45}_n$-*consistent basic formula* α *of maximal* a, b-*depth* k, j *is satisfiable wrt some* (k, j)-*model.*

Proof. Every $\mathbf{K45}_n$-consistent basic formula is satisfiable wrt the canonical model. By Theorem 15, the correspondence (k, j)-model also satisfies the formula. ∎

We now demonstrate the soundness of $\mathbf{AX}_n$. We use $\vdash$ to denote provability.

THEOREM 17. *For all $\alpha \in \mathcal{ONL}_n^-$, if $\mathbf{AX}_n \vdash \alpha$ then $\models \alpha$.*

Proof. The soundness is easily shown to hold for $\mathbf{A1}_n - \mathbf{A4}_n$. For $\mathbf{A5}_n$, let $\neg\alpha$ be a $\mathbf{K45}_n$-consistent basic a-objective formula. Suppose α has a max b-depth k. By Lemma 16 there is $(\{\}, e^{*k}_b, w^*)$ such that $\{\}, e^{*k}_b, w^* \models \neg\alpha$, *i.e.* we ignore the epistemic state for a since α is a-objective. For an arbitrary e_a^{k+1}, if $(w^*, e_b^{*k}) \in e_a^{k+1}$ then $e_a^{k+1}, \{\}, w \models \neg K_a\alpha$. Otherwise $e_a^{k+1}, \{\}, w \models \neg N_a\alpha$. Thus $e_a^{k+1}, \{\}, w \models K_a\alpha \supset N_a\alpha$. Analogously for $\mathbf{K45}_n$-consistent basic b-objective formulas. ∎

The proof that $\mathbf{AX}_n$ is also complete for $\mathcal{ONL}_n^-$ is long and tedious. This and other proofs in the remaining parts of the chapter are presented in [Belle and Lakemeyer 2010]. The argument is based on a certain normal form introduced by HL. The normal form allows us to reduce any arbitrary $\alpha \in \mathcal{ONL}_n^-$ to ones where strictly i-objective basic formulas appear in the scope of K_i and N_i. Using this simplification, it is possible to show that every $\mathbf{AX}_n$-consistent formula of maximal a, b-depth k, j is satisfiable in some (k, j)-model.

THEOREM 18. *For all formulas $\alpha \in \mathcal{ONL}_n^-$, if $\models \alpha$ then $\mathbf{AX}_n \vdash \alpha$.*

4.2 $\mathbf{AX}_n$ is not complete for $\mathcal{ONL}_n$

It is straightforward to extend the soundness result for $\mathbf{AX}_n$ in Theorem 17 to the full language $\mathcal{ONL}_n$. To prove that it is an incomplete axiomatization, it suffices to obtain a formula $\alpha \in \mathcal{ONL}_n$ such that α is valid but not provable using $\mathbf{AX}_n$. The sentence $K_a(\mathit{false}) \supset \neg N_a\neg O_b\neg O_a p$ from Theorem 10 is such a formula. In fact, $\mathbf{AX}_n$ is sound for $\mathcal{ONL}_n$ wrt the canonical model and the i-set approach as well. Therefore the validity of non-provable formulas $\neg O_a\neg O_b p$ and $K_a(\mathit{false}) \supset \neg N_a\neg O_b\neg O_a p$ (from Section 3.3) wrt these approaches respectively demonstrates the incompleteness of $\mathbf{AX}_n$. Intuitively, part of the problem is $\mathbf{A5}_n$. It has to somehow go beyond basic formulas. However, as we mentioned earlier, the issue is one of circularity: to reason about *consistent* i-objective formulas we have to first specify an axiomatization.

The approach taken in HL is to introduce *validity*, and its dual satisfiability, directly into the language as modal operators. The motivation is that by representing satisfiability in the object language, the notion of consistency can be inductively defined from propositional formulas to modal ones. Not surprisingly, a new set of axioms is needed to characterize this feature, by way of which the axiomatization is significantly different from Levesque's formulation for the single agent case. We give the details in Section 5 but for now let us refer to the proposal as $\mathbf{AX}_n'$.

HL demonstrate that the axioms $\mathbf{AX}_n'$ characterize an infinite model, very much in the spirit of the canonical model from Definition 12, but where worlds correspond to maximally $\mathbf{AX}_n'$-consistent sets of formulas. They also handle K_i and N_i as two separate operators, by which we mean that the model is defined using two separate accessibility relations. Owing to these differences, they term the structure as an *extended canonical model*.

One approach towards an axiomatization for our semantics is to perhaps show that the set of valid formulas coincide in the extended canonical model and k-structures for the new language. But then, as we argued, axiomatizing validity is not natural. Moreover, the proof theory in HL is difficult to use. And in the end, we would still understand the axioms to characterize a semantics bridged on proof-theoretic elements.

4.3 A Proof Theory for $\mathcal{ONL}_n$

What is really desired is a generalization of Levesque's axiom **A5**, and nothing more. To this end, we propose a new axiom system, that is subtly related to the structure of formulas. Formally, let us begin by defining a sequence of languages:

DEFINITION 19. Let $\mathcal{ONL}_n^1 = \mathcal{ONL}_n^-$. Let $\mathcal{ONL}_n^t$ be all Boolean combinations of formulas of $\mathcal{ONL}_n^{t-1}$ and formulas of the form $K_i\alpha$ and $N_i\alpha$ for $\alpha \in \mathcal{ONL}_n^{t-1}$.

Clearly $\mathcal{ONL}_n^t \supsetneq \mathcal{ONL}_n^{t-1}$, e.g. N_aN_bp is not in $\mathcal{ONL}_n^1$ but it is in $\mathcal{ONL}_n^t, t \geq 2$.

We have already established that $\mathbf{AX}_n$ characterize formulas from $\mathcal{ONL}_n^1$. The axiom system that characterizes formulas from $\mathcal{ONL}_n^t$ is given as:

Axioms:

$\mathbf{A1}_n - \mathbf{A4}_n,$

$\mathbf{A5}_n^1. \ N_i\alpha \supset \neg K_i\alpha$ if $\neg\alpha$ is a $\mathbf{K45}_n$-consistent i-objective basic formula,

$\mathbf{A5}_n^t. \ N_i\alpha \supset \neg K_i\alpha$, if $\neg\alpha \in \mathcal{ONL}_n^{t-1}$ is i-objective

and consistent wrt $\mathbf{A1}_n - \mathbf{A4}_n, \mathbf{A5}_n^1 - \mathbf{A5}_n^{t-1}$.

Inference Rules:

MP and **NEC** (as before).

Let us refer to this set of axioms as $\mathbf{AX}_n^t$. That is, for a given t, the proof theory has t axioms in addition to $\mathbf{A1}_n - \mathbf{A4}_n$. It is worth taking note that we resolve the circularity regarding consistency by carefully limiting the nesting of *at most* operators. By this we mean that each *language* allows another level of nesting of only-knowing with varying agent indices. So if we restrict our attention to formulas from $\mathcal{ONL}_n^1$, it is clear that Lakemeyer's proof theory suffices. But to reason about formulas of the form N_aN_bp, we have to consider the next language, *viz.* $\mathcal{ONL}_n^2$ (or higher).[13]

To show that $\mathbf{AX}_n^t$ is sound and complete for all $\alpha \in \mathcal{ONL}_n^t$ wrt our semantics, we do an induction on the parameter t. The base case corresponds to when t is 1, the proof of which is Theorem 17 and Theorem 18.

For the induction step, the normal form reduction by HL (as discussed in Section 4.1) is once again helpful. This reduction is applicable for $\mathcal{ONL}_n$ in the sense that it allows us to reduce any $\alpha \in \mathcal{ONL}_n$ to one where strictly i-objective (but not necessarily basic) formulas appear in the scope of K_i and N_i. This enables the use of similar techniques from Theorem 18 to obtain the following:

THEOREM 20. *For all $\alpha \in \mathcal{ONL}_n^t$, $\mathbf{AX}_n^t \vdash \alpha$ iff $\models \alpha$.*

Thus, we have a sound and complete axiomatization for propositional $\mathcal{ONL}_n$. In comparison to L93 the axiomatization goes beyond a language that restricts the nesting of N_i. In contrast to HL, the axiomatization does not necessitate the use of semantic notions in the proof theory. A third axiomatization by Waaler [2004] considers an interesting alternative to deal with the circularity in a generalized **A5**. The idea is to first define consistency by

---

[13]The idea was also suggested by a reviewer in [Halpern and Lakemeyer 2001] for an axiomatic characterization of the extended canonical model, although its completeness was left open.

formulating a fragment of the axiom system in the sequent calculus. Quite analogous to having t-axioms, they allow us to apply $\mathbf{A5}_n$ on i-objective formulas of a lower depth, thereby avoiding circularity without the need to appeal to satisfiability as in HL. Waaler and Solhaug [2005] also define a semantics for multi-agent only-knowing which does not appeal to canonical models. Instead, they define a class of Kripke structures which need to satisfy certain constraints. Unfortunately, these constraints are quite involved and, as the authors admit, the nature of these models "is complex and hard to penetrate."

4.4 Formal Derivations of Nonmonotonic Inferences

To get a feel for the axiomatization, let us consider examples, adapted from HL. Suppose Alice assumes the following default:

> *unless I know that Bob knows my secret then he does not know it.*

If the default is all that she knows, then she *nonmonotonically* comes to believe that Bob does not know her secret. Let $p \in \Phi$ denote Alice's secret. Then we can prove that

$$\vdash O_a \alpha \supset K_a \neg K_b p$$

where $\alpha = \neg K_a K_b p \supset \neg K_b p$. The justifications are below, and they indicate a previous line or the axioms utilized to prove the current line. We write **Def** to mean the equivalence $O_i \alpha \equiv K_i \alpha \wedge N_i \neg \alpha$. Further, we freely reason with propositional logic (**PL**) or $\mathbf{K45}_n$.

1. $O_a \alpha \supset (K_a \neg K_a K_b p \supset K_a \neg K_b p)$ **Def, PL, A2**$_n$

2. $O_a \alpha \supset (N_a \neg K_a K_b p \wedge N_a K_b p)$ **Def, PL, K45**$_n$

3. $N_a K_b p \supset \neg K_a K_b p$ $\mathbf{A5}_n^1$

4. $\neg K_a K_b p \supset K_a \neg K_a K_b p$ $\mathbf{A4}_n$

5. $O_a \alpha \supset K_a \neg K_a K_b p$ $2, 3, 4,$ **PL**

6. $O_a \alpha \supset K_a \neg K_b p$ $1, 5,$ **PL**

Most of the lines involve standard $\mathbf{K45}_n$ reasoning. Line (3) is the only place where we need to make use of the special connection between N_i and K_i. In particular, we use $\mathbf{A5}_n^1$ and it is applicable because $\neg K_b \gamma$ is a-objective, basic and $\mathbf{K45}_n$-consistent. However, we could have equivalently restricted ourselves to $\mathbf{AX}_n$ to derive the result.

To consider an example where $\mathbf{AX}_n$ is not enough, simply change the default to $\beta = \neg K_a O_b p \supset \neg O_b p$. Intuitively, we could imagine a rather cautious Alice who assumes:

> *if I do not believe that p is all that Bob knows, then he usually knows more.*

By analogy to the above, we can prove that Alice believes that p is not all that Bob knows:

$$\vdash O_a \beta \supset K_a \neg O_b p.$$

Here, on the one hand, $\beta \notin \mathcal{ONL}_n^-$ and on the other, $O_b p$ is not a basic formula. Nevertheless, the proof is essentially similar except for line (3) where we prove $N_a O_b p \supset \neg K_a O_b p$ by means of the axiom $\mathbf{A5}_n^2$. This is applicable because $\neg O_b p \in \mathcal{ONL}_n^1$ is a-objective and $\mathbf{AX}_n^1$-consistent. We remark that the latter proof requires reasoning with the satisfiability modal operator in HL.

5 Axiomatizing Validity

We pointed out that both L93 and H93 fail to satisfy the properties of Levesque's semantics. Besides, L93's proof theory is restricted to formulas in $\mathcal{ONL}_n^-$. Extending this work, HL proposed a formalism that does capture the desiderata of multi-agent only-knowing. But as discussed, there are two undesirable features. The first is a semantics based on canonical models, and the second is a proof theory that axiomatizes validity. Although such a construction is far from natural, we show now that they do indeed capture the desired properties of only-knowing. This also instructs us that our axiomatization avoids such problems in a reasonable manner.

We begin by presenting the main formal features of their approach. Mainly, this involves enriching $\mathcal{ONL}_n$ with a modality V, for validity. Let $\mathcal{ONL}_n^+$ be the addition of V to $\mathcal{ONL}_n$. A modality S, for satisfiability, is used freely such that $V(\alpha)$ denotes $\neg S(\neg\alpha)$.

To handle the circularity of consistency, HL propose an axiom system, which we will denote as $\mathbf{AX}_n'$. It has the following schemas:

Axioms:

$\mathbf{A1}_n - \mathbf{A4}_n,$

$\mathbf{A5}_n'.$ $S(\neg\alpha) \supset (N_i\alpha \supset \neg K_i\alpha)$ if α is i-objective,

$\mathbf{V1}.$ $V(\alpha) \wedge V(\alpha \supset \beta) \supset V(\beta),$

$\mathbf{V2}.$ $S(\alpha)$ if α is a satisfiable propositional formula,

$\mathbf{V3}.$ $\bigwedge S(\alpha \wedge \beta_k) \wedge \bigwedge S(\gamma \wedge \delta_z) \wedge V(\alpha \vee \gamma) \supset$
$$S(K_i\alpha \wedge \bigwedge \neg K_i \neg \beta_k \wedge N_i\gamma \wedge \bigwedge \neg N_i \neg \delta_z)$$
if $\alpha, \beta_k, \gamma, \delta_z$ are i-objective formulas,

$\mathbf{V4}.$ $S(\alpha) \wedge S(\beta) \supset S(\alpha \wedge \beta)$ if α is i-objective and β is i-subjective.

Inference Rules:

MP and NEC,

$\mathbf{NEC_V}.$ From α infer $V(\alpha)$.[14]

While the axioms seem somewhat mysterious, intuitively, they allow us to extend the notion of consistency to modal formulas. To see this, consider the axiom $\mathbf{V2}$ which allows us to include satisfiable (and hence consistent) propositional formulas in the scope of S. $\mathbf{V3}$ allows us to derive consistent i-subjective formulas from consistent i-objective ones. The generalized $\mathbf{A5}$ in the proof theory is $\mathbf{A5}_n'$, which allows us to invoke the N_i vs. K_i relationship on falsifiable i-objective formulas, $i.e.$ formulas α for which $S(\neg\alpha)$ is provable from the given premises.

In order to enable comparisons to this approach, it seems obvious that we need to extend our logic to handle the enriched language $\mathcal{ONL}_n^+$.[15] However, it turns out that there is a much simpler alternative, which rests on the following result proved by HL:

---

[14]The axiom $\mathbf{V1}$ and the rule $\mathbf{NEC_V}$ makes the modality V a *normal modal operator*, like K_i and N_i.

[15]This is indeed the direction pursued in [Belle and Lakemeyer 2010]. But the methodology considered in the sequel is cleaner and more direct.

THEOREM 21. *[HL] Every formula $\alpha \in \mathcal{ONL}_n^+$ is provably equivalent (wrt $\mathbf{AX}_n'$) to some formula $\alpha' \in \mathcal{ONL}_n$.*

The theorem essentially tell us that as far as we are concerned regarding derivations of $\mathbf{AX}_n'$, it suffices to restrict our attention to the language $\mathcal{ONL}_n$. With this in hand, we are able to show that the provability of $\mathbf{AX}_n'$ and $\mathbf{AX}_n^t$ coincide in the following sense:

THEOREM 22. *For all $\alpha \in \mathcal{ONL}_n^t$, $\mathbf{AX}_n' \vdash \alpha$ iff $\mathbf{AX}_n^t \vdash \alpha$.*

We mentioned earlier that $\mathbf{AX}_n'$ is sound and complete for the language $\mathcal{ONL}_n^+$ wrt the extended canonical model. Therefore, as a corollary we obtain:

COROLLARY 23. *For all $\alpha \in \mathcal{ONL}_n^t$, $\models \alpha$ iff α is valid in the extended canonical model.*

In this sense, we establish an exact correspondence between HL and our approach. The proofs are presented in [Belle 2011].

6 Autoepistemic Logic

Levesque related only-knowing to the notion of *stable expansions*, which is the basis for AEL. Stable expansions, in a certain sense, rationally reconstruct perfect introspection via meta-theoretic constructions. So besides the independent semantical account to nonmonotonicity in a monotonic framework, he also generalizes Moore's account to a quantificational language with equality.

While L93, H93, and HL propose a many agent extension to Moore's AEL, they remain propositional. In the sequel, we consider an analogue to Levesque's result, and propose a multi-agent quantificational generalization of AEL. We begin with the definition of an i-stable expansion, which can be seen as the generalization of a stable expansion in the presence of multiple agents.

DEFINITION 24. A set of sentences Γ is the *i-stable expansion* of a set of basic sentences A iff Γ satisfies the fixed-point equation:

$$\Gamma = \{\beta \mid \beta \text{ is basic and } A \cup \{\mathbf{K}_i\alpha \mid \alpha \in \Gamma\} \cup \{\neg\mathbf{K}_i\alpha \mid \alpha \notin \Gamma\} \models \beta\}.$$

It is worth noting that our definition differs from the one due to Moore [1985] (and Levesque [1990]) in the following way. We consider logical consequence wrt $\models$ instead of propositional (and first-order) consequence. This is necessary in the many agent case to capture the intuition that an agent attributes perfect introspective capabilities to the other agents. For instance, if a believes $\neg\mathbf{K}_b\alpha$ then we would also require a to believe $\mathbf{K}_b\neg\mathbf{K}_b\alpha$.

Our main result regarding i-stable expansions and what i only-knows is the following:

THEOREM 25. *Suppose α and β are basic formulas. Then β is an element of all i-stable expansions of α iff $\models \mathbf{O}_i\alpha \supset \mathbf{K}_i\beta$.*

This result is weaker than Levesque's, who showed, for example, that for a given stable expansion Γ of $\{\alpha\}$ there is a unique e whose basic beliefs are Γ and which only-knows α. We cannot obtain such a result since, as discussed earlier, our models are defined only for formulas of bounded depth, yet Γ is unbounded. However, if we restrict Γ to the subset Γ_k of formulas of maximal i-depth k, where $k \geq i$-depth of α, then there is again a unique k-structure for agent i where i only-knows α and whose basic beliefs up to i-depth k coincides with Γ_k.

7 Conclusions

We have presented the following results. We proposed a first-order modal logic for multi-agent only-knowing that we show, for the first time, generalizes Levesque's semantics. Unlike all attempts so far, we neither make use of proof-theoretic notions of maximal consistency nor Kripke structures [Waaler and Solhaug 2005]. The benefit is that the semantic proofs are straightforward, and we understand possible worlds precisely as Levesque meant.

We then analyzed a propositional subset, and showed first that the axiom system from Lakemeyer [1993] is sound and complete for a restricted language. We used this result to devise a new proof theory that does not require us axiomatize any semantic notions [Halpern and Lakemeyer 2001]. Our axiomatization was shown to be sound and complete for the semantics, and its use is straightforward on formulas involving the nesting of *at most* operators. In the process, we revisited the features of only-knowing and compared the semantical framework to other approaches. Its behavior seems to coincide with our intuitions, and it also captures a multi-agent generalization of Moore's AEL.

Finally, although the axiomatization of Halpern and Lakemeyer [2001] is not natural, we showed that they essentially capture the desired properties of multi-agent only-knowing, but at much expense.

References

Belle, V. [2011]. Ph.D. Dissertation. Dept. of Computer Science, RWTH Aachen University. *Forthcoming.*

Belle, V. and G. Lakemeyer [2010]. Multi-agent Only-knowing Revisited. In *Proc. KR*, pp. 49–60. AAAI Press.

Ben-David, S. and Y. Gafni [1989]. All we believe fails in impossible worlds. *Manuscript.*

Brachman, R.J. and Levesque, H.J. [2004]. *Knowledge Representation and Reasoning.* Morgan Kaufmann Pub.

Fagin, R., J. Y. Halpern, Y. Moses, and M. Y. Vardi [1995]. *Reasoning About Knowledge.* The MIT Press.

Halpern, J. and G. Lakemeyer [2001]. Multi-agent only knowing. *Journal of Logic and Computation 11*(1), 41.

Halpern, J. Y. [1993]. Reasoning about only knowing with many agents. In *AAAI*, pp. 655–661.

Halpern, J. Y. and G. Lakemeyer [1995]. Levesque's axiomatization of only knowing is incomplete. *Artif. Intell. 74*(2), 381–387.

Halpern, J. Y. and Y. Moses [1984]. Towards a theory of knowledge and ignorance: Preliminary report. In *NMR*, pp. 125–143.

Hughes, G. E. and M. J. Cresswell [1972]. *An introduction to modal logic.* Methuen London.

Lakemeyer, G. [1993]. All they know: A study in multi-agent autoepistemic reasoning. In *Proc. IJCAI*, pp. 376–381.

Levesque, H. and G. Lakemeyer [2001]. *The logic of knowledge bases.* The MIT Press.

Levesque, H. J. [1990]. All I know: a study in autoepistemic logic. *Artif. Intell.* *42*(2-3), 263–309.

Moore, R. C. [1985]. Semantical considerations on nonmonotonic logic. *Artif. Intell.* *25*(1), 75–94.

Rosati, R. [2000]. On the decidability and complexity of reasoning about only knowing. *Artif. Intell.* *116*(1-2), 193–215.

Waaler, A. [2004]. Consistency proofs for systems of multi-agent only knowing. In *Advances in Modal Logic*, pp. 347–366. King's College Publications.

Waaler, A. and B. Solhaug [2005]. Semantics for multi-agent only knowing: extended abstract. In *TARK*, pp. 109–125.

5

Description Logics vs. First Order Logic as KBMS Interaction Languages

ALEXANDER BORGIDA

ABSTRACT. Description Logics (DLs) and First Order Predicate Calculus (FOPC) are among the best known languages in the KR community. FOPC is also very well known in the information management community. When one adopts Levesque's functional view of systems that manage information [23], they can both be considered as candidates for languages associated with various operators for telling information and querying/receiving answers. We compare their suitability for these tasks in two ways: First, by relying on their common formal semantics (model theory), we can compare their expressive power. Second, we consider the advantages that have accrued to DLs due to their variable-free term-like structure, built with constructors that have long been found empirically useful for modeling the natural world.

1 Introduction

Hector Levesque's seminal paper [23] made a number of contributions that deeply influenced those of us interested in data and knowledge management using computers.

1. It reformulated the problem of interacting with knowledge management systems as one of interacting with a data abstraction equipped with two essential kinds of operators: TELL, for adding information, and ASK, for querying. To support this view, one in fact needs to associate with these operators a series of languages: $\mathcal{L}_{\text{tell}}$, $\mathcal{L}_{\text{query}}$, and $\mathcal{L}_{\text{answer}}$. And in fact one can distinguish several variants of TELL and ASK for any particular system.

2. It introduced the notion of *query answering* specified at a "knowledge level", rather than an implementation one, using the information that has been told to the system, and the semantics of the query.

3. It introduced an epistemic language which was provably *more expressive* than standard FOPC as an $\mathcal{L}_{\text{query}}$, allowing one to make distinctions that are highly important in databases for example, but which were a puzzle until then. For example, although one could assert that *anna* had an age as $\exists x.hasAge(anna, x)$, the relational database requirement that this value is not allowed to be null could not be expressed until one had sentences such as $\exists x.\mathbf{K}hasAge(anna, x)$ ("the age of Anna is known"), which is distinguished from $\mathbf{K}\exists x.hasAge(anna, x)$ ("it is known that Anna has an age").

The notion of multiple operators and languages, as well as greater expressive power are not tied to the specific epistemic language and logic introduced and investigated by Levesque. It is the purpose of this chapter to consider the use of a family of languages and logics called *Description Logics* (DLs) as interaction languages with a KBMS, and to compare them with standard FOPC.

Description Languages/Logics (DLs) are descendants of the KL-ONE [8] knowledge representation system, and have been the object of study for a number of decades, forming at this moment the foundation of the OWL 2 language proposed for defining ontologies to be used on the Semantic Web [25].

We propose to first provide an informal guide to DLs and their distinctive features, and then, after giving more precise definitions, contrast the use of DLs and of FOPC fragments in the roles they can play as $\mathcal{L}_{query}$, $\mathcal{L}_{tell}$, and $\mathcal{L}_{answer}$ languages respectively. Among the interesting results we will find is that even the most expressive DLs proposed have mostly features that can be expressed in FOPC with a small number of variables, and, quite surprisingly the converse will also hold.

2   Aspects of Description Logics

An initial distinctive feature of the original KL-ONE pproposal by Brachman was the idea that in addition to *primitive* concepts, such as "Person", one can also *define* concepts such as "Persons with at least three friends" — the definition providing both necessary and sufficient conditions for membership in this new concept. The question is how can one define concepts.

2.1   DL Descriptions, Axioms and Reasoning - an informal review

As other conceptual modeling notations, DLs view the world as being populated by *individuals*, which can be related to each other by binary relations called *roles*; individuals are also grouped into *concepts*. Concepts and roles are specified by *descriptions*. For example, consider the concept description in Figure 1. It is constructed from identifiers denoting binary relations (e.g., bornIn, players), individuals (e.g., Toronto) and other concepts (e.g., TEAM, CITY) using *description constructors* **intersection**, **allValuesFrom**, **minCardinality**, and **hasValue**. The description in Figure 1 has as intended meaning *"Teams which play in a city"* (their playsIn role's value must be an instance of concept CITY), *involving at least 11 players* (the players role has at least 11 values/fillers), *all of whom were born in Toronto* (all players values are restricted to have value Toronto for role bornIn).

Anther key distinction of DLs is that unlike other conceptual modeling languages, roles need not be atomic identifiers since there are constructors which can be used to build composite descriptions of roles. For example, the term **roleChain**(parents, **roleRestrict**(siblings, MALE))   corresponds to the notion of *"uncles"* in English, since **roleRestrict** takes a binary relation such as siblings and derives the one where all elements in its range are instances of concept MALE (hence brothers), while **roleChain**

> **intersection**(TEAM
>   **allValuesFrom**(playsIn,CITY)
>   **minCardinality**(11,players)
>   **allValuesFrom**(players, **hasValue**(bornIn,Toronto))

Figure 1. Composite description for a concept

corresponds to binary relation composition.

Descriptions of concepts are essentially noun-phrases, or "terms" in the terminology of predicate calculus. They are therefore by themselves not sentences that can be said to be true or false. To make sentences (assertions, axioms), one uses descriptions as arguments to predicates. Since the list of such predicates is also one of the specific features of each modern DL, we shall refer to them as "axiom constructors". In DLs, the *subsumption relationship* $\sqsubseteq$ holds pride of place, and is used to create sentences of the form $D \sqsubseteq C$; these are to be read as "every individual satisfying description D also satisfies description C"; for example, one could *assert* that MALE $\sqsubseteq$ PERSON; or *deduce* that **intersection**(PLAYER,**allValuesFrom**(bornIn,France) $\sqsubseteq$ PLAYER, or brothers $\sqsubseteq$ siblings. An example of what we call here an "axiom constructor" might be the assertions that role R is transitive or symmetric, say. In DLs, such generic assertions are gathered into a so-called "terminology" (TBox), which essentially specifies the meaning of terms. *Reasoning* in DLs is then often a matter of deciding whether the above kinds of relationships hold in the presence of (potentially empty) TBoxes. However, other, related and important reasoning tasks include deciding (un)satisfiability of a description (whether it can ever have individual instances),

Other classical and essential axioms in DLs concern the membership of an individual b in (the extent) of concept C, and whether a pair of individuals is related by a role or not. Such assertions are collected into a so-called ABox, which in principle is supposed to describe the state of the world. Deducing membership in a concept in the presence of an ABox and TBox is another general form of reasoning that DL systems are called upon to perform.

As observable from the above examples, a notable distinguishing feature of DLs is the the variable-free term-like notation of descriptions, and the absence of variables in axioms. Any particular description logic is then characterized by (i) the set of concept and role constructors from which new descriptions can be constructed according to its syntax, and (ii) the axiom constructors. The set of constructors is chosen with two goals in mind: empirical utility in applications, and the complexity of reasoning in the resulting logic, which should always be decidable.

2.2 The language and semantics of descriptions

Formally, the language of descriptions is obtained recursively by starting from a schema $S = (\mathcal{N_C}, \mathcal{N_R}, \mathcal{N_I})$ of atomic names for concepts, roles, and individuals, and recursively building from them more *complex terms* using description constructors.

Over the years, a considerable variety of DLs have been investigated. Table 1 presents a relatively comprehensive list of the constructors (with their syntax and semantics) based on early survey papers [2; 26] as well as more recent developments, such as the OWL-2 proposal [25].[1] The set of constructors in Table 1 is not minimal, since considerable work in the field has been devoted to finding subsets of constructors for which the various reasoning problems have good computational complexity properties. Also, the precise name of each constructor may vary, just like the surface syntax of programming languages; we have chosen to adopt wherever possible the "functional syntax" used in the recent OWL-2 web ontology language proposed by W3C [25].

---

[1] Some constructors have been omitted here and will be discussed in a separate section.

In the table, and elsewhere, we use the symbols A, B, C to range over concept descriptions, $p, q, \ldots$ to range over role descriptions, $a, b, \ldots$ for individual names, and D, E, F to denote descriptions in general.

The semantics of description terms is given denotationally using the notion of an *interpretation* $\mathcal{I} = \langle \Delta^{\mathcal{I}}, (\cdot)^{\mathcal{I}} \rangle$, which starts with a domain (non-empty universe) of values $\Delta^{\mathcal{I}}$, and a mapping $(\cdot)^{\mathcal{I}}$ from atomic concepts in $\mathcal{N}_C$ to subsets of the domain, and atomic roles in $\mathcal{N}_R$ to sets of 2-tuples over the domain; the mapping also associates with every individual name $\mathcal{N}_I$ some value in $\Delta^{\mathcal{I}}$. For simplicity, in this paper we will require that distinct names denote distinct values, following the Unique Name Assumption, and in fact will use the Horn-domain, equating $\mathcal{N}_I$ with $\Delta^{\mathcal{I}}$. We note however that recent proposal no longer follow this restriction. The interpretation function $(\cdot)^{\mathcal{I}}$ is extended recursively to composite descriptions in Table 1, where the interpretation of roles is viewed, in the obvious ways, as a function from an individual to the set of individuals related to it by the role (i.e., as $\mathcal{N}_R \rightarrow (\Delta^{\mathcal{I}} \rightarrow 2^{\Delta^{\mathcal{I}}})$).

For the purposes of the results to be presented below, we will distinguish several subsets of DL constructors: the complete set, $\mathcal{DCN}$, in Table 1, $\mathcal{DC}$, which omits the cardinality constraints **minCardinality** and **maxCardinality**, and $\mathcal{D}$, which also omits role composition **roleChain**.

2.3 The language and semantics of FOPC

Because we plan to compare DLs with FOPC, we quickly review FOPC and its semantics, with formulas having free variables being the only non-standard material.

We start, as usual, from a set of predicates and individual constants – a schema $S = (\mathcal{N}_P, \mathcal{N}_I)$ of predicate and constant names. From these, as well as the equality predicate and variable symbols in the set $\mathcal{N}_V$ (which is assumed to have a lexicographic ordering on it), atomic and composite formulas are built as usual, using connectives $\neg, \wedge, \exists$, with the other logical symbols defined as abbreviations in terms of these in the usual manner. The notion of "free variable" is defined as usual, and we use the notation $\Psi(x, y)$ to refer to a formula that has as free variables exactly x and y.

The semantics of FOPC is based on the same kind of interpretation $\mathcal{I}$ as we used for DLs, in this case starting with denotations for $\mathcal{N}_P$ and $\mathcal{N}_I$. However, in this case one traditionally also has to deal with variable symbols, which is done by means of a partial function $\nu : \mathcal{N}_V \rightarrow \Delta^{\mathcal{I}}$ that provides a substitution for the variables. An interpretation $\mathcal{I}$ and a substitution ν define a function $[\![\cdot]\!]^{\mathcal{I}, \nu}$ from formulas to truth values $\{\ True, False\ \}$ in the usual way; for example, for the formula $P(x, a)$, we have $[\![P(x, a)]\!]^{\mathcal{I}, \nu} = True$ iff $(\nu(x), a^{\mathcal{I}}) \in P^{\mathcal{I}}$.

3 Descriptions vs. FOPC as query languages for databases

We are driven by the dominant paradigm of comparing database query languages, which started with Codd's paper on the theory of relational databases [14] and is now standard in every textbook: one exhibits several query languages, for example the relational algebra RA and the relational tuple calculus RTC, and then shows that although superficially they look quite different, the RA is in fact "as expressive" as RTC. This meant that for every RTC query q_T there exists an RA query q_A such that for *every database state*, the two queries return the same answers. Moreover, this proof is accomplished by showing an explicit translation which constructs q_T from q_A.

TERM	INTERPRETATION
THING	$\Delta^{\mathcal{I}}$
NOTHING	$\emptyset$
intersection(C,D)	$C^{\mathcal{I}} \cap D^{\mathcal{I}}$
union(C,D)	$C^{\mathcal{I}} \cup D^{\mathcal{I}}$
complement(C)	$\Delta^{\mathcal{I}} \setminus C^{\mathcal{I}}$
allValuesFrom(p,C)	$\{ \delta \in \Delta^{\mathcal{I}} \mid p^{\mathcal{I}}(\delta) \subseteq C^{\mathcal{I}} \}$
someValuesFrom(p,C)	$\{ \delta \in \Delta^{\mathcal{I}} \mid p^{\mathcal{I}}(\delta) \cap C^{\mathcal{I}} \neq \emptyset \}$
sameAs(p,q)	$\{ \delta \in \Delta^{\mathcal{I}} \mid p^{\mathcal{I}}(\delta) = q^{\mathcal{I}}(\delta) \}$
subset(p,q)	$\{ \delta \in \Delta^{\mathcal{I}} \mid p^{\mathcal{I}}(\delta) \subseteq q^{\mathcal{I}}(\delta) \}$
notSameAs(p,q)	$\{ \delta \in \Delta^{\mathcal{I}} \mid p^{\mathcal{I}}(\delta) \neq q^{\mathcal{I}}(\delta) \}$
hasValue(p,b)	$\{ \delta \in dom^{\mathcal{I}} \mid b^{\mathcal{I}} \in p^{\mathcal{I}}(\delta) \}$
oneOf(b_1,...,b_m)	$\{ b_1^{\mathcal{I}}, \ldots, b_m^{\mathcal{I}} \}$
hasSelf(p)	$\{ \delta \in \Delta^{\mathcal{I}} \mid p^{\mathcal{I}}(\delta, \delta) \}$
minCardinality(n,p,C)	$\{ \delta \in \Delta^{\mathcal{I}} \mid \lvert p^{\mathcal{I}}(\delta) \cap C^{\mathcal{I}} \rvert \geq n \}$
maxCardinality(n,p,C)	$\{ \delta \in \Delta^{\mathcal{I}} \mid \lvert p^{\mathcal{I}}(\delta) \cap C^{\mathcal{I}} \rvert \leq n \}$

TERM	INTERPRETATION
TOPROLE	$\Delta^{\mathcal{I}} \times \Delta^{\mathcal{I}}$
IDENTITY	$\{ (\delta, \delta) \mid \delta \in \Delta^{\mathcal{I}} \}$
roleIntersection(p,q)	$p^{\mathcal{I}} \cap q^{\mathcal{I}}$
roleUnion(p,q)	$p^{\mathcal{I}} \cup q^{\mathcal{I}}$
roleComplement(p)	$\Delta^{\mathcal{I}} \times \Delta^{\mathcal{I}} \setminus R^{\mathcal{I}}$
inverseOf(p)	$\{ (\delta, \delta') \mid (\delta', \delta) \in R^{\mathcal{I}} \}$
roleRestrict(p,C)	$\{ (\delta, \delta') \in p^{\mathcal{I}} \mid \delta' \in C^{\mathcal{I}} \}$
product(C,D)	$C^{\mathcal{I}} \times D^{\mathcal{I}}$
roleChain(p,q)	$p^{\mathcal{I}} \circ q^{\mathcal{I}}$

Table 1. *A collection of DL concept and role constructors*

We face a similar situation with DLs and FOPC: they look very different superficially, but we want to show that in fact there is quite a close relationship between them: not only can "queries" in $\mathcal{D}$ be expressed in (a special subset of) FOPC, which we will call $\mathcal{FOL}^2$, but also the converse is true. So that we have equally expressive languages.

To achieve this we need to specify

- what is a database;

- how a description in a DL is considered a query, and what the answer to such a DL query is for a database;

- how a formula in FOPC is viewed as a query, and what is the answer to an FOPC query for a database.

In general, the above can be difficult if DLs and FOPC view databases in a different way (e.g., relational databases based on FOPC have finite domains). We sidestep this difficulty by making a key simplification, based on the fact that both DL and FOPC have Tarskian semantics: the information captured by a complete database is simply taken to be an interpretation $\mathcal{I}$.

Concerning answering DL queries D in a database $\mathcal{I}$, we will use a model theoretic definition indicating that the values returned by a concept query, say, are just the individuals in its interpretations.

DEFINITION 1.
$$\mathbf{Answ}(D, \mathcal{I}) = D^{\mathcal{I}}$$

Turning to FOPC, as traditional in databases, queries are formulas $\Psi(x_1, ..., x_k)$ with free variables, and we define the answer to such queries in a database $\mathcal{I}$ as substitutions that make the query evaluate to true (the so-called "model theoretic" interpretation of databases):

DEFINITION 2.
$$\mathbf{Answ}(\Psi(x_1, ..., x_k), \mathcal{I}) = \{\langle \alpha_1, ..., \alpha_k \rangle \mid \alpha_i \in \Delta^{\mathcal{I}}, [\![\Psi(x_1, ..., x_k)]\!]^{\mathcal{I}, \nu} = True\}$$ for any substitution ν mapping x_i to α_i for $1 \leq i \leq k$.

Since the tuples are ordered, to obtain a unique answer to a query we will require the free variables $x_1, x_2, ...$ to appear in lexicographic order. Note that queries with no free variables have either empty interpretation, or one containing the 0-tuple. Closed formulas with no free variables have $True$ or $False$ as interpretation.

We will then say that some language $\mathcal{L}_2$ is as expressive as language $\mathcal{L}_1$, if there is a total function $transl$ from $\mathcal{L}_1$ to $\mathcal{L}_2$ such that for every formulas L in $\mathcal{L}_1$, $transl$(L) returns the same answers as L. Two languages are equally expressive if each is as expressive as the other.

For example, Schmolze and Israel [27] show informally that FOPC is as expressive as the KL-ONE DL by essentially defining a translation function $\tau\langle . \rangle$, which maps concepts to formulas with free variable x, and roles to formula with free variables x and y. For example, $\tau\langle$**allValuesFrom** $(\text{p}, \text{C})\rangle$ is "$\lambda x. \forall w. p(x, w) \Rightarrow C(w)$", while the translation of **roleChain**(p,q) is "$\lambda x, y. \exists z. p(x, z) \wedge q(z, y)$."

In the next section, we will explore in some detail the relationship between DLs and FOPC as query languages. Afterwards, we will return to consider the consequences of the results we discovered.

4   DLs and possibly limited FOPC

Note that the translation from descriptions to FOPC mentioned in the previous section introduces new variables whenever a new quantifier appears, in order to avoid spurious capture of variables. For example, without this precaution, the translation of **roleChain** (p, **roleChain** (p, p)) would yield $\exists z.p(x, z) \wedge \exists z.(p(z, z) \wedge p(z, y))$, which is clearly wrong — one wants $\exists z_2.p(x, z_2) \wedge \exists z_1.(p(z_2, z_1) \wedge p(z_1, y))$.

An important result in this paper is that one does not need arbitrarily many variable symbols in order to translate into FOPC the constructors in Table 1. Specifically, let $\mathcal{FOL}^k$ be the set of all FOPC formulas with equality which use at most k variable symbols. Note that $\mathcal{FOL}^k$ does not limit the number of nested quantifiers in a formula since the same variable may be reused in nested subformulas. Properties of such language families have been studied, among others, in [24; 19; 11]. In our case, since we are dealing with roles and concepts, we will be interested only in those formulas that (i) have one or two free variables (though they may have closed subformulas), and (ii) have only monadic and dyadic predicates.

4.1   Translating $\mathcal{DC}$ into FOPC with 2 or 3 variables

Our first result shows that everything that can be said with $\mathcal{D}$, can be said with just 2 variables.

THEOREM 3. *The language* $\mathcal{FOL}^2$ *is as expressive as* $\mathcal{D}$.

Proof The proof relies on a more careful encoding of the constructors into predicate calculus, where the same variable is reused as much as possible. The translation function comes in several variants that behave as follows: $\mathcal{T}^x \langle \rangle$ makes x be the free variable of the monadic predicate it will produce for its argument concept, while $\mathcal{T}^y \langle \rangle$ makes the free variable be y. So, for a concept C in $\mathcal{D}$, $\mathcal{T}^x \langle C \rangle = C(x)$, while $\mathcal{T}^y \langle C \rangle = C(y)$. In the case of roles R, $\mathcal{T}^{x,y} \langle R \rangle$ produces a predicate $R(x, y)$, while $\mathcal{T}^{y,x} \langle R \rangle$ produces predicate $R(y, x)$. The translation functions $\mathcal{T}^x \langle \rangle$ and $\mathcal{T}^{x,y} \langle \rangle$ are presented in the following two tables. $\mathcal{T}^y \langle \rangle$ and $\mathcal{T}^{y,x} \langle \rangle$ are obtained from $\mathcal{T}^x \langle \rangle$ and $\mathcal{T}^{x,y} \langle \rangle$ respectively by simultaneously exchanging *all* occurrences of x and y (whether free or bound).

TERM C	$\mathcal{T}^x \langle C \rangle$
THING	$x = x$
NOTHING	$\neg(x = x)$
intersection(C,D)	$\mathcal{T}^x \langle C \rangle \wedge \mathcal{T}^x \langle D \rangle$
union(C,D)	$\mathcal{T}^x \langle C \rangle \vee \mathcal{T}^x \langle D \rangle$
complement(C)	$\neg \mathcal{T}^x \langle C \rangle$
allValuesFrom(p,C)	$\forall y. \mathcal{T}^{x,y} \langle p \rangle \Rightarrow \mathcal{T}^y \langle C \rangle$
someValuesFrom(p,C)	$\exists y. \mathcal{T}^{x,y} \langle p \rangle \wedge \mathcal{T}^y \langle C \rangle$
subset(p,q)	$\forall y. \mathcal{T}^{x,y} \langle p \rangle \Rightarrow \mathcal{T}^{x,y} \langle q \rangle$
sameAs(p,q)	$\forall y. \mathcal{T}^{x,y} \langle p \rangle \Leftrightarrow \mathcal{T}^{x,y} \langle q \rangle$
notSameAs(p,q)	$\exists y. \neg(\mathcal{T}^{x,y} \langle p \rangle \Leftrightarrow \mathcal{T}^{x,y} \langle q \rangle)$
hasValue(p,b)	$\exists y.(y = b) \wedge \mathcal{T}^{x,y} \langle p \rangle$
oneOf(b$_1$,...,b$_m$)	$x = b_1 \vee \ldots \vee x = b_m$

TERM R	TRANSLATION $\mathcal{T}^{x,y}\langle \mathrm{R}\rangle$
TOPROLE	$x = x \land y = y$
IDENTITY	$x = y$
roleIntersection(p,q)	$\mathcal{T}^{x,y}\langle \mathrm{p}\rangle \land \mathcal{T}^{x,y}\langle \mathrm{q}\rangle$
roleUnion(p,q)	$\mathcal{T}^{x,y}\langle \mathrm{p}\rangle \lor \mathcal{T}^{x,y}\langle \mathrm{q}\rangle$
roleComplement(p)	$\neg \mathcal{T}^{x,y}\langle \mathrm{p}\rangle$
inverseOf(p)	$\mathcal{T}^{y,x}\langle \mathrm{p}\rangle$
roleRestrict(p,C)	$\mathcal{T}^{x,y}\langle \mathrm{p}\rangle \land \mathcal{T}^{y}\langle \mathrm{C}\rangle$
product(C,D)	$\mathcal{T}^{x}\langle \mathrm{C}\rangle \land \mathcal{T}^{y}\langle \mathrm{D}\rangle$

The only non-trivial part of the proof relies on the alternating use of $\mathcal{T}^{x}\langle\rangle$ and $\mathcal{T}^{y}\langle\rangle$ in the translation of **allValuesFrom** and **someValuesFrom**, which ensure that $\mathbf{Answ}(C,\mathcal{I}) = \mathbf{Answ}(\mathcal{T}^{x}\langle \mathrm{C}\rangle, \mathcal{I})$.

∎

The second theorem extends this result to the case of the **roleChain** constructor

THEOREM 4. *The language $\mathcal{FOL}^3$ is as expressive as $\mathcal{DC}$.*

The proof relies simply (but essentially) on the use the equalities $x = z$ and $y = z$ in the translation of **roleChain**

$$\mathcal{T}^{x}\langle \textbf{roleChain}\,(\mathrm{p},\mathrm{q})\rangle = \exists z.(\exists y.y = z \land \mathcal{T}^{x,y}\langle \mathrm{p}\rangle) \land \exists z.(\exists y.y = z \land \mathcal{T}^{x,y}\langle \mathrm{p}\rangle)$$

4.2 Translating FOPC with 2 or 3 variables into $\mathcal{DC}$

The more surprising, and more complex result is that the converses of the above theorems also hold: we can simulate any FOPC formula with unary and binary predicates and one or two free variables using descriptions. To begin with, we have

THEOREM 5. *$\mathcal{D}$ is as expressive as $\mathcal{FOL}^2$.*

Proof Suppose the two variables we can use in $\mathcal{FOL}^2$ are x and y. We shall proceed by structural recursion on the syntax of formulas with up to two free variables.

The following table lists the various kinds of formulas $\Upsilon(x)$ that have a single free variable x, and shows how each kind is translated into a concept description $\tau_c\langle\Upsilon\rangle$. ($C$ is an atomic concept name from $\mathcal{N}_C$, and P is in $\mathcal{N}_\mathcal{R}$.)

$\Upsilon(x)$	$\tau_c\langle\Upsilon\rangle$
$C(x)$	C
$P(x,b)$	**hasValue**(P,b)
$P(b,x)$	**hasValue**(inverseOf(P),b)
$P(x,x)$	**someValuesFrom**(roleIntersection(P,IDENTITY),THING)
$x = b$	**oneOf**(b)
$x = x$	THING
$\neg\Psi_1(x)$	**complement**($\tau_c\langle\Psi_1\rangle$)
$\Psi() \land \Phi(x)$	**intersection**($\tau_c\langle\Psi()\rangle,\tau_c\langle\Phi\rangle$)
$\Psi(x) \land \Phi(x)$	**intersection**($\tau_c\langle\Psi\rangle,\tau_c\langle\Phi\rangle$)
$\exists y.\Psi(x,y)$	**someValuesFrom**($\tau_r\langle\Psi\rangle$, THING)
$\exists y.\Psi(x)$	$\tau_c\langle\Psi\rangle$

The translation of formulas with a single free variable y is identical, except for the case when $\Upsilon(y)$ is of the form $\exists x.\Psi(x,y)$, when we need to invert the relationship represented by Ψ, so it is translated as **someValuesFrom(inverseOf($\tau_c\langle\Psi\rangle$)), THING)**

Formulae of the form $\Psi(x,y)$ are translated to roles relating x to y according to the following table

$\Upsilon(x,y)$	$\tau_r\langle\Upsilon\rangle$
$P(x,y), P \in \mathcal{N_R}$	P
$P(y,x)$	**inverseOf(P)**
$x = y$	IDENTITY
$\neg\Psi(x,y)$	**roleComplement**$(\tau_r\langle\Psi(x,y)\rangle)$
$\Psi(x) \wedge \Phi(y)$	**product**$(\tau_c\langle\Psi\rangle,\tau_c\langle\Phi\rangle)$
$\Psi(x,y) \wedge \Phi()$	**roleIntersection**$(\tau_r\langle\Psi\rangle,\tau_r\langle\Phi()\rangle)$
$\Psi(x,y) \wedge \Phi(x)$	**roleIntersection**$(\tau_r\langle\Psi\rangle,$ **product**$(\tau_c\langle\Phi\rangle,$THING$)$ $)$
$\Psi(x,y) \wedge \Phi(y)$	**roleIntersection**$(\tau_r\langle\Psi\rangle,$ **product**$($THING$, \tau_c\langle\Phi\rangle)$ $)$
$\Psi(x,y) \wedge \Phi(x,y)$	**roleIntersection**$(\tau_r\langle\Phi\rangle,\tau_r\langle\Psi\rangle)$

The novelty is the need to translate formulas $\Upsilon()$ without free variables. However, these can occur only as a conjunct, and the number of free variables (1 or 2) in its context determines its translation: a concept or a role. For the case when a concept is desired, we need a description D_Υ with the property that for any interpretation $\mathcal{I}$, if $\Upsilon()^\mathcal{I} = True$ then $D_\Upsilon^\mathcal{I} = \Delta^\mathcal{I}$, and if $\Upsilon()^\mathcal{I} = False$ then $D_\Upsilon^\mathcal{I} = \emptyset$. This essentially "gates" the meaning of the other conjunct. The following table provides such translations:

$\Upsilon()$	$\tau_c\langle\Upsilon()\rangle$
$C(b)$	**allValuesFrom(product(**THING,**oneOf**(b)**)**, C)
$P(b,b)$	**allValuesFrom(product(**THING,**oneOf**(b)**)**, **hasValue**(P,b)**)**
$P(a,b)$	**allValuesFrom(product(**THING,**oneOf**(a)**)**, **hasValue**(P,b)**)**
$b = b$	THING
$a = b$	NOTHING
$\neg\Psi()$	**complement**$(\tau_c\langle\Psi()\rangle)$
$\Psi() \wedge \Phi()$	**intersection**$(\tau_c\langle\Psi()\rangle,\tau_c\langle\Phi()\rangle)$
$\exists x.\Psi(x)$	**someValuesFrom**(TOPROLE, $\tau_c\langle\Psi\rangle$)
$\exists y.\Psi(y)$	**someValuesFrom**(TOPROLE, $\tau_c\langle\Psi\rangle$)
$\exists x.\Psi()$	$\tau_c\langle\Psi\rangle$

In contexts where we require roles, the translation is just $\tau_r\langle\Upsilon()\rangle$ = **product**$(\tau_c\langle\Upsilon()\rangle, \tau_c\langle\Upsilon()\rangle)$.

■

Matching Theorem 5 we have

THEOREM 6. *The description language* $\mathcal{DC}$ *is as expressive as* $\mathcal{FOL}^3$.

Proof In this case we deal with formulas having possibly three variables: x, y, and z. Once again we define a recursive procedure for translating formulas into descriptions.

The main novelty therefore lies in the translation of subformulas $\Psi(x,y,z)$ with three free variables. Since the final formula can have no more than two free variables, every subformula $\Psi(x,y,z)$ must in fact be part of a larger subformula of the form $\exists\gamma.\Phi(x,y,z)$, where γ is either x, y or z. Without loss of generality, suppose it is $\exists z.\Phi(x,y,z)$. Because all atomic formulas involve only monadic and dyadic predicates, we will be able to prove that it is sufficient to consider the case when $\exists z\Phi(x,y,z)$ is of the form $\exists z.\Phi_1(x,z)\wedge\Phi_2(y,z)$, which can be translated as the role description **roleChain**$(\tau_r\langle\Phi_1\rangle,$ **inverseOf**$(\tau_r\langle\Phi_2\rangle))$.

It therefore remains to show that every formula of the form $\exists z\Phi(x,y,z)$ can be reduced to the form $\exists z.\Phi_1(x,z) \wedge \Phi_2(y,z)$. This is accomplished by massaging $\Phi(x,y,z)$ into a normal form that allows the quantifier to be moved in. Specifically, let Φ_i be the *maximal subformulas* of $\Phi(x, y, z)$ that have at most two free variables. Since there are only three variables and all predicates are of arity at most 2, $\Phi(x, y, z)$ must be the boolean combination these Φ_i — any intervening quantifier would reduce the number of variables to 2, and hence would be part of some Φ_j. Using de Morgan's laws, it is therefore possible to rewrite $\Phi(x, y, z)$ into disjunctive normal form $\bigvee_j (\bigwedge_k \Phi_{j,k})$, where each $\Phi_{j,k}$ is either some Φ_i or its negation. Since an existential quantifier can be moved in past disjunctions $(\exists z.(\alpha \vee \beta) \equiv (\exists z.\alpha) \vee (\exists z.\beta))$, $\exists z.\Phi(x, y, z)$ is therefore logically equivalent to $\bigvee_j \Theta_j$ where $\Theta_j = \exists z.(\bigwedge_k \Phi_{j,k})$, and note that each Θ_j has at most two free variables (z is being quantified over).

Therefore we need only consider formulas $\exists z.\Phi(x, y, z)$ where $\Phi(x, y, z)$ is the conjunction of subformulas $\Phi_{j,k}$, each with at most two free variables. By associativity and commutativity of conjunction, group together the subformulas that have the same free variables, thereby obtaining that $\Phi(x, y, z)$ is in the most general case of the form $\Psi_0() \wedge \Psi_1(x) \wedge \Psi_2(y) \wedge \Psi_3(z) \wedge \Psi_4(x, y) \wedge \Psi_5(x, z) \wedge \Psi_6(y, z)$. But then we can move the subformulas not containing z outside the quantifier, rewriting $\exists z \Phi(x, y, z)$ in the form $\beta(x, y) \wedge \exists z.((\Psi_3(z) \wedge \Psi_5(x, z)) \wedge \Psi_6(y, z))$. Therefore, in the end the formula in the scope of $\exists z$ does have the desired restricted form $\exists z.\Psi_7(x, z) \wedge \Psi_5(y, z)$, establishing our claim. (Note that if subformulas Ψ_5 or Ψ_6 are empty then we are left inside the scope of $\exists z$ with a formula with at most two free variables, whose translation had already been provided earlier.)

∎

Combining the preceding theorems we get

COROLLARY 7. *(i) The description language $\mathcal{D}$ and $\mathcal{FOL}^2$ are equally expressive. (ii) The description language $\mathcal{DC}$ and $\mathcal{FOL}^3$ are equally expressive.*

4.3 Number Restrictions in DLs

Probably the best known DL constructors we have omitted so far from discussion are **minCardinality** and **maxCardinality**. This is mostly because they no longer obey the beautiful symmetry of the preceding results. In particular, although there is an obvious translation of concepts involving these constructors to FOPC

$$\mathcal{T}^x \langle \; \textbf{minCardinality}\,(\texttt{n}, \texttt{p}) \; \rangle = \exists z_1, ..., \exists z_n \; p(x, z_1) \wedge ... \wedge p(x, z_n) \wedge z_i \neq z_j$$

it seems to use n variable symbols in an essential way. The solution is to use a variant of FOPC with so called "counting quantifiers" (e.g., [20]), which allows one to say, for example, $\exists_6 y.\texttt{hasPlayers}(x, y)$, predicating the existence of six distinct values for y for which the formula is satisfied. Note that this is not treated as an abbreviation because we will wish to say that the above formula has only two variables, x and y! Theorems 3 and 4 do extend to this case. Unfortunately we have been unable to show that $\exists_n z.p(x, z) \wedge q(z, y)$ can be expressed in DLs, so the full equivalence breaks down since we cannot show that formulas with 1 or 2 free variables but with up to 3 counting quantifiers can be expressed in $\mathcal{DCN}$.

4.4 Other DL constructors

Of great interest to database research is the notion of "key" – a set of a properties whose values uniquely identify an object. Usually, keys are introduced as *constraints* or axioms on database states, but in the spirit of DLs it is more interesting to consider them as concept constructors. In fact, Toman and Weddell [28] have done exactly this, using a generalization of the notion of key, called a functional dependency in databases. For example, **functionallyDetermine**(EMPLOYEE, (hasLevel, inDept), hasSalary) denotes individuals (whether employees or not) which have the property that if they share the same hasLevel and inDept values with an employee, they will also have the same hasSalary values. Formally, the interpretation of **functionallyDetermine**$(A, (p_1, ..., p_k), p)$ is $\{ \delta \in \Delta^{\mathcal{I}} \mid \forall \alpha \in A^{\mathcal{I}}.(\bigwedge_i p_i^{\mathcal{I}}(\delta) = p_i^{\mathcal{I}}(\alpha)) \Rightarrow p^{\mathcal{I}}(\delta) = p^{\mathcal{I}}(\alpha) \}$. By definition, keys are then functional dependencies with the identity relationship being used at the end: **keyFor**(EMPLOYEE, (hasName, inDept)) $\equiv$ **functionallyDetermine**(EMPLOYEE, (hasName, inDept), IDENTITY)

The **functionallyDetermine** construct can in fact be translated into FOPC with 3 variables. The translation is based on the observation that a logical formula of the form $(b_1 \wedge ... \wedge b_k) \Rightarrow c$ is equivalent to $b_1 \Rightarrow (... \Rightarrow (b_k \Rightarrow c)...)$. We therefore first represent FOPC expressions of the form $q^{\mathcal{I}}(x) = q^{\mathcal{I}}(y)$, with x, y free variables, by using an additional variable z: $\forall z.(T^{x,y}\langle q \rangle \wedge y = z \Leftrightarrow T^{y,x}\langle q \rangle \wedge x = z)$. Then we add $\forall y.T^y\langle A \rangle \Rightarrow$ at the beginning of the chain of implications, leaving x as the only free variable in the formula, as expected for a concept. Note therefore that functional dependencies also require three variables to represent in FOPC in our translation, and incidentally showing that **functionallyDetermine** is definable in terms of the constructors in Table 1.

There are several other extensions of Description Logics, many discussed in the Description Logic Handbook [3], which turn out to be problematic not just for FOPC with limited numbers of variables but also FOPC in general.

For example, many investigators have considered constructing composite roles using regular expressions. For this, one needs not just role union and intersection but also transitive closure. For example, ancestor $\equiv$ **transClosure**(parentOf) is a very useful term to define in an ontology about genealogy. However, it is well known that First Order Logic cannot represent the transitive closure of a relationship, although it can assert that a relationship is transitive. More powerful extensions in the same direction involve adding *fixed point constructors* which allow the expression of inductive definitions, such as that of a LISP list.

A rather different way to change the language of DLs is to allow n-ary rather than just binary relationships. Calvanese et al. [12] conducted a thorough investigation of such a description logic, $\mathcal{DLR}$. Clearly, an n-ary role corresponds to an n-ary predicate, and it is not hard to show that $\mathcal{DLR}$ role and concept expressions can be translated into FOPC. Now n-ary roles allow n-tuples to be returned as answers to roles treated as queries, so clearly in this sense $\mathcal{DLR}$ is more expressive than $\mathcal{DCN}$. Calvanese et al. show that it is possible to *reify* an n-ary relationship R, turning it into a concept R_c, whose instances are connected via n functional roles $f_1, ..., f_n$ to the components of the corresponding n-tuple which is an instance of the original n-ary relation R. To make the correspondence perfect one needs to make sure that every tuple is represented by exactly one individual in R_c, which in $\mathcal{DLR}$ is ensured by model-theoretic properties of the particular set of concept constructor, but in our more general case could also be enforced using **keyFor**. So we are in a situation where $\mathcal{DLR}$ descriptions can be *encoded* into the standard description logic $\mathcal{DN}$, but only thru the addition of new concept and role names, and especially axioms which express their intended meaning to simulate R. This allows reasoning about $\mathcal{DLR}$, including subsumption, to be reduced to reasoning in a (decidable) subset of $\mathcal{DN}$. However, as far as databases and queries, interpretations $\mathcal{I}$ of the original $\mathcal{DLR}$ are quite different from the interpretations $\mathcal{I}'$ of its encoding. Therefore one is left in a situation where it is hard to say that $\mathcal{DN}$ is as expressive as even the concept sublanguage of $\mathcal{DLR}$ according to our original definition. One possible avenue to explore is that initiated by Baader [1], who was the first to consider the comparative expressiveness of description logics, and who explicitly considered the introduction of new predicate names and axioms during an encoding.

An even more complex situation is presented by so-called *epistemic description logics* [15] which add modalities to the language of descriptions, along the same way in which Levesque added epistemic modality to FOPC, as mentioned in the introduction. In this case the model-theory is quite different from that of standard FOPC so one is hard-pressed to even compare them in the framework we have here, which relies on the use of the same model theory to compare the formalisms.

Higher order DLs [17] are another extension of DLs, which preserve the basic variable-free term-like notation, by introducing axiom constructors. Once again the problem is that the model theory is not the one based on standard FOPC interpretations, which we have been using for comparisons. However, as with $\mathcal{DLR}$, it is shown that satisfiability reasoning over a theory in this particular language can be reduced to reasonsing over an encoding of it in a subset of $\mathcal{DN}$. This leaves us again with the question of how encodings affect the expressiveness of query languages.

5 DL vs FOPC as interaction languages

5.1 DLs vs FOPC in $\mathcal{L}_{\text{query}}$ revisited

Let us now return to our original consideration of using DLs in $\mathcal{L}_{\text{query}}$ for simple databases (interpretations), especially in contrast to the use of FOPC, which is the golden standard in the database community.

First, for decades, researchers have seen benefits in having queris be objects about which one can reason [7]. These include

- (Decidable) reasoning about a description D can tell us if it always returns the empty answer (it is unsatisfiable), which is almost always a sign that the user has made some mistake in stating it. Such feedback is obviously highly useful.

- Queries (and possibly retrieved answered) can be organized with respect to each other in a hierarchy according to the subsumption relationship $\sqsubseteq$. Therefore, when one asks a new query Q, one can see related queries that have been asked previously, together with even the answers returned to them. This is useful in situations where many users are asking questions independently of each other, in data exploration.

  One can also use this in query optimization, by retrieving only individuals not already found by the subsumed queries.

- When queries are classified near other existing concepts in the conceptual schema describing the KBMS contents, the roles and concepts involved in them may suggest ways in which to refine the current query. This is useful in situations where the user is unfamiliar with the schema of the data source.

On the other hand, the preceding results concerning the equivalence of most DLs with FOPC having limited number of variables are quite pessimstic because Immerman [19] has shown that there are no formulas in $\mathcal{FOL}^{k-1}$ over a schema involving predicates { NODE,Edge } that allow the expression of formulas stating such graph-theoretic properties as the existence of a k-subclique. And similar results have been extended by Cai et al [11] to $\mathcal{FOL}^{k-1}$ with counting qunatifiers. Yet such properties can be expressed even in the simplest so-called "conjunctive queries" studied in databases.

Therefore it is pretty clear that, on their own, DLs are not well suited as query languages for standard databases. For this reason, much of the research on query languages in the presence of DLs has been focused on unions of conjunctive queries, which have the form $\{ \vec{x} \mid \exists \vec{y}.conj(\vec{x}, \vec{y}) \}$ where $conj(\vec{x}, \vec{y})$ is a conjunction of atoms which have as predicates concepts or roles, and as arguments constants or variables from $\vec{x}$ and/or $\vec{y}$. There are often restrictions on what kinds of concepts and roles (primitive or composite) are allowed in the query. Actually, for standard relational databases one would want the most expressive possible query language, FOPC, *except* that it would deny the benefits of using subsumption on queries, listed at the beginning of this section, since subsumption would be undecidable.

5.2 DL vs FOPC in $\mathcal{L}_{answer}$

In the work above, as in almost all research on query answering, we have been content to use as $\mathcal{L}_{answer}$ just an enumeration of the individuals that satisfy the query. However, the term-like syntax of descriptions, especially in expressively limited languages — ones with few constructors — raises the possibility that individuals would come accompanied by *"descriptive information"* consisting of a concept they belong to which best characterizes them. Importantly, this concept would not have to be just a previously defined identifier, but could be a complex description. DL researchers have studied exactly such non-standard inference problems, where funtions in the system API return descriptions, rather than true/false. For example, Baader and Kuesters [4] summarize results concerning the finding of the *most specific concept* describing an individual, the *least common subsumer* of a set of concepts, and *rewriting a concept using definitions from a TBox*. All of these operations provide for the opportunity to either augment individuals with descriptions about them, or replace an answer set by the most specific description of its elements. These techniques are mostly applicable to DLs which use so-called "structural subsumption" to reason about concepts — an approach where descriptions can be put into an expanded normal form that makes all information explicitly available. But, in principle, there is nothing to prevent one from approximating (from "above" and "below") descriptions from a richer language using such a weaker language. Also, for such applications it does not seem essential to have perfectly sound and complete algorithms.

In this context, DLs appear to hold an advantage over FOPC, where every individual can be described by an infinite set of formulas, and it is hard to decide a priori in what way they should be restricted to obtain a manageable set.

5.3 Telling things to the system

There are at least three kinds of things that one can consider telling a database management system.

Telling extensional facts

For standard databases these are atomic facts, and the important aspect is that there is an automatic Closed World Assumption. As a result, one does indeed get a single interpretation that satisfies the resulting set of told facts — and this is what we have been imagining working with so far.

For description logics, we have mentioned that an ABox contains assertions about individuals which describe the state of the world. The following two short tables summarize the kinds of assertions supported by OWL-2 (modulo assumptions we have made earlier), and their translation to FOPC formulas.

DL ASSERTION	INTERPRETATION
classAssertion(C a)	$\mathcal{T}^x\langle$C$\rangle$ with a substituted for x
propertyAssertion(r a b)	$\mathcal{T}^{x,y}\langle$r$\rangle$ with a, b substituted for x, y
negativePropAssertion(r a b)	$\mathcal{T}^{x,y}\langle$**roleComplement** r$\rangle$ with a, b substituted for x, y

DL AXIOM φ	$\mathcal{FOL}^3$ AXIOM $\sigma(\varphi)$
classAssertion(C a)	$\mathcal{T}^x\langle$C$\rangle$ with a substituted for x
propertyAssertion(r a b)	$\mathcal{T}^{x,y}\langle$r$\rangle$ with a, b substituted for x, y
negativePropAssertion(r a b)	$\mathcal{T}^{x,y}\langle$**roleComplement** r$\rangle$ with a, b substituted for x, y

DLs work with the Open World Assumption, mostly because they are monotonic and can be used to infer new things. Therefore, the result of telling a system a series of facts of the above form will result in a **set** of interpretations $\mathcal{S}$ that satisfy it, and therefore we are in a situation where we need to define the meaning of **Answ**(D, $\mathcal{S}$), for a possibly infinite set of interpretations. The usual approach is to accept as answers $\bigcap_{\mathcal{I} \in \mathcal{S}} D^{\mathcal{I}}$ — the set of "certain answers", which hold in all possible interpretations.

In DLs supporting **oneOf** constructor and cardinality constraints, one can express certain localized Closed World Assumptions, e.g., FAVE_CHARACTER $\sqsubseteq$ **oneOf**(calvin, hobbes). But in general, one of the benefits of using DLs is exactly the ability to provide incomplete information, and have the logic of constructors reasons with it. For example if **classAssertion(minCardinality**(2,hasSiblings), cindy), then the query **minCardinality**(1,hasSiblings) will return cindy as an answer, without knowing cindy's siblings.

FOPC acts in the same way as DLs in expressing information about individuals. In fact, disjunction and existential quantifiers are well-known ways in which information can be incomplete. Because of the expressive limitations noted for DLs in Section 4, there are clearly ways in which incomplete knowledge cannot be expressed in $\mathcal{DCN}$ but can be told via FOPC. However, this does not appear to have arisen as a significant practical limitation.

Telling schemas and constraints

Information sources usually have schemas which help the user understand what information is available, by providing at least a vocabulary of terms to use in queries. It is also a source of certain (integrity) constraints that can be used to reject information being told to the system in order to guard against ever-present data entry errors.

The expression of such schemas at the user's conceptual level (so-called conceptual models), is one of the greatest strengths of DLs. The following is a subset of the rather rich set of axiom constructors, from the OWL-2 proposal[2].

---

[2]Most of the others can be reconstructed using other axiom constructors or concept constructors in $\mathcal{DN}$. Also note that the OWL 2 functional syntax uses **subClassOf** for the infix $\sqsubseteq$ we use throughout the paper.

DL AXIOM	INTERPRETATION
subClassOf(C D)	$C^{\mathcal{I}} \subseteq D^{\mathcal{I}}$
subPropertyOf(p q)	$p^{\mathcal{I}} \subseteq q^{\mathcal{I}}$
propertyDomain(p C)	$\forall x, y.(x, y) \in p^{\mathcal{I}} \Rightarrow x \in C^{\mathcal{I}}$
functionaProperty(p)	$\forall x, y, z.(x, y) \in p^{\mathcal{I}} \wedge (x, z) \in p^{\mathcal{I}} \Rightarrow y = z$
reflexiveProperty(p)	$\forall x \in \Delta \Rightarrow (x, x) \in p^{\mathcal{I}}$
symmetricProperty(p)	$\forall x, y.(x, y) \in p^{\mathcal{I}} \Rightarrow (y, x) \in p^{\mathcal{I}}$
transitiveProperty(p)	$\forall x, y, z.(x, y) \in p^{\mathcal{I}} \wedge (y, z) \in p^{\mathcal{I}} \Rightarrow (x, z) \in p^{\mathcal{I}}$

It is relatively easy to translate the DL axioms into (closed) FOPC formulas with at most 3 variables such that an interpretation $\mathcal{I}$ makes one true iff it makes the other true. During the translation below, we make use of the $\mathcal{T}^x \langle . \rangle$ functions defined earlier.

DL AXIOM φ	$\mathcal{FOL}^3$ AXIOM $\sigma(\varphi)$
subClassOf(C D)	$\forall x.\mathcal{T}^x \langle \mathrm{C} \rangle \Rightarrow \mathcal{T}^x \langle \mathrm{D} \rangle$
subPropertyOf(p q)	$\forall x, y.\mathcal{T}^{x,y} \langle \mathrm{p} \rangle \Rightarrow \mathcal{T}^{x,y} \langle \mathrm{q} \rangle$
propertyDomain(p C)	$\forall x, y.\mathcal{T}^{x,y} \langle \mathrm{p} \rangle \Rightarrow \mathcal{T}^x \langle \mathrm{C} \rangle$
functionalProperty(p)	$\forall x, y, z.\mathcal{T}^{x,y} \langle \mathrm{p} \rangle \wedge y = z \Rightarrow (\mathcal{T}^{x,y} \langle \mathrm{p} \rangle \Rightarrow x = z)$
reflexiveProperty(p)	$\forall x.\mathcal{T}^{x,y} \langle \mathrm{p} \rangle \Rightarrow x = y$
symmetricProperty(p)	$\forall x, y.\mathcal{T}^{x,y} \langle \mathrm{p} \rangle \Rightarrow \mathcal{T}^{y,x} \langle \mathrm{p} \rangle$
transitiveProperty(p)	$\forall x, y, z.\mathcal{T}^{x,y} \langle \mathrm{p} \rangle \wedge y = z \Rightarrow$ $[\mathcal{T}^{x,y} \langle \mathrm{p} \rangle \wedge x = z \Rightarrow \mathcal{T}^{x,y} \langle \mathrm{p} \rangle]$

Traditional relational databases have usually been designed starting from so-called Extended Entity Relationship diagrams or, more recently, UML diagrams as conceptual models. It has been shown that using only **subClassOf** and **subPropertyOf**, standard DLs can express these formalisms [13][5], especially with the addition of keys, thus providing additional evidence of the suitability of DLs for this task.

Since, in principle, constraints are only supposed to reject updates that would lead to errors, they can be used as queries, which are supposed to return true in valid states.

Of course, since FOPC is more expressive than $\mathcal{DC}$, clearly such constraints can be stated as FOPC theories. It is unclear though where there are gains in the extra expressiveness.

Telling general knowledge

It may also be desired to add generic information to the system in order to allow it to deduce new facts, which corresponds to eliminating some of the interpretations from S in a case when there is incomplete knowledge of the state of the world.

The standard way to do so is to tell the system FOPC axioms forming a theory Σ, and then perform question answering with respect to this theory, as well as the extensional information. In other words, if the extensional information is given through axioms DB, then we look for answers $\vec{a}$ to query $Q(\vec{x})$ which satisfy the entailment $\Sigma, DB \models Q(\vec{a})$.

In the case of description logics, this corresponds to using the full richness of the DL constructors to create a TBox, and then perform query answering with respect to the combination of the ABox and TBox. As with query languages, more power seems to be needed for the axioms one would want to add to an information manager. Evidence for this comes from the number of efforts for adding *deductive rules* to DLs [22] [16], including the SWRL rule addition to OWL [18]. An even more powerful approach was explored a long time ago by the Krypton system [9], which integrated a regular (resolution) theorem prover with a TBox that defined the predicates.

6 Discussion and Conclusions

This chapter attempted to compare the capabilities of Description Logics and First Order Predicate Calculus as languages for interacting with databases, and more generally knowledge bases. We used the framework of a common model theory, based on the interpretation $\mathcal{I} = <\Delta^{\mathcal{I}}, (\cdot)^{\mathcal{I}}>$, to consider the expressive power of the two kinds of formalisms, because it is used to provide the semantics of both kinds of logics. The comparison on this level playing field revealed that with a few exceptions, Description Logics proposed until today are much less expressive than FOPC, since they can be translated into formulas of $\mathcal{FOL}^3$ — FOPC with at most three variable symbols, possibly augmented with counting quantifiers to express role cardinality constraints. (The exceptions to this are constructors for role transitive closure, concept fixed points, and variants that express modal and higher order logic features.) Since for increasing k, $\mathcal{FOL}^k$ is known to be strictly increasingly expressive, this does indeed provide a limitation on what can be said with DLs, though this statement is empirical in the sense that new constructors may be added in the future that do not have this property. We consider it interesting that several constructors added more recently, such as **functionallyDetermine**, maintain this property. We have also shown that *any* formula in $\mathcal{FOL}^2$ (resp.$\mathcal{FOL}^3$) can be captured either as an axiom or a description in the "universal" description logic $\mathcal{D}$ (resp. $\mathcal{DC}$).

This has been a view taken from a very high level – one that essentially contrasted $\mathcal{D}(\mathcal{N})(\mathcal{C})$ with FOPC. Within the field of DL, the interest however has been and continues to be the selection of subsets of concept, role and axiom constructors, and even finer restrictions on their use (e.g., **subPropertyOf** applies only on certain kinds of roles), which yield languages for which reasoning tasks such as satisfiability or axiom entailment are *decidable*, and characterizing the precise computational complexity of these tasks. So one approach at comparing the expressive power of different DLs can be based on the computational complexity of reasoning with them.

In contrast, two general schemes for comparing the expressive power of descriptions in different DLs which, similar to ours, are based on model theory deserve mention here. First, the pioneering work of Franz Baader [1] took into account the fact that one could encode certain language features (e.g., concept negation) using others (e.g., cardinality constraints), if one is allowed to introduce new elements of the vocabulary, and add axioms to the TBox. For this purpose, it provided an embedding function ψ. We have mentioned already that the effect of encodings seems crucial to the expressive power of several DLs, including $\mathcal{DLR}$. A second approach, presented by Kurtonina and de Rijke [21], relies on considering the sets of models for any particular description logic $\mathcal{L}$ (built with its specific set of constructors) and then showing that all and only these models are preserved under a specific notion of "(bi)-simulation" characteristic of $\mathcal{L}$. This results in obtaining much finer distinctions than were given in either the present work or in Baader's [1], separating DLs as one adds new constructors.

A second question considered in the paper, inspired by Levesque's functional view of knowledge bases, has been the contrasting use of DLs and FOPC as languages involved with the TELL and ASK operators. In this respect, the expressive limitations of DLs with respect to FOPC are most problematic for the $\mathcal{L}_{\text{query}}$ language, and explain in part the necessity to consider (unions of) conjunctive queries, which are currently the focus of research in the Description Logics community. On the other hand, as far as $\mathcal{L}_{\text{answer}}$ is concerned, the availability of operations to find such things as the "most specific concept" describing an individual or collection of them, makes it possible to consider principled notions of non-standard, descriptive answers that go beyond the standard language of individual enumerations which we started from for FOPC. Finally, concerning $\mathcal{L}_{\text{tell}}$, we found it useful to distinguish certain variants of the TELL operator: describing the (possibly incomplete) knowledge of the state of the world, declaring a schema describing/constraining the kinds of data/knowledge known, and providing general knowledge to deduce new information. In these respects, DLs seemed very well suited for the second task, did not have obvious practical deficiencies for the first task, but were once again hampered by expressive limitations w.r.t. FOPC for the third task.

To summarize, one must first acknowledge that reasoning with full FOPC is undecidable, while reasoning with $\mathcal{D}$ (though not $\mathcal{DC}$) is actually known to be decidable [24], although the best current DL reasoners implement a subset of $\mathcal{DN}$, with some limited use of **roleChain**. To some, this is would trump all other issues. In other respects, one can say that FOPC holds the edge in those facets of interacting with information bases where expressiveness is at a premium (e.g., querying, possibly inference rules), while DLs have an advantage because of the greater richness of description/formula constructors, whose utility has been demonstrated empirically over decades of applications in knowledge representation, and therefore provide guidance to the user in the kinds of information to capture for conceptual models or provide an opportunity to give "impoverished" but useful descriptions of answers. The weaknesses of DLs are being addressed by considering "hybrid" systems that combine often Horn rules with concepts and roles from a DL acting as a source of predicates names, about which additional terminological knowledge is available.

Acknowledgment This chapter contains some technical material from the article "On the expressive power of description logics and predicate logics", *Artificial Intelligence*, 1996, pp.353-367. I am particularly indebted to Franz Baader for insight in that matter.

References

[1] F. Baader, "A formal definition for expressive power of Knowledge Representation Languages", *Proc. ECAI-90*, Stockholm, 1990, pp.53-58.

[2] F. Baader, H-J. Bürckert, J. Heinsohn, B. Hollunder, J. Müller, B. Nebel, W. Nutt, H. Profitlich, *Terminological Knowledge Representation: A Proposal for a Terminological Logic*, DFKI Report, DFKI, Saarbrucken, GERMANY, October 1992.

[3] Franz Baader, Diego Calvanese, Deborah McGuinness, Daniele Nardi, and Peter Patel-Schneider, *The Description Logic Handbook: Theory, Implementation and Applications*, Cambridge University Press, 2nd edition (2007)

[4] Franz Baader and Ralf Ksters, "Nonstandard Inferences in Description Logics: The Story So Far", *Mathematical Problems from Applied Logic I, vol.4*, 2006, pp. 1-75

[5] D. Berardi, D. Calvanese, G. De Giacomo: "Reasoning on UML class diagrams". *Artificial Intelligence 168(1-2)* pp. 70-118 (2005)

[6] A. Borgida and R. Brachman, "Loading data into description reasoners", *Proc. ACM SIGMOD Conf. on Data Management*, 1993, Washington, DC, pp. 217 – 226.

[7] A. Borgida, "Description Logics in Data Management", *IEEE Trans. on Knowledge and Data Engineering*, 7(5), October 1995, pp.671–682.

[8] R. J. Brachman, "A Structural Paradigm for Representing Knowledge," Ph.D. Thesis, Harvard University, Division of Engineering and Applied Physics, 1977.

[9] R. J. Brachman, V. P. Gilbert, H. J. Levesque: "An Essential Hybrid Reasoning System: Knowledge and Symbol Level Accounts of KRYPTON" *Proc. IJCAI*, 1985, pp. 532-539

[10] M.Buchheit, M. Jeusfeld, W. Nutt, and M. Staudt, "Subsumption between queries in object-oriented databases", *Information Systems* 19(1), pp.33-54, 1994.

[11] J. Cai, M. Fürer, and N. Immerman, "An optimal lower bound on the number of variables for graph identification", *Proc. 30th IEEE Symp. on Foundations of Computer Science*, 1989, pp.612–617.

[12] D. Calvanese, G. De Giacomo, M. Lenzerini, "Conjunctive Query Containment in Description Logics with n-ary Relations. *Proc. 1997 Description Logics Workshop (DL97)*, pp.5–9, 1997

[13] D. Calvanese, M. Lenzerini, D. Nardi. "Unifying Class-Based Representation Formalisms" *J. Artif. Intell. Res. (JAIR) 11* pp. 199-240 (1999)

[14] E. F. Codd. "Relational Completeness of Data Base Sublanguages". In: R. Rustin (ed.): *Database Systems*, Prentice Hall, pp.65-98, 1972

[15] F.Donini, M. Lenzerini, D.Nardi , A. Schaerf, and W. Nutt. "An epistemic operator for description logics" *Articial Intelligence 100(1-2)*: pp. 225274, 1998.

[16] F.M.Donini, M.Lenzerini, D.Nardi, A.Schaerf, "A Hybrid System with Datalog and Concept Languages" *Trends in Artificial Intelligence*, 1991, Springer, LNAI 549, pp.88-97.

[17] Giuseppe De Giacomo, Maurizio Lenzerini, Riccardo Rosati: "On Higher-Order Description Logics", *Proc. 2009 Description Logics Workshop (DL'09)*

[18] I. Horrocks, P.F. Patel-schneider, S. Bechhofer and D. Tsarkov, "OWL Rules: A Proposal and Prototype Implementation", *Journal of Web Semantics 3*, 2005, pp.23-40.

[19] N. Immerman, "Upper and lower bounds for first-order expressibility", *J. Comp. Syst. Sciences*, **25**, 1982, pp. 76-98.

[20] N. Immerman, "Relational queries computable in polynomial time", *Information and Control*, **68**, 1986, pp. 86-104.

[21] N. Kurtonina and M. de Rijke, "Expressiveness of concept expressions in first-order description logics" *Artificial Intelligence 107*, 1999, pp.303-333

[22] A. Y. Levy and M.-C. Rousset. Combining Horn rules and description logics in CARIN. *Articial Intelligence, 104(12)*, 1998, pp.165209.

[23] Hector J. Levesque, "Foundations of a Functional Approach to Knowledge Representation", *Artificial Intelligence 23(2)*: 1984, pp.155-212

[24] M. Mortimer, "On languages with two variables", *Zeitschr. f. Math. Logik und Grundlagen d. Math.*, 21, pp.135–140, 1975.

[25] *OWL 2 Web Ontology Language Primer*, W3C Recommendation, 27 October 2009 ; http://www.w3.org/TR/owl2-primer/,

[26] P.F. Patel-Schneider, B. Swartout, *Description Logic Knowledge Representation System Specification from the KRSS Group of the ARPA Knowledge Sharing Effort*, AT&T Bell Laboratories Report, 1994.

[27] J.G. Schmolze and D. Israel, "KL-ONE: semantics and classification", in *Research in Knowledge Represetnation for NL Understanding*, Tech Report 5421, BBN Laboratories, 1983.

[28] D. Toman and G. Weddell, "On Keys and Functional Dependencies as First-Class Citizens in Description Logics," *J. Automated Reasoning 40(2-3)*, pp. 117-132 (2008)

[29] W. A. Woods and J. G. Schmolze, "The KL-ONE family," *Computers and Mathematics with Applications 23*(2-5), Special Issue on Semantic Networks in Artificial Intelligence.

6

On Continual Planning with Runtime Variables

MICHAEL BRENNER AND BERNHARD NEBEL

ABSTRACT. This article discusses the problem of planning and acting in partially observable environments. In many such domains conditional planning for all contingencies is prohibitively hard. Therefore we advocate a continual planning approach, where decisions can, by means of so-called *assertions*, be deferred until execution time when more information is available. Additionally, we formalize the notion of *runtime variables* as functional fluents, which can act as placeholders for sensing results unknown at planning time. Using runtime variables and assertions we show how a *series of sequential plans* can solve planning tasks that in non-continual planning would necessitate plans with conditional branching and loops.

1 Introduction

In his 1996 article "What is Planning in the Presence of Sensing?" Hector Levesque discusses the *Airport* example, in which a traveler must find a way onto flight 123 at the local airport. To do this, she must find out the boarding gate for the flight and move there. In such a scenario execution-time sensing must be taken into account already during planning because the agent's behavior will be influenced by the outcome of the sensing action. It is widely accepted that solutions for this or similar problems must be conditional plans that branch over the possible sensor outcomes or even more general structures like policies or finite-state controllers. In 1996 Hector noted: "There clearly is no sequence of actions that can be shown to achieve the desired goal; which gate to go to depends on the (runtime) result of checking the departure screen".

The above statement is obviously true—if by "actions" we mean fully instantiated actions as typically used in classical propositional planning. However, consider how a human would probably represent her "recipe" for getting onto flight 123 in some arbitrary imperative programming language:

```
gate = read-departure-screen(flight123)
move-to(gate)
board(flight123)
```

This *is* a sequence of actions; however, it makes use of one the most basic concepts of imperative programming, namely a *runtime variable* `gate`. If our planning model allowed for such runtime variables, it would be possible to find a sequential plan to achieve the goal. It is true, however, that not all problems of planning with sensing can be solved using sequential plans and runtime variables—sometimes different sensing results will require different reactions. For instance, big international airports, such as Frankfurt or Toronto, usually have several terminals, so that getting to the gate for flight 123 may or may not involve a transfer to another terminal. This surely cannot be expressed with a sequential plan, can it?

Our approach to Levesque's Airport Example is based on the observation that when people find their way around an airport they do act in a goal-directed, planful way, yet they do not seem to branch over all possible conditions in advance. This does not mean that they randomly explore the airport or make guesses about the departure gate that they must revoke when they find them violated. Rather, they rely on additional domain knowledge stating, e.g., that there will be signs to guide them to any gate—even if this includes transfer to another terminal. As a result, a typical passenger at Frankfurt airport with little information about the airport's terminals, gates, and layout may still confidently follow the following plan:

---

gate = read-departure-screen(flight123)
follow-signs(gate)
board(flight123)

This plan (or program) not only features the runtime variable `gate`, but also a call to a *subroutine* `follow-signs()`. Execution of the subroutine, i.e. actually following the signs will most likely involve moving around the airport and reading signs repeatedly. Interestingly, those details are usually not planned for in advance by humans and indeed often *cannot* be, because the planning domain is incompletely specified at planning time. In the Airport example, a traveler usually neither knows the number of possible gates nor the layout of an airport before getting there.

Instead, at planning time the human relies on the "contract"[1] of the subroutine, i.e. the yet unspecified subplan `follow-signs(gate)`: she knows that, once she is at the airport and has found out her gate, she will be able to find her way to that gate. In other words, although the subplan named `follow-signs(gate)` is not yet concretized, it is already characterized by a *Hoare triple* $\{P\}$`follow-signs(gate)`$\{Q\}$[Hoare 1969]: If certain preconditions P are guaranteed to hold before the subplan, it guarantees that certain postconditions Q can be achieved.

We will call such unspecified subplans *assertions*. The notion is inspired by the use of the same term in computer programming, where a programmer asserts that certain conditions must hold at a certain point in the program execution. An assertion is characterized by its pre- and postconditions, exactly like normal actions in the domain. However, like subroutines in a program or methods in Hierarchical Task Networks it will be replaced by a concretized plan before being executed. Note, however, that in both programming and hierarchical planning this concretization is assumed to be given in advance. In our approach, as in the Airport example, it is the agent itself that will fill in the details of the plan as soon as the missing information becomes available. The task is thus not decomposed hierarchically, but *temporally*: parts of the planning problem that cannot be solved *yet* are postponed and plan execution is started early in order to gather additional knowledge. Planning and execution are thus integrated and interleaved deliberately by the agent—a process that we call *Continual Planning*.

Interestingly, the assertion `follow-signs(gate)` is used like any normal action in the above plan, which therefore remains sequential. Likewise, when the assertion is *expanded*,

---

[1]This terminology is borrowed from the programming language Eiffel and its principle of "Design by Contract" [Meyer 1992].

i.e. replaced by a more concrete subplan[2] once the agent reads the departure screen, this new plan is again a sequential one. By then the agent will know the concrete departure gate and must only find *one* plan for how to reach it. It is the great advantage of this form of continual planning that branching over possible contingencies is thus avoided. Instead, a *series* of sequential plans is generated and executed.

The remainder of the paper is structured as follows. In the following section, we will discuss related approaches to planning with partial observability. Section 3 presents our continual planning formalism and algorithm. In Section 4 we show how an extended version of Levesque's Airport example can be solved by continual sequential planning, which otherwise would require plans with conditional branches and loops. We conclude with a brief discussion of future work.

2 Related Work

Planning with sensing actions has been extensively studied in the planning literature [Levesque 1996; Golden and Weld 1996; Weld, Anderson, and Smith 1998; Bonet and Geffner 2000; Petrick and Bacchus 2002; Petrick and Bacchus 2004]. Usually, this is done in a conditional planning setting. Unfortunately, conditional planning with partial observability is of prohibitively high computational complexity, even in relation to classical planning [Rintanen 2004].

Additionally, it is widely acknowledged that, as stated in the textbook by Russell and Norvig, "even the best-laid plans of mice, men and conditional planning agents frequently fail" [Russell and Norvig 2003]. Practical planning agents must therefore be extended with capabilities for execution monitoring and plan adaption or replanning in order to be able to react quickly to unexpected circumstances and events [Fritz and McIlraith 2007]. Such agents are sometimes called continual planning agents, because they continually switch between plan execution and replanning until they have reached their goal. In accordance with Russell and Norvig's textbook we will, however, refer to such agents as *replanning agents* and reserve the term *continual planning agent* to agents that can deliberately switch to plan execution during the planning process, i.e. even before a plan has fully been elaborated.

Continual planning agents try to evade the complexity of conditional planning by not only planning for information gathering, but by actually performing information gathering before planning for all future contingencies. Possibly the earliest approach so tightly integrating planning, monitoring, execution and information gathering was the IPEM system [Ambros-Ingerson and Steel 1988]. In this, as well as in later systems like Sage [Knoblock 1995], execution is integrated into a partial-order planning algorithm by treating unexecuted actions in the plan as a special kind of "flaw" that the planner has to resolve by executing the action at some point. It is not clear in these earlier approaches, however, if or how a planner can decide that it is necessary or helpful to start execution before a complete plan has been found.

In the present work, assertions are used to ensure the planner that it can safely switch to execution before having a detailed plan in order if necessary to proactively gather relevant information. The notion of assertions was developed in our previous work [Brenner and Nebel 2009]. However, there we used a different model for the symbolic effects of sensing actions which, in a nutshell, just stated that the value of a fluent would be *known* after a sensing action, similarly to corresponding models in epistemic logic [Fagin, Halpern,

---

[2]This plan may also contain assertions, but must have at least an executable prefix (cf. Section 3).

Moses, and Vardi 1995]. Assertions then used knowledge preconditions to make state-ments about the planning domain such as "If I knew my gate, I could find a plan to get on my flight". However, normal operators do not have knowledge preconditions, but refer to concrete facts in the current situation for testing applicability. When sensing operators are only modeled in terms of their epistemic effects, their results cannot be used to make these preconditions true. Therefore, in this work, we use a different model for sensing, based on runtime variables.

Etzioni and colleagues first used *runtime variables* to refer to sensing results in a plan that will only become known at execution time [Etzioni, Hanks, Weld, Draper, Lesh, and Williamson 1992; Etzioni, Golden, and Weld 1997; Golden 1998]. Similarly, Petrick and Bacchus used 0-ary functional fluents as runtime variables [Petrick and Bacchus 2002; Petrick and Bacchus 2004]. While these approaches successfully made use of runtime variables, they did not clearly define the semantics of planning with them. Here, we provide such a definition in the context of the Functional STRIPS language developed by Geffner [Geffner 2000].

Assertions are similar to schemata or "methods" in Hierarchical Task Networks (HTNs) [Yang 1997; Erol, Hendler, and Nau 1996; Nau, Cao, Lotem, and Munoz-Avila 1999; Nau, Au, Ilghami, Kuter, Murdock, Wu, and Yaman 2003] in that they will be decomposed into more concrete subplans until the goal can be reached. The purpose of both approaches is different, though: While HTNs essentially provide search guidance for a planner by explicitly decomposing a problem into (ideally) independent subtasks, assertions enable the planner to decompose the problem temporally into different planning and execution phases. As a result, where HTN planners generate abstraction hierarchies, continual planning with assertions generates a series of non-hierarchical plans.

While HTN domain designers need to explicitly provide method decompositions, no pre-defined abstraction hierarchies are required in continual planning. Rather, the designer only specifies conditions for the existence of (sub-)plans in a particular domain, much like a programmer specifying the "contract" of a procedure in programming languages like Eiffel [Meyer 1992]. The planner itself then finds expansions for assertions, i.e. it synthesizes concrete subprograms once the preconditions of the "contract" are satisfied.

Our approach is also related to work on integrating planning into Golog-like action lan-guages [Giacomo, Lesperance, and Levesque 2000; De Giacomo, Lesprance, Levesque, and Sardiña 2002], in particular to our previous work on integrating a planner with In-diGolog [Classen, Eyerich, Lakemeyer, and Nebel 2007]. IndiGolog programs, like contin-ual planning agents, can interleave planning, acting and information gathering, so that the results of sensing actions can be reacted to immediately. Previously, we have shown how planning subproblems can be extracted from an IndiGolog program at runtime and solved by a planning system [Classen, Eyerich, Lakemeyer, and Nebel 2007]. These calls to an external planner replace the generic forward-search operator *achieve* (or *search*), which is used to refer to some yet unspecified solution to a subproblem in an IndiGolog program— which is exactly the role of an assertion in a continual plan. Thus, where *achieve* enables an IndiGolog programmer to hand over some control to an autonomous planner, assertions can be said to enable the domain designer of a planning domain to specify some additional domain information, namely about the solvability of subproblems, to the planner.

3 Planning Model

Our planning language is a variation of Geffner's Functional STRIPS [Geffner 2000], a typed first-order language with *name*, *type* and *function* symbols without quantifiers or variables. Relations are not explicit in the language, but can be expressed by using functions with a Boolean codomain. To simplify the presentation, we will use standard relational notation in our examples anyway, i.e. we will write, e.g., $\neg connected(pos(), dest)$ instead of $connected(pos(), dest) = \perp$ when comparing or $connected(pos(), dest) := \perp$ when assigning.

Name symbols are assumed to be non-fluent, i.e. they are fixed names of objects interpreted *under the unique name assumption*; therefore we will simply equate name symbols with their denotations in the following. Function symbols are *fluent*, i.e. their denotation depends on the state they are interpreted in. Terms and formulas must obey the usual formation rules ("well-formedness"). For a term t and formula f, we write t^s and $f^s(t^s)$ to denote their interpretations in state s (but omit the superscript s if it is irrelevant or obvious from context).

3.1 Tasks

We can now define the tasks a continual planning agent is supposed to solve.

DEFINITION 1. A **continual planning task** is a tuple $T = \langle \mathcal{T}, F, \mathcal{O}, s_0, s_* \rangle$ where

- $\mathcal{T}$ is a set of **types**, where each type t is associated with a set of names, called its **domain** $\mathcal{D}^t$. $\mathcal{C} = \cup_{t \in \mathcal{T}} \mathcal{D}^t$ is the set of all **names**.

- $\mathcal{F}$ is a set of **function symbols**. Each $f \in \mathcal{F}$ has an associated signature $f : \mathcal{D}^{t_1} \times \cdots \times \mathcal{D}^{t_n} \to \mathcal{D}^{t_{n+1}}$ (with $n \geq 0$ and $t_i \in \mathcal{T}$). $\mathcal{D}^f = \mathcal{D}^{t_1} \times \cdots \times \mathcal{D}^{t_n}$ is called the domain of f, and $\mathcal{D}^{t_{n+1}}$ is called the codomain of f and referred to by $codomain(f)$.

- A **state** s is an interpretation of the language that is defined by the names $\mathcal{C}$ and function symbols $\mathcal{F}$ over $\mathcal{C}$. Since, as stated above, we treat names as their own interpretations, the state s is represented by the interpretations of each function symbol f over its domain $\mathcal{D}^f$, i.e. the values $w \in codomain(f)$.

- A **belief state** b is a set of states representing the states the agent assumes to be possible. A set of belief states B will be called **belief set** and represents the possible beliefs the agent can have - given the sensing actions the agent performed.

- $\mathcal{O}$ is a set of **operators** of the form $\langle param, pre, eff, sense \rangle$.

    - *param* is a list of typed schema variables. They can be used in the rest of the operator definition as place holders for fluents.

    - In analogy to Functional STRIPS [Geffner 2000], the precondition *pre* is a conjunction of conditions of the form $t = w$. With t and w be well-formed terms[3] .

    - The effect *eff* is a list of conditional updates $(c \to e)$ where c is an effect condition with the same restrictions as a precondition and e is an atomic update of the form $t := w$, and t and w are terms of the same type.

---

[3] *Terms* are built in the usual manner using function and name symbols of compatible types and arities [Geffner 2000].

- *sense* is the sensed fluent in a sensing action, otherwise it is unspecified.

Every operator o is declared to belong to one of the following three disjoint sets:

- the set of **sensing operators** $\mathcal{O}^s$ ($o \in \mathcal{O}^s$ iff *sense* is specified for o)
- the set of **assertions** $\mathcal{O}^a$,
- the set of **standard operators** $\mathcal{O}^o$,

The induced set of actions $\mathcal{A}$ is the set of parameter-free operators generated by substituting all possible functional fluent expressions for the scheme variables in the operators. This set is potentially infinite. The sets $\mathcal{A}^o$, $\mathcal{A}^a$, and $\mathcal{A}^s$ are defined analogously to the corresponding sets of operators.

- s_0 is a singleton belief set, called the **initial belief set**. It is specified by a conjunction of expressions of the kind $t = c$, with t being some term of our language and c a name. Since s_0 is singleton, it denotes the unique *belief state*, where for each fluent either the value is known or it can be any value of the domain of the fluent.

- s_* is a formula describing the **goal condition**, which we will assume to have a form like a precondition of an action, i.e., a conjunction of fluent equations.

Since continual planning alternates between planning and plan execution, we will need to distinguish between the semantics of the physical and the symbolic execution of an action, the latter describing the state transitions reasoned about in the planning process under incomplete knowledge, the former modeling the actual results of executing the action in a particular world state.

Furthermore, for the symbolic execution, we will distinguish between execution in the *full belief set space* and a *simplified model*, which is defined below.

In order to do so, we have to define what it means that a condition φ of an action a is satisfied by a belief set B, a belief state b and a classical state s. A state s satisfies φ, in symbols $s \models \varphi$, if s satisfies φ classically. A belief state b satisfies φ, symbolically $b \| \models \varphi$, if for all $s \in b$ we have $s \models \varphi$. That is, regardless of what we consider as possible, the condition is true. Finally, a condition ϕ is **satisfied by the current belief set** B, symbolically $B \| \models \varphi$, iff all $b \in B$ satisfy φ.

DEFINITION 2. The **symbolic execution over the full belief set space** is defined as follows:

- Standard actions and assertions $a \in \mathcal{A}^o \cup \mathcal{A}^a$ can only be executed if the preconditions are satisfied by B. The next belief set consists of all belief states that result from executing a in each of the belief states. Executing an action a in a belief state b is simply the execution on each state $s \in b$, which means the application of all conditional effects of a in s, i.e., if the effect condition is satisfied in s, then the effect is made true in the resulting state.

- A sensing action follows the same execution model except that right in the beginning, before the effects of a are applied, the sensing of the fluent f of type t in *sense* takes place. This results in splitting each belief state into $\| \mathcal{D}^t \|$ belief states, such that for each $c_i \in \mathcal{D}^t$ that is possible for f in the belief state b, there is a belief state b_i that satisfies $f = c_i$ and is otherwise identical to the original belief state b. After that the effects of the sensing action are applied. The resulting belief set contains all belief states generated in this way.

Based on these definitions, we say that a plan P is a successful plan for a planning task $T = \langle \mathcal{T}, F, \mathcal{O}, s_0, s_* \rangle$ if the actions in P sequentially executed on the belief set s_0 lead to a belief set B that satisfies s_*.

Domains:	*Egg: egg0, egg1*
	Pen: pen1, pen2, pen3
	Object; egg0, pen1, sample
	Color: red, blue, green
Fluents:	*color : Object $\rightarrow$ Color*
	sampleColor : $\emptyset \rightarrow$ Color
Action:	**paint-egg**(*egg:Egg, pen:Pen*)
Prec:	$\emptyset$
Post:	*color(egg) := color(pen)*
Sensor:	**sense-color**(*obj:Object, col():Color*)
Sense:	*color(obj)*
Prec:	$\emptyset$
Post:	*col() := color(obj)*
Init:	*color(pen1)=green, color(pen2)=blue, color(pen3)=red*
Goal:	*color(egg0)=color(sample), color(egg1)=color(sample)*

Figure 1. *Easter eggs* continual planning domain

An example planning domain and task is shown in Figure 1 (syntactically, we follow Geffner's Functional Strips [Geffner 2000]). In this task, easter eggs must be painted in some sample color which must be determined by sensing first. In our framework, the goal is achieved without branching by the following plan:

```
sense-color(sample, sampleColor())
paint-egg(egg0, sampleColor())
paint-egg(egg1, sampleColor())
```

A more elaborate example will be discussed in Section 4.

3.2 A Simplified Execution Model

Our planning language gives us only limited ways of expressing uncertainty and knowledge gathering. In particular, because we stick to linear plans, we cannot branch on sensed values in the plan (we will later discuss how assertions can mitigate this restriction). On the positive side, a more compact representation of belief sets seems possible in our framework.

DEFINITION 3. The special value u does not belong to a type and denotes the **unknown** value. For any type t, $\mathcal{D}_u^t = \mathcal{D}^t \cup \{u\}$ and $\mathcal{C}_u = \cup_{t \in \mathcal{T}} \mathcal{D}_u^t$

A **simplified belief state** is a singleton set containing one state that is an interpretation of all the fluents over $\mathcal{C}_u$ such that for all fluents that are interpreted as u, all values of the codomain of the fluent are possible.

A condition φ is satisfied by a simplified belief state $b = \{s\}$ iff none of the subterms in φ are interpreted as u in s and φ is classically satisfied by the state s.

Using this notion of simplified belief set, we can express the initial belief set that contains only one belief state as one interpretation. All fluents that have a value assigned have that value, all others have the value u.

Without sensing and without conditional effects, such a simple execution model would be actually equivalent to one that is executed over full belief sets. This can be shown by a simple compilation to standard basic STRIPS [Nebel 2000].

Domains:	*Type1: 1, 2*
Fluents:	*right, wrong, do, done:* $\emptyset \rightarrow$ *Bool*
	a: $\emptyset \rightarrow$ *Type1*
Action:	**case()**
Prec:	$\emptyset$
Post:	$a() = 1 \rightarrow right(), a() = 2 \rightarrow right(), a() = 1 \rightarrow \neg wrong(),$
	$\top \rightarrow do()$
Action:	**unsound()**
Prec:	*wrong(), do()*
Post:	*done()*
Init:	$\emptyset$
Goal:	*right()*

Figure 2. Small example demonstrating incompleteness and unsoundness

When the planning language contains conditional effects, planning with respect to simplified belief sets becomes incomplete with respect to planning over the full belief set space, as is demonstrated by the artificial planning task in Figure 2. In order to achieve the goal *right()*, the action **case** is sufficient in the full belief set space. Under the simplified model, however, the conditional effects are not activated, because the conditions are not satisfied.

Furthermore, one has to be careful to avoid the traps of being unsound with respect to the belief set model. For example, if we consider the initial description $\{wrong\}$ and the goal $\{done\}$, then there is no plan for the full belief set space. However, the plan $\langle$**case, unsound**$\rangle$ seems to achieve the goal under the simplified belief set space because none of the conditional effects in the action **case** are executed.

In order to avoid unsoundness, uncertainty has to be propagated over conditional effects. So, **execution over simplified belief sets** differs as follows: if a fluent has the value u and this fluent is part of an effect condition, then the effect fluent will become unknown as well.

This leads in general to the fact that less actions can be applied and that less fluents will have a known value.

PROPOSITION 4 (Soundness of the simplified model). *Any successful plan that can be executed over the simplified belief set space is a successful plan over the full belief set space.*

While this simplification reduces the number of states we have to store exponentially, the belief set still contains a number of states that is exponential in the number of sensing actions. However, do we really need to track all of these states? Wouldn't it be enough to simply remember that the value of a fluent is *known* after a sensing action has been executed on this fluent? For instance, we could introduce another special domain element k that could be (implicitly) assigned when a fluent is sensed and tested afterwards, e.g., with an expression such as *departure(flight123)* = k. In fact, this was the intended semantics of the approach described in an earlier paper of ours [Brenner and Nebel 2009] (where we did not deal with functional fluents, though).

It is to be expected that this move, again, introduces incompleteness. However, even worse, it does not even allow to execute our examples, because they rely on the equality between two fluents, e.g., *gate()* equaling *departure(flight123)*, of which the concrete value is unknown at planning time. Similarly, our plan for the easter eggs example of Figure 1 would not ensure that eggs and sample object share the same color in the final state.

Another idea might be to introduce a new anonymous object for each value sensed. Such a move would make our examples executable. Again, such a model would not be complete with respect to the simplified belief model. For instance, in some task there may be a number of different persons, with each person's age being known. If we sense a particular person, we should know the age of the person as well. This is true for the semantics of Definition 2. However, if during the simplified symbolic execution a new anonymous object is introduced for the person seen, the value for her age will be unknown.

Things are even worse. Namely, introducing anonymous objects can make planning unsound with respect to the full belief set model, provided we can introduce unlimited many new anonymous objects. As sketched in Figure 3, we could simulate a Turing machine and solve the Halting Problem on the empty tape by planning in the execution model with anonymous objects.

Of course, the number of new anonymous objects could be restricted, but it is not clear whether this will lead to soundness. For this reason, as a compromise between efficiency and completeness, we adapted the simplified belief set space as described above. While it seems obvious that this simplified model is easier to implement, the question of whether there is indeed a reduction in computational complexity for planning has still to be resolved.

3.3 Continual Planning with Assertions

We can now describe how planning, i.e. reasoning about the symbolic execution of actions, can lead to physically achieving a goal.

In the following, we will refer to the **symbolic execution** of an action a or a plan P in the simplified belief set state B as $project(B, a)$ respectively as $project(B, P)$. This is to be contrasted with the *physical execution* of actions. These always take place in a single simplified belief state b and consist of checking the preconditions and executing the (conditional) effects. In case of *sensing* actions, one particular value is nondeterministically chosen. We denote the **physical execution** of an action a in a simplified belief state b by $execute(s, a)$ and, respectively, the physical execution of a plan P by $execute(s, P)$.

Domains:	*Cell: cell0*
	Char: 1, 0, *
	State: q0, q1, q2, ..., qf

Fluents:	*next : Cell $\rightarrow$ Cell*
	prev : Cell $\rightarrow$ Cell
	cont : Cell $\rightarrow$ Char
	head: $\emptyset \rightarrow$ Cell
	state: $\emptyset \rightarrow$ State
	init: $\emptyset \rightarrow$ Bool

Sensor:	**fresh(*cell:Cell*)**
Sense:	*next(cell)*
Prec:	*init()*
Post:	*prev(next(cell)) := cell, char(next(cell)) := **

Action:	**switch-to-computation()**
Prec:	*init()*
Post:	*$\neg$ init()*

Action:	**rule$_1$()**
Prec:	*cont(head()) = 1, state() = qi*
Post:	*cont(head()) := 0, head() := next(head()), state() = qk*

$$\vdots$$

| Init: | *cont(cell0) = *, init(), head() = cell0, state() = q0* |

| Goal: | *state() = qf* |

Figure 3. Turing machine encoding

Assertions *cannot* be executed physically, i.e. *execute*(s, a) is undefined for any assertion a, even if its preconditions are satisfied in s. This crucial semantic difference between *project*() and *execute* will lead to the intended "expansion" of assertions during continual planning: CP algorithms must take into account that assertions can never be selected for physical execution, therefore they must make sure that assertions never appear first in a plan (in the case of partially ordered plans, they must never be ordered immediately after the *init* action). Additionally, after an action is physically executed and removed from the plan *monitoring* must check if this constraint is violated in the updated plan. In that case, a new planning phase is triggered in which the assertion is not allowed to be used at the beginning of the plan any more.

Algorithm 3.3 shows the basic continual planning algorithm. Here, PLANNER is a generic planning algorithm computing valid, i.e. symbolically executable, plans and ensuring that every plan P returned by PLANNER($\{b\}, G$) does not start with an expandable

---

Algorithm 1 Generic continual planning algorithm.

$P \leftarrow \langle \rangle$
while $b \not\models s_*$ **do**
   $P \leftarrow$ PLANNER($\{b\}, G$)
   if $P =$ failure **then**
      return "No solution from b"
   while PLANISVALID(b, P) **do**
      $a \leftarrow$ POP(P)
      $b \leftarrow$ EXECUTEACTION(a, b)
return "Goal reached"

---

assertion a, i.e. $\{b\}$ does not satisfy $pre(a)$. Similarly, PLANISVALID is a generic plan monitoring procedure [Russell and Norvig 2003]. It checks if the plan P updated after its first action has been executed does not begin with an expandable assertion and, if that is not the case, computes $project(\{b\}, P)$ to verify whether the state resulting from symbolic execution of the updated plan P still satisfies the goal.

Algorithm 3.3 will produce a sequence of belief states and plans that we call a *CP trace*.

DEFINITION 5 (CP trace). Let $\langle \mathcal{T}, F, \mathcal{O}, s_0, s_* \rangle$ be a continual planning task. $T = \langle b_0, P_0, \ldots, P_{n-1}, b_n \rangle$ is a sequence alternating belief states and plans and is a called a **CP trace** if

- $project(\{b_i\}, P_i)\|\models s_*$, i.e. P_i symbolically executed in $\{b_i\}$ achieves s_*

- $b_{i+1} = execute(s_i, a)$ where $a = P[0]$, i.e. b_{i+1} results from physically executing the first action of P in b_i.

While it would be desirable that all CP traces for a goal G generated by Algorithm 3.3 always end in a goal state, in general this can be guaranteed only if assertions are expanded in a side-effect free manner.

PROPOSITION 6 (Soundness). *Let $\langle \mathcal{T}, F, \mathcal{O}, s_0, s_* \rangle$ be a continual planning task and b the simple belief state representing s_0. Then Algorithm 3.3 generates a CP trace $T = \langle b_0, P_0, \ldots, P_{n-1}, b_n \rangle$ such that b_n satisfies s_*, provided for each assertion it is always possible to find a plan that makes the effects of the assertions in all plan P_i true and does not change anything else.*

Obviously, enforcing complete absence of side effects is a rather strong limitation of applicability. However, in practice we can mitigate the restriction: We can determine those potential side effects that would be harmful to the plan suffix by regressing from the goal, and then force the planner to not achieve these when expanding assertions. How this affects soundness is a topic of future work. In particular, we are interested in describing structural properties of *domains* that guarantee soundness of planning with assertions.

4 Worked Example

Let's step through Hector Levesque's *Airport* example once more, modeled as a continual planning problem, as shown in Figure 4. In the beginning, the agent is situated at *gate0*, not knowing at which gate her flight, *flight123*, departs. At *gate0* there is a big screen indicating the departing gates for all flights, as indicated by the precondition of action

Domains:	*Gate : gate0, gate1, gate2, gate3, ...*
	Flight: flight123, flight456, flight789, ...
Fluents:	*departure : Flight → Gate*
	pos : ∅ → Gate
	connected : Gate × Gate → Bool
	direction : Gate × Gate → Gate
	gateA, viaA : ∅ → Gate
Action:	**move-to**(*dest:Gate*)
Prec:	*connected(pos(), dest)*
Post:	*pos() := dest*
Action:	**board**(*flight:Flight*)
Prec:	*pos() = departure(flight)*
Post:	*pos() := flight*
Assertion:	**follow-signs**(*dest:Gate, via:Gate*)
Prec:	*connected(pos(), via), direction(pos(), dest) = via*
Post:	*pos() := dest*
Sensor:	**read-departure-screen**(*flight:Flight,gate():Gate*)
Sense:	*departure(flight)*
Prec:	*pos() = gate0*
Post:	*gate() := departure(flight)*
Sensor:	**read-sign**(*dest:Gate,via():Gate*)
Sense:	*direction(pos(), dest)*
Prec:	*∅*
Post:	*via() := direction(pos(), dest),*
	connected(pos(), via()), connected(via(), pos())
Init:	*pos() := gate0*
Goal:	*pos() = flight123*

Figure 4. *Airport* continual planning domain

read-departure-screen. This situation is described by the *Init:* statement of Figure 4, which defines the initial belief set $\{b_0\}$.

Using the PLANNER of Algorithm 3.3 the agent comes up with the following plan P_0:

```
read-departure-screen(flight123, gateA())
read-sign(gateA(), viaA())
follow-signs(gateA(), viaA())
board(flight123, gateA())
```

The first two actions in this plan are sensing actions, with fluents *gateA()* and *viaA()* acting as runtime variables that represent the values sensed in the remainder of the plan. The first action, *read-departure-screen(flight123, gate())* equates *gate()* with the departing gate for flight 123. Next, the agent will sense where to go next in direction of *gate()*. To enable this, the planning domain models "signs" with the function fluent *direction*. It is assumed to map pairs *(src,dest)* to the position closest to *src* on the way to *dest*. Thus, the second action, *read-sign(gateA(), viaA())*, determines the gate *via()* closest to the current position on the path to *gateA()*.

After the two sensing actions, *follow-signs(gateA(), viaA())* asserts that the agent will be able to find a path to *gateA()*, i.e. a subplan that will be continually developed after the agent has found out her concrete gate and while she is following the signs around the airport. Based on the postcondition of this assertion, the agent can safely plan to *board(flight123, gateA())* in the last step.

Having produced an initial plan that "hides" all contingencies and future plan variants, the agent can follow Algorithm 3.3 and switch to plan execution. When she physically executes *read-departure-screen(flight123)* she has reached belief state b_1 in the CP trace, in which she will know the actual departure gate. Let's assume this is *gate37*. The fluents *departure(flight123)* and *gateA()* both have this value in b_1. Likewise, when *follow-signs(gateA(), viaA())* is executed, *via()* is assigned a value, e.g. *gate1*, that corresponds to the one of *direction(gate0,gate37)*.

After their physical execution both sensing actions have been removed from the plan. We have reached belief state b_2 in the CP trace and the current plan looks like this (to clarify the presentation, the runtime variable fluents have been replaced with their values):

---

follow-signs(gate37, gate1)
board(flight123, gate37)

This plan, however, is no longer accepted by PLANISVALID, because it starts with an assertion. Thus, the agent enters a new planning phase, which produces the plan P_2:

---

move-to(gate1)
read-sign(gate37, viaA())
follow-signs(gate37, viaA())
board(flight123, gate37)

As can be seen, the original assertion *follow-signs(gate37, gate1)* has been expanded into a new subplan that starts with executable actions again. Note also that by deferring choices until the value of *departure(flight123)* was known, the continual planning agent avoided branching and can immediately commit to one specific more detailed plan now, namely the one having *gate37* as the boarding gate.

Since P_2 contains an assertion again, it can obviously still not be the final, fully executable plan in the CP trace. Most interestingly, P_2 uses the same assertion, *follow-signs*, as the previous plans *again*. It is obvious that the continual planner will *loop* through a cycle of producing similar plans and executing their prefixes until the agent has moved to a gate adjacent to *gate37*. This loop, however, is not explicit in the plan, as would be the case, e.g., in the approaches to planning with loops developed by Hector Levesque [Levesque 2005; Hu and Levesque 2009]. Instead, it is continually created by the continual planning algorithm through the interleaving of planning with acting. Again, this phenomenon can be described with an analogy from computer programming: The continual planning algorithm creates looping behavior through *recursion*.

5 Discussion and Future Work

The conventional approach to incomplete knowledge and sensing is to employ conditional plans (or policies) in order to deal with contingencies after sensing actions. Since this is computationally infeasible in many applications, we instead proposed to rely on linear plans and defer decisions as much as possible to the execution time when more information is available [Brenner and Nebel 2009]. The tool to delay such planning time decisions are *assertions*, which are basically abstract actions that can be expanded into subplans at execution time. On top of that, we propose here to employ *runtime variables* to deal with values that only become known at runtime. This devices permits us to deal with a number of scenarios introduced in the literature dealing with incomplete knowledge and sensing.

In order to formalize the relationship between plans interpreted on the symbolic level and plans that are executed and replanned while being executed, we developed formal execution models for the symbolic and physical execution and showed that the planning process is sound under some severe restrictions.

A number of questions remain unanswered at this point. In particular,

- we would like to determine the precise computational complexity for verifying and generating plans under the different symbolic execution models;

- we are interested in devising efficient data structures for representing the simplified belief set space;

- we intend to identify more properties of domain structures that guarantee soundness of abstract plans;

- and finally, we want to evaluate the proposed approach empirically..

Acknowledgements

This work has been supported by the European Commission as part of the Integrated Project *CogX* (FP7-ICT-2xo15181-CogX), by the German Research Foundation (DFG) as part of the collaborative research center SFB/TR-14 *AVACS*, and by the German Space Agency (DLR) as part of the project *KontiPlan*.

References

Ambros-Ingerson, J. A. and S. Steel [1988, August). Integrating planning, execution and monitoring. In *Proceedings of the 7th National Conference of the American Association for Artificial Intelligence (AAAI-88)*, Saint Paul, MI, pp. 83–88.

Bonet, B. and H. Geffner [2000]. Planning with incomplete information as heuristic search in belief space. In *Proceedings of the 5th International Conference on Artificial Intelligence Planning Systems (AIPS-00)*, pp. 52–61. AAAI Press, Menlo Park.

Brenner, M. and B. Nebel [2009]. Continual planning and acting in dynamic multiagent environments. *Journal of Autonomous Agents and Multiagent Systems 19*(3), 297–331.

Classen, J., P. Eyerich, G. Lakemeyer, and B. Nebel [2007]. Towards an integration of Golog and planning. In *Proceedings of the 20th International Joint Conference on Artificial Intelligence (IJCAI-07)*, pp. 1846–1851. AAAI Press.

De Giacomo, G., Y. Lesprance, H. Levesque, and S. Sardiña [2002, April]. On the semantics of deliberation in indigolog – from theory to implementation. In D. Fensel, F. Giunchiglia, D. McGuinness, and M. A. Williams (Eds.), *Proceedings of Eighth International Conference in Principles of Knowledge Representation and Reasoning (KR-2002)*, Toulouse, France, pp. 603–614. Morgan Kaufmann.

Erol, K., J. Hendler, and D. Nau [1996]. Complexity results for hierarchical task-network planning. *Annals of Mathematics and Artificial Intelligence 18*, 69–93.

Etzioni, O., K. Golden, and D. S. Weld [1997]. Sound and efficient closed-world reasoning for planning. *Artificial Intelligence 89*(1-2), 113–148.

Etzioni, O., S. Hanks, D. Weld, D. Draper, N. Lesh, and M. Williamson [1992]. An approach to planning with incomplete information. In *Principles of Knowledge Representation and Reasoning: Proceedings of the 3rd International Conference (KR-92)*, Cambridge, MA, pp. 115–125. Morgan Kaufmann.

Fagin, R., J. Y. Halpern, Y. Moses, and M. Y. Vardi [1995]. *Reasoning About Knowledge*. MIT Press.

Fritz, C. and S. A. McIlraith [2007]. Monitoring plan optimality during execution. In *Proceedings of the 17th International Conference on Automated Planning and Scheduling (ICAPS-07)*, Providence, Rhode Island, USA. Morgan Kaufmann.

Geffner, H. [2000]. Functional Strips: A more flexible language for planning and problem solving. In J. Minker (Ed.), *Logic-Based Artificial Intelligence*. Dordrecht, Holland: Kluwer.

Giacomo, G. D., Y. Lesperance, and H. J. Levesque [2000]. Congolog, a concurrent programming language based on the situation calculus. *Artificial Intelligence 121*(1-2), 109–169.

Golden, K. [1998]. Leap before you look: Information gathering in the PUCCINI planner. In *Proceedings of the 4th International Conference on Artificial Intelligence Planning Systems (AIPS-98)*, pp. 70–77.

Golden, K. and D. Weld [1996]. Representing sensing actions: The middle ground revisited. In *Principles of Knowledge Representation and Reasoning: Proceedings of the 5th International Conference (KR-96)*. Morgan Kaufmann.

Hoare, C. A. R. [1969, October]. An axiomatic basis for computer programming. *Communications of the ACM 12*, 576–580.

Hu, Y. and H. J. Levesque [2009]. Planning with loops: Some new results. In *Proceedings of the ICAPS 2009 Workshop on Generalized Planning: Macros, Loops, Domain Control. September 20th, 2009, Thessaloniki, Greece.*

Knoblock, C. A. [1995]. Planning, executing, sensing, and replanning for information gathering. In C. Mellish (Ed.), *Proc. the Fourteenth International Joint Conference on Artificial Intelligence*, San Francisco, pp. 1686–1693. Morgan Kaufmann.

Levesque, H. J. [1996]. What is planning in the presence of sensing? In *Proceedings of the 13th National Conference of the American Association for Artificial Intelligence (AAAI-96)*, pp. 1139–1146. MIT Press.

Levesque, H. J. [2005]. Planning with loops. In *Proceedings of the 19th International Joint Conference on Artificial Intelligence (IJCAI-05)*, Edinburgh, Scotland, UK, pp. 509–515. Professional Book Center.

Meyer, B. [1992, October). Applying "Design by Contract". *Computer 25*, 40–51.

Nau, D., Y. Cao, A. Lotem, and H. Munoz-Avila [1999, August). SHOP: Simple hierarchical ordered planner. In T. Dean (Ed.), *Proceedings of the 16th International Joint Conference on Artificial Intelligence (IJCAI-99)*, Stockholm, Sweden. Morgan Kaufmann.

Nau, D. S., T.-C. Au, O. Ilghami, U. Kuter, J. W. Murdock, D. Wu, and F. Yaman [2003]. SHOP2: An HTN planning system. *Journal of Artificial Intelligence Research 20*, 379–404.

Nebel, B. [2000]. On the compilability and expressive power of propositional planning formalisms. *Journal of Artificial Intelligence Research 12*, 271–315.

Petrick, R. and F. Bacchus [2002]. A knowledge-based approach to planning with incomplete information and sensing. In M. Ghallab, J. Hertzberg, and P. Traverso (Eds.), *Proceedings of the Sixth International Conference on Artificial Intelligence Planning Systems*, Toulouse, France. AAAI Press.

Petrick, R. P. A. and F. Bacchus [2004]. Extending the knowledge-based approach to planning with incomplete information and sensing. In S. Zilberstein, J. Koehler, and S. Koenig (Eds.), *Proceedings of the Fourteenth International Conference on Automated Planning and Scheduling (ICAPS 2004), June 3-7, 2004, Whistler, British Columbia, Canada*, pp. 2–11. AAAI Press.

Rintanen, J. [2004]. Complexity of planning with partial observability. In S. Zilberstein, J. Koehler, and S. Koenig (Eds.), *Proceedings of the Fourteenth International Conference on Automated Planning and Scheduling (ICAPS 2004), June 3-7, 2004, Whistler, British Columbia, Canada*, pp. 345–354. AAAI Press.

Russell, S. and P. Norvig [2003]. *Artificial Intelligence: A Modern Approach* (Second ed.). Englewood Cliffs, NJ: Prentice-Hall.

Weld, D. S., C. R. Anderson, and D. E. Smith [1998]. Extending Graphplan to handle uncertainty and sensing actions. In *Proceedings of the 15th National Conference of the American Association for Artificial Intelligence (AAAI-98)*, Madison, WI, pp. 897–904. MIT Press.

Yang, Q. [1997]. *Intelligent Planning: A decomposition and abstraction based approach*. Springer-Verlag.

7

Actions and Programs over Description Logic Knowledge Bases: A Functional Approach

DIEGO CALVANESE, GIUSEPPE DE GIACOMO, MAURIZIO LENZERINI, AND RICCARDO ROSATI

ABSTRACT. We aim at reasoning about actions and about high-level programs over knowledge bases (KBs) expressed in Description Logics (DLs). This is a critical issue that has resisted good, robust solutions for a long time. In particular, while well-developed theories of actions and high-level programs exist in AI, e.g., the ones based on the Situation Calculus, these theories do not apply to DL KBs, since these impose very rich state constraints (all the intensional part of the ontology itself). Here we propose a radical solution: we assume a Levesque's functional view of KBs and see them as systems that allow for two kinds of operations: ASK, which returns the (certain) answer to a query, and TELL, which produces a new knowledge base as the result of the application of an atomic action. In particular, we consider DL KBs formed by two components: a TBox, providing the intensional knowledge about the domain of interest, which we assume to be immutable over time; and an ABox, providing (incomplete) information at the extensional level, which we assume to be changed by atomic actions based on generic forms of instance level updates. The only requirement that we pose on such updates is that the resulting ABox is still expressible in the original DL language. We demonstrate the effectiveness of the approach by introducing Golog/ConGolog-like high-level programs on DL KBs, characterizing the notion of single-step executability of such programs, and devising nice methods for reasoning about sequences of actions generated by such programs. All our basic results are parametric wrt the specific DL language. Though for concreteness we present them using a particularly well behaved DL, namely $DL\text{-}Lite_{A,id}$.

1 Introduction

In this paper we look at reasoning about actions over Description Logic (DL) knowledge bases (KBs). In doing so, we merge two areas to which Hector Levesque has profoundly contributed: that of DLs, where the Brachman & Levesque's seminal paper on the trade-off between expressiveness and computational complexity in DLs, presented at AAAI'84 [Brachman and Levesque 1984], has shaped nearly all successive research in the field; and that of high-level programs directly based on logics, in particular Golog and ConGolog based on the Situation Calculus (SitCalc), where the work of Levesque and Reiter has demonstrated, possibly for the first time, the maturity of the area of reasoning about actions [Levesque et al. 1997; De Giacomo et al. 2000; Reiter 2001]. Interestingly, in order to make reasoning about actions over DL KBs feasible, we resort to a third foundational contribution by Hector Levesque: the so-called "functional view of knowledge bases" presented in [Levesque 1984], in which KBs are seen as sophisticated objects whose basic operation are "ASK", to extract knowledge from the KB, and "TELL", to update the knowledge in the KB, and where both operation are based on well characterized logical reasoning tasks.

In fact, research on reasoning about actions over DL KBs is currently of particular interest. Indeed, DL KBs [Baader et al. 2003] are generally advocated as the right tool to express ontologies, and this belief is one of the cornerstones of the Semantic Web [Smith et al. 2004; Horrocks et al. 2003]. Notably, semantic web services [Martin et al. 2004] constitute another cornerstone of the Semantic Web. These are essentially high-level descriptions of computations that abstract from the technological issues of the actual programs that realize them. An obvious concern is to combine in some way the static descriptions of the information provided by ontologies with the dynamic descriptions of the computations provided by semantic web services. However, such a critical issue has resisted good solutions for a long time, and even big efforts such as OWL-S [Martin et al. 2004] have not really succeeded.

In AI, the importance of combining static and dynamic knowledge has been recognized early [McCarthy 1962; McCarthy and Hayes 1969]. By now, well developed theories of actions and high-level programs, such as Levesque and others' Golog/ConGolog, exist. Note that high-level programs share with semantic web services the emphasis on abstracting from the technological issues of actual programs, and are indeed abstract descriptions of computations over a domain of interest. Unfortunately, these theories do not apply easily to general DL KBs, which impose a very rich kind of state constraints.

DL KBs are formed by two components, a TBox and an ABox. The TBox provides intensional knowledge about the domain of interest, expressed in terms of assertions about concepts, denoting sets (or classes) of individuals, and roles, denoting binary relationships (or associations) between individuals in such classes. The ABox provides facts asserting the membership of single individuals to classes or of pairs of individuals to roles. Such facts constitute an incomplete description of the information about the domain of interest at the extensional level. The TBox assertions in a DL KB typically do not provide *definitions* of concepts and roles, but only interrelations between them (cf. cyclic TBox interpreted according to the so-called descriptive semantics [Baader et al. 2003]). Such non-definitorial nature of DL KBs makes them one of the most difficult kinds of domain descriptions for reasoning about actions, since to come with them means to come with complex forms of state constraints that must be maintained over time [Baader et al. 2005; Liu et al. 2006a].

As mentioned, here we propose a radical solution: we assume a functional view [Levesque 1984] of KBs and see them as systems that allow for two kinds of operations: ASK, which returns the (certain) answer to a query over teh KB, and TELL, which produces a new KB as a result of the application of an atomic action. Observe that this approach, whose origins come from [De Giacomo et al. 1996; Giuseppe and Rosati 1999; Petrick and Bacchus 2004; van Riemsdijk et al. 2006], has some subtle limitations, due to the fact that we lose the possibility of distinguishing between "knowledge" and "truth" as pointed out in [Sardiña et al. 2006]. On the other hand, it has a major advantage: it decouples reasoning on the static knowledge from reasoning on the dynamics of the computations over such knowledge. As a result, we gain the ability of lifting to DLs many of the results developed in reasoning about actions in the years.

We demonstrate such an approach in this paper. Specifically, we assume the TBox of a KB to be immutable over time, while the ABox might be changed by atomic actions used by the TELL operation and based on generic forms of instance level updates. The only requirement that we pose on such updates is that the resulting ABox is still expressible in the original KB language. Building on this functional view, we introduce Golog/ConGolog-like high level programs over DL KBs, we characterize the notion of single-step executability of such programs, and we devise methods for reasoning about sequences of actions generated by such programs. All our basic results are parametric with respect to the specific DL language. Though for concreteness we present them using a particularly well behaved DL, namely *DL-Lite$_{A,id}$*, which is an expressive member of the *DL-Lite* family [Calvanese et al. 2007b], a family of DLs that enjoys particularly nice computational properties when reasoning about knowledge at the instance level. We stress that this paper is really an illustration of what a functional view on KBs can bring about in combining static and dynamic aspects in the context of DL KBs, and that many extensions of this work can be investigated (we will mention some of them in the conclusions).

2 Preliminaries

DL ontologies. Description Logics (DLs) [Baader et al. 2003] are knowledge representation formalisms that are tailored for representing the domain of interest in terms of *concepts* (or classes), which denote sets of objects, and *roles* (or relations), which denote denote binary relations between objects. DLs *knowledge bases* (KBs) are based on an alphabet of object, concept, and role symbols, and are formed by two distinct parts: the so-called *TBox*, which represents the *intensional level* of the KB, and contains an intensional description of the domain of interest; and the so-called *ABox*, which represents the *instance level* of the KB, and contains extensional information.

We give the semantics of a DL KB in terms of interpretations over a fixed infinite domain Δ of objects. We assume to have one constant in the alphabet for each object in Δ denoting exactly that object. In this way we blur the distinction between constants and objects, so that we can use them interchangeably (with a little abuse of notation), without causing confusion (cf. standard names [Levesque and Lakemeyer 2001]).

An *interpretation* $\mathcal{I} = \langle \Delta, \cdot^{\mathcal{I}} \rangle$ consists of a first order structure over Δ, and an interpretation function $\cdot^{\mathcal{I}}$, mapping each concept to a subset of Δ and each role to a subset of $\Delta \times \Delta$. We say that $\mathcal{I}$ is a *model of a (TBox or ABox) assertion* α, or also that $\mathcal{I}$ *satisfies* α, if α is true in $\mathcal{I}$. We say that $\mathcal{I}$ is a *model of the KB* $\mathcal{K} = \langle \mathcal{T}, \mathcal{A} \rangle$, or also that $\mathcal{I}$ *satisfies* $\mathcal{K}$, if $\mathcal{I}$ is a model of all the assertions in $\mathcal{T}$ and $\mathcal{A}$. Given a set $\mathcal{S}$ of (TBox or ABox) assertions, we denote by $Mod(\mathcal{S})$ the set of interpretations that are models of all assertions in $\mathcal{S}$. In particular, the set of *models of* $\mathcal{K}$, denoted $Mod(\mathcal{K})$, is the set of models of all assertions in $\mathcal{T}$ and $\mathcal{A}$, i.e., $Mod(\mathcal{K}) = Mod(\langle \mathcal{T}, \mathcal{A} \rangle) = Mod(\mathcal{T} \cup \mathcal{A})$. A KB $\mathcal{K}$ is *consistent* if $Mod(\mathcal{K}) \neq \emptyset$, i.e., it has at least one model. We say that a KB $\mathcal{K}$ *logically implies* an expression α (e.g., an assertion, an instantiated conjunctive query (CQ), or an instantiated union of conjunctive queries (UCQ), etc.), written $\mathcal{K} \models \alpha$, if for every interpretation $\mathcal{I} \in Mod(\mathcal{K})$, we have that $\mathcal{I} \in Mod(\alpha)$, i.e., all the models of $\mathcal{K}$ are also models of α.

When dealing with queries, we are interested in *query answering* (for CQs and UCQs): given a KB $\mathcal{K}$ and a query $q(\vec{x})$ over $\mathcal{K}$, return the *certain answers* to $q(\vec{x})$ over $\mathcal{K}$, i.e., all tuples $\vec{t}$ of elements of Δ such that $\mathcal{K} \models q(\vec{t})$, where $q(\vec{t})$ denotes the query obtained from $q(\vec{x})$ by substituting $\vec{x}$ with $\vec{t}$.

DL-Lite$_{A,id}$. The *DL-Lite* family [Calvanese et al. 2007b] is a family of low complexity DLs particularly suited for dealing with KBs with very large ABoxes, and forms the basis of OWL 2 QL, one of the profiles of OWL 2, the official ontology specification language of the World-Wide-Web Consortium (W3C)[1].

We now present the DL *DL-Lite*$_{A,id}$ [Calvanese et al. 2008], which is the most expressive logic in the family. Expressions in *DL-Lite*$_{A,id}$ are formed according to the following syntax:

$$
\begin{aligned}
B &\longrightarrow A \mid \exists Q \mid \delta(U) & E &\longrightarrow \rho(U) \\
C &\longrightarrow B \mid \neg B & F &\longrightarrow \top_D \mid T_1 \mid \cdots \mid T_n \\
Q &\longrightarrow P \mid P^- & V &\longrightarrow U \mid \neg U \\
R &\longrightarrow Q \mid \neg Q &&
\end{aligned}
$$

where A, P, and U denote respectively an atomic concept name, an atomic role name, and an attribute name, $T_1, \ldots, T_n$ are the value-domains allowed in the logic (which correspond to the data types adopted by Resource Description Framework (RDF)[2]), $\top_D$ denotes the union of all domain values, P^- denotes the inverse of P, $\exists Q$ denotes the objects related to some object by the role Q, $\neg$ denotes negation of concepts, roles, or attributes, $\delta(U)$ denotes the *domain* of U, i.e., the set of objects that U relates to values, and $\rho(U)$ denotes the *range* of U, i.e., the set of values to which U relates objects.

A *DL-Lite*$_{A,id}$ TBox $\mathcal{T}$ contains intensional assertions of three types, namely inclusion assertions, functionality assertions, and identification assertions (IDs) [Calvanese et al. 2008]. More precisely, *DL-Lite*$_{A,id}$ assertions are of the form:

$$
\begin{aligned}
& B \sqsubseteq C && \textit{concept inclusion assertion} \\
& E \sqsubseteq F && \textit{value-domain inclusion assertion} \\
& Q \sqsubseteq R && \textit{role inclusion assertion} \\
& (\mathsf{funct}\ Q) && \textit{role functionality assertion} \\
& (\mathsf{funct}\ U) && \textit{attribute functionality assertion} \\
& (\mathsf{id}\ B\ \pi_1, ..., \pi_n) && \textit{identification assertions}
\end{aligned}
$$

In the identification assertions, π denotes a *path*, which is an expression built according to the following syntax rule:

$$
\pi \longrightarrow S \mid B? \mid \pi_1 \circ \pi_2
$$

where S denotes an atomic role, the inverse of an atomic role, or an atomic attribute, $\pi_1 \circ \pi_2$ denotes the composition of the paths π_1 and π_2, and $B?$, called *test relation*, represents the identity relation on instances of the concept B. The *length* of a path is inductively defined as follows: the length of a path whose form is S or $B?$ is 1; the length of a path of the form $\pi_1 \circ \pi_2$ is the sum of the lengths of π_1 and of π_2. In *DL-Lite*$_{A,id}$, identification assertions are *local*, i.e., at least one $\pi_i \in \{\pi_1, ..., \pi_n\}$ has length 1. In what follows, we only refer to IDs which are local.

---

[1] http://www.w3.org/TR/2008/WD-owl2-profiles-20081008/
[2] http://www.w3.org/RDF/

A concept inclusion assertion specifies that a (basic) concept B is subsumed by a (general) concept C. Analogously for the other types of inclusion assertions. Attribute functionality assertions are used to impose that attributes are actually functions from objects to domain values. Finally, an ID (id B $\pi_1, ..., \pi_n$) asserts that for any two different instances a, b of B, there is at least on π_i such that a and b differ in the set of their π_i-fillers.

In order to guarantee the good computational properties of the DLs of the *DL-Lite* family, a *DL-Lite*$_{A,id}$ TBox $\mathcal{T}$ has to satisfy the following conditions:

- for each atomic role P, if either (funct P) or (funct P^-) occur in $\mathcal{T}$, then $\mathcal{T}$ does not contain assertions of the form $Q \sqsubseteq P$ or $Q \sqsubseteq P^-$, where Q is a basic role;

- for each ID α in $\mathcal{T}$, every role that occurs (in either direct or inverse direction) in a path of α, does not appear in the right-hand side of assertions of the form $Q \sqsubseteq Q'$.

Intuitively, these conditions say that, in *DL-Lite*$_{A,id}$ TBoxes, roles occurring in functionality or identification assertions cannot be specialized.

We observe that *DL-Lite*$_{A,id}$ is able to capture all essential features of conceptual modeling formalisms, such as UML Class Diagrams or Entity-Relationship schemas, with the notable exception of *covering* constraints in generalization hierarchies, which would require the introduction of disjunction. Indeed, it can be shown that, if disjunction was added to *DL-Lite*$_{A,id}$, then query answering would become intractable with respect to the size of the ABox [Calvanese et al. 2006].

A *DL-Lite*$_{A,id}$ ABox $\mathcal{A}$ is a finite set of assertions of the form $A(a)$, $P(a,b)$, and $U(a,v)$, where A, P, and U are as above, a and b are object constants, and v is a value constant.

Answering *EQL-Lite*(UCQ) queries over *DL-Lite*$_{A,id}$ knowledge bases. As query language, here we consider *EQL-Lite*(UCQ) [Calvanese et al. 2007a]. This language is essentially formed by full (domain-independent) FOL query expressions built on top of atoms that have the form $\mathbf{K}\alpha$, where α is a union of conjunctive queries[3]. The operator $\mathbf{K}$ is a minimal knowledge operator [Levesque 1984; Reiter 1990; Levesque and Lakemeyer 2001], which is used to formalize the epistemic state of the KB. Informally, the formula $\mathbf{K}\alpha$ is read as "α is known to hold" or "α is logically implied by the knowledge base".

Note that answering *EQL-Lite*(UCQ) queries over *DL-Lite*$_{A,id}$ KBs $\mathcal{K} = \langle \mathcal{T}, \mathcal{A} \rangle$ is LOGSPACE with respect to the size of $\mathcal{A}$. Notably, query answering can be reduced to evaluating (pure) FOL queries over the ABox, considered as a database. We refer to [Calvanese et al. 2007a] for more details.

DL instance-level update and erasure. Besides answering queries, Description Logic systems should be able to cope with the *evolution of the KB*. There are two types of evolution operators, corresponding to inserting and deleting chunks of knowledge, respectively. In the case of insertion, the aim is to incorporate new knowledge into the KB, and the corresponding operator should be defined in such a way to compute a consistent KB that supports the new knowledge. In the case of deletion, the aim is to come up with a consistent KB where the retracted knowledge is not valid. Many recent papers demonstrate that the interest towards a well-defined approach to KB evolution is growing significantly [Flouris et al. 2008; Liu et al. 2006b; De Giacomo et al. 2009; Wang et al. 2010; Calvanese et al. 2010].

Following the tradition of the work on knowledge revision and update [Katsuno and Mendelzon 1991], all the above papers advocate some minimality criterion in the changes

---

[3]For queries consisting of only one atom $\mathbf{K}\alpha$, the $\mathbf{K}$ operator is omitted.

of the KB that must be undertaken to realize the evolution operations. In other words, the need is commonly perceived of keeping the distance between the original KB and the KB resulting from the application of an evolution operator minimal. There are two main approaches to define such a distance, called *model-based* and *formula-based*, respectively. In the model-based approaches, the result of an evolution operation applied to the KB $\mathcal{K}$ is defined in terms of a set of models, with the idea that such a set should be as close as possible to the models of $\mathcal{K}$. One basic problem with this approach is to characterize the language needed to express the KB that exactly captures the resulting set of models. Conversely, in the formula-based approaches, the result is explicitly defined in terms of a formula, by resorting to some minimality criterion with respect to the formula expressing $\mathcal{K}$. Here, the basic problem is that the formula constituting the result of an evolution operation is not unique in general.

Virtually all model-based approaches suffer from the expressibility problem (see, e.g., [Liu et al. 2006b; De Giacomo et al. 2009; Calvanese et al. 2010]). For this reason, we adopt here a formula-based approach, inspired in particular by the work developed in [Fagin et al. 1983] for updating logical theories. As in [Fagin et al. 1983], we consider both insertions and deletions, but we limit our attention to insertions and deletions of ABox assertions. In other words, we consider the evolution of the ABox of a KB under an invariant TBox, on the basis of the fact that, in many applications, the TBox represents a stable representation of the intensional knowledge about the domain. The specific approach we take is inspired by two recent methods proposed for the evolution of KBs expressed in the *DL-Lite* family [Calvanese et al. 2010; Lenzerini and Savo 2011]. Both methods follow the formula-based approach, and guarantee that the result of the evolution of a KB is always expressible in the DL used to specify the original KB.

3 Atomic actions

Under Levesque's functional view, KBs are seen as systems that are able to perform two basic kinds of operations, namely ASK and TELL operations [Levesque 1984; Levesque and Lakemeyer 2001]:

- ASK: given a KB and a *query* (in the query language recognized by the KB), returns a *finite* set of tuples of objects (constituting the answers to the query over the KB).

- TELL: given a KB and an *atomic action*, returns a new KB resulting from executing the action, if the action is executable wrt the given KB.

We consider KBs expressed in arbitrary DLs languages, except that we require that they are able to deal with *EQL-Lite*(UCQ) as query language. This implies that they need to be able to handle certain answers of unions of conjunctive queries in a decidable/effective way. With this assumption in place, we base ASK on certain answers to *EQL-Lite*(UCQ). Specifically, we denote by $q(\vec{x})$ an (*EQL-Lite*(UCQ)) query with distinguished variables $\vec{x}$. For a KB $\mathcal{K}$, we define $\text{ASK}(q(\vec{x}), \mathcal{K}) = \{\vec{t} \mid \mathcal{K} \models q(\vec{t})\}$, where $\vec{t}$ denotes a tuple of constants of the same arity as $\vec{x}$. We denote by ϕ queries with no distinguished variables. Such queries are called *boolean* queries and return either *true* (i.e., the empty tuple) or *false* (i.e., no tuples at all).

As for TELL, we base atomic actions on instance level update and erasure [De Giacomo et al. 2006; De Giacomo et al. 2007]. Specifically, we allow for *atomic actions* of the form

$$\textbf{update}_{op} \; L(\vec{x}) \; \textbf{where} \; q(\vec{x})$$

where $q(\vec{x})$ stands for a query with $\vec{x}$ as distinguished variables, $L(\vec{x})$ stands for a set of membership assertions on constants and variables in $\vec{x}$, and **update**$_{op}$ is an update operator that makes use of $L(\vec{x})$ to update the KB.

We allow for several update operators, for various forms of update and erasure [Katsuno and Mendelzon 1991; Eiter and Gottlob 1992]. However, we require that by applying one such update operator **update**$_{op}$ to a KB $\mathcal{K} = (\mathcal{T}, \mathcal{A})$ we get a single new KB $\mathcal{K}' = (\mathcal{T}, \mathcal{A}')$ with $\mathcal{A}'$ still in the same DL language of $\mathcal{K}$, or we fail. The new ABox is defined as $\mathcal{A}' = f_{op}(L(\vec{x}), q(\vec{x}), \mathcal{K})$, where f_{op} characterizes the semantics of the update operator **update**$_{op}$. If for any reason $f_{op}(L(\vec{x}), q(\vec{x}), \mathcal{K})$ is not defined, then the update fails, and the action is not executable.

Requiring that the result of the update is still a single KB in the same language as the original one is somehow a severe restriction, since we know that most classical knowledge update operators [Katsuno and Mendelzon 1991; Eiter and Gottlob 1992] applied to DL KBs produce results that are not expressible in the DL of the original KB [De Giacomo et al. 2006; De Giacomo et al. 2007]. On the other hand, from a pragmatical point of view, such an assumption is essential, since it guarantees that by applying the update we can still use the reasoning techniques/algorithms that we used for the original KB. Dropping such assumption would have a disruptive effect on the system: the reasoning techniques in the various states of our system would need to be different.[4]

Let $\mathcal{K} = (\mathcal{T}, \mathcal{A})$, then we define:

$$\text{TELL}([\textbf{update}_{op} \, L(\vec{x}) \, \textbf{where} \, q(\vec{x})], \mathcal{K}) = \begin{cases} \bot, & \text{if } f_{op}(L(\vec{x}), q(\vec{x}), \mathcal{K}) \text{ is undefined} \\ (\mathcal{T}, f_{op}(L(\vec{x}), q(\vec{x}), \mathcal{K})), & \text{otherwise} \end{cases}$$

If $\bot$ is returned by the TELL operation, we say that the atomic action a is *not executable* in $\mathcal{K}$. We extend ASK with expressions of the form $\text{ASK}([executable(a)], \mathcal{K})$, so as to be able to check executability of actions. Observe that the executability of actions as defined above can indeed be checked on the KB.

We close the section by discussing these notions on *DL-Lite*$_{A,id}$ KBs. First *DL-Lite*$_{A,id}$ allows for computing certain answers of *EQL-Lite*(UCQ) in LOGSPACE in data complexity (as in relational databases) and PTIME in KB complexity, making ASK particularly effective (we assume the size of the query fixed). As for TELL we consider two forms of updates in *DL-Lite*$_{A,id}$: **update**$_{add}$ is based on the notion of update mentioned above, and **update**$_{erase}$ is based on the notion of erasure. Both these operation can be done in polynomial time wrt the KB, and hence also TELL is polynomial (again we assume the size of the query fixed).

4 Programs

We now consider how atomic actions can be organized within a program. In particular, we focus on a variant of Golog [Levesque et al. 1997; De Giacomo et al. 2000; Sardiña et al. 2004] tailored to work on KBs. Instead of situations, we consider KBs, or, to be more precise, KB states. We recall that when considering KBs we assume the TBox to be invariant, so the only part of the KB that can change as a result of an action (or a program) is the ABox.

---

[4]In fact one aspect of this assumption could be dropped: the fact that the update results in a *single* resulting ABox. We could possibly assume that the resulting ABoxes could be many. Now if they could be finitely many, then the results here could be easily extended, if they could be infinitely many, then more work needs to be done.

While all constructs of the original Golog/ConGolog have a counterpart in our variant, here for brevity we concentrate on a core fragment only, namely:

a	atomic actions
ϵ	the empty sequence of actions
$\delta_1; \delta_2$	sequential composition
if ϕ **then** δ_1 **else** δ_2	if-then-else
while ϕ **do** δ	while
pick $q(\vec{x}).\delta[\vec{x}]$	pick

where a is an atomic instruction that corresponds to the execution of the atomic action a; ϵ is an empty sequence of instructions (needed for technical reasons); **if** ϕ **then** δ_1 **else** δ_2 and **while** ϕ **do** δ are the standard constructs for conditional choice and iteration, where the test condition is a boolean query (or an executability check) to be asked to the current KB; finally, **pick** $q(\vec{x}).\delta[\vec{x}]$ picks a tuple $\vec{t}$ in the answer to $q(\vec{x})$, instantiates the rest of the program δ by substituting $\vec{x}$ with $\vec{t}$ and executes δ. The latter construct is a variant of the pick construct in Golog: the main difference being that $\vec{t}$ is bounded by a query to the KB. Also, while in Golog such a choice is nondeterministic, here we think of it as possibly made *interactively*, see below.

The general approach we follow is the *structural operational semantics* approach based on defining a single step of program execution [Plotkin 1981; De Giacomo et al. 2000]. This single-step semantics is often called *transition semantics* or *computation semantics*. Namely, to formally define the semantics of our programs we make use of a *transition relation*, named *Trans*, and denoted by "⟶":

$$(\delta, \mathcal{K}) \xrightarrow{a} (\delta', \mathcal{K}')$$

where δ is a program, $\mathcal{K}$ is a KB in which the program is executed, a is the executed atomic action, $\mathcal{K}'$ is the KB obtained by executing a in $\mathcal{K}$, and δ' is what remains to be executed of δ after having executed a.

We also make use of a *final predicate*, named *Final*, and denoted by "$\cdot\sqrt{}$":

$$(\delta, \mathcal{K})^{\sqrt{}}$$

where δ is a program that can be considered (successfully) terminated with the KB $\mathcal{K}$.

Such a relation and predicate can be defined inductively in a standard way, using the so called *transition (structural) rules*. The structural rules for defining the transition relation and the final predicate are given in Figure 1 and Figure 2 respectively. All structural rules have the following schema:

$$\frac{\text{CONSEQUENT}}{\text{ANTECEDENT}} \text{ if SIDE-CONDITION}$$

which is to be interpreted logically as:

$$\forall(\text{ANTECEDENT} \wedge \text{SIDE-CONDITION} \rightarrow \text{CONSEQUENT})$$

$$act: \quad \frac{(a, \mathcal{K}) \xrightarrow{a} (\epsilon, \text{TELL}(a, \mathcal{K}))}{true} \quad \text{if } a \text{ is executable in } \mathcal{K}$$

$$seq: \quad \frac{(\delta_1; \delta_2, \mathcal{K}) \xrightarrow{a} (\delta_1'; \delta_2, \mathcal{K}')}{(\delta_1, \mathcal{K}) \xrightarrow{a} (\delta_1'; \mathcal{K}')} \qquad \frac{(\delta_1; \delta_2, \mathcal{K}) \xrightarrow{a} (\delta_2', \mathcal{K}')}{(\delta_2, \mathcal{K}) \xrightarrow{a} (\delta_2'; \mathcal{K}')} \quad \text{if } (\delta_1, \mathcal{K})^\checkmark$$

$$if: \quad \frac{(\textbf{if } \phi \textbf{ then } \delta_1 \textbf{else } \delta_2, \mathcal{K}) \xrightarrow{a} (\delta_1', \mathcal{K}')}{(\delta_1, \mathcal{K}) \xrightarrow{a} (\delta_1', \mathcal{K}')} \quad \text{if } \text{ASK}(\phi, \mathcal{K}) = true$$

$$\frac{(\textbf{if } \phi \textbf{ then } \delta_1 \textbf{else } \delta_2, \mathcal{K}) \xrightarrow{a} (\delta_2', \mathcal{K}')}{(\delta_2, \mathcal{K}) \xrightarrow{a} (\delta_2', \mathcal{K}')} \quad \text{if } \text{ASK}(\phi, \mathcal{K}) = false$$

$$while: \quad \frac{(\textbf{while } \phi \textbf{ do } \delta, \mathcal{K}) \xrightarrow{a} (\delta'; \textbf{while } \phi \textbf{ do } \delta, \mathcal{K}')}{(\delta, \mathcal{K}) \xrightarrow{a} (\delta', \mathcal{K}')} \quad \text{if } \text{ASK}(\phi, \mathcal{K}) = true$$

$$pick: \quad \frac{(\textbf{pick } q(\vec{x}).\ \delta[x], \mathcal{K}) \xrightarrow{a} (\delta'[\vec{t}], \mathcal{K}')}{(\delta[\vec{t}], \mathcal{K}) \xrightarrow{a} (\delta'[\vec{t}], \mathcal{K}')} \quad \textit{(for } \vec{t} = \text{CHOICE}[\text{ASK}(q(\vec{x}), \mathcal{K})])$$

Figure 1. Transition rules

where $\forall Q$ stands for the universal closure of all free variables occurring in Q, and, typically, ANTECEDENT, SIDE-CONDITION, and CONSEQUENT share free variables. The structural rules define inductively a relation, namely *the smallest relation satisfying the rules*.

Observe the use of the parameter CHOICE, which denotes a choice function, to determine the tuple to be picked in executing the pick constructs of programs. More precisely, CHOICE stands for any function, depending on an arbitrary number of parameters, returning a tuple from the set $\text{ASK}(q(\vec{x}), \mathcal{K})$. In the original Golog/ConGolog proposal [Levesque et al. 1997; De Giacomo et al. 2000] such a choice function (there also extended to other nondeterministic constructs) is implicit, the idea there being that Golog executions use a choice function that would lead to the termination of the program (angelic nondeterminism). In [Sardiña et al. 2004], a choice function is also implicit, but based on the idea that choices are done randomly (devilish nondeterminism). Here, we make use of choice functions explicitly, so as to have control on nondeterministic choices. Indeed, one interesting use of CHOICE is to model the delegation of choices to the client of the program, with the idea that the pick construct is interactive: it presents the result of the query to the client, who chooses the tuple s/he is interested in. For example, if the query is about hotels that are available in Rome, the client sees the list of available hotels resulting from the query and chooses the one s/he likes most.[5] We say that a program is *deterministic* when no pick instructions are present or a fixed choice function for CHOICE is considered.

---

[5]Note that, since the query q is an *EQL-Lite*(UCQ) query, it is range restricted by definition (cf. [Calvanese et al. 2007a]).

$$\epsilon : \quad \frac{(\epsilon, \mathcal{K})^{\checkmark}}{true} \qquad\qquad seq : \quad \frac{(\delta_1; \delta_2, \mathcal{K})^{\checkmark}}{(\delta_1, \mathcal{K})^{\checkmark} \wedge (\delta_2; \mathcal{K})^{\checkmark}}$$

$$if : \quad \frac{(\mathbf{if}\ \phi\ \mathbf{then}\ \delta_1 \mathbf{else}\ \delta_2, \mathcal{K})^{\checkmark}}{(\delta_1, \mathcal{K})^{\checkmark}} \quad \text{if } \mathrm{ASK}(\phi, \mathcal{K}) = true$$

$$\frac{(\mathbf{if}\ \phi\ \mathbf{then}\ \delta_1 \mathbf{else}\ \delta_2, \mathcal{K})^{\checkmark}}{(\delta_2, \mathcal{K})^{\checkmark}} \quad \text{if } \mathrm{ASK}(\phi, \mathcal{K}) = false$$

$$while : \quad \frac{(\mathbf{while}\ \phi\ \mathbf{do}\ \delta, \mathcal{K})^{\checkmark}}{true} \quad \text{if } \mathrm{ASK}(\phi, \mathcal{K}) = false$$

$$\frac{(\mathbf{while}\ \phi\ \mathbf{do}\ \delta, \mathcal{K})^{\checkmark}}{(\delta, \mathcal{K})^{\checkmark}} \quad \text{if } \mathrm{ASK}(\phi, \mathcal{K}) = true$$

$$pick : \quad \frac{(\mathbf{pick}\ q(\vec{x}).\ \delta[\vec{x}], \mathcal{K})^{\checkmark}}{(\delta[\vec{t}], \mathcal{K})^{\checkmark}} \quad (for\ \vec{t} = \mathrm{CHOICE}[\mathrm{ASK}(q(\vec{x}), \mathcal{K})])$$

Figure 2. Final rules

Examples. Let us look at some simple examples of programs. Consider the KB on companies and grants shown in Figure 3, also depicted graphically in Figure 4.

The first program we write aims at populating the concept *IllegalOwner* with those companies that own themselves, either directly or indirectly. We assume *temp* to be an additional role in the alphabet of the TBox. Then, the following deterministic program `ComputeIllegalOwners` can be used to populate *IllegalOwner*:

```
ComputeIllegalOwners =
  UPDATEerase temp(x1,x2) where q(x1,x2) <- temp(x1,x2);
  UPDATEerase IllegalOwner(x) where q(x) <- IllegalOwner(x);
  UPDATEadd temp(x1,x2) where q(x1,x2) <- owns(x1,x2);
  while (q() <- K(temp(y1,z), owns(z,y2)), not K(temp(y1,y2))) do (
    UPDATEadd temp(x1,x2) where
      q(x1,x2) <- K(temp(x1,z), owns(z,x2)), not K(temp(x1,x2))
  );
  UPDATEadd IllegalOwner(x) where q(x) <- temp(x,x)
```

The second program we look at is a program that, given a research group r and a company c, interactively—through a suitable choice function for CHOICE—selects a public company owned by c to ask a grant to; if c does not own public companies, then it selects the company c itself:

```
askNewGrant(r,c) =
  if (q() <- owns(c,y), PublicCompany(y)) then (
    pick (q(x) <- owns(c,x), PublicCompany(x)). (
```

$$
\begin{aligned}
\exists owns &\sqsubseteq Company \\
\exists owns^- &\sqsubseteq Company \\
(\text{funct } owns^-) & \\
PublicCompany &\sqsubseteq Company \\
PrivateCompany &\sqsubseteq Company \\
PublicCompany &\sqsubseteq \neg PrivateCompany \\
IllegalOwner &\sqsubseteq Company \\
\exists grantAsked &\sqsubseteq ResearchGroup \\
\exists grantAsked^- &\sqsubseteq Company \\
ResearchGroup &\sqsubseteq \exists grantAsked \\
\exists belongsTo &\sqsubseteq ResearchGroup \\
\exists belongsTo^- &\sqsubseteq ResearchDept \\
ResearchGroup &\sqsubseteq \exists belongsTo \\
(\text{funct } belongsTo) & \\
\delta(rid) &\sqsubseteq ResearchGroup \\
\rho(rid) &\sqsubseteq String \\
(\text{id } ResearchGroup\ rid, belongsTo) &
\end{aligned}
$$

Figure 3. A simple *DL-Lite*$_{A,id}$ TBox

```
UPDATEadd grantAsked(r,x) where true
  )
)
else UPDATEadd grantAsked(r,c) where true
```

Finally, consider the case of a program that erases a pair (g, d) from the *belongsTo* role, where g is an instance of *ResearchGroup* and d an instance of *ResearchDept*. Since *belongsTo* is a functional role, in the models of $\mathcal{K}$ before the erasure operation, there cannot be any other object d' connected to g via *belongsTo*. However, the inclusion assertion *ResearchGroup* $\sqsubseteq \exists belongsTo$ in the TBox "enforces" in all models the existence of such an object, possibly an existentially implied one. Notice also that this does not lead in any way to a violation of the ID (id *ResearchGroup rid, belongsTo*).

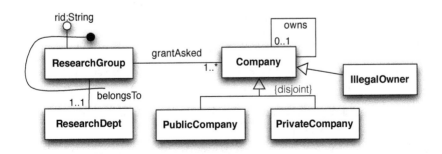

Figure 4. The *DL-Lite*$_{A,id}$ TBox of Figure 3 rendered as a UML Class Diagram

5   Results

In this section, we assume that KBs are expressed in *DL-Lite*$_{A,id}$ and that the ASK and TELL operations are defined for *DL-Lite*$_{A,id}$ using query answering, update (add), and erasure discussed in Section 2 and Section 3.

Given a KB $\mathcal{K}$ and a program δ, we define the set *next step*, denoted by *Next*, as:

$$Next(\delta, \mathcal{K}) = \{\langle a, \delta', \mathcal{K}' \rangle \mid (\delta, \mathcal{K}) \xrightarrow{\ a\ } (\delta', \mathcal{K}')\}$$

The following two theorems tell us that programs are indeed computable.

THEOREM 1. *Let $\mathcal{K}$ be a KB and δ a program. Then, the set $Next(\delta, \mathcal{K})$ has a finite cardinality, and can be computed in polynomial time in $\mathcal{K}$ and δ (considering the size of the queries in δ fixed). Moreover, if δ is deterministic then, for each action a, the number of tuples $\langle a, \delta', \mathcal{K}' \rangle \in Next(\delta, \mathcal{K})$ is at most one (one if a is executable, zero otherwise).*

THEOREM 2. *Let $\mathcal{K}$ be a KB and δ a program. Then, checking $(\delta, \mathcal{K})^{\surd}$ can be done in polynomial time in $\mathcal{K}$ and δ (considering the size of the queries in δ fixed).*

Given a KB $\mathcal{K}_0$ and a sequence $\rho = a_1 \cdots a_n$ of actions, we say that ρ is a *run* of a program δ_0 over the KB $\mathcal{K}_0$ if there are $(\delta_i, \mathcal{K}_i)$, for $i = 1, \ldots, n$, such that

$$(\delta_0, \mathcal{K}_0) \xrightarrow{\ a_1\ } (\delta_1, \mathcal{K}_1) \xrightarrow{\ a_2\ } \cdots \xrightarrow{\ a_n\ } (\delta_n, \mathcal{K}_n)$$

We call δ_n and $\mathcal{K}_n$ above respectively the program and the KB resulting from the run ρ. If $(\delta_n, \mathcal{K}_n)$ is final (i.e., $(\delta_n, \mathcal{K}_n)^{\surd}$), then we say that ρ is a *terminating run*. Note that, if the program δ_0 is deterministic, then $(\delta_n, \mathcal{K}_n)$ is functionally determined by $(\delta_0, \mathcal{K}_0)$ and ρ.

THEOREM 3. *Let $\mathcal{K}_0$ be a KB, δ_0 a deterministic program, and $\rho = a_1 \cdots a_n$ a sequence of actions. Then checking whether ρ is a run of δ_0 starting from $\mathcal{K}_0$ can be done in polynomial time in the size of $\mathcal{K}_0$, ρ, and δ_0 (considering the size of the queries in δ_0 fixed)*

THEOREM 4. *Let $\mathcal{K}_0$ be a KB, δ_0 a deterministic program, and ρ a run of δ_0 starting from $\mathcal{K}_0$. Then, computing the resulting program δ_n and the resulting KB $\mathcal{K}_n$, as well as checking $(\delta_n, \mathcal{K}_n)^{\surd}$ and computing a query $q(\vec{x})$ over $\mathcal{K}_n$, can be done in polynomial time in the size of $\mathcal{K}_0$, ρ, and δ_0 (considering the size of the queries in δ_0 fixed).*

For nondeterministic programs, i.e., when we do not fix a choice function for CHOICE, Theorems 3 and 4 do not hold anymore. Indeed, it can be shown the problems in the theorems become NP-complete.

We conclude this section by turning to the two classical problem in reasoning about actions, namely the executability problem and the projection problem [Reiter 2001]. In our setting such problems are phrased as follows:

- *executability problem*: check whether a sequence of actions is executable in a KB;

- *projection problem*: compute the result of a query in the KB obtained by executing a sequence of actions in an initial KB.

Now, considering that a sequence of actions can be seen as a simple deterministic program, from the theorems above we get the following result:

THEOREM 5. *Let $\mathcal{K}_0$ be a KB and ρ a sequence of actions. Then, checking the executability of ρ in $\mathcal{K}_0$, and computing the result of a query $q(\vec{x})$ over the KB obtained by executing ρ in $\mathcal{K}_0$, can both be done in polynomial time in the size of $\mathcal{K}_0$ and ρ.*

In fact, all the above results can be immediately extended (with different complexity bounds) to virtually every DL and associated ASK and TELL operations, as long as ASK and TELL are decidable and conform to the requirements mentioned in Section 3.

6 Conclusion

In this paper we have laid the foundations for an effective approach to reasoning about actions and programs over KBs, based on Levesque's functional view of the KB. Namely, the KB is seen as a system that can perform two kinds of operations: ASK and TELL. We have focused on *DL-Lite*, but the approach applies to more expressive DLs. It suffices to have a decidable ASK, i.e., decidable query answering on the chosen query and KB languages, and a decidable TELL, i.e., define atomic actions so that, through their effects, they produce one successor KB (or, in fact, a finite number of successor KBs) and such that their executability can be decided. Works such as those reported in [Baader et al. 2005; Liu et al. 2006a; Gu and Soutchanski 2007] are certainly relevant.

Our approach (and the results for *DL-Lite$_{A,id}$*) can be extended to all other programming constructs studied within Golog (i.e., non determinism, procedures) [Levesque et al. 1997], ConGolog (i.e., concurrency, prioritized interrupts) [De Giacomo et al. 2000] and, with some care –see the discussion on analysis and synthesis below– even to those in IndiGolog (search) [Sardiña et al. 2004].

Also, the works on forms of execution developed within Golog/ConGolog/IndiGolog can be lifted to DL KBs by applying the proposed approach. Specifically, notions like online execution [Sardiña et al. 2004], offline execution [Levesque et al. 1997; De Giacomo et al. 2000], monitored execution [De Giacomo et al. 1998], can all be lifted to the setting studied here.

Golog/ConGolog-like programs do not have a store to keep memory of previous results of queries to the KB. An interesting extension would be to introduce such a store, i.e., variables for storing results of queries or partial computations. Notice that this would make also the program infinite state in general (the KB is already infinite state). Also, this would make such programs much more alike programs in standard procedural languages such as C or *Java*, which manipulate global data structures—in our case the KB—and local data structures, in our case the information stored in the variables of the program.

Finally, we can adopt the functional view of KBs also to specify interactive and nonterminating processes acting on them, similarly to what is done when specifying web services on relational databases [Berardi et al. 2005; Deutsch et al. 2009].

We close the paper by noticing that the KB is not finite state if we allow for introducing new individuals in its updates. This infinite state nature makes tasks related to automated analysis and automated synthesis of programs (e.g., verifying executability on every KB, verifying termination, synthesizing a plan that achieves a goal, or synthesizing a service that fulfills a certain specification) difficult in general. This difficulty is shared with Situation Calculus based and Golog/ConGolog-like high-level programs. On the other hand if we bound the number of new individuals that can be introduced by updates then the KB becomes finite state. It is not obvious whether we can in practice bound the numbers of individual a priori, but if we cannot, we could still make use of forms of *abstraction*, studied in verification (see e.g.,. [Zuck and Pnueli 2004]), to force finite states.

References

Baader, F., D. Calvanese, D. McGuinness, D. Nardi, and P. F. Patel-Schneider (Eds.) [2003]. *The Description Logic Handbook: Theory, Implementation and Applications*. Cambridge University Press.

Baader, F., C. Lutz, M. Milicic, U. Sattler, and F. Wolter [2005]. Integrating description logics and action formalisms: First results. In *Proc. of the 20th Nat. Conf. on Artificial Intelligence (AAAI 2005)*, pp. 572–577.

Berardi, D., D. Calvanese, G. De Giacomo, R. Hull, and M. Mecella [2005]. Automatic composition of transition-based Semantic Web services with messaging. In *Proc. of the 31st Int. Conf. on Very Large Data Bases (VLDB 2005)*, pp. 613–624.

Brachman, R. J. and H. J. Levesque [1984]. The tractability of subsumption in frame-based description languages. In *Proc. of the 4th Nat. Conf. on Artificial Intelligence (AAAI'84)*, pp. 34–37.

Calvanese, D., G. De Giacomo, D. Lembo, M. Lenzerini, and R. Rosati [2006]. Data complexity of query answering in description logics. In *Proc. of the 10th Int. Conf. on the Principles of Knowledge Representation and Reasoning (KR 2006)*, pp. 260–270.

Calvanese, D., G. De Giacomo, D. Lembo, M. Lenzerini, and R. Rosati [2007a]. EQL-Lite: Effective first-order query processing in description logics. In *Proc. of the 20th Int. Joint Conf. on Artificial Intelligence (IJCAI 2007)*, pp. 274–279.

Calvanese, D., G. De Giacomo, D. Lembo, M. Lenzerini, and R. Rosati [2007b]. Tractable reasoning and efficient query answering in description logics: The *DL-Lite* family. *J. of Automated Reasoning 39*(3), 385–429.

Calvanese, D., G. De Giacomo, D. Lembo, M. Lenzerini, and R. Rosati [2008]. Path-based identification constraints in description logics. In *Proc. of the 11th Int. Conf. on the Principles of Knowledge Representation and Reasoning (KR 2008)*, pp. 231–241.

Calvanese, D., E. Kharlamov, W. Nutt, and D. Zheleznyakov [2010]. Evolution of *DL-Lite* knowledge bases. In *Proc. of the 9th Int. Semantic Web Conf. (ISWC 2010)*, Volume 6496 of *Lecture Notes in Computer Science*, pp. 112–128. Springer.

De Giacomo, G., L. Iocchi, D. Nardi, and R. Rosati [1996]. Moving a robot: the KR&R approach at work. In *Proc. of the 5th Int. Conf. on the Principles of Knowledge Representation and Reasoning (KR'96)*, pp. 198–209.

De Giacomo, G., M. Lenzerini, A. Poggi, and R. Rosati [2006]. On the update of description logic ontologies at the instance level. In *Proc. of the 21st Nat. Conf. on Artificial Intelligence (AAAI 2006)*, pp. 1271–1276.

De Giacomo, G., M. Lenzerini, A. Poggi, and R. Rosati [2007]. On the approximation of instance level update and erasure in description logics. In *Proc. of the 22nd AAAI Conf. on Artificial Intelligence (AAAI 2007)*, pp. 403–408.

De Giacomo, G., M. Lenzerini, A. Poggi, and R. Rosati [2009]. On instance-level update and erasure in description logic ontologies. *J. of Logic and Computation, Special Issue on Ontology Dynamics 19*(5), 745–770.

De Giacomo, G., Y. Lespérance, and H. J. Levesque [2000]. ConGolog, a concurrent programming language based on the situation calculus. *Artificial Intelligence 121*(1–2), 109–169.

De Giacomo, G., R. Reiter, and M. Soutchanski [1998]. Execution monitoring of high-level robot programs. In *Proc. of the 6th Int. Conf. on the Principles of Knowledge Representation and Reasoning (KR'98)*, pp. 453–465.

Deutsch, A., R. Hull, F. Patrizi, and V. Vianu [2009]. Automatic verification of data-centric business processes. In *Proc. of the 12th Int. Conf. on Database Theory (ICDT 2009)*, pp. 252–267.

Eiter, T. and G. Gottlob [1992]. On the complexity of propositional knowledge base revision, updates and counterfactuals. *Artificial Intelligence 57*, 227–270.

Fagin, R., J. D. Ullman, and M. Y. Vardi [1983]. On the semantics of updates in databases. In *Proc. of the 2nd ACM SIGACT SIGMOD Symp. on Principles of Database Systems (PODS'83)*, pp. 352–365.

Flouris, G., D. Manakanatas, H. Kondylakis, D. Plexousakis, and G. Antoniou [2008]. Ontology change: Classification and survey. *Knowledge Engineering Review 23*(2), 117–152.

Giuseppe, D. G. and R. Rosati [1999]. Minimal knowledge approach to reasoning about actions and sensing. *Electronic Trans. on Artificial Intelligence 3*(C), 1–18.

Gu, Y. and M. Soutchanski [2007]. Decidable reasoning in a modified situation calculus. In *Proc. of the 20th Int. Joint Conf. on Artificial Intelligence (IJCAI 2007)*, pp. 1891–1897.

Horrocks, I., P. F. Patel-Schneider, and F. van Harmelen [2003]. From $\mathcal{SHIQ}$ and RDF to OWL: The making of a web ontology language. *J. of Web Semantics 1*(1), 7–26.

Katsuno, H. and A. Mendelzon [1991]. On the difference between updating a knowledge base and revising it. In *Proc. of the 2nd Int. Conf. on the Principles of Knowledge Representation and Reasoning (KR'91)*, pp. 387–394.

Lenzerini, M. and F. Savo [2011]. On the evolution of the instance level of *DL-Lite* knowledge bases. Submitted for publication.

Levesque, H. J. [1984]. Foundations of a functional approach to knowledge representation. *Artificial Intelligence 23*, 155–212.

Levesque, H. J. and G. Lakemeyer [2001]. *The Logic of Knowledge Bases*. The MIT Press.

Levesque, H. J., R. Reiter, Y. Lesperance, F. Lin, and R. Scherl [1997]. GOLOG: A logic programming language for dynamic domains. *J. of Logic Programming 31*, 59–84.

Liu, H., C. Lutz, M. Milicic, and F. Wolter [2006a]. Reasoning about actions using description logics with general TBoxes. In *Proc. of the 10th Eur. Conference on Logics in Artificial Intelligence (JELIA 2006)*, Volume 4160 of *Lecture Notes in Computer Science*. Springer.

Liu, H., C. Lutz, M. Milicic, and F. Wolter [2006b]. Updating description logic ABoxes. In *Proc. of the 10th Int. Conf. on the Principles of Knowledge Representation and Reasoning (KR 2006)*, pp. 46–56.

Martin, D., M. Paolucci, S. McIlraith, M. Burstein, D. McDermott, D. McGuinness, B. Parsia, T. Payne, M. Sabou, Solanki, N. Srinivasan, and K. Sycara [2004]. Bringing semantics to web services: The OWL-S approach. In *Proc. of the 1st Int. Workshop on Semantic Web Services and Web Process Composition (SWSWPC 2004)*.

McCarthy, J. [1962]. Towards a mathematical science of computation. In *Proc. of the IFIP Congress*, pp. 21–28.

McCarthy, J. and P. J. Hayes [1969]. Some philosophical problems from the standpoint of aritificial intelligence. *Machine Intelligence 4*, 463–502.

Petrick, R. P. A. and F. Bacchus [2004]. Extending the knowledge-based approach to planning with incomplete information and sensing. In *Proc. of the 9th Int. Conf. on the Principles of Knowledge Representation and Reasoning (KR 2004)*, pp. 613–622.

Plotkin, G. D. [1981]. A structural approach to operational semantics. Technical Report DAIMI FN-19, University of Aarhus.

Reiter, R. [1990]. What should a database know? *J. of Logic Programming 14*, 127–153.

Reiter, R. [2001]. *Knowledge in Action: Logical Foundations for Specifying and Implementing Dynamical Systems*. The MIT Press.

Sardiña, S., G. De Giacomo, Y. Lespérance, and H. J. Levesque [2004]. On the semantics of deliberation in IndiGolog - from theory to implementation. *Ann. of Mathematics and Artificial Intelligence 41*(2–4), 259–299.

Sardiña, S., G. De Giacomo, Y. Lespérance, and H. J. Levesque [2006]. On the limits of planning over belief states under strict uncertainty. In *Proc. of the 10th Int. Conf. on the Principles of Knowledge Representation and Reasoning (KR 2006)*, pp. 463–471.

Smith, M. K., C. Welty, and D. L. McGuiness [2004, February]. OWL Web Ontology Language guide. W3C Recommendation, World Wide Web Consortium. Available at http://www.w3.org/TR/owl-guide/.

van Riemsdijk, M. B., F. S. de Boer, M. Dastani, and J.-J. C. Meyer [2006]. Prototyping 3APL in the Maude term rewriting language. In *Proc. of 5th Int. Joint Conf. on Autonomous Agents and Multiagent Systems (AAMAS 2006)*, pp. 1279–1281.

Wang, Z., K. Wang, and R. W. Topor [2010]. A new approach to knowledge base revision in *DL-Lite*. In *Proc. of the 24th AAAI Conf. on Artificial Intelligence (AAAI 2010)*.

Zuck, L. D. and A. Pnueli [2004]. Model checking and abstraction to the aid of parameterized systems (a survey). *Computer Languages, Systems & Structures 30*(3–4), 139–169.

8

Teamwork

PHILIP R. COHEN AND HECTOR J. LEVESQUE

Teamwork Retrospective by Philip Cohen

I am pleased to offer this paper[1] to the *Hecfest* as a representative of the research that Hector, I and colleagues accomplished between 1987 and 2000. During the early portion of that period, we worked fertile ground at the Center for the Study of Language and Information at Stanford University, where numerous researchers were investigating belief-desire-intention theories and architectures. Much of our research was thoroughly interdisciplinary, examining concepts at the intersection of artificial intelligence, philosophy, and linguistics. Because of this, we published Teamwork in a philosophy journal, Noûs, whose editor (Héctor-Neri Castañeda) had commissioned a special issue on artificial intelligence and cognitive science. We deliberately wrote in a style familiar to the readers of the journal by expressing our theory in a precisely formal use of English, rather than via a logical language. But, since the concepts we presented were based on our prior formal theory of intention and commitment [5], it should have been clear how the logical rendition of the joint action theory (as first discussed in [15]) would play out.

Regarding the paper itself, the theory of teamwork presented here was inspired initially by analyses of 'we-intentions' ([21]; [22]) and of Shared Plans [10]. Our previous research on the theory of intention and commitment had provided a set of basic concepts that we believed could be used to express notions of joint commitment, joint intention, and teamwork, without the need to postulate new underlying semantic entities. Thus, the more basic commitment/intention theory laid the formal foundations for the joint commitment/ intention analysis, enabling a set of entailments that met a variety of reasonable desiderata. Moreover, the joint action theory provided a basis for analyzing the communicative interactions needed to form and maintain a team [23], predicting the kinds of mental states and communications that would need to occur (using speech acts or via mutual observation). As such, the resulting communication theory enabled a more comprehensive analysis of speech acts and dialogue than our earlier ones (e.g., [6]), in fact describing both of these concepts in terms of joint commitments and joint intentions. It also led to the development of fault tolerant multi-agent architectures using teams of broker agents who maintained joint intentions [24].

These theories influenced numerous researchers working on individual and joint intention, speech acts, and multiagent systems. A number strove (often successfully) to overcome limitations in our formalism, proposing alternative logics (e.g, using temporal logic rather than dynamic logics of action.) Others built formal and computational models of human-human, computer-computer, and human-computer and human-robot dialogue. Numerous individual agent and multiagent system architectures were built as a result of the theoretical specifications we provided. Overall, our personal joint commitment was to start research conversations on these topics, not to finish them. In recognition of this work, I am

---

[1]Original appeared in *Noûs 25(4), Special Issue on Cognitive Science and Artificial Intelligence 1991.*

proud that the International Foundation for Autonomous Agents and Multiagent Systems honored us with one of their two inaugural Influential Paper awards for the work on the formal theory of intention.

1 Introduction

What is involved when a group of agents decide to do something together? Joint action by a team appears to involve more than just the union of simultaneous individual actions, even when those actions are coordinated. We would not say that there is any teamwork involved in ordinary automobile traffic, even though the drivers act simultaneously and are coordinated (one hopes) by the traffic signs and rules of the road. But when a group of drivers decide to do something together, such as driving somewhere as a convoy, it appears that the group acts more like a single agent with beliefs, goals, and intentions of its own, over and above the individual ones.

But given that actions are performed by individuals, and that it is individuals who ultimately have the beliefs and goals that engender action, what motivates agents to form teams and act together? In some cases, the answer is obviously the inherent value in doing something together, such as playing tennis, performing a duet, or dancing. These are examples of activities that simply cannot be performed alone. But in many cases, team activity is only one way among many of achieving the goals of the individuals. What benefits do agents expect to derive from their participation in a group effort?

In this paper, we attempt to provide an answer to these questions. In particular, we argue that a joint activity is one that is performed by individuals sharing certain specific mental properties. We show how these properties affect and are affected by properties of the participants. Regarding the benefits of teamwork, we show that in return for the overhead involved in participating in a joint activity, an agent expects to be able to share the load in achieving a goal in a way that is robust against certain possible failures and misunderstandings.

In the next section, we sketch our methodology and the adequacy criteria that have guided us. In section 3, we motivate certain aspects of our account by looking in detail at the example of a convoy. We then review the notion of an individual intention, and build on it for the joint case in sections 4 and 5. Then, in sections 7 and 8 , we discuss how our account satisfies the adequacy criteria we have laid out, and how this account relates to others. Finally, we draw some general conclusions.

2 Approach

The account of joint action presented here should probably not be regarded as a descriptive theory. We are primarily concerned with the design of artificial agents, under the assumption that these agents may need to interact with other agents (including people) having very different constitutions. At this stage, what we seek are reasonable *specifications*, that is, properties that a design should satisfy, and that would then lead to desirable behaviour. Thus we are not so much interested in characterizing some natural concept of joint activity; rather, we want to specify an idealized concept that has appropriate consequences. From this point of view, our discussion is in terms of what a specification guarantees, those properties an agent or group of agents satisfying the specification must have, as well as what the specification allows, those properties not ruled out by the specification. We attempt to guard against specifications that are too weak, in that they would fail to guarantee intuitively appropriate outcomes, as well as specifications that are too strong, in that they would place unreasonable demands on agents.

In our previous work [5], we have presented a belief-goal-commitment model of the mental states of individuals in which intentions are specified not as primitive mental features, but as internal commitments to perform an action while in a certain mental state. Our notion of commitment, in turn, was specified as a goal that persists over time. A primary concern of the present research is to investigate in what ways a team is in fact similar to an aggregate agent, and to what extent our previous work on individual intention can be carried over to the joint case. Hence, we continue our earlier development and argue for a notion of joint intention, which is formulated as a joint commitment to perform a collective action while in a certain shared mental state, as the glue that binds team members together.

To achieve a degree of realism required for successful autonomous behaviour, we model individual agents as situated in a dynamic, multi-agent world, as possessing neither complete nor correct beliefs, as having changeable goals and fallible actions, and as subject to interruption from external events. Furthermore, we assume that the beliefs and goals of agents need not be known to other agents, and that even if agents start out in a state where certain beliefs or goals are shared, this situation can change as time passes.

This potential divergence of mental state clearly complicates our task. If we could limit ourselves to cases where every agent knew what the others were doing, for instance, by only considering joint actions that can be performed publicly, it would be much simpler to see how a collection of agents could behave as a single agent, because so much of their relevant beliefs would be shared.

On the other hand, it is precisely this potential divergence that makes joint activity so interesting: agents will not necessarily operate in lock step or always be mutually co-present, so there will be tension in trying to keep the team acting as a unit. Indeed, a primary goal of this research is to discover what would hold the team together, while still allowing the members to arrive at private beliefs about the status of the shared activity. In other words, even if we are willing to assume that everything is progressing smoothly during some shared activity, we will still be concerned with cases where, for example, one of the agents no longer has the belief that some other agent intends to do her share.

Moreover, it is this divergence among the agents that makes communication necessary. Whereas the model of individual intention in our earlier work [5, 6] was sufficient to show how communicative acts were defined in terms of beliefs and intentions, and could be used to achieve various goals, it did so only from the perspective of each individual agent, by constraining the rational balance that agents maintain among their own beliefs, goals, commitments, intentions, and actions. But special communicative demands are placed on agents involved in joint activities, and we wish to examine how these arise as a function of more general constraints on team behaviour.

Before looking at an example of the sort of joint activity we have in mind and possible specifications of the underlying team behaviour, we briefly list further questions that we expect our theory to address, in addition to those cited above:

Joint intentions leading to individual ones: As we said above, ultimately, it is agents that act based on their beliefs, goals, and intentions. How then do the joint beliefs, goals, and intentions of teams lead to those of the individuals, so that anything gets done? Typically, teams will be involved in joint activities that consist of many parts performed concurrently or in sequence. How do joint intentions to perform complex actions lead to appropriate intentions to perform the pieces? Assuming that an agent will only intend to do her own actions, what is her attitude towards the others' share?

The functional role of joint intentions: Bratman [2] has argued that in the case of individuals, intentions play certain functional roles: they pose problems for agents, which can be solved by means-end analysis; they rule out the adoption of intentions that conflict with existing ones; they dispose agents to monitor their attempts to achieve them; and, barring major changes, they tend to persist. Which of these roles have analogues for teams?

Communication required: Any theory of joint action should indicate when communication is necessary. What do agents need to know (and when) about the overall activity, about their own part, and about the other agents' shares? Should agents communicate when the joint action is to begin, when one agent's turn is over, when the joint action is finished, when the joint action is no longer needed? How does communication facilitate the monitoring of joint intentions?

3   A Convoy Example

If team behaviour is more than coordinated individual behaviour, how does it work? This question is perhaps best answered by considering what would happen in the case of a convoy example without the right sort of joint intention.

Suppose we have two agents Alice and Bob; Bob wants to go home, but does not know his way, but knows both that Alice is going near there also and that she does know the way. Clearly Alice and Bob do not have to do anything together for Bob to get home; Bob need only follow Alice. In many circumstances, this plan would be quite adequate.

But it does have problems. For example, Alice might decide to drive very quickly through traffic, and Bob may be unable to keep up with her. It would be much better, from Bob's point of view, if Alice knew that he intended to follow her until he finds his way, counting on the fact that Alice, being a kind soul, would plan on keeping him in sight. Let us say that Bob arranges for Carl to tell her what he is going to do. Then, assuming she is helpfully disposed, she would not speed away. However, there is no reason for her to know that Bob is the one who sent Carl. As far as she is concerned, Bob might not know that she knows what he is up to. In particular, she would not expect Bob to signal her when he knows his way. So if, for example, Bob starts having car trouble and needs to pull over, Alice may very well speed off, believing that all is now well.

Realizing this is a possibility, Bob might try to get around it. He might get Carl also to tell Alice that he (Bob) is the one who asked Carl to talk to Alice. This would ensure that Alice was aware of the fact that Bob knew that she was being told. Assuming now that all goes well, at this point, both Bob and Alice would have appropriate intentions, both would know that the other had such intentions, and both would know that the other knew that they had such intentions.

However, there is still room for misunderstanding. Alice might say to herself: "Carl told me that Bob sent him to talk to me. So Bob now knows that I know what his driving plans are. But does he know that Carl mentioned that it was Bob who sent him? I think Carl just decided on the spot to say that, and so Bob doesn't realize that I know that it was him. So although Bob knows that I know his driving plans, he thinks I found out more or less accidentally, and so he thinks I won't expect him to signal me when he finds his way." Such reasoning might not happen, of course, but if it did, again Alice might speed off when Bob runs into car trouble. In fact, the situation is slightly worse than this, since even if this incorrect reasoning does not take place, Bob could still believe that it has, and not want to pull over for fear of being misunderstood.

This is clearly not the kind of robustness one expects from a convoy. The whole point of driving together is precisely to be able to deal better with problems that occur en route. The kind of misunderstanding that is taking place hereand it could go on to deeper levels − is due to the fact that although both parties have the right intentions and the right beliefs about each other (at whatever level), they lack mutual belief of what they have agreed to. This suggests that Bob should approach Alice directly and get her to agree to the convoy, so that the agreement would be common knowledge between both parties.

Without being too precise about what exactly this means at this stage, we can nonetheless think of this as a rough first proposal for a concept of joint intention, that is, the property that will hold the group together in a shared activity. In other words, we expect agents to first form future-directed joint intentions to act, keep those joint intentions over time, and then jointly act.

> **Proposal 1.** x and y jointly intend to do some collective action iff it is mutually known between x and y that they each intend that the collective action occur, and it is mutually known that they each intend to do their share (as long as the other does theirs).

As we will discuss later in section 8, something very much like this has been proposed in the literature. As above, and assuming a tight connection between intention and commitment, it does indeed guarantee that the two agents commit to achieving the goal. Moreover, it is common knowledge that they are committed in this way and that neither party will change their mind about the desirability of the activity. In addition, we can assume that there are no hidden obstacles, in that if both parties did their share, then Bob would indeed get home. But even with these strong assumptions, the specification by itself is still too weak, once we allow for a divergence of mental states.

To see this, consider two appropriate reasons for dropping participation in the convoy: first, Bob could come to realize that he now knows his way, and so the intended action has successfully terminated; second, Alice could realize that she does not know where Bob lives after all, and so the intended action cannot be performed. We assume that in each case the agent in question has no choice but to give up the intention to act, terminating the convoy. The problem is that while Bob and Alice are driving together, Alice may come to believe that Bob now knows his way, that the convoy is over, and then speed off. Or Bob may come to believe that Alice does not know the way, that the convoy is over, and plan to get home some other way. As above, even if neither party comes to such an erroneous conclusion, they could suspect that something similar is happening with the other, and again the convoy would fall apart. Although both parties still have the right intentions and start with the right mutual knowledge, there is nothing to prevent this mutual knowledge from dissipating as doubt enters either agent about the private beliefs of the other regarding the status of the activity. But, these are potential troubles we would expect a joint activity to overcome.

More precisely, the problem with the first proposal is that although it guarantees goals and intentions that will persist suitably in time, it does not guarantee that the mutual knowledge of these goals and intentions will persist. So a second proposal is this:

> **Proposal 2.** x and y jointly intend to do some action iff it is mutually known between x and y that they each intend that the collective action occur, and also that they each

intend to do their share as long as the other does likewise, and this mutual knowledge persists until it is mutually known that the activity is over (successful, unachievable, irrelevant).

This is certainly strong enough to rule out doubt-induced unraveling of the team effort, since both parties will know exactly where they stand until they arrive at a mutual understanding that they are done.

The trouble with this specification is that, allowing for the divergence of mental states, it is too strong. To see this, suppose that at some point, Alice comes to realize privately that she does not know where Bob lives after all. The intention to lead Bob home is untenable at that point, and so there is no longer mutual belief that both parties are engaged in the activity. But to have been involved in a joint intention (in proposal 2) meant keeping that intention until it was mutually believed to be over. Since under these circumstances, it is not now mutually believed to be over, we are led to the counterintuitive conclusion that there was not really a joint intention to start with. The specification is too strong because it stipulates at the outset that the agents must mutually believe that they will each have their respective intentions until it is mutually known that they do not. It therefore does not allow for private beliefs that the activity has terminated successfully or is unachievable.

In section 5, we will propose more precisely a third specification for joint intention that lies between these two in strength and avoids the drawbacks of each. Roughly speaking we consider what one agent should be thinking about the other during the execution of some shared activity:

- The other agent is working on it (the normal case), or

- The other agent has discovered it to be over (for some good reason).

We then simply stipulate that for participation in a team, there is a certain *team overhead* to be expended, in that, in the second case above, it is not sufficient for a team member to come to this realization privately, she must make this fact mutually known to the team as a whole. As we will see, if we ensure that mutual knowledge of this condition persists, we do get desirable properties.

To see this in detail, we first briefly describe our analysis of individual commitment and intention, and then discuss the joint case.

4 Individual Commitment and Intention

Our formal account of individual and joint commitments and intentions [5, 15] is given in terms of beliefs, mutual beliefs, goals, and events. In this paper, we will not present the formal language, but simply describe its features in general terms. At the very lowest level, our account is formulated in a modal quantificational language with a possible-world semantics built out of the following primitive elements.

> **Events:** We assume that possible worlds are temporally extended into the past and future, and that each such world consists of an infinite sequence of primitive events, each of which is of a type and can have an agent.[2]

---

[2]Currently, we picture these events as occurring in a discrete synchronized way, but there is no reason not to generalize the notion to a continuous asynchronous mode, modeled perhaps by a function from the real numbers to the set of event types occurring at that point.

Belief: We take belief to be what an agent is sure of, after competing opinions and wishful thinking are eliminated. This is formalized in terms of an accessibility relation over possible worlds in the usual way: the accessible worlds are those the agent has ruled capable of being the actual one. Beliefs are the propositions that are true in all these worlds. Although beliefs will normally change over time, we assume that agents correctly remember what their past beliefs were.

Goal: We have formalized the notion of goal also as accessibility over possible worlds, where the accessible worlds have become those the agent has selected as *most desirable*. Goals are the propositions that are true in all these worlds. As with belief, we presume that conflicts among choices and beliefs have been resolved. Thus, we assume that these chosen worlds are a subset of the belief-accessible ones, meaning that anything believed to be currently true must be chosen, since the agent must rationally accept what cannot be changed. However, one can have a belief that something is false now and a goal that it be true later, which is what we call an *achievement goal*. Finally, we assume agents always know what their goals are.

Mutual belief: The concept of mutual belief among members of a group will be taken to be the usual infinite conjunction of beliefs about other agents' beliefs about other agents' beliefs (and so on to any depth) about some proposition. Analogous to the individual case, we assume that groups of agents correctly remember what their past mutual beliefs were.

This account of the attitudes suffers from the usual possible-world problem of logical omniscience (see [14], for example), but we will ignore that difficulty here. Moreover, we will take knowledge simply (and simplistically) to be true belief, and mutual knowledge to be true mutual belief.

To talk about actions, we will build a language of action expressions inductively out of primitive events, and complex expressions created by action-forming operators for sequential, repetitive, concurrent, disjunctive, and contextual actions, where contextual actions are those executed when a given condition holds, or resulting in a given condition's holding. These dynamic logic primitives are sufficient to form a significant class of complex actions, such as the "if-then-else" and "while- loops" familiar from computer science [12]. In all cases, the agents of the action in question are taken to be the set of agents of any of the primitive events that constitute the performance of the action. To ground the earlier definition of collective action in the formal framework, we note that although a complex collective action may involve the performance by one agent of individual actions sequentially, repetitively, disjunctively, or concurrently with the performance of other individual actions by other agents, the collection of agents are not necessarily performing the action together, in the sense being explained in this paper.

For our purposes, it is not necessary to talk about actions with respect to arbitrary intervals (and thus have variables ranging over time points), but merely to have the ability to say that an action is happening, has just happened, and will happen next, with the implicit quantification that implies. It is also useful to define (linear) temporal expressions from these action expressions, such as a proposition's being *eventually, always,* or *never* true henceforth; similar expressions can be defined for the past. Finally, we say that a proposition remains true *until* another is true, with the obvious interpretation: if at some point in the future the former proposition is false, there must be an earlier future point where the latter is true.

4.1 Individual Commitment

Based on these primitives, we define a notion of individual commitment called persistent goal:[3]

DEFINITION 1. An agent has a persistent goal relative to q to achieve p if

1. she believes that p is currently false;

2. she wants p to be true eventually;

3. it is true (and she knows it) that (2) will continue to hold until she comes to believe either that p is true, or that it will never be true, or that q is false.

Some important points to observe about individual commitments are as follows: once adopted, an agent cannot drop them freely; the agent must keep the goal at least until certain conditions arise; moreover, other goals and commitments need to be consistent with them; and, agents will try again to achieve them should initial attempts fail. Clause 3 states that the agent will keep the goal, subject to the aforementioned conditions, in the face of errors and uncertainties that may arise from the time of adoption of the persistent goal to that of discharge. Condition q is an irrelevance or "escape" clause, which we will frequently omit for brevity, against which the agent has relativized her persistent goal. Should the agent come to believe it is false, she can drop the goal. Frequently, the escape clause will encode the network of reasons why the agent has adopted the commitment. For example, with it we can turn a commitment into a subgoal, either of the agent's own supergoal, or of a (believed) goal of another agent. That is, an agent can have a persistent goal to achieve p relative to her having the goal of achieving something else. Note that q could in principle be quite vague, allowing disjunctions, quantifiers, and the like. Thus, we need not specify precisely the reasons for dropping a commitment. In particular, it could be possible to have a commitment to p relative to p being the most favored of a set of desires; when those rankings change, the commitment could be dropped. However, most observers would be reluctant to say that an agent is committed to p if the q in question is sufficiently broad, for example, such as that the agent could not think of anything better to do.

Finally, it is crucial to notice that an agent can be committed to another agent's acting. For example, an agent x can have a persistent goal to its being the case that some other agent y has just done some action. Just as with committing to her own actions, x would not adopt other goals inconsistent with y's doing the action, would monitor y's success, might request y to do it, or help y if need be. Although agents can commit to other's actions, they do not intend them, as we will see shortly.

4.2 Individual Intention

We adopt Bratman's [2] methodological concern for treating the future-directed properties of intention as primary, and the intention-in-action properties as secondary, contra Searle [20, 21]. By doing so, we avoid the notoriously difficult issue of how an intention self-referentially causes an agent to act, as discussed in [20], although many of those properties are captured by our account. Rather, we are concerned with how adopting an intention constrains the agents' adoption of other mental states.

An intention is defined to be a commitment to act in a certain mental state:

---

[3]This definition differs slightly from that presented in our earlier work [5], but that difference is immaterial here.

DEFINITION 2. An agent intends relative to some condition to do an action just in case she has a persistent goal (relative to that condition) of having done the action and, moreover, having done it, believing throughout that she is doing it.

Intentions inherit all the properties of commitments (e.g., tracking, consistency with beliefs and other goals) and also, because the agent knows she is executing the action, intention inherits properties that emerge from the interaction of belief and action. For example, if an agent intends to perform a conditional action, for which the actions on the branches of the conditional are different, then one can show that, provided the intention is not dropped for reasons of impossibility or irrelevance, eventually the agent will have to come to a belief about the truth or falsity of the condition. In our earlier paper [5], we also show how this analysis of intention satisfies Bratman's [2, 3] functional roles for intentions and solves his "package deal" problem, by not requiring agents also to intend the known side-effects of their intended actions, despite our possible-world account of belief and goal.

Typically, an intention would arise within a subgoal-supergoal chain as a decision to do an action to achieve some effect. For example, here is one way to come to intend to do an action to achieve a goal. Initially the agent commits to p's becoming true, without concern for who would achieve it or how it would be accomplished. This commitment is relative to q, so if the agent comes to believe q is false, she can abandon the commitment to p. Second, the agent commits to a or b as the way to achieve p, relative to the goal of p being true. Thus, she is committing to one means of achieving the goal that p be true. Third, the agent chooses one of the actions (say, a) and forms the intention to do it, that is, commits to doing a knowingly. The intention could be given up if the agent discovers that she has achieved p without realizing it, or if any other goal higher in the chain was achieved. For example, the intention might be given up if she learns that some other agent has done something to achieve q.[4] This example of intention formation illustrates the pivotal role of the relativization condition that structures the agent's network of commitments and intentions. We now turn to the joint case.

5 Joint Commitment

How should the definition of persistent goal and intention be generalized to the case where a group is supposed to act like a single agent? As we said earlier in the discussion of Proposal 2, joint commitment cannot be simply a version of individual commitment where a team is taken to be the agent, for the reason that the team members may diverge in their beliefs. If an agent comes to think a goal is impossible, then she must give up the goal, and fortunately knows enough to do so, since she believes it is impossible. But when a member of a team finds out a goal is impossible, the team as a whole must again give up the goal, but the team does not necessarily know enough to do so. Although there will no longer be mutual belief that the goal is achievable, there need not be mutual belief that it is unachievable. Moreover, we cannot simply stipulate that a goal can be dropped when there is no longer mutual belief, since that would allow agreements to be dissolved as soon as there was uncertainty about the state of the other team members. This is precisely the problem with the failed convoy discussed above. Rather, any team member who discovers privately that a goal is impossible (has been achieved, or is irrelevant) should be left with a goal to make this fact known to the team as a whole. We will specify that before this commitment can be discharged, the agents must in fact arrive at the mutual belief that a

---

[4]Of course, the agent may still intend to achieve p again if she is committed to doing so herself.

termination condition holds; this, in effect, is what introspection achieves in the individual case.

We therefore define the state of a team member nominally working on a goal as follows:

DEFINITION 3. An agent has a weak achievement goal relative to q and with respect to a team to bring about p if either of these conditions holds:

- The agent has a normal achievement goal to bring about p, that is, the agent does not yet believe that p is true and has p eventually being true as a goal.

- The agent believes that p is true, will never be true, or is irrelevant (that is, q is false), but has as a goal that the status of p be mutually believed by all the team members.

So this form of weak goal involves four cases: either she has a real goal, or she thinks that p is true and wants to make that mutually believed,[5] or similarly for p never being true, or q being false.

A further possibility, that we deal with only in passing, is for an agent to discover that it is impossible to make the status of p known to the group as a whole, when for example, communication is impossible. For simplicity, we assume that it is always possible to attain mutual belief and that once an agent comes to think the goal is finished, she never changes her mind.[6] Among other things, this restricts joint persistent goals to conditions where there will eventually be agreement among the team members regarding its achievement or impossibility.[7]

The definition of joint persistent goal replaces the "mutual goal" in Proposal 2 by this weaker version:

DEFINITION 4. A team of agents have a joint persistent goal relative to q to achieve p just in case

1. they mutually believe that p is currently false;

2. they mutually know they all want p to eventually be true;

3. it is true (and mutual knowledge) that until they come to mutually believe either that p is true, that p will never be true, or that q is false, they will continue to mutually believe that they each have p as a weak achievement goal relative to q and with respect to the team.

Thus, if a team is jointly committed to achieving p, they mutually believed initially that they each have p as an achievement goal. However, as time passes, the team members cannot conclude about each other that they still have p as an achievement goal, but only that they have it as a weak achievement goal; each member allows that any other member may have discovered privately that the goal is finished (true, impossible, or irrelevant) and

---

[5]More accurately, we should say here that her goal is making it mutually believed that p had been true, in case p can become false again.

[6]For readers familiar with the results in distributed systems theory [11] in which it is shown that mutual knowledge is impossible to obtain for computers by simply passing messages, we point out that those results do not hold for mutual beliefs acquired by default, nor for agents that can be co-present or communicate instantly.

[7]Actually, agents do have the option of using the escape clause q to get around this difficulty. For example, $\sim$q could say that there was an unresolvable disagreement of some sort, or just claim that an expiry date had been reached, or that the agents each no longer wants to have the joint intention. In such cases, mutual belief in $\sim$q amounts to an agreement to dissolve the commitment regardless of the status of p.

be in the process of making that known to the team as a whole. If at some point, it is no longer mutually believed that everyone still has the normal achievement goal, then the condition for a joint persistent goal no longer holds, even though a mutual belief in a weak achievement goal will continue to persist. This is as it should be: if some team member privately believes that p is impossible, even though the team members continue to share certain beliefs and goals, we would not want to say that the team is still committed to achieving p.

The first thing to observe about this definition is that it correctly generalizes the concept of individual persistent goal, in that it reduces to the individual case when there is a single agent involved.

THEOREM 5. *If a team consists of a single member, then the team has a joint persistent goal if that agent has an individual persistent goal.*

The proof is that if an agent has a weak goal that persists until she believes it to be true or impossible, she must also have an ordinary goal that persists. It can also be shown that this definition of joint commitment implies individual commitments from the team.

THEOREM 6. *If a team has a joint persistent goal to achieve p, then each member has p as an individual persistent goal.*

To see why an individual must have p as a persistent goal, imagine that at some point in the future the agent does not believe that p is true or impossible to achieve. Then there is no mutual belief among the whole team either that p is true or that p is impossible, and so p must still be a weak goal. But under these circumstances, it must still be a normal goal for the agent. Consequently, p persists as a goal until the agent believes it to be satisfied or impossible to achieve. A similar argument also shows that if a team is jointly committed to p, then any subteam is also jointly committed. This generalization will also apply to other theorems about intention presented below.

So if agents form a joint commitment, they are each individually committed to the same proposition p (relative to the same escape condition q). If p is the proposition that the agents in question have done some collective action constructed with the action-formation operators discussed above, then each is committed to the entire action's being done, including the others' individual actions that comprise the collective. Thus, one can immediately conclude that agents will take care to not foil each other's actions, to track their success, and to help each other if required.

Furthermore, according to this definition, if there is a joint commitment, agents can count on the commitment of the other members, first to the goal in question and then, if necessary, to the mutual belief of the status of the goal. This property is captured by the following theorem, taken from our earlier work [15].

THEOREM 7. *If a team is jointly committed to some goal, then under certain conditions, until the team as a whole is finished, if one of the members comes to believe that the goal is finished but that this is not yet mutually known, she will be left with a persistent goal to make the status of the goal mutually known.*

In other words, once a team is committed to some goal, then any team member that comes to believe privately that the goal is finished is left with a commitment to make that fact known to the whole team. So, in normal circumstances,[8] a joint persistent goal to

---

[8]The normality conditions referred to here are merely that once the agent comes to a belief about the final status of the goal, she does not change her mind before arriving at a mutual belief with the others.

achieve some condition will lead to a private commitment to make something mutually believed. Thus, although joint persistent goal was defined only in terms of a weak goal — a concept that does not by itself incorporate a commitment — a persistent goal does indeed follow.

This acquisition of a commitment to attain mutual belief can be thought of as the team overhead that accompanies a joint persistent goal. A very important consequence is that it predicts that communication will take place, as this is typically how mutual belief is attained, unless there is co-presence during the activity. Thus, at a minimum, the team members will need to engage in communicative acts to attain mutual belief that a shared goal has been achieved.

6 Joint Intention

Just as individual intention is defined to be a commitment to having done an action knowingly, joint intention is defined to be a joint commitment to the agents' having done a collective action, with the agents of the primitive events as the team members in question, and with the team acting in a joint mental state.

DEFINITION 8. (and Proposal 3.) A team of agents jointly intends, relative to some escape condition, to do an action if the members have a joint persistent goal relative to that condition of their having done the action and, moreover, having done it mutually believing throughout that they were doing it.[9]

That is, the agents are jointly committed to its being the case that throughout the doing of the action, the agents mutually believe they are doing it.

Next, we examine some of the important properties of joint intention.

6.1 Properties of Joint Intention

Given that joint intention is a property of a group of agents, but that only individual agents act, what is the origin of the individual intentions that lead those agents to perform their share? We have shown that joint persistent goals imply individual goals among the team members. We now wish to show a similar property for joint intentions.

First, observe that joint intention implies individual intention when one agent is the only actor.

THEOREM 9. *If a team jointly intends to do an action, and one member believes that she is the only agent of that action, then she privately intends to do the action.*

This holds because joint commitment entails individual commitment, and mutual belief entails individual belief. Of importance is the added condition that the agent must believe herself to be the only agent of the action. As desired, we do not allow agents to intend to perform other agents' actions, although they can be committed to them.

In the case of multi-agent actions, we will only consider two types: those that arise from more basic actions performed concurrently, and those that are formed from a sequence of more basic actions.

6.2 Jointly Intending Concurrent Actions

Consider the case of two agents pushing or lifting a heavy object, or one bracing an object while the other acts upon it. First, we need the following property of individual intention.

---

[9] A more precise version of this definition [15] also requires that they mutually know when they started.

THEOREM 10. *An individual who intends to perform actions a and b concurrently intends to perform a (resp. b) relative to the broader intention.*

The proof of this depends on the treatment of concurrency as a conjunction of actions performed over the same time interval. Hence, the conjuncts can be detached and treated separately. Note that the intention to perform a is only relative to the intention to do both parts together; should the agent come to believe that it is impossible to do b at all, she may very well not want to do a alone.

Analogously, for joint intention, the following holds:

THEOREM 11. *If a team jointly intends to do a complex action consisting of the team members concurrently doing individual actions, then the individuals will privately intend to do their share relative to the joint intention.*

In other words, agents who jointly intend concurrent actions also individually intend to do their parts as long as the joint intention is still operative. The proof of this parallels the proof that joint intention leads to individual intention in the case of single-agent actions. Individual intentions thus persist at least as long as the joint intention does. But the commitment can be dropped if a team member discovers, for example, that some other team member cannot do her share.

Thus, an unrestricted individual intention, that is, an intention that is not relative to the larger intention, does not follow from a joint intention. Still, as with any joint persistent goal, even if one agent discovers privately that the joint intention is terminated, there will remain residual commitments to attain mutual belief of the termination conditions.

Notice also that agents are supposed to mutually believe, throughout the concurrent action, that they are performing it together. Thus, while the agents are performing their individual actions, they also each believe that together they are performing the group action.

6.3 Jointly Intending Sequential Actions

Next, we need to ascertain how joint intentions for sequential actions result in the agents acquiring their own individual intentions. This case is more complex, since temporal properties and execution strategies need to be considered.

Stepwise Execution

Consider first individual intention and action. Processors for programming languages in computer science are usually designed to step through a program "deliberately," by keeping track of what part of the action is being executed and, if there are conditions (such as for if-then-else actions), by ascertaining the truth or falsity of those conditions before proceeding with the computation or execution. However, the framework we have adopted allows individual agents to be considerably more flexible in executing action expressions. For example, though an agent may know she is executing a repetitive or sequential action, she need not know where she is in the sequence. For example, an agent can click on a phone receiver a number of times and know that one of those clicks disconnects the line and produces a dial tone without ever having to know which click was the one that did it. Similarly, an agent need not know the truth value of a condition on an if-then-else action if the two branches share an initial sequence of events. So, for instance, to execute an action expressed as "if it is raining, then bring all your rain gear, otherwise just bring the umbrella" it is sufficient to get an umbrella before checking the weather, since that is required in either case. Only at the point at which those execution paths diverge will it be necessary for the agent to have a belief about the truth of the past condition.

This freedom may seem like unnecessary generality, although, as we will see, it plays an important role in the case of joint activity. However, one consequence it has is that an agent who intends to do a sequential action does not necessarily intend to perform the first step in the sequence, even relative to the larger intention. It is consistent with our specification that an agent can intend to do the sequence without expecting to know when the first part of that sequence is over. Thus, the reasons for dropping the commitments entailed in having an intention would not be present. Moreover, the agent need not intend to do the remainder of the sequence either: since she might not know when the first part has been completed, she might not know she is doing the second part. In other words, because one may not know when subactions start and stop, it is possible to execute a sequence of actions knowingly without knowingly executing the individual steps.

However, it is possible to stipulate a condition on the execution of a sequence that would guarantee the distribution of intention throughout the sequence: we can require the agent to believe after each step both that the step was just done and that she is doing the remainder. We call this stepwise execution. That is, in the stepwise execution of a sequence, each step becomes a contextual action: it must be performed in a context where the agent has certain beliefs. In effect, this forces the agent to execute the action like a traditional programming language processor and leads to the following theorem.

THEOREM 12. *If an agent intends to do a sequential action in a stepwise fashion, the agent also intends to do each of the steps, relative to the larger intention.*

The proof of this theorem is that if an agent believes she has done the entire sequence in a stepwise fashion, she must believe that she had the belief at the relevant times about having done each step, and by the memory assumption of section 4, these beliefs cannot simply be after-the-fact reconstructions.[10]

Joint Stepwise Execution

Given that to obtain a seemingly desirable property of intending sequences which most agent designers implicitly assume, the agent must explicitly intend to execute the sequence in stepwise fashion, the freedom offered in our formulation of sequential action execution may seem like a dubious advantage. However, it has considerable merit when one considers joint action. Recall that one of our principles has been to maximize the similarity of joint commitments and intentions to their individual counterparts. If stepwise execution of actions were the only way to execute actions, and if, following the similarity principle, we applied that strategy to a team, we would thereby enforce a joint stepwise execution strategy, requiring the attainment of mutual belief after each step that the step had been accomplished and that the agents were embarking on the remainder.

But we do not want to require a team to always execute complex actions in lock step. There are many types of joint actions where such team overhead would be undesirable. Consider, for example, an expert and an apprentice performing a sequence together, where the expert has to do something immediately after the apprentice. The two may separate from one another, with only the expert being able to sense the apprentice, and hence know when it is her turn to act. In fact, it is possible that only the expert will know when the apprentice has successfully done his part. She may be free to continue the sequence, and

---

[10]Another way to obtain a similar result might be to change the definition of persistent goal to say that an agent can drop her goal that p if she comes to believe that p has been made true, rather than is currently true. However, this introduces additional complexity, since one must be careful not to consider times when p was true before the adoption of the goal.

report success of the whole enterprise, without reporting intermediate results. Thus, we want to allow for individuals to contribute privately to a sequence, when that is compatible with the performance of the overall activity. So, to allow for such actions, the joint intention to do a sequence must not require the agents to come to a mutual belief that each step has just been done successfully.

How, then, do team members get individual intentions in such cases? Essentially, all that is needed is that each agent know that it is her turn and what she is doing.

THEOREM 13. *If a team jointly intends to do a sequential action, then the agent of any part will intend to do that part relative to the larger intention, provided that she will always know when the antecedent part is over, when she is doing her share, and when she has done it.*

As always, the individual intentions are formed relative to the larger joint intention.

However, many joint activities, such as games and dialogue, are supposed to be performed in a joint stepwise fashion. For example, agents who jointly intend to play a set of tennis jointly intend to play the first point. After the point, the agents must agree that it is over (and who won it) before proceeding. So, we need to allow for both forms of joint execution of a sequential action. Fortunately, our earlier analysis of individual action provides just the right kind of generalization and offers an immediate analogue for the joint case.

THEOREM 14. *If a team intends to do a sequence of actions in a joint stepwise fashion, the agents of any of the steps will jointly intend to do the step relative to the larger intention.*

As before, appropriate individual intentions and commitments will then follow from the joint intentions.

7 Meeting the Adequacy Criteria

In characterizing joint commitments and intentions, we have specified a notion of weak achievement goal as the property that persists and holds the group together. Given this, we have addressed our first adequacy criterion by showing the conditions under which joint intentions to perform simple actions, concurrent actions, and sequential actions entail the team members forming the relevant individual intentions. Joint intentions embody a precise notion of commitment, yet are not defined in terms of the individual intentions. Instead, both are defined in terms of the same primitives, and the individual intentions follow from the joint ones.

As we have seen, joint commitments give rise to individual commitments relative to the overarching joint commitment. Thus, the individual commitments are subsidiary to the joint ones and can be abandoned if the joint ones are given up. Moreover, because a jointly intended action requires the agents to mutually believe they are acting together, an agent does not merely believe she is acting alone. Rather, the agents believe their actions are part of and depend on the group's commitment and efforts.

Turning now to the functional role of joint intentions, our discussion of execution strategies implies that the adoption of a joint intention need not always lead to a process of joint problem-solving that culminates in a mutual belief among the team members regarding what each is to do. Rather, this property would only hold if it were necessary for the execution of the action or if the agents agreed to perform their actions in a more deliberate way, such as in a joint stepwise fashion.

However, joint intentions do form a "screen of admissibility" [2], analogous to those of individual commitments, because joint commitments and, hence, intentions must be consistent with the individual commitments. Just as agents cannot knowingly and intentionally act to make their individual commitments and intentions impossible to achieve, they similarly cannot act to make their joint commitments and joint intentions impossible. In particular, they will not knowingly and intentionally act to foil their team members' actions. On the other hand, if it is mutually known that one team member requires the assistance of another, our account predicts that the other will intend to help. All of these properties follow immediately from the fact that joint commitments entail individual commitments, and that these must be mutually consistent. For more specific analyses of the kinds of consistency predicted by our analysis of individual commitment and intention, see our more comprehensive paper [5].

In addition, as in the individual case, a group will monitor the success or failure of the joint effort, and, in particular, with joint stepwise execution, it will monitor the intermediate results as well. These results follow from the facts that agents who have jointly intended to do some collective action are jointly committed to mutually believing that they are performing the action, and that they must ultimately come to mutually believe that they have done it or that it is impossible or is irrelevant.

As for the communication criterion, by showing that agents who have adopted a joint intention commit themselves to attaining mutual belief about the status of that intention, we derive commitments that may lead to communication. For example, in other of our papers on dialogues about a task [7, 8], we have analyzed how joint intentions to engage in the task lead to the discourse goals that underlie various speech acts.

8 Comparison with Other Analyses of Joint Intention

Numerous analyses of concepts similar to joint intention have been given. Tuomela and Miller [22] propose a conceptual analysis of an individual agent's "we-intending" a group action. Essentially, that agent must intend to do her part of the action and believe it is mutually believed that the other members of the team will do their parts as well. Power [18] is perhaps the earliest researcher within the artificial intelligence research community to be concerned with modeling joint activity. He defines a mutual intention to be each agent's having the intention to do her part and there being a mutual assumption that each agent has such intentions. Grosz and Sidner [10] propose a concept of shared plans, using Pollack's [17] analysis of plans and Goldman's [9] analysis of action. In their model, two agents have a shared plan if those agents mutually know that each agent intends to do her own part to achieve the jointly done action, and that each agent will do her part if and only if the other agent does likewise.

Though differing in detail, these analyses share a number of disadvantages as compared to the analysis proposed here. First, they do not make clear how, if at all, the agents are committed to a joint activity or to its parts, although Grosz and Sidner's come closest with their use of the biconditional relating agents' intentions. Specifically, they do not show how one agent can be committed to the other's acting, without stating that the agent intends the other agent's actions, an expression that would be ill-formed in most analyses of intention. Such commitment to the others' actions are important, since they would lead one agent to help another, to stay out of her way, etc., as we have described.

Second, even granting some notion of commitment inherent in their uses of the term 'intention', these analyses all possess the defects of Proposal 1: though the agents' intentions to do their parts may persist, there is no constraint on the persistence of the agents' mutual beliefs about those intentions. Hence, such analyses are dissolved by doubt. Finally, because there is no requirement to start or terminate joint actions with mutual belief, these analyses make no predictions for communication.

Searle [21] provides a different argument against approaches such as these, claiming that collective intentions are not reducible to individual intentions, even when supplemented with mutual beliefs. He claims to provide a counterexample of a group of students who have been jointly educated to be selfish capitalists, and mutually know their classmates have been similarly indoctrinated to compete vigorously, with the collective goal of serving humanity and themselves. The students are claimed to satisfy Tuomela and Miller's definition (and, by extension, Power's), but are not acting collectively.[11] On the other hand, Searle argues that had the students made a pact on graduation day to compete vigorously, their subsequent actions would constitute a joint activity.

Instead of reducing collective intention to some combination of individual intention and mutual belief, Searle proposes a primitive construct, one not defined in terms of other concepts, for "we- intending" in which individual agents we-intend to do an action by means of the individual agents doing their share. By using a new primitive construct, Searle attempts to solve the problem addressed earlier, namely, how a group's collective intention leads the individual agents to form their own intentions.[12] Rather than propose a primitive construct for collective intention, we have shown that we can derive reasonable entailments and meet a substantial set of adequacy criteria by defining both joint and individual intentions in terms of the same set of primitive elements.

A major concern of the present paper has been to characterize joint commitment suitably so that it keeps a group together long enough to take action. Thus, it is crucial to our understanding that joint intention be regarded as a future-directed joint commitment. Although Searle's examples are motivated by cases of future-directed collective intention, Searle's analysis extends only his notion of intention-in-action [20] to the collective case. Thus, the analysis is silent about how a group could plan to do some action in the future, and about how such collective future-directed intentions could eventually result in the formation of a collective present-directed intentions.

9 Conclusions

At this point, we have exhibited some of the consequences of a group's adopting joint commitments and intentions. Once adopted, agents should be able to build other forms of interaction upon them. Here, we only have space to remark in passing on how this might work by looking briefly at contracts and agreements, speech acts, and dialogue.

First, an interesting extension of our analysis would be to describe how the properties that result from the adoption of joint commitments and intentions compare with those inherent in the formulation of valid contracts [1]. We suspect that informal versions of many of the properties of contracts can be found in our notion of a joint commitment, especially in cases where there can be disagreement about the final contractual outcome. Historically, contracts (in British contract law) were regarded as formalized agreements. Hence, if our account were to bear some relation to contract law, it would be through some understanding of what constitutes an agreement.

---

[11]Whether Searle's example also counters Grosz and Sidner's analysis, as claimed by Hobbs [13], is arguable. They may escape the example's force because of the biconditional in their definition: there must be mutual belief that each agent intends to do his part iff the other agent does likewise.

[12]In Tuomela and Miller's, Power's, and Grosz and Sidner's analyses, the means by which we-intentions, mutual intentions, and shared plans, respectively, lead agents to have individual intentions is no mystery: they are simply defined in terms of individual intention.

The clearest cases of joint activity are ones in which either an explicit or implicit agreement to act is operative, by which we mean some sort of mental state that agents enter into and the speech acts by which they do so. Although there surely is a complex interrelationship between having a joint intention and there being such an agreement in force, we have taken the concept of joint intention simply to be present in all agreements. For the purposes of this paper, the two concepts have been treated as one.

Future work will examine how speech acts of various kinds might be used to create agreements and, hence, joint commitments. Currently, our theory of speech acts [6] argues that the intended effect of a request is to get the addressee to form an individual commitment to do the requested action relative to the speaker's desire that he do so. Although the addressee may be individually committed, nothing in our account prevents the speaker from changing her mind, not notifying the addressee, and then deliberately making the requested action impossible. This would be a clear violation of tacit rules of social behaviour, but nothing in an individualistic account of the commitments entailed by acceding to a request would prevent it. The question remains, for designing artificial agents: Should we augment the semantics of their speech acts to somehow make mutual promises or requests followed by acknowledgments yield joint commitments? And if not, where do the joint commitments come from?

One can also now imagine developing a more general account of dialogue, in which a theorist formally analyses the social contract implicit in dialogue in terms of the conversants' jointly intending to make themselves understood and to understand the other. From our perspective, the signals of understanding and requests for them, which are so pervasive in ongoing discourse [4, 16, 19], would thus be predictable as the means to attain the states of mutual belief that discharge this joint intention [8, 7]. More generally, if such an account of dialogue were successful, it might then be possible to formalize cooperative conversation in a way that leads to the derivation of Gricean maxims.

Finally, let us return to one of our original motivations, designing agents that can work together in groups. Research in artificial intelligence has in the main concentrated on the design of individual agents. If that work is successful (a big "if" indeed), there will undoubtedly be many agents constructed and let loose on the world. Without consideration of how they will cooperate and communicate with other agents, perhaps of dissimilar design, and with people, we risk a kind of "sorcerer's apprentice" scenario — once let loose, they cannot be controlled, and will compete with the other agents for resources in achieving their selfish aims. Joint commitments, we claim, can form the basis for a social order of agents, specifying how groups remain together in the face of unexpected events and the fallible and changeable nature of the agents' attitudes. If built according to our specifications, once such agents agree to cooperate, they will do their best to follow through.

Acknowledgments

This research was supported by a grant from the National Aeronautics and Space Administration to Stanford University (subcontracted to SRI International) for work on "Intelligent Communicating Agents," by a contract from ATR International to SRI International, and by a gift from the System Development Foundation. The second author was supported in part by a grant from the Natural Sciences and Engineering Research Council of Canada. This is a slightly revised version of a paper that appears in Nous 25/4, 1991.

Many thanks go to Michael Bratman, David Israel, Henry Kautz, Kurt Konolige, Joe Nunes, Sharon Oviatt, Martha Pollack, William Rapaport, Yoav Shoham, and Bonnie Webber for their valuable comments. The second author also wishes to acknowledge the Center for the Study of Language and Information and the Department of Computer Science of Stanford University, where he was a visitor during the preparation of this paper.

References

[1] P. S. Atiyah. *An Introduction to the Law of Contract.* Oxford University Press, Oxford, U. K., 1989.

[2] M. Bratman. *Intentions, Plans, and Practical Reason.* Harvard University Press, 1987.

[3] M. Bratman. What is intention? In P. R. Cohen, J. Morgan, and M. E. Pollack, editors, *Intentions in Communication.* MIT Press, Cambridge, Massachusetts, 1990.

[4] H. H. Clark and D. Wilkes-Gibbs. Referring as a collaborative process. *Cognition*, 22:1-39, 1986.

[5] P. R. Cohen and H. J. Levesque. Intention is choice with commitment. *Artificial Intelligence*, 42(3), 1990.

[6] P. R. Cohen and H. J. Levesque. Rational interaction as the basis for communication. In P. R. Cohen, J. Morgan, and M. E. Pollack, editors, *Intentions in Communication.* MIT Press, Cambridge, Massachusetts, 1990.

[7] P. R. Cohen and H. J. Levesque. Confirmations and joint action. In *Proceedings of the 12th International Joint Conference on Artificial Intelligence*, Sydney, Australia, August 1991. Morgan Kaufmann Publishers, Inc.

[8] P. R. Cohen, H. J. Levesque, J. Nunes, and S. L. Oviatt. Task-oriented dialogue as a consequence of joint activity. In *Proceedings of the Pacific Rim International Conference on Artificial Intelligence*, Nagoya, Japan, November 1990.

[9] A. I. Goldman. *A Theory of Human Action.* Princeton University Press, Princeton, New Jersey, 1970.

[10] B. Grosz and C. Sidner. Plans for discourse. In P. R. Cohen, J. Morgan, and M. E. Pollack, editors, *Intentions in Communication.* MIT Press, Cambridge, Massachusetts, 1990.

[11] J. Y. Halpern and Y. O. Moses. Knowledge and common knowledge in a distributed environment. In *Proceedings of the 3rd ACM Conference on Principles of Distributed Computing*, New York City, New York, 1984. Association for Computing Machinery.

[12] D. Harel. *First-Order Dynamic Logic.* Springer:Verlag, New York City, New York, 1979.

[13] J. R. Hobbs. Artificial intelligence and collective intentionality: Comments on Searle and on Grosz and Sidner. In P. R. Cohen, J. Morgan, and M. E. Pollack, editors, *Intentions*

in Communication. MIT Press, Cambridge, Massachusetts, 1990.

[14] H. J. Levesque. A logic of implicit and explicit belief. In *Proceedings of the National Conference of the American Association for Artificial Intelligence*, Austin, Texas, 1984.

[15] H. J. Levesque, P. R. Cohen, and J. Nunes. On acting together. In *Proceedings of AAAI-90*, San Mateo, California, July 1990. Morgan Kaufmann Publishers, Inc.

[16] S. L. Oviatt and P. R. Cohen. *Discourse structure and performance efficiency in interactive and noninteractive spoken modalities.* Computer Speech and Language, 1991, in press.

[17] M. E. Pollack. Plans as complex mental attitudes. In P. R. Cohen, J. Morgan, and M. E. Pollack, editors, *Intentions in Communication*. M.I.T. Press, Cambridge, Massachusetts, 1990.

[18] R. Power. Mutual intention. *Journal for the Theory of Social Behavior*, 14(1):85-102, March 1984.

[19] E. A. Schegloff. Discourse as an interactional achievement: Some uses of unh-huh and other things that come between sentences. In D. Tannen, editor, *Analyzing discourse: Text and talk*. Georgetown University Roundtable on Languages and Linguistics, Georgetown University Press, Washington, D.C., 1981.

[20] J. R. Searle. *Intentionality: An Essay in the Philosophy of Mind*. Cambridge University Press, New York, New York, 1983.

[21] J. R. Searle. Collective intentionality. In P. R. Cohen, J. Morgan, and M. E. Pollack, editors, *Intentions in Communication*. M.I.T. Press, Cambridge, Massachusetts, 1990.

[22] R. Tuomela and K. Miller. We-intentions. *Philosophical Studies*, 53:367-389,1988.

[23] Cohen, P. R., Levesque, H. J., and Smith, I. On Team Formation, in Tuomela, R. and Hintikka, G., *Social Action Theory, Synthese Series*, North Holland, 1997.

[24] Kumar, S., Cohen, P. R., and Levesque, H. J., The Adaptive Agent Architecture: Achieving fault-tolerance using persistent broker teams, *Proceedings of the Fourth International Conference on Multiagent Systems*, Boston, July, 2000.

9

Honouring Sydney J. Hurtubise
Amanuensis to Hector J. Levesque

ROGATIEN "G." CUMBERBATCH

ABSTRACT. This abstract is a summary of a tribute to Sydney J. Hurtubise, which first appeared as [Cumberbatch 2011].

1 Introduction

The accomplishments of Hector Levesque are now legendary. But, there is a man behind the legend. His name is Sydney J. Hurtubise. Wherever Hector has gone, Sydney J. has not been too far away, lurking in the background. Whatever Hector has done, Sydney J. probably deserves some of the credit. Insights into Sydney J. will help to provide insights into Hector J. Let's look more closely at this important figure, a man now justly ignored despite incredible accomplishments of his own.

Sydney J. Hurtubise first came to the attention of the artificial intelligence community with the publication of his paper entitled "A Conversation Between Some Man and a Smart-Aleck Computer" [Hurtubise 1976]. In this paper, he outlined some important modeling techniques for early dialogue research, in particular introducing the concepts of "belief cloud" representation and "buck-passing" control, ideas that have never been surpassed (nor in fact ever used). This paper got Dr. Hurtubise tenure at the University of North Bay (Callander Campus), and launched his career, but didn't bring him the international fame he so desired. This fame was not long in coming, however. Working with his first Ph.D. student, Rogatien "G." Cumberbatch, he produced the world's first "syllogistic" reasoning system, thus unifying the wisdom of the ages with the new truths being discovered by modern computer science.

The SILLI system, as it was labeled, was a brilliant system that drew inspiration for its knowledge representation primitives from both computational complexity and linguistics. Able to leap tall syllogisms in a single bound, and with superb error detection capabilities, it proved to be a highly effective reasoning system, for anybody (or indeed any system) wishing to reason in syllogisms. A seminal paper appeared in the 1978 Canadian AI conference [Hurtubise and Cumberbatch 1978], describing this foundational system to an amazed world. They were amazed not only by the paper's discoveries, but also were amazed to discover the paper at all, since it did not appear either in the table of contents of the Proceedings of the conference, nor in the program. Perhaps this in some ways explains the current deserved obscurity of Sydney J. Hurtubise: nobody could find the paper. To rectify this, the original SILLI paper is reproduced here in exact facsimile (complete with its IBM Selectric type font and in its old-fashioned 1970's English idiom). Readers I'm sure will agree that it truly is a SILLI system, fully deserving of the accolades it did not receive. Most importantly, however, its methodologies consolidated an emerging new school of knowledge representation, ignorance-based reasoning, which has now spun off the applied area of ignorance engineering. Ignorance-based reasoning and ignorance engineering are now much in evidence, in both the research literature and in many systems in actual use in the "real world".

Although this SILLI paper got Sydney J. Hurtubise promoted to full professor at UNB (CC), he did not stop pursuing creative research directions merely because he had reached the pinnacle of his academic career. A scientist is not motivated by mere earthly rewards, but seeks eternal truth (or in Sydney J. Hurtubise's case, ignorance). Thus, Sydney J. continued to show up at international conferences, constantly taking part in discussions after talks, or over a beer later in the bar, intervening with many ignorance-based opinions. I know, because I was often there backing him up. He also continued to create new research ideas. One notable example was his 1996 invited panel presentation at the Canadian artificial intelligence conference in which he showed how the Dr. Seuss classic story "Sam I Am" could be represented in the situation calculus, a new use for logic-based AI that only somebody as orthogonally orientated as Sydney J. Hurtubise could have devised. Another notable discovery was the "Reductio Ad Absurdum" argumentative design pattern, now in wide use throughout science. Although now of great age, Sydney J. Hurtubise continues to be active, always thinking of newly ignorant things, always trying to influence all those around him, still a seminal influence on Hector J. Levesque.

2 Brief Biography of Sydney J. Hurtubise

Born in North Bay, Ontario, Canada, at just about the same time as Hector J. Levesque, Sydney J. Hurtubise grew up there, right in the heart of the "shad fly capital of Canada". It is rumoured that the young Sydney (known to all those around him as "Syd the Kyd") and the young Hector might even have met on occasion, although there is no documentation to that effect. Sydney J. Hurtubise was educated at the University of Toronto right through his Ph.D., where his thesis explored territory on the boundary of Classics and Computer Science, in particular looking into whether the structure of ancient Latin poems could prove insightful for the design of programming languages. His development of the notion of programming rhythms, in particular the "iambic pentameter" coding rhythm, proved innovative, but has never been followed up. Nor has his proposal that programs should have a rhyming scheme, even though this idea was promoted as a contribution to the then important area of structured programming, and even though it was by no means the least useful idea to come out of that school of software engineering.

After his doctorate, Sydney J. Hurtubise took up a position in the United States at the Slum Burger fast food chain in silicon valley. It wasn't long, however, before he was lured back to Canada to the University of Toronto as an assistant professor. His subsequent discovery that Toronto's CN Tower was not the world's tallest structure, but merely its tallest free standing structure, so disillusioned him that it wasn't long before he made the move to the University of North Bay (Callander Campus), where his career blossomed. Here is a very brief list of some of his career accomplishments:

- appointed Canada Research Chair in Ignorance-Based Reasoning at UNB (CC)

- winner Lack of Thought Award for early career contributions to formalizing ignorance-based representation techniques

- winner Lifetime Non-Achievement Award of the International Ignorance and Idiocy Association (IIIA) for contributions to ignorance-based reasoning

- elected Jolly Good Fellow of the IIIA

- winner of the "Purity of Essence" award, one of Canada's highest scientific honours, for his discovery of the "Reductio Ad Absurdum" argumentative design pattern

3 Brief Biography of Rogatien "Gatemouth" Cumberbatch

Born in the Canadian wild west of Alberta in the frontier town of Edmonton, Rogatien Cumberbatch was the most garrulous in a very talkative family, earning him his lifelong nickname. After an undergraduate degree at the University of Alberta (in Non-Statistical Reasoning) and a M.Sc degree there too (in Confusing Science), Rogatien headed out to the University of British Columbia to do his Ph.D. His Ph.D. work was on natural language understanding, where he began development of the D-FENDER system to answer questions posed on the BC Driver's Written Examination, a precursor to the ignorance-based approaches he and Hurtubise later made infamous. However, when an astute committee at UBC determined that Cumberbatch had help in developing the system (by noticing that the paper describing D-FENDER [Cumberbatch and Hopozopen 1974] was, in fact, co-authored), he was asked to leave the University. He moved to the wilderness of Ottawa, the world's coldest capital city (working "for the man" as he put it, although which man isn't precisely clear). While there, he happened to run into Sydney J. Hurtubise, who was on his annual shopping trip to Ottawa for provisions to tide him over the long North Bay winter. Professor Hurtubise convinced the no longer so young Cumberbatch to come to UNB (CC) to resume his Ph.D. under Hurtubise's supervision, leading eventually to the SILLI system and a mutual commitment by them both to pursue ignorance. A life long fruitful collaboration was launched.

After graduation, Cumberbatch moved west, to Floral University, because as he put it, he "missed the wheat". With the broad perspectives of Canada's rectangular province to inspire him, Rogatien "G." Cumberbatch went on to fame of his own, as the founder of the area of ignorance engineering, the applied "yin" to Sydney J. Hurtubise's theoretical "yang". Here are some career highlights for Rogatien "G." Cumberbatch:

- appointed Senior Fellow of the Society for the Encouragement of Ignorance, Illogic, and Illiteracy in Computational Systems (EIII-CS)

- winner EIII-CS Lifetime Non-Achievement Award in Ignorance Engineering for his seminal role in creating the field of ignorance engineering

- induction into the Tabula Rasa Society of Scholars for his commitment to the importance of not evaluating computer systems

- elected First President, Society of Professional Ignorance Engineers (SPIEs)

- certified as a P.Ig.

Acknowledgements

Professor Cumberbatch would like to be able to acknowledge his research funding sources.

References

Cumberbatch, R. "G." and H. P. Hopozopen [1975], "An Extensible Feature-Based Procedural Question Answering System to Handle the Written Section of the British Columbia Driver's Examination", *Second Newsletter of the Canadian Society for Computational Studies of Intelligence*.

Cumberbatch, R. "G." [2011], "Honouring Sydney J. Hurtubise, Amenuensis to Hector Levesque", G. Lakemeyer and S. McIlraith (eds.), *Knowing, Reasoning, and Acting. Essays in Honour of Hector J. Levesque*, this volume.

Hurtubise, S. J. [1976], "A Model and Stack Implementation of a Conversation Between Some Man and a Smart-Aleck Computer", *Third Newsletter of the Canadian Society for Computational Studies of Intelligence*, pp. 46-50.

Hurtubise, S. J. and R. "G." Cumberbatch [1978], "In Defence of Syllogisms", *Proceedings of the Second National Conference of the Canadian Society for Computational Studies of Intelligence*, Toronto, Ontario, July, pp. 311-314.

In Defence of Syllogisms

by

Sydney J. Hurtubise
and
Rogatien "Gatemouth" Cumberbatch

Department of Issues, Distinctions, Controversies, and Puzzles
University of Artificial Intelligence
North Bay, Ontario*

Abstract

In this paper we present a serious study of syllogistic reasoning. We believe that powerful new developments in computer technology (such as the process metaphor, the type / token distinction, procedural semantics, pattern recognition, and the INTEL 8080 chip) add significantly to man's understanding of the universe and hence make it possible to break new ground in this classic philosophical endeavour. The paper itself provides the details, and we strongly urge all progressive AIers to read it - it would be a shame to miss out on one of the greatest pieces of research carried out since the advent of the space age.

1. Introduction

Recently there has been a renewal of interest in using tried and true logical methods for solving problems in artificial intelligence. What we propose here is going back to the original tried and true logical method: syllogistic deduction. There are a number of reasons for our decision. First, syllogisms are a nicely limited domain, approachable and understandable by anybody, even the common man. Second, syllogisms have been around so long they are probably by now cognitive primitives underlying all other reasoning. Third, we believe that researchers in AI should not look at more esoteric logics (e.g. propositional logic, predicate calculus, CONNIVER) until there has been a full understanding of earlier approaches. Moreover, much of the so-called power of advanced logical reasoning systems (e.g. resolution theorem proving) could trivially be achieved syllogistically.

----

*This research was supported in part by charitable donations.

In this paper we describe a model called SILLI which accepts as input two premises stated in natural language and produces a syllogistically valid natural language conclusion. We have chosen to use natural language rather than a logical notation because for the common man natural language is very clear cut, but logic, full as it is of brackets and squiggly little symbols, is extremely ambiguous, even incomprehensible at times (brackets are unnatural (at least that's what we think (and we are not alone, either)))).

The SILLI model is comprised of four main components:

(i)   a natural language front end that translates users' premises into internal notation;

(ii) a deductive component to actually carry out the syllogistic reasoning;

(iii) a semantic network to encode knowledge of the world;

and (iv) a natural language back end which produces the conclusion derived from the premises.

Lets look at these components in more detail.

2. Components of the Model

2.1 The Natural Language Front End

This component must interpret natural language sentences of the form "A verb B" or "All A verb B". Using the revolutionary new picture theory of meaning ("a picture is worth a thousand words"), we convert an input sentence into an analogic representation called a picture.+ Obvious efficiencies

----

+A picture is a high level gestalt representation that completely circumvents the problems of the usual low-level pixels. See Hurtubise (1976) for more details.

in storage can be achieved by this method. Thus, the sentence

"Pick up the big red block and hit Mary with it."

would take up only .011 of a picture. Another advantage of using pictures is that there is absolutely no need to do pragmatics (this is a well known result from computer vision research). Because all this is done by efficient parallel procedures, we call our approach to language the procedural approach.

2.2 The Deductive Component

This component of the SILLI system takes a couple of interpreted premises and deduces a conclusion from them. To do this it uses two main logical postulates:

(i) The Active / Passive Postulate

If the first (or active) premise is of the form "All A something-or-other" and the second (or passive) premise is of the form "B is A", then the conclusion "B something-or-other" can be derived. Thus, this syllogism is valid:

active: "All Saints Cathedral has
            stained glass windows."
passive: "Saint James is a saint."

conclusion: "Saint James Cathedral has
            stained glass windows."

(ii) The EQ-NP Deletion Postulate

If there is a premise of the form "NP1 is NP2", the phrases NP1 and NP2 are equivalent. Either can be deleted and replaced by the other anywhere in any premise. For example, in the syllogism

active: "The temperature is 90°F."
passive: "The temperature is 32.2222°C."

conclusion: "90°F is 32.2222°C."

we have used the EQ-NP deletion postulate to delete "the temperature" from the passive premise and replace it with "90°F" from the active premise.

2.3 The Semantic Network

The semantic network component of the model is used to check user's statements of the form "A is B" for real world validity. The network consists of a bunch of nodes and arcs connected up into a generalization hierarchy. There are several kinds of arcs in this hierarchy: IS-A, A-K-O, IS, SUP, and ISA. The distinctions separating these arc types are subtle and are unjustly ignored in most semantic networks.

Figure 1 shows a sample network:

Figure 1 - A Sample Network

To see how the semantic network component is used by SILLI, lets make the simplistic assumption that the above sample net is the model's entire knowledge base. Now, if the user were to say something like

"John is a graduate student."

then this would not check out with the facts and the error message

"You have made an error."

would be printed out. Note that error messages, too, are in natural language, a nice piece of generality.

2.4 The Natural Language Back End

This works just like the natural language front end, only in reverse.

3. Implementation

Unlike many AI systems, our model has been implemented. Our programs are all encoded in a structured production system, consisting of structured productions of the form

IF ξ THEN ψ ELSE θ ;

where ξ, ψ, and θ are LISP procedures. Structured productions are better than normal productions because they have a natural syntax, they have two right-hand sides (ψ, θ) for each left-hand side (ξ), and they give the power of arbitrary LISP procedures to productions.

Our data structures are different from our programs because data and programs are quite distinct in real life (as opposed to the cloistered academic environment where sometimes data and programs can be the same). Data are stored in LISP lists because these are in widespread use throughout the AI community and also because they are flexible enough to store anything if you really try.

We have found that this combination of structured productions and LISP lists is very easy to use and, moreover, allows the construction of programs that are not only

162

efficient in time and storage, but are also readily understandable to the user. In fact our SILLI program has the following complexity bounds:

$space:$ $lower\ bound:$ $o(n^3 \log n)$

$\qquad upper\ bound:$ $o(\log (\log (\log)))$

$time:$ $lower\ bound:$ $o(n \log (\log n^2))$

$\qquad upper\ bound:$ $o((\log n) \div (\log n))$

We are still trying to precisely work out lower and upper bounds on understandability, although we suspect polysyllabic bounds in both cases.

A note of interest here is that these bounds seem to be related to the bounds discovered by numerical analysts for Runge-Kutta methods of order 6 that solve systems of stiff partial differential equations (PDEs). In fact for any bound "B" of our SILLI program, the equivalent bound for the Runge-Kutta method is "mB". What we would like to know is: with suitable analysis of round-off and truncation error (perhaps using backward error analysis), will we ever be able to show that B = mB?

4. Sample Run

At last we arrive at the moment of truth (or at least not falsity). Lets look at how SILLI handles a set of real syllogisms. We think the following actual run speaks for itself, although we will annotate the output (using comments surrounded by /* --- */) to indicate interesting points, much as is done in chess protocols (see any SIGART Newsletter for more details).

```
          ^C
          .LOGIN 1978,0721
          ENTER PASSWORD
          ▮▮▮▮▮▮▮▮        /* The password is hidden from the reading public to protect */
          READY           /* our computer dollar budget.                              */
          .R LISP
          *(INVOKE STRUCTURED_PRODUCTION_SYSTEM_PROGRAMS_AND_DATA_STRUCTURES)
          NIL
          *(START THE SYSTEM UP NOW PLEASE)    /* The SILLI system is started up.        */
          WELCOME TO THE SILLI WORLD           /* The syllogisms have been numbered for  */
          ENTER PREMISES                       /* the purposes of identification.        */
     1.   #MARY HAS BROWN HAIR.
          #MARY IS STANDING.                   /* These are the user's premises.         */
          THE CONCLUSION IS
          ¢STANDING HAS BROWN HAIR.            /* This illustrates the EQ-NP deletion    */
                                               /* postulate.                             */

          ENTER PREMISES
          #+§√ ¶E∫ (FAST_MODE_NOW_PLEASE)      /* Here we use the macro characters       */
          OKEY DOKEY                           /* +§√ ¶E∫ to temporarily call     LISP   */
                                               /* from ° within SILLI.  In this case     */
     2.   #ALL MEN ARE CREATED EQUAL.          /* we have SILLI stop printing "ENTER     */
          #JOHN IS A MAN.                      /* PREMISES" and "THE CONCLUSION IS"      */
          ¢JOHN IS CREATED EQUAL.              /* and instead just use # and ¢.          */
                                       /* This illustrates the active / passive postulate. */

     3.   #ALL STUDENTS LAID END TO END WOULD STRETCH AT LEAST
               FROM HERE TO NORTH BAY.
          #JOHN IS A STUDENT.                       /* Premises and conclusion        */
          ¢JOHN LAID END TO END WOULD STRETCH AT LEAST  /* can be more than one line. */
               FROM HERE TO NORTH BAY.

     4.   #JOHN WANTS TO BE THE PRESENT KING OF FRANCE.
          #THE PRESENT KING OF FRANCE IS BALD.
          ¢JOHN WANTS TO BE BALD.                   /* A classic problem, handled */
                                                    /* well by SILLI.             */

     5.   #MARY REALLY THINKS SHE IS IT.
          #IT IS RAINING OUTSIDE.                   /* SILLI handles this without */
          ¢MARY REALLY THINKS SHE IS RAINING OUTSIDE.  /* needing to resort to com- */
                                                    /* plicated structures such as*/
     6.   #PICK IT UP.                              /* belief spaces.             */
          #IT IS SUCH A PRETTY WORLD TODAY.
          ¢PICK SUCH A PRETTY WORLD TODAY UP.  /* Note the displacement of the particle */
                                               /* "up" in the conclusion.  This flaw    */
          ^C                                   /* is discussed in section 5.            */
          .REE
```

```
7.  #ALL MEN SHOULD LAY DOWN THEIR ARMS
        AND LIVE IN PEACE AND BROTHERHOOD.
    #WHAT IS A MAN?                        /* The "?" in the conclusion is preserved*/
    ¢WHAT SHOULD LAY DOWN THEIR ARMS       /* because of our sophisticated natural  */
        AND LIVE IN PEACE AND BROTHERHOOD? /* language back end.                    */

8.  #NO MAN IS AN ISLAND.
    #HE IS A REAL NOWHERE MAN.             /* SILLI can handle negation. */
    ¢HE IS A REAL ANWHERE ISLAND.

9.  #ALL MEN ARE MORTAL.
    #SOCRATES IS A MAN.         /* SILLI picks up the user's mistake - since      */
    ¢YOU HAVE MADE AN ERROR.    /* Socrates is dead, how can he still be considered*/
                               /* to be a man?                                   */
    ^C
    .KJOB      /* This ends the sample run. We do not show our final statistics   */
               /* on the run due to acute embarrassment.                         */
```

5. Evaluation

Anybody looking at the performance and competence of SILLI on these examples has to be impressed. The deductive component handles a wide variety of syllogisms, usually with great aplomb and amazing grace and without combinatorial explosion. The natural language front end not only does what the user wants but also what he means to want; and vice versa for the back end. The system is very robust, able to overcome erroneous input (e.g. syllogism 9) with ease. Finally, SILLI seems to be well debugged. In fact, the only bug, if you can call it that, which manifested itself in this run was caused when we inadvertently hit ^C after syllogism 6. This just indicates that we are still human and haven't become dehumanized and depersonalized, an ever present danger in AI research that (unlike ours) attempts to down grade mankind by simulating aspects of humanity that are inherently so warm, moist, and fuzzy that no computer could ever do them.

Of course there are still a few unsolved problems. For instance the syllogism

active: "All students laid end to end
             would stretch at least from
             here to North Bay."
passive: "Students are people."

conclusion: "All people laid end to end
             would stretch at least from
             here to North Bay."

does not work (and in fact will send SILLI into an infinite recursive descent) despite its obvious real world validity. There are also minor difficulties in the natural language processing, as illustrated by the particle displacement in syllogism 6. A final criticism that nit-picking nay-sayers (such as the referees) could use to denigrate our approach is that "a few of the syllogisms in the sample run are a bit suspect" (emphasis mostly mine). Even if this is true, it must be admitted that they do exist (otherwise we wouldn't have been able to use them) and they must therefore be explained. The test of any theory is that it

account for all the facts! Subtle problems notwithstanding, we believe our theory does just that.

6. Conclusion

Interesting as it is from a theoretical standpoint, this research as it now stands has only a limited number of practical applications. However, we have plans to vastly expand the usefulness of SILLI. We believe that methods similar to those used for syllogisms will also work well for the closely related areas of limericks and sonnets, and may be applicable to less structured domains such as blank verse, free verse, or even computer vision. The long term future of this research is assured.

Acknowledgements

The authors would like to gratefully acknowledge the Greeks. Without their invention of both the syllogism and the Greek alphabet, this research would have taken somewhat longer to complete.

Bibliography

Hurtubise (1976). S. J. Hurtubise, "A
    Model and Stack Implementation of a
    Conversation Between Some Man and a
    Smart Aleck Computer", Third CSCSI/
    SCEIO Newsletter, July 1976.

Hurtubise and Cumberbatch (1978).
    S. J. Hurtubise, R. "G." Cumberbatch,
    "In Defence of Syllogisms", see this
    volume.

Hopozopen (1975). Horace P. Hopozopen,
    "An Extensible Feature-Based Pro-
    cedural Question Answering System to
    Handle the Written Section of the
    British Columbia Driver's Examination",
    Second CSCSI/SCEIO Newsletter, 1975.

10

From Intentions to Social Commitments: Adaptation in Multiagent Systems

FABIANO DALPIAZ, AMIT K. CHOPRA, JOHN MYLOPOULOS, AND PAOLO GIORGINI

ABSTRACT. Runtime adaptation of software systems is an area of software engineering that is gaining increasing attention from researchers. Adaptive software can successfully cope with failures and the volatility of the environment it operates in. Researchers in software requirements and architecture are especially interested in model-driven adaptation. The idea is that software would reflect upon its own model in order to reason about adaptation.

This chapter represents a summary of our recent work for supporting adaptation in multiagent systems. Our key contribution is to exploit the notion of *social commitment* in order to reason about adaptation in multiagent systems. While other approaches for adaptation assume cooperation or some form of logical centralization, our approach works for open multiagent systems wherein agents are autonomous and heterogeneously constructed. We conceive adaptation from the perspective of an intentional agent, who needs to interact with other agents in the multiagent system—by engaging in social commitments—to achieve its goals.

0 An Allegory.[1]

Imagine! You are an enterprise architect with HyperTech Industries Unlimited ("Hype", for short) and a client comes along wanting to re-engineer all business processes for her multinational publications organization, Home Periodicals Inc. (hereafter "Hope") to eliminate redundant positions while improving performance. Your mission—should you decide to accept it—is to re-engineer these business processes in consultation with Hope staff and, most importantly, keep them happy and off your boss' back.

Now, you happen to be a recent graduate from a Computer Science program and you are aching to use all these great ideas you learned back in college. So, you dig up your course notes from your home basement and proclaim at the next meeting with your boss and your client that you intend to go about this using the latest work from the best minds in the field. Ignoring the anxious looks on their faces—and the sneers behind your back among your fellow architects the following day—you begin your task going over all your material, reading up references and taking notes.

At the end of your search, you proudly present your boss with your findings. There seems to be unanimous agreement among experts, you note, that existing business process specification techniques—based on old and well-tried system modeling concepts such as finite state machines, interaction diagrams, and Petri nets—lead to inflexible, procedural descriptions of what's to be done. Since human activity is intrinsically situated and emergent, such specifications lead to serious discrepancies between what is specified and what

---

[1] Adapted loosely from [Mylopoulos 1992]

actually happens. Such specifications also fail with respect to flexibility, as they are difficult to customize to user preferences, or adapt as the environment changes. While your boss waits patiently, you point out that there seems to be no consensus on how to deal with these fundamental problems. Some organizations fall back on informal specifications of their systems, whereas others adopt simpler diagrammatic techniques (flow charts and so on) that leave out important details. Sketches of processes (formal or informal) are fine sometimes, you remark, but they often lead to misunderstandings among Hope employees, and suboptimal performance in the execution of its business processes. Your boss (who, you just notice, bears remarkable resemblance to Dilbert's Pointy-Haired boss[2]) shakes his head knowingly. Following your carefully laid out script, you draw an analogy between what current business process specifications practices do and the proverbial drunk, who late one night is looking under a street lamp for his keys, lost elsewhere, because there he can at least see.

As your boss becomes restless, you get to the punchline. According to your recently completed multiagent systems course, the key concepts for talking about multiagent systems (which business processes surely are) are *agent* and *social commitment*. A social commitment consists of one agent x committing to another agent y to fulfill proposition q if p (often, p would be something that y would have to bring about). Such atomic contracts can be used to model complex interactions, such as those involving the execution of a multiparty process. To specify such processes, we need to abstract away from the concept of agent to that of a *role*. Then, system specifications consist of generic commitments among roles. These can be instantiated and enacted whenever concrete agents are bound to all the roles that are part of a specification. With this approach, you conclude while waiving didactically your longest finger, we avoid the pitfalls of current process specification techniques. Multiagent system specifications based on commitments are neither procedural nor do they talk about tasks, activities, and plans. Instead, they specify the social expectations (via social commitments) that would arise from the agents' interactions.

Your boss stops looking at the ceiling and is now staring directly into your eyes (and yes, he does look Pointy-Haired!). "Could it be that this guy is on to something?", he wonders. The tools offered for commitment-based specifications, you continue, are based on ideas from multiagent systems. The key concern is to change the nature of multiagent interaction specifications so that they are social models, rather than procedural ones.

Your boss is ecstatic. He has recently watched the movie *The Social Network* and is fleetingly favorable towards all things social. "Here is another wacky idea", he thinks, "but it actually sounds better than all the others we have been peddling to our customers. Let's try it!" He gives you his blessing and you give him a good book on multiagent systems ([Singh and Huhns 2005]) for background reading before embarking on the Hope project. For the rest of the week you are definitely on the good side of your boss' balance sheet as you rock-and-roll between Hype and Hope. And it's all thanks to multiagent systems...

1 Introduction

In the age when the department of Computer Science at the University of Toronto was in its infancy, Hector Levesque had a trail-blazing career that took him from junior undergraduate to a faculty position within the span of a little more than a decade. Hector was the undergraduate who asked—by correspondence—Marvin Minsky when he had questions about Artificial Intelligence (AI). And it was him who put the instructor of the AI course to task

---

[2]http://www.dilbert.com/

for not using Lisp as the programming language taught in the course[3]. His MSc thesis was inspired by Lisp in defining a self-descriptive knowledge representation language[4]. And as a PhD student he taught his supervisor things (modal logics, and more), rather than the other way around, as he was working towards an epistemic account of knowledge bases. Among other ideas, that thesis used a functional account of knowledge bases [Levesque 1981] that was later adopted, first by KRYPTON [Brachman et al. 1983] and then by Description Logics. And all that was a preface to a sparkling research career.

Multiagent interactions among human, organizational and artificial agents account for much of the social activity of modern world, be they business processes, socio-technical systems or collaborative efforts. To better manage and support such multiagent interactions, computer scientists have used a variety of specification techniques—often based on Petri nets, finite state machines and the like. Unfortunately, such techniques are too rigid, prescriptive and inflexible to describe human activity in two orthogonal ways. Firstly, they don't account for agent autonomy. Secondly, they are static and difficult to dynamically adapt in the face of varying user preferences and environmental settings.

A multiparty interaction (business process or otherwise) is above all else a social activity. During such an interaction, participants care about whom they interact with; what contractual relationships arise from their interactions and if they have recourse in case others violate their contractual obligations; whether their goals are likely to be met by participating in the activity; whether they can interact flexibly and adapt in appropriate ways when necessary.

This paper adopts concepts and ideas from the literature on multiagent systems to address the problem of *agent adaptation*—an area Levesque has explored in the past [Kumar et al. 2000]. Levesque's work relies upon the notion of *internal commitment* to intentions. By contrast, our proposal builds upon ongoing work by Munindar Singh and his group on the specification of multiagent interactions in terms of *social commitments*. Social commitments, as we show later, are distinct from internal commitments.

Organization. Section 2 contrasts cognitive abstractions against social abstractions. It introduces a key social primitive, that of commitment among agents. Section 3 introduces a conceptual model of adaptation that takes into account social commitments. Section 4 introduces an architecture for adaptive agents. Section 5 summarizes the paper and concludes with directions for future research.

2 From Intentions to Social Commitments

Broadly, we understand an agent as an active autonomous entity that acts according to its own motivations. Typically, an agent would not be able to achieve all of its goals by interacting with inactive objects in the environment. Rather, it would need to interact with other agents and enter into social relationships with them to achieve its goals. The interactions give rise to a multiagent system.

Multiagent systems may be specified a priori. Such specifications would prescribe the legal interactions among the agents, not with reference to specific agents however, but with reference to *roles*. We call such multiagent system specifications *protocols*. An agent could then adopt a role in a protocol if doing so suited its goals. For example, if Barbara wanted to sell her cellphone, she may consider adopting the role *seller* in some auction protocol.

---

[3]The University of Toronto didn't offer any form of interactive computing at the time, so Lisp was out of the question.

[4]Elements of that thesis were published in [Levesque and Mylopoulos 1979].

Researchers in multiagent systems are keenly concerned with protocols and agents and the abstractions in terms of which they can be specified. Generally speaking, computer science has tended to address more the challenges of high-level agent specification rather than high-level protocol specification. This is likely due to traditions from AI, where agents are viewed as intelligent computational entities capable of reasoning in terms of cognitive (mental) abstractions such as beliefs, goals, intentions, and obligations [Bratman 1987].

Cohen and Levesque's work [1990] on cognitive abstractions has been especially influential over the decades. They formulated a rich theory of rational action based on the concepts of *intention* and *internal commitment* to intention. For an agent to succeed with its intentions, it must be internally committed to realizing them. Further, the agent should not be overcommitted or undercommitted to the intentions as that would produce undesirable results.

Another widely influential strand of research, following research in the philosophy of language, concerns itself with the semantics of communication. Searle's account of speech act theory [1969] categorizes communications into types such as *assertives, commissives*, and so on, and formalizes each in terms of the mental states of the communicating agents. For example, if Barbara asserts to Alice "It is raining outside", it would be taken to mean that Barbara actually believes it is raining outside. If Barbara commits to all bidders to honor the winning bid for her cellphone, it would be taken to mean that she intends to honor the winning bid. In effect, speech act theory gives meaning to communication—an observable social phenomenon—in terms of the private mental states of agents. In a similar vein, Cohen and Levesque [1990] extrapolated internal commitment as being foundational for understanding communication among agents. We shall refer to this class of approaches as *mentalist*. The mentalist approach to communication influenced greatly the agent-oriented programming paradigm [Shoham 1993] and the attempts to standardize agent communication languages (KQML [Finin et al. 1994] and FIPA ACL [FIPA 2003] being the prominent ones).

Meanwhile, another strand of research focused on a convention-based view of communication, where agents interact with each other on the basis of public conventions [Winograd and Flores 1986]. Singh [1991] argues that *social commitments* are distinct and orthogonal to internal commitments, and they are intimately tied to public convention. By contrast, internal commitments, as their name suggests, are private to an agent. Let us return to our example to illustrate the difference. On the one hand, Barbara may be internally committed to honor the winning bid, but may not have communicated so to the bidders; in other words, even though Barbara is internally committed, she is not socially committed. On the other hand, Barbara may be socially committed to honor the winning bid (because she communicated that to the bidders) but not internally committed to do so (maybe because she has a minimum price in mind). Singh pointed the importance of keeping the two notions separate: while social commitments could be used to formalize convention and communication, internal commitments could be used to design specific agents.

Singh [1998] expanded on the argument against mentalism to show that speech act theory could not form the foundations of agent communication. He argued against the approach taken by KQML and FIPA ACL by showing that they led to unrealistic assumptions about the nature of multiagent systems. For example, assumptions about the sincerity of agents were unrealistic because one agent could not possibly know what another intended. In particular, such assumptions violated the autonomy of agents, and limited multiagent systems only to systems of cooperative agents. A recent joint manifesto [Chopra et al.

2011] by researchers in agent communication

- *affirms* the necessity of a social semantics for communication,

- *affirms* social commitments as a key part of any such social semantics, and

- *rejects* mentalist semantics and approaches based thereupon such as the FIPA ACL.

The distinction between social and internal commitments is crucially relevant to the design of multiagent interactions, where agents could be organizations, humans, or their software surrogates, with distinct, private, and often competitive motivations. These entities interact with each other on the basis of public conventions, not on the basis of internal commitments or any other mentalist notion. These public conventions are in fact protocols.

Cognitive abstractions have a place. However, this place is not in formalizing communication; it is in capturing agent intentions. Such intentions lead to communication and account for its purpose. For example, Barbara's desire to buy a cruise ticket may be the motivation for selling her cellphone, but her desires cannot explain the social effects of actually communicating the offer to others. Chopra et al. [2010] offer an account of the reasoning that connects goals with social commitments.

A social commitment is of the form C(debtor, creditor, antecedent, consequent), where debtor and creditor are agents, and antecedent and consequent are propositions [Singh 1999]. From now on, we will simply use *commitment* to refer to social commitments. Other kinds of commitments, for example internal, will be appropriately qualified. $C(x, y, r, u)$ means that x is committed to y that if r holds, then it will bring about u. If r holds, then $C(x, y, r, u)$ is *detached*, and the commitment $C(x, y, \top, u)$ holds ($\top$ being the constant for truth). If u holds, then the commitment is *discharged* and doesn't hold any longer. All commitments are *conditional*; an unconditional commitment is merely a special case where the antecedent equals $\top$. (For general rules of commitment reasoning, see [Singh 2008].)

For example, C(*Barbara, Alice, paid, deliverPhone*) means that Barbara commits to Alice that if payment is made, the phone will be delivered. When the payment is made, then the unconditional commitment C(*Barbara, Alice,* $\top$, *deliverPhone*) holds.

Commitments are rooted in communication. A commitment can be created or otherwise manipulated only by explicit communication among agents. The primitive messages for manipulating commitments are *Create, Cancel, Release, Delegate, Assign* [Singh 1999].

The notion of commitment protocols [Yolum and Singh 2002] has been highly influential in multiagent systems. The basic idea there is that application-specific messages can be assigned meanings (by a protocol designer or collectively by the application community) in terms of commitment-specific messages. Thus for example, in a particular community of fruit-sellers, a quote from a merchant to a customer may be formalized as creating the corresponding commitment:

Quote(m,c,item,price) counts as *Create(m,c,item,price)*

In another community, a quote may not mean any such commitment. The community provides the *context* of the commitment, which we omit here.

3 Conceptualizing Agent Adaptation

Adaptation in agents has largely been confined to two kinds of settings. One considers only a single agent; in other words, it does not consider interaction. The other considers multiple agents but in the context of a cooperative setting. Levesque, in joint work with others [Kumar et al. 2000], formulated adaptation in *team* settings. Essentially, a team of brokers has to ensure the fault tolerant operation of a system of agents (via services such as locating services, routing, and so on). The key feature of their approach is the establishment of *team commitments* and individual internal commitments among the brokers. For example, the brokers adopt the team commitment that if a client loses connection with a broker, the brokers will attempt to reestablish connection with the client. Further, if a registered client is disconnected, each broker would have the internal commitment to make that believed among the team, and so on.

Our approach to adaptation instead relies on *social* commitments. As such, it applies to cooperative as well as competitive systems. We characterize an adaptive agent in terms of both intentional concepts—the *goals* it wants to achieve—as well as social concepts—the social commitments it makes to or takes from other agents. We aim to ensure that a specific agent achieves its goals in a dynamic environment where failures and under-performance are common events. Although we do not talk explicitly about teams, extensions of our approach should be able to support teamwork as a special case. In general, our approach for agent adaptation should work in any open multiagent system, that is, a system where the agents are autonomous and independently designed.

We conceptualize adaption in terms of changing strategies in order to achieve a goal. Hence, the adaptive agent in our approach is goal-driven. Essentially, the agent monitors his goals, and depending on environmental conditions and interactions with other agents, adopts new strategies to achieve the goal if necessary.

We illustrate our approach on a scenario concerning firefighting, where Jim, a fire chief, has to decide how to efficiently respond to a fire. Jim's goal model is shown in Figure 1 using a subset of the concepts of Tropos [Bresciani et al. 2004]. Jim's top-level goal is "fire extinguished". This goal is OR-decomposed to indicate that there are alternative ways to achieve it. Jim can either use a fire hydrant or rely on a water tanker truck. Both goals are AND-decomposed. In order to use a fire hydrant, Jim has to notify this need and get authorized. To fight the fire with a tanker truck, the tanker service should be paid, the fire should be reached by a truck, and a water pipe should be connected to the truck. Jim's capabilities—goals he can achieve without interacting with other agents—are "hydrant need notified" and "pipe connected".

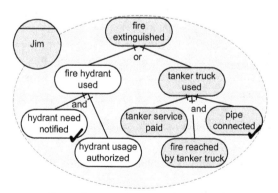

Commitments in the variant: C_2, C_3

Figure 1. Jim's goal model, emphasizing his current variant to extinguish the fire

Table 1 lists some possible commitments that may arise in the firefighting scenario. C_1 is made by fire brigade Brigade1 to Jim. Brigade1 commits that, if hydrant need is notified, then hydrant usage will be authorized. Jim commits (C_2) to the fire brigade that, if the tanker service is paid, then the water tanker truck will be used. Commitments C_3 and C_4 are made by different water tankers to Jim. Both commitments tell that, if the tanker service is paid, fire will be reached by the tanker truck.

C_1	C(Brigade1, Jim, hydrant need notified, hydrant usage authorized)
C_2	C(Jim, Brigade1, tanker service paid, tanker truck used)
C_3	C(Tanker1, Jim, tanker service paid, fire reached by tanker truck)
C_4	C(Tanker2, Jim, tanker service paid, fire reached by tanker truck)

Table 1. Some example commitments in the fire response scenario

The notion of *variant* is fundamental in our framework as it defines a common semantic substrate that characterizes a goal-oriented agent operating in a multiagent setting. The intuition is that, given a certain goal to achieve, a variant is a strategy that could lead to the achievement of the goal—if successfully carried out at runtime.

We provide the basic intuition behind the notion of variant. More technical details can be found in [Dalpiaz et al. 2010]. A variant refers to one or more goals—typically, top-level goals—the agent aims to attain. A variant consists of a set of goals $\mathcal{G}$ the agent wants to achieve, a set of commitments $\mathcal{P}$ the agent intends to make or take, and a set of capabilities $\mathcal{C}$ the agent plans to exploit. Roughly, an agent's variant $\langle \mathcal{G}, \mathcal{P}, \mathcal{C} \rangle$ supports a goal g if:

- the agent is capable of goal g, i.e. $g \in \mathcal{C}$;

- the agent takes a commitment (in $\mathcal{P}$) from another agent; the commitments' antecedent is supported by the variant and the consequent entails g;

- the agent makes a commitment (in $\mathcal{P}$) to another agent; the commitment's antecedent entails g;

- a goal supported by the variant contributes positively to g;

- g is AND- (OR-) decomposed and all (at least one of) the subgoals is supported.

EXAMPLE 1. Figure 1 shows the active variant for Jim's goal fire extinguished (the goals in the current variant are grayed, and the commitments in the variant are shown below the goal model). The leaf-level goals in the variant are all supported: pipe connected via Jim's capability for that goal, tanker service paid by commitment C_2 made to Brigade1, fire reached by tanker truck via commitment C_3 Jim takes from Tanker1. Thus, the AND-decomposed goal tanker truck used is supported. In turn, it supports the OR-decomposed goal fire extinguished.

4   Adaptive Agent Architecture

Based on our conceptualization of agent adaptation, we propose here an architecture for adaptive agents. Our architecture operates in accordance with the Monitor-Diagnose-Reconcile-Compensate (MDRC) control loop:

(M) **Monitor** collects data about the state of the environment and the agents participating in the system from a variety of sources;

(D) **Diagnose** interprets the data with respect to goal models to determine if all is well. If not, the problem-at-hand is diagnosed;

(R) **Reconcile** searches for a different variant that best deals with the problem-at-hand;

(C) **Compensate** defines and executes a plan that enacts the new variant.

We exemplify how the MDRC cycle works in the fire response scenario where agent Jim is playing role fire chief.

EXAMPLE 2. Jim monitors fire severity and evolution, wind conditions, traffic in nearby areas, as well as retrieving messages from other agents, e.g., firefighters and water tanker truck providers. Jim's diagnosis would involve verifying whether the collected information threatens his goals. For example, if a firefighter notifies he cannot come, Jim's goal to respond to the emergency might be in danger. Jim can reconcile this situation by identifying a new strategy, e.g., relying on water tanker helicopters. Ultimately, Jim may compensate the nonavailability of firefighters by calling upon the helicopters.

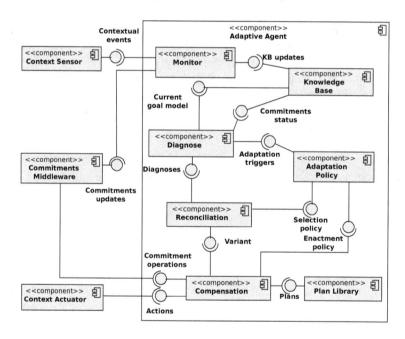

Figure 2. Logical view of the architecture for adaptive agents

The logical view of our architecture is shown in Figure 2. The architecture includes a number of external components the agent interacts with. *Context sensors* are computational entities providing raw data about the environment the agent runs in (e.g. to retrieve wind conditions, temperature, and traffic status). *Context actuators* represent effectors in the environment that can receive commands that modify the environment itself, e.g., sirens, loudspeakers, and door openers.

An essential external component of the architecture is a *Commitments Middleware*, the communication infrastructure that enables agents to interact in terms of commitments. The component allows for asynchronous message exchange between agents through an API that triggers a call-back method for every received message. Within this method, the agent defines how the message is handled according to its internal policy. Specifically, such middleware has to support the standard commitment manipulation operations [Singh 1998]. Below we list some messages the agents would typically exchange.

- *Create*(C_3): Tanker1 sends a message to Jim whereby C_3 is created.

- *Declare*(tanker service paid): Jim sends a message to Tanker1. As a result C_3 is detached and C(Tanker1, Jim, $\top$, fire reached by tanker truck) holds.

- *Release*(C_3): Jim sends a message to Tanker1 telling that the truck is not needed anymore, thus releasing the provider from its commitment;

- *Cancel*(C_3): Tanker1 sends a message to Jim telling that the truck will not be able to reach the fire, as the truck was involved in a pile-up;

- *Delegate*(C_3): Tanker1 sends a message to another truck provider of a nearby town, thereby delegating its commitment to reach the fire.

In the next subsections, we detail the core components of the architecture that deliver the MDRC adaptive control loop.

4.1 Monitor and Diagnose

The purpose of the *Monitor* component is to detect relevant changes in the physical and social context. To collect events, this component relies on two sources: context sensors provide changes in the environment, whereas the commitments middleware furnishes the messages sent by other agents—in terms of commitments operations.

EXAMPLE 3. Jim's monitor receives *Cancel*(C_3) from Tanker1. Such message informs Jim that Tanker1 is canceling its commitment C_3, presumably due to an unforeseen traffic jam. Further, Jim's monitor has received data from a context sensor notifying it that there has been a car accident in the city center.

The collected data is then provided to the *Knowledge Base* component, which represents the agent's current knowledge. This corresponds to correlating collected data against the agent's goal model and the current commitments. The knowledge base provides these information through interfaces *Current goal model* and *Commitments status*.

The knowledge base does not analyze whether the agent's goals are at risk. Such activity is conducted by the *Diagnose* component. In particular, its role is to determine whether the current *variant* is adequate to achieve the agent's goals, if enacted correctly at runtime. While evaluating the adequacy of the current variant, the diagnose component takes into consideration the *adaptation triggers* specified in the agent's adaptation policy.

Adaptation triggers specify under what circumstances the agent should search for an alternative variant to achieve its goals. Various types of adaptation triggers exist [Dalpiaz et al. 2010] including the following:

- *Failure.* An agent may fail in using its capability (because the corresponding low-level procedure failed) or another agent may violate a commitment. For example, Jim's plan to deliver capability "pipe connected" may fail if the pipe socket is incompatible with the plug. The violation of a commitment is shown in Example 3, wherein Tanker1 cancels C_3 (by sending a message), which was part of Jim's adopted variant to extinguish fire.

- *Threat.* Threats are identified by specific agents on the basis of their own policies. A possible way to detect threats is to exploit risk analysis techniques. For example, if Tanker1 sends a message saying it could be late (it has not violated its commitment yet), Jim may interpret that as a threat to C_3. Whereas commitment failure is publicly observable, the evaluation of threats may be a matter of agent policy, i.e., one agent may consider a commitment threatened when another does not.

- *Opportunity.* Here monitored data suggests an opportunity to improve current performance. This kind of trigger constitutes an example of the proactivity of agents, as adaptation is stimulated by new opportunities rather than as reaction to failures or underperformance. For instance, if a new water tanker truck provider Tanker3 comes into play and commits to reach the fire in less time—its headquarters are close to the fire—Jim may check if this opportunity is exploitable. Deciding whether or not to adopt a new variant is up to the reconciliation component.

4.2 Reconcile and Compensate

Once the problem at-hand is identified and the root cause diagnosed, the architecture begins the reconciliation phase. The main activity here is to identify a new variant that is more likely to succeed than the current one. Specifically, the *Reconciliation* component takes as input the diagnoses as well as an agent variant selection policy, and returns the variant to be adopted.

After identifying possible variants, the agent has to select among them. Several criteria can be exploited and combined, such as (i) minimize cost expressed as money, resources, or time; (ii) preserve stability by minimizing change from the status quo; (iii) maximize quality, e.g. by considering soft-goals such as performance, security, and risk; (iv) choose preferred goals and commitments; (v) apply redundancy to ensure achievement of critical goals.

Element	AC	CC
hydrant need notified	6	4
pipe connected	17	8
C_1	15	7
C_2	8	0
C_3	13	20
C_4	31	16

Table 2. Activation (AC) and compensation cost (CC) for capabilities and commitments

Our architecture exploits a variant selection algorithm that takes into account cost and stability [Dalpiaz 2011]. As shown in Table 2, we associate cost values to capabilities and commitments. Activation cost refers to the effort required to exploit a capability and to make/take a commitment. Compensation cost represents the effort required to nullify/mitigate the effects of a capability and to cancel/release a commitment. Our selection algorithm considers the adaptation policy of the agent, which includes several factors:

- *Variant selection criteria* specify how to determine variant costs that determine selection. Two algorithms are supported: (i) *overall-cost* considers the overall variant cost, and selects the variant having minimal total cost; (ii) *delta* computes delta costs between a candidate and the current variant.

- *Variant exclusion factors* provide reasons why some variants should be excluded. An agent might want a different solution to avoid threatened capabilities and commitments. Also, failed capabilities and violated commitments might be excluded.

- *Compensation cost* represents how expensive it is for an agent to revert/mitigate the effects of a capability that is currently exploited and that will not be in the next variant. Compensation cost is considered when determining variant cost, if such option is in the policy.

- *Opportunity threshold* determines when an opportunity should be adopted. If the current variant is not at risk, the agent needs a clear incentive to switch to another variant. If the variant selection strategy is chosen, this threshold is how less the new variant would cost, in percentage. If the delta strategy is chosen, opportunities are taken only if their delta is lower than a fixed value.

We illustrate how our cost-based variant selection algorithm works at runtime to support agent Jim playing role fire chief (as in Figure 1) in the scenario depicted in Table 1 with respect to the costs in Table 2.

EXAMPLE 4. Suppose C_3 made by Tanker1 is threatened. This happens because Tanker1 notifies it will be late due to a traffic jam. C_3 is part of Jim's current variant for goal fire extinguished and the variant is threatened. In response, an adaptation process is triggered wherein variant selection is based on the delta criteria, and threatened/violated commitments should be avoided.

Jim can currently choose between three variants to extinguish the fire:

- V_1: Exploit his capability for hydrant need notified and get C_1 from Brigade1 to support goal hydrant usage authorized. Bringing about the antecedent of C_1 will make Brigade1 unconditionally committed to authorize hydrant usage.

- V_2: Make C_2 to Brigade1 to support tanker service paid, chain the commitment to C_3 so that Tanker1 is unconditionally committed to fire reached by tanker truck, and use his capability for pipe connected. Notice that this is the current variant, which is already partially enacted.

- V_3: The same strategy as the previous one, but relies on C_4 instead of C_3. This corresponds to making Tanker2 unconditionally committed to reach the fire.

Variant V_1 involves C_1 and Jim's capability for hydrant need notification. The variant supports hydrant need notified, hydrant usage authorized, fire hydrant used, and fire extinguished. To compute the variant cost, activation and compensation costs should be considered. The activation cost of hydrant need notified is 6; that of C_1 is 15. Compensation cost should be considered for capability pipe connected (8), C_2 (0), and C_3 (20). The variant overall cost is 49.

Variant V_2 (C_3, C_2, pipe connected) supports states of affairs tanker service paid, fire reached by tanker truck, pipe connected, tanker truck used, and fire extinguished. Unfortunately, this variant violates the agent's adaptation policy. Indeed C_3 in the variant is threatened.

Variant V_3 (C_4, C_2, pipe connected) supports the same states of affairs as V_2. Cost computation for capabilities is also the same and adds no cost. Cost computation for commitments adds the activation cost of C_4 (31). The only compensation cost to consider is that for commitment C_3 (20). Indeed, C_3 is in the current variant but not in the analyzed variant. The variant overall cost is 51.

Jim will therefore choose—according to his adaptation strategy—V_1. The next step is deciding on how to enact this variant, i.e. defining the course of actions to switch from the current variant to V_1. This is the role of the *Compensation* component, which uses the selected variant, the agent's enactment policy, and the agent's plans—taken from the plan library—to determine a course of action. Two types of atomic operations are possible: (i) actions that use actuators in the context and (ii) operations on commitments performed by sending messages through the commitments middleware.

To enact V_1, the architecture needs to carry out several steps. Some are intended to reverse the effects of the current variant, others to enact the new plan.

1. Cancel C_2 made to Brigade1, and release Tanker1 from commitment C_3. These operations are executed by sending messages through the commitments middleware;

2. Compensate capability pipe connected by using context actuators, such as an engine to retract the water pipe that was already ready to be plugged;

3. Carry out some actions to adopt the new variant: use capability for hydrant need notified (via an actuator such as a phone) to detach C_1 and make Brigade1 unconditionally committed to hydrant usage authorized by sending a message through the commitments middleware.

5 Discussion

In this chapter we have proposed an approach to agent adaptation. Our proposal applies to different settings than Kumar et al.'s adaptive agent architecture [2000]. Whereas their work applies only to multiagent settings where agents are collaborative, ours supports also settings where agents are designed by different stakeholders and might be competitive. The core message of our chapter is that, in multiagent systems where agents are weakly controllable, *social abstractions*—here, commitments—are the mechanisms that make the system work. In fact, relying on the intentions of other agents is risky, due to their autonomy and heterogeneity.

A distinguishing objective of our work is to exploit high-level models to represent both the agent purposes and the social relationships with other agents (social commitments). The combined usage of these abstractions makes our proposal very flexible. By focusing on the *purpose* of the agent and the *meaning* of interaction, adaptation guarantees that the agent meets its strategic interests (its purpose) and acts in compliance with its commitments to other agents. Central to our framework is the notion of variability, the existence of multiple strategies (variants) to achieve an agent's goals. We have shown how variants are constructed and presented a cost-based framework that enables an agent to select the best variant to adopt.

Future work comprises different research lines: (i) early detection of failures, i.e. the capability of agents to anticipate failures by adapting; (ii) considering quality-of-service attributes to choose between alternatives; (iii) efficient variant generation algorithms, e.g. heuristics that generate good-enough variants; (iv) adaptation policies that guide the choices made by an agent throughout its adaptation control loop.

References

Brachman, R. J., R. E. Fikes, and H. J. Levesque [1983, October]. Krypton: A functional approach to knowledge representation. *Computer 16*, 67–73.

Bratman, M. E. [1987]. *Intention, Plans, and Practical Reason*. Cambridge, MA: Harvard University Press.

Bresciani, P., A. Perini, P. Giorgini, F. Giunchiglia, and J. Mylopoulos [2004]. Tropos: An agent-oriented software development methodology. *Autonomous Agents and Multi-Agent Systems 8*(3), 203–236.

Chopra, A. K., A. Artikis, J. Bentahar, M. Colombetti, F. Dignum, N. Fornara, A. J. I. Jones, M. P. Singh, and P. Yolum [2011]. Research directions in agent commmunication. *ACM Transactions on Intelligent Systems*. To appear.

Chopra, A. K., F. Dalpiaz, P. Giorgini, and J. Mylopoulos [2010]. Reasoning about agents and protocols via goals and commitments. In *Proceedings of the Ninth International Conference on Autonomous Agents and Multiagent Systems (AAMAS)*, pp. 457–464.

Cohen, P. R. and H. J. Levesque [1990]. Intention is choice with commitment. *Artificial Intelligence 42*, 213–261.

Dalpiaz, F. [2011]. *Exploiting Contextual and Social Variability for Software Adaptation*. Ph.D. thesis, Information and Communication Technology, University of Trento.

Dalpiaz, F., A. K. Chopra, P. Giorgini, and J. Mylopoulos [2010]. Adaptation in open systems: Giving interaction its rightful place. In *Proceedings of the 29th International Conference on Conceptual Modeling*, Volume 6412 of *LNCS*, pp. 31–45. Springer.

Finin, T., R. Fritzson, D. McKay, and R. McEntire [1994]. KQML as an agent communication language. In *Proceedings of the International Conference on Information and Knowledge Management*, pp. 456–463. ACM Press.

FIPA [2003]. FIPA interaction protocol specifications. FIPA: The Foundation for Intelligent Physical Agents, http://www.fipa.org/repository/ips.html.

Kumar, S., P. R. Cohen, and H. J. Levesque [2000]. The adaptive agent architecture: Achieving fault-tolerance using persistent broker teams. In *Proceedings of the Fourth International Conference on Multi-Agent Systems*, pp. 159–166.

Levesque, H. J. [1981]. The interaction with incomplete knowledge bases: A formal treatment. In *Proceedings of the 7th International Joint Conference on Artificial Intelligence (IJCAI'81)*, pp. 240–245. AAAI Press.

Levesque, H. J. and J. Mylopoulos [1979]. A procedural semantics for semantic networks. In N. V. Findler (Ed.), *Associative Networks: The Representation and Use of Knowledge by Computers*, pp. 93–120. Academic Press.

Mylopoulos, J. [1992]. Conceptual modelling and Telos. In P. Loucopoulos and R. Zicari (Eds.), *Conceptual Modelling, Databases and CASE: An Integrated View of Information System Development*. McGraw Hill.

Searle, J. R. [1969]. *Speech Acts*. Cambridge, UK: Cambridge University Press.

Shoham, Y. [1993]. Agent-oriented programming. *Artificial Intelligence 60*(1), 51–92.

Singh, M. P. [1991]. Social and psychological commitments in multiagent systems. In *AAAI Fall Symposium on Knowledge and Action at Social and Organizational Levels*, pp. 104–106.

Singh, M. P. [1998, December). Agent communication languages: Rethinking the principles. *IEEE Computer 31*(12), 40–47.

Singh, M. P. [1999]. An ontology for commitments in multiagent systems: Toward a unification of normative concepts. *Artificial Intelligence and Law 7*, 97–113.

Singh, M. P. [2008]. Semantical considerations on dialectical and practical commitments. In *Proceedings of the 23rd Conference on Artificial Intelligence*, pp. 176–181.

Singh, M. P. and M. N. Huhns [2005]. *Service-Oriented Computing: Semantics, Processes, Agents*. Chichester, UK: John Wiley & Sons.

Winograd, T. and F. Flores [1986]. *Understanding Computers and Cognition: A New Foundation for Design*. Norwood, New Jersey: Ablex Publishing.

Yolum, P. and M. P. Singh [2002]. Flexible protocol specification and execution: Applying event calculus planning using commitments. In *Proceedings of the 1st International Joint Conference on Autonomous Agents and MultiAgent Systems*, pp. 527–534. ACM Press.

11

Beliefs, Belief Revision, and Noisy Sensors

JAMES DELGRANDE

ABSTRACT. In logical AI, an agent's beliefs are typically categorical, in that they are specified by a set of formulas. An agent may change its beliefs as a result of being informed in one fashion or another about some aspect of the world, or following the execution of some action. The areas of belief revision and reasoning about action deal with just such change in belief. However, most information about the real world is not categorical. While there are well-established accounts for accommodating non-categorical information via probability theory, it is worth asking whether probabilistic information may be reconciled with the logical accounts of belief change. We present such an account in this paper. An agent receives uncertain information as input and its probabilities, expressed as probabilities on possible worlds, are updated via Bayesian conditioning. A set of formulas among the (noncategorical) beliefs is identified as the agent's (categorical) belief set. This set is defined in terms of the most probable worlds such that the summed probability of these worlds exceeds a given threshold. The effect of this updating on the belief set is examined with respect to its appropriateness as a revision operator. It proves to be the case that a subset of the classical AGM belief revision postulates are satisfied. Most significantly, the success postulate is not guaranteed to hold. However it does hold after a sufficient number of iterations. Not is it the case that in revising by a formula consistent with the agent's beliefs, revision corresponds to expansion. On the other hand, limiting cases of the presented approach correspond to specific approaches to revision that have appeared in the literature.

It is a great pleasure to dedicate this paper to Hector Levesque on the occasion of his 60th birthday. While Hector's work has broadly focussed on representational aspects of an agent's beliefs together with accounts of reasoning – whether epistemic, nonmonotonic, limited, in a theory of action, or otherwise – it has certainly touched on many other areas over the years. This paper outlines a possible linking of two such areas, that of reasoning about noisy sensors on the one hand [Bacchus, Halpern, and Levesque 1999], and revision in the presence of (categorical) observations on the other [Shapiro, Pagnucco, Lespérance, and Levesque 2011].

1 Introduction

In logical AI, an agent is generally regarded as holding, or *believing*, some set of formulas to be true. As well, an agent's knowledge of a domain will most often be incomplete and inaccurate. Consequently, an agent must change its beliefs in response to receiving new information. *Belief revision* addresses the problem of how an agent may incorporate a new formula into its set of beliefs. That is, the the agent has some set of beliefs K which are accepted as holding in the domain of application, and the agent is given a new formula ϕ which it is to incorporate into the set of beliefs. If ϕ is consistent but conflicts with K, some beliefs will have to be dropped from K before ϕ can be added. The original and best-known approach to belief revision is called the *AGM approach* [Alchourrón, Gärdenfors, and Makinson 1985; Gärdenfors 1988], named after the developers of this framework. Subsequently, the area of belief revision has developed into an active area of research in KR&R [Peppas 2007].

However, information about the world is often not categorical, but is received only with a certain level of confidence. For example, an agent may make observations about the world via sensors, but such sensors may be inaccurate or may provide incorrect information. There are of course approaches for modifying an agent's beliefs in such a situation, most obviously via probability theory and using Bayesian conditioning (e.g. [Pearl 1988]). However, in this case formulas are held with attached (subjective) probabilities, and are not generally held as being absolutely true or false.

In one sense, these approaches seem to be addressing the same problem, since they both consider the change in an agent's beliefs in the presence of new information. Yet the approaches also seem to be quite different. In the case of belief revision, an agent *accepts* a certain set of beliefs as categorically holding, and another categorical belief is to be consistently incorporated into this set. This approach is essentially *qualitative*, since the sentences making up the agent's knowledge base are (simply) believed to be true. In the case of updating via Bayesian conditioning, beliefs are generally not held with certainty, but rather with varying levels or degrees of confidence. The task then is to modify these degrees of confidence, expressed as probabilities, as new evidence is received. Hence this approach is fundamentally *quantitative*.

This division into qualitative and quantitative approaches to belief represents two fundamentally different ways of dealing with uncertain information, often referred to as the *logicist* and *probabilist* camps. The distinction can be paraphrased as concerning on the one hand those approaches that adopt a proposition as holding, but hedged in the sense that one is prepared to give it, as opposed to accepting a proposition in a hedged fashion, in that it isn't held with certainty but just with some level of confidence [Kyburg 1994].

However, it can also be observed that categorical beliefs arise from noncategorical, hedged claims: in fact, it can be argued that *none* of our knowledge about the world is certain, yet we often – perhaps most often – act as if it were. Hence, not only do people act assuming that the sun will rise tomorrow, they will usually act as if their car is guaranteed to be where they parked it. Consequently, it is an interesting question to ask how one may move from a noncategorical account of a domain to a categorical account. As alluded to above, Kyburg, among others, has been occupied with this question. In this paper we take a different tack and ask whether an underlying non-categorical approach, based on subjective probabilities, may have something to say about the categorical approach of belief revision. That is, evidence about the real world is generally uncertain, and it is of interest to examine how such a setting may be reconciled with the assumptions underlying belief revision.

We begin with a simple model of an agent's beliefs, in which probabilities are associated with possible worlds which in turn characterise the agent's subjective knowledge. The agent's accepted, categorical beliefs are characterised by the set of worlds with highest probability such that the sum of the probabilities over those worlds exceeds a certain threshold. As new, uncertain information is received, the probabilities attached to worlds are modified and the set of accepted beliefs consequently changes. One can then examine the dynamics of these accepted beliefs to see how it accords with accounts of belief revision. Perhaps not surprisingly, only a subset of the AGM revision postulates are satisfied. Most notably, if a formula ϕ is received with an attached probability, it does not necessarily appear among the set of accepted beliefs. However, it proves to be the case that after some number of iterations of revision by ϕ, ϕ will come to be believed. This makes intuitive sense: if one is informed of a formula with probability $> .5$, one may still not immediately believe ϕ. However after repeated such reports one eventually accepts the formula. We also examine variants of the approach. It proves to be the case that two extant approaches to belief revision are closely related to instances of the approach developed here.

The next section reviews background material. Section 3 reviews the updating of probabilities first, by way of motivation, in terms of formulas and second in terms of probabilities on possible worlds. The following section motivates and defines the notion of *epistemic state* as used in the paper. Section 5 describes belief revision in this framework, including properties of the resulting revision operator and a comparison to related work. Section 6 gives a brief summary.

2   Background

2.1   Formal Preliminaries

Let $\mathcal{P} = \{a, b, \ldots\}$ be a finite set of atomic sentences, and let $\mathcal{L}$ be the language over $\mathcal{P}$ closed under the usual connectives $\neg$, $\wedge$, $\vee$, and $\supset$. The classical consequence relation is denoted $\vdash$; $Cn(A)$ is the set of logical consequences of a formula or set of formulas A; that is $Cn(A) = \{\phi \in \mathcal{L} \mid A \vdash \phi\}$. $\top$ stands for some arbitrary tautology and $\bot$ is defined to be $\neg\top$. Given two sets of formulas A and B, $A + B$ denotes the *expansion* of A by B, that is $A + B = Cn(A \cup B)$. Expansion of a set of formulas A by a formula ϕ is defined analogously. Sentences ϕ and ψ are *logically equivalent*, $\phi \equiv \psi$, iff $\phi \vdash \psi$ and $\psi \vdash \phi$. This also extends to sets of formulas. A propositional *interpretation* (or *possible world*) is a mapping from $\mathcal{P}$ to {true, false}. The set of interpretations of $\mathcal{L}$ is denoted $W_{\mathcal{L}}$. A *model* of a sentence ϕ is an interpretation w that makes ϕ true according to the usual definition of truth, and is denoted by $w \models \phi$. We also write $W \models \phi$ if $w \models \phi$ for every $w \in W$. $Mod(A)$ is the set of models of the set of formulas A. $Mod(\{\phi\})$ is also written as $Mod(\phi)$. For $W \subseteq W_{\mathcal{L}}$, we denote by $T(W)$ the set of sentences which are true in all elements of W; that is $T(W) = \{\phi \in \mathcal{L} \mid w \models \phi \text{ for every } w \in W\}$.

A total preorder $\preceq$ is a reflexive, transitive binary relation, such that either $w_1 \preceq w_2$ or $w_2 \preceq w_1$ for every w_1, w_2. As well, $w_1 \prec w_2$ iff $w_1 \preceq w_2$ and $w_2 \not\preceq w_1$. $w_1 = w_2$ abbreviates $w_1 \preceq w_2$ and $w_2 \preceq w_1$. Given a set S and total preorder $\preceq$ defined on members of S, we denote by $\min(S, \preceq)$ the set of minimal elements of S in $\preceq$.

Let $\rho : W_{\mathcal{L}} \mapsto [0, 1]$ be a function such that $0 \leq \rho(w) \leq 1$ and $\sum_{w \in W_{\mathcal{L}}} \rho(w) = 1$. ρ is a *probability assignment* to worlds. We distinguish the function $\rho_\top$ where $\rho_\top(w) = \frac{1}{|W_{\mathcal{L}}|}$ for every world w, and we use $\rho_\bot$ to denote the (non-probability) assignment where $\rho_\bot(w) = 1$ for every world w. $\rho_\top$ can be used to characterise a state of ignorance for an agent, while $\rho_\bot$ is a technical convenience that will be used to characterise an inconsistent set of beliefs. Mention of probability assignments will include $\rho_\bot$ as a special case. These functions are extended to subsets of $W_{\mathcal{L}}$ by, for $W \subseteq W_{\mathcal{L}}$, $\rho(W) = \sum_{w \in W} \rho(w)$. Informally, $\rho(w)$ is the (subjective) probability that, as far as the agent knows, w is the actual world being modelled; and for $W \subseteq W_{\mathcal{L}}$, $\rho(W)$ is the probability that the real world is a member of W. As will be later described, the function ρ can be taken as comprising the major part of an agent's *epistemic state* [Darwiche and Pearl 1997; Peppas 2007]. The probability of a formula ϕ then is given by: $Pr_\rho(\phi) = \sum_{w \models \phi} \rho(w) = \rho(Mod(\phi))$. *Conditional probability* is defined, as usual, by $Pr_\rho(\phi|\psi) = Pr_\rho(\phi \wedge \psi)/Pr_\rho(\psi)$ and is undefined when $Pr_\rho(\psi) = 0$.

2.2 Belief revision

The *AGM approach* [Gärdenfors 1988] provides the best-known approach to belief revision. Belief change is described at the *knowledge level*, that is on an abstract level, independent of how beliefs are represented and manipulated. An agent's beliefs are modelled by a set of sentences, or *belief set*, closed under the logical consequence operator of a logic that includes classical propositional logic. Thus a belief set K satisfies the constraint: $\phi \in K$ if and only if $K \vdash \phi$. K can be understood as a partial theory of the world. $K_\perp$ is the inconsistent belief set (i.e. $K_\perp = \mathcal{L}$).

In the revision of K by a formula ϕ, the intent is that ϕ is to be incorporated into K so that the resulting belief set is consistent whenever ϕ is consistent. If ϕ is inconsistent with K, revision will require the removal of beliefs from K in order to retain consistency. In this approach, revision is a function from belief sets and formulas to belief sets. However, various researchers have argued that it is more appropriate to consider *epistemic states* (also called *belief states*) as objects of revision. An epistemic state $\mathcal{K}$ includes information regarding how the revision function itself changes following a revision. The belief set corresponding to belief state $\mathcal{K}$ is denoted $Bel(\mathcal{K})$. As well, we will use the notation $Mod(\mathcal{K})$ to mean $Mod(Bel(\mathcal{K}))$. Formally, a revision operator $*$ maps an epistemic state $\mathcal{K}$ and new information ϕ to a revised epistemic state $\mathcal{K} * \phi$. Then, in the spirit of [Darwiche and Pearl 1997], the AGM postulates for revision can be reformulated as follows:

$(\mathcal{K} * 1)$ $Bel(\mathcal{K} * \phi) = \mathcal{C}n(Bel(\mathcal{K} * \phi))$

$(\mathcal{K} * 2)$ $\phi \in Bel(\mathcal{K} * \phi)$

$(\mathcal{K} * 3)$ $Bel(\mathcal{K} * \phi) \subseteq Bel(\mathcal{K}) + \phi$

$(\mathcal{K} * 4)$ If $\neg\phi \notin Bel(\mathcal{K})$ then $Bel(\mathcal{K}) + \phi \subseteq Bel(\mathcal{K} * \phi)$

$(\mathcal{K} * 5)$ $Bel(\mathcal{K} * \phi)$ is inconsistent, only if $\nvdash \neg\phi$

$(\mathcal{K} * 6)$ If $\phi \equiv \psi$ then $Bel(\mathcal{K} * \phi) \equiv Bel(\mathcal{K} * \psi)$

$(\mathcal{K} * 7)$ $Bel(\mathcal{K} * (\phi \wedge \psi)) \subseteq Bel(\mathcal{K} * \phi) + \psi$

$(\mathcal{K} * 8)$ If $\neg\psi \notin Bel(\mathcal{K} * \phi)$ then $Bel(\mathcal{K} * \phi) + \psi \subseteq Bel(\mathcal{K} * (\phi \wedge \psi))$

The postulates express very basic properties for revision. Thus, the result of revising $\mathcal{K}$ by ϕ is an epistemic state in which ϕ is believed in the corresponding belief set $((\mathcal{K} * 1)$, $(\mathcal{K} * 2))$; whenever the result is consistent, the revised belief set consists of the expansion of $Bel(\mathcal{K})$ by ϕ $((\mathcal{K} * 3)$, $(\mathcal{K} * 4))$; the only time that $Bel(\mathcal{K})$ is inconsistent is when ϕ is inconsistent $((\mathcal{K} * 5))$; and revision is independent of the syntactic form of the formula for revision $((\mathcal{K} * 6))$. The last two postulates state that whenever consistent, revision by a conjunction corresponds to revision by one conjunct and expansion by the other.

Various constructions have been proposed to characterise belief revision. Katsuno and Mendelzon [1991] have shown that a revision can be characterised in terms of a total preorder on the set of possible worlds. For epistemic state $\mathcal{K}$, a *faithful ranking* on $\mathcal{K}$ is a total preorder $\preceq_{\mathcal{K}}$ on the possible worlds $W_{\mathcal{L}}$, such that for any possible worlds $w_1, w_2 \in W_{\mathcal{L}}$:

1. If $w_1, w_2 \models Bel(\mathcal{K})$ then $w_1 =_{\mathcal{K}} w_2$

2. If $w_1 \models Bel(\mathcal{K})$ and $w_2 \not\models Bel(\mathcal{K})$, then $w_1 \prec_{\mathcal{K}} w_2$

Intuitively, $w_1 \preceq_{\mathcal{K}} w_2$ if w_1 is at least as plausible as w_2 according to the agent. The first condition asserts that all models of the agent's knowledge are ranked equally, while the second states that the models of the agent's knowledge are lowest in the ranking. It follows directly from the results of [Katsuno and Mendelzon 1991] that a revision operator $*$ satisfies $(\mathcal{K} * 1)$–$(\mathcal{K} * 8)$ iff there exists a faithful ranking $\preceq_{\mathcal{K}}$ for an arbitrary belief state $\mathcal{K}$, such that for any sentence ϕ:

$$Bel(\mathcal{K} * \phi) = \begin{cases} \mathcal{L} & \text{if } \vdash \neg\phi \\ \mathcal{T}(\min(Mod(\phi), \preceq_{\mathcal{K}})) & \text{otherwise} \end{cases}$$

Thus when ϕ is satisfiable, the belief set corresponding to $\mathcal{K} * \phi$ is characterised by the least ϕ models in the ranking $\preceq_{\mathcal{K}}$.

The AGM postulates do not address properties of iterated belief revision. This has led to the development of additional postulates for iterated revision; the best-known approach is that of Darwiche and Pearl [1997]. They propose the following postulates, adapted according to our notation:

C1 If $\psi \vdash \phi$, then $Bel((\mathcal{K} * \phi) * \psi) = Bel(\mathcal{K} * \psi)$

C2 If $\psi \vdash \neg\phi$, then $Bel((\mathcal{K} * \phi) * \psi) = Bel(\mathcal{K} * \psi)$

C3 If $\phi \in Bel(\mathcal{K} * \psi)$, then $\phi \in Bel((\mathcal{K} * \phi) * \psi)$

C4 If $\neg\phi \notin Bel(\mathcal{K} * \psi)$, then $\neg\phi \notin Bel((\mathcal{K} * \phi) * \psi)$

Darwiche and Pearl show that an AGM revision operator satisfies each of the Postulates (C1)–(C4) iff the way it revises faithful rankings satisfies the respective conditions:

CR1 If $w_1, w_2 \models \phi$, then $w_1 \preceq_{\mathcal{K}} w_2$ iff $w_1 \preceq_{\mathcal{K}*\phi} w_2$

CR2 If $w_1, w_2 \not\models \phi$, then $w_1 \preceq_{\mathcal{K}} w_2$ iff $w_1 \preceq_{\mathcal{K}*\phi} w_2$

CR3 If $w_1 \models \phi$ and $w_2 \not\models \phi$, then $w_1 \prec_{\mathcal{K}} w_2$ implies $w_1 \prec_{\mathcal{K}*\phi} w_2$

CR4 If $w_1 \models \phi$ and $w_2 \not\models \phi$, then $w_1 \preceq_{\mathcal{K}} w_2$ implies $w_1 \preceq_{\mathcal{K}*\phi} w_2$

Thus postulate (C1) asserts that revising by a formula and then by a logically stronger formula yields the same belief set as simply revising by the stronger formula at the outset. The corresponding semantic condition (CR1) asserts that in revising by a formula ϕ, the relative ranking of ϕ worlds remains unchanged. The other postulates and semantic conditions can be interpreted similarly. Subsequently, other approaches for iterated revision have been proposed, including [Boutilier 1996; Nayak, Pagnucco, and Peppas 2003; Jin and Thielscher 2007]. While interesting, we do not consider them further since they add little to the exposition.

2.3 Related Work

In probability theory and related approaches, there has of course been work on incorporating new evidence to produce a new probability distribution. The simplest means of updating probabilities is via conditionalisation: If an agent holds ϕ with probability q, and so $Pr(\phi) = q$, and the agent learns ψ with certainty, then one can define the updated probability $Pr'(\phi)$ via

$$Pr'(\phi) = Pr(\phi|\psi) = Pr(\phi \wedge \psi)/Pr_\rho(\psi).$$

Of course an agent may not learn ψ with certainty, but rather may change its probability assignment to ψ from $Pr(\psi)$ to a new value $Pr'(\psi)$. The question then is how probabilities assigned to other variables should be modified. Jeffrey [1983] proposes that for proposition ϕ, the new probability should be given by what has come to be known as Jeffrey's Rule for updating probabilities:

$$Pr'(\phi) \; = \; Pr(\phi|\psi)Pr'(\psi) + Pr(\phi|\neg\psi)Pr'(\neg\psi).$$

So $Pr'(\psi) = q$ means that the agent has learned that the probability of ψ is q. In particular, if the probability of ψ is further updated to $Pr''(\psi)$ but it turns out that $Pr''(\psi) = Pr'(\psi)$, then the distributions Pr' and Pr'' will coincide.

This is orthogonal to our goals here. Instead, we are interested in the case where we have some underlying proposition, say that a light is on, represented by on, and we are given an observation Obs_{on}, where Obs_{on} has an attached probability. Then if the agent receives repeated observations that the light is on, the agent's confidence that on is true will increase with each positive observation. Details are given in the next section; the main point here is that Bayes' Rule will be more appropriate in this case, where Bayes' Rule is given by:

$$Pr(\phi|\psi) \; = \; Pr(\psi|\phi)Pr(\phi)/Pr(\psi).$$

Previous research dealing with the intersection of belief revision and probability is generally concerned with revising a probability function. In such approaches, an agent's belief set K is given by those formulas that have probability 1.0. These formulas with probability 1.0 are referred to as the *top* of the probability function. For a revision $K * \phi$, the probability function is revised by ϕ, and the belief set corresponding to $K * \phi$ is given by the top of the resulting probability function. So such approaches allow the characterisation of not just the agent's beliefs, but also allow probabilities to be attached to non-beliefs. As will be subsequently described, this is in contrast to the present approach, in which an agent's categorical beliefs will generally have probability less than 1.

One difficulty with revising probability functions is the *non-uniqueness problem*, that there are many different probability functions that have K as their top. Lindström and Rabinowicz [1989] consider various ways of dealing with this problem. Boutilier [1995] considers the same general framework, but rather focuses on issues of iterated belief revision. However the approach described herein addresses a different problem: a means of incorporating uncertain information into a given probability function is assumed, and the question addressed is how such an approach may be reconciled with AGM revision, or alternatively, how such an approach may be considered as an instance (or proto-instance) of revision. To this end, Gärdenfors [1988, Ch. 5] has also considered an extension of the AGM approach to the revision of probability functions; we discuss this work in detail after our approach has been presented.

With respect to qualitative, AGM-style belief revision, the approach at hand might seem to be an instance of an *improvement operator* [Konieczny and Pino Pérez 2008]. An improvement operator according to Konieczny and Pino Pérez is a belief change operator where new information isn't necessarily immediately accepted. However plausibility is increased and, after a sufficient number of iterations, the information will come to be believed. Interestingly, as we discuss later, the approach described here differs in significant ways from those of [Konieczny and Pino Pérez 2008].

The setting adopted here is similar to that of [Bacchus, Halpern, and Levesque 1999]: Agents receive uncertain information, and as a result alter their (probabilistic) beliefs about the world. However, the goals are quite different. [Bacchus, Halpern, and Levesque 1999] is concerned with an extension of the situation calculus [Levesque, Pirri, and Reiter 1998] to deal with noisy sensors. Consequently their focus is on a version of the situation calculus in which the agent doesn't hold just categorical beliefs, but also probabilistic beliefs. The main issue then is how to revise these probabilities in the presence of sensing and non-sensing actions. In contrast, the present paper is concerned with the possible role of probabilistic beliefs with respect to a (classical AGM-style) belief revision operator. We further discuss this and other related work once the approach has been presented.

3 Unreliable Observations and Updating Probabilities

An agent will make observations concerning a domain. These observations may be unreliable, in that a value may be incorrectly sensed or reported. We wish to update the probability assignment to possible worlds appropriately, given such a possibly-erroneous observation. Consider first a situation in which an agent observes or senses ϕ with a given probability $q > .5$. Our interpretation of this event is that ϕ is reported as being true, but that the probability is $1 - q$ that the sensing is incorrect (and so the probability is $1 - q$ that $\neg\phi$ is in fact the case).[1] Since $q > .5$, the agent's confidence in ϕ will increase. We can write $Pr(Obs_\phi|\phi) = q$ for the probability of observing that ϕ is true given that ϕ is in fact true. As well, the agent will also have some prior probability $Pr(\phi)$ that ϕ is true. We wish to compute the probability that ϕ is true given the new piece of evidence, $Pr(\phi|Obs_\phi)$. This can be determined by Bayes' Rule:

$$(1) \quad Pr(\phi|Obs_\phi) = \frac{Pr(Obs_\phi|\phi)Pr(\phi)}{Pr(Obs_\phi)} = \frac{Pr(Obs_\phi|\phi)Pr(\phi)}{Pr(Obs_\phi|\phi)Pr(\phi) + Pr(Obs_\phi|\neg\phi)Pr(\neg\phi)}$$

For example assume that the agent has no prior information about a light being on or not, and so $Pr(on) = .5$. As well, the light sensor is correct 80% of the time, and so $Pr(Obs_{on}|on) = .8$ while $Pr(Obs_{on}|\neg on) = .2$. Hence following an observation that the light is on, we would obtain:

$$
\begin{aligned}
Pr(on|Obs_{on}) &= \frac{Pr(Obs_{on}|on)Pr(on)}{Pr(Obs_{on}|on)Pr(on) + Pr(Obs_{on}|\neg on)Pr(\neg on)} \\
&= (.8 \times .5)/[(.8 \times .5) + (.2 \times .5)] \\
&= .8.
\end{aligned}
$$

Thus on observing that the light was on, the agent's (subjective) probability that the light was on would increase from .5 to .8. If the agent was to subsequently re-sense the light, and again sense that the light was on, its degree of belief would then be given by:

$$
\begin{aligned}
Pr(on|Obs_{on}) &= \frac{Pr(Obs_{on}|on)Pr(on)}{Pr(Obs_{on}|on)Pr(on) + Pr(Obs_{on}|\neg on)Pr(\neg on)} \\
&= (.8 \times .8)/[(.8 \times .8) + (.2 \times .2)] \\
&\approx .94.
\end{aligned}
$$

---

[1] The case of non-binary valued sensing is straightforward and adds nothing of additional interest with respect to the problem at hand; see for example [Bacchus, Halpern, and Levesque 1999] for how this can be handled.

Observations and possible worlds: The preceding discussion reviews how a probability assignment to formulas may be updated given new information. Since we have a finite language and a finite set of possible worlds, it is straightforward to extend this to updating probabilities attached to worlds, and hence updating a probability function ρ. Since a world may be associated with the conjunction of literals true at that world, we can repeat the steps in the preceding section, but with respect to worlds.

Consider a situation in which we observe ϕ with probability q. As before, Obs_ϕ is true if ϕ is observed and false otherwise. We wish to update the probability that a world w is the real world given this additional piece of information. That is, if ϕ is true at w then the probability attached to ρ will increase, and decrease if ϕ is false at w. For a world $w \in W_\mathcal{L}$, we have the prior probability assignment $\rho(w)$. We can use this in probability expressions by letting $Pr_\rho(w)$ be understood such that the occurrence of w in $Pr_\rho(\cdot)$ stands for the (finite) conjunction of literals true in w, and thus $Pr_\rho(w) = \rho(w)$. Then, again with Bayes' rule we have:

$$(2) \quad Pr_\rho(w|Obs_\phi) = \frac{Pr_\rho(Obs_\phi|w)Pr_\rho(w)}{Pr_\rho(Obs_\phi)}$$

Thus on the left side of the equality, we wish to determine the (updated) probability of w, given that ϕ is observed. For the numerator on the right hand side, $Pr_\rho(Obs_\phi|w)$ is the probability of observing ϕ given that one is in w; this is just q if $w \models \phi$ and $1 - q$ otherwise. $Pr_\rho(w)$ is just $\rho(w)$. For the denominator, we have that

$$
\begin{aligned}
Pr_\rho(Obs_\phi) &= Pr_\rho(Obs_\phi|\phi)Pr_\rho(\phi) + Pr_\rho(Obs_\phi|\neg\phi)Pr_\rho(\neg\phi) \\
&= q \times \rho(Mod(\phi)) + (1 - q) \times \rho(Mod(\neg\phi)).
\end{aligned}
$$

This justifies the following definition.

DEFINITION 1. Let ρ be a probability assignment to worlds. Let $\phi \in \mathcal{L}$ and $q \in [0, 1]$. Let $\eta = \rho(Mod(\phi)) \times q + \rho(Mod(\neg\phi)) \times (1 - q)$.

Define the probability assignment $\rho(\phi, q)$ by:

$$
\begin{aligned}
\rho(\phi, q) &= \rho_\perp && \text{if } \eta = 0; \quad \text{otherwise:} \\
\rho(\phi, q)(w) &= \begin{cases} (\rho(w) \times q)/\eta & \text{if } w \models \phi \\ (\rho(w) \times (1 - q))/\eta & \text{if } w \not\models \phi \end{cases}
\end{aligned}
$$

Thus, for probability function ρ, a new probability function $\rho(\phi, q)$ results after sensing ϕ with probability q. Observe that $\rho(\phi, q)(w)$ in Definition 1 corresponds to $Pr_\rho(w|Obs_\phi)$ in (2). If $\eta = 0$, the updated probability assignment involves accepting with certainty (i.e. $q = 1$) an impossible proposition ($\rho(Mod(\phi)) = 0$), or rejecting with certainty a necessarily true proposition. In either case, an incoherence state of affairs ($\rho_\perp$) results.

Example: Consider Table 1. The first column lists possible worlds in terms of an assignment of truth values to atoms, where $\bar{a}$ stands for $\neg a$. The second column gives an initial probability function, while the next three columns show how ρ changes under different updates. At the outset $Pr(a) = .5$, $Pr(b) = .6$, and $Pr(c) = .5$. Following an observation of a with reliability .8, we obtain that $Pr(a) = .8$, $Pr(b) = .6$, and $Pr(c) = .5$. If we iterate the process and again observe a with the same reliability, the probabilities become $Pr(a) = .9412$, $Pr(b) = .6$, and $Pr(c) = .5$. Thus the probability of a increases, and the probability of b and c varies depending on the probabilities assigned to individual worlds;

Worlds	ρ	$\rho(a, .8)$	$\rho(a, .8)(a, .8)$	$\rho(a, .8)(b, .8)$
a, b, c	.150	.240	.2824	.3333
$a, b, \overline{c}$	.150	.240	.2824	.3333
$a, \overline{b}, c$	.100	.160	.1882	.0556
$a, \overline{b}, \overline{c}$	.100	.160	.1882	.0556
$\overline{a}, b, c$	.150	.060	.0176	.0972
$\overline{a}, b, \overline{c}$	.150	.060	.0176	.0972
$\overline{a}, \overline{b}, c$	.100	.040	.0118	.0139
$\overline{a}, \overline{b}, \overline{c}$	.100	.040	.0118	.0139

Table 1. Example of Updating Probabilities of Worlds

in the example they happen to be unchanged. Last if we sense a and then b, in both cases with reliability .8 we obtain that $Pr(a) = .7778$, $Pr(b) = .8610$, and $Pr(c) = .5$.

4 Epistemic States: A Model of Categorical Belief based on Noncategorical Belief

This section presents the notion of *epistemic state* as it is used in the present approach. We first discuss intuitions then give the formal details.

4.1 Intuitions

An agent's epistemic state $\mathcal{K}$ is given by a pair (ρ, c), where ρ is a probability assignment over possible worlds and c is a *confidence level*. The probability function captures what the agent knows about the world. We wish to say that an agent *accepts* a belief represented by a formula just if, in some fashion, the probability of the formula exceeds the confidence level c. Thus, the agent accepts a formula if its probability is "sufficiently high". This notion of *acceptance* is nonstandard, in that an accepted belief will be categorical yet its associated probability may be less than 1. This is in contrast to the approaches combining probability and revision described in Section 2.3, where the agent's categorical beliefs have probability 1. This also is in contrast with [Bacchus, Halpern, and Levesque 1999], where non-beliefs have probability 0. In any case, for us an accepted belief is one that is categorical, in that the agent may act under the assumption that it is true, yet it is also noncategorical, in that its probability is less than 1, and hence it can be given up following a revision. The issue then becomes one of suitably defining the worlds characterising an agent's accepted beliefs.

The most straightforward way of defining acceptance is to say that a formula ϕ is accepted just if $\rho(Mod(\phi)) \geq c$. This leads immediately to the *lottery paradox* [Kyburg 1961]. This problem is that for any $c < 1.0$ one can construct a scenario where $p_1, \ldots, p_n$, along with $\neg p_1 \vee \ldots \vee \neg p_n$ are all accepted. But the set consisting of these formulas is of course inconsistent. The resolution proposed here is to focus instead on the set of possible worlds characterising an agent's beliefs. That is, the agent's (categorical) beliefs will be identified with a subset of possible worlds in which the set of beliefs is true. The issue then is to determine the appropriate subset of possible worlds. To this end, we suggest that the appropriate set of worlds is comprised of those worlds of greatest probability such that the probability of the set exceeds c. Since the agent's accepted beliefs are characterised by a unique set of worlds, the lottery paradox doesn't arise.

The assumption that worlds with higher probability are to be preferred to those with lower probability for characterising an agent's beliefs can be justified by (at least) two arguments. First, if an agent had to commit to a single world being the real world, then it would choose a world w for which the probability $\rho(w)$ was maximum; if it had to commit to n worlds, then it would choose the n worlds with highest probability. Similarly, if it were to choose the most likely set of worlds containing the real world, such that the probability of the set exceeded a certain bound (here c), then it would choose the set of worlds of maximal probability that meets or exceeds c. Since there is nothing that distinguishes worlds beyond their probability, if $\rho(w) = \rho(w')$ then if w is in this set then so is w'.

A second argument is related to a principle of *informational economy*: It seems reasonable to assume that, given a set of candidate belief sets, an agent will prefer a set that gives more information over one that gives less. This is the case here. In general there will be more than one set of worlds where the probability of the set exceeds c. The set composed of worlds of maximal probability is generally also the set with the least number of worlds, which in turn will correspond to the belief set with the maximum number of (logically distinct) formulas. So this approach commits the agent to the maximum set of accepted beliefs, where the overall probability of the set exceeds c.[2] Such a set may be said to have the greatest *epistemic content* among the candidate belief sets.

Thus an epistemic state consists principally of a probability function on possible worlds. Via an assumption of maximality of beliefs (or maximal epistemic content), and given the confidence level c, a set of accepted beliefs is defined. So this differs significantly from prior work, in that an accepted formula will generally have an associated probability that is less than 1.0. Arguably this makes sense: for example, I believe that my car is where I left it this morning, in that I act as if this was a true fact even though I don't hold that it is an absolute certainty that the car is where I left it. If pressed, I would be happy to attach a probability to the possibility of my car not being where I left it, but I would continue to act as if it were (simply) true that my car was where I left it. Moreover, of course, I am prepared to revise this belief if I receive information to the contrary.

4.2 Epistemic States: Formal Details

In this subsection we define our notion of epistemic state, and relate it to *faithful rankings* that have been used to characterise AGM revision.

DEFINITION 2. $\mathcal{K} = (\rho, c)$ is an *epistemic state*, where:

- ρ is probability assignment to possible worlds and

- $c \in (0, 1]$ is a *confidence level*.

As described, an epistemic state characterises the state of knowledge of the agent, both its (contingent) beliefs as well as, implicitly, those beliefs that it would adopt or abandon in the presence of new information. We need to also define the agent's *belief set* or beliefs about the world at hand. This is most easily done by first defining the worlds that characterise the agent's belief set, and then defining the belief set in terms of these worlds.

DEFINITION 3. For epistemic state $\mathcal{K} = (\rho, c)$, the set of worlds characterising the agent's belief set, $Mod(\mathcal{K}) \subseteq W_{\mathcal{L}}$, is the least set such that:

  If $\rho = \rho_{\perp}$ then $Mod(\mathcal{K}) = \emptyset$; otherwise:

---

[2]These notions of course make sense only in a finite (under equivalence classes) language, which was assumed at the outset.

1. $\rho(Mod(\mathcal{K})) \geq c$,

2. If $w \in Mod(\mathcal{K})$ and $w' \notin Mod(\mathcal{K})$ then $\rho(w) > \rho(w')$.

$Mod(\cdot)$ is uniquely characterised; in particular we have that if $\rho(w) = \rho(w')$ then $w \in Mod(\mathcal{K})$ iff $w' \in Mod(\mathcal{K})$.

DEFINITION 4. For epistemic state $\mathcal{K}$, the agent's accepted (categorical) beliefs, $Bel(\mathcal{K})$, are given by
$$Bel(\mathcal{K}) = \{\phi \mid Mod(\mathcal{K}) \models \phi\} = \mathcal{T}(Mod(\mathcal{K})).$$

Thus, an agent accepts a sentence if it is sufficiently likely, and a sentence is "sufficiently likely" if it is true in the set of most plausible worlds such that the probability of the set exceeds the given confidence level. Since an agent's beliefs are characterised by a single set of possible worlds, the lottery paradox doesn't arise.

This then describes the static aspects of an epistemic state. For the dynamic aspects (i.e. revision) it will be useful to distinguish those formulas that are *possible*, in the sense that they are *conceivable*, which is to say they have a non-zero probability. We use $Poss_{\mathcal{K}}(\phi)$ to indicate that, according to the agent involved, ϕ is possible; that is, there is possible world w such that $w \models \phi$ and $\rho(w) > 0$. $Poss_{\mathcal{K}}(\cdot)$ can be axiomatised as the modality $\Diamond$ in the modal logic S5 [Hughes and Cresswell 1996]. We have the simple consequence:

PROPOSITION 5. *If not* $Poss_{\mathcal{K}}(\phi)$ *then* $\neg\phi \in Bel(\mathcal{K})$

The probability assignment to possible worlds defines a ranking on worlds, where worlds with higher probability are lower in the ranking:

DEFINITION 6. For given ρ, define $rank_\rho(w)$ for every $w \in W_{\mathcal{L}}$ by:

1. $rank_\rho(w) = 0$ if $\nexists w'$ such that $\rho(w') > \rho(w)$

2. Otherwise, $rank_\rho(w) = 1 + \max\{rank_\rho(w') : \rho(w') > \rho(w)\}$.

Lastly, we can define a *faithful ranking* (as given in Section 2) to relate the ranking defined here to rankings used in belief revision:

DEFINITION 7. The *faithful ranking* $\preceq_{\mathcal{K}}$ is given by:

1. If $w_1, w_2 \models Bel(\mathcal{K})$ then $w_1 =_{\mathcal{K}} w_2$

2. If $w_1 \models Bel(\mathcal{K})$ and $w_2 \not\models Bel(\mathcal{K})$ then $w_1 \prec_{\mathcal{K}} w_2$

3. Otherwise if $rank_\rho(w_1) \leq rank_\rho(w_2)$ then $w_1 \preceq_{\mathcal{K}} w_2$.

Thus, a faithful ranking on worlds can be defined in a straightforward manner from an epistemic state as given in Definition 2. The first two conditions stipulate that we have a faithful ranking. The third condition ensures that we have total preorder that conforms to the probability assignment for those worlds not in $Mod(\mathcal{K})$. Clearly this faithful ranking suppresses detail found in ρ. First, quantitative information is lost in going from Definition 2 to Definition 6. Second, gradations in an agent's beliefs are lost: worlds in $Mod(\mathcal{K})$ may have varying probabilities, yet in the corresponding faithful ranking given in Definition 7, all worlds in $Mod(\mathcal{K})$ are ranked equally. Consequently, the notion of epistemic state as defined here is a richer structure than that of a faithful ordering.

5 Belief Dynamics in a Probabilistic Framework

We next consider how this approach fits with work in belief revision. A natural way to define the revision of an epistemic state $\mathcal{K} = (\rho, c)$ by ϕ with reliability q is to set $\mathcal{K} * (\phi, q) = (\rho(\phi, q), c)$. Of course, revision so defined is a ternary function, as opposed to the usual expression of revision as a binary function, $\mathcal{K} * \phi$. There are various ways in which this mismatch may be resolved. First, we could simply regard revision in a probabilistic framework as a ternary function, with the extra argument giving the reliability of the observation. This is problematic, with regards to our aims, since a ternary operator represents a *quantitative* approach, where the degree of support q of ϕ is taken into account. In contrast, AGM revision is *qualitative*, in that for a revision $\mathcal{K} * \phi$, it is the (unqualified) formula ϕ that is a subject of revision. This clash then highlights the main issue of this paper: a probabilistic approach is intrinsically quantitative, while standard approaches to belief revision are inherently qualitative.

In re-considering revision $*$ as a binary function, the intent is that in expressing $\mathcal{K} * (\phi, q)$ as a binary function $\mathcal{K} * \phi$, we want to study *properties* of the function $*$ without regard to specific values assigned to q. Consequently, we assume that the reliability of a revision is some fixed probability q. Given that the reliability is fixed, we can drop the probability argument from a statement of revision, and simply write $\mathcal{K} * \phi$. We later also consider the situation where the reliability of observations may vary.

Revision by ϕ is intended to *increase* the agent's confidence in ϕ, and so for $\mathcal{K} * \phi$ it is understood that the probability of ϕ is greater than 0.5. Since revision corresponds to the incorporation of *contingent* information, it is reasonable to assume that nothing can be learned with certainty, and so we further assume that $q < 1$.[3] Consequently, in what follows, we assume that the reliability of a revision is a fixed number q in the range $(0.5, 1.0)$.

DEFINITION 8. Let $q \in (0.5, 1.0)$ be fixed. Let $\mathcal{K} = (\rho, c)$ be an epistemic state and $\phi \in \mathcal{L}$. Define the revision of $\mathcal{K}$ by ϕ by:

$$\mathcal{K} * \phi = (\rho(\phi, q), c)$$

Clearly one needs to know the value of q (along with $\mathcal{K}$ and ϕ) before being able to determine $\mathcal{K} * \phi$. However, without knowing the value of q, one can still investigate properties of the class of revision functions, which is our goal here. Other aspects of the definition are discussed below, in the discussion of postulates. We first revisit our previous example.

Example (continued): Consider again Table 1, and assume that our initial epistemic state is given by $\mathcal{K} = (\rho, 0.9)$. At the outset, $Bel(\mathcal{K}) = Cn(\top)$. If the probability associated with the world given by $\{\bar{a}, \bar{b}, \bar{c}\}$ was .05, with the balance distributed uniformly across other possible worlds, we would have $Bel(\mathcal{K}) = Cn(a \vee b \vee c)$.

We obtain that $Bel(\mathcal{K} * a) = Cn(a \vee b)$, and $Bel(\mathcal{K} * a * a) = Cn(a)$. We also obtain $Bel(\mathcal{K}*a*b) = Cn(a \vee b)$ and (not illustrated in Table 1) $Bel(\mathcal{K}*a*b*b) = Cn(b)$. So, not unexpectedly, for repeated iterations, the resulting belief set "converges" toward accepting the iterated formula, with the results being biased by the initial probability distribution.

---

[3]It might be pointed out that a tautology can be learned with absolute certainty. However, it can be pointed out in return that a tautology is in fact *known* with certainty, so the probability being 1 or less makes no difference. In any case, we later examine the situation where $q = 1$.

5.1 Properties of Probability-Based Belief Revision

Recall that $Poss_{\mathcal{K}}(\phi)$ indicates that, according to the agent, ϕ is possible, in that there is w such that $w \models \phi$ and $\rho(w) > 0$. $\mathcal{K} *^n \phi$ stands for the n-fold iteration of $\mathcal{K} * \phi$, that is:

$$\mathcal{K} *^n \phi = \begin{cases} \mathcal{K} * \phi & \text{if } n = 1 \\ (\mathcal{K} *^{n-1} \phi) * \phi & \text{otherwise} \end{cases}$$

We obtain the following results; numbering corresponds to the AGM revision postulates.

THEOREM 9. *Let $\mathcal{K}$ be an epistemic state and $\phi, \psi \in \mathcal{L}$.*

$(\mathcal{K} * 1)$ $Bel(\mathcal{K} * \phi) = Cn(Bel(\mathcal{K} * \phi))$

$(\mathcal{K} * 2a)$ *If $Poss_{\mathcal{K}}(\phi)$ then $\phi \in Bel(\mathcal{K} *^n \phi)$ for some $n > 0$*

$(\mathcal{K} * 2b)$ *If $\mathcal{K} \neq \mathcal{K}_{\perp}$ and $\phi \in Bel(\mathcal{K})$ then $\phi \in Bel(\mathcal{K} * \phi)$*

$(\mathcal{K} * 2c)$ *If $\mathcal{K} \neq \mathcal{K}_{\perp}$ and not $Poss_{\mathcal{K}}(\phi)$ then $Bel(\mathcal{K} * \phi) = Bel(\mathcal{K})$*

$(\mathcal{K} * 5)$ *$Bel(\mathcal{K} * \phi)$ is consistent.*

$(\mathcal{K} * 6)$ *If $\phi \equiv \psi$ then $Bel(\mathcal{K} * \phi) \equiv Bel(\mathcal{K} * \psi)$*

Proof: $(\mathcal{K} * 1)$ follows directly from Definition 4. For $(\mathcal{K} * 2a)$, it follows from Definition 1 that if $0 < Pr_{\rho}(\phi) \leq 1$ then $Pr_{\rho}(\phi) < Pr_{\rho(\phi,q)}(\phi) \leq 1.0$, and so if we iterate a revision by ϕ, the probability of ϕ monotonically increases, with upper bound 1.0. It follows that for some n, $Mod(\mathcal{K} *^n \phi) \subseteq Mod(\phi)$, and so for some $n > 0$, $\phi \in Bel(\mathcal{K} *^n \phi)$. $(\mathcal{K} * 2b)$ and $(\mathcal{K} * 2c)$ have prerequisite condition that $\mathcal{K}$ is not the incoherent epistemic state. $(\mathcal{K} * 2b)$ is obvious. For $(\mathcal{K} * 2c)$, if $\mathcal{K} \neq \mathcal{K}_{\perp}$ then if there are no ϕ-worlds with non-zero probability, then Definition 1 can be seen to leave the probability function unchanged. For $(\mathcal{K} * 5)$ it can be seen from the definitions that if $\mathcal{K} \neq \mathcal{K}_{\perp}$, then there will be worlds with a non-zero probability, and so $Mod(\mathcal{K}) \neq \emptyset$ in Definition 3, and so $Bel(\mathcal{K})$ is well defined in Definition 4 and specifically $Bel(\mathcal{K}) \neq \mathcal{L}$. In particular, in the case of a revision by an inconsistent formula ϕ, we get that $\mathcal{K} * \phi = \mathcal{K}$: All ϕ worlds (of which there are none) share in the probability q, and all $\neg\phi$ worlds share in the probability $1 - q$. The result is normalised, leaving the probabilities unchanged. If $\mathcal{K} = \mathcal{K}_{\perp}$, then in Definition 1 we get that $\eta \neq 0$, $0 < q < 1$, and so the probability assignment $\rho(\phi,q) \neq \rho_{\perp}$, and so in Definition 3 we obtain that $Mod(\mathcal{K}) \neq \emptyset$. Postulate $(\mathcal{K} * 6)$ holds trivially, but by virtue of the fact that the reliability of an observation of ϕ is the same as that of ψ. ■

The weaker version of postulate $(\mathcal{K} * 2)$, given by $(\mathcal{K} * 2a)$, means that an agent will accept that ϕ is true after a sufficient number of iterations (or "reports") of ϕ. Hence, despite the absence of other AGM postulates, the operator $*$ counts as a revision operator, since the formula ϕ will eventually be accepted, provided that it is possible. Note that if a formula ϕ is not possible, then from our earlier (non-revision) result

If not $Poss_{\mathcal{K}}(\phi)$ then $\neg\phi \in Bel(\mathcal{K})$

together with $(\mathcal{K} * 5)$, we have that ϕ can never be accepted. As well, if ϕ is accepted, it will continue to be accepted following revisions by ϕ $(\mathcal{K} * 2b)$. This last point would seem to be obvious, but is necessitated by the absence of a postulate of success and the absence of $(K * 4)$. If a formula ϕ is deemed to be not possible, but the agent is not in the incoherent state $K_\perp$, $(K * 2c)$ shows that revising by ϕ leaves the agent's belief set unchanged. While this may seem noncontentious, [Makinson 2011] discusses the case where it may be meaningful to condition on a contingent formula whose probability is zero. However, such a situation appears to rely on an underlying infinite domain.

It can be noted that $K_\perp$ plays no interesting role in revisions; this is reflected by $(K * 5)$, which asserts that no revision can yield $K_\perp$. Hence an epistemic state can be inconsistent only if the original assignment of probabilities to worlds is the absurd probability assignment $\rho_\perp$. Any subsequent revision will have $Bel(K_\perp * \phi) \neq L$. In particular if ϕ is $\perp$ then $\rho_\perp(\phi, q) = \rho_\top$ and so $Bel(K_\perp * \perp) = Cn(\top)$.

Postulate $(K*5)$ is quite strong, in that it imposes no conditions on the original epistemic state K or the formula for revision. If one begins with the inconsistent epistemic state $K_\perp$, then revision is defined as being the same as a revision of the epistemic state of complete ignorance $\rho_\top$. This is pragmatically useful: from $K_\perp$, if one revises by a formula ϕ where $Pr(\phi) \neq 0$, then analogous to the AGM approach, one arrives at a consistent belief state. This also goes beyond the AGM postulate $(K * 5)$, since if ϕ is held to be impossible (i.e. there are no worlds with nonzero probability in which ϕ is true), then there will be worlds in which $\neg\phi$ is true and with nonzero probability, and so revision yields meaningful results, in particular yielding the epistemic state with probability function $\rho_\top$.

Postulate $(K*6)$ holds trivially, given the assumption that the reliability of an observation of ϕ is the same as that of ψ. This assumption is, of course, limiting, and in the case where observations may be made with differing degrees of reliability, the postulate would not hold. It can be noted that in the case where observations may be made with differing degrees of reliability, the postulate can be replaced by the weaker version:

If $\phi \equiv \psi$ then $Bel(K * \phi) \subseteq Bel(K * \psi)$ or $Bel(K * \psi) \subseteq Bel(K * \phi)$.

We next consider those postulates that don't hold, and why they fail to hold. For a counterexample for $(K * 3)$, let $\mathcal{P} = \{a, b\}$, $K = (\rho, 0.97)$, and ρ is given as follows:

$$\rho(\{a, b\}) = .96 \qquad \rho(\{a, \neg b\}) = .02,$$
$$\rho(\{\neg a, b\}) = .01 \qquad \rho(\{\neg a, \neg b\}) = .01$$

$Bel(K)$ is characterised by $\{a, b\}, \{a, \neg b\}$, $Bel(K) = Cn(a)$, and so $Bel(K) + a = Cn(a)$. If we revise by a with confidence .8, we get

$$\rho(a, .9)(\{a, b\}) \approx .9746 \qquad \rho(a, .9)(\{a, \neg b\}) \approx .0203$$
$$\rho(a, .9)(\{\neg a, b\}) \approx .0025 \qquad \rho(a, .9)(\{\neg a, \neg b\}) \approx .0025$$

Thus $Bel(K * a)$ is characterised by $\{a, b\}$, i.e. $Bel(K * a) = Cn(a \wedge b) \neq Cn(a) = Bel(K) + a$. This illustrates a curious phenomenon: In the counterexample we have that $Bel(K) = Cn(a)$ yet $Bel(K * a) = Cn(a \wedge b)$. In revising by a, the probability of worlds given by $\{a, b\}, \{a, \neg b\}$ both increase, but that of $\{a, b\}$ increases so that its probability exceeds the confidence level c, and so it alone characterises the agent's set of accepted beliefs. We discuss this behaviour later, once we have presented the approach as a whole.

For $(K * 4)$, it is possible to have formulas ϕ and ψ such that ϕ and ψ are logically independent, $Bel(K) = Cn(\phi)$ and $Bel(K * \psi) = Cn(\psi)$, thus contradicting the postulate. To see this, consider where $\mathcal{P} = \{a, b\}$, and $K = (\rho, 0.9)$ and where:

$$\rho(\{a, b\}) = .46 \qquad \rho(\{a, \neg b\}) = .46,$$
$$\rho(\{\neg a, b\}) = .06 \qquad \rho(\{\neg a, \neg b\}) = .02$$

$Bel(\mathcal{K})$ is characterised by $\{a, b\}, \{a, \neg b\}$, i.e. $Bel(\mathcal{K}) = Cn(a)$, and so $Bel(\mathcal{K}) + b = Cn(a \wedge b)$. If we revise by b with confidence .9, we get

$$\rho(b, .9)(\{a, b\}) \approx .802 \qquad \rho(b, .9)(\{a, \neg b\}) \approx .089$$
$$\rho(b, .9)(\{\neg a, b\}) \approx .105 \qquad \rho(b, .9)(\{\neg a, \neg b\}) \approx .004$$

Since $\rho(b, .9)(\{a, b\}) + \rho(b, .9)(\{\neg a, b\}) > .9 = c$, we get that $Bel(\mathcal{K} * b)$ is characterised by $\{a, b\}, \{\neg a, b\}$ and so $Bel(\mathcal{K}) = Cn(b)$.

This illustrates another interesting point: not only does the postulate fail but, unlike $(\mathcal{K} * 2)$, it may fail over any number of iterations. For the example given, the probability of the world given by $\{a, b\}$ will converge to something just over .88, which is below the given confidence level of $c = 0.9$. Since $Cn(a, b)$ is the result of expansion in the example, this shows that $Cn(a, b)$ will never come to be accepted. Similar remarks hold for $(\mathcal{K} * 8)$.

$(\mathcal{K} * 7)$ doesn't hold for the same reason $(\mathcal{K} * 3)$ doesn't. Substituting $\top$ for ϕ in $(\mathcal{K} * 7)$ in fact yields $(\mathcal{K} * 3)$. Similarly, $(\mathcal{K} * 8)$ doesn't hold for the same reason that $(\mathcal{K} * 4)$ doesn't. Substituting $\top$ for ϕ in $(\mathcal{K} * 8)$ yields $(\mathcal{K} * 4)$.

5.2 Variants of the Approach

We next examine three variants of the approach. In the first, observations are made with certainty. This variation coincides with an extant approach in belief revision. As well it has close relations to Gärdenfors' revision of probability functions; a discussion of the relation with this latter work is deferred to the next section. In the second variant, observations are made with near certainty; again this variant corresponds with an extant approach in belief revision. In the last variant, informally possible worlds that characterise an agent's beliefs are retained after a revision if there is no reason to eliminate them.

Certain Observations Consider where observations are certain, and so the (binary) revision $\mathcal{K} * \phi$ corresponds to $\mathcal{K} * (\phi, 1.0)$. Clearly, if $\rho(w) = 0$, then $\rho(\phi, 1.0)(w) = 0$ for any ϕ; that is, if a world had probability 0, then no observation is going to alter this probability. As well, if $w \models \neg\phi$ then $\rho(\phi, 1.0)(w) = 0$. So in a revision by ϕ with certainty, any $\neg\phi$ world will receive probability 0, and by the previous observation, this probability of 0 will remain fixed after subsequent revisions.

Thus in this case, revision is analogous to a form of *expansion*, but with respect to the epistemic state $\mathcal{K}$. So following a revision by ϕ, all $\neg\phi$ worlds are discarded from the derived faithful ranking. This corresponds to revision in the approach of [Shapiro, Pagnucco, Lespérance, and Levesque 2011], where their account of revision is embedded in an account of reasoning about action. For their approach, a plausibility ordering over worlds is given at each world. Observations are assumed to be correct; thus an observation of ϕ means that $\neg\phi$ is impossible in the current world, and so all $\neg\phi$ worlds are discarded. This also means that an observation of ϕ followed by $\neg\phi$ yields the inconsistent epistemic state. This result can be justified by the argument that, if ϕ is observed with certainty, then if the world does not change, then it is *impossible* for $\neg\phi$ to be observed. In this approach, postulates $(K * 1) - (K*4)$, and $(K * 6)$ are shown to be satisfied.

Near-Certain Observations Consider where the binary revision $\mathcal{K} * \phi$ is defined to be $\mathcal{K} * (\phi, 1.0 - \epsilon)$, where ϵ is "sufficiently small" compared to the probabilities assigned by ρ. If the minimum and maximum values in the range of ρ are min_ρ and max_ρ, then "sufficiently small" would mean that

$$\max\{\rho(w) \mid w \in Mod(\neg\phi)\} \times \epsilon \; < \; \min\{\rho(w) \mid w \in Mod(\phi)\} \times (1.0 - \epsilon).$$

Thus for $\rho' = \rho(\phi, 1.0 - \epsilon)$ we would have for $w \models \phi$, $w' \not\models \phi$ that $\rho'(w) > \rho'(w')$. This yields *lexicographic revision* [Nayak 1994] in which, in revising by ϕ, every ϕ world is ranked below every $\neg\phi$ world, but the relative ranking of ϕ worlds (resp. $\neg\phi$ worlds) is retained. In this approach, all AGM revision postulates hold.

Retaining Confirmed Possible Worlds The present approach clearly falls within belief revision, since under reasonable conditions a formula will become accepted. However, it has notable weaknesses compared to the AGM approach; in particular the postulates $(\mathcal{K} * 3)$, $(\mathcal{K} * 4)$, $(\mathcal{K} * 7)$, and $(\mathcal{K} * 8)$ all fail. In the case of $(\mathcal{K} * 4)$ and $(\mathcal{K} * 8)$ this seems unavoidable. However, an examination of $(\mathcal{K} * 3)$ and $(\mathcal{K} * 7)$ shows a curious phenomenon. Consider $(\mathcal{K} * 3)$: In the counterexample presented, the agent believed that the real world was among the set of worlds $\{\{a, b\}, \{a, \neg b\}\}$. On revising by a, the agent believed that the real world was among the set of worlds $\{\{a, b\}\}$, which is to say, that $\{a, b\}$ was the real world. But this means is that $\{a, \neg b\}$ was considered to be possibly be the real world according to the agent, but on receiving confirmatory evidence (viz. revision by a), this world was dropped from the characterising set. But arguably if w may be the actual world according to the agent, and the agent learns ϕ where $w \models \phi$, then it seems that the agent should still consider w as possibly being the actual world.

The reason for this phenomenon is clear: The probability of other worlds (in the example, given by $\{a, b\}$) becomes large enough following revision so that the "dropped" world isn't required in making up $Mod(\mathcal{K})$. To counteract this phenomenon, it seems reasonable to assume that if an agent considers a world to be possible, then it remains possible after confirmatory evidence. To this end, the approach can be modified so that one keeps track of worlds considered possible by the agent, where these are the worlds characterising the agent's contingent beliefs. An epistemic state now would be a triple (ρ, c, B) where $B \subseteq W_{\mathcal{L}}$ characterises the agent's belief set following a revision by ϕ with probability q. Thus after revising by ϕ, the new value of B would be given by:

$$Mod((\rho(\phi, q), c, B)) \cup (Mod((\rho, c.B)) \cap Mod(\phi)).$$

In this case postulates $(K * 3)$ and $(K * 7)$ also hold.

5.3   Iterated Belief Revision

Turning to iterated revision, it proves to be the case that three of the Darwiche-Pearl postulates fail to hold. However, the reason that these postulates do not hold is not a result of the probabilistic approach per se, but rather is a result of the expression of a belief set in terms of possible worlds.

THEOREM 10. *Let $\mathcal{K}$ be an epistemic state with associated revision operator $*$. Then $\mathcal{K}$ satisfies* **C3**.

Proof: For **C3**, we obtain that the semantic condition **CR3** holds: If $w_1 \models \phi$ and $w_2 \not\models \phi$, then $w_1 \prec_{\mathcal{K}} w_2$ implies that $\rho(w_1) < \rho(w_2)$ from which it follows that $\rho(\phi, q)(w_1) < \rho(\phi, q)(w_2)$ and so $w_1 \prec_{\mathcal{K}*\phi} w_2$. By the same argument as [Darwiche and Pearl 1997, Theorem 13], we get that **C3** is satisfied. ∎

$\mathcal{K}$ does not necessarily satisfy **C1**, **C3**, and **C4**. Consider **C1**, and let $\mathcal{P} = \{a, b\}$, $\mathcal{K} = (\rho, 0.9)$, and where:

$$\rho(\{a, b\}) = .85 \qquad \rho(\{a, \neg b\}) = .06,$$
$$\rho(\{\neg a, b\}) = .05 \qquad \rho(\{\neg a, \neg b\}) = .04$$

$Bel(\mathcal{K})$ is characterised by $\{a, b\}$, $\{a, \neg b\}$ and so $Bel(\mathcal{K}) = Cn(a)$. If we revise by a with confidence .7, we get

$$\rho(a, .7)(\{a, b\}) \approx .891 \qquad \rho(a, .7)(\{a, \neg b\}) \approx .063$$
$$\rho(a, .7)(\{\neg a, b\}) \approx .023 \qquad \rho(a, .7)(\{\neg a, \neg b\}) \approx .018$$

Thus $Bel(\mathcal{K}) = Bel(\mathcal{K} * a)$. If we again revise by a with confidence .7, we get

$$\rho(a, .7)(\{a, b\}) \approx .917 \qquad \rho(a, .7)(\{a, \neg b\}) \approx .065$$
$$\rho(a, .7)(\{\neg a, b\}) \approx .010 \qquad \rho(a, .7)(\{\neg a, \neg b\}) \approx .008$$

Since $\rho(a, .9)(\{a, b\}) > .9 = c$, so $Mod(\mathcal{K} * a * a) = \{a, b\}$. Hence $Bel(\mathcal{K} * a) \neq Bel(\mathcal{K} * a * a)$, thereby violating **C1**.[4] Other postulates fail for analogous reasons.

It is worth considering why most of the iteration postulates fail. Interestingly, for the semantic conditions, **CR1 – CR4**, if expressions of the form $w_1 \prec_{\mathcal{K}} w_2$ are replaced by expressions of the form $\rho(w_1) \leq \rho(w_2)$, then the modified conditions hold in the current approach. That is, it is easily verified that all of the following hold:

THEOREM 11.

PCR1 *If* $w_1, w_2 \models \phi$, *then* $\rho(w_1) \leq \rho(w_2)$ *iff* $\rho(\phi, q)(w_1) \leq \rho(\phi, q)(w_2)$.

PCR2 *If* $w_1, w_2 \not\models \phi$, *then* $\rho(w_1) \leq \rho(w_2)$ *iff* $\rho(\phi, q)(w_1) \leq \rho(\phi, q)(w_2)$.

PCR3 *If* $w_1 \models \phi$ *and* $w_2 \not\models \phi$, *then* $\rho(w_1) < \rho(w_2)$ *implies* $\rho(\phi, q)(w_1) < \rho(\phi, q)(w_2)$.

PCR4 *If* $w_1 \models \phi$ *and* $w_2 \not\models \phi$, *then* $\rho(w_1) \leq \rho(w_2)$ *implies* $\rho(\phi, q)(w_1) \leq \rho(\phi, q)(w_2)$.

Proof: Straightforward from Definition 1. ∎

The problem is that our faithful ranking (Definition 7) doesn't preserve the ordering given by ρ. In particular, if $w_1, w_2 \in Mod(\mathcal{K})$ then $w_1 =_{\mathcal{K}} w_2$ in the derived faithful ranking, while most often we will have $\rho(w_1) \neq \rho(w_2)$. Essentially, in moving from values assigned via ρ to the faithful ranking, gradations (given by probabilities) among worlds in $Mod(\mathcal{K})$ are lost. That is, in a sense, the probabilistic approach provides a finer-grained account of an epistemic state than is given by a faithful ranking on worlds, in that models of the agent's belief set also come with gradations of belief.

5.4 Relation with Other Work

Other Approaches to Revision We have already discussed the relation of the approach to [Shapiro, Pagnucco, Lespérance, and Levesque 2011] and [Nayak 1994].

The work in belief change that is closest to that described here is that of *improvement operators* [Konieczny and Pino Pérez 2008], where an an improvement operator is a belief change operator in which new information isn't necessarily immediately accepted, but where the plausibility is increased. Thus after a sufficient number of iterations, the information will come to be believed. The general idea of this approach then is similar to the present approach. As well, in both approaches the success postulate does not necessarily hold, so new information is not necessarily immediately accepted. However, beyond failure of the success postulate, the approaches have quite different characteristics.

---

[4]In terms of **CR1**, we have $\{a, b\} =_{\mathcal{K}*a} \{a, \neg b\}$ but $\{a, b\} \prec_{\mathcal{K}*a*a} \{a, \neg b\}$, thereby violating **CR1**.

In the postulate set following,[5] $\circ$ is an improvement operator, and $\times$ is defined by: $\mathcal{K} \times \phi = \mathcal{K} \circ^n \phi$ where n is the first integer such that $\phi \in Bel(\mathcal{K} \circ^n \phi)$.

(I1) There exists n such that $\phi \in Bel(\mathcal{K} \circ^n \phi)$

(I2) If $\neg\phi \notin Bel(\mathcal{K})$ then $Bel(\mathcal{K} \times \phi) \equiv Bel(\mathcal{K}) + \phi$

(I3) $Bel(\mathcal{K} \circ \phi)$ is inconsistent, only if $\nvdash \neg\phi$

(I4) If $\phi_i \equiv \psi_i$ for $1 \leq i \leq n$ then $Bel(\mathcal{K} \circ \phi_1 \circ \ldots \circ \phi_n) \equiv Bel(\mathcal{K} \circ \psi_1 \circ \ldots \circ \psi_n)$

(I5) $Bel(\mathcal{K} \times (\phi \wedge \psi)) \subseteq Bel(\mathcal{K} \times \phi) + \psi$

(I6) If $\neg\psi \notin Bel(\mathcal{K} \times \phi)$ then $Bel(\mathcal{K} \times \phi) + \psi \subseteq Bel(\mathcal{K} \times (\phi \wedge \psi))$

To show the approaches are independent, it suffices to compare $(\mathcal{K}*3)/(\mathcal{K}*4)$ with **(I2)**. According to **(I2)**, after some number of iterations of an improvement operator, the resulting belief set will correspond to expansion of the original belief set by the formula in question. However, there are cases in which neither $(K * 3)$ nor $(K * 4)$ are satisfied regardless of the number of iterations. Similar comments apply to $(K * 7)$ and $(K * 8)$, and **(I5)** and **(I6)**, respectively. The need for the extended postulate for irrelevance of syntax for epistemic states **I4** was noted in [Booth and Meyer 2006]. In the present approach $(K * 6)$ suffices.

Other Related Work As discussed in Section 2, earlier work dealing specifically with revision and probability has been concerned with revising probability functions. Thus, [Gärdenfors 1988; Lindström and Rabinowicz 1989; Boutilier 1995] deal with extensions to the AGM approach for revising probability functions. In these approaches there is a probability function associated with possible worlds, but where the agent's belief set is characterised by worlds with probability 1.0. For a revision $K * \phi$, ϕ represents new *evidence*, and the probability function is revised by ϕ. The belief set corresponding to $K * \phi$ then is the set of propositions with probability 1.0. In contrast, in the approach at hand, an agent's *accepted* beliefs are characterised by a set of possible worlds whose overall probability in the general case will be less than 1.0. In a sense then there is finer granularity with regards the present approach, since the worlds characterising a belief set may have varying probability. As well, for us if a formula ϕ has probability of 1.0, then it cannot be removed by subsequent revisions; a formula is accepted as true if its probability is sufficiently high, although it may potentially be revised away. This arguably confirms to intuitions, in that if a formula is held with complete certainty then it *should* be immune from revisions.

It was noted that [Bacchus, Halpern, and Levesque 1999] presents the same general setting in which an agent receives possibly-unreliable observations. However, the concern in this paper was to update probabilities associated with worlds and then to use this for reasoning about dynamic domains expressed via the situation calculus. The approach at hand employs the same method for updating probabilities but addresses the question of how this may be regarded as, or used to formulate, an approach to belief revision. The present approach also has finer granularity, in that in [Bacchus, Halpern, and Levesque 1999] non-beliefs are given by worlds with probability 0; in the approach at hand, non-beliefs are those that fall outside the set of accepted beliefs, and may have non-zero probability. Again, arguably the present approach conforms to intuitions, since if a formula is held to be impossible then it seems it should forever remain outside the realm of revision.

---

[5][Konieczny and Pino Pérez 2008] follow [Katsuno and Mendelzon 1991], where the result of revision is a formula, not a belief set. We rephrased the [Konieczny and Pino Pérez 2008] postulates in terms of belief sets. In particular, $(\mathcal{K}*3)$ and $(\mathcal{K}*4)$ correspond to **(I2)**, while $(\mathcal{K}*7)$ and $(\mathcal{K}*8)$ correspond to **(I5)** and **(I6)** respectively.

6 Conclusion

We have explored an approach to beliefs and belief revision, based on an underlying model of uncertain reasoning. With few exceptions, research in belief revision has dealt with categorical information in which an agent has some set of beliefs and the goal is to incorporate a formula into this set of beliefs. However, most information about the real world is not categorical, and arguably no non-tautological belief may be held with complete certainty. To accommodate this, one alternative is to adopt a purely probabilistic framework for belief change. However, such a framework ignores the fact that an agent may well *accept* a formula as being true, even if this acceptance is tentative, or hedged in some fashion. So another alternative, and the one followed here, is to begin with a probabilistic framework, but also define a set of formulas that the agent accepts. Revision can then be defined in this framework, and the effect of revision on the agent's accepted beliefs examined.

To this end we assumed that an agent receives uncertain information as input, and the agent's probabilities on possible worlds are updated via Bayesian conditioning. A set of formulas among the (noncategorical) beliefs is then identified as the agent's (categorical) belief set. We show that a subset of the AGM belief revision postulates are satisfied by this approach. Most significantly, though not surprisingly, the success postulate is not guaranteed to hold, though it is after a sufficient number of iterations. As well, it proves to be the case that in revising by a formula consistent with the agent's beliefs, revision does not necessarily correspond to expansion. As another point of interest, of the postulates for iterated revision that we considered, only **C3** holds. This is because, even though the updating of the probability assignment ρ satisfies all of the corresponding semantic conditions, the induced faithful ordering $\prec_\kappa$ does not. Last, although the approach shares motivation and intuitions with improvement operators, these approaches have different properties.

There are two ways that these results may be viewed with respect to classical AGM-style belief revision. On the one hand, it could be suggested that the current approach simply provides a revision operator that is substantially weaker than given in the AGM approach and approaches to iterated revision. On the other hand, the AGM approach and approaches to iterated revision have been justified by appeals to rationality, in that it is claimed that *any* rational agent should conform to the AGM postulates and, say, the Darwiche/Pearl iteration postulates. Thus, to the extent that the presented approach is rational, this would appear to undermine the rationale of these approaches, at least in the case of uncertain information.

References

Alchourrón, C., P. Gärdenfors, and D. Makinson [1985]. On the logic of theory change: Partial meet functions for contraction and revision. *Journal of Symbolic Logic 50*(2), 510–530.

Bacchus, F., J. Halpern, and H. Levesque [1999]. Reasoning about noisy sensors and effectors in the situation calculus. *Artificial Intelligence 111*(1-2), 171–208.

Booth, R. and T. A. Meyer [2006]. Admissible and restrained revision. *Journal of Artificial Intelligence Research 26*, 127–151.

Boutilier, C. [1995]. On the revision of probabilistic belief states. *Notre Dame Journal of Formal Logic 36*(1), 158–183.

Boutilier, C. [1996]. Iterated revision and minimal revision of conditional beliefs. *Journal of Logic and Computation 25*, 262–305.

Darwiche, A. and J. Pearl [1997]. On the logic of iterated belief revision. *Artificial Intelligence 89*, 1–29.

Gärdenfors, P. [1988]. *Knowledge in Flux: Modelling the Dynamics of Epistemic States*. Cambridge, MA: The MIT Press.

Hughes, G. and M. Cresswell [1996]. *A New Introduction to Modal Logic*. London and New York: Routledge.

Jeffrey, R. [1983]. *The Logic of Decision* (second ed.). University of Chicago Press.

Jin, Y. and M. Thielscher [2007]. Iterated belief revision, revised. *Artificial Intelligence 171*(1), 1–18.

Katsuno, H. and A. Mendelzon [1991]. Propositional knowledge base revision and minimal change. *Artificial Intelligence 52*(3), 263–294.

Konieczny, S. and R. Pino Pérez [2008]. Improvement operators. In G. Brewka and J. Lang (Eds.), *Proceedings of the Eleventh International Conference on the Principles of Knowledge Representation and Reasoning*, Sydney, Australia, pp. 177–186. AAAI Press.

Kyburg, Jr., H. [1961]. *Probability and the Logic of Rational Belief*. Wesleyan University Press.

Kyburg, Jr., H. [1994]. Believing on the basis of evidence. *Computational Intelligence 10*(1), 3–20.

Levesque, H., F. Pirri, and R. Reiter [1998]. Foundations for the situation calculus. *Linköping Electronic Articles in Computer and Information Science 3*(18).

Lindström, S. and W. Rabinowicz [1989]. On probabilistic representation of non-probabilistic belief revision. *Journal of Philosophical Logic 11*(1), 69–101.

Makinson, D. [2011]. Conditional probability in the light of qualitative belief change. *Journal of Philosophical Logic*. To appear.

Nayak, A. [1994]. Iterated belief change based on epistemic entrenchment. *Erkenntnis 41*, 353–390.

Nayak, A. C., M. Pagnucco, and P. Peppas [2003]. Dynamic belief revision operators. *Artificial Intelligence 146*(2), 193–228.

Pearl, J. [1988]. *Probabilistic Reasoning in Intelligent Systems: Networks of Plausible Inference*. San Mateo, CA: Morgan Kaufman.

Peppas, P. [2007]. Belief revision. In F. van Harmelen, V. Lifschitz, and B. Porter (Eds.), *Handbook of Knowledge Representation*, pp. 317–359. San Diego, USA: Elsevier Science.

Shapiro, S., M. Pagnucco, Y. Lespérance, and H. J. Levesque [2011]. Iterated belief change in the situation calculus. *Artificial Intelligence 175*(1), 165–192.

12

Agent Trust and Reputation – A Prospectus

NORMAN FOO, ABHAYA NAYAK, AND PAVLOS PEPPAS

ABSTRACT. Trust and reputation are central to interacting systems of agents. There are numerous approaches to work in this area but no dominant paradigm has emerged. We report here recent as well as ongoing work by us and our colleagues, and outline a prospectus for amalgamating the external with the internal view of agents that promises dividends in insight and engagement with the very active simulation community in this area.

1 Introduction

An engaging aspect of Hector Levesque's work is his constant return to *epistemological issues* – how agents think about the world and about other agents, and what is meant by agent knowledge. His early seminal work on shallow reasoning, and his much later work in incorporating epistemic predicates into the situation calculus, are examples. Our paper here is dedicated to him in celebration of his myriad fundamental contributions to knowledge representation and reasoning, in appreciation of a friend who is unmatched in achievement, famous for his self-efacement and loved for his wit. Thanks Hector, for the comradeship!

Our presentation here is a prospectus for sone promising approaches to formalizing how multiagents evaluate ond interact with one another dynamically. The importance of this area is obvious. The internet is replete with such evaluations – formal, explicit, informal, implicit, open, cladestine. Well known examples of explicit and open systems are in *eBay* ranking and *Amazon* customer reviews. Implicit systems are in use by some networks to assess and modify the effevtiveness of nodes in data packet routing; the "agents" here are the nodes. *Google* uses its privileged algorithms to score sites, and the *WOT* application scores sites for safety and reputation. Away from the internet, real human societies are readily viewed as highly dynamic interacting communities of agents in which trust and reputation are powerful determinants of activity. Indeed historically it was the sociologists who first investigated the constituents of trust and reputation, and how they evolved. Issues pertaining to social trust in virtual societies were identified by [Falcone and Castelfranchi 2001].

The well-known *Prisoners' Dilemma (PD)* (see any standard book on game theory, e.g., [Watson 2008] stimulated interest in agents' reasoning about other agents rational decisions in formal two-person complete information games. For good reason it has become the paradigmatic basis for work on trust and reputation, so it is worthwhile to recall its main features. Its setting is a game matrix in figure 1.

The two agents know the payoffs for their choice of strategies, with the payoff for agent 1 being the first component of the pair and that for agent 2 being the second in each matrix entry identified by the strategies chosen by each. Thus, if agent 1 chooses to cooperate while agent 2 defects, the payoff for agent 1 is 0 and that for agent 2 is 3. If the agents have no reason to believe that they can trust each other to cooperate, then rationally they

Agent 2

		cooperate	defect
Agent 1	cooperate	(2,2)	(0,3)
	defect	(3,0)	(1,1)

Figure 1. The Prisoners' Dilemma Payoff Martix

will choose to defect and end up in the bottom right entry where both receive sub-optimal payoffs of 1. This is the celebrated *Nash Equilibrium* (Watson, op.cit.) of the PD, meaning a paired choice in which each is an optimal response to the other. However, if the game is repeated without the agents knowing when it will end, mutual cooperation may emerge as a dynamic equilibrium and both agents benefit from higher payoffs. This was first demonstrated in a tournament run by Axelrod [Axelrod 1984]. The winning entry was submitted by A. Rapaport using a strategy now called *Tit-for-Tat* which cooperated with its opponent until the latter defects, in which case it retaliates by defecting the next round, and reverts to cooperting until its opponent again defects. Thus it defects once for opponent defection and then it forgives. Axelrod succinctly describes *Tit-for-Tat*'s algorithm as follows: it begins by cooperating, thence it copies its opponents previous move, hence its name Tit-for-Tat. Thus, a continuing sequence of cooperation strategies by both agents will redound to the mutual optimal payoff for both. This, Axelrod argued, is evidence for the evolution of cooperation. Others have argued that in real life it may also be interpreted as the *emergence of mutual trust*. Unfortunately another iterated PD equilibrium is one where both agents simply iterate mutual defection to ther mutual detriment. The almost canonical status of the PD and its variants is largely due to its effectiveness in modeling animal and human behavior, e.g., *Tit-for-Tat*'s success has been paraphrased as "assume people are good, punish swiftly if they behave badly, then forgive and assume goodness again".

2   Trust and Reputation – External View

In this section we view agents objectively, i.e., not considering their internal or cognitive reasons for their perception or evaluation of other agents, nor of their personalities or social norms [Savarimuthu, Purvis, and Purvis 2008]. We return in section 4 to these considerations.

The problems we address are about a community of interacting agents engaging in a class of transactions, and each agent attempts to rank other agents for trustworthiness and reputation. An agent may not rank itself — this is modelled by a self-rank with neutral trust value. The reputation of an agent arises from some form of aggregation of of its trust ranks by other agents. Both trust and reputation ranks may have to be revised to reflect the informal meanings of these ranks.

How do trust ranks arise? The view we accept is that an agent i will assign a degree of trustworthiness to another agent j by observing how j behaves in transactions with other agents (possibly also with i). This is described in detail in [Jonker and Treur 1999] which formalized the history of such observations as an *experience sequence* and the assignment of a trust rank to it as a function that satisfies certain desirable intuitive properties. An experience sequence of agent i of agent j is a function over discrete time with values that represent i's judgments of the quality of j's behavior in transactions at those points in time. Technically time points are identifiable with the set N of natural numbers, while in practice agent interactions are defined at most within some initial finite interval ("segment") of N. In [Foo and Renz 2008] it is shown how to collect all such initial segments into an experience sequence space ES by introducing the notion of *concatenation* of two segments, whence $s_1 \circ s_2$ is informally the appending of segment s_2 to segment s_1. An example of an intuitive property is *monotonicity* which means the following. Suppose agent i's experience sequences e_j and e_k of two agents j and k are such that at each time point the value of e_j is no better than that of e_k; then a constraint is that i maps e_j to a trust rank no better than that of e_k. Foo and Renz (op.cit.) observed that the main results of the Jonker and Treur paper (op.cit.) can be subsumed in mathematical systems theory. This is continued by ongoing work of Ignatović and his colleagues [Ignjatovic, Foo, and Lee 2008] in refining the desirable properties of functions from experience to trust that accord both with existing systems and intuition. The following notation is used in their postulates. Experience sequences are indexed by both the evaluating and the evaluated agents; hence, if we need to be specific, the above example of agent i evaluating agent j at time t is denoted by $e_{i,j}(t)$. Let $F : ES \rightarrow T$ be the function that maps experience sequences into a trust space T. These are some of their postulates.

Congruence For all $e_1, e_2.e_3$ if $F(e_1) = F(e_2)$ then $F(e_1 \circ e_3) = F(e_2 \circ e_3)$

Monotonicity If $e_1(t) \leq e_2(t)$ for all t then $F(e_1) \leq F(e_2)$

Continuity This requires respective metrics d_{ES} and d_T to be defined over ES and T. For all $\epsilon > 0$ there is a $\delta > 0$ such that if $d_{ES}(e_1, e_2) < \delta$ then $d_T(F(e_1), F(e_2)) < \epsilon$

Averaging If there are orderings on the experience values of sequences in ES and the values in T, the mononicity above coupled with its dual yields a form of averaging, i.e.:
Let $min(e)$ (resp. $max(e)$) be the minimum (resp. maximum) experience value in e. Let e_{min} (resp. e_{max}) be the sequence obtained from e by replacing all its values by $min(e)$ (resp. e_{max}). Then $F(e_{min}) \leq F(e) \leq F(e_{max})$.

Discounting This notion embodies the common disclaimer "past performance is no guarantee of future performance", i.e. how an agent has behaved in the past should influence its assigned trust rank less than more recent behavior. This is captured as the following.
Let $v1 < v2$ be two experience values, e_1 and e_2 be experience sequences with respective values $v1$ and $v2$ uniformly. Let $e_a = e_1 \circ e_2$ and $e_b = e_2 \circ e_1$. Then $e_2 < e_1$.

Ignjatovic et.al. (op.cit) showed that a representation exists for these postulates. Their notion of trust assignment was further refined to reflect the intuitive property that while it should be hard to achieve high trust rank, once that is achieved it only takes a few bad experiences to lose a lot of that trust. Their work so far has not taken into account stochasticity, but they have initiated simulation studies of the theory to see if trust can be regarded as an *expectation* of future behavior. This is also considered in the context of agent reputation in subsection 3.3 below.

3 Reputation

Social norms as viewed by researchers such as Savarimuthu (op.cit.) are community *expectations* of individual behavior, subject to either formal or informal penalties for breach, the ancient Greek *ostrakismos* - ostracism - being a famous one. Bad behavior often leads to loss of reputation. The difference between *trust* and *reputation* is that the former arises from pairwise interactions between agents, whereas the latter is a *community* aggregation of perceptions of the trustworthiness of agents. To say of an agent i that it has high reputation is to say that most other agents j have assigned it high trust values. Inasmuch as new experiences with an agent cause its trust values to be revised, the changing trust values in turn affect its reputation. This section addresses some of the issues of computing and revising reputation. The very fruitful research in *belief revision* [Peppas 2008] and that which it spawned, e.g. belief merging, suggests that postulational approaches may provide guidance on that computation. Tennenholtz [Tennenholtz 2004] and Nayak [Nayak 2010] exemplify two strands of postulational vuews. The former has features of Google's page ranking, with trust values being strengthened if they come from agents which already have high trustworthiness. The latter has rationality features of belief revsion's *AGM* postulates (see Peppas, op.cit.) that served it so well. We return to Nayak's work below.

We propose a concrete setting that is useful for reifying postulational approaches to trust and reputation updates and revision. First let us give the trust space T above an explicit order structure, viz., $\langle T, < \rangle$, For $h1$ and $h2$ in T, $h1 < h2$ means that the trust level of $h1$ is lower than that of $h2$. The particular choice $T = \{-2, -1, 0, 1, 2\}$ is used here to illustrate the key ideas. There are n interacting agents whose behavior is observable by all. What constitutes the relevant attributes of the behavior of an agent j are specific to the other agents observing it, and these individual perpectives are reflected by the individual evaluations of agent j in their assigned trust ranks as follows. At any point in time t agent i ranks agent j for trustworthiness. Agent i does this for all other (including itself) agents, so let us model this as a function with value $tr_i(t, j) \in T$. meaning at time t agent i ranks agent j with trust value $tr_i(t, j)$. A trust value $tr_i(t, j) = 0$ is interpreted to mean that at time t agent i is neutral about the trustworthiness of agent j; $+2$ is the best and -2 is the worst.

There are n such functions, one for each agent. Hence at time t any agent j has associated with it a *trust vector* $tv_j(t) = \langle tr_1(t, j), tr_2(t, j), \ldots, tr_n(t, j) \rangle$ of trust rank values assigned to it by the other agents. It is not assumed that this is visible to all. Denote the space of all such vectors by $TV(t)$.

The trust vectors are provided to some central unit to compute the *reputation* $\rho_i(t)$ of each agent at time t. We assume that reputations also take values in some finite ordered structure, e.g., $\{-k, -(k-1), \ldots, 0, \ldots, k-1, k\}$. There are two issues: (1) how this should be computed; (2) once computed, is any revision necessary. Why is (2) even considered? The reason is that if the computed $\rho_i(t)$ has value $-k$, which is the worst reputation, then agent i may be very unreliable and its judgments highly questionable. Hence its trust assignments $tr_i(t, j)$ are suspect and should be neutralized by changing their values to 0. This would result in re-computing $\rho_i(t)$ — which is *reputation revision*. Note that this all takes place for a fixed time point t, so to simplify notation in this context we drop the t

parameter.

3.1 Trust Matrix Revision

We introduce a convenient notation to discuss the reputation revision problem.

Each agent j has a trust vector $tv_j = \langle tr_1(j), tr_2(j), \ldots, tr_n(j) \rangle$. Let us re-write $tr_i(j)$ as $tr(j, i)$. Doing this, the vector for agent j is correspondingly re-written as $\langle tr(j, 1) tr(j, 2), \ldots, tr(j, n) \rangle$. We can then collect all the vectors as rows of a matrix M:

$$\begin{pmatrix} tr(1,1) & tr(1,2) & \ldots & tr(1,n) \\ tr(2,1) & tr(2,2) & \ldots & tr(2,n) \\ & \ldots & & \\ tr(n,1) & tr(n,2) & \ldots & tr(n,n) \end{pmatrix}$$

An agent i's trust scores for all agents is therefore the column $M(\cdot, i)$, and the trust vector for agent i is the row $M(i, \cdot)$.

There are two approaches to the reputation revision problem, one operational or procedural and the other semantic.

3.2 Operational

The *operational* view is the following. Suppose Π is an algorithm that computes reputations $\rho(i)$. M is the input for Π, and intermediate revision of the results will be reflected its entries, i.e., Π starts with M as the initial data and changes it as the revision progresses. If there are stages in Π's revisions, the k-th stage matrix value is denoted by M_k. The reputations $\rho(i)$ are output only when Π terminates. The intention is along the following lines. Π identifies from M a set L of agents of low reputation and adjusts their trust scores of other agents to reflect the unreliability of the agents in L, yielding a new M_1. This in turn could render other agents unreliable, so the process repeats to give M_2, etc. Π stops when no new unreliable agents are generated, and a *fixed point* M_k is reached.

Here are a number of examples to sharpen intuition. Let the trust assignments have possible values $-1, 0, +1$ with $+1$ being good, 0 being neutral and -1 bring bad. Suppose Π computes the (tentative) reputation ρ'_i using $\sum_{j=1}^{4} M(i, j)$, and i is considred unreliable if $\rho'_i \leq -2$. Also, the adjustment in that case is to change i's scores for other agents to 0, neutralizing the assignments of i to reflect its unreliabilty..

Example 1. Consider the matrix:

$$M_1 = \begin{pmatrix} 0 & -1 & -1 & 0 \\ 1 & 0 & -1 & -1 \\ 0 & 1 & 0 & 1 \\ 1 & 1 & 0 & 0 \end{pmatrix}$$

The agent 1 is unreliable, giving a new matrix M_2.

$$M_2 = \begin{pmatrix} 0 & -1 & -1 & 0 \\ 0 & 0 & -1 & -1 \\ 0 & 1 & 0 & 1 \\ 0 & 1 & 0 & 0 \end{pmatrix}$$

This now renders agent 2 unreliable, and so its score for 1 is changed to 0 in the next matrix.

$$M_3 = \begin{pmatrix} 0 & 0 & -1 & 0 \\ 0 & 0 & -1 & -1 \\ 0 & 0 & 0 & 1 \\ 0 & 0 & 0 & 0 \end{pmatrix}$$

And now agent 1 is reliable again! The question is this: should we now restore its original scores, i.e., those in $M_1(\cdot, 1)$? If we do that agent 2 becomes reliable again. If restoration of original scores is part of the reputation revision this example shows that a cycle results and there is no convergence. However, if the revision never restores the original scores, viz. once a score is changed to 0 it is "sticky", then convergence is guaranteed because the number of columns that change to 0 values is non-decreasing.

The next example shows that even if convergence happens, the answer may not be unique unless a kind of "parallel" revision is used.

Example 2. Consider the matrix:

$$M_1 = \begin{pmatrix} 0 & -1 & -1 & 0 \\ -1 & 0 & -1 & 0 \\ 1 & 0 & 0 & 1 \\ 0 & 1 & 1 & 0 \end{pmatrix}$$

Agents 1 and 2 are both unreliable. Choosing to change M_1 by neutralizing 1's scores for other agents first renders agent 2 reliable and the revision terminates. On the other hand, choosing to adjust the scores given by agent 2 first renders agent 1 reliable and the revision terminates. There are two different solutions. But choosing to neutralize *both* their scorings at once results in a unique conclusion:

$$M_{stop} = \begin{pmatrix} 0 & 0 & -1 & -1 \\ 0 & 0 & -1 & -1 \\ 0 & 0 & 0 & 1 \\ 0 & 0 & 1 & 0 \end{pmatrix}$$

It is not hard to see that if this parallel revision is adopted for all (new) unreliable agents in the phases of Π there will be a unique terminating matrix.

3.3 Behaviors

Given some reputation revision algorithm Π can we say what are the possible kinds of behavior Π can exhibit? To answer this we first note that the space $\mathcal{M}$ of trust matrices is *finite*, i.e. there are only finitely manny distinct trust matrices. Let us call each such matrix a *state* of $\mathcal{M}$. Any reputation revision function Π induces a sequence of trust matrix transformations (as in the examples above) whose final effect is a function $F_\Pi : \mathcal{M} \to \mathcal{M}$. This F_Π may be regarded as a *state transition function* on an input-free finite automaton $\mathcal{A}_\mathcal{M}$ whose states are $\mathcal{M}$. As a corollary we see that there are at most $|\mathcal{M}|^{|\mathcal{M}|}$ distinct revision functions. It is well-known (and easy to see via the pigeon hole principle) that for a fixed transition (reputation revision) function of an input-free automaton, the only possible long-term behavior is a cylcle of states, a trivial cycle being one in which a state is mapped back to itself. The non-trivial cycles are undesirable as they precisely the cases where Π does not converge. On the other hand the "self loops" of the trivial cycles are the fixed points of Π.

The question that suggests itself is this. Are the fixed points of Π meaningful in terms of agent trustworthiness and reputation? There are several ways to look at this. One intuition is that if the change in behavior is small the change in computed reputation is also small. The second intuition is that "stable behavior" should not perturb the fixed points, i.e. given a fixed point and the evaluations that led to it, if the agents do not vary much in their future behavior the reputations represented by the fixed point should not change. Let us see if these can be made more formal.

The first intuition has an analogy in analysis with various notions of "smoothness" which are formalized as different strengths of *continuity*. The second intuition has an analogy in stochastic process in the notion of *stationarity*. A stochastic process is a family of random variables, and it is stationary if the statistics (meanns, variance, higher moments) are time invariant despite varying random variables. Both analogs can usefully drive our desired formalizations of Π with good fixed points.

Here is an attempt at the first intuition, bearing in mind that we are in a discrete space and therefore the classical notions of continuity do not apply. Consider a function $F : \mathcal{M} \to \mathcal{M}$ arising from an algorithm Π for revising trust matrices prior to computing reputation. The smoothness analog here assumes a metric d on $\mathcal{M}$ (of which many are possible). Denote by F^+ the iterated application of F as many times as as needed to either get to a cycle or a trivial self-loop. We say that F is *(p,q)-bounded* if for all M_1, M_2, $d(M_1, M_2) \leq p$ implies $d(F^+(M_1), F^+(M_2)) \leq q$, noting that the number of times F may have iterated may differ for M_1 and M_2. If F^+ always ends in self loops the boundedness condition is a realization of the smoothness intuition. Suppose otherwise. There are two possibilities. First consider the case when $F^+(M_1)$ results in a trivial self-loop at M_3 while $F^+(M_2)$ is a non-trivial cycle of states. If the original distance between M_1 and M_2 is p-bounded, then all the states in the cycle of $F^+(M_2)$ are within a q distance from M_3. The second possibility is when both the iterates result in non-trivial cycles. Nevertheless all states in one cycle are no further than q from all states in the other cycle. If the Π computation of reputation is based on values of $F^+(M)$ and is itself subject to a boundedness condition, it is plausible that (p,q)-boundedness provides a way of "averaging" the reputations of states "near" one another. If F^+ always ends up in self-loops there is no need to average and therefore it is the ideal case.

3.4 Postulates

Reputation revison algorithms should satisfy certain rationality properties. The notations for the postulates below are these. There are two (hypothetically alternative) trust vectors $v_a = \langle tr_a(j, 1), tr_a(j, 2), \ldots, tr_a(j, n) \rangle$ and $v_b = \langle tr_b(j, 1), tr_b(j, 2), \ldots, tr_b(j, n) \rangle$. We write $v_a^j(i)$ for $tr_a(j, i)$ and similarly $v_b^j(i)$ for $tr_b(j, i)$. Then by $v_a^j \leq v_b^j$ we mean that for all i $v_a^j(i) \leq v_b^j b(i)$. We assume that there is a metric d_T on the trust vectors and also one d_ρ on the reputation space. Limits based on these metrics are denoted as usual by $x_n \to x$.

Monotonicity For all j $v_a^j \leq v_b^j \Rightarrow \Pi(v_a^j) \leq \Pi(v_b^j)$

Continuity If $v_n^j \to v$ then $\Pi(v_n^j) \to \Pi(v)$

(p,q) boundedness For all i, j $d_T(v_a^i, v_b^j) \leq p \Rightarrow d_\rho(\rho(i), \rho(j)) \leq q$

A simnple realization that satisfies these postulates is presented in (Ignjatovic, op.cit.). For an approach more akin to belief revision Nayak (op.cit) has proposed a view of trust that is informally described by "an agent a trusts agent b if a is suspicious not of b but of the enemies of b". His ensuing formalism relies on an agent a's perception of who are the friends or foes of any agent b, and also which agents c whom it regards as suspicious. Suspicion carries ordinal degrees and its value depends on a's evaluation of c. Using this framework Nayak proposes trust updates as perceptual changes are reported, and he has a version of group trust that is similat to what we call reputation here.

4 Trust ans Reputation – Internal View

Human agents, i.e., *people* have *beliefs* about one another, they *reason* with those beliefs, and they *act* accordingly in their interactions.. A community of interacting people have internal structure with subcommunities comprising people who not only transact with each other more frequently than with other people, but who also trust one another much higher. Information flow is asymmetric – "insiders" have privileged knowledge. The framework in section 2 is inadequate for these, one limitation being its lack of representation of how a person's beliefs influences its behavior. This is where the formalism of the situation calculus is appropriate.

We now consider the stucture needed to model how agents *reason* about other agents beyond simply assigning them trust ranks. The preceding considerations are "external" in the sense that each agent k can be regarded as a blackbox, a "mere" function F_k that produces a trust value $T(e)$ given an input experience sequence e. To analyze the details of interactive agent behavior including how it reasons we use the PD in figure 1 as the example.

That a logic-based formalism is suitable for modeling agent choice of strategies is unsurprising because concepts like Nash equilibria in game theory are essentially about rational agents reasoning about one another, with the implicit reasoning buried in fixed point characterizations of the equilibria. Moreover, it has been shown [Foo, Meyer, and Brewka 2005] that they can alternatively be represented as answer sets of a dialect of logic programs in which the rules encode agent strategy preferences in response to those of other agents. However, even in such programs the agents' beliefs are not made explicit but have been "built-in", and are therefore inflexible with regard to issues like changing contexts, traumatic experience (e.g. the recent international finance), and adaptive agent attitudes. The situation calculus, enhanced with epistemic predicates, is not only sufficiently expressive to meet these requirements but also provide a machinery for reasoning.

Here is a situation calculus specification of two Tit-for-Tat agents playing the PD game. It is convenient to let a situation s be a pair $\langle s_1, s_2 \rangle$ to make explicit the simultaneous strategy choices of the respective agents 1 and 2. So, a typical situation records the history of a sequence of $do_1(\_), do_2(\_)$ terms, and the status of the $believe_i(\_)$ fluents evqluated accordingly. We also need a binary predicte $perform(b,s)$ that abbreviates the formula $\forall s'[do(a, s) = s' \leftrightarrow a = b]$ which says b is the only admissible action in situation s.

For $i = 1$ and $j = 2$, or $i - 2$ and $j = 1$:

$$perform(do_i(cooperate, s)) \leftarrow holds(believe_i(j, good), s) \tag{1}$$

$$perform(do_i(defect, s)) \leftarrow holds(believe_i(j, bad), s) \tag{2}$$

$$holds(believe_i(j, good), do_j(coopperate, s)) \tag{3}$$

$$holds(believe_i(j, bad), do_j(defect, s)) \tag{4}$$

$$believe_i(j, good, s) \leftarrow \neg believe_i(j, bad, s) \tag{5}$$

In these formulas $believe_i(j, v)$ is interpreted as agent i believing that agent j has the quality x (good or bad). The term $do_i(a, s)$ means that in situation s agent i chooses the strategy a (cooperate or defect). Formulas 3 and 4 encode Tit-for-Tat's copying of its opponent's last strategy. The last formula is a contraint saying an agent has to beleive the other is either good or bad in any situation s, with the unique names assumption separating *good* from *bad*. To complete the specification for these Tit-for-Tat agents we launch them off in the initial situation s_0 with their mutually charitable beliefs, again with $i = 1$ and $j = 2$, or $j = 1$ and $i = 2$.:

$$holds(believe_i(j, good, s_0)) \tag{6}$$

It then follows using progression that the agents will cooperate in every situation. It is also easy to model a Tit-for-Tat agent's forgiveness after retaliating once for bad behavior by supplementing formula 2 with a companion formula

$$believe_i(j, good, do_i(defect, s)) \leftarrow holds(believe_i(j, bad), s) \tag{7}$$

where agent i, after punishing agent j, reverts to believing by default that agent j is good again. The awful equilibrium of two mutually distrusting and punishing agents is also easy to encode by an initial situation of mutual belief in the badness of the other agent, and never forgiving. This is the sad "cycle of revenge" seen today in too many societies and countries. Further, encoding an exploitative agent i which takes advantage of agent j's (mistaken) belief that i is good can be made explicit thus:

$$do_i(defect, s) \leftarrow holds(believe_i(believe_j(i, good)), s))) \tag{8}$$

Strictly speaking the nesting of *believe* in formula 8 requires it to be a modal operator with an attendant Kripke semantics, but we will let that pass in this prospectus. It suffices to see that the situation calculus offers an expressive modeling and reasoning framework for agent trust and reputation.

5 Elaborations

Some simple elaborations of the preceding sections can be made. As above, let $e_{i,j}(t)$ be the experience of agent i in repeated PD games with agent j. Agent i can use this encoding to rank agent j: if j cooperated at time t then encode that experience as 1, otherwise it is 0. Let δ be a number less than 1. A trust assignment by agent i for j that respects a discount factor of δ is $\delta \sum_1^t \delta^{t-i} e_{i,j}(i)$. This reflects the intuition that recent experiences are more relevant than those in the remote past. Now consider the case of an agent j assessed by agent i to have the nearly maximal trust rank of 1, meaning that j has been overwhelmingly cooperative up to current time, say t. Then in a sequence of plays after t it defects. What is a rational policy for i to revise it's trust rank for j? Ignjotvic, et.al. (op.cit) argue that the penalty for j should be proportionately greater than if its trust rank had been much lower than 1. Their intuition, reflected in their postulates, is that unexpected betrayal is much worse than routine bad behavior insofar as trust is concerned. In the situation calculus this can be captured with an elaboration of the *good* and *bad* trust values into the real interval $[0, 1]$, and a short history of agent strategy choices as a parameter. Another easy elaboration is that investigated by Nowak and his team [Ohtsuki, Hauert, Lieberman, and Nowak 2006] where the agent community is not randomly interacting but proximity (space, kinship, hierarchies, etc) enhances the frequency of interaction. The trust matrices of section 3.1 can be supplemented with matrices of interaction probabilities that encode the proximities. Nowak's team has used analytical methods as well as simulations for their work. Their view of agents is what we have called external. The interesting question is whether an internal view using the logic of the situation calculus would yield even more insight.

References

Axelrod, R. M. [1984]. *The evolution of cooperation.* New York: Basic Books.

Falcone, R. and C. Castelfranchi [2001]. Social trust: a cognitive approach. In C. Castelfranchi and Y.-H. Tan (Eds.), *Trust and deception in virtual societies*, Chapter 3, pp. 55–90. Dordrecht, The Netherlands.: Kluwer Academic Publishers.

Foo, N., T. Meyer, and G. Brewka [2005]. Lpod answer sets and nash equilibria. In M. Maher (Ed.), *Advances in Computer Science - ASIAN 2004*, Volume 3321 of *Lecture Notes in Computer Science*, pp. 343–351. Springer-Verlag.

Foo, N. and J. Renz [2008]. Experience and trust —a systems-theoretic approach. In *Proceeding of the 2008 conference on ECAI 2008: 18th European Conference on Artificial Intelligence*, Amsterdam, The Netherlands, The Netherlands, pp. 867–868. IOS Press.

Ignjatovic, A., N. Foo, and C. T. Lee [2008]. An analytic approach to reputation ranking of participants in online transactions. *Web Intelligence and Intelligent Agent Technology, IEEE/WIC/ACM International Conference on 1*, 587–590.

Jonker, C. M. and J. Treur [1999]. Formal analysis of models for the dynamics of trust based on experiences. In *Multi-Agent System Engineering*, Volume 1647 of *Lecture Notes in Computer Science*, pp. 221–231. Springer-Verlag.

Nayak, A. C. [2010]. The deficit and dynamics of trust. *Embedded and Ubiquitous Computing, IEEE/IFIP International Conference on 0*, 517–522.

Ohtsuki, H., C. Hauert, E. Lieberman, and M. A. Nowak [2006, 05]. A simple rule for the evolution of cooperation on graphs and social networks. *Nature 441*(7092), 502–505.

Peppas, P. [2008]. Chapter 8 belief revision. In V. L. Frank van Harmelen and B. Porter (Eds.), *Handbook of Knowledge Representation*, Volume 3 of *Foundations of Artificial Intelligence*, pp. 317 – 359. Elsevier.

Savarimuthu, B. T. R., M. Purvis, and M. Purvis [2008]. Social norm emergence in virtual agent societies. In *Proceedings of the 7th international joint conference on Autonomous agents and multiagent systems - Volume 3*, AAMAS '08, Richland, SC, pp. 1521–1524. International Foundation for Autonomous Agents and Multiagent Systems.

Tennenholtz, M. [2004]. Reputation systems: an axiomatic approach. In *AUAI '04: Proceedings of the 20th conference on Uncertainty in artificial intelligence*, Arlington, Virginia, United States, pp. 544–551. AUAI Press.

Watson, J. [2008]. *Strategy: an introduction to game theory* (2nd ed ed.). New York: W.W. Norton.

13

Programming Hierarchical Task Networks in the Situation Calculus

ALFREDO GABALDON

ABSTRACT. Hierarchical Task Networks (HTNs) is a successful approach for planning where domain specific knowledge is used in order to make the search for plans more efficient. Similarly, high-level action programming languages like Golog provide a way to use procedural domain knowledge as a search control mechanism in planning. In this work, we present a systematic encoding of HTNs into Golog/ConGolog, allowing one to use both forms of search control under the same framework and also shedding some light on the relationship between the two formalisms.

Foreword

Some years ago, on Hector's suggestion[1] a research effort was begun that has resulted in the Golog family of action programming languages and the new field of *Cognitive Robotics*. It also led to the creation of the Cognitive Robotics Group where as a student I had the fortune of interacting with and learning from Hector. I am pleased to make a contribution to this collection in honor of Hector's 60th birthday.

1 Introduction

Recognizing the intractability of domain-independent planning, researchers have proposed various approaches to take advantage of frequently available domain specific knowledge, to speed up the search for a plan. One of these approaches is Hierarchical Task Network planning, which originated in the work of Sacerdoti [Sacerdoti 1974] and Tate [Tate 1977]. This approach, to which we will refer to as HTN-planning, computes a set of primitive actions or operators just as in (classical) planning. However, instead of a goal condition to be achieved by the actions, in HTN-planning problem is described as a set of *tasks* to be performed. The domain specific knowledge is provided in the form of task decomposition directives, that is, the HTN-planning system is given a set of *methods* that tell it how a high-level task can be decomposed into lower-level tasks. The HTN-planning problem is solved when a sequence of primitive tasks is found that corresponds to performing the original set of tasks.

The intractability of classical planning, in particular for high-level robotic control, was one of the motivations [Levesque and Reiter 1998] behind the introduction of the action programming language Golog [Levesque, Reiter, Lespérance, Lin, and Scherl 1997]. In this approach, instead of relying on classical planning, high-level control is driven by programs written in a language with procedural constructs like conditionals, loops and procedure definitions and calls, and whose primitive statements are *actions* from an underlying action

---
[1] as recounted in [Reiter 2001].

theory. Although there is more to these programs, one way to see them is as providing a means to give the robot a more specific idea of how to do what it needs to do.

Yet another approach was advanced by Bacchus and Kabanza [2000] in their TLPlan planning system. Their approach consists in providing a planning system with domain specific knowledge in the form of Linear Temporal Logic formulae, to help the system discard early some of the unpromising partial plans. Here we will not consider this approach, but see Section 5 for some comments on related work.

Our aim here is to look at the relationship between HTN-planning and high-level action programming languages such as Golog. More specifically, we present an encoding of HTN-planning problems in Golog and ConGolog [De Giacomo, Lesperance, and Levesque 2000]—an extension of Golog with concurrency and interrupts. In addition to the formal result itself showing that HTN-planning can be embedded into Golog, there are several practical benefits from the embedding. One obvious benefit is that instead of having to manually re-encode knowledge given in the form of an HTN, one can use the systematic encoding introduced here to incorporate or mix it with procedural knowledge given in the form of Golog programs. Second, the encoding brings to HTN-planning a considerable set of new features which have been formalized and are available in the Golog family of languages. For instance: explicit time, sensing actions, exogenous events, execution monitoring, incomplete information about the initial state and stochastic actions, to mention a few, all within a single, coherent logical framework. Thus the range of problems that can be tackled potentially becomes much larger. As an illustration of this, we will use the ConGolog encoding of a logistics domain HTN, and add to it run-time package delivery requests (exogenous actions).

2 Preliminaries

2.1 The Situation Calculus

The Situation Calculus [McCarthy 1963] is a logical language for describing and reasoning about dynamic worlds. It has three main components which are used to talk about change in a dynamic world: actions, which are assumed to be the cause of all change in the world and are treated as first-class objects in the language; situations, which correspond to possible states the world may evolve into and are also treated as fist-class objects; and fluents, relations or functions representing the properties of the world that change as the world evolves due to the occurrence of actions.

In this work we use the particular axiomatization of the Situation Calculus that has been developed by Levesque, Reiter and their colleagues and students in the Cognitive Robotics group at the U. of Toronto [Reiter 1991; Levesque, Pirri, and Reiter 1998; Reiter 2001]. This axiomatization uses a classical logic language with sorts $action$, $situation$ and $object$. A special constant S_0 is used to denote the initial situation, and the function symbol do of sort $(action, situation) \mapsto situation$ is used to represent situations that result from executing actions, that is, a term $do(\alpha, \sigma)$, where α is an action term and σ a situation term, represents the situation that results from doing α in situation σ. By nesting function do, it is possible to build sequences of actions. For instance, a sequence consisting of actions a_1, a_2, a_3 is represented by the term $do(a_3, do(a_2, do(a_1, S_0)))$.

Fluents are represented by means of relations $F(\vec{x}, s)$ where $\vec{x}$ is a tuple of arguments of sorts $object$ or $action$ and the last argument, s, is always of sort $situation$. For example, a fluent $atTruck(trk, loc, s)$ could be used to represent the location loc of a tuck trk in situation s.

Function symbols of sort $object \mapsto action$ are used for terms $A(\vec{x})$ that represent action types. For instance, a function $loadTruck(obj, trk)$ could be used to represent the action of loading an object obj onto a truck trk. They are called action *types* because a single function symbol can be used to create multiple *instances* of an action, e.g. the instances $loadTruck(Box_1, Trk_1)$, $loadTruck(Box_2, Trk_2)$, etc. We will use $a_1, a_2, \ldots$ to denote action variables, $\alpha_1, \alpha_2, \ldots$ to denote action terms, and $s_1, s_2, \ldots$ to denote situation variables.

A Basic Action Theory $\mathcal{D}$ consists of the following sets of axioms (variables that appear free are implicitly universally quantified. $\vec{x}$ denotes a tuple of variables $x_1, \ldots, x_n$):

1. For each action type $A(\vec{x})$, there is exactly one **Action Precondition Axiom** (APA) of the form:

$$Poss(A(\vec{x}), s) \equiv \Pi_A(\vec{x}, s)$$

   where variable s is the only term of sort situation in formula $\Pi_A(\vec{x}, s)$. The latter formula represents the conditions under which an action $A(\vec{x})$ is executable. The restriction that the only situation term mentioned in this formula be s intuitively means that these preconditions depend only on the situation where the action would be executed.

2. For each fluent $F(\vec{x}, s)$, there is exactly one **Successor State Axiom** (SSA) of the form:

$$F(\vec{x}, do(a, s)) \equiv \Phi_F(\vec{x}, a, s)$$

   where s is the only term of sort situation in formula $\Phi_F(\vec{x}, a, s)$. This formula represents all and the only conditions under which executing an action a in a situation s causes the fluent to hold in situation $do(a, s)$. These axioms embody Reiter's solution to the frame problem [Reiter 1991; Reiter 2001].

3. The **Initial Database** $\mathcal{D}_{S_0}$ which is a finite set of sentences whose only situation term is S_0 and describe the initial state of the domain, i.e. before any actions have occurred. Any sentence is allowed as long as the only situation term that appears in it is S_0, so one can write sentences such as $(\exists box)atLoc(box, L, S_0)$, reflecting incomplete information about the initial state of a domain.

4. The **Foundational Axioms** Σ which define situations in terms of the constant S_0 and the function do. Intuitively, these axioms define a tree-like structure for situations with S_0 as the root of the tree. They also define relation $\sqsubset$ on situations. Intuitively, $s \sqsubset s'$ means that the sequence of actions s is a prefix of sequence s'.

5. A set of unique names axioms (UNA) for actions. For example,

$$loadTruck(o, t) \neq driveTruck(t, l_1, l_2),$$
$$loadTruck(o, t) = loadTruck(o', t') \supset (o = o' \wedge t = t')$$

EXAMPLE 1. Through out this chapter we will use the well known logistics domain as an example. In this domain, there are objects that need to be moved between locations by truck or plane. Cities contain different locations some of which are airports. Primitive actions include loading/unloading an object onto a truck or plane, driving a truck and flying a plane. A basic action theory for this domain is comprised of a set of axioms that includes the following:

Action Precondition Axioms:

$$Poss(loadTruck(o, tr), s) \equiv (\exists l).atTruck(tr, l, s) \wedge atObj(o, l, s),$$

$$Poss(unloadTruck(o, tr), s) \equiv inTruck(o, tr, s),$$

$$Poss(loadAirplane(o, p), s) \equiv (\exists l).atObj(o, l, s) \wedge atAirplane(p, l, s),$$

$$Poss(unloadAirplane(o, p), s) \equiv inAirplane(o, p, s),$$

$$Poss(driveTruck(tr, or, de), s) \equiv atTruck(tr, or, s) \wedge (\exists c).inCity(or, c) \wedge \\ inCity(de, c).$$

$$Poss(fly(p, or, de), s) \equiv atAirplane(p, or, s) \wedge airport(de).$$

Successor State Axioms:

$$atObj(o, l, do(a, s)) \equiv (\exists tr)[a = unloadTruck(o, tr) \wedge atTruck(tr, l, s)] \vee \\ (\exists p)[a = unloadAirplane(o, p) \wedge atAirplane(p, l, s)] \vee \\ atObj(o, l, s) \wedge \neg(\exists tr)a = loadTruck(o, tr) \wedge \\ \neg(\exists p)a = loadAirplane(o, p).$$

$$atTruck(tr, l, do(a, s)) \equiv (\exists or)a = driveTruck(tr, or, l) \vee \\ atTruck(tr, l, s) \wedge \neg(\exists or, de)a = driveTruck(tr, or, de),$$

$$atAirplane(p, ap, do(a, s)) \equiv (\exists oap)a = fly(p, oap, ap) \vee \\ atAirplane(p, ap, s) \wedge \neg(\exists oap, dap)a = fly(p, oap, dap),$$

$$inTruck(o, tr, do(a, s)) \equiv a = loadTruck(o, tr) \vee \\ inTruck(o, tr, s) \wedge a \neq unloadTruck(o, tr),$$

$$inAirplane(o, p, do(a, s)) \equiv a = loadAirplane(o, p) \vee \\ inAirplane(o, p, s) \wedge a \neq unloadAirplane(o, p).$$

Initial situation. These would include sentences such as:

$atAirplane(p, l, S_0) \equiv p = Plane_1 \wedge l = Loc_{5,1} \vee p = Plane_2 \wedge l = Loc_{2,1}.$

$atTruck(t, l, S_0) \equiv t = Truck_{1,1} \wedge l = Loc_{1,1} \vee t = Truck_{2,1} \wedge l = Loc_{2,1} \vee \ldots$

$airport(loc) \equiv loc = Loc_{1,1} \vee loc = Loc_{2,1} \vee loc = Loc_{3,1} \vee \ldots$

$inCity(l, c) \equiv l = Loc_{1,1} \wedge c = City_1 \vee l = Loc_{2,1} \wedge c = City_2 \vee \ldots$

$atObj(p, l, S_0) \equiv p = Package_1 \wedge l = Loc_{3,3} \vee p = Package_2 \wedge l = Loc_{3,1} \vee \ldots$

The above set of axioms is almost a complete basic action theory for the logistics domain. The only axioms missing are the domain independent foundational axioms and the unique names axioms for actions. Note that the initial situation includes non-fluent relations such as $airport(loc)$ and $inCity(l, c)$. It may also include other "utility" sentences such as formulae specifying that different constants denote different objects, e.g. $Loc_{1,1} \neq Loc_{1,2}$.

Basic action theories in the Situation Calculus is one of the established formalisms for reasoning about actions. It has been used to formalize various reasoning problems in terms of logical deduction, including the classical planning problem: given a basic action theory, $\mathcal{D}$, describing the planning domain and including a description of the initial state, and given also a goal formula $G(s)$, the planning problem is specified in terms of logical entailment as follows:

$$\mathcal{D} \models (\exists s).executable(s) \wedge G(s)$$

where $executable(s)$ intuitively means that the situation s, i.e. the plan, includes only actions whose preconditions are satisfied.

2.2 Golog and ConGolog

The high-level action programming languages Golog [Levesque, Reiter, Lespérance, Lin, and Scherl 1997] and ConGolog [De Giacomo, Lesperance, and Levesque 2000] are defined on top of a basic action theory and provide a set of programming constructs that allow one to define compound actions in terms of simpler ones. These constructs are the typical constructs of an imperative programming language. Golog includes the following:

- Test condition: $\phi?$. Test whether ϕ is true in the current situation.

- Sequence: $\delta_1; \delta_2$. Execute δ_1 followed by δ_2.

- Non-deterministic action choice: $\delta_1 | \delta_2$. Execute δ_1 or δ_2.

- Non-deterministic choice of arguments: $(\pi x)\delta$. Choose a value for x and execute δ for that value.

- Non-deterministic iteration: δ^*. Execute δ zero or more times.

- Procedure definitions: **proc** $P(\vec{x})$ δ **endProc** . $P(\vec{x})$ is the name of the procedure, $\vec{x}$ its parameters, and δ is the body.

ConGolog includes the above constructs plus the following:

- Synchronized conditional: **if** ϕ **then** δ_1 **else** δ_2.

- Synchronized loop: **while** ϕ **do** δ.

- Concurrent execution: $\delta_1 \parallel \delta_2$.

- Prioritized concurrency: $\delta_1 \rangle\rangle \delta_2$. Execute δ_1 and δ_2 concurrently but δ_2 executes only when δ_1 is blocked or done.

- Concurrent iteration: $\delta^{\parallel}$. Execute δ zero or more times in parallel.

- Interrupt: $\phi \rightarrow \delta$. Execute δ whenever condition ϕ is true.

EXAMPLE 2. In the logistics domain, moving an object to a new location may be achieved by transporting it within a city, or it may require flying it to another city. So it may be useful to define a complex action $moveObj(o, loc)$ that invokes the required actions to move o to the new location loc.

Such a complex action can be defined as a Golog procedure as follows:

$$
\begin{aligned}
&\textbf{proc } moveObj(o, loc) \\
&\quad (\pi\ oloc, ocity). \\
&\qquad [atObj(o, oloc) \wedge inCity(oloc, ocity)]?\ ; \\
&\qquad \textbf{if } inCity(loc, ocity) \textbf{ then} \\
&\qquad\qquad inCityDeliver(o, oloc, loc) \\
&\qquad \textbf{else} \\
&\qquad\qquad (\pi\ dcity). \\
&\qquad\qquad\ inCity(loc, dcity)?\ ; \\
&\qquad\qquad (\pi\ oap, dap). \\
&\qquad\qquad\quad [inCity(oap, ocity) \wedge inCity(dap, dcity)]?\ ; \\
&\qquad\qquad\quad inCityDeliver(o, oloc, oaprt)\ ; \\
&\qquad\qquad\quad airDeliver(o, oaprt, daprt)\ ; \\
&\qquad\qquad\quad inCityDeliver(o, daprt, loc) \\
&\textbf{endProc}
\end{aligned}
$$

This procedure calls procedures $inCityDeliver(\cdots)$ and $airDeliver(\cdots)$ according to the origin and destination of the object being moved.

ConGolog is clearly a more complex language than Golog, and that is fairly evident in the logical formalization of the semantics. In fact, the semantics of Golog is defined through a set of 'macros' that stand for standard Situation Calculus formulae [Levesque, Reiter, Lespérance, Lin, and Scherl 1997; Reiter 2001]. However, since we will need the features of ConGolog to encode HTNs with partially ordered tasks and ConGolog includes all the constructs of Golog, we will only give an overview of the ConGolog semantics.

The formal semantics of ConGolog is defined in terms of a relation $Trans(\delta, s, \delta', s')$ that defines single computation steps of a program. Intuitively, the relation defines a single step transition from a *configuration* δ, s into a configuration δ', s' that results from executing one step of program δ in situation s. An additional relation $Final(\delta, s)$ defines those configurations that are terminating, i.e. where the computation may end successfully.

The full axiomatization of $Trans(\delta, s, \delta', s')$ and $Final(\delta, s)$ can be found in [De Giacomo, Lesperance, and Levesque 2000]. Below we show a sample of the axioms, which provide some idea of how it captures the semantics of each construct.

$$Trans(nil, s, \delta', s') \equiv False,$$
$$Trans(a, s, \delta', s') \equiv Poss(a, s) \wedge \delta' = nil \wedge s' = do(a, s),$$
$$Trans(\phi?, s, \delta', s') \equiv \phi[s] \wedge \delta' = nil \wedge s' = s,$$
$$Trans(\delta_1; \delta_2, s, \delta', s') \equiv$$
$$(\exists \gamma)\delta' = (\gamma; \delta_2) \wedge Trans(\delta_1, s, \gamma, s') \vee Final(\delta_1, s) \wedge Trans(\delta_2, s, \delta', s'),$$

$$Trans((\pi v)\delta, s, \delta', s') \equiv (\exists x)Trans(\delta_x^v, s, \delta', s'),$$
$$Trans(\textbf{if } \phi \textbf{ then } \delta_1 \textbf{ else } \delta_2, s, \delta', s') \equiv$$
$$\phi[s] \wedge Trans(\delta_1, s, \delta', s') \vee \neg\phi[s] \wedge Trans(\delta_2, s, \delta', s'),$$
$$Trans(\textbf{while } \phi \textbf{ do } \delta, s, \delta', s') \equiv$$
$$(\exists \gamma).(\delta' = \gamma \; ; \; \textbf{while } \phi \textbf{ do } \delta) \wedge \phi[s] \wedge Trans(\delta, s, \gamma, s'),$$
$$Trans(\delta_1 \parallel \delta_2, s, \delta', s') \equiv$$
$$(\exists \gamma)[\delta' = (\gamma \parallel \delta_2) \wedge Trans(\delta_1, s, \gamma, s')] \vee (\exists \gamma)[\delta' = (\delta_1 \parallel \gamma) \wedge Trans(\delta_2, s, \gamma, s')].$$

$$Final(nil, s) \equiv True,$$
$$Final(a, s) \equiv False,$$
$$Final(\phi?, s) \equiv False,$$
$$Final(\delta_1; \delta_2, s) \equiv Final(\delta_1, s) \wedge Final(\delta_2, s),$$
$$Final((\pi x)\delta, s) \equiv (\exists x)Final(\delta, s),$$
$$Final(\textbf{if } \phi \textbf{ then } \delta_1 \textbf{ else } \delta_2, s) \equiv \phi[s] \wedge Final(\delta_1, s) \vee \neg\phi[s] \wedge Final(\delta_2, s),$$
$$Final(\textbf{while } \phi \textbf{ do } \delta, s) \equiv \neg\phi[s] \vee Final(\delta, s),$$
$$Final(\delta_1 \parallel \delta_2, s) \equiv Final(\delta_1, s) \wedge Final(\delta_2, s).$$

An abbreviation $Do(\delta, s, s')$, meaning that executing δ in situation s is possible and it legally terminates in situation s', can then be defined in terms of the transitive closure of $Trans$ and of relation $Final$:

$$Do(\delta, s, s') \stackrel{\text{def}}{=} (\exists \delta').Trans^*(\delta, s, \delta', s') \wedge Final(\delta', s').$$

The axiom defining $Trans^*$, as well as one of the foundational axioms which recursively defines sequences of actions, requires second order quantification (for details see [De Giacomo, Lesperance, and Levesque 2000]). The intention with these axiomatizations is not to use them directly in implementations. The are meant to semantically characterize dynamic wolds and programs. Nevertheless, under suitable assumptions, various practical interpreters have been devised. In Section 4 we shall use on of these interpreters to demonstrate some sample runs with our HTN encoding in ConGolog.

2.3 HTN Planning

In this section we give a brief overview of HTN-planning. Our discussion is based on the definitions of HTN-planning from [Erol, Hendler, and Nau 1996], where an *operational semantics* of HTN-planning is given. HTNs are defined over a first-order language with a vocabulary consisting of sets symbols for variables, constants, predicates, primitive tasks, compound tasks, and some other symbols. Here we will take the language from a basic action theory and just add symbols for compound tasks. The only assumption we need to make is that the set of constants is finite and denote the elements of the domain. We will then talk about ground instances of tasks with the usual meaning. Furthermore, instead of new symbols for primitive tasks, we will use action terms from an underlying basic action theory as the primitive task symbols. Finally, we use situations instead of state symbols.

A *primitive task* is an action term $A(\vec{x})$. A *compound task* is a term of the form $tname(\vec{x})$ where $tname$ is a compound task (function) symbol. A *task network* is a pair (T, ϕ) where T is a list of tasks and ϕ a boolean formula of constraints of the forms $(t \prec t')$, (t, l), (l, t), (t, l, t'), $(v = v')$ and $(v = c)$ where t, t' are tasks from T, l is a fluent literal, v, v' are variables and c is a constant. A task network consisting only of primitive tasks is called a *primitive task network*. An HTN *method* is a pair (h, d) where h is a compound task and d is a task network. Methods are the HTN construct for building complex tasks from primitive ones.

An HTN *planning problem* is a tuple (d, s, D) where d is a task network, s is a situation, and D is a *planning domain* consisting of a set of methods plus a basic action theory (which includes an initial database $\mathcal{D}_{S_0}$). The parameter s is the the situation from where planning starts and in general it can be any situation. If a situation different from the initial situation, S_0, is given, then the problem involves planning after some actions have already occurred. Nevertheless, for brevity, we will only consider problems where the situation parameter is S_0. A *plan* is a sequence of ground primitive tasks.

Let d be a primitive task network, s a situation, and D a planning domain. A sequence of primitive tasks σ is a *completion* of d in s, denoted by $\sigma \in comp(d, s, D)$, if σ is a total ordering of a ground instance of the primitive task network d, it is executable in s and satisfies the constraint formula in d. For d containing a non-primitive task, $comp(d, s, D)$ is defined to be the empty set.

Let d be a task network that contains a compound task t and $m = (h, d')$ be a method such that θ is a most general unifier of t and h. Define $reduce(d, t, m)$ to be the task network obtained from $d\theta$ by replacing $t\theta$ with the tasks $t_1\theta, \ldots, t_k\theta$ from d' and incorporating (see [Erol, Hendler, and Nau 1996] for details) the constraints in d' with those in d. Define $red(d, D)$ as the set of all reductions of d by methods of D, that is $red(d, D) = \{d' \mid t$ is a task in d, m is a method in d and $d' = reduce(d, t, m)\}$.

A solution, $sol(d, s, D)$, of a planning problem (d, s, D) is defined recursively as follows:

$$sol_1(d, s, D) = comp(d, s, D)$$
$$sol_{n+1}(d, s, D) = sol_n(d, s, D) \cup \bigcup_{d' \in red(d, D)} sol_n(d', s, D)$$
$$sol(d, s, D) = \bigcup_{n < \omega} sol_n(d, s, D)$$

The set $sol(d, s, D)$ contains all plans that can be derived in a finite number of steps.

EXAMPLE 3. The following are methods for a task $moveObj(o, loc)$ for moving an object in the logistics domain as described in Example 2. The first method works for moving an object within the same city. The second is for moving an object between cities.[2]

---

[2]For improved readability, we use the labelling notation $t_i = tname(\vec{x})$ and labels t_i to refer to tasks.

$(moveObj(o, loc)$
$[t = inCityDeliver(o, oloc, loc)]$
$(atObj(o, oloc), t) \land (inCity(oloc, ocity), t) \land (inCity(loc, ocity), t)$
$)$

$(moveObj(o, loc)$
$[t_1 = inCityDeliver(o, oloc, oaprt),$
$t_2 = airDeliver(o, oaprt, daprt),$
$t_3 = inCityDeliver(o, daprt, loc)]$
$(atObj(o, oloc), t_1) \land (inCity(oloc, ocity), t_1) \land (inCity(loc, dcity), t_1) \land$
$(inCity(oaprt, ocity), t_1) \land (inCity(daprt, dcity), t_1) \land$
$(t_1 \prec t_2) \land (t_2 \prec t_3)$
$)$

3 Programming HTNs in Golog/ConGolog

In this section we show how HTN-planning problems can be encoded in Golog/ConGolog. We will considered two versions of the HTN-planning: totally ordered and partially ordered. In totally ordered HTNs, the task networks of all methods must be totally ordered. Considering totally ordered and partially ordered HTNs separately is interesting because they have different computational complexity which will indeed allow us to use the simpler language Golog to encode totally ordered HTNs. Moreover, implemented planning systems exist for each type of problem, namely, systems SHOP [Nau, Cao, Lotem, and Munoz-Avila 1999] for totally ordered HTNs and SHOP2 [Nau, Munoz-Avila, Cao, Lotem, and Mitchell 2001] for partially ordered HTNs. These systems have been very successfully applied to real world domains. Also, it is probably useful in practice to have separate solutions for each class of problem.

3.1 Totally ordered task networks

Let us then first consider totally ordered HTNs. A task network (T, ϕ) is totally ordered if the boolean formula ϕ includes precedence constraints $(t_1 \prec t_2)$, $(t_2 \prec t_3), \ldots, (t_{n-1} \prec t_n)$ on the tasks in T. We will further assume that ϕ is a conjunction of the precedence constraints $t_i \prec t_j$ and of a set of constraints of the form (l, t) representing task preconditions. An HTN-planning problem (d, s, D) is *totally ordered* if d and all the task networks in the methods of D are totally ordered. This corresponds to the form of task network that the SHOP system handles.

We show an encoding of totally ordered HTN-planning problems in Golog. Consider an HTN-planning problem $P = (d, S_0, D)$. For each compound task h, with a set of methods $(h, d_1), (h, d_2), \ldots, (h, d_k)$ in D, we define the following Golog procedure:

> **proc** h
> $(\pi \vec{x}_1)[(L_{1,1})? \; ; \; t_{1,1} \; ; \ldots ; (L_{1,i_1})? \; ; \; t_{1,i_1}] \; |$
> $(\pi \vec{x}_2)[(L_{2,1})? \; ; \; t_{2,1} \; ; \ldots ; (L_{2,i_2})? \; ; \; t_{2,i_2}] \; |$
> $\ldots$
> $(\pi \vec{x}_k)[(L_{k,1})? \; ; \; t_{k,1} \; ; \ldots ; (L_{k,i_k})? \; ; \; t_{k,i_k}]$
> **endProc**

where $t_{i,j}$ is the jth task in d_i according to the ordering, $L_{i,j}$ is the conjunction of the literals l such that $(l, t_{i,j})$ is a constraint in d_i and $\vec{x}_i$ are the variables that appear free in the constrains and tasks of method d_i. Intuitively, the procedure non-deterministically chooses one of the methods to execute. Each line in the procedure non-deterministically chooses values for the free variables and then executes the sequence of precondition tests and tasks.

Let Δ_P denote the resulting set of Golog procedure declarations. To complete the encoding of the HTN-planning problem P we include a Golog program δ_d obtained from the task network d. This program has the same form as the subprogram for a single method:

$$(\pi\vec{x}).(L_1)? \; ; \; t_1 \; ; \ldots; \; (L_n)? \; ; \; t_n.$$

EXAMPLE 4. The procedure in Example 2 is in fact a simplified version of the encoding of the methods in Example 3. The procedure that results by following exactly the encoding described above is as follows.

> **proc** $moveObj(o, loc)$
> $(\pi \, oloc, ocity)[$
> $(atObj(o, oloc) \land inCity(oloc, ocity) \land inCity(loc, ocity))? \; ;$
> $inCityDeliver(o, oloc, loc)] \mid$
> $(\pi \, oloc, ocity, dcity, oap, dap)[$
> $(atObj(o, oloc) \land inCity(oloc, ocity) \land inCity(loc, ocity) \land$
> $inCity(loc, dcity) \land inCity(oap, ocity) \land inCity(dap, dcity))? \; ;$
> $inCityDeliver(o, oloc, oaprt) \; ;$
> $airDeliver(o, oaprt, daprt) \; ;$
> $inCityDeliver(o, daprt, loc)]$
> **endProc**

Given the above encoding of (totally ordered) HTNs in Golog, we can formulate a specification of the HTN-planning problem in terms of the axiomatization of primitive tasks (a basic action theory) and Golog, and of logical entailment, as follows. Let P be a totally ordered HTN-planning problem, then

$$\mathcal{D}_P \models (\exists s) Do(\Delta_P \; ; \; \delta_d, S_0, s). \tag{1}$$

Here, $\mathcal{D}_P$ is the basic action theory of P and it would also include the axioms defining the semantics of the Golog constructs shown in Section 2.2. However, we would like to remark again that the logical formalization of Golog is much simpler than that of ConGolog and the axiomatization of $Trans$ and $Final$ is not really necessary. This means that, unlike the case of partially ordered HTN-planning, the formalization of totally ordered HTN-planning requires only a standard basic action theory in the Situation Calculus plus a set of macros for the Golog constructs.

The following theorem formalizes the relationship between the operational semantics of totally ordered HTN-planning and the Golog encoding described above. For a sequence σ of ground primitive tasks, i.e. ground action terms $\alpha_1, \ldots, \alpha_n$, let $do(\sigma, s)$ denote the situation term $do(\alpha_n, do(\alpha_{n-1}, \ldots, do(\alpha_1, s) \ldots))$.

THEOREM 5. *Let $P = (d, S_0, D)$ be a totally ordered HTN-planning problem, Δ_P and δ_d be the corresponding Golog procedures and program. Then*

$$\sigma \in sol(d, S_0, D) \text{ iff } \mathcal{D}_P \models Do(\Delta_P \; ; \; \delta_d, \; S_0, do(\sigma, s)).$$

3.2 Partially ordered task networks

Let us next considered partially ordered HTNs. The only difference with the HTNs of the previous subsection is that the boolean formulae ϕ in task networks specify through constraints $t \prec t'$ a partial instead of a total ordering on the tasks. Thus, formulae ϕ will be assumed to be arbitrary conjunctions of constraints of the form $t \prec t'$ and (l, t).

This form of HTN corresponds to the HTNs the SHOP2 system handles, except that SHOP2 additionally allows so-called *protection requests* and *protection cancellations*. These are used, for example, to ensure that some condition be maintained from the time one task finishes executing and another starts. These constraints are not difficult to add to our encoding, however, they are a side issue which for clarity of presentation we prefer not to address here.

The main motivation behind allowing the tasks to be partially ordered is the possibility of allowing multiple tasks to execute concurrently, that is, to allow their subtasks to interleave when there is no precedence ordering specified between them. Interleaved concurrency is exactly the type of concurrency that the construct ‖ of ConGolog provides.

Let us then consider encoding partially ordered HTN-planning problems in ConGolog. As before, for each compound task we will define a procedure. But in this case we will need to introduce some auxiliary fluents and actions that will help enforce the task precedence constraints.

First, we need a fluent $terminated_{tname}(\vec{x}, s)$ for every compound task $tname(\vec{x})$ or primitive task (action) $A(\vec{x})$. Given a task, let τ stand for symbol $tname$ if a compound task and for A if a primitive task. Intuitively, $terminated_\tau(\vec{x}, s)$ holds when task $\tau(\vec{x})$ has already executed and terminated in situation s. We also need to introduce an auxiliary primitive task $end_\tau(\vec{x})$. For compound tasks, $end_\tau(\vec{x})$ will be added as a sub-task of $\tau(\vec{x})$ and will be the last sub-task to execute. The intention obviously is that this task will set the fluent $terminated_\tau(\vec{x}, s)$ to true when $\tau(\vec{x})$ finishes executing.

Formally, the successor state axioms for fluents $terminated_\tau(\vec{x}, s)$ are as follows:

$$terminated_\tau(\vec{x}, do(a, s)) \equiv \quad a = \tau(\vec{x}) \vee a = end_\tau(\vec{x}) \vee$$
$$terminated_\tau(\vec{x}, s).$$

These fluents are all initially false, i.e. $(\forall x)\neg terminated_\tau(\vec{x}, S_0)$ is included in the initial database $\mathcal{D}_{S_0}$. Since these fluents and the actions $end_\tau(\vec{x})$ are auxiliary "system actions," so to speak, they can be kept transparent from the "user".

Before we introduce the ConGolog encoding, we need to introduce one last notation. Let (T, ϕ) be a task network and $\tau(\vec{x})$ be one of its tasks. We define $pred_\tau(\vec{x}, s)$ as the following conjunction:

$$pred_\tau(\vec{x}, s) \stackrel{\text{def}}{=} \bigwedge_{\{\tau' : (\tau'(\vec{x}') \prec \tau(\vec{x})) \in \phi\}} terminated_{\tau'}(\vec{x}', s).$$

If there is no constraint $(\tau'(\vec{x}') \prec \tau(\vec{x}))$ in ϕ then $pred_\tau(\vec{x}, s) \stackrel{\text{def}}{=} True$. Intuitively, $pred_\tau(\vec{x}, s)$ holds in s if all tasks which precede τ according to the constraints have already executed.

We are now ready to present the encoding. For the sake of readability we omit term arguments. The ConGolog procedure that encodes the methods $(h, d_1), (h, d_2), \ldots, (h, d_k)$ for a compound task h is:

$$\textbf{proc } h$$
$$(\delta_1 | \delta_2 | \ldots | \delta_k) ;$$
$$end_h$$
$$\textbf{endProc}$$

where each δ_i stands for a program as follows:

$$\delta_i \begin{cases} (\pi\vec{v})\{ & [(pred_{t_{i,1}})? \; ; \; (L_{i,1})? \; ; \; t_{i,1}] \; \| \\ & [(pred_{t_{i,2}})? \; ; \; (L_{i,2})? \; ; \; t_{i,2}] \; \| \\ & \ldots \\ & [(pred_{t_{i,k_i}})? \; ; \; (L_{i,k_i})? \; ; \; t_{i,k_i}] \quad \} \end{cases}$$

The $t_{i,j}$ are the tasks in d_i and $L_{i,j}$ is as before the conjunction of the literals l such that $(l, t_{i,j})$ is a constraint in d_i. Intuitively, each δ_i encodes the task network d_i of the corresponding h method. The program δ_i essentially executes all the subtasks concurrently. But each subtask must first successfully test that all the subtasks that precedes it have already terminated, before it can execute. The semantics of construct $\|$ captures the intended meaning that the execution of a task is suspended until the preceding tasks, according to the ordering constraints, have executed.

One aspect of the above encoding is not completely faithful to the HTN semantics of [Erol, Hendler, and Nau 1996]. The intended meaning of a constraint (l, t) is that the literal l must hold immediately before task t executes. In the above ConGolog encoding, the corresponding code is l? ; t. This encoding and interleaved concurrency with other tasks means that some other tasks may execute after the test l? and before the execution of t. It would not be difficult to add to ConGolog a new programming construct for synchronized *test-and-execute* or use interrupts as in [Gabaldon 2002]. However, requiring a condition to hold immediately before a task begins executing seems to us to make sense only for primitive tasks, since the subtasks of a compound task would be interleaved anyway. Then in the case of a primitive task, a condition required to hold before the task executes can be encoded in the action precondition axiom instead, since primitive tasks are actions from the basic action theory. We will therefore proceed with the relaxed form of the constraint.

EXAMPLE 6. This is a simple blocks world example method for moving a block v_1 from on top a block v_2 onto a block v_3:

$$(move(v_1, v_2, v_3)$$
$$clear(v_1), \; clear(v_3), \; unstack(v_1, v_2), stack(v_1, v_3)$$
$$(clear(v_1) \prec unstack(v_1, v_2)) \wedge (clear(v_3) \prec unstack(v_1, v_2)) \wedge$$
$$(unstack(v_1, v_2) \prec stack(v_1, v_3))$$
$$)$$

The encoding as a ConGolog procedure is as follows. Assuming the above method is the only method for task $move(v_1, v_2, v_3)$, the procedure declaration is:

$$\textbf{proc } move(v_1, v_2, v_3)$$
$$\delta_1;$$
$$end_{move}(v_1, v_2, v_3)$$
$$\textbf{endProc}$$

where δ_1 is the following program (the tests $(True)?$ can obviously be removed):

$$(True)? \; ; \; clear(v_1) \; \|$$
$$(True)? \; ; \; clear(v_3) \; \|$$
$$(terminated_{clear}(v_1) \wedge terminated_{clear}(v_3))? \; ; \; unstack(v_1, v_2) \; \|$$
$$(terminated_{unstack}(v_1, v_2))? \; ; \; stack(v_1, v_3)$$

If the options of writing task networks or ConGolog programs are both available, in some cases it may be easier to write ConGolog directly instead of an HTN and then translate. Writing ConGolog directly had an additional advantage that the direct encoding may require a simpler theory compared to the result of an automatic translation of an HTN into ConGolog. The reason is that there may be less overhead in the form of auxiliary fluents and actions. For instance, consider again the method for $move(v_1, v_2, v_3)$. This could have been encoded as the following much simpler ConGolog program:

$$(clear(v_1) \; \| \; clear(v_3)) \; ;$$
$$unstack(v_1, v_2) \; ;$$
$$stack(v_1, v_3)$$

On the other hand, for more complex methods it may be much easier and more natural to write instead a task network with partial ordering constraints. In that case, ConGolog procedures can be obtained automatically by applying the above encoding.

The logical specification (1) of HTN-planning is the same in the case of partially ordered HTNs encoded in ConGolog as shown above. The only difference is that in this case, the theory $\mathcal{D}_P$ must include the axiomatization of the $Trans$ and $Final$ relations, so it is necessarily a more complex theory.

A similar result on the correspondence between the operational semantics of HTNs and the ConGolog encoding can be obtained.

THEOREM 7. *Let $P = (d, S_0, D)$ be a partially ordered HTN-planning problem, Δ_P and δ_d be the corresponding ConGolog procedures and program. Then*

$$\sigma \in sol(d, S_0, D) \; \textit{iff} \; \mathcal{D}_P \models Do(\Delta_P \; ; \; \delta_d, \; S_0, do(\sigma, s)).$$

4   On-line Execution with Exogenous Actions

As we mentioned earlier, one of the benefits of the encoding is that it brings to HTNs a number of additional features that already exist for the Golog family of languages. The purpose of this section is 1) to show some sample runs obtained with the Prolog interpreters of ConGolog and some translated HTNs shown earlier, and 2) to demonstrate one particular feature of the Golog family of languages that is normally not available in HTN systems, namely, online execution with exogenous events. To this end, we will resort again to the logistics domain example and its ConGolog encoding, and extend it for modeling online execution with exogenous events in the form of run-time package delivery requests. Of course, once a set of HTNs has been translated into ConGolog, features like exogenous actions become an orthogonal issue. Nevertheless, this section also illustrates the use of the translated logistics domain HTNs discussed earlier and shows a sample run obtained with the implementation.

Online execution of a ConGolog program means that once the next primitive action to execute is determined according to the control structure of the program, which due to non-determinism may involve randomly choosing one, this action is actually executed "in the world." This entails that once such an action is chosen, the ConGolog interpreter cannot backtrack since it is not possible to backtrack from actually executing an action in the phys-ical world. This behaviour is in fact very easy to model with the interpreters by means of the Prolog cut operator. The offline interpreter from [De Giacomo, Reiter, and Soutchanski 1998] includes the rule:

```
offline(Prog,S0,Sf):- final(Prog,S0), S0=Sf ;
                      trans(Prog,S0,Prog1,S1),
                      offline(Prog1,S1,Sf).
```

To prevent the interpreter from backtracking on primitive actions, including exogenous ones, De Giacomo et al. simply add a cut after a one-step transition of the program. This one-step transition involves, among other things, choosing the next primitive action to ex-ecute, according to the program and the current situation. Hence, an online interpreter should not backtrack after such a step. The modified rule of the online interpreter is as follows:

```
online(Prog,S0,Sf):- final(Prog,S0), S0=Sf ;
                     trans(Prog,S0,Prog1,S1), !,
                     online(Prog1,S1,Sf).
```

This rule results in an interpreter called *brave* because after choosing an action it im-mediately commits and executes it. Alternatively, a *cautious* online interpreter may check, *offline*, before committing to execute an action, whether it is possible for the remainder of the program to terminate successfully. This behavior is captured by the following rule:

```
online(Prog,S0,Sf):- final(Prog,S0), S0=Sf ;
                     trans(Prog,S0,Prog1,S1),
                     offline(Prog1,S1,Soff), !,
                     online(Prog1,S1,Sf).
```

Online vs offline execution interpreters are further discussed in [De Giacomo, Reiter, and Soutchanski 1998; Reiter 2001]. An extension of ConGolog that includes a programming construct for offline deliberation was introduced in [De Giacomo and Levesque 1999].

Let us now turn to exogenous actions. Although an agent, or in our case the logistics program, does not have control over when exogenous actions occur, the standard assump-tion is that the agent has complete knowledge about the possible exogenous actions that can occur and what their effects are. In other words, the background basic action theory includes precondition and successor state axioms for exogenous actions too, but the agent does not control those actions. In our logistics example, we will consider one exogenous action: $requestDelivery(obj, loc)$, intuitively meaning that a request to deliver obj to loc has been issued. In our Prolog implementation, we simulate the occurrence of these exogenous requests by having the interpreter prompt the user to input them. Instead of this interactive approach to exogenous actions, another possibility would be to have the exogenous actions generated at random.

Following [De Giacomo, Lesperance, and Levesque 2000], a special procedure, $exoProg$ models the "observation" of exogenous actions. This procedure executes concurrently with the main logistics procedure shown below. The $exoProg$ procedure is defined as follows using the ConGolog construct for interrupts:

> **proc** $exoProg$
> $(\pi e)(exoActionOccurred(e) \rightarrow e)$
> **endProc**

The condition $exoActionOccurred(e)$ always succeeds when evaluated and it comes back with a user supplied value for e, which can be an exogenous action, nil, a dummy action with no effects meaning that no exogenous action occurred, or $endSim$ which has no effect either but instructs the interpreter to stop requesting the user to enter exogenous actions.

The main logistics procedure, called $deliveryDaemon$, is a recursive program that reacts to the occurrence of exogenous actions by triggering the execution of a $moveObj(obj, loc)$ task. An exogenous action $requestDelivery(obj, loc)$ causes fluent $deliveryReq(pck, loc)$ to become true, which in turn causes an interrupt to fire.

> **proc** $deliveryDaemon$
> $(\exists pck, loc)deliveryReq(pck, loc) \rightarrow$
> $\qquad (\pi pck, loc)\{deliveryReq(pck, loc)?\ ;$
> $\qquad\qquad startDelivery(pck, loc)\ ;$
> $\qquad\qquad moveObj(pck, loc)\ ;$
> $\qquad\qquad endDelivery(pck, loc)\}$
> $\quad \|\ deliveryDaemon$
> **endProc**

Note that the number of delivery requests that will need to be served concurrently cannot be anticipated. What the above procedure does when a delivery request arrives (the interrupt fires) is to concurrently start serving the request and recursively invoke itself. Through the recursive call the procedure continues to wait for the arrival of new requests. This is admittedly a bit of a hack. What we really would like to use is a fork construct, which is not (yet) available in the Golog family.

Finally, the main ConGolog program for running the simulation of the logistics domain with run-time delivery requests consists of the parallel execution of the logistics procedure and the exogenous actions procedure:

$$exoProg\ \|\ deliveryDaemon.$$

223

The full Prolog implementation is available online.[3] Here is a sample run:

```
[eclipse 2]: runSim.
startSim
    Enter an exogenous action:  requestDelivery(package1, loc5_1).
requestDelivery(package1, loc5_1)
startDelivery(package1, loc5_1)
    Enter an exogenous action:  nil.
driveTruck(truck3_1, loc3_1, loc3_3)
    Enter an exogenous action:  nil.
loadTruck(package1, truck3_1)
    Enter an exogenous action:  nil.
driveTruck(truck3_1, loc3_3, loc3_1)
unloadTruck(package1, truck3_1)
    Enter an exogenous action:  nil.
fly(plane1, loc5_1, loc3_1)
    Enter an exogenous action:  requestDelivery(package2, loc3_2).
requestDelivery(package2, loc3_2)
loadAirplane(package1, plane1)
fly(plane1, loc3_1, loc5_1)
unloadAirplane(package1, plane1)
startDelivery(package2, loc3_2)
    Enter an exogenous action:  nil.
endDelivery(package1, loc5_1)
loadTruck(package2, truck3_1)
driveTruck(truck3_1, loc3_1, loc3_2)
unloadTruck(package2, truck3_1)
    Enter an exogenous action:  requestDelivery(package3, loc1_3).
requestDelivery(package3, loc1_3)
    Enter an exogenous action:  nil.
startDelivery(package3, loc1_3)
endDelivery(package2, loc3_2)
driveTruck(truck2_1, loc2_1, loc2_3)
    Enter an exogenous action:  nil.
loadTruck(package3, truck2_1)
driveTruck(truck2_1, loc2_3, loc2_1)
unloadTruck(package3, truck2_1)
    Enter an exogenous action:  nil.
loadAirplane(package3, plane2)
    Enter an exogenous action:  nil.
fly(plane2, loc2_1, loc1_1)
    Enter an exogenous action:  nil.
    Enter an exogenous action:  endSim.
endSim
unloadAirplane(package3, plane2)
loadTruck(package3, truck1_1)
driveTruck(truck1_1, loc1_1, loc1_3)
unloadTruck(package3, truck1_1)
endDelivery(package3, loc1_3)

Plan length: 32 More?  n.
```

The non-indented lines above are primitive tasks listed in the order they occur. The user is prompted for an exogenous action every time the condition $exoActionOccurred(e)$ is evaluated. This happens every time the interpreter computes a transition for the $exoProg$ procedure.

___
[3]http://centria.di.fct.unl.pt/~ag/exologistics/

5 Conclusion

Our main goal in this chapter was to look at the relationship of HTN-planning and the Golog family of high-level action programming languages. We have shown that it is possible to encode HTN-planning problems into these languages without adding new constructs. In particular, totally ordered HTNs in Golog and partially ordered HTNs, which allow tasks to execute concurrently, in ConGolog. We showed that the operational semantics of an HTN-planning problem, i.e. the set of solutions, corresponds to the set of execution traces derived from the Situation Calculus formalization and (Con)Golog. We also showed a logical specification of the HTN-planning problem in terms of logical consequence from a basic action theory and the (Con)Golog axioms. Furthermore, we illustrated how through the translation one can combine HTNs with other features of the Golog family by taking the logistics domain example, adding exogenous delivery requests and running it on an online interpreter.

As we discussed in the introduction, HTNs and high-level action languages like Golog are two complementary ways of incorporating domain specific knowledge into automated planning. We also mentioned a third alternative that has been introduced by Bacchus and Kabanza [2000]. In their approach, domain specific knowledge is expressed in terms of Linear Temporal Logic, and is used to instruct their planning system, TLPlan, what alternatives *not* to explore. There is a consensus that this type of control knowledge and procedural knowledge as in HTNs and ConGolog, are both useful. Although we did not consider TLPlan style control knowledge here, we have considered it elsewhere [Gabaldon 2003; Gabaldon 2004]. In that work, we introduced a procedure for "compiling" TLPlan style control knowledge into a Situation Calculus basic action theory in a way that achieves the same pruning of unpromising plans as in TLPlan. That work together with our encoding of HTNs presented here, amount to a combination of both types of control knowledge into the single formal framework of the Situation Calculus. This combination is much further explored and developed in [Sohrabi, Baier, and McIlraith 2009].

References

Bacchus, F. and F. Kabanza [2000]. Using temporal logics to express search control knowledge for planning. *Artificial Intelligence 116*, 123–191.

De Giacomo, G., Y. Lesperance, and H. Levesque [2000]. ConGolog, a concurrent programming language based on the situation calculus. *Artificial Intelligence 121*, 109–169.

De Giacomo, G. and H. J. Levesque [1999]. An incremental interpreter for high-level programs with sensing. In *Logical Foundations for Cognitive Agents: Contributions in Honor of Ray Reiter*, pp. 86–102. Springer.

De Giacomo, G., R. Reiter, and M. Soutchanski [1998]. Execution monitoring of high-level robot programs. In A. Cohn and L. Schubert (Eds.), *6th International Conference on Principles of Knowledge Representation and Reasoning (KR'98)*, pp. 453–465.

Erol, K., J. A. Hendler, and D. S. Nau [1996]. Complexity results for hierarchical task-network planning. *Annals of Mathematics and Artificial Intelligence 18*, 69–93.

Gabaldon, A. [2002]. Programming hierarchical task networks in the situation calculus. In *AIPS'02 Workshop on On-line Planning and Scheduling*, Toulouse, France.

Gabaldon, A. [2003]. Compiling control knowledge into preconditions for planning in the situation calculus. In G. Gottlob and T. Walsh (Eds.), *18th International Joint Conference on Artificial Intelligence (IJCAI'03)*, pp. 1061–1066.

Gabaldon, A. [2004]. Precondition control and the progression algorithm. In D. Dubois, C. Welty, and M.-A. Williams (Eds.), *9th International Conference on Principles of Knowledge Representation and Reasoning (KR'04)*, pp. 634–643.

Levesque, H., F. Pirri, and R. Reiter [1998]. Foundations for the situation calculus. *Electronic Transactions on Artificial Intelligence* 2(3-4), 159–178. http://www.ep.liu.se/ej/etai/1998/005/.

Levesque, H. and R. Reiter [1998]. High-level robotic control: beyond planning. Position paper. In *Cognitive Robotics AAAI Fall Symposium*, pp. 106–108.

Levesque, H., R. Reiter, Y. Lespérance, F. Lin, and R. B. Scherl [1997]. Golog: A logic programming language for dynamic domains. *Journal of Logic Programming 31*(1–3), 59–83.

McCarthy, J. [1963]. Situations, actions and causal laws. Technical report, Stanford University. Reprinted in Semantic Information Processing (M. Minsky ed.), MIT Press, Cambridge, Mass., 1968, pp. 410–417.

Nau, D., H. Munoz-Avila, Y. Cao, A. Lotem, and S. Mitchell [2001]. Total-order planning with partially ordered subtasks. In B. Nebel (Ed.), *17th International Joint Conference on Artificial Intelligence (IJCAI'01)*, pp. 425–430.

Nau, D. S., Y. Cao, A. Lotem, and H. Munoz-Avila [1999]. SHOP: Simple hierarchical ordered planner. In *Proceedings of the 16th International Joint Conference on Artificial Intelligence (IJCAI'99)*, pp. 968–975.

Reiter, R. [1991]. The frame problem in the situation calculus: A simple solution (sometimes) and a completeness result for goal regression. In V. Lifschitz (Ed.), *Artificial Intelligence and Mathematical Theory of Computation*, pp. 359–380. Academic Press.

Reiter, R. [2001]. *Knowledge in Action: Logical Foundations for Describing and Implementing Dynamical Systems*. Cambridge, MA: MIT Press.

Sacerdoti, E. [1974]. Planning in a hierarchy of abstraction spaces. *Artificial Intelligence 5*, 115–135.

Sohrabi, S., J. A. Baier, and S. A. McIlraith [2009]. HTN planning with preferences. In C. Boutilier (Ed.), *21st International Joint Conference on Artificial Intelligence (IJCAI'09)*, pp. 1790–1797.

Tate, A. [1977]. Generating project networks. In R. Reddy (Ed.), *5th International Joint Conference on Artificial Intelligence (IJCAI'77)*, pp. 888–893.

14

A Logical Theory of Coordination and Joint Ability

Hojjat Ghaderi, Hector J. Levesque, and Yves Lespérance

ABSTRACT. A team of agents is jointly able to achieve a goal if despite any incomplete knowledge they may have about the world or each other, they still know enough to be able to get to a goal state. Unlike in the single-agent case, the mere existence of a working plan is not enough as there may be several incompatible working plans and the agents may not be able to choose a share that coordinates with those of the others. Some formalizations of joint ability ignore this issue of coordination within a coalition. Others, including those based on game theory, deal with coordination, but require a complete specification of what the agents believe. Such a complete specification is often not available. Here we present a new formalization of joint ability based on logical entailment in the situation calculus that avoids both of these pitfalls.[1]

Foreword

The first author (Hojjat Ghaderi) was fortunate to have Hector as one of his Ph.D. supervisors. The problem we tackled, reasoning about the coordination of agents under incomplete specifications and without assuming communication, turned out to be very complex. Hector's broad knowledge, deep technical mastery, brilliant insight and intuition, and generous support were instrumental in our success in tackling this hard problem. Every research meeting with him was enlightening and rewarding. Hector, it has been a pleasure and an honor to work under you guidance, and to know you as a friend.

1 Introduction

The coordination of teams of cooperating but autonomous agents is a core problem in multiagent systems research. A team of agents is *jointly able* to achieve a goal if despite any incomplete knowledge or even false beliefs that they may have about the world or each other, they still know enough to be able to get to a goal state, should they choose to do so. Unlike in the single-agent case, the mere existence of a working plan is not sufficient since there may be several incompatible working plans and the agents may not be able to choose a share that coordinates with those of the others.

There is a large body of work in game theory [Osborne and Rubinstein 1999] dealing with coordination and strategic reasoning for agents. The classical game theory framework has been very successful in dealing with many problems in this area. However, a major limitation of the framework is that it assumes that there is a *complete specification* of the structure of the game including the beliefs of the agents. It is also often assumed that this

[1]This article originally appeared in the proceedings of the 22nd Conference on Artificial Intelligence (AAAI'07). Hojjat Ghaderi acknowledges the financial support he received from the Department of Computer Science, University of Toronto.

structure is common knowledge among the agents. These assumptions often do not hold for the team members, let alone for a third party attempting to reason about what the team members can do.

In recent years, there has been a lot of work aimed at developing symbolic logics of games so that more incomplete and qualitative specifications can be dealt with. This can also lead to faster algorithms as sets of states that satisfy a property can be abstracted over in reasoning. However, this work has often incorporated very strong assumptions of its own. Many logics of games like Coalition Logic [Pauly 2002] and ATEL [van der Hoek and Wooldridge 2003] ignore the issue of coordination within a coalition and assume that the coalition can achieve a goal if there exists a strategy profile that achieves the goal. This is only sufficient if we assume that the agents can communicate arbitrarily to agree on a joint plan / strategy profile. In addition, most logics of games are propositional, which limits expressiveness.

In this paper, we develop a new first-order (with some higher-order features) logic framework to model the coordination of coalitions of agents based on the situation calculus. Our formalization of joint ability avoids both of the pitfalls mentioned above: it supports reasoning on the basis of very incomplete specifications about the belief states of the team members and it ensures that team members do not have incompatible strategies. The formalization involves iterated elimination of dominated strategies [Osborne and Rubinstein 1999]. Each agent compares her strategies based on her private beliefs. Initially, they consider all strategies possible. Then they eliminate strategies that are not as good as others given their beliefs about what strategies the other agents have kept. This elimination process is repeated until it converges to a set of preferred strategies for each agent. Joint ability holds if all combinations of preferred strategies succeed in achieving the goal.

In the next section, we describe a simple game setup that we use to generate example games, and test our account of joint ability. Then, we present our formalization of joint ability in the situation calculus. We show some examples of the kind of ability results we can obtain in this logic. This includes examples where we prove that joint ability holds given very weak assumptions about the agents. Then, we discuss related work and summarize our contributions.

2   A simple game setup

To illustrate our formalization of joint ability, we will employ a simple game setup that incorporates several simplifying assumptions. Many of them (*e.g.* only two agents, public actions, no communicative acts, goals that can be achieved with just two actions) are either not required by our formalization or are easy to circumvent; others are harder to get around. We will return to these in the Discussion section.

In our examples, there are two agents, P and Q, one distinguished fluent F, and one distinguished action A. The agents act synchronously and in turn: P acts first and then they alternate. There is at least one other action A', and possibly more. All actions are public (observed by both agents) and can always be executed. There are no preestablished conventions that would allow agents to rule out or prefer strategies to others or to use actions as signals for coordination (e.g. similar to those used in the game of bridge). The sorts of goals we will consider will only depend on whether or not the fluent F held initially, whether or not P did action A first, and whether or not Q then did action A.[2] Since there

---

[2] We use the term "goal" not in the sense of an agent's attitude, but as a label for the state of affairs that we ask whether the agents have enough information to achieve, *should they choose to do so.*

are $2 \times 2 \times 2$ options, and since a goal can be satisfied by any subset of these options, there are $2^8 = 256$ possible goals to consider.

This does not mean, however, that there are only 256 possible games. We assume the agents can have beliefs about F and about each other. Since they may have beliefs about the other's beliefs about their beliefs and so on, there are, in fact, an infinite number of games. At one extreme, we may choose not to stipulate anything about the beliefs of the agents; at the other extreme, we may specify completely what each agent believes. In between, we may specify some beliefs or disbeliefs and leave the rest of their internal state open. For each such specification, and for each of the 256 goals, we may ask if the agents can jointly achieve the goal.[3]

Example 1: Suppose nothing is specified about the beliefs of P and Q. Consider a goal that is satisfied by P doing A and Q not doing A regardless of F. In this case, P and Q can jointly achieve the goal, since they do not need to know anything about F or each other to do so. Had we stipulated that P believed that F was true and Q believed that F was false, we would still say that they could achieve the goal despite the false belief that one of them has.

Example 2: Suppose we stipulate that Q knows that P knows whether or not F holds. Consider a goal that is satisfied by P doing A and Q not doing A if F is true and P not doing A and Q doing A if F is false. In this case, the two agents can achieve the goal: P will do the right thing since he knows whether F is true; Q will then do the opposite of P since he knows that P knows what to do. The action of P in this case behaves like a signal to Q. Interestingly, if we merely require Q to *believe* that P knows whether or not F holds, then even if this belief is true, it would not be sufficient to imply joint ability (specifically, in the case where it is true for the wrong reason; we will return to this).

Example 3: Suppose again we stipulate that Q knows that P knows whether or not F holds. Consider a goal that is satisfied by P doing anything and Q not doing A if F is true and P doing anything and Q doing A if F is false. In a sense this is a goal that is easier to achieve than the one in Example 2, since it does not require any specific action from P. Yet, in this case, it would not follow that they can achieve the goal. Had we additionally stipulated that Q did not know whether F held, we could be more definite and say that they definitely cannot jointly achieve this goal as there is nothing P can do to help Q figure out what to do.

Example 4: Suppose again we stipulate that Q knows that P knows whether or not F holds. Consider a goal that is like in Example 3 but easier, in that it also holds if both agents do not do A when F is false. In this case, they can achieve the goal. The reason, however, is quite subtle and depends on looking at the various cases according to what P and Q believe. Similar to Example 2, requiring Q to have true belief about P knowing whether F holds is not sufficient.

To the best of our knowledge, there is no existing formal account where examples like these and their variants can be formulated. We will return to this in the Related Work section. In the next section, we present a formalization of joint ability that handles the game setup above and much more based on entailment in the situation calculus.

---

[3]We may also ask whether the agents *believe* or *mutually believe* that they have joint ability, but we defer this to later.

3 The formal framework

The basis of our framework for joint ability is the situation calculus [McCarthy and Hayes 1969; Levesque, Pirri, and Reiter 1998]. The situation calculus is a predicate calculus language for representing dynamically changing domains. A *situation* represents a possible state of the domain. There is a set of initial situations corresponding to the ways the domain might be initially. The actual initial state of the domain is represented by the distinguished initial situation constant, S_0. The term $do(a, s)$ denotes the unique situation that results from an agent doing action a in situation s. Initial situations are defined as those that do not have a predecessor: $Init(s) \doteq \neg \exists a \exists s'. s = do(a, s')$. In general, the situations can be structured into a set of trees, where the root of each tree is an initial situation and the arcs are actions. The formula $s \sqsubseteq s'$ is used to state that there is a path from situation s to situation s'. Our account of joint ability will require some second-order features of the situation calculus, including quantifying over certain functions from situations to actions, that we call *strategies*.

Predicates and functions whose values may change from situation to situation (and whose last argument is a situation) are called *fluents*. The effects of actions on fluents are defined using successor state axioms [Reiter 2001], which provide a succinct representation for both effect and frame axioms [McCarthy and Hayes 1969]. To axiomatize a dynamic domain in the situation calculus, we use action theories [Reiter 2001] consisting of (1) successor state axioms; (2) initial state axioms, which describe the initial states of the domain including the initial beliefs of the agents; (3) precondition axioms, which specify the conditions under which each action can be executed (we assume here that all actions are always possible); (4) unique names axioms for the actions, and (5) domain-independent foundational axioms (we adopt the ones given in [Levesque, Pirri, and Reiter 1998] which accommodate multiple initial situations).

For our examples, we only need three fluents: the fluent F mentioned in the previous section in terms of which goals are formulated, a fluent *turn* which says whose turn it is to act, and a fluent B to deal with the beliefs of the agents.

Moore [Moore 1985] defined a possible-worlds semantics for a logic of knowledge in the situation calculus by treating situations as possible worlds. Scherl and Levesque [Scherl and Levesque 2003] adapted this to Reiter's action theories and gave a successor state axiom for B that states how actions, including sensing actions, affect knowledge. Shapiro et al. [Shapiro, Lespérance, and Levesque 1998] adapted this to handle the beliefs of multiple agents, and we adopt their account here. $B(x, s', s)$ will be used to denote that in situation s, agent x thinks that situation s' might be the actual situation. Note that the order of the situation arguments is reversed from the convention in modal logic for accessibility relations. Belief is then defined as an abbreviation:[4]

$$Bel(x, \phi[now], s) \doteq \forall s'. B(x, s', s) \supset \phi[s'].$$

We will also use

$$TBel(x, \phi[now], s) \doteq Bel(x, \phi[now], s) \wedge \phi[s]$$

as an abbreviation for true belief (which we distinguish from knowledge formalized as a *KT45* operator, for reasons alluded to above in Example 2). Whenever we need knowledge

---

[4]Free variables are assumed to be universally quantified from outside. If ϕ is a formula with a single free situation variable, $\phi[t]$ denotes ϕ with that variable replaced by situation term t. Instead of $\phi[now]$ we occasionally omit the situation argument completely.

and not merely true belief, we can simply add the following initial reflexivity axiom (called **IBR**) to the theory:

$$Init(s) \supset B(x, s, s).$$

Mutual belief among the agents, denoted by *MBel*, can be defined either as a fix-point or by introducing a new accessibility relation using a second-order definition. Common knowledge is then *MBel* under the **IBR** assumption.

Our examples use the following successor state axioms:

- $F(do(a, s)) \equiv F(s)$.
  The fluent F is unaffected by any action.

- $turn(do(a, s)) = x \ \equiv x = P \wedge turn(s) = Q \ \vee \ x = Q \wedge turn(s) = P$.
  Whose turn it is to act alternates between P and Q.

- $B(x, s', do(a, s)) \equiv \exists s''.B(x, s'', s) \wedge s' = do(a, s'')$.
  This is a simplified version of the successor state axiom proposed by Scherl and Levesque. See the Discussion section for how it can be generalized.

The examples also include the following initial state axioms:

- $Init(s) \supset turn(s) = P$. So, agent P gets to act first.

- $Init(s) \wedge B(x, s', s) \supset Init(s')$.
  Each agent initially knows that it is in an initial situation.

- $Init(s) \supset \exists s' B(x, s', s)$.
  Each agent initially has consistent beliefs.

- $Init(s) \wedge B(x, s', s) \supset \forall s''. B(x, s'', s') \equiv B(x, s'', s)$.
  Each agent initially has introspection of her beliefs.

The last two properties of belief can be shown to hold for all situations using the successor state axiom for B so that belief satisfies the modal system *KD45* [Chellas 1980]. If we include the **IBR** axiom, belief will satisfy the modal system *KT45*. Since the axioms above are universally quantified, they are known to all agents, and in fact are common knowledge. We will let Σ_e denote the action theory containing the successor and initial state axioms above. All the examples in the next section will use Σ_e with additional conditions.

3.1 Our definition of joint ability

We assume there are N agents named 1 to N. We use the following abbreviations for representing strategy[5] profiles:

- A vector of size N is used to denote a complete strategy profile consisting of one strategy per agent, e.g. $\vec{\sigma}$ for $\sigma_1, \sigma_2, \cdots, \sigma_N$.

- $\vec{\sigma}_{-i}$ represents an incomplete profile with strategies for everyone except player i, i.e. $\sigma_1, \cdots, \sigma_{i-1}, \sigma_{i+1} \cdots, \sigma_N$.

---

[5]Strictly speaking, the σ_i's are second-order variables ranging over functions from situations to actions. We use $Strategy(i, \sigma_i)$ to restrict them to valid strategies.

- $\oplus_i$ is used to insert a strategy for player i into an incomplete profile, e.g. $\vec{\sigma}_{-i} \oplus_i \delta$: $\sigma_1, \cdots, \sigma_{i-1}, \delta, \sigma_{i+1} \cdots, \sigma_N$.

- $|_i$ is used to substitute the ith player's strategy in a complete profile, for example, $\vec{\sigma}|_i \delta : \sigma_1, \cdots, \sigma_{i-1}, \delta, \sigma_{i+1} \cdots, \sigma_N$.

All of the definitions below are abbreviations for formulas in the language of the situation calculus presented above. The joint ability of N agents to achieve ϕ is defined as follows:[6]

- $JCan(\phi, s) \doteq$
  $\forall \vec{\sigma}. [\bigwedge_{i=1}^{N} Pref(i, \sigma_i, \phi, s)] \supset Works(\vec{\sigma}, \phi, s).$

  Agents $1 \cdots N$ can jointly achieve ϕ iff all combinations of their preferred strategies work together.

- $Works(\vec{\sigma}, \phi, s) \doteq \exists s''. s \sqsubseteq s'' \wedge \phi[s''] \wedge$
  $\forall s'. s \sqsubseteq s' \sqsubseteq s'' \supset$
  $\qquad \bigwedge_{i=1}^{N}(turn(s') = i \supset do(\sigma_i(s'), s') \sqsubseteq s'').$

  Strategy profile $\vec{\sigma}$ works if there is a future situation where ϕ holds and the strategies in the profile prescribe the actions to get there according to whose turn it is.

- $Pref(i, \sigma_i, \phi, s) \doteq \forall n. Keep(i, n, \sigma_i, \phi, s)$

  Agent i prefers strategy σ_i if it is kept for all levels n.[7]

- $Keep$ is defined inductively:[8]

  - $Keep(i, 0, \sigma_i, \phi, s) \doteq Strategy(i, \sigma_i).$
    At level 0, all strategies are kept.

  - $Keep(i, n+1, \sigma_i, \phi, s) \doteq Keep(i, n, \sigma_i, \phi, s) \wedge$
    $\neg \exists \sigma_i'. Keep(i, n, \sigma_i', \phi, s) \wedge GTE(i, n, \sigma_i', \sigma_i, \phi, s) \wedge \neg GTE(i, n, \sigma_i, \sigma_i', \phi, s).$
    For each agent i, the strategies kept at level $n+1$ are those kept at level n for which there is not a better one (σ_i' is better than σ_i if it is as good as, i.e. greater than or equal to, σ_i while σ_i is not as good as it).

- $Strategy(i, \sigma_i) \doteq$
  $\forall s. turn(s) = i \supset \exists a. TBel(i, \sigma_i(now) = a, s).$

  Strategies for an agent are functions from situations to actions such that the required action is known to the agent whenever it is the agent's turn to act.

- $GTE(i, n, \sigma_i, \sigma_i', \phi, s) \doteq$
  $Bel(i, \forall \vec{\sigma}_{-i}. ([\bigwedge_{j \neq i} Keep(j, n, \sigma_j, \phi, now) \wedge Works(\vec{\sigma}_{-i} \oplus_i \sigma_i', \phi, now)]$
  $\qquad\qquad\qquad\qquad\qquad \supset Works(\vec{\sigma}_{-i} \oplus_i \sigma_i, \phi, now)), s).$

  Strategy σ_i is as good as (Greater Than or Equal to) σ_i' for agent i if i believes that whenever σ_i' works with strategies kept by the rest of the agents so does σ_i.

---

[6]In the Discussion section, we consider the case where there may be agents outside of the coalition of N agents.

[7]The quantification is over the sort natural number.

[8]Strictly speaking, the definition we propose here is ill-formed. We want to use it with the second argument universally quantified (as in $Pref$). $Keep$ and GTE actually need to be defined using second-order logic, from which the definitions here emerge as consequences. We omit the details for space reasons.

These formulas define joint ability in a way that resembles the iterative elimination of weakly dominated strategies of game theory [Osborne and Rubinstein 1999] (see the Related Work section). As we will see in the examples to follow, the mere *existence* of a working strategy profile is not enough; the definition requires coordination among the agents in that *all* preferred strategies must work together.

4 Formalizing the examples

As we mentioned, for each of the 256 possible goals we can consider various assumptions about the beliefs of the agents. In this section, we provide theorems for the goals of the four examples mentioned earlier. Due to lack of space we omit the proofs. Since there are only two agents, the previous definitions can be simplified. For better exposition, we use g (possibly superscripted) to refer to strategies of the first agent (called P) and h for those of the second (called Q).

4.1 Example 1

For this example, the goal is defined as follows:

$$\phi_1(s) \doteq \exists s'. \, Init(s') \land \exists a. \, a \neq A \land s = do(a, do(A, s'))$$

Because the goal in this example (and other examples) is only satisfied after exactly two actions, we can prove that $Works(g, h, \phi_1, s)$ depends only on the first action prescribed by g and the response prescribed by h:

LEMMA 1. $\Sigma_e \models Init(s) \supset$
$$Works(g, h, \phi_1, s) \equiv g(s) = A \land h(do(A, s)) \neq A.$$

We can also show that P will only prefer to do A, and Q will only prefer to do a non-A action. So, we have the following:

THEOREM 2. $\Sigma_e \models Init(s) \supset JCan(\phi_1, s)$.

It then trivially follows that the agents can achieve ϕ_1 despite having false beliefs about F:

COROLLARY 3.
$$\Sigma_e \models [Init(s) \land Bel(P, \neg F, s) \land Bel(Q, F, s)] \supset JCan(\phi_1, s).$$

We can also trivially show that the agents have mutual belief that joint ability holds:

COROLLARY 4. $\Sigma_e \models Init(s) \supset MBel(JCan(\phi_1, now), s)$.

4.2 Example 2

For this example, the goal is defined as follows:
$$\phi_2(s) \doteq \exists s', a. \, Init(s') \land a \neq A \land$$
$$[F(s') \land s = do(a, do(A, s')) \lor \neg F(s') \land s = do(A, do(a, s'))].$$

LEMMA 5. $\Sigma_e \models Init(s) \supset$
$$Works(g, h, \phi_2, s) \equiv [F(s) \land g(s) = A \land h(do(A, s)) \neq A \lor$$
$$\neg F(s) \land g(s) \neq A \land h(do(g(s), s)) = A].$$

We will also use the following definitions:

- $BW(x, \phi, s) \doteq Bel(x, \phi, s) \lor Bel(x, \neg\phi, s)$

  the agent believes whether ϕ holds.

- $TBW(x, \phi, s) \doteq TBel(x, \phi, s) \lor TBel(x, \neg\phi, s)$.

As mentioned in Example 2, Q's having true belief about P truly believing whether F is not sufficient for joint ability:

THEOREM 6. $\Sigma_e \cup \{TBel(Q, TBW(P, F, now), S_0)\} \not\models JCan(\phi_2, S_0)$.

This is because $TBel(Q, TBW(P, F, now), s)$ does not preclude Q having a false belief about P, namely believing that P believes that F is false when in fact P believes correctly that F is true. To resolve this, we can simply add the **IBR** axiom. Another approach is to remain in the *KD45* logic but assert that Q's belief about P's belief of F is correct:

$$BTBel(Q, P, F, s) \doteq$$
$$[Bel(Q, TBel(P, F, now), s) \supset TBel(P, F, s)] \wedge$$
$$[Bel(Q, TBel(P, \neg F, now), s) \supset TBel(P, \neg F, s)]$$

To keep our framework as general as possible, we take the second approach and add $BTBel(Q, P, F, s)$ whenever needed. With this, we have the following theorem:

THEOREM 7. $\Sigma_e \models [Init(s) \wedge BTBel(Q, P, F, s) \wedge TBel(Q, TBW(P, F, now), s)] \supset$
$$JCan(\phi_2, s).$$

It follows from the theorem that Q's *knowing* that P *knows* whether F holds is sufficient to get joint ability. More interestingly, it follows immediately that common knowledge of the fact that P knows whether F holds implies common knowledge of joint ability.

4.3 Example 3

The goal for this example is easier to satisfy than the one in Example 2 (in the sense that $\Sigma_e \models \phi_2(s) \supset \phi_3(s)$):

$$\phi_3(s) \doteq \exists s', a, b. Init(s') \wedge$$
$$[F(s') \wedge b \neq A \wedge s = do(b, do(a, s')) \vee$$
$$\neg F(s') \wedge b = A \wedge s = do(b, do(a, s'))].$$

Nonetheless, unlike in Example 2, it does not follow that the agents can achieve the goal (cf. Theorem 7):

THEOREM 8. $\Sigma_e \not\models [BTBel(Q, P, F, S_0) \wedge TBel(Q, TBW(P, F, now), S_0)] \supset JCan(\phi_3, S_0)$.

In fact, we can prove a stronger result that if Q does not believe whether F holds they cannot jointly achieve ϕ_3:

THEOREM 9. $\Sigma_e \models \neg BW(Q, F, S_0) \supset \neg JCan(\phi_3, S_0)$.

Note that there are two strategy profiles that the agents believe achieve ϕ_3: (1) P does A when F holds and a non-A action otherwise, and Q does the opposite of P's action; (2) P does a non-A action when F holds and A otherwise, and Q does the same action as P. However, Q does not know which strategy P will follow and hence might choose an incompatible strategy. Therefore, although there are working profiles, the existence of at least two incompatible kept profiles results in the lack of joint ability. We did not have this problem in Example 2 since profile (2) did not work there.

We can prove that if Q truly believes whether F holds they can achieve the goal as Q will know exactly what to do:

THEOREM 10. $\Sigma_e \models Init(s) \wedge TBW(Q, F, s) \supset JCan(\phi_3, s)$.

4.4 Example 4

The goal here is easier than the ones in Examples 2 and 3:

$$\phi_4(s) \doteq \exists s', a, b. \, Init(s') \wedge$$
$$F(s') \wedge b \neq A \wedge s = do(b, do(a, s')) \vee$$
$$\neg F(s') \wedge [s = do(A, do(A, s')) \vee b \neq A \wedge s = do(a, do(b, s'))].$$

Similarly to Example 2, we can show that if Q has true belief about P truly believing whether F holds, then assuming $BTBel(Q, P, F, s)$, the agents can achieve the goal:

THEOREM 11. $\Sigma_e \models [Init(s) \wedge BTBel(Q, P, F, s) \wedge TBel(Q, TBW(P, F, now), s)] \supset$
$$JCan(\phi_4, s).$$

Note that there are many profiles that achieve ϕ_4 (including profiles (1) and (2) mentioned in Example 3). Nonetheless, unlike in Example 3, we can prove by looking at various cases that the agents can coordinate (even if $\neg BW(Q, F, s)$ holds) by eliminating their dominated strategies.

From the above theorem, it follows that Q's *knowing* that P knows whether F holds is sufficient to get joint ability. More interestingly, common knowledge of the fact that P knows whether F holds implies common knowledge of joint ability even though Q may have incomplete beliefs about F.

5 Properties of the definition

We now present several properties of our definition to show its plausibility in general terms. Let Σ be an arbitrary action theory with a *KD45* logic for the beliefs of N agents.

Our definition of ability is quite general and can be nested within beliefs. The consequential closure property of belief can be used to prove various subjective properties about joint ability. For example, to prove $Bel(i, Bel(j, JCan(\phi, now), now), S_0)$, it is sufficient to find a formula γ such that $\Sigma \models \forall s. \, \gamma[s] \supset JCan(now, s)$ and $\Sigma \models Bel(i, Bel(j, \gamma, now), S_0)$.

One simple case where we can show that an agent believes that joint ability holds is when there is no need to coordinate. In particular, if agent i has a strategy that she believes achieves the goal regardless of choices of other team members, then she believes that joint ability holds:[9]

THEOREM 12. $\Sigma \models [\exists \sigma_i \forall \vec{\sigma}_{-i}. \, Bel(i, Works(\vec{\sigma}_{-i} \oplus_i \sigma_i, \phi, now), S_0)] \supset$
$$Bel(i, JCan(\phi, now), S_0).[10]$$

(Example 3 with $BW(Q, F, s)$ is such a case: Q believes that a strategy that says do A when F is false and do a non-A action otherwise achieves the goal whatever P chooses.) However, there are theories Σ' such that even though agent i has a strategy that always achieves the goal (regardless of choices of others) joint ability does not actually follow:

THEOREM 13. *There are Σ' and ϕ such that*
$$\Sigma' \cup \{\exists \sigma_i \forall \vec{\sigma}_{-i} Works(\vec{\sigma}_{-i} \oplus_i \sigma_i, \phi, S_0)\} \not\models JCan(\phi, S_0).$$

This is because agent i might not know that σ_i always works, and hence might keep other incompatible strategies as well.

Another simple case where joint ability holds is when there exists a strategy profile that every agent truly believes works, and moreover everyone believes it is impossible to achieve the goal if someone deviates from this profile:[11]

---

[9]Note that, however, this does not imply that joint ability holds in the real world since i's beliefs might be wrong.

[10]From here on, σ's are intended to range over strategies.

[11]$EBel(\gamma, s) \doteq \bigwedge_{i=1}^{N} Bel(i, \gamma, s)$.

THEOREM 14. $\Sigma \models [\exists\vec{\sigma}. ETBel(Works(\vec{\sigma}, \phi, now), S_0) \land$
$$\forall\vec{\sigma}' \neq \vec{\sigma}. EBel(\neg Works(\vec{\sigma}', \phi, now), S_0)] \supset JCan(\phi, S_0).$$

It turns out that joint ability can be proved from even weaker conditions. In particular, $ETBel(Works(\vec{\sigma}, \phi, now), S_0)$ in the antecedent of Theorem 14 can be replaced by $Works(\vec{\sigma}, \phi, S_0) \land \bigwedge_{i=1}^{N} \neg Bel(i, \neg Works(\vec{\sigma}, \phi, now), S_0)$, i.e. $Works(\vec{\sigma}, \phi, S_0)$ holds and is consistent with the beliefs of the agents. This is because each agent i will only prefer her share of $\vec{\sigma}$ (i.e. σ_i).

We can generalize the result in Theorem 14 if we assume there exists a strategy profile that is *known* by everyone to achieve the goal. Then, it is sufficient for every agent to *know* that their share in the profile is at least as good as any other available strategy to them, for *JCan* to hold:

THEOREM 15. $\Sigma \cup \{\textbf{\textit{IBR}}\} \models$
$[\exists\vec{\sigma}. EBel(Works(\vec{\sigma}, \phi, now), S_0) \land$
$$\forall\vec{\sigma}'. \bigwedge_{i=1}^{N} Bel(i, Works(\vec{\sigma}', \phi, now) \supset Works(\vec{\sigma}'|_i \sigma_i, \phi, now), S_0)] \supset JCan(\phi, S_0).$$

Another important property of joint ability is that it is nonmonotonic w.r.t. the goal. Unlike in the single agent case, it might be the case that a team is able to achieve a strong goal while it is unable to achieve a weaker one (the goals in Examples 3 and 4 are an instance of this):

THEOREM 16. *There are Σ', ϕ, and ϕ' such that*
$$\Sigma' \models \phi(s) \supset \phi'(s) \text{ but } \Sigma' \not\models JCan(\phi, S_0) \supset JCan(\phi', S_0).$$

6   Discussion

Although the game in this paper makes several simplifying assumptions for illustration purposes, many of them are not actually required by our account of joint ability or are easy to circumvent. For example, the formalization as presented here supports more than two agents and goals that are achieved after more than two actions. Although the game does not involve any sensing or communicative acts, our definition stays the same should we include such actions. We only need to revise the successor state axiom for belief accessibility as in [Scherl and Levesque 2003; Shapiro, Lespérance, and Levesque 1998]. Similarly, actions that are not public can be handled with a variant axiom making the action known only to the agent who performed it. However, handling concurrent actions and actions with preconditions is more challenging.

Also, the definition of joint ability presented here assumes all N agents are in the same coalition. It can be straightforwardly generalized to allow some agents to be outside of the coalition. Let C be a coalition (i.e. a subset of agents $\{1, \cdots, N\}$). Since each agent $j \notin C$ might conceivably choose any of her strategies, agents inside the coalition C must coordinate to make sure their choices achieve the goal regardless of the choices of the agents outside C. It turns out that a very slight modification to the definition of *Keep* is sufficient for this purpose. In particular, the definition of *Keep* for agents inside C remains unchanged while for every agent $j \notin C$, we define $Keep(j, n, \sigma_j, s) \doteq Strategy(j, \sigma_j)$. Therefore, for every agent j outside the coalition we have $Pref(j, \sigma_j, s) \equiv Strategy(j, \sigma_j)$.

7   Related work

As mentioned, there has been much recent work on developing symbolic logics of cooperation. In [Wooldridge and Jennings 1999] the authors propose a model of cooperative problem solving and define joint ability by simply adapting the definition of single-agent ability, i.e. they take the existence of a joint plan that the agents mutually believe achieves the goal as sufficient for joint ability. They address coordination in the plan formation phase where agents negotiate to agree on a promising plan before starting to act. Coalition logic, introduced in [Pauly 2002], formalizes reasoning about the power of coalitions in strategic settings. It has modal operators corresponding to a coalition being able to enforce various outcomes. The framework is propositional and also ignores the issue of coordination *inside* the coalition. In a similar vein, van der Hoek and Wooldridge propose ATEL, a variant of alternating-time temporal logic enriched with epistemic relations in [van der Hoek and Wooldridge 2003]. Their framework also ignores the issue of coordination inside a coalition. In [Jamroga and van der Hoek 2004], the authors acknowledge this shortcoming and address it by enriching the framework with extra cooperation operators. These operators nonetheless require either communication among coalition members, or a third-party choosing a plan for the coalition.

The issue of coordination using domination-based solution concepts has been thoroughly explored in game theory [Osborne and Rubinstein 1999]. Our framework differs from these approaches, however, in a number of ways. Foremost, our framework not only handles agents with incomplete information [Harsanyi 1967], but also it handles incomplete *specifications* where some aspects of the world or agents including belief/disbelief are left unspecified. Since our proofs are based on entailment, they remain valid should we add more detail to the theory. Second, rather than assigning utility functions, our focus is on the achievability of a state of affairs by a team. Moreover, we consider strict uncertainty where probabilistic information may be unavailable. Our framework supports a weaker form of belief (as in *KD45* logic) and allows for false belief. Our definition of joint ability resembles the notion of admissibility and iterated weak dominance in game theory [Osborne and Rubinstein 1999; Brandenburger and Keisler 2001]. Our work can be related to these by noting that every *model* of our theory with a *KT45* logic of belief can be considered as a partial extensive form game with incomplete information represented by a set of infinite trees each of which is rooted at an initial situation. We can add Nature as a player who decides which tree will be chosen as the real world and is indifferent among all her choices. Also, for all agents (other than Nature) we assign utility 1 to any situation that has a situation in its past history where ϕ is satisfied, and utility 0 to all other situations. However, since there are neither terminal nodes nor probabilistic information, the traditional definition of weak dominance cannot be used and an alternative approach for comparing strategies (as described in this paper) is needed, one that is based on the private beliefs of each agent about the world and other agents and their beliefs.

8 Conclusion

In this paper, we proposed a logical framework based on the situation calculus for reasoning about the coordination of teams of agents. We developed a formalization of joint ability that supports reasoning on the basis of very incomplete specifications of the belief states of the team members, something that classical game theory does not allow. In contrast to other game logics, our formalization ensures that team members are properly coordinated. We showed how one can obtain proofs of the presence or lack of joint ability for various examples involving incomplete specifications of the beliefs of the agents. We also proved several general properties about our definitions.

In future work, we will consider some of the generalizations of the framework noted in the Discussion section. We will also examine how different ways of comparing strategies (the *GTE* order) lead to different notions of joint ability, and try to identify the best. We will also evaluate our framework on more complex game settings. Finally, we will look at how our framework could be used in automated verification and multiagent planning.

References

Brandenburger, A. and H. J. Keisler [2001]. Epistemic conditions for iterated admissibility. In *TARK '01*, San Francisco, pp. 31–37. Morgan Kaufmann.

Chellas, B. [1980]. *Modal logic: an introduction*. United Kingdom: Cambridge University Press.

Harsanyi, J. C. [1967]. Games with incomplete information played by Bayesian players. *Management Science 14*, 59–182.

Jamroga, W. and W. van der Hoek [2004]. Agents that know how to play. *Fundamenta Informaticae 63*(2-3), 185–219.

Levesque, H., F. Pirri, and R. Reiter [1998]. Foundations for the situation calculus. *Electronic Transactions on Artificial Intelligence 2*(3-4), 159–178.

McCarthy, J. and P. J. Hayes [1969]. Some philosophical problems from the standpoint of artificial intelligence. In B. Meltzer and D. Michie (Eds.), *Machine Intelligence 4*, pp. 463–502.

Moore, R. [1985]. A formal theory of knowledge and action. In J. Hobbs and R. Moore (Eds.), *Formal Theories of the Commonsense World*, pp. 319–358.

Osborne, M. J. and A. Rubinstein [1999]. *A Course in Game Theory*. MIT Press.

Pauly, M. [2002]. A modal logic for coalitional power in games. *J. of Logic and Computation 12*(1), 149–166.

Reiter, R. [2001]. *Knowledge in Action: Logical Foundations for Specifying & Implementing Dynamical Systems*. The MIT Press.

Scherl, R. and H. Levesque [2003]. Knowledge, action, and the frame problem. *Artificial Intelligence 144*, 1–39.

Shapiro, S., Y. Lespérance, and H. Levesque [1998]. Specifying communicative multiagent systems. In W. Wobcke, M. Pagnucco, and C. Zhang (Eds.), *Agents and Multiagent systems*, pp. 1–14. Springer-Verlag.

van der Hoek, W. and M. Wooldridge [2003]. Cooperation, knowledge, and time: Alternating-time temporal epistemic logic and its applications. *Studia Logica 75*(1), 125–157.

Wooldridge, M. and N. R. Jennings [1999]. The cooperative problem-solving process. *Journal of Logic and Computation 9*(4), 563–592.

15

Chronolog: It's about Time for Golog

GIUSEPPE DE GIACOMO AND MAURICE PAGNUCCO

ABSTRACT. In this paper we introduce a notion of discrete time into the cognitive robotics languages of the Golog family. Our variant of Golog, named *Chronolog*, is achieved in a rather straightforward manner but we show that it allows one to express a rich variety of temporal concepts. It is based on exogenous `tick` actions generated by an external "clock" and a special functional fluent to keep track of the passing of time. Moreover, we consider on-line and off-line versions of our proposal as well as real-time and reactive variants and the ability to deal with exogenous actions.

1 Introduction

In this paper we introduce a notion of discrete time into the cognitive robotics language Golog [Levesque, Reiter, Lespérance, Lin, and Scherl 1997; Reiter 2001]. The main advantages of our approach are that it is relatively straightforward yet novel and surprisingly expressive. We consider that an external "clock" will generate discrete clock "ticks" and that their arrival will be kept track of by a special functional fluent for time.

The introduction of time into Golog and the situation calculus is not new. It has already been considered by several authors. Grosskreutz and Lakemeyer in describing their Golog variant cc-Golog [Grosskreutz and Lakemeyer 2003] suggest the use of an explicit clock with discrete clock tick actions but object to it due to the problem of determining the correct granularity of clock ticks. They also object to this on the basis that the execution traces would be "glutted with irrelevant 'clock tick' actions". Gans *et al.* [2003] also use the notion of clock ticks in Golog but do not develop the idea formally. Reiter [1996, 1998, 2001] considers instantaneous actions with a temporal argument $A(\bar{x}, t)$ denoting that the action occurred at time t. He is able to deal with continuous actions, concurrent actions and actions with duration by allowing processes (or events) to have a start action that initiates the process and an end action that terminates it. This approach to actions with duration is based on one developed by Pinto [1997]. Davis [1990] also discusses time in the situation calculus and draws his account mainly on the work of McDermott [1982]. Here a fluent *clock_time* associates a time with situations. Situations can also belong to intervals (sets of situations belonging to the interval). This allows for continuous actions and branching time.

In this paper, we give a nice and simple framework to deal with discrete time, under certain assumptions. We consider our agent's behavior to be controlled by (variants of) Golog programs, and we consider the discrete passing of time (i.e., clock ticks) to be outside the control of the agent. We then make the simplifying assumption that program execution is orders of magnitudes quicker than the external ticking. And we equip agents' programs with simple abilities for testing time flow and synchronize with time ticks when needed. In this setting we show that we can actually go surprisingly far in modeling sophisticated time related properties within programs.

Again we stress that the simplicity and the effectiveness of the approach rely strongly on the assumption that the computation of the agent is so fast that it can execute any amount of computation in between time ticks. This is obviously an assumption that we cannot hope to hold in every context. However there are contexts where it can be thought of as a good approximation of reality. Obvious examples are robots that perform physical movements directed by a high level control program.

The remainder of the paper is organized as follows. In the next section we provide some background to the situation calculus and the cognitive robotics language Golog and some of its variants. In Section 3 we introduce our notion of time and the axioms required to capture it while in Section 4 we introduce some additional syntactic constructs to enhance readability. Section 4 also illustrates the expressiveness of our approach with a list of temporal notions that are captured. Section 5 discusses the execution of Chronolog programs. We provide some examples in Section 6. In Section 7 we show the modifications required of the IndiGolog [De Giacomo and Levesque 1999] interpreter to implement Chronolog and in Section 8 we show how to implement a timed search construct.

2 Preliminaries

The basis for Golog and its variants is provided by the situation calculus [McCarthy 1963; McCarthy and Hayes 1969; Reiter 2001]. We will not go over the language in detail here except to note the following components: there is a special constant S_0 used to denote the *initial situation*, namely that situation in which no actions have yet occurred; there is a distinguished binary function symbol do where $do(a, s)$ denotes the situation resulting from performing the action a at situation s; relations whose truth values vary from situation to situation, are called (relational) *fluents*, and are denoted by predicate symbols taking a situation term as their last argument; and, there is a special predicate $Poss(a, s)$ used to state that action a is executable in situation s.

An action theory of the following form is a common scenario [Reiter 2001]:

- Axioms describing the initial situation, S_0;

- Action precondition axioms, one for each primitive action a, that are used to characterize $Poss(a, s)$—when it is possible to perform action a at situation s;

- Successor state axioms, one for each fluent F, stating the conditions under which $F(\vec{x}, do(a, s))$ holds as a function of what holds in situation s; these take the place of effect axioms, but also provide a solution to the frame problem;

- Unique names axioms for the primitive actions; and,

- Some foundational, domain independent axioms [Levesque and Lakemeyer 2000; Reiter 2001].

Next we turn to programs. The programs we consider here are based on the ConGolog language defined in [De Giacomo, Lespérance, and Levesque 2000], which provides a rich set of programming constructs summarized below:

α	primitive action
ϕ	wait for a condition
$\delta_1; \delta_2$	sequence
$\delta_1 \mid \delta_2$	nondeterministic branch

$\pi \, x. \, \delta$	nondeterministic choice of argument
δ^*	nondeterministic iteration
if ϕ **then** δ_1 **else** δ_2 **endIf**	conditional
while ϕ **do** δ **endWhile**	while loop
$\delta_1 \parallel \delta_2$	concurrency with equal priority
$\delta_1 \rangle\!\rangle \, \delta_2$	concurrency with δ_1 at a higher priority
$\delta^\parallel$	concurrent iteration
$\langle \, \phi \rightarrow \delta \, \rangle$	interrupt
$p(\vec{\theta})$	procedure call[1]

Among these constructs, we notice the presence of nondeterministic constructs. These include $(\delta_1 \mid \delta_2)$, which nondeterministically chooses between programs δ_1 and δ_2, $\pi \, x. \, \delta$, which nondeterministically picks a binding for the variable x and performs the program δ for this binding of x, and δ^*, which performs δ zero or more times. It should be noted that these forms of nondeterminism represent "reasoned" choices; during execution any choices to be made are carried out on the basis of what will guarantee termination of the program. Also notice that ConGolog includes constructs for dealing with concurrency. In particular $(\delta_1 \parallel \delta_2)$ expresses the concurrent execution (interpreted as interleaving) of the programs δ_1 and δ_2. Observe that a program may become blocked when it reaches a primitive action whose preconditions are false or a wait action ϕ? whose condition ϕ is false. Then, execution of $(\delta_1 \parallel \delta_2)$ may continue provided another program executes next. Beside $(\delta_1 \parallel \delta_2)$ ConGolog includes other constructs for dealing with concurrency, such as prioritized concurrency $(\delta_1 \rangle\!\rangle \, \delta_2)$, and interrupts $\langle \phi \rightarrow \delta \rangle$. In $(\delta_1 \rangle\!\rangle \, \delta_2)$, δ_1 has higher priority than δ_2, and δ_2 may only execute when δ_1 is done or blocked. $\delta^\parallel$ is like nondeterministic iteration δ^*, but the instances of δ are executed concurrently rather than in sequence. Finally, an interrupt $< \vec{x} : \phi \rightarrow \delta >$ has variables $\vec{x}$, a trigger condition ϕ, and a body δ. If the interrupt gets control from higher priority processes and the condition ϕ is true for some binding of the variables, the interrupt triggers and the body is executed with the variables taking these values. Once the body completes execution, the interrupt may trigger again. We refer the reader to [De Giacomo, Lespérance, and Levesque 2000] for a detailed account of ConGolog.

In [De Giacomo, Lespérance, and Levesque 2000], a single step transition semantics in the style of [Plotkin 1981] is defined for ConGolog programs. Two special predicates $Trans$ and $Final$ are introduced. $Trans(p, s, p', s')$ means that by executing program p starting in situation s, one can get to situation s' in one elementary step with the program p' remaining to be executed, that is, there is a possible transition from the configuration (p, s) to the configuration (p', s'). $Final(p, s)$ means that program p may successfully terminate in situation s, i.e., the configuration (p, s) is final.[2]

Offline executions of programs, which are the kind of executions originally proposed for Golog and ConGolog [Levesque, Reiter, Lespérance, Lin, and Scherl 1997; De Giacomo,

---

[1] For the sake of simplicity, we will not consider procedures in this paper.

[2] For example, the transition requirements for sequence are

$$Trans([p_1; p_2], s, p', s') \equiv$$
$$Final(p_1, s) \wedge Trans(p_2, s, p', s') \vee$$
$$\exists q'. \, Trans(p_1, s, q', s') \wedge p' = (q'; p_2)$$

i.e., to single-step the program $(p_1; p_2)$, either p_1 terminates and we single-step p_2, or we single-step p_1 leaving some q', and $(q'; p_2)$ is what is left of the sequence.

Lespérance, and Levesque 2000], are characterized using the $Do(p, s, s')$ predicate, which means that there is an execution of program p that starts in situation s and terminates in situation s':

$$Do(p, s, s') \stackrel{\text{def}}{=} \exists p'.Trans^*(p, s, p', s') \wedge Final(p', s'),$$

where $Trans^*$ is the reflexive transitive closure of $Trans$. An offline execution of program p from situation s is a sequence of actions $a_1, \ldots, a_n$ such that:

$$Axioms \models Do(p, s, do(a_n, \ldots, do(a_1, s))).$$

Observe that an offline executor is in fact similar to a planner that given a program, a starting situation, and a theory describing the domain, produces a sequence of actions to execute in the environment. In doing this, it has no access to sensing results, which will only be available at runtime. See [De Giacomo, Lespérance, and Levesque 2000] for more details.

In [De Giacomo and Levesque 1999], IndiGolog, an extension of ConGolog that deals with online executions with sensing is developed. The semantics defines an *online execution* of an IndiGolog program p starting from a history σ, as a sequence of *online configurations* $(p_0 = p, \sigma_0 = \sigma), \ldots, (p_n, \sigma_n)$ such that for $i = 0, \ldots, n-1$:

$$Axioms \cup \{Sensed[\sigma_i]\} \models$$
$$Trans(p_i, end[\sigma_i], p_{i+1}, end[\sigma_{i+1}]),$$

$$\sigma_{i+1} = \begin{cases} \sigma_i & \text{if } end[\sigma_{i+1}] = end[\sigma_i], \\ \sigma_i \cdot (a, x) & \text{if } end[\sigma_{i+1}] = do(a, end[\sigma_i]) \\ & \text{and } a \text{ returns } x. \end{cases}$$

An *online execution successfully terminates* if

$$Axioms \cup \{Sensed[\sigma_n]\} \models Final(p_n, end[\sigma_n]).$$

There is no automatic lookahead in IndiGolog. Instead, a *search* operator $\Sigma(p)$ is introduced to allow the programmer to specify when lookahead should be performed [De Giacomo and Levesque 1999; De Giacomo, Lespérance, Levesque, and Sardina 2002].

3 Time in our Framework

We presume the existence of an external "clock" that generates exogenous `tick` actions. All actions, including `tick`, are instantaneous however and have no duration. While an action may occur after a certain number of `tick`s t have occurred and so can be considered to have occurred at t, it takes no time for the action itself to be performed. We shall discuss how actions with duration may be handled later. No further assumptions are made about the clock, or the regularity of `tick` actions, etc.

Chronolog introduces the special functional fluent *Time* and exogenous action `tick`. In the initial situation, *Time* has the value 0 and is incremented by one each time an exogenous `tick` occurs. In other words, *Time* serves to keep track of the number of exogenous `tick`s that have occurred. Given these assumptions we can proceed as follows.

The initial state axiom for time:

$$Time(S_0) = 0$$

The preconditions for `tick` are:

$$Poss(\texttt{tick}, s) \equiv true$$

The successor state axiom for *Time* is:

$$Time(do(a, s)) = t \equiv$$
$$a = \texttt{tick} \land Time(s) = t - 1 \lor$$
$$a \neq \texttt{tick} \land Time(s) = t$$

Observe that these axioms conform to the requirements for Reiter's Action Theories and hence we can use regression and, more generally, all results on Basic Action Theories apply to them as well.

These axioms can be added to the standard Golog, ConGolog and IndiGolog axioms in a straightforward way as described in Section 2. The repercussions of doing so, we shall return to shortly. First however, it will be convenient to introduce some additional syntactic constructs that will simplify our presentation. These will be defined in terms of the notions introduced above.

4 Synchronization

Now we turn to *Chronolog* programs. These are simply IndiGolog programs that *access the time ticking* by testing the fluent $Time(s)$. However, here we make the fundamental assumption that no time ticks occur during the execution of a Chronolog program, unless it explicitly *synchronizes* with time. In practice we are requiring that Chronolog programs can be executed order of magnitudes quicker than time ticking.

Let us look at how synchronization can be performed in a Chronolog program. The most basic synchronization facility is to wait for the passing of a certain number of clock ticks. We introduce the construct $wait(t)$ where t is an integer number of `ticks` and define it as follows.

$$wait(t) \equiv \pi t'.[(Time = t')?; (Time = t' + t)?]$$

or, equivalently,

$$wait(t) \equiv \pi t'.[(t' = Time + t)?; (Time = t')?]$$

In other words, $wait(t)$ in a Chronolog program causes the program to wait for t exogenous `tick` actions to occur.

The definitions above make use of Chronolog's blocking wait construct ?. The first instance is used to determine the current time (the number of clock ticks that have occurred since the system was started in the initial situation S_0). The second instance causes the Golog program to block until an additional t exogenous `tick` actions have occurred. It can be easily seen that $wait(1)$ can be used to synchronize with the next clock tick. $wait(0)$ essentially does nothing.

Notice that here the assumption that Chronolog execution is much quicker than time ticking plays an important role. If this was not the case, and an exogenous tick occurs after the execution of the first blocking wait construct ? of a $wait(0)$ but before the second, the program will block forever and never terminate.

Other basic synchronization constructs that can be immediately expressed in Chronolog as well include the following.

wait until time t:
$$(Time = t)?$$

wait until after time t:
$$\pi u.[(u \geq t)?; \ (Time = u)]$$

wait no more than time t:
$$\pi u.[(u \leq t)?; \ (Time = u)]$$

Using these notions it turns out that we can express a surprisingly rich set of constructs. To illustrate the range of possibilities we list a number of useful temporal concepts that our approach can provide. To simplify notation we make the assumptions that unless otherwise stated $t \geq Time$ and also, $t \leq t'$.

perform action a at t:
$$\pi u.[(u = t - Time)?; wait(u); a]$$

perform action a before t:
$$\pi u.[(u \leq t - Time)?; wait(u); a]$$

perform action a after t:
$$\pi u.[(u \geq t - Time)?; wait(u); a]$$

In the following constructs we consider actions with duration. $start\_a$ indicates the initiation of the action and end_a indicates its termination. We consider actions with duration further in the discussion.

perform action a between t and t':
$$\pi u.[(u = t - Time)?; wait(u); start\_a; wait(t' - t); end\_a]$$

perform action a for time t:
$$start\_a; \ wait(t); \ end\_a$$

perform action a for at least time t:
$$\pi u.(u \geq t)?; \ start\_a; \ wait(u); \ end\_a$$

perform action a for no more than time t:
$$\pi u.(u \leq t)?; \ start\_a; \ wait(u); \ end\_a$$

perform action a at t for time t':
$$(Time = t)?; start\_a; \ wait(t'); \ end\_a$$

perform action a at t for at least time t':
$$(Time = t)?; \pi u.(u \geq t'); start\_a; \ wait(u); \ end\_a$$

perform action a at t for no more than time t':
$$(Time = t)?; \pi u.(u \leq t'); start\_a; \ wait(u); \ end\_a$$

perform action a start before t for time t':
$$\pi u.(u \leq t - Time)?; \ wait(u); \ start\_a; \ wait(t); \ end\_a$$

perform action a start before t for at least time t':
$$\pi u.(u \leq t - Time)?; \ wait(u); \ \pi u'.(u' \geq t')?; \ start\_a;$$
$$wait(u'); \ end\_a$$

perform action a start before t for no more than t':

$\pi u.(u \leq t - Time)?;\ wait(u);\ \pi u'.(u' \leq t')?;\ start\_a;$
$wait(u');\ end\_a$

perform action a start after t for time t':

$\pi u.(u \geq t - Time)?;\ wait(u);\ start\_a;\ wait(t);\ end\_a$

perform action a start after t for at least time t':

$\pi u.(u \geq t - Time)?;\ wait(u);\ \pi u'.(u' \geq t')?;\ start\_a;$
$wait(u');\ end\_a$

perform action a start after t for no more than time t':

$\pi u.(u \geq t - Time)?;\ wait(u);\ \pi u'.(u' \leq t')?;\ start\_a;$
$wait(u');\ end\_a$

start action a before t and complete it at t'

$\pi u.(u \leq t - Time)?;\ wait(u);\ start\_a;$
$\pi u'.(u' = t' - Time)?;\ wait(u');\ end\_a$

start action a before t and complete it before t'

$\pi u.(u \leq t - Time)?;\ wait(u);\ start\_a;$
$\pi u'.(u' \leq t' - Time \wedge u' \geq u)?;\ wait(u');\ end\_a$

start action a before t and complete it after t'

$\pi u.(u \leq t - Time)?;\ wait(u);\ start\_a;$
$\pi u'.(u' \geq t' - Time)?;\ wait(u');\ end\_a$

start action a after t and complete it before t'

$\pi u.(u \geq t - Time \wedge u \leq t' - Time)?;\ wait(u);\ start\_a;$
$\pi u'.(u' \leq t' - Time \wedge u' \geq u)?;\ wait(u');\ end\_a$

start action a after t and complete it after t'

$\pi u.(u \geq t - Time \wedge u \leq t' - Time)?;\ wait(u);\ start\_a;$
$\pi u'.(u' \geq t' - Time \wedge u' \geq u)?;\ wait(u');\ end\_a$

start action a at t and complete it after t'

$\pi u.(u = t - Time)?;\ wait(u);\ start\_a;$
$\pi u'.(u' \geq t' - Time)?;\ wait(u');\ end\_a$

4.1 Triggers and Exceptions

By using ConGolog's interrupt construct, we can arrange for certain actions to occur at a particular time or on a regular basis while executing a program (denoted by δ here).

trigger action a at time t:

$\langle (Time = t) \rightarrow a \rangle \,\rangle\!\rangle \delta$

trigger action a at every t ticks:

$\langle ((Time\%t) = 0) \rightarrow a \rangle \,\rangle\!\rangle \delta$

Where $\%$ is the modulus operator.

5 Chronolog execution

To understand the semantics of a Chronolog program, one has to keep in mind that a Chronolog program is not executable in isolation in general, but needs to be executed concurrently with a process emitting ticks in a continuous way. Also the ticking should be slow enough to allow for complete execution of the parts of a Chronolog program that do not include synchronization with time.

Observe that even if our program does not contain concurrency explicitly, in order to understand the semantics we need to resort to concurrency. Hence the semantics, even for the offline version, must be single-step. We use notions from ConGolog [De Giacomo, Lespérance, and Levesque 2000] to discuss the execution of Chronolog programs.

The next question to settle is how we model the ticking process. The ticking must go on forever in principle, but in practice it is sufficient to have enough ticks to complete the execution of the ConGolog program: i.e., a finite but unbounded number of ticks. Hence we can render in ConGolog the ticking process simply as:

$$tick^*$$

Finally we must formalize the execution of a ConGolog program δ concurrently with the ticking process, and since we said that the Chronolog program executes freely unless a synchronization with the ticking is required, we can formalize the execution of the program δ together with the ticking process $tick^*$ as:

$$\delta \,\rangle\!\rangle\, tick^*$$

In this way if the Chronolog program can execute (make a transition) it does so, and only when it synchronizes with a tick (i.e., waits for a given tick) does it stop and allow the ticking to go on.

Observe that when we turn to the online semantics, we must pay special attention to the search operator. Indeed, we can retain IndiGolog's normal search $search(\delta)$ that will not take time synchronization into account. But can also introduce a "timed" search for parts of Chronolog programs that require time synchronization. This last variant of search can be defined on the basis of the original one as follows:

$$timed\_search(\delta) \equiv search(\delta \,\rangle\!\rangle\, tick^*)$$

That is, we search as normal but simulate the occurrence of $tick$s as required by the program.

6 Examples

In the following examples we shall omit the specification of action precondition axioms and successor state axioms for fluents as they do not add anything to what we're trying to demonstrate here. Furthermore, action and fluent names will give an indication of their purpose.

One common use for temporal notions is to arrange the scheduling of regular actions. For instance, consider the cron daemon under Unix. It's purpose is to schedule the regular execution of certain programs for effecting system maintenance tasks. As we have seen above when discussing triggers, this can be easily implemented in Chronolog through the use of ConGolog's interrupt mechanism. Here is an example of how we might schedule regular maintenance tasks in Chronolog.

$$\langle\langle((Time\%3600) = 0) \rightarrow hourly\_task\rangle\,\rangle\rangle$$
$$\langle\langle((Time\%86400) = 0) \rightarrow daily\_task\rangle\,\rangle\rangle$$
$$\langle\langle((Time\%604800) = 0) \rightarrow weekly\_task\rangle\,\rangle\rangle$$
$$\langle\langle((Time\%2592000) = 0) \rightarrow monthly\_task\rangle\,\rangle\rangle$$
$$\langle true \rightarrow wait(1)\rangle$$

Note that in this example we would need to use action preconditions to ensure that an action is only executed once each time the interrupt's condition is true. However, an advantage of our execution model is that more than one rule may fire in between clock ticks. For example, when the time is a multiple of 86400, we can be sure that the *hourly_task* will also be performed since 86400 is divisible by 3600. If we wanted to ensure that only one interrupt was triggered between any two clock ticks (and assuming no other exogenous actions), we could add a $wait(1)$ after all other actions that are to be executed in the body of the interrupt. Another thing to be noted in this example is that we have not attempted to model Unix' multitasking abilities.

Of course, it is also possible to use preconditions that involve time. For example, suppose you wish to serve coffee to three people given when they will be available (in this case, between when they arrive and when they leave). A robot r can serve a person x provided that the person is around and that they are thirsty. Once x is served they are not thirsty anymore. The precondition can be specified in Chronolog as follows.

$$Poss(serve(x), s) \equiv$$
$$Time \le arrive(x, s) \wedge Time \ge leave(x, s) \wedge thirsty(x, s))$$

We omit the successor axiom, and just say that they model the fact that once a person has been served they are not thirsty anymore, while arrival and leaving time remain fixed (situation independent).

A possible initial situation could be described as follows:

$$arrive(giuseppe, S_0) = 10, leave(giuseppe, S_0) = 20$$
$$arrive(maurice, S_0) = 15, leave(maurice, S_0) = 30$$
$$arrive(bob, S_0) = 1, leave(bob, S_0) = 50$$
$$thirsty(giuseppe, S_0), thirsty(maurice, S_0),$$
$$thirsty(bob, S_0)$$

A Chronolog program to decide the ordering of the serving of those that want coffee is

$$timed_s earch($$
$$(\pi x.serve(x) \mid wait(1))^*; (\forall x.\neg thirsty(x))?$$
$$)$$

This program will attempt to find a sequence of actions (a plan) to serve everyone while they are around. More complex variants where we take into account the time the robots need to serve someone can also be easily modeled.

7 On-line Real-Time Interpreter

Implementing Chronolog requires a fairly straightforward modification to the IndiGolog interpreter [De Giacomo and Levesque 1999] by modifying the `indigo(E, H)` predicate and adding the `exec_inst(E, H, E1, H1)` predicate. The `wait_for_a_tick(H1, H2)` predicate blocks until an exogenous `tick` action occurs. The `exec_inst` predicate executes as many primitive (instantaneous) actions as possible. In other words, the idea is to execute as much of the program as possible (since primitive actions in the situation calculus are considered to be instantaneous) and then check for the occurrence of exogenous `ticks`.

```
indigo(E, H) :-
    exec_inst(E, H, E1, H1),
    (final(E1, H1) ->
        true;
        (wait_for_a_tick(H1, H2), indigo(E1, H2)).

exec_inst(E, H, E, H) :-
    final(E, H).

exec_inst(E, H, E2, H2) :-
    trans(E, H, E1, H1), !,
    exec_inst(E1, H1, E2, H2).

exec_inst(E, H, E, H) :-
    not trans(E, H, _ , _),
    not final(E, H).
```

Modifying IndiGolog in this way means that the program essentially behaves in the following manner (where δ is a Chronolog program) $\delta\rangle\rangle$`tick`* as desired. Notice that the underlying transition semantics, in terms of $trans$ and $final$, is on-line.

8 Conclusions

We have introduced a notion of discrete time into the situation calculus and the cognitive robotics language Golog. Our Golog variant, called Chronolog, achieves this through the introduction of a special functional fluent $Time$ to keep track of the passage of time and exogenous `tick` actions assumed to be generated by an external clock. All we require is a small number of additional axioms, yet the resulting framework allows us to express a surprising number of temporal notions. We have also described the execution of a Chronolog program and introduced the notion of a timed search.

Several extensions that follow in the same spirit are possible. It would be possible to introduce a construct $wait(action)$ that waits until action a occurs and implement it in a similar way to $wait(t)$. Similarly we could introduce $wait$; wait for an exogenous action (any exogenous action) to occur.

Our introduction of time into Golog would allow one to easily deal with actions with duration. These would need to be "clipped" along the lines that Pinto [Pinto 1997] deals with such actions, so that if we have an action a, it will be clipped into $start\_a$ and end_a. See also the discussion regarding ConGolog [De Giacomo, Lespérance, and Levesque 2000]. Note that it may not be easy to split all actions in this way.

We may also want to introduce constructs into Golog to explicitly allow for such actions. For example:

$Clipped(a, start\_a, end\_a)$

means that action a can be clipped and has $start\_a$ as the initiation of action a and end_a as the termination of action a. Note that end_a can be specified as exogenous or not. In some cases we do not want an exogenous termination but want to explicitly terminate the action at a particular time. It might be interesting to investigate the possibility of allowing end_a to be both exogenous and primitive so that you can do things like: "perform action a and if it has not terminated before a particular condition is true, (explicitly) terminate the action."

Precondition axioms and effect axioms (equivalently, successor state axioms for fluents) would need to be supplied for these actions. What is the relationship between the preconditions and effect/successor state axioms of the three actions? Here is one possibility:

$$Poss(a, s) \equiv Poss(start\_a, s)$$
$$Poss(end\_a, s) \equiv true$$
$$effects(do(a, s)) \equiv effects(do(end\_a, do(start\_a, s)))$$

Moreover, supposing that we wish to perform a instantaneously, that would be the same as the action sequence $start\_a;\ end\_a$.

Acknowledgments

This paper is dedicated to Hector J. Levesque; a colleague, a mentor, an inspiration and, above all, a friend.

This work was completed while the second author was on Special Studies Program leave at the Università di Roma "La Sapienza".

References

Davis, E. [1990]. *Representations of Commonsense Knowledge*. San Francisco: Morgan Kaufmann.

De Giacomo, G., Y. Lespérance, and H. J. Levesque [2000]. ConGolog, a concurrent programming language based on the situation calculus. *Artificial Intelligence 121*(1-2), 109–169.

De Giacomo, G., Y. Lespérance, H. J. Levesque, and S. Sardina [2002]. On the semantics of deliberation in indigolg – from theory to implementation. In *Proceedings of the 8th International Conference on the Principles of Knowledge Representation and Reasoning (KR'02)*, pp. 603–614. Morgan Kaufmann Publishers.

De Giacomo, G. and H. J. Levesque [1999]. An incremental interpreter for high-level programs with sensing. In H. J. Levesque and F. Pirri (Eds.), *Logical Foundations for Cognitive Agents*, pp. 86–102. Springer-Verlag.

Gans, G., G. Lakemeyer, M. Jarke, and T. Vits [2003]. Snet: A modeling and simulation environment for agent networks based on i* and congolog. In *Proceedings of the Conference on Advanced Information Systems Engineering (CAiSE'03)*.

Grosskreutz, H. and G. Lakemeyer [2003, March]. ccgolog – a logical language dealing with continuous change. *Logic Journal of the IGPL 11*(2), 179–221.

Levesque, H. J. and G. Lakemeyer [2000]. *The Logic of Knowledge Bases*. Cambridge, Massachusetts: MIT Press.

Levesque, H. J., R. Reiter, Y. Lespérance, F. Lin, and R. Scherl [1997]. GOLOG: A logic programming language for dynamic domains. *Journal of Logic Programming 31*, 59–84.

McCarthy, J. [1963]. Situations, actions, and causal laws. Technical report, Stanford University Artificial Intelligence Project.

McCarthy, J. and P. Hayes [1969]. Some philosophical problems from the standpoint of artificial intelligence. In D. Michie and B. Meltzer (Eds.), *Machine Intelligence 4*, pp. 463–502. University of Edinburgh Press.

McDermott, D. [1982]. A temporal logic for reasoning about processes and plans. *Cognitive Science 6*, 101–155.

Pinto, J. [1997]. Integrating discrete and continuous change in a logical framework. *Computational Intelligence 14*(1).

Plotkin, G. [1981]. A structural approach to operational semantics. Technical Report DAIMI-FN-19, Computer Science Dept., Aarhus University, Denmark.

Reiter, R. [1996]. Natural actions, concurrency and continuous change in the situation calculus. In *Proceedings of the Fifth International Conference on Principles of Knowledge Representation and Reasoning*, pp. 2–13.

Reiter, R. [1998]. Sequential, temporal golog. In *Proceedings of the Sixth International Conference on Principles of Knowledge Representation and Reasoning*.

Reiter, R. [2001]. *Knowledge in Action: Logical Foundations for Describing and Implementing Dynamical Systems*. MIT Press.

16

Model Counting: A New Strategy for Obtaining Good Bounds[1]

CARLA P. GOMES, ASHISH SABHARWAL, AND BART SELMAN

ABSTRACT. Model counting is the classical problem of computing the number of solutions of a given propositional formula. It vastly generalizes the NP-complete problem of propositional satisfiability, and hence is both highly useful and extremely expensive to solve in practice. We present a new approach to model counting that is based on adding a carefully chosen number of so-called streamlining constraints to the input formula in order to cut down the size of its solution space in a controlled manner. Each of the additional constraints is a randomly chosen XOR or parity constraint on the problem variables, represented either directly or in the standard CNF form. Inspired by a related yet quite different theoretical study of the properties of XOR constraints, we provide a formal proof that with high probability, the number of XOR constraints added in order to bring the formula to the boundary of being unsatisfiable determines with high precision its model count. Experimentally, we demonstrate that this approach can be used to obtain good bounds on the model counts for formulas that are far beyond the reach of exact counting methods. In fact, we obtain the first non-trivial solution counts for very hard, highly structured combinatorial problem instances. Note that unlike other counting techniques, such as Markov Chain Monte Carlo methods, we are able to provide high-confidence guarantees on the quality of the counts obtained.

Preface by Bart Selman

For our Festschrift contribution, I selected a paper describing some of our recent work. I believe the paper is a good illustration of the continued profound influence Hector has had on my research, ever since my Ph.D. studies. In this preface, I will briefly share some of my personal anecdotes and experiences with Hector as my Ph.D. advisor. I still remember vividly — a reference for those in the know — when Hector joined the Computer Science department at UofT. It was a much anticipated event with significant "buzz" because the department had managed to bring a true star home. The event was even covered in the general media; Hector came with a scientific "rock star" status that even impressed my non-CS friends. Having just finished my M.Sc. work, I was eager to sign up with Hector for my Ph.D. I got lucky: when I met with Hector he had room for one more student! Quite uncharacteristically for me, I scheduled a morning meeting; a fellow student found out that afternoon that Hector could not take on any more students. As they say, timing is everything. After my excitement of having signed up with Hector, I also immediately came to appreciate Hector's sense of humor. From looking over Hector's own Ph.D. thesis, I noticed an alarmingly large number of proofs and theorems. Having been trained as a physicist, I was not a great fan of formal proofs, so I casually suggested to Hector that,

---

[1]Based on a paper that appeared in the Proceedings of AAAI-06, the 21st National Conference on Artificial Intelligence, Boston, MA, July 2006.

surely, a good CS Ph.D. thesis did not necessarily need to contain a large number of formal proofs. Fortunately, Hector confirmed that "lots of proofs" was indeed not a requirement for a good CS thesis, but then, after a brief pause added, "of course, it *is* for a thesis of any of *my* students. . . " At least, I knew right then what my thesis needed to look like, and, over the years, I have come to appreciate that Hector's insistence on mathematical rigor and precision in research provided the best graduate training any student could have wished for.

The need to make precise verifiable claims about the properties of proposed formalisms and methods is now well-accepted among most AI researchers. However, it was Hector's research program in the late seventies and early eighties that provided the key impetus to raise the level of rigor in AI. It was great experience to be part of this movement within our field. It is not even so much the need for mathematical proofs per se, but rather the implied need to rigorously formulate claims and properties about AI systems and methods that helped move our field forward. Note that before these developments many approaches in AI were somewhat ad hoc and often ill-specified. This made it difficult to validate and replicate research results. In fact, the lack of precise formulations made it hard to even properly discuss certain research directions and claims.

Working as a student with Hector provided a great example of the apprenticeship style model for graduate studies. As his student, you were fully engaged in a joint project with him right from the start, and learned by doing; you learned what it meant to "ask the right question," to search for the right methodology, and to continually question your methods and insights. You also learned to work on the "big questions," which is exactly what makes a scientific career exciting. Any student of Hector will remember with great fondness the lively and vigorous scientific discussions that would take place in Hector's research meetings. Hector literally elevates the intellectual level of discourse of those around him in an infectious manner. I have come to realize that only a few top researchers consistently manage to do so.

The paper in this chapter highlights some of the spirit of Hector's research program and how it continues to influence and shape my own research agenda. The paper introduces a new approach to counting the number of models of a logical theory. The basic idea was inspired by work done in the theory community on so-called Unique SAT, a formal problem studied in computational complexity. I first encountered Unique SAT in work with Hector when we considered the question as to whether theories with an unique model are somehow easier to compute with than general theories. (The short answer is "most likely No.") Counting models is a #P complete problem, and for interesting theories, one generally resorts to Monte Carlo style methods to get estimates of model counts. Unfortunately, these approximate counts come without any guarantee on the quality of the approximation. In fact, it is not difficult to find in the literature examples of both over-counts and under-counts by many orders of magnitude. This a level of imprecision that Hector certainly would not settle for. In the paper, we introduce the first model counting technique that provides good lower-bounds on the true model count with concrete guarantees on the confidence of the lower-bound. This confidence can be boosted arbitrarily high with more computation. (The method, in principle, also provides a way for obtaining upper-bounds but the computation becomes quite inefficient.)

This work is also a nice example of research on the Boolean Satisfiability (SAT) problem. During the last two decades, we have seen an explosion of approaches in AI and computer science in general (in particular in the verification community) that translate various NP-complete computational tasks into large SAT instances. These instances can often be solved quite effectively with modern SAT solvers, thereby providing a solution to the original problem. In the early nineties, in joint work with Hector, Henry Kautz, and David Mitchell, we provided some of the original impetus for this line of research. It is good to see that this community is still going strong.

1 Introduction

Propositional model counting is the problem of computing the number of models for a given propositional formula, *i.e.*, the number of distinct variable assignments for which the formula evaluates to TRUE. This problem generalizes the well-known NP-complete problem of propositional satisfiability, SAT, which has played a key role in complexity theory as well as in automated reasoning. Indeed, computing the exact model count is a #P-complete problem, which means that it is no easier than solving a propositional formula with an unbounded number of "there exist" and "forall" quantifiers in its variables [Toda 1989]. For comparison, recall that SAT can be thought of as a propositional formula with exactly one level of "there exist" quantification.

Effective model counting procedures would open up a range of new applications. For example, various probabilistic inference problems, such as Bayesian net reasoning, can be effectively translated into model counting problems (cf. Roth 1996; Littman, Majercik, and Pitassi 2001; Darwiche 2005). Another application is in the study of hard combinatorial problems, such as combinatorial designs, where the number of solutions provides further insights into the problem. Even finding a single solution can be a challenge for such problems: counting the number of solutions is much harder yet. Using our counting method, we will obtain the first non-trivial lower bounds on the number of solutions of several complex combinatorial problems.

The earliest practical approach for counting models is based on an extension of systematic DPLL-based SAT solvers. By using appropriate multiplication factors and continuing the search after a single solution is found, Relsat [Bayardo Jr. and Pehoushek 2000] is able to provide incremental lower bounds on the model count as it proceeds, and finally computes the exact model count. Newer tools such as Cachet [Sang, Bacchus, Beame, Kautz, and Pitassi 2004] often improve upon this by using techniques such as component caching.

All exact counting methods, including Relsat and Cachet, essentially attack a #P-complete problem "head on" — by searching the raw combinatorial search space. Consequently, these algorithms often have difficulty scaling up to larger problem sizes. We should point out that problems with a higher solution count are not necessarily harder to determine the model count of. In fact, Relsat can compute the exact model count of highly under-constrained problems with many "don't care" variables and a lot of models by exploiting big clusters in the solution space. The model counting problem is instead much harder for more intricate combinatorial problems where the solutions are spread much more finely throughout the combinatorial space. We consider examples of such problems in our experiments.

A relatively new approach introduced by Wei and Selman [2005] is to use Markov Chain Monte Carlo sampling to compute an approximation of the exact model count. Their tool, ApproxCount, is able to solve several instances quite accurately, while scaling much better than both Relsat and Cachet as problem size increases. The drawback of ApproxCount is that one is not able to provide any hard guarantees on the model count it computes. To output a number close to the exact count, the counting strategy of Wei and Selman requires *uniform sampling from the set of solutions*, which is generally very difficult to achieve. Uniform sampling from the solution space is much harder than just generating a single solution. MCMC methods can provide theoretical convergence guarantees but only in the limit, generally after an exponential number of steps.

Interestingly, the inherent strength of most state-of-the-art SAT solvers comes actually from the ability to quickly narrow down to a certain portion of the search space the solver is designed to handle best. Such solvers therefore sample solutions in a highly non-uniform manner, making them seemingly ill-suited for model counting, unless one forces the solver to explore the full combinatorial space. An interesting question is whether there is a way around this apparent limitation of the use of state-of-the-art SAT solvers for model counting. Our key contribution is *a new method for model counting, which uses a state-of-the-art SAT solver "as is"*. It follows immediately that the more efficient the SAT solver used, the more powerful our counting strategy becomes.

Our approach is inspired by recent work on so-called "streamlining constraints" [Gomes and Sellmann 2004], in which additional, non-redundant constraints are added to the original problem to increase constraint propagation and to focus the search on a small part of the subspace, (hopefully) still containing solutions. This strategy was shown to be successful in solving very hard combinatorial design problems, with carefully created, domain-specific streamlining constraints. In contrast, in this work, we introduce a domain-independent streamlining technique.

Streamlining could potentially also be used to obtain an accurate estimate of the total solution count, if the solution density in the remaining (streamlined) search space was similar to that of the overall solution density. In that case, one could count the solutions remaining in the subspace and multiply this by the relative size of the subspace to the overall search space. Interestingly, there exist *generic* constraints that can be used to probabilistically streamline the search space sufficiently uniformly. These are so-called parity or XOR constraints, represented by logical XOR formulas.

The central idea of our approach is to repeatedly add randomly chosen XOR or parity constraints on the problem variables to the input formula and feed the result to a state-of-the-art SAT solver. We will discuss the technical details below, but at a very high level, our approach works as follows. Each random XOR constraint will cut the search space approximately in half. So, intuitively, if after the addition of s XOR's the formula is still satisfiable, the original formula must have at least on the order of 2^s models. More rigorously, we will show that if we perform t experiments of adding s random XOR constraints and our formula remains satisfiable in each case, then with probability at least $1 - 2^{-\alpha t}$, our original formula will have at least $2^{s-\alpha}$ satisfying assignments for any $\alpha \geq 1$. So, by repeated experiments or by weakening the claimed bound, one can arbitrarily boost the confidence in the lower bound count. We also give results for the upper bound, and formalize two variants of this approach as algorithms `MBound` and `Hybrid-MBound`.

Of course, the above argument might raise suspicion, because it does not depend at all on the how the solutions are distributed throughout the search space. This however is the surprising feature of the approach. We rely on the very special properties of random parity constraints, which in effect provide a good hash function, randomly dividing the solutions into two near-equal sets. Such constraints were first used by Valiant and Vazirani [1986] in a randomized reduction from SAT to so-called unique SAT. They provided evidence that unique SAT problems (formulas with at most one satisfying assignment) are essentially as hard as general SAT problems. In this work, we show a different, more positive, use of

XOR constraints, allowing us to count assignments of hard combinatorial problems.

In the theoretical section of the paper, we give much more specific details on the bounds obtained from XOR-streamlining. To demonstrate that this strategy is not just of theoretical interest, we also provide experimental results. Specifically, we applied our technique to three hard combinatorial problems, the Ramsey problem, the Schur problem, and the clique coloring problem. Our technique provides the first good lower bounds on solution counts for these problems. For both problems, we compute lower bounds with 99% confidence. For the Ramsey problems, we obtained a lower bound of $2^{64} \approx 1.8 \times 10^{19}$ solutions, in under two hours of computation. By comparison, `Relsat` found only 194,127 models in over 12 hours (`Cachet` does not provide partial counts and timed out, and `ApproxCount` does not converge to a solution). For the Schur problem, we obtained a lower bound of $2^{26} \approx 6.7 \times 10^{7}$ solutions, in under 5 hours of computation. `Relsat`, `Cachet`, and `ApproxCount` could not find any solutions in over 12 hours. For the clique coloring problem, we found over 10^{40} solutions in only a few minutes while other methods didn't finish in 12 hours.

In summary, we provide a new approach to model counting. Our method is unique in that it can use any state-of-the-art SAT solver without any modifications. Our approach uses randomized streamlining XOR constraints and gives concrete bounds with high probability (desired confidence level is under control of the user). Our experiments provide the first non-trivial lower bounds on solution counts for three highly combinatorial problems. In each case these bounds dramatically improve upon existing methods.

2 Preliminaries

For the rest of this paper, fix the set of propositional variables in all formulas to be V, $|V| = n$. A *variable assignment* $\sigma : V \to \{0,1\}$ is a function that assigns a value in $\{0,1\}$ to each variable in V. We may think of the value 0 as FALSE and the value 1 as TRUE. We will often abuse notation and write $\sigma(i)$ for valuations of entities $i \notin V$ when the intended meaning is either already defined or is clear from the context. In particular, $\sigma(1) = 1$ and $\sigma(0) = 0$. When $\sigma(i) = 1$, we say that σ *satisfies* i. For $x \in V$, $\neg x$ denotes the corresponding *negated* variable; $\sigma(\neg x) = 1 - \sigma(x)$.

Let F be a formula over the set V of variables and let σ be a variable assignment. $\sigma(F)$ denotes the valuation of F under σ. If σ satisfies F, i.e., $\sigma(F) = 1$, then σ is a *model, solution,* or *satisfying assignment* for F. The *model count* of F, denoted $MC(F)$, is the number of models of F. The *(propositional) model counting problem* is to compute $MC(F)$ given a (propositional) formula F.

Although our theoretical results hold for propositional formulas in general, we present our work in the context of formulas in the standard conjunctive normal form or CNF, on which our experimental results rely. A *clause* (also called a CNF *constraint*) C is a logical disjunction of a set of possibly negated variables; σ satisfies C if it satisfies at least one signed variable of C. A formula F is in the CNF *form* if it is a logical conjunction of a set of clauses; σ satisfies F if it satisfies all clauses of F.

An XOR *constraint* D over variables V is the logical "xor" or parity of a subset of $V \cup \{1\}$; σ satisfies D if it satisfies an *odd number* of elements in D. The value 1 allows us to express even parity. For instance, $D = \{a, b, c, 1\}$ represents the xor constraint $a \oplus b \oplus c \oplus 1$, which is TRUE when an even number of a, b, c are TRUE. Note that it suffices to use only positive variables. E.g., $\neg a \oplus b \oplus \neg c$ and $\neg a \oplus b$ are equivalent to $D = \{a, b, c\}$ and $D = \{a, b, 1\}$, respectively.

Our focus will be on formulas in the CNFXOR *form*, i.e., a logical conjunction of clauses and XOR constraints. Note that for every two complementary XOR constraints involving the same subset of V (e.g., $c \oplus d$ and $c \oplus d \oplus 1$), any assignment σ satisfies exactly one of them. This simple property will be crucial for our analysis.

Let $\mathbb{X}$ denote the set of all XOR constraints over V. For $1 \leq k \leq n = |V|$, let $\mathbb{X}^k$ denote the subset of $\mathbb{X}$ containing only those constraints that involve exactly k variables (and possibly the element 1). For simplicity, we will assume in the rest of the paper that n is even, and will be interested only in $1 \leq k \leq n/2$. This assumption can be avoided, for instance, by adding a dummy variable with a fixed TRUE/FALSE value or by defining $\mathbb{X}^{n/2} = \mathbb{X}^{\lfloor n/2 \rfloor} \cup \mathbb{X}^{\lceil n/2 \rceil}$ when n is odd.

3 A Simple Model Counting Algorithm

Our main algorithm is MBound, described below as Algorithm 1. It provides bounds on the model count of an input formula F by adding s random XOR constraints to F, solving the resulting CNFXOR formula using an arbitrary SAT solver as a subroutine, repeating this process t times, and looking at the observed satisfiable vs. unsatisfiable (sat-unsat) distribution of the t CNFXOR formulas. If this observed distribution is biased away from half-and-half, a bound on $MC(F)$ is reported. Specifically, for a slack factor α, if most instances are satisfiable, $2^{s-\alpha}$ is reported as a lower bound, and if most instances are unsatisfiable, $2^{s+\alpha}$ is reported as an upper bound. The correctness of these bounds depends on various factors and is quantified in the next section.

<u>Parameters</u>: MBound has five parameters: (1) the *size k* of the XORs used, (2) the *number s* of the XORs used, (3) the number t of repetitions or *trials*, (4) the *deviation $\delta \in (0, \frac{1}{2}]$* from the 50-50 sat-unsat ratio, and (5) the *precision slack $\alpha \geq 1$*. s and t will be the most crucial parameters, and we will often use $k \ll n/2$ for our experiments.

<u>Output</u>: MBound has three modes of termination: (1) return a lower bound on $MC(F)$, (2) return an upper bound on $MC(F)$, or (3) return "Failure" without reporting any bound whatsoever. A Failure happens when the observed sat-unsat ratio is less than δ away from 50-50.

We say that MBound *makes an error* if it reports an incorrect lower or upper bound on $MC(F)$, and that it *fails* if it reports Failure. We expect the probability of MBound making an error to go down as k, t, δ, and α increase.

MBound is based on the following central idea. As observed by Valiant and Vazirani [1986], the effect of adding a random XOR from $\mathbb{X}$ to F is to cut down the number of models by approximately a half. The same holds also when using $\mathbb{X}^{n/2}$ instead of $\mathbb{X}$. Somewhat surprisingly, this works *no matter how* solutions are structured in the space of all variable assignments. This is because constraints in $\mathbb{X}$ and $\mathbb{X}^{n/2}$ act as pairwise-independent uniform hash functions — uniformity allowing them to accept each model with probability exactly a half, and pairwise-independence making them oblivious to their acceptance or rejection of another model.

As the solution space is being iteratively cut down into halves, the number of XOR constraints one expects to add to F to bring it to the boundary of being unsatisfiable is roughly $s^* = \log_2 MC(F)$. This is the key property MBound uses to approximate $MC(F)$. Of course, this is only the expected behavior. To make the algorithm robust, we give a detailed probabilistic analysis to show that MBound is unlikely to deviate significantly from its expected behavior. This analysis forms the core of the technical contribution of this paper on the theoretical side and extends to provable guarantees for our second algorithm,

---

Algorithm 1: MBound

Params: $k, s, t, \delta, \alpha :$ k, s, t positive integers, $k \leq n/2$, $0 < \delta \leq \frac{1}{2}$, $\alpha \geq 1$

Input : A CNF formula F

Output: A lower or upper bound on $MC(F)$, or Failure

begin

$\quad$ | $\quad numSat \leftarrow 0$

$\quad$ | $\quad$ **for** $i \leftarrow 1$ **to** t **do**

$\quad$ | $\quad\quad$ | $\quad Q_s \leftarrow \{s \text{ random constraints from } \mathbb{X}^k\}$

$\quad$ | $\quad\quad$ | $\quad F_s^k \leftarrow F \cup Q_s$

$\quad$ | $\quad\quad$ | $\quad result \leftarrow \text{SATSolve}(F_s^k)$

$\quad$ | $\quad\quad$ | $\quad$ **if** $result = \text{TRUE}$ **then** $numSat \leftarrow numSat + 1$

$\quad$ | $\quad$ **if** $numSat \geq t \cdot (\frac{1}{2} + \delta)$ **then**

$\quad$ | $\quad\quad$ | $\quad$ **return** Lower bound: $MC(F) > 2^{s-\alpha}$

$\quad$ | $\quad$ **else if** $numSat \leq t \cdot (\frac{1}{2} - \delta)$ **then**

$\quad$ | $\quad\quad$ | $\quad$ **return** Upper bound: $MC(F) < 2^{s+\alpha}$

$\quad$ | $\quad$ **else return** Failure

end

Hybrid-MBound, as well.

In practice, adding XOR constraints from $\mathbb{X}^{n/2}$ (i.e., *large* XORs) can make the underlying SAT solver quite inefficient, and one is forced to consider the spaces $\mathbb{X}^k, k \ll n/2$, of *small* XORs. Such small XORs, however, do not necessarily act pairwise-independently on the solution space, resulting in an algorithm of not as high a quality as with large XORs. Interestingly, small XORs turn out to be sufficient to obtain guaranteed *lower* bounds. Moreover, as k increases, one provably approaches the truly pairwise-independent random behavior of large XORs. In fact, in several domains, fairly small XORs work quite well in practice. Further, our preliminary results suggest that this fact can be used to our advantage in order to provide key insights into the clustering structure of the solution space. This, however, is beyond the scope of this paper.

4 Analysis of Algorithm MBound

For $1 \leq k \leq n/2$, let Q_s^k denote a set of s XOR constraints chosen independently and uniformly at random from $\mathbb{X}^k$, and let F_s^k denote the random CNFXOR formula obtained by adding the constraints Q_s^k to F. In our analysis, the probability will be over the random choice of Q_s^k.

Our technical arguments have the following flavor. We show that when one adds too many XORs, even small ones, the resulting CNFXOR formula is quite unlikely to remain satisfiable. Further, if one conducts a *meta*-experiment by repeating this experiment with a fixed number of (too many) XORs several times, one is extremely unlikely to see many satisfiable formulas. By focusing on a meta-experiment that produces enough satisfiable formulas, one is thus able to probabilistically conclude that the number of XORs added was indeed *not* too many, providing a lower bound on the model count. A similar reasoning with too few XORs provides an upper bound, though with some complications arising from the lack of pairwise-independence of small XORs.

We will use standard bounds on the concentration of moments of a probability distribution, namely, Markov's inequality, Chebyshev's inequality, and the Chernoff bound (cf. Motwani and Raghavan 1994).

4.1 The Lower Bound

We begin by arguing using Markov's inequality that as more and more XOR constraints are added at random from $\mathbb{X}^k$ for *any* k to a formula F, its model count, $MC(F) = 2^{s^*}$, is quite likely to go down *at least* nearly as fast as expected. Later, in the upper bound section, we will argue that $MC(F)$ is likely to go down *at most* nearly as fast as expected when $k = n/2$. The precision slack α captures the notion of "nearly" as fast.

LEMMA 1. *For* $1 \leq k \leq n/2$ *and* $\alpha \geq 1$, $\Pr[MC(F_s^k) \geq MC(F)/2^{s-\alpha}] \leq 2^{-\alpha}$.

Proof. Let S be the set of satisfying assignments for F; $|S| = MC(F)$. For each $\sigma \in S$, let $Y_\sigma = \sigma(F_s^k)$ be a 0-1 random variable. The expected value of Y_σ is the probability that σ satisfies all of the s XOR constraints in Q_s^k. Recall that σ satisfies exactly half the constraints in $\mathbb{X}^k$. Since the s constraints in Q_s^k are chosen uniformly and independently from $\mathbb{X}^k$, the probability that σ satisfies all of them is 2^{-s}, implying $\mathbb{E}[Y_\sigma] = 2^{-s}$.

Let $Y = \sum_\sigma Y_\sigma$. The random variable Y equals $MC(F_s^k)$, and we have $\mathbb{E}[Y] = \mathbb{E}[\sum_\sigma Y_\sigma] = \sum_\sigma \mathbb{E}[Y_\sigma] = \sum_\sigma 2^{-s} = MC(F)/2^s$. It follows that $\Pr[MC(F_s^k) \geq MC(F)/2^{s-\alpha}] = \Pr[Y \geq 2^\alpha \mathbb{E}[Y]] \leq 2^{-\alpha}$ by Markov's inequality. □

COROLLARY 2. *For* $1 \leq k \leq n/2, \alpha \geq 1$, *and* $s \geq s^* + \alpha$, $\Pr[F_s^k$ *is satisfiable*$] \leq 2^{-\alpha}$.

Proof. Observe that $MC(F)/2^{s-\alpha} = 2^{s^*-(s-\alpha)} \leq 1$. Therefore, $\Pr[F_s^k$ is satisfiable$] = \Pr[MC(F_s^k) \geq 1] \leq \Pr[MC(F_s^k) \geq MC(F)/2^{s-\alpha}] \leq 2^{-\alpha}$. □

We use this result along with the Chernoff bound to show that after adding $s \geq s^* + \alpha$ XOR constraints from $\mathbb{X}_s^k$ *several times*, the fraction of instances F_s^k that are satisfiable is unlikely to be much more than $2^{-\alpha}$. Consequently, if one does see a significantly larger fraction of satisfiable instances than $2^{-\alpha}$ in this meta-experiment, then s is very likely to be less than $(s^* + \alpha)$, providing a high probability lower bound of $(s - \alpha)$ on s^*. Clearly, a weaker bound with a large α holds with a higher probability than a stronger one with a small α.

Formally, let $F_s^{k,(i)}, 1 \leq i \leq t$, denote t random formulas obtained by independently adding s random XOR constraints from $\mathbb{X}^k$ to F. The probability in what follows is on the collective choice of these formulas. In particular, whether $F_s^{k,(i)}$ is satisfiable or not is a random event of interest.

For convenience, we define the following quantity related to the Chernoff bound.

DEFINITION 3. *For any positive integer* $t, 0 < \delta \leq \frac{1}{2}$, *and* $\alpha \geq 1$, *let* $\beta = 2^\alpha(\frac{1}{2}+\delta)-1$ *and define*

$$p(t, \delta, \alpha) = \begin{cases} 2^{-\alpha t} & \text{if } \delta = \frac{1}{2} \\ \left(\frac{e^\beta}{(1+\beta)^{1+\beta}}\right)^{t/2^\alpha} & \text{if } \delta < \frac{1}{2}. \end{cases}$$

When $\delta < \frac{1}{2}$, we can simplify the above expression for $\alpha \in \{1, 2\}$ to get $p(t, \delta, 1) \leq e^{-0.77\delta^2 t}$ and $p(t, \delta, 2) \leq e^{-0.07(1+4\delta)^2 t} \leq e^{-1.12\delta^2 t - 0.07t}$. This demonstrates that $p(t, \delta, \alpha)$ decreases exponentially as δ and t increase, and is significantly smaller for larger α. This will help in qualitatively understanding the correctness guarantees we provide.

LEMMA 4. *For* $s \geq s^* + \alpha$ *and* $0 < \delta \leq \frac{1}{2}$, *the probability that at least a* $(\frac{1}{2} + \delta)$ *fraction of the* t *formulas* $F_s^{k,(i)}, 1 \leq i \leq t$, *is satisfiable is at most* $p(t, \delta, \alpha)$.

Proof. Let $Z_i, 1 \leq i \leq t$, be a random variable whose value is 1 if $F_s^{k,(i)}$ is satisfiable and 0 otherwise. Let $Z = \sum_i Z_i$ be the random variable that equals the number of satisfiable formulas $F_s^{k,(i)}, 1 \leq i \leq t$. Note that Z is the sum of *independent* 0-1 random variables, and, by Chernoff bound, is highly concentrated around its expected value.

By Corollary 2, $\Pr[F_s^{k,(i)}$ is satisfiable$] \leq 2^{-\alpha}$. Therefore, $\mathbb{E}[Z_i] = \Pr[Z_i = 1] = \Pr[F_s^{k,(i)}$ is satisfiable$] \leq 2^{-\alpha}$ so that $\mathbb{E}[Z] = \sum_i \mathbb{E}[Z_i] \leq t2^{-\alpha}$.

The probability that at least a ($\frac{1}{2} + \delta$) fraction of these t random formulas is satisfiable equals $\Pr[Z \geq t \cdot (\frac{1}{2} + \delta)]$. For $\delta = \frac{1}{2}$, this equals $\Pr[Z \geq t] = \Pr[Z_i = 1$ for all $i] \leq 2^{-\alpha t} = p(t, \frac{1}{2}, \alpha)$ because the random variables Z_i are independent. For $\delta < \frac{1}{2}$, $\Pr[Z \geq t \cdot (\frac{1}{2} + \delta)] \leq \Pr[Z \geq 2^{\alpha}(\frac{1}{2} + \delta)\mathbb{E}[Z]]$. Using the Chernoff bound, this probability is bounded above by $p(t, \delta, \alpha)$. □

Recall that $p(t, \delta, \alpha)$ is a quantity that decreases exponentially with t and δ.

THEOREM 5 (Main Result). *For $1 \leq k \leq n/2$, the lower bound of $2^{s-\alpha}$ reported by* MBound *with parameters $(k, s, t, \delta, \alpha)$ is correct with probability at least $1 - p(t, \delta, \alpha)$.*

Proof. Suppose MBound with parameters $(k, s, t, \delta, \alpha)$ makes a lower bound error on input F. Let r denote the final value of the variable *numSat* in that run of MBound, i.e., r is the number of satisfiable formulas amongst the t random formulas generated by the algorithm.

Since MBound reported a (wrong) lower bound, it must be that $r \geq t \cdot (\frac{1}{2} + \delta)$. Further, since it made an error, it must be that $\log_2 MC(F) \leq s - \alpha$ in reality, i.e., $s \geq s^* + \alpha$. In this case, by Lemma 4, the probability of the algorithm encountering $r \geq t \cdot (\frac{1}{2} + \delta)$ satisfiable formulas amongst the t random ones is bounded above by $p(t, \delta, \alpha)$. □

In particular, when *all* formulas encountered by MBound are satisfiable, we have a simpler correctness guarantee.

COROLLARY 6 (Simplified Result). *When* MBound *finds all t CNFXOR formulas to be satisfiable, the lower bound $2^{s-\alpha}$ it reports is correct with probability at least $1 - 2^{-\alpha t}$.*

EXAMPLE 7. Consider an experiment with $t = 20$ runs and δ set to $1/4$. Then $t \cdot (\frac{1}{2} + \delta) = 15, p(t, \delta, 1) \leq 0.34$, and $p(t, \delta, 2) \leq 0.002$. It follows that if we observe at least 15 out of 20 runs to be satisfiable, the lower bound of 2^{s-1} is correct with probability at least 0.66, and that of 2^{s-2} is correct with probability at least 0.998.

This shows that the probability of MBound making a lower-bound error indeed goes down *exponentially* as the number of trials t increase, making the algorithm quite robust for providing lower bounds on the model count. Further, when the precision slack α is increased from 1 to 2, the error probability decreases dramatically.

4.2 The Upper Bound: Ideal Case, Large XORs

For the upper bound, Markov's inequality is insufficient. We instead argue using Chebyshev's inequality that as more and more XOR constraints are added at random from $\mathbb{X}^{n/2}$ to a formula F, its model count is quite likely to go down *at most* nearly as fast as expected. In particular, F is quite unlikely to become unsatisfiable with too few XORs. Note that this result as such *does not hold* when small XORs are used.

LEMMA 8. *For $\alpha \geq 1$ and $s \leq s^*$, (A) $\Pr[MC(F_s^{n/2}) \leq MC(F)/2^{s+\alpha}] \leq 1/((1 - 2^{-\alpha})^2 2^{s^*-s})$ and (B) $\Pr[F_s^{n/2}$ is unsatisfiable$] \leq 1/2^{s^*-s}$.*

Proof. (proof-sketch, see Appendix for details) As in the proof of Lemma 1, let S be the set of satisfying assignments for F. For each $\sigma \in S$, let $Y_\sigma = \sigma(F_s^{n/2})$ be a 0-1 random variable. The expected value of Y_σ is, as before, given by $\mathbb{E}[Y_\sigma] = 2^{-s}$. Further, its variance is $\mathrm{Var}[Y_\sigma] = \mathbb{E}[Y_\sigma^2] - \mathbb{E}[Y_\sigma]^2$. Ignoring the negative term and using the fact that Y_σ is a 0-1 variable, $\mathrm{Var}[Y_\sigma] \leq \mathbb{E}[Y_\sigma]$.

A key thing to observe here is that the random variables Y_σ for various σ are *pairwise-independent* because of an argument that relies on both the fact that we are dealing with XOR constraints (as opposed to, say, CNF constraints) and that they are chosen uniformly from $\mathbb{X}^{n/2}$ rather than from $\mathbb{X}^k$ for $k < n$. The result then follows from a variance computation and an application of Chebyshev's inequality. □

COROLLARY 9. *For $\alpha \geq 1$ and $s \leq s^* - \alpha$, $\Pr[F_s^{n/2}$ is unsatisfiable$] \leq 2^{-\alpha}$.*

The meta-experiment providing an upper bound on s^* works essentially the same as Lemma 4 (see Appendix).

LEMMA 10. *For $s \leq s^* - \alpha$ and $0 < \delta \leq \frac{1}{2}$, the probability that at least a $(\frac{1}{2} + \delta)$ fraction of the t formulas $F_s^{n/2,(i)}, 1 \leq i \leq t$, is unsatisfiable is at most $p(t, \delta, \alpha)$.*

From this follows our main upper bound result for large XORs, in a fashion very similar to the proof of Theorem 5.

THEOREM 11. *An upper bound of $2^{s+\alpha}$ reported by MBound with parameters $(n/2, s, t, \delta, \alpha)$ is correct with probability at least $1 - p(t, \delta, \alpha)$.*

4.3 The Upper Bound: Practical Case, Small XORs

As mentioned earlier, computation with large XORs is quite expensive in practice. While the *correctness* of the lower bound reported by MBound does not depend on the length of the XORs, that of the upper bound does. When the solution space is highly structured, small XORs do not act pairwise-independently on various variable assignments.

After adding $s \leq s^* - 2$ small random XORs to F, while one still expects $2^{s^* - s}$ solutions of F to survive on average, the *variance* in this number could be quite high. In the worst case, one could have a tiny number of resulting formulas $F_s^{k,(i)}$ be satisfiable with an enormous number of solutions, and a huge number of such formulas be unsatisfiable. This would still maintain the expected number of surviving solutions, but would make the sat-unsat distribution of $F_s^{k,(i)}$ highly skewed towards unsat even with too few XORs.

On the positive side, as k increases and approaches $n/2$, one expects random XORs from $\mathbb{X}^k$ to act on different variable assignments in a more and more pairwise independent manner. This can be proved formally using a straightforward variance calculation. We omit the proof for lack of space.

PROPOSITION 12. *As the length k of XORs increases, the variance in the number of satisfiable CNFXOR formulas observed by MBound decreases.*

5 A Hybrid Model Counting Algorithm

By using a good SAT solver as a subroutine, MBound is already able to provide high quality lower bounds in practice. Its performance can be boosted even further by combining it with an exact model counting algorithm. Algorithm 2, Hybrid-MBound, that we present in this section does precisely this. The idea is to add randomly chosen XORs as before, but solve

---

Algorithm 2: `Hybrid-MBound`

Params: k, s, t, α : k, s, t positive integers, $k \leq n/2$, $\alpha \geq 1$
Mode : Conservative, Moderate, or Aggressive
Input : A CNF formula F
Output : A lower and an upper bound on $MC(F)$
begin

    $numSeq \leftarrow empty$
    for $i \leftarrow 1$ **to** t **do**
        $Q_s \leftarrow \{s \text{ random constraints from } \mathbb{X}^k\}$
        $F_s^{k,(i)} \leftarrow F \cup Q_s$
        $numModels \leftarrow \texttt{ExactModelCount} (F_s^{k,(i)})$
        $numSeq.\texttt{PushBack} (numModels)$
    $minModels \leftarrow \texttt{Min} (numSeq)$
    $maxModels \leftarrow \texttt{Max} (numSeq)$
    $avgModels \leftarrow \texttt{Average} (numSeq)$
    if $mode = $ Conservative **then**
        return Lower bound: $MC(F) > 2^{s-\alpha} \cdot minModels$,
                    Upper bound: $MC(F) < 2^{s+\alpha} \cdot maxModels$
    else if $mode = $ Moderate **then**
        return Lower bound: $MC(F) > 2^{s-\alpha} \cdot avgModels$,
                    Upper bound: $MC(F) < 2^{s+\alpha} \cdot avgModels$
    else   `/* mode = Aggressive */`
        return Lower bound: $MC(F) > 2^{s-\alpha} \cdot maxModels$,
                    Upper bound: $MC(F) < 2^{s+\alpha} \cdot minModels$

end

---

the resulting formula using an *exact model counting algorithm* as a subroutine instead of a SAT solver.

Two key factors make the hybrid approach work well in practice. First, by streamlining hard-to-count formulas with random XORs, we bring them within the reach of exact counting methods while maintaining the accuracy of the overall bound. Second, by throwing in relatively fewer XORs and relying on an exact counting method for the residual formula, the quality of the obtained bound is improved.

`Hybrid-MBound` has a subset of the parameters of `MBound` and has only a single mode of termination: return both a lower and an upper bound on $MC(F)$, within a factor of $2^{2\alpha+1}$ of each other. Once `Hybrid-MBound` generates exact counts for t streamlined formulas, the overall bound it reports can naturally be based on the minimum, average, or maximum of the t residual counts. We call this the *mode* of operation. The correctness of `Hybrid-MBound` is captured by the following results (see Appendix for proofs).

THEOREM 13 (Main Hybrid Result). *For* $1 \leq k \leq n/2$, *when* `Hybrid-MBound` *is run with parameters* (k, s, t, α), *then*
$$\Pr[\text{Conservative lower bound is correct}] \geq 1 - 2^{-\alpha t},$$
$$\Pr[\text{Moderate lower bound is correct}] \geq 1 - 2^{-\alpha}, \text{ and}$$
$$\Pr[\text{Aggressive lower bound is correct}] \geq (1 - 2^{-\alpha})^t.$$

THEOREM 14. *The lower and upper bounds reported by* `Hybrid-MBound` *with parame-*

ters $(n/2, s, t, \alpha)$ *are correct with probability at least* $1 - 1/2^{s^*-s-2}$ *independent of* t, α.

6   Experimental Results

To demonstrate the practical relevance of our approach, we considered the model counting problem for three hard combinatorial domains: the Ramsey problem, the Schur problem, and the clique coloring problem. All three problems deal with the question of the existence of certain intricate combinatorial objects.

In the *Ramsey domain*, one considers all possible two-colorings (red and blue) of the edges of the complete graph on n nodes. Ramsey showed that when n gets sufficiently large, certain structures of red or blue edges will be found in every coloring. In particular, $R(k, l)$ denotes the minimum value of n such that every coloring has at least one red clique of k vertices *or* one blue clique of l vertices. It is known that $R(4, 5) = 25$. So, if we consider a complete graph of 23 vertices, we are guaranteed to have solutions that neither contain a red clique of size 4 nor a blue clique of size 5. However, this is a highly non-trivial coloring problem. We can translate this problem into a SAT formula with 253 variables and 42,504 clauses. Finding a single solution using the fastest available SAT solver for this problem (MiniSAT) takes approximately 30 seconds on a 1 GHz machine. Our challenge is to find an interesting lower bound on the number of solutions.

We used MBound with small XOR constraints and MiniSAT as the subsolver. The parameters were chosen so as to make the streamlined formula easy for MiniSAT. In particular, when the XORs are too large, they do not provide enough constraint propagation for MiniSAT. (Note that the XORs are converted into CNF using auxiliary variables and clauses.) We found that we could streamline with 65 random XOR constraints with 4 to 5 variables in each constraint.

In the *Schur problem*, we are given the set of integers $\{1, 2, \ldots, n\}$. The question is whether this set can be divided into k sum-free subsets. A set S is sum-free if the sum of any pair of numbers in S is not in S. For each value of k, there is a certain value of n such that no partition into k sum-free subsets exists. For given values of n and k, we can again construct a SAT problem representing the formula. It is known that for $k = 5$ and $n = 140$, sum-free partitions still exist. The corresponding SAT problem has 700 variables and 51,600 clauses. This formula is already beyond the reach of current state-of-the-art SAT solvers (i.e., we could not find a single model in approximately 12 hours of CPU time). With random XOR streamlining with good parameters, we were able to solve the instance using MiniSAT by adding 27 XOR constraints containing an average of 9 variables.

In the *clique coloring problem* with parameters n, m, and k, the task is to construct a graph on n nodes such that it can be colored with m colors and also contains a clique of size k. This problem has interesting properties that make it very useful in proof complexity research on exponential lower bounds for powerful proof systems [Pudlák 1997]. We experimented with instances that had 600-750 variables and 20,000-35,000 clauses. When satisfiable, finding a single solution to these is quite easy. However, counting all solutions turns out to be extremely challenging even for approximate methods. By streamlining with XORs of size 6-8, we obtained lower bounds of 10^{40} and higher within minutes.

Table 1 summarizes the results obtained[1] on a 550 MHz 8 processor Intel Pentium III machine with 4 GB shared memory. All reported lower bounds are based on $t = 7$ and

---

[1] The code and complete data are available from the authors.

Table 1. MBound on problems beyond the reach of exact counting methods (99% confidence). Note that ApproxCount does not provide any guarantee on correctness or accuracy.

instance	MBound		Relsat		Cachet		ApproxCount	
	models	time	models	time	models	time	models	time
Ramsey-20-4-5	$\geq 1.2 \times 10^{30}$	$<$ 2 hr	$\geq 7.1 \times 10^{8}$	12 hr	—	12 hr	$\approx 1.8 \times 10^{19}$	4 hr
Ramsey-23-4-5	$\geq 1.8 \times 10^{19}$	$<$ 2 hr	$\geq 1.9 \times 10^{5}$	12 hr	—	12 hr	$\approx 7.7 \times 10^{12}$	5 hr
Schur-5-100	$\geq 2.8 \times 10^{14}$	$<$ 2 hr	—	12 hr	—	12 hr	$\approx 2.3 \times 10^{11}$	7 hr
Schur-5-140	$\geq 6.7 \times 10^{7}$	$<$ 5 hr	—	12 hr	—	12 hr	—	12 hr
fclqcol-18-14-11	$\geq 2.1 \times 10^{40}$	3 min	$\geq 2.8 \times 10^{26}$	12 hr	—	12 hr	—	12 hr
fclqcol-20-15-12	$\geq 2.2 \times 10^{46}$	9 min	$\geq 2.3 \times 10^{20}$	12 hr	—	12 hr	—	12 hr

$\alpha = 1$ so that Corollary 6 guarantees a $1 - 2^{-7} \geq 99\%$ confidence. The confidence level can, of course, be boosted by simply doing more XOR streamlined runs with MiniSAT (higher t) or reducing the reported bound by a factor of 2 (higher α).

We see that we obtain non-trivial lower bounds on the model counts. We also see that these counting problems are effectively beyond the reach of other state-of-the-art model counting approaches. Both Relsat and Cachet do not finish in 12 hours. Relsat gives very low partial counts while Cachet is not designed to report partial counts. ApproxCount computes a medium quality approximate count without any guarantees. *These results show that counting using randomized XOR streamlining provides a powerful new approach for obtaining lower bounds on model counts of hard combinatorial problems.*

Often lower bounds obtained from MBound can be made stronger with Hybrid-MBound. With Cachet as a subsolver and 30 XORs, we could boost the lower bound model count for Schur-5-140 to 1.8×10^{12}. Similarly, the lower bound for fclqcolor-18-14-11 was improved to 4.1×10^{45}. Note that the ability of Hybrid-MBound to boost MBound relies partly on exact model counting technology. The latter generally lags quite a bit behind SAT solvers, which are sufficient for MBound.

Finally, the results summarized in Table 2 confirm that, in practice, lower bounds reported by MBound and Hybrid-MBound can come quite close to exact counts even with very small XORs. The instance bitmax is a circuit synthesis problem and log_a is a logistics planning problem. The exact counts for these are obtained using Relsat. The last two instances are pigeonhole-type problems, with n pigeons and k holes, $n \leq k$. Relsat timed out on these instances after 12 hours. However, the exact count can be analytically computed to be $k!/(k-n)!$.

In all four cases, the average length of XORs used was less than 5% of the number of problem variables. Nevertheless, MBound came within a factor of 20 of the exact counts, which exceed 10^{11} and sometimes even 10^{28}. Hybrid-MBound typically requires shorter XORs than MBound in order to make the streamlined formula solvable using an exact counting method. Despite this, it further boosted the results to within a factor of 2 of the ex-

Table 2. Comparison of lower bounds with exact counts. The XOR size column reports the average.

prob.	num vars	exact count	XOR size	MBound	Hybrid-MBound
bitmax	252	21.0×10^{28}	9	$\geq 1.9 \times 10^{28}$	$\geq 9.2 \times 10^{28}$
log_a	1719	26.0×10^{15}	36	$\geq 1.1 \times 10^{15}$	$\geq 11.0 \times 10^{15}$
php.10.20	200	6.7×10^{11}	17	$\geq 1.3 \times 10^{11}$	$\geq 2.9 \times 10^{11}$
php.15.20	300	20.0×10^{15}	20	$\geq 1.1 \times 10^{15}$	—

act counts in three cases; in the fourth, even the streamlined problem remained hard for Relsat. Of course, one could yet further improve these bounds using somewhat larger XORs and additional computational resources.

7 Conclusion

Current techniques for model counting are based on either an exact counting paradigm or an approximate counting approach (e.g., MCMC methods), both of which have their limitations. We propose a third alternative based on randomized streamlining using XOR constraints. Our approach has two key strengths: it can generate model counts using any state-of-the-art SAT solver off-the-shelf, and it provides concrete bounds along with a high probability correctness guarantee that can be easily boosted by repetition. The model count lower bounds obtained using our algorithm MBound dramatically improve upon the results of existing techniques on three very difficult combinatorial problems. Our algorithm Hybrid-MBound combines the strength of existing exact counting methods with our XOR streamlining approach, boosting the results even further.

Acknowledgments: This research was supported by Intelligent Information Systems Institute (IISI), Cornell University (AFOSR grant F49620-01-1-0076) and DARPA (REAL grant FA8750-04-2-0216).

References

Bayardo Jr., R. J. and J. D. Pehoushek [2000, July]. Counting models using connected components. In *17th AAAI*, Austin, TX, pp. 157–162.

Darwiche, A. [2005, July]. The quest for efficient probabilistic inference. Invited Talk, IJCAI-05.

Gomes, C. P. and M. Sellmann [2004, October]. Streamlined constraint reasoning. In *10th CP*, Volume 3258 of *LNCS*, Toronto, Canada, pp. 274–289.

Littman, M. L., S. M. Majercik, and T. Pitassi [2001]. Stochastic boolean satisfiability. *J. Auto. Reasoning 27*(3), 251–296.

Motwani, R. and P. Raghavan [1994]. *Randomized Algorithms*. Cambridge University Press.

Pudlák, P. [1997, September]. Lower bounds for resolution & cutting plane proofs & monotone computations. *J. Symb. Logic 62*(3), 981–998.

Roth, D. [1996]. On the hardness of approximate reasoning. *J. AI* 82(1-2), 273–302.

Sang, T., F. Bacchus, P. Beame, H. A. Kautz, and T. Pitassi [2004, May). Combining component caching and clause learning for effective model counting. In *7th SAT*, Vancouver, B.C., Canada. Online Proceedings.

Toda, S. [1989]. On the computational power of PP and ⊕P. In *30th FOCS*, pp. 514–519.

Valiant, L. G. and V. V. Vazirani [1986]. NP is as easy as detecting unique solutions. *Theoretical Comput. Sci.* 47(3), 85–93.

Wei, W. and B. Selman [2005, June). A new approach to model counting. In *8th SAT*, Volume 3569 of *LNCS*, St. Andrews, U.K., pp. 324–339.

A Appendix: Proofs

Proof. (Lemma 8) As in the proof of Lemma 1, let S be the set of satisfying assignments for F; $|S| = MC(F)$. For each $\sigma \in S$, let Y_σ be the 0-1 random variable whose value is $\sigma(F_s^{n/2})$. The expected value of Y_σ is, as before, given by $\mathbb{E}[Y_\sigma] = 1/2^s$. Further, its variance is $\text{Var}[Y_\sigma] = \mathbb{E}[Y_\sigma^2] - \mathbb{E}[Y_\sigma]^2$. Ignoring the negative term and using the fact that Y_σ is a 0-1 variable, $\text{Var}[Y_\sigma] \leq \mathbb{E}[Y_\sigma]$.

A key point to observe here is that the random variables Y_σ for various σ are *pairwise-independent* because of the following argument which relies on both the fact that we are dealing with XOR constraints (as opposed to, say, CNF constraints) and that they are chosen uniformly from $\mathbb{X}^{n/2}$ rather than from $\mathbb{X}^k$ for $k < n$. Consider two random variables, Y_{σ_1} and Y_{σ_2}. The question we need to answer is: what can we say about the value of Y_{σ_2} when we know the value of Y_{σ_1}? The answer is determined by the behavior of σ_1 and σ_2 on *any single* XOR constraint D. We will show that for D chosen uniformly from $\mathbb{X}^{n/2}$, $\sigma_1(D)$ differs from $\sigma_2(D)$ with probability exactly a $1/2$. This will imply that knowing the value of Y_{σ_1} does not tell us anything about the value of Y_{σ_2}.

To this end, let $V' \subseteq V$ be the non-empty set of variables on which σ_1 and σ_2 differ, and D be an random XOR constraint from $\mathbb{X}^{n/2}$. Since D is an XOR constraint, the probability that $\sigma_1(D)$ differs from $\sigma_2(D)$ equals the probability that D involves an odd number of variables in V'. Since D is chosen at random from $\mathbb{X}^{n/2}$ and exactly half the constraints in $\mathbb{X}^{n/2}$ involve an odd number of variables in any subset of V, this probability is exactly $1/2$. As argued above, this implies that Y_{σ_1} and Y_{σ_2} are pairwise-independent.

Getting back to the main argument, let $Y = \sum_\sigma Y_\sigma$. The random variable Y equals the number of models of the formula $F_s^{n/2}$. The expected value of Y is given by $\mathbb{E}[Y] = \mathbb{E}[\sum_\sigma Y_\sigma] = \sum_\sigma \mathbb{E}[Y_\sigma] = \sum_\sigma 1/2^s = MC(F)/2^s$. Further, since Y is the sum of pairwise-independent random variables, its variance is given by $\text{Var}[Y] = \sum_\sigma \text{Var}[Y_\sigma]$. Recall that $\text{Var}[Y_\sigma] \leq \mathbb{E}[Y_\sigma]$. Hence, $\text{Var}[Y] \leq \sum_\sigma \mathbb{E}[Y_\sigma] = \mathbb{E}[Y]$.

For part (A), $\Pr[MC(F_s^{n/2}) \leq MC(F)/2^{s+\alpha}] = \Pr[Y \leq \mathbb{E}[Y]/2^\alpha] \leq \Pr[|Y - \mathbb{E}[Y]| \geq (1 - 2^{-\alpha})\mathbb{E}[Y]]$. By Chebyshev's inequality, this last expression is at most $\text{Var}[Y]/((1 - 2^{-\alpha})^2\mathbb{E}[Y]^2) \leq 1/((1 - 2^{-\alpha})^2\mathbb{E}[Y])$. For part (B), $\Pr[F_s^{n/2}$ is unsatisfiable$] = \Pr[Y = 0] = \Pr[|Y - \mathbb{E}[Y]| \geq \mathbb{E}[Y]]$. Again, by Chebyshev's inequality, this last expression is at most $\text{Var}[Y]/\mathbb{E}[Y]^2 \leq 1/\mathbb{E}[Y] \leq 1/2^{s^*-s}$. $\qquad\square$

Proof. (Lemma 10) Let $Z_i, 1 \leq i \leq t$, be a random variable whose value is 1 if $F_s^{n/2,(i)}$ is *un*satisfiable and 0 otherwise. Let $Z = \sum_i Z_i$ be the random variable that equals the number of unsatisfiable formulas $F_s^{n/2,(i)}, 1 \leq i \leq t$. Note that Z is the sum of *independent*

0-1 random variables, and, by Chernoff bound, is highly concentrated around its expected value.

By Corollary 9, $\Pr[F_s^{n/2,(i)}$ is unsatisfiable$] \leq 1/2^\alpha$. Therefore, $\mathbb{E}[Z_i] = \Pr[Z_i = 1] = \Pr[F_s^{n/2,(i)}$ is unsatisfiable$] \leq 1/2^\alpha$ so that $\mathbb{E}[Z] = \sum_i \mathbb{E}[Z_i] \leq t/2^\alpha$.

The probability that at least a $(1/2 + \delta)$ fraction of these t random formulas is unsatisfiable equals $\Pr[Z \geq t \cdot (1/2 + \delta)]$. For $\delta = 1/2$, this is $\Pr[Z \geq t] = \Pr[Z_i = 1$ for all $i] \leq 1/2^{\alpha t} = p(t, 1/2, \alpha)$ because the Z_i are independent. For $\delta < 1/2$, $\Pr[Z \geq t \cdot (1/2 + \delta)] \leq \Pr[Z \geq 2^\alpha (1/2 + \delta) \mathbb{E}[Z]]$. Using the Chernoff bound separately for $\alpha = 1, 2$ as in Lemma 4, this probability is bounded above by $p(t, \delta, \alpha)$. $\qquad\square$

Proof. (Theorem 13) Suppose HybridMC makes a lower bound error in Conservative mode, that is, $MC(F) \leq 2^{s-\alpha} \cdot minModels$. Equivalently, $minModels \geq MC(F)/2^{s-\alpha}$ so that *all* t of the random formulas $F_s^{k,(i)}, 1 \leq i \leq t$, have model counts at least $MC(F)/2^{s-\alpha}$. By Lemma 1, this happens for a single such formula with probability at most $2^{-\alpha}$, and, by independence, for all of them with probability at most $2^{-\alpha t}$.

From the opposite perspective, the lower bound reported in Aggressive mode is correct iff *all* t random formulas $F_s^{k,(i)}$ correctly have model counts less than $MC(F)/2^{s-\alpha}$. By Lemma 1, this happens for a single such formula with probability at least $1 - 2^{-\alpha}$, and, by independence, for all of them with probability at least $(1 - 2^{-\alpha})^t$.

The result for Moderate mode can be proved using a variant of Lemma 1. Specifically, for the ith trial, we have a 0-1 random variable Y_σ^i instead of Y_σ in the proof of Lemma 1, and now $Y = \sum_i \sum_\sigma Y_\sigma^i$ is the total number of solutions of all t random formulas. $\mathbb{E}[Y] = t \cdot MC(F)/2^s$ and the result follows from Markov's inequality as before. $\qquad\square$

Proof. (Theorem 14) We will focus on the hardest case, namely, $t = \alpha = 1$. Let $F_s^{n/2}$ denote the CNFXOR formula generated by HybridMC on input F, so that $minModels = maxModels = avgModels = MC(F_s^{n/2})$. Suppose the algorithm makes an error, that is, either $MC(F) \leq 2^{s-1} MC(F_s^{n/2})$ or $MC(F) \geq 2^{s+1} MC(F_s^{n/2})$. We will show that these events occur with a low probability.

The proof uses a slight generalization of Lemma 8. Assume the same setup as in the proof of that lemma, namely, a random variable Y that equals $MC(F_s^{n/2})$ and whose mean and variance are $\mathbb{E}[Y] = MC(F)/2^s$ and $\text{Var}[Y] \leq \mathbb{E}[Y]$, respectively. We then have $\Pr[MC(F) \geq 2^{s+1} MC(F_s^{n/2})] = \Pr[MC(F_s^{n/2}) \leq MC(F)/2^{s+1}] = \Pr[Y \leq \mathbb{E}[Y]/2]$. Similarly, $\Pr[MC(F) \leq 2^{s-1} MC(F_s^{n/2})] = \Pr[MC(F_s^{n/2}) \geq MC(F)/2^{s-1}] = \Pr[Y \geq 2\mathbb{E}[Y]]$.

Considering the two error modes of the algorithm, $\Pr[$HybridMC makes an error$] = \Pr[Y \leq \mathbb{E}[Y]/2$ or $Y \geq 2\mathbb{E}[Y]] \leq \Pr[|Y - \mathbb{E}[Y]| \geq \mathbb{E}[Y]/2]$. By Chebyshev's inequality, this probability is at most $\text{Var}[Y]/(\mathbb{E}[Y]/2)^2 \leq 4/\mathbb{E}[Y] = 1/2^{s^*-s-2}$. $\qquad\square$

17

The Typed Situation Calculus[1]

YILAN GU AND MIKHAIL SOUTCHANSKI

ABSTRACT. We propose a theory for reasoning about actions based on order-sorted predicate logic where one can consider an elaborate taxonomy of objects. We are interested in the projection problem: whether a statement is true after executing a sequence of actions. To solve it we design a regression operator takes advantage of well-sorted unification between terms. We show that answering projection queries in our logical theories is sound and complete wrt answering similar queries in Reiter's basic action theories. This proves correctness of our approach. Moreover, we demonstrate that our regression operator based on order-sorted logic can provide significant computational advantages in comparison to Reiter's regression operator.

Hector Levesque's research about structured representation of knowledge and about computationally tractable reasoning over this knowledge makes long-standing and important contributions to AI. He contributed to the idea that assertional knowledge (ABox) and terminological knowledge (TBox) should be represented separately, invented the terms ABox and TBox, and together with Ron Brachman founded a research area that later became known as Description Logics [Brachman and Levesque 1982; Brachman, Levesque, and Fikes 1983]. He summarized some of his ideas in a well-known lecture presented upon receipt of the 1985 Computers and Thought Award [Levesque 1986]. A few years later, he also made significant contributions to the situation calculus and to Cognitive Robotics [Levesque 1994; Levesque, Reiter, Lespérance, Lin, and Scherl 1997; Bacchus, Halpern, and Levesque 1999; Scherl and Levesque 2003; Levesque and Lakemeyer 2008; Lakemeyer and Levesque 2011]. As Hector's graduate students we had an opportunity to learn about his research first hand. Not surprisingly, our own research has been influenced by his ideas. The results of one of our research exercises are presented below. Our main departure from more traditional themes in Hector's work is in exploring how many-sorted (or more precisely, order-sorted) representations can be useful to provide an internal structure for objects, while Hector himself explored different forms for representing knowledge. Nevertheless, our research benefited from discussions with him, and from his comments.

1 Introduction

Starting from 1970s, many-sorted reasoning and taxonomies gained in popularity in deductive databases and automated reasoning. In particular, [McSkimin 1976] subdivides a domain into semantic categories, uses them to build a semantic category graph and argues that this semantic world would provide computational advantages in a query answering system. These ideas have been implemented in MRPPS (the Maryland Refutation Proof Procedure System) described in in [McSkimin and Minker 1977]. In deductive databases,

---

[1] A preliminary shorter version of this paper originally appeared in the Proceedings of the Ninth International Symposium on Logical Formalizations of Commonsense Reasoning, Toronto, Canada, June 1-3, 2009.

[Reiter 1977a; Reiter 1977b] also use boolean combination of monadic predicates to express taxonomies of types (each simple type is represented by a monadic predicate). Reiter develops a typed unification algorithm and argues that his approach is more suitable for real world databases than the approach in deductive question-answering research that deals with unrestricted first-order databases.

In his influential paper [Hayes 1971] titled "A Logic of Actions", P. Hayes proposed an outline of a logical theory for reasoning about actions based on many-sorted logic with equality. His paper inspired subsequent work on many-sorted logics in AI. In particular, A. Cohn [Cohn 1987; Cohn 1989] developed expressive many-sorted logic and reviewed all previous work in this area. Reasoning about action and change based on the situation calculus (SC) has been extensively developed in [Reiter 2001]. However, it considers a logical language with sorts for actions, situations and just one catch-all sort $Object$ for the rest that remains unelaborated. Surprisingly, even if the idea proposed by Hayes seems straightforward, there is still no formal study of logical and computational properties of a version of the SC with many related sorts for objects in the domain. Perhaps, this is because mathematical proofs of these properties are not straightforward although the intuition is. There are other action formalisms that permit sorted objects; such representations have also been used in planning formalisms, such as the standard PDDL language. However, to the best of our knowledge, none of the previous work investigated logical foundations and semantics of reasoning about actions over typed domains. We undertake this study and demonstrate that reasoning about actions with elaborated sorts has significant computational advantages in comparison to reasoning without them. In contrast to an approach to many-sorted reasoning [Schmidt 1938; Wang 1952; Herbrand 1971] where variables of different sorts range over unrelated universes, we consider a case when sorts are related to each other, so that one can construct an elaborated taxonomy. This is often convenient for representation of common-sense knowledge about a domain.

Generally speaking, we are usually interested in a comprehensive taxonomic structure for sorts, where sorts may inherit from each other and may have non-empty intersections. Note that even if both many-sorted logic and order-sorted logic can be translated to unsorted logic, using sorted ones can bring about significant computational advantages, for example in deduction, comparing to unsorted logic. This was a primary driving force for [Walther 1987; Cohn 1987]. Hence, we consider formulating the SC in an order-sorted logic to describe taxonomic information about objects. Based on the newly formulated language, we consider solving the projection problem via regression. We show that regression in the order-sorted SC can benefit from well-sorted unification. One can gain computational efficiency by terminating regression steps earlier when objects of incommensurable sorts are involved. However, before we can address issues of computational advantages, we have to investigae formally the relations between the proposed typed version and the situation calculus developed in [Reiter 2001], since the latter became de-facto standard. It turns out that proving soundness and completness of our new version with respect to Reiter's formulation is non-trivial matter.

2 Background

In general, order-sorted logic (OSL) [Oberschelp 1962; Walther 1987; Oberschelp 1990; Schmidt-Schauβ 1989; Bierle, Hedtstück, Pletat, Schmitt, and Siekmann 1992; Weidenbach 1996] restricts the domain of variables to subsets of the universe (i.e., *sorts*). Notation $x : Q$ means that variable x is of sort Q and $\mathbf{V}_Q$ is the set of variables of sort Q. For any

n, sort cross-product $Q_1 \times \cdots \times Q_n$ is abbreviated as $\vec{Q}_{1..n}$; term vector $t_1, \ldots, t_n$ is abbreviated as $\vec{t}_{1..n}$; variable vector $x_1, \ldots, x_n$ is abbreviated as $\vec{x}_{1..n}$; and, variable declaration sequence $x_1 : Q_1, \ldots, x_n : Q_n$ is abbreviated as $\vec{x}_{1..n} : \vec{Q}_{1..n}$.

A theory in OSL includes a set of declarations (called *sort theory*) to describe the hierarchical relationships among sorts and the restrictions on ranges of the arguments of predicates and functions. In particular, a sort theory $\mathbb{T}$ is a set of *term declarations* of the form $t : Q$ representing that term t is of sort Q, *subsort declarations* of the form $Q_1 \leq Q_2$ representing that sort Q_1 is a (direct) subsort of sort Q_2 (i.e., every object of sort Q_1 is also of sort Q_2), and *predicate declarations* of the form $P : \vec{Q}_{1..n}$ representing that the i-th argument of the n-ary predicate P is of sort Q_i for $i = 1..n$. In particular, when a logic has more than one sort symbol and there are no subsort declarations (any two sorts are disjoint), it is called *many-sorted* logic. A *function declaration* is a special term declaration where term t is a function with distinct variables as arguments: for each n-ary function f, the abbreviation of its function declaration is of the form $f : \vec{Q}_{1..n} \to Q$, where Q_i is the sort of the i-th argument of f and Q is the sort of the value of f. $c : Q$ is a special function declaration, representing that constant c is of sort Q. Arguments of equality "$=$" can be of any sort. Below, we consider a *finite simple* sort theory only, in which there are finitely many sorts and declarations, the term declarations are all function declarations, and for each function there is one and only one declaration (i.e., no polymorphism is allowed).

For any sort theory $\mathbb{T}$, subsort relation $\leq_{\mathbb{T}}$ is a partial ordering defined by the reflexive and transitive closure of the subsort declarations. Then, following the standard terminology of lattice theory, if each pair of sort symbols in $\mathbb{T}$ has greatest lower bound (g.l.b.), we say that *the sort hierarchy of $\mathbb{T}$ is a meet semi-lattice* [Walther 1987]. Moreover, a *well-sorted term* (wrt $\mathbb{T}$) is either a sorted variable, or a constant declared in $\mathbb{T}$, or a functional term $f(\vec{t}_{1..n})$, in which each t_i is well-sorted and the sort of t_i is a subsort of Q_i, given that $f : \vec{Q}_{1..n} \to Q$ is in $\mathbb{T}$. A *well-sorted atom* (wrt $\mathbb{T}$) is an atom $P(\vec{t}_{1..n})$ (can be $t_1 = t_2$), where each t_i is a well-sorted term of sort Q'_i, and $Q'_i \leq_{\mathbb{T}} Q_i$, given that $P : \vec{Q}_{1..n}$ is in $\mathbb{T}$. A *well-sorted formula* (wrt $\mathbb{T}$) is a formula in which all terms (including variables) and atoms are well-sorted. Any term or formula that is not well-sorted is called *ill-sorted*. A formal definition of *well-sorted formulas* is given in [Bierle, Hedtstück, Pletat, Schmitt, and Siekmann 1992]. A *well-sorted substitution* (wrt $\mathbb{T}$) is a substitution ρ s.t. for any variable $x : Q$, ρx (the result of applying ρ to x) is a well-sorted term and its sort is a (non-empty) subsort of Q. Given any set $E = \{\langle t_{1,1}, t_{1,2} \rangle, \ldots, \langle t_{n,1}, t_{n,2} \rangle\}$, where each $t_{i,j}$ ($i = 1..n, j = 1..2$) is a well-sorted term, a *well-sorted most general unifier* (well-sorted mgu) of E is a well-sorted substitution that is an mgu of E. It is important that in comparison to an mgu in unsorted logic (i.e., predicate logic without sorts), an mgu in OSL can include new weakened variables of sorts which are subsorts of the sorts of unified terms. For example, assume that $E = \{\langle x, y \rangle\}$, $x \in \mathbf{V}_{Q_1}$, $y \in \mathbf{V}_{Q_2}$ and the g.l.b. of $\{Q_1, Q_2\}$ is a non-empty sort Q_3. Then, $\mu = [x/z, y/z]$ (x is substituted by z, y is substituted by z) for some new variable $z \in \mathbf{V}_{Q_3}$ is a well-sorted mgu of E. A well-sorted mgu neither always exists nor it is unique. However, it is proved that the well-sorted mgu of unifiable sorted terms is unique up to variable renaming when the sort hierarchy of $\mathbb{T}$ is a meet semi-lattice [Walther 1987].

The semantics of OSL is defined similar to unsorted logic. Note that the definition of interpretations for well-sorted terms and formulas is similar to unsorted logic, but the semantics is not defined for ill-sorted terms and formulas. For any well-sorted formula ϕ, a $\mathbb{T}$-interpretation $\mathbb{I} = \langle \mathcal{M}, I \rangle$ is a tuple with a structure $\mathcal{M}$ and an assignment I from the set of free variables to the universe of $\mathcal{M}$, s.t. it satisfies the following conditions: (1) For each sort Q, $Q^{\mathbb{I}}$ is a subset of the whole universe $\mathbf{U}$. In particular, $\top^{\mathbb{I}} = \mathbf{U}$, $\bot^{\mathbb{I}} = \emptyset$, and $Q_1^{\mathbb{I}} \subseteq Q_2^{\mathbb{I}}$ for any $Q_1 \leq_{\mathbb{T}} Q_2$. (2) For any predicate declaration $P : \vec{Q}_{1..n}$, $P^{\mathbb{I}} \subseteq Q_1^{\mathbb{I}} \times \cdots \times Q_n^{\mathbb{I}}$ is a relation in $\mathcal{M}$. (3) For any function declaration $f : \vec{Q}_{1..n} \to Q$, $f^{\mathbb{I}} : Q_1^{\mathbb{I}} \times \cdots \times Q_n^{\mathbb{I}} \to Q^{\mathbb{I}}$ is a function in $\mathcal{M}$. (4) $x^{\mathbb{I}} = I(x)$ is in $Q^{\mathbb{I}}$ for any variable $x \in \mathbf{V}_Q$, $c^{\mathbb{I}} \in Q^{\mathbb{I}}$ for any constant declaration $c : Q$, and $(f(\vec{t}_{1..n}))^{\mathbb{I}} \stackrel{def}{=} f^{\mathbb{I}}(t_1^{\mathbb{I}}, \ldots, t_n^{\mathbb{I}})$ for any well-sorted term $f(\vec{t}_{1..n})$. $\mathbb{I}$ is not defined for ill-sorted terms and formulas. (5) If $\mathbb{T}$ includes a declaration for equality symbol "=", then $=^{\mathbb{I}}$ must be defined as a set $\{(d, d) \mid d \in \mathbf{U}\}$. For any sort theory $\mathbb{T}$ and a well-sorted formula ϕ, a structure $\mathcal{M}$ is a $\mathbb{T}$-*model* of ϕ, written as $\mathcal{M} \models_{\mathbb{T}}^{os} \phi$ iff for every $\mathbb{T}$-interpretation $\mathbb{I} = \langle \mathcal{M}, I \rangle$, $\mathbb{I}$ satisfies ϕ. In particular, when ϕ is a sentence, this does not depend on any variable assignment and $\mathbb{I} = \mathcal{M}$. Moreover, the notion a $\mathbb{T}$-interpretation $\mathbb{I} = \langle \mathcal{M}, I \rangle$ satisfies ϕ, written as $\mathbb{I} \models_{\mathbb{T}}^{os} \phi$, is defined by structural induction on formulas as follows. (a) $\mathbb{I} \models_{\mathbb{T}}^{os} P(\vec{t}_{1..n})$ iff $(t_1^{\mathbb{I}}, \ldots, t_n^{\mathbb{I}}) \in P^{\mathbb{I}}$. (b) $\mathbb{I} \models_{\mathbb{T}}^{os} \neg \phi$ iff $\mathbb{I} \models_{\mathbb{T}}^{os} \phi$ does not hold. (c) $\mathbb{I} \models_{\mathbb{T}}^{os} \phi_1 \wedge \phi_2$ iff $\mathbb{I} \models_{\mathbb{T}}^{os} \phi_1$ and $\mathbb{I} \models_{\mathbb{T}}^{os} \phi_2$. (d) $\mathbb{I} \models_{\mathbb{T}}^{os} \phi_1 \vee \phi_2$ iff $\mathbb{I} \models_{\mathbb{T}}^{os} \phi_1$ or $\mathbb{I} \models_{\mathbb{T}}^{os} \phi_2$. (e) $\mathbb{I} \models_{\mathbb{T}}^{os} \phi_1 \supset \phi_2$ iff $\mathbb{I} \models_{\mathbb{T}}^{os} \neg\phi_1 \vee \phi_2$. (f) $\mathbb{I} \models_{\mathbb{T}}^{os} \forall x : Q.\phi$ iff for every $d \in Q^{\mathbb{I}}$, $\mathbb{I} \models_{\mathbb{T}}^{os} \phi[x/d]$, where $\phi[x/o]$ represent the formula obtained by substituting x with o. (g) $\mathbb{I} \models_{\mathbb{T}}^{os} \exists x : Q.\phi$ iff there is some $d \in Q^{\mathbb{I}}$ s.t. $\mathbb{I} \models_{\mathbb{T}}^{os} \phi[x/d]$. Given a sort theory $\mathbb{T}$ as the background, a theory Φ including well-sorted sentences only satisfies a well-sorted sentence ϕ, written as $\Phi \models_{\mathbb{T}}^{os} \phi$, iff every model of Φ is a model of ϕ.

Due to the space limitations, we do not introduce the SC. Details can be found in [Reiter 2001] and we refer to this language as Reiter's SC $\mathcal{L}_{sc}$ below. Also note that we use $\models_{\mathbb{T}}^{os}$ to represent the logical entailment wrt a sort theory $\mathbb{T}$ in order-sorted logic, $\models^{ms}$ to represent the logical entailment in Reiter's SC (a many-sorted logic), and $\models^{fo}$ to represent the logical entailment in unsorted predicate logic.

3   An Order-Sorted Situation Calculus

Here, we consider a modified SC based on order-sorted logic, called *order-sorted SC* and denoted as $\mathcal{L}^{OS}$ below. $\mathcal{L}^{OS}$ includes a set of sorts $\mathbf{Sort} = \mathbf{Sort}_{obj} \cup \{\top, \bot, Act, Sit\}$, where $\top$ represents the whole universe, $\bot$ is the empty sort, Act is the sort for all actions, Sit is the sort for all situations, and $\mathbf{Sort}_{obj}$ is a set of sub-sorts of $Object$ including sort $Object$ itself. We assume that for every sort (except $\bot$) there is at least one ground term (constant) of this sort to avoid the problem with "empty sorts" [Goguen and Meseguer 1987]. Moreover, the number of individual variable symbols of each sort in $\mathbf{Sort}$ is infinitely countable. For the sake of simplicity, we do not consider functional fluents here.

Here, we will consider dynamical systems that can be described using *order-sorted basic action theories* (order-sorted BATs). An order-sorted BAT $\mathcal{D} = (\mathbb{T}, \mathbf{D})$ includes the following two parts of theories.

- $\mathbb{T}$ is a sort theory based on a finite set of sorts $\mathbf{Q}_{\mathcal{D}}$ s.t. $\mathbf{Q}_{\mathcal{D}} \subseteq \mathbf{Sort}$ and $\{\bot, \top, Object, Act, Sit\} \subseteq \mathbf{Q}_{\mathcal{D}}$. Moreover, the sort theory includes the following declarations for finitely many predicates and functions:

  1. Subsort declarations of the form $Q_1 \leq Q_2$ for $Q_1, Q_2 \in \mathbf{Q}_{\mathcal{D}} - \{\top, Act, Sit\}$, and subsort declarations: $Object \leq \top$, $Act \leq \top$, $Sit \leq \top$, $\bot \leq Act$, $\bot \leq Sit$. Here, we only consider those sort theories whose sort hierarchies are meet semi-lattices.

  2. One and only one predicate declaration of the form $F : \vec{Q}_{1..n}$ for each n-ary relational fluent F in the system, where $Q_i \leq_{\mathbb{T}} Object$ and $Q_i \neq \bot$ for $i = 1..(n-1)$, and Q_n is Sit.

3. One and only one predicate declaration for the special predicate $Poss$, that is, $Poss$: $Act \times Sit$.

4. One and only one predicate declaration of the form $P : \vec{Q}_{1..n}$ for each n-ary situation independent predicate P in the system, where $Q_i \leq_\mathbb{T} Object$ and $Q_i \neq \perp$ for $i = 1..n$.

5. A special declaration for equality symbol $=: \top \times \top$.

6. One and only one function declaration of the form $A : \vec{Q}_{1..n} \to Act$ for each n-ary action function A in the system, where $Q_i \leq_\mathbb{T} Object$ and $Q_i \neq \perp$ for $i = 1..n$. Note that, when $n = 0$, the declaration is $A : Act$ for constant action function A.

7. One and only one function declaration of the form $f : \vec{Q}_{1..n} \to Q_{n+1}$ for each n-ary ($n \geq 0$) situation independent function f (other than action functions), where each $Q_i \leq_\mathbb{T} Object$ and $Q_i \neq \perp$ for each $i = 1..(n+1)$. When $n = 0$, the function declaration is of the form $c : Q$ for constant c of sort Q.

8. One and only one function declaration $do : Act \times Sit \to Sit$, and $S_0 : Sit$ for the initial situation S_0.

• **D** is a set of axioms represented using well-sorted sentences wrt $\mathbb{T}$, which includes the following subsets of axioms.

1. Foundational axioms Σ for situations, which are the same as those in [Reiter 2001].

2. A set $\mathcal{D}_{una}$ of unique name axioms for actions: for any two distinct action function symbols A and B with declarations $A : \vec{Q}_{1..n} \to Act$ and $B : \vec{Q}'_{1..m} \to Act$, we have
$$(\forall \vec{x}_{1..n} : \vec{Q}_{1..n}, \vec{y}_{1..m} : \vec{Q}'_{1..m}). A(\vec{x}_{1..n}) \neq B(\vec{y}_{1..m}).$$
Moreover, for each action function symbol A, we have
$$(\forall \vec{x}_{1..n} : \vec{Q}_{1..n}, \vec{y}_{1..n} : \vec{Q}_{1..n}). A(\vec{x}_{1..n}) = A(\vec{y}_{1..n}) \supset \bigwedge_{i=1}^{n} x_i = y_i.$$

3. The initial theory $\mathcal{D}_{S_0}$, which includes well-sorted (first-order) sentences that are uniform in S_0.

4. A set $\mathcal{D}_{ap}$ of precondition axioms for actions represented using well-sorted formulas: for each action symbol A, whose sort declaration is $A : \vec{Q}_{1..n} \to Act$, its precondition axiom is
$$(\forall \vec{x}_{1..n} : \vec{Q}_{1..n}, s : Sit). Poss(A(\vec{x}_{1..n}), s) \equiv \Pi_A(\vec{x}_{1..n}, s), \tag{1}$$
where $\Pi_A(\vec{x}_{1..n}, s)$ is a well-sorted formula uniform in s, whose free variables are at most among $\vec{x}_{1..n}$ and s.

5. A set $\mathcal{D}_{ss}$ of successor state axioms (SSAs) for fluents represented using well-sorted formulas: for each fluent F with declaration $F : \vec{Q}_{1..n} \times Sit$, its SSA is of the form
$$(\forall \vec{x}_{1..n} : \vec{Q}_{1..n}, a : Act, s : Sit). F(\vec{x}_{1..n}, do(a, s)) \equiv \phi_F(\vec{x}_{1..n}, a, s), \tag{2}$$
where $\phi_F(\vec{x}_{1..n}, a, s)$ is a well-sorted formula uniform in s, whose free variables are at most among $\vec{x}_{1..n}$ and a, s.

Here is a simple example of an order-sorted BAT.

EXAMPLE 1 (Transport Logistics) We present an order-sorted BAT $\mathcal{D}$ of a simplified example of logistics. $\mathbb{T}$ includes the following subsort declarations:

$MovObj \leq Object,\ \bot \leq City,\ \bot \leq Box,\ \bot \leq Truck,$

$Truck \leq MovObj,\ City \leq Object,\ Box \leq MovObj,$

where $MovObj$ is the sort of movable objects, and other sorts are self-explanatory. The predicate declarations are $\quad InCity : MovObj \times City \times Sit,\ On : Box \times Truck \times Sit$ for the fluents $InCity(o, l, s)$ and $On(o, t, s)$. An example of function declaration can be $twinCity : City \rightarrow City$. The function declarations for actions $load(b, t)$, $unload(b, t)$ and $drive(t, c_1, c_2)$ are obvious. For instance, $\quad drive : Truck \times City \times City \rightarrow Act$. Besides $S_0 : Sit$, the constant declarations may include:

$B_1 : Box,\qquad B_2 : Box,\qquad\quad T_1 : Truck,$

$T_2 : Truck,\quad Toronto : City,\quad Boston : City.$

Axioms in $\mathcal{D}_{S_0}$ can be:

$\exists x : Box.\ InCity(x, Boston, S_0),$

$(\forall x : Box, t : Truck).\ \neg On(x, t, S_0),$

$InCity(T_1, Boston, S_0) \vee InCity(T_2, Boston, S_0).$

As an example, the precondition axiom for $load$ is:

$(\forall x : Box, t : Truck, s : Sit).\ Poss(load(x, t), s) \equiv$
$\qquad\qquad \neg On(x, t, s) \wedge \exists y : City.InCity(x, y, s) \wedge InCity(t, y, s),$

and the preconditions for $unload$ and $drive$ are obvious. As an example, the SSA of fluent $InCity$ is:

$(\forall d : MovObj, c : City, a : Act, s : Sit).InCity(d, c, do(a, s)) \equiv$
$\qquad (\exists t : Truck, c_1 : City).a = drive(t, c_1, c) \wedge$
$\qquad (d = t \vee \exists b : Box.b = d \wedge On(b, t, s))) \vee$
$\qquad InCity(d, c, s) \wedge \neg(\exists t : Truck, c_1 : City.a = drive(t, c, c_1) \wedge$
$\qquad (d = t \vee \exists b : Box.b = d \wedge On(b, t, s))),$

and the SSA of fluent On is obvious.

4 Order-Sorted Regression and Reasoning

We now consider the central reasoning mechanism in the order-sorted SC. The definition of a regressable formula of $\mathcal{L}^{OS}$ is the same as the definition of a regressable formula of $\mathcal{L}_{sc}$ except that instead of being stated for a formula in $\mathcal{L}_{sc}$, it is formulated for a well-sorted formula in $\mathcal{L}^{OS}$.

A formula W of $\mathcal{L}^{OS}$ is *regressable* (wrt an order-sorted BAT $\mathcal{D}$) iff (1) W is a well-sorted first-order formula wrt $\mathbb{T}$; (2) every term of sort Sit in W starts from S_0 and has the syntactic form $do([\alpha_1, \cdots, \alpha_n], S_0)$, where each α_i is of sort Act; (3) for every atom of the form $Poss(\alpha, \sigma)$ in W, α has the syntactic form $A(\vec{t}_{1..n})$ for some n-ary action function symbol A; and (4) W does not quantify over situations, and does not mention the relation symbols "$\sqsubset$" or "$=$" between terms of sort Sit. A *query* is a regressable sentence.

EXAMPLE 2 Consider the BAT $\mathcal{D}$ from Example 1. Let W be

$$\exists d : Box.\ d = Boston \wedge On(d, T_1, do(load(B_1, T_1), S_0)).$$

W is a (well-sorted) regressable sentence (wrt $\mathcal{D}$); while

$$On(Boston, T_1, do(load(B_1, T_1), S_0))$$

is ill-sorted and therefore is not regressable.

The regression operator $\mathcal{R}^{os}$ in $\mathcal{L}^{OS}$ is defined recursively similar to the regression operator in [Reiter 2001]. Moreover, we take advantages of the sort theory during regression: when there is no well-sorted mgu for equalities between terms that occur in a conjunctive sub-formula of a query, this sub-formula is logically equivalent to false and it should not be regressed any further. We will see that this key idea helps eliminate useless sub-trees of a regression tree.

DEFINITION 3 Consider a regressable formula W in $\mathcal{L}^{OS}$ with respect to a background order-sorted BAT $\mathcal{D} = (\mathbb{T}, \mathbf{D})$. The *regression* of W, $\mathcal{R}^{os}[W]$, is recursively defined as follows. In what follows, $\vec{t}$ and $\vec{\tau}$ are tuples of terms, α and α' are terms of sort Act, σ and σ' are terms of sort Sit, and W is a regressable formula of $\mathcal{L}^{OS}$.

1. If W is a non-atomic formula and is of the form $\neg W_1$, $W_1 \vee W_2$, $(\exists v : Q).W_1$ or $(\forall v : Q).W_1$, for some regressable formulas W_1, W_2 in $\mathcal{L}^{OS}$, then

$$\mathcal{R}^{os}[\circ W_1] = \circ \mathcal{R}^{os}[W_1] \text{ for constructor } \circ \in \{\neg, (\exists x : Q), (\forall x : Q)\}$$
$$\mathcal{R}^{os}[W_1 \vee W_2] = \mathcal{R}^{os}[W_1] \vee \mathcal{R}^{os}[W_2].$$

2. Else, if W is a non-atomic formula, W is not of the form $\neg W_1$, $W_1 \vee W_2$, $(\exists v : Q)W_1$ or $(\forall v : Q)W_1$, but of the form $W_1 \wedge W_2 \wedge \cdots \wedge W_n$ $(n \geq 2)$, where each W_i $(i = 1..n)$ is not of the form $W_{i,1} \wedge W_{i,2}$ for some sub-formulas $W_{i,1}, W_{i,2}$ in W_i. After using commutative law for $\wedge$, without loss of generality, there are two sub-cases:

2(a) Suppose that for some j, $j = 1..n$, each W_i $(i = 1..j)$ is of the form $t_{i,1} = t_{i,2}$ for some (well-sorted) terms $t_{i,1}, t_{i,2}$, and none of W_k, $k = (j + 1)..n$, is an equality between terms. In particular, when $j = n$, $\bigwedge_{k=j+1}^{n} W_k \stackrel{def}{=} true$. Then,

$$\mathcal{R}^{os}[W] = \begin{cases} W_1 \wedge W_2 \wedge \cdots \wedge W_j \wedge \mathcal{R}^{os}[W_0'] & \text{if there is a well-sorted mgu } \mu \\ & \text{for } \{\langle t_{i,1}, t_{i,2} \rangle \,|\, i = 1..j\}; \\ false & \text{otherwise.} \end{cases}$$

Here, W_0' is a new formula obtained by applying mgu μ to $\bigwedge_{k=j+1}^{n} W_k$ and it is existentially-quantified at front for every newly introduced sort weakened variable in μ. Moreover, note that based on the assumption that we consider meet semi-lattice sort hierarchies only, such mgu is unique if it exists. Notice that $W_1 \wedge \cdots \wedge W_j$ needs to be kept, because it carries unification information between terms $\{\langle t_{i,1}, t_{i,2} \rangle | i = 1..j\}$ that cannot be omitted, and the unifiability of these terms does not mean $W_1 \wedge \cdots \wedge W_j \equiv true$.

2(b) Otherwise, $\mathcal{R}^{os}[W] = \mathcal{R}^{os}[W_1] \wedge \cdots \wedge \mathcal{R}^{os}[W_n]$.

3. Otherwise, W is atomic. There are four sub-cases.

3(a) Suppose that W is of the form $Poss(A(\vec{t}), \sigma)$ for an action term $A(\vec{t})$ and a situation term σ, and the action precondition axiom for A is of the form (1). Without loss of generality, assume that all variables in Axiom (1) have had been renamed (with variables of the same sorts) to be distinct from the free variables (if any) of W. Then,

$$\mathcal{R}^{os}[W] = \mathcal{R}^{os}[\Pi_A(\vec{t}, \sigma)].$$

3(b) Suppose that W is of the form $F(\vec{t}, do(\alpha, \sigma))$ for some relational fluent F. Let F's SSA be of the form (2). Without loss of generality, assume that all variables in Axiom (2) have had been renamed (with variables of the same sorts) to be distinct from the free variables (if any) of W. Then,

$$\mathcal{R}^{os}[W] = \mathcal{R}^{os}[\phi_F(\vec{t}, \alpha, \sigma)].$$

3(c) Suppose that atom W is of the form $t_1 = t_2$. for some well-sorted terms t_1, t_2. Then,

$$\mathcal{R}^{os}[W] = \begin{cases} W & \text{if there is a well-sorted mgu } \mu \text{ for } \langle t_1, t_2 \rangle; \\ false & \text{otherwise.} \end{cases}$$

3(d) Otherwise, if atom W has S_0 as its only situation term, then

$$\mathcal{R}^{os}[W] = W.$$

Notice that although the definition seems to depend on syntactic form of a formula, we prove below that for any regressable formulas W_1 and W_2 in $\mathcal{L}^{OS}$ that are logically equivalent, their regressed results are still equivalent wrt $\mathcal{D}$ (see Corollary 7). Here are some examples.

EXAMPLE 4 Consider the order-sorted BAT $\mathcal{D}$ from Example 1 and the query W from Example 2. Then, it is easy to see that $\mathcal{R}^{os}[W] = false$, since there is no well-sorted mgu for $\langle d, Boston \rangle$, where $d : Box$. Now, let W_1 be

$$\neg \forall d : Box.\, d \neq Boston \vee \neg On(d, T_1, do(load(B_1, T_1), S_0)).$$

W_1 is a sentence that is equivalent to W. It is easy to check that $\mathcal{R}^{os}[W_1]$ is a formula equivalent to $false$ (wrt $\mathcal{D}$).

Here is another example to illustrate the necessity of keeping $W_1 \wedge \cdots \wedge W_j$ in **2(a)** of Def. 3. We consider the regression of a well-sorted formula

$$InCity(B_1, city, do(drive(T_1, Boston, Toronto), S_0)),$$

where $city$ is a free variable of sort $City$.

$\mathcal{R}^{os}\left[InCity(B_1, city, do(drive(T_1, Boston, Toronto), S_0))\right]$

$\quad = \quad \mathcal{R}^{os}[(\exists t : Truck, c_1 : City).drive(T_1, Boston, Toronto) = drive(t, c_1, city)$
$\qquad \wedge (B_1 = t \vee \exists b : Box.b = B_1 \wedge On(b, t, S_0))) \vee InCity(B_1, city, S_0)$
$\qquad \wedge \neg (\exists t : Truck, c_1 : City.\, drive(T_1, Boston, Toronto) = drive(t, city, c_1)$
$\qquad \wedge (B_1 = t \vee \exists b : Box.b = B_1 \wedge On(b, t, S_0)))]$

$\quad = \quad \cdots \cdots$

$\quad = \quad (\exists t : Truck, c_1 : City).drive(T_1, Boston, Toronto) = drive(t, c_1, city)$
$\qquad \wedge \exists b : Box.b = B_1 \wedge On(B_1, T_1, S_0) \vee InCity(B_1, city, S_0)$
$\qquad \wedge \neg (\exists t : Truck, c_1 : City.drive(T_1, Boston, Toronto) = drive(t, city, c_1)$
$\qquad \wedge \exists b : Box.b = B_1 \wedge On(b, T_1, S_0)).$

In the above formula, for instance, if we omit the condition $drive(T_1, Boston, Toronto) = drive(t, c_1, city)$ (an example of the component $W_1 \wedge \cdots \wedge W_j$ in **2(a)** of Def. 3), we will lose unification information between the variable $city$ and the constant $Toronto$, and won't be able to maintain logical equivalence between the original formula and its regression.

Given an order-sorted BAT $\mathcal{D} = (\mathbb{T}, \mathbf{D})$ and the order-sorted regression operator defined above, to show the correctness of the newly defined regression operator, we prove the following theorems similar to the theorems in [Reiter 2001].

THEOREM 5 *If W is a regressable formula wrt $\mathcal{D}$, then $\mathcal{R}^{os}[W]$ is a well-sorted $\mathcal{L}^{OS}$ formula (including false) that is uniform in S_0. Moreover,*
$$\mathbf{D} \models_{\mathbb{T}}^{os} W \equiv \mathcal{R}^{os}[W].$$

THEOREM 6 *If W is a regressable formula wrt $\mathcal{D}$, then*
$$\mathbf{D} \models_{\mathbb{T}}^{os} W \text{ iff } \mathcal{D}_{S_0} \cup \mathcal{D}_{una} \models_{\mathbb{T}}^{os} \mathcal{R}^{os}[W].$$

Hence, to reason whether $\mathbf{D} \models_{\mathbb{T}}^{os} W$ is the same as to compute $\mathcal{R}^{os}[W]$ first and then to reason whether $\mathcal{D}_{S_0} \cup \mathcal{D}_{una} \models_{\mathbb{T}}^{os} \mathcal{R}^{os}[W]$. Besides, according to Th. 5, it is easy to see that the following consequence holds.

COROLLARY 7 *If W_1 and W_2 are regressable formulas in $\mathcal{L}^{OS}$ s.t. $\models_{\mathbb{T}}^{os} W_1 \equiv W_2$, then*
$$\mathbf{D} \models_{\mathbb{T}}^{os} \mathcal{R}^{os}[W_1] \equiv \mathcal{R}^{os}[W_2].$$

5 Order-Sorted Situation Calculus v.s. Reiter's Situation Calculus

Although BATs and regressable formulas in $\mathcal{L}^{OS}$ are based on order-sorted logic, they can be related to BATs and regressable formulas in Reiter's SC $\mathcal{L}_{sc}$.

First, given a well-sorted formula W wrt the sort theory $\mathbb{T}$ of some order-sorted BAT $\mathcal{D}$ (or simply say, wrt $\mathcal{D}$) in $\mathcal{L}^{OS}$, we define what is a *translation* of W in Reiter's SC $\mathcal{L}_{sc}$. Some concepts are introduced here for later convenience. For any sort Q in the language of $\mathcal{L}^{OS}$, we introduce a unary predicate $Q(x)$, which will be true iff x is of sort Q in $\mathcal{L}^{OS}$. Note that we can use same symbols for both sorts and their corresponding unary predicates based on the assumption that all sort symbols are distinct from usual predicate symbols in $\mathcal{L}^{OS}$.

DEFINITION 8 Consider any well-sorted formula ϕ in $\mathcal{L}^{OS}$. A *translation* of ϕ to a (well-sorted) sentence in Reiter's SC, denoted as $tr(\phi)$, is defined recursively as follows:

For every atom $P(\vec{t})$, $tr(P(\vec{t})) \overset{def}{=} P(\vec{t})$;

$tr(\neg\phi) \overset{def}{=} \neg tr(\phi)$;

$tr((\exists x : \bot)\phi) \overset{def}{=} false$;

$tr((\forall x : Q)\phi) \overset{def}{=} \neg tr((\exists x : Q. \neg\phi))$;

$tr((\exists x : Q)\phi) \overset{def}{=} (\exists x : Q)tr(\phi)$, if $Q \in \{Object, Act, Sit\}$;

$tr((\exists x : \top)\phi) \overset{def}{=} (\exists x : Object)tr(\phi) \vee (\exists x : Act)tr(\phi) \vee \exists x : Sit)tr(\phi)$;

$tr((\exists x : Q)\phi) \overset{def}{=} (\exists y : Object)[Q(y) \wedge tr(\phi(x/y))]$, if $Q \notin \{\top, \bot, Object, Act, Sit\}$;

$tr(\phi \circ \psi) \overset{def}{=} tr(\phi) \circ tr(\psi)$ for $\circ \in \{\supset, \wedge, \vee, \supset, \equiv\}$.

The intuition behind the definition above is obvious, for any well-sorted formula W in $\mathcal{L}^{OS}$, we can always find an "equivalent" formula in Reiter's format. The meaning of equivalence between W and $tr(W)$ is formally given in Lemma 11 below.

We would like to show that the order-sorted situation calculus $\mathcal{L}^{OS}$ is correct, or *sound*, in the sense that for any BAT $\mathcal{D}$ in $\mathcal{L}^{OS}$ we can always find a way to represent the BAT in Reiter's situation calculus $\mathcal{L}_{sc}$ (known as the corresponding BAT $\mathcal{D}'$ of $\mathcal{D}$ in $\mathcal{L}_{sc}$) such that for any regressable formula W, it can be entailed by $\mathcal{D}$ iff the translation of W in Reiter's situation calculus $\mathcal{L}_{sc}$ can be entailed by the corresponding BAT $\mathcal{D}'$ in $\mathcal{L}_{sc}$. Later, the corresponding BAT $\mathcal{D}'$ of $\mathcal{D}$ is denoted as $TR(\mathcal{D})$ to remind that it is constructed out of $\mathcal{D}$. That is,

THEOREM 9 (Soundness) *For any order-sorted BAT* $\mathcal{D} = (\mathbb{T}, \mathbf{D})$ *in* $\mathcal{L}^{OS}$, *there exists a corresponding BAT* $\mathcal{D}'$ *(denoted as* $TR(\mathcal{D})$ *below), such that*

$$\mathbf{D} \models^{os}_{\mathbb{T}} W \ \text{iff} \ TR(\mathcal{D}) \models^{ms} tr(W)$$

for any regressable sentence (i.e., a query) W.

It is hard to prove Th. 9 directly. Inspired by the *standard relativization* of order-sorted logic to unsorted predicate logic, our general idea of proving Th. 9 is as follows (see the diagram in Fig. 1). In Step 1, we construct a BAT $TR(\mathcal{D})$ (called the *corresponding Reiter's BAT of* $\mathcal{D}$ above) in Reiter's situation calculus. In Step 2, we prove that there is an unsorted theory $\mathcal{D}''$ (*strong relativization* of $\mathcal{D}$) and an unsorted first-order sentence W'' (*relativization* of W) such that $\mathbf{D} \models^{os}_{\mathbb{T}} W$ iff $\mathcal{D}'' \models^{fo} W''$. In Step 3, we show that $TR(\mathcal{D}) \models^{ms} tr(W)$ iff $\mathcal{D}''' \models^{fo} W'''$, for some unsorted theory $\mathcal{D}'''$ (standard relativization of $TR(\mathcal{D})$) and first-order sentence W''' (relativization of $tr(W)$). Finally, in Step 4, we show that $\mathcal{D}''' \models^{fo} W'''$ iff $\mathcal{D}'' \models^{fo} W''$.

$$\mathbf{D} \models^{os}_{\mathbb{T}} W \qquad \overset{\text{(Step 2)}}{\Longleftrightarrow} \qquad \mathcal{D}'' \models^{fo} W''$$
$$\Updownarrow \text{(Step 4)}$$
$$TR(\mathcal{D}) \models^{ms} tr(W) \ \text{(Step 1)} \qquad \overset{\text{(Step 3)}}{\Longleftrightarrow} \qquad \mathcal{D}''' \models^{fo} W'''$$

Figure 1. Diagram of the outline for proving Theorem 9

The following definition of relativization of order-sorted logic to unsorted predicate logic (Def. 10) and the bridge axioms (Def. 12) were given in [Oberschelp 1962; Schmidt 1951; Walther 1987; Schmidt-Schauß 1989].

DEFINITION 10 For any well-sorted formula ϕ in $\mathcal{L}^{OS}$, $rel(\phi)$, a *relativization* of ϕ, is an unsorted formula defined as:

For every atom $P(\vec{t})$, $rel(P(\vec{t})) \overset{def}{=} P(\vec{t})$;

$rel(\neg\phi) \overset{def}{=} \neg rel(\phi)$;

$rel(\phi \circ \psi) \overset{def}{=} rel(\phi) \circ rel(\psi)$ for $\circ \in \{\wedge, \vee, \supset\}$;

$rel((\forall x : Q)\phi) \overset{def}{=} (\forall y)[Q(y) \supset rel(\phi[x/y])]$;

$rel((\exists x : Q)\phi) \overset{def}{=} (\exists y)[Q(y) \wedge rel(\phi[x/y])]$.

Moreover, for any set Set of well-sorted formulas, $rel(Set) = \{rel(\phi) \mid \phi \in Set\}$.

Note that Reiter's SC $\mathcal{L}_{sc}$ is in fact based on many-sorted logic, which is a special case of order-sorted logic, with three disjoint sorts (Act, Sit and $Object$). All formulas in $\mathcal{L}_{sc}$ are well-sorted wrt the sort theory of $\mathcal{L}_{sc}$ with all quantified variables restricted to suitable sorts by default. Hence, the definition of rel can also be applied to any formula in Reiter's SC.

Now, it is straightforward to prove the following lemma (Lemma 11) for rel and tr by structural induction, which shows the equivalence relationship between a well-sorted formula W in $\mathcal{L}^{OS}$ and its translation $tr(W)$ in $\mathcal{L}_{sc}$.

LEMMA 11 *Consider any well-sorted formula W in $\mathcal{L}^{OS}$. Then, given the default assumption that everything in the universe is either an action, or a situation, or an object, we have $\models^{fo} rel(tr(W)) \equiv rel(W)$.*

DEFINITION 12 For any sort theory $\mathbb{T}$, which includes predicate declarations, function declarations and/or subsort declarations, *the set of bridge axioms* of $\mathbb{T}$, $BA(\mathbb{T})$, is a set of unsorted formulas as follows:

(a) $(\forall x).\, Q_2(x) \supset Q_1(x)$ for each $Q_2 \leq Q_1$ in $\mathbb{T}$;

(b) $Q(c)$ for each $c:Q$ in $\mathbb{T}$;

(c) $(\forall \vec{x}_{1..n}).\, \bigwedge_{i=1}^{n} Q_i(x_i) \supset Q(f(\vec{x}_{1..n}))$ for each function $f:\vec{Q}_{1..n} \to Q$ in $\mathbb{T}$.

Note that in particular, when we compute the bridge axioms for a sort theory $\mathbb{T}$ in a given order-sorted BAT $\mathcal{D}$, $Sit(S_0)$ is always included in $BA(\mathbb{T})$ for $S_0:Sit$ in $\mathbb{T}$, and the axioms of the form (c) are introduced for all functions, including action functions and the special situation function $do(a, s)$.

Based on the definition of relativization and the bridge axioms, the following lemma has been proved in [Schmidt-Schauβ 1989; Walther 1987; Bierle, Hedtstück, Pletat, Schmitt, and Siekmann 1992]: *For any well-sorted sentence ϕ wrt a sort theory $\mathbb{T}$, we have that $\models^{os}_{\mathbb{T}} \phi$ iff $BA(\mathbb{T}) \models^{fo} rel(\phi)$.*

We then define the standard relativization as follows.

DEFINITION 13 Consider a sort theory $\mathbb{T}$ in an order-sorted (or many-sorted) logic and a set of well-sorted axioms $\mathcal{K}$ wrt the given sort theory. Then, the *standard relativization of $\mathcal{K}$*, an unsorted theory, is defined as

$$REL(\mathcal{K}) \overset{def}{=} rel(\mathcal{K}) \cup BA(\mathbb{T}).$$

In particular, for any BAT $\mathcal{D}_1$ in Reiter's SC $\mathcal{L}_{sc}$ that has a finite set $\mathbb{T}_{\mathcal{D}_1}$ of function declarations and predicate declarations for all predicates and functions appeared in $\mathcal{D}_1$, the *standard relativization of $\mathcal{D}_1$* is

$$REL(\mathcal{D}_1) \overset{def}{=} rel(\mathcal{D}_1 - \phi_\Sigma) \cup BA(\mathbb{T}_{\mathcal{D}_1}) \cup \{rel(\phi_\Sigma)\},$$

where ϕ_Σ is one of the foundational axioms representing the second-order induction axiom

$$(\forall P).P(S_0) \wedge (\forall a, s)[P(s) \supset P(do(a, s))] \supset (\forall s)P(s), \qquad (3)$$

and the relativization of ϕ_Σ, $rel(\phi_\Sigma)$, is defined as

$$(\forall P).P(S_0) \wedge (\forall a, s)[Act(a) \wedge Sit(s) \supset (P(s) \supset P(do(a, s)))]$$
$$\supset (\forall s)Sit(s) \supset P(s). \qquad (4)$$

It is easy to see that the standard relativization of a BAT of Reiter's SC is a very slight extension of the standard relativization of a set of well-sorted (first-order) formulas by applying the (standard) relativization function to a (second-order) well-sorted formula. Therefore, similar to the Relativization Theorem proved in [Schmidt-Schauβ 1989], we have:

LEMMA 14 *Consider any regressable formula W with a background BAT $\mathcal{D}$ in Reiter's SC $\mathcal{L}_{sc}$. Then,*

$$\mathcal{D} \models^{\mathrm{ms}} W \text{ iff } REL(\mathcal{D}) \models^{\mathrm{fo}} rel(W).$$

Now we proceed to Step 1 mentioned in Fig. 1 of the outline. Consider any order-sorted BAT $\mathcal{D}$. We construct the *corresponding Reiter's BAT of $\mathcal{D}$*, denoted as $TR(\mathcal{D})$, that will be the Reiter's BAT we are looking for in Th. 9. In $TR(\mathcal{D})$, we introduce three new special predicates $SortedObj(x)$ $SortedAct(a)$ and $SortedSit(s)$. Intuitively, for any term t (a, or s, respectively) of sort object (action, or situation, respectively), $SortedObj(t)$ ($SortedAct(a)$, or $SortedSit(s)$, respectively) means that t (a, or s, respectively) needs to be well-sorte with respect to the given sort theory $\mathbb{T}$ in the order-sorted BAT $\mathcal{D}$. Note that the reason why we introduce three different special predicates for well-sorte terms ($SortedObj(x)$, $SortedAct(a)$ and $SortedSit(s)$) is because Reiter's Situation Calculus is a many-sort logic with three sorts only and his BATs have a particular syntactic format. For instance, every formula in an initial theory needs to be uniform in the initial situation S_0, and every SSA has to be of the form $F(\vec{x}, do(a,s)) \equiv \phi_F(\vec{x}, a, s)$. In order to construct a BAT in Reiter' Situation Calculus that satisfies Th. 9, we need to "encode" information about the well-sortedness of terms into the constructed BAT. However, if we introduce onl one predicate to describe well-sortedness, say $sorted(x)$ for any x (including objects, actions and situations) representin x is well-sorted then it would be problematic when we want to axiomatize the property of $sorted$ – it neither can b considered as an axiom in the initial theory (since it is not uniform in S_0), nor can be considered as an SSA (since its last argument is not of sort situation). Notice that in [Reiter 2001], sorted quantifiers are omitted as a convention, because their sorts are always obvious from context. Hence, when we construct the BAT $TR(\mathcal{D})$ in Reiter's SC below, all free variables are implicitly universally sorted-quantified according to their obvious sorts. The declarations for functions and predicates (including for predicates $SortedObj$, $SortedAct$ and $SortedSit$) are always standard, hence are not mentioned here.

• $TR(\mathcal{D})$ includes the standard foundational axioms and the set of unique name axioms for action functions in Reiter's SC.

• The initial theory of $TR(\mathcal{D})$, say $\mathcal{D}'_{S_0}$, includes the following axioms.

1. For any well-sorted sentence $\phi \in \mathcal{D}_{S_0}$, $tr(\phi)$ is in $\mathcal{D}'_{S_0}$.

2. For each declaration $Q_2 \leq Q_1$ in $\mathbb{T}$, add an axiom

$$tr((\forall x : \top).(\exists y_2 : Q_2.x = y_2) \supset (\exists y_1 : Q_1.x = y_1)).$$

3. For any constant declaration $c : Q$ where $Q \leq_{\mathbb{T}} Object$ and $Q \neq Object$, add an axiom $Q(c)$. Note that other constant declarations will still be kept in the sort theory of $TR(\mathcal{D})$ in language $\mathcal{L}_{sc}$ by default. For example, $S_0 : Sit$, $C : Object$ for any constant object C appeared in $TR(\mathcal{D})$ and $A : Act$ for any constant action function A.

4. For each (situation-independent) function f (including action function) whose declaration is $f : \vec{Q}_{1..n} \to Q$ in $\mathbb{T}$ ($n \geq 1$), add an axiom $tr((\forall \vec{x}_{1..n} : \vec{Q}_{1..n}).(\exists y : Q).y = f(\vec{x}_{1..n}))$.

5. The axioms related to $SortedObj$, $SortedAct$, and $SortedSit$:

(a) $SortedObj(y) \equiv tr(\bigvee_{i=1}^{k}(\exists x_{i,1}:Q_{i,1},\ldots,x_{i,n_i}:Q_{i,n_i}).y = f_i(x_{i,1},\ldots,x_{i,n_i}) \wedge \bigwedge_{j=1}^{n_i} SortedObj(x_{i,j}))$, where $f_1,\ldots,f_k$ (including constant Objects) are all functions other than action functions and do function included in $\mathcal{D}$, and the function declaration for each f_i in $\mathbb{T}$ is $f_i:Q_{i,1} \times \cdots \times Q_{i,n_i} \rightarrow Q_{i,1+n_i}$ (each $Q_{i,j} \leq_{\mathbb{T}} Object$);

(b) $SortedAct(a) \equiv tr(\bigvee_{i=1}^{m}(\exists x_{i,1}:Q_{i,1},\ldots,x_{i,n_i}:Q_{i,n_i}).a = A_i(x_{i,1},\ldots x_{i,n_i}) \wedge \bigwedge_{j=1}^{n_i} SortedObj(x_{i,j}))$, where $A_1,\ldots,A_m$ (including constant action functions) are all action functions included in $\mathcal{D}$, and the function declaration for each A_i in $\mathbb{T}$ is $A_i:Q_{i,1} \times \cdots \times Q_{i,n_i} \rightarrow Act$ (each $Q_{i,j} \leq_{\mathbb{T}} Object$);

(c) $tr(P(\vec{x}_{1..n}) \supset \bigwedge_{i=1}^{n}(\exists y_i:Q_i.y_i = x_i \wedge SortedObj(x_i)))$ for each situation-independent predicate $P:\vec{Q}_{1..n}$ in $\mathbb{T}$;

(d) $SortedSit(S_0)$;

(e) $tr(F(\vec{x}_{1..n},S_0) \supset \bigwedge_{i=1}^{n}(\exists y_i:Q_i.y_i = x_i \wedge SortedObj(x_i)))$ for each fluent $F:\vec{Q}_{1..n} \times Sit$ in $\mathbb{T}$. Here, all y_i's are distinct auxiliary variables never appearing in $\vec{x}_{1..n}$.

- For action $A(\vec{x}_{1..n})$ whose precondition axiom in $\mathcal{D}_{ap}$ has the form Eq.(1), we replace it with a precondition axiom in the format of Reiter's SC:

$$Poss(A(\vec{x}_{1..n}),s) \equiv \Pi'_A(\vec{x}_{1..n},s),$$

where $\Pi'_A(\vec{x}_{1..n},s)$ is uniform in s, resulting from

$$tr((\exists \vec{y}_{1..n}:\vec{Q}_{1..n}).(\bigwedge_{i=1}^{n} x_i = y_i \wedge SortedObj(x_i)) \wedge \Pi_A(\vec{y}_{1..n},s)).$$

Here, all y_i's are distinct auxiliary variables never appearing in $\Pi_A(\vec{x}_{1..n},s)$.
- The set of successor state axioms of $TR(\mathcal{D})$ now includes the following axioms:

1. For each relational fluent $F(\vec{x}_{1..n},s)$, whose SSA in $\mathcal{D}_{ss}$ is of the form Eq.(2), we replace it with SSA in the format of Reiter's SC:

$$F(\vec{x}_{1..n},do(a,s)) \equiv \phi'_F(\vec{x}_{1..n},a,s),$$

where $\phi'_F(\vec{x}_{1..n},a,s)$ is uniform in s, resulting from

$tr(SortAct(a) \wedge SortedSit(s) \wedge (\exists \vec{y}_{1..n}:\vec{Q}_{1..n}).\bigwedge_{i=1}^{n}(x_i = y_i \wedge SortedObj(x_i)) \wedge \phi_F(\vec{y}_{1..n},a,s)).$

Here, all y_i's are distinct auxiliary variables never appearing in $\phi_F(\vec{x}_{1..n},s)$.

2. $SortedSit(do(a,s)) \equiv SortedAct(a) \wedge SortedSit(s)$.

Now, we define a different relativization, the *strong relativization*, for BATs in order-sorted SC $\mathcal{L}^{OS}$ (Def. 15) to help us prove Th. 9 because of the following reasons: (1) We include the sort theory in each order-sorted BAT in the language of $\mathcal{L}^{OS}$, while Reiter's SC mentions sort declarations generally in the signature of $\mathcal{L}_{sc}$. (2) We are not able to use standard relativization to relate order-sorted BATs to BATs in Reiter's SC directly because of the particular syntactic formats of BATs in Reiter's SC. (3) We expect that any predicates (including fluents) will be true only if they are for "reasonable" types of objects in unsorted logic, i.e., for well-sorted terms wrt a given sort theory in order-sorted logic.

DEFINITION 15 For any order-sorted BAT $\mathcal{D} = (\mathbb{T},\mathbf{D})$ in $\mathcal{L}^{OS}$, besides introducing unary predicates that correspond to sorts in $\mathbb{T}$, same as the special new predicates introduced in the corresponding Reiter's BAT of $\mathcal{D}$, $TR(\mathcal{D})$, we also use $SortedObj(x)$,

($SortedAct(a)$ and $SortedSit(s)$, respectively) to represent that t (a, or s, respectively) is well-sorted with respect to the given sort theory $\mathbb{T}$ in the order-sorted BAT $\mathcal{D}$.

The *strong relativization of* $\mathcal{D}$ is an unsorted theory defined as

$$REL_S(\mathcal{D}) \overset{def}{=} rel_S(\mathbf{D}) \cup BA(\mathbb{T}),$$

where $rel_S(\mathbf{D})$ is a set of axioms including the following axioms.

(a) $(\forall \vec{x}_{1..n}). \bigwedge_{i=1}^{n} Q'_i(x_i) \supset Q'_{n+1}(f(\vec{x}_{1..n}))$, where each Q'_j ($j = 1..n+1$) is a predicate in $\{Action, Situation, Object\}$ and its corresponding sort Q'_j satisfies $Q_j \leq_\mathbb{T} Q'_j$, for any function $f : \vec{Q}_{1..n} \to Q_{n+1}$ in $\mathbb{T}$ (including constant functions, action functions and $do(a, s))^1$.

(b) all axioms in $rel(\mathcal{D}_{S_0} \cup \Sigma - \{\phi_\Sigma\})$, where ϕ_Σ is Axiom (3).

(c) the relativization of Axiom (3), i.e., Axiom (4).

(d) $(\forall \vec{x}_{1..n}, \vec{y}_{1..n}). \bigwedge_{i=1}^{n} \big(Object(x_i) \wedge Object(y_i)\big) \supset$
$$\big(A(\vec{x}_{1..n}) = A(\vec{y}_{1..n}) \supset \bigwedge_{i=1}^{n} x_i = y_i\big)$$
for each action function symbol A.

(e) $(\forall \vec{x}_{1..n}, \vec{y}_{1..m}). \bigwedge_{i=1}^{n} Object(x_i) \wedge \bigwedge_{j=1}^{m} Object(y_i) \supset A(\vec{x}_{1..n}) \neq B(\vec{y}_{1..m})$ for any two distinct action function symbols A and B.

(f) $(\forall y).Object(y) \supset [SortedObj(y) \equiv \bigvee_{i=1}^{k} (\exists x_{i,1}, \ldots, x_{i,n_i}).y = f_i(x_{i,1}, \ldots, x_{i,n_i}) \wedge (\bigwedge_{j=1}^{n_i} Q_{i,j}(x_{i,j}) \wedge SortedObj(x_{i,j}))]$, where $f_1, \ldots, f_k$ (including constant Objects) are *all* functions other than action functions and do function included in $\mathcal{D}$, and the function declaration for each f_i in $\mathbb{T}$ is $f_i : Q_{i,1} \times \cdots \times Q_{i,n_i} \to Q_{i,n_i+1}$ (each $Q_{i,j} \leq_\mathbb{T} Object$). Note that for any i, if $n_i = 0$ (i.e., f_i is a constant object), there are no quantifiers for variables $x_{i,1}, \ldots, x_{i,n_i}$ at the front and $\bigwedge_{j=1}^{n_i} Q_{i,j}(x_{i,j}) \wedge SortedObj(x_{i,j})) \equiv true$.

(g) $(\forall a).Action(a) \supset [SortedAct(a) \equiv \bigvee_{i=1}^{m} (\exists x_{i,1}, \ldots, x_{i,n_i}).a = A_i(x_{i,1}, \ldots, x_{i,n_i}) \wedge (\bigwedge_{j=1}^{n_i} Q_{i,j}(x_{i,j}) \wedge SortedObj(x_{i,j}))]$, where $A_1, \ldots, A_m$ (including constant action functions) are *all* action functions included in $\mathcal{D}$, and the function declaration for each A_i in $\mathbb{T}$ is $A_i : Q_{i,1} \times \cdots \times Q_{i,n_i} \to Action$ (each $Q_{i,j} \leq_\mathbb{T} Object$). Note that for any i, if $n_i = 0$ (i.e., A_i is a constant action function), there are no quantifiers for variables $x_{i,1}, \ldots, x_{i,n_i}$ at the front and $\bigwedge_{j=1}^{n_i} Q_{i,j}(x_{i,j}) \wedge SortedObj(x_{i,j})) \equiv true$.

(h) $(\forall \vec{x}_{1..n}). \bigwedge_{i=1}^{n} Object(x_i) \supset [P(\vec{x}_{1..n}) \supset \bigwedge_{i=1}^{n} Q_i(x_i) \wedge SortedObj(x_i)]$ for each situation-independent predicate $P : \vec{Q}_{1..n}$ in $\mathbb{T}$.

(i) $SortedSit(S_0)$.

(j) $(\forall a, s).Action(a) \wedge Situation(s) \supset$
$$[SortedSit(do(a, s)) \equiv SortedAct(a) \wedge SortedSit(s)].$$

---

[1] In particular, when $n = 0$, $f(\vec{x}_{1..n})$ is a constant function c, and we have $Q'(f)$, where Q' is a predicate in $\{Action, Situation, Object\}$ and its corresponding sort Q' satisfies $Q \leq_\mathbb{T} Q'$, for $c : Q$ in $\mathbb{T}$ (including constant action functions and the initial situation S_0).

(k) $(\forall \vec{x}_{1..n}, a, s). \bigwedge_{i=1}^{n} Object(x_i) \wedge Action(a) \wedge Situation(s) \supset [F(\vec{x}_{1..n}, do(a, s)) \equiv$
$\bigwedge_{i=1}^{n} Q_i(x_i) \wedge SortedObj(x_i) \wedge SortedAct(a) \wedge SortedSit(s) \wedge rel(\phi_F(\vec{x}_{1..n}, a, s))]$
for each fluent F, whose SSA in $\mathcal{D}$ is of the form Axiom (2).

(l) $(\forall \vec{x}_{1..n}). \bigwedge_{i=1}^{n} Object(x_i) \supset [F(\vec{x}_{1..n}, S_0) \supset \bigwedge_{i=1}^{n} Q_i(x_i) \wedge SortedObj(x_i)]$ for each
fluent $F : \vec{Q}_{1..n} \times Situation$ in $\mathbb{T}$.

(m) $(\forall \vec{x}_{1..n}, s). \bigwedge_{i=1}^{n} Object(x_i) \wedge Situation(s) \supset [Poss(A(\vec{x}_{1..n}), s) \equiv$
$\bigwedge_{i=1}^{n} (Q_i(x_i) \wedge SortedObj(x_i)) \wedge SortedSit(s) \wedge rel(\Pi_A(\vec{x}_{1..n}, s))]$ for each
n-ary action function A, whose precondition axiom in $\mathcal{D}$ is of the form Axiom (1).

We can also prove a relativization theorem as follows for the strong relativization similar to the Relativization Theorem proved in [Schmidt-Schauß 1989].

LEMMA 16 *Consider any regressable formula W with a background BAT $\mathcal{D} = (\mathbb{T}, \mathbf{D})$ in order-sorted SC $\mathcal{L}^{OS}$. Then,*

$$\mathbf{D} \models_{\mathbb{T}}^{os} W \text{ iff } REL_S(\mathcal{D}) \models^{fo} rel(W).$$

We can prove Step 2 in Fig. 1 using Lemma 16. Because Reiter's SC is a many-sorted logical language that has particular syntactic formats for precondition axioms and SSAs, we cannot use rel to relate $\mathcal{D}$ in $\mathcal{L}^{OS}$ with a Reiter's BAT directly. It is also the reason why strong relativization is introduced.

Proof of Th. 9. Overall, in Fig. 1, in Step 1, $TR(\mathcal{D})$ is constructed as above. Let $\mathcal{D}'' = REL_S(\mathcal{D})$ and $W'' = rel(W)$, and Step 2 can be proved by using Lemma 16. Let $W''' = rel(tr(W))$ and $\mathcal{D}''' = REL(\mathcal{D}')$ (see Def. 13), and Step 3 can be proved by using Lemma 14. Finally, Step 4 is true according to Lemma 11.

It is important to notice that any query in $\mathcal{L}^{OS}$ has to be well-sorted wrt a given background order-sorted BAT $\mathcal{D}$; while, in general, a query that can be answered in the corresponding Reiter's BAT of $\mathcal{D}$ are not necessarily well-sorted wrt $\mathcal{D}$. Below, Th. 17 shows that for any query that can be answered in $TR(\mathcal{D})$, it can be answered in $\mathcal{D}$ in a "well-sorted" way too. Proof details are omitted due to the space limitations. But, we provide some examples below to illustrate the statement.

THEOREM 17 (Completeness) *Let $\mathcal{D}$ be an order-sorted BAT in $\mathcal{L}^{OS}$, and $TR(\mathcal{D})$ be its corresponding Reiter's BAT. Consider any regressable formula W in Reiter's SC, in which there is no appearance of special predicates $SortedObj$, $SortedAct$ or $SortedSit$, W can be translated to a (well-sorted) formula wrt $\mathcal{D}$, denoted as $os(W)$ below, such that*

$$TR(\mathcal{D}) \models^{ms} tr(os(W)) \equiv W.$$

Furthermore, we have

$$TR(\mathcal{D}) \models^{ms} W \text{ iff } \mathbf{D} \models_{\mathbb{T}}^{os} os(W)$$

when W is a regressable sentence wrt $TR(\mathcal{D})$.

We provide an example to illustrate some axioms in the corresponding Reiter's BAT of an order-sorted BAT in Th. 9. We also give some examples to illustrate the idea of Th. 17.

EXAMPLE 18 Consider the BAT $\mathcal{D}$ from Example 1. Most of the axioms in $TR(\mathcal{D})$ are obvious and we just provide examples of the axiom of $SortedObj$, a precondition axiom and an SSA:

$$SortedObj(x) \equiv x = B_1 \lor x = B_2 \lor x = Boston \lor x = Toronto$$
$$\lor x = T_1 \lor x = T_2 \lor \exists y.City(y) \land x = twinCity(y),$$
$$Poss(load(x,t),s) \equiv Box(x) \land Truck(t) \land \neg On(x,t,s) \land$$
$$(\exists y.City(y) \land InCity(x,y,s) \land InCity(t,y,s)),$$
$$InCity(d,c,do(a,s)) \equiv MovObj(d) \land City(c) \land$$
$$[(\exists t, c_1.Truck(t) \land City(c_1) \land a = drive(t,c_1,c)$$
$$\land (d = t \lor \exists b.Box(b) \land b = d \land On(b,t,s))) \lor InCity(d,c,s) \land$$
$$\neg (\exists t, c_1.Truck(t) \land City(c_1) \land a = drive(t,c,c_1)$$
$$\land (d = t \lor \exists b.Box(b) \land b = d \land On(b,t,s)))].$$

Now, let $On(twinCity(Boston), T_1, s)$ (denoted as W_3) be a regressable formula in $\mathcal{L}_{sc}$, where s is a variable of sort situation. According to the way $TR(\mathcal{D})$ is constructed, we have $TR(\mathcal{D}) \models^{ms} On(o,t,s) \supset Box(o)$. Then, for any situation s, if $TR(\mathcal{D}) \models^{ms}$ $On(twinCity(Boston), T_1, s)$, we need to have $TR(\mathcal{D}) \models^{ms} Box(twinCity(Boston))$, which in fact does not hold according to the axioms in $TR(\mathcal{D})$. Hence, $TR(\mathcal{D}) \models^{ms} W_3 \equiv os(W_3)$, where $os(W_3) = false$.

Let W_4 be $\forall s.\exists c. \neg InCity(B_1, twinCity(c), s)$, which is a regressable sentence in $\mathcal{L}_{sc}$, where $c : Object$ and $s : Situation$ hold by default. Then, $os(W_4)$ is $\forall s : Sit.\exists c : Object.\neg(\exists c_1 : City.c_1 = c \land InCity(B_1, twinCity(c_1), s))$. Since $TR(\mathcal{D}) \models^{ms}$ $InCity(B_1, twinCity(c), s) \supset City(c)$, it is easy to see $TR(\mathcal{D}) \models^{ms} W_4 \equiv tr(os(W_4))$.

6   Computational Advantages of $\mathcal{L}^{OS}$

Given any BAT $\mathcal{D}$ in $\mathcal{L}^{OS}$, it is easy to see that Reiter's regression operator $\mathcal{R}$ [Reiter 2001] still can be applied to (well-sorted) regressable formulas (wrt $\mathcal{D}$). Moreover, one can prove that $\mathcal{R}[W]$ is a formula in $\mathcal{L}^{OS}$ uniform in S_0 and $\mathbf{D} \models^{os}_T W \equiv \mathcal{R}[W]$. However, using the order-sorted regression operator $\mathcal{R}^{os}$ sometimes can give us computational advantages in comparison to using Reiter's regression operator $\mathcal{R}$. But first of all, we show that the computational complexity of using $\mathcal{R}^{os}$ is no worse than that of $\mathcal{R}$.

For the regression operator $\mathcal{R}$ that can be used either in $\mathcal{L}^{OS}$ or in $\mathcal{L}_{sc}$ ($\mathcal{R}^{os}$ used in $\mathcal{L}^{OS}$, respectively), we can construct a *regression tree* rooted at W for any regressable query W in either language. Each node in a regression tree of $\mathcal{R}[W]$ ($\mathcal{R}^{os}[W]$, respectively) corresponds to a sub-formula computed by regression, and each edge corresponds to one step of regression according to the definition of the regression operator. In the worst case scenario, for any query W in $\mathcal{L}^{OS}$, the regression tree of $\mathcal{R}^{os}[W]$ will have the same number of nodes as the regression tree of $\mathcal{R}[W]$ (and linear to the number of nodes in the regression tree of $\mathcal{R}[tr(W)]$ wrt $TR(\mathcal{D})$). Moreover, based on the assumption that our sort theory of $\mathcal{D}$ is simple with empty equational theory, whose corresponding sort hierarchy is a meet semi-lattice, finding a unique (well-sorted) mgu takes the same time as in the unsorted case [Schmidt-Schauß 1989; Jouannaud and Kirchner 1991; Weidenbach 1996]. Hence, the overall computational complexity of building the regression tree of $\mathcal{R}^{os}[W]$ is at most linear to the size of Reiter's regression tree.

THEOREM 19 *Consider any regressable sentence W with a background BAT $\mathcal{D}$ in order-sorted SC $\mathcal{L}^{OS}$. Then, in the worst case scenario, the complexity of computing $\mathcal{R}^{os}[W]$ is the same as that of computing $\mathcal{R}[W]$, which is also the same as the complexity of computing $\mathcal{R}[tr(W)]$ in $TR(\mathcal{D})$, the corresponding Reiter's BAT.*

On the other hand, under some circumstances, the regression of a query in $\mathcal{L}^{OS}$ using $\mathcal{R}^{os}$ instead of $\mathcal{R}$ will give us computational advantages. Consider any query (i.e., a regressable sentence) W with a background BAT $\mathcal{D}$ in $\mathcal{L}^{OS}$. Then, the computation of $\mathcal{R}^{os}[W]$ wrt $\mathcal{D}$ can sometimes terminate earlier than that of $\mathcal{R}[W]$ wrt $\mathcal{D}$, and also earlier than the computation of $\mathcal{R}[tr(W)]$ wrt $TR(\mathcal{D})$. In particular, we have:

PROPERTY 20 *Consider an order-sorted BAT $\mathcal{D}$ in $\mathcal{L}^{OS}$, at least one of the SSAs in $\mathcal{D}$ is not context-free, and any regressable formula W of the syntactic form $t_{1,1} = t_{1,2} \wedge \ldots \wedge t_{m,1} = t_{m,2} \wedge W_1$. Let the size of W (including the length of the terms in W) be n. If there is no well-sorted mgu for equalities between terms $\{\langle t_{i,1}, t_{i,2} \rangle \mid i = 1..m\}$, then in the worst-case scenario, computing $\mathcal{R}^{os}[W]$ runs in $O(n)$, while computing $\mathcal{R}[W]$ with respect to $\mathcal{D}$ ($\mathcal{R}[tr(W)]$ with respect to $TR(\mathcal{D})$) runs in time $O(2^n)$. Moreover, the size of the resulting formula of $\mathcal{R}^{os}[W]$, which is false, is always constant, while the size of the resulting formula using $\mathcal{R}$ is $O(2^n)$.*

According to the definition of Reiter's regression operator, the equalities will be kept and regression will be further performed on W_1 (or on $tr(W_1)$ in $TR(\mathcal{D})$, respectively), which in general takes exponential time wrt the length of W_1 and causes exponential blow-up in the size of the formula. Once Reiter's regression has terminated, a theorem prover will find that the resulting formula is false either because there is no mgu for terms when reasoning is performed in $\mathcal{L}^{OS}$ (or, due to the clash between sort related predicates when reasoning in $\mathcal{L}_{sc}$, respectively). Hence, using the order-sorted regression operator can sometimes prune brunches of the regression tree built by $\mathcal{R}$ exponentially (wrt the size of the regressed formula), and therefore make regression terminated exponentially faster.

We provide an example below to show the computational advantage of using $\mathcal{R}^{os}$. This example also illustrates the the class of conjunctive queries in Property 20 is common in regression and leads to significant savings if regression trees are pruned earlier on.

EXAMPLE 21 Consider the BAT $\mathcal{D}$ from Example 1. Let W_5 be a $\mathcal{L}^{OS}$ query, i.e., a (well-sorted) regressable sentence,

$$InCity(T_1, Toronto, do(drive(T_1, Boston, Toronto), S_1)),$$

where S_1 is a well-sorted ground situation term that involves a long sequence of actions. According to the SSA of $InCity$, at the branch of computing $\mathcal{R}^{os}[\exists b : Box.b = T_1 \wedge On(b, t, S_1)]$ in the regression tree, since there is no well-sorted mgu for (b, T_1), the application of order-sorted regression equals to $false$ immediately. However, using Reiter's regression operator (no matter in $\mathcal{D}$ or in $TR(\mathcal{D})$), his operator will keep doing useless regression on $On(b, t, S_1)$ until getting (a potentially huge) sub-formula uniform in S_0. Once his regression has terminated, such sub-formula will also be proved equivalent to $false$ wrt the initial theory ($\mathcal{D}_{S_0}$ or $TR(\mathcal{D})_{S_0}$, respectively) using a theorem prover, for the same reason as above.

In addition, since our sort theory of a BAT $\mathcal{D}$ in $\mathcal{L}^{OS}$ is finite and it has one and only one declaration for each function and predicate symbol, for any query W (wrt $TR(\mathcal{D})$) in $\mathcal{L}_{sc}$, it takes linear time (wrt the length of the query) to find a well-sorted formula $os(W)$ in $\mathcal{L}^{OS}$ that satisfies Th. 17. But, reasoning whether $\mathbf{D} \models_{\mathbb{T}}^{os} os(W)$ (starting from finding $os(W)$) sometimes can terminate exponentially earlier than finding whether $TR(\mathcal{D}) \models^{ms} W$. In particular, we study a certain class of formulas.

DEFINITION 22 Let $\mathcal{D}$ be a BAT in the order-sorted SC $\mathcal{L}^{OS}$, and $TR(\mathcal{D})$ be its corresponding Reiter's BAT. Any term t in Reiter's SC is a *possibly sortable term wrt* $\mathcal{D}$, if one of the following conditions holds:
(1) t is a variable of sort Act, $Object$ or Sit in $\mathcal{L}_{sc}$;
(2) t is a constant c, and $c{:}Q$ in $\mathbb{T}$ (we say that the sort of c is Q wrt $\mathcal{D}$); or,
(3) t is of form $f(\vec{x}_{1..n})$, function declaration $f : \vec{Q}_{1..n} \rightarrow Q$ in $\mathbb{T}$, for every i ($i = 1..n$), t_i either is a variable or is a non-variable possibly sortable term of sort Q_i' wrt $\mathcal{D}$ and $Q_i' \leq_{\mathbb{T}} Q_i$ in $\mathbb{T}$ (we say that the sort of $f(\vec{t}_{1..n})$ is Q wrt $\mathcal{D}$).

Similarly, any atom $P(\vec{t}_{1..n})$ in Reiter's SC (well-sorted wrt $TR(\mathcal{D})$) and P is not $SortedObj$, $SortedAct$ or $SortedSit$, is a *possibly sortable atom wrt* $\mathcal{D}$, if for every i, t_i either is a variable or is a non-variable term of sort Q_i' wrt $\mathcal{D}$ satisfying that:
(a) it is a possibly sortable term wrt $\mathcal{D}$; and
(b) $P{:}\vec{Q}_{1..n}$ is in $\mathbb{T}$ and $Q_i' \leq_{\mathbb{T}} Q_i$ wrt $\mathcal{D}$.

Any regressable formula W in Reiter's SC (well-sorted wrt $TR(\mathcal{D})$) is *possibly sortable wrt* $\mathcal{D}$ if every atom in W is possibly sortable wrt $\mathcal{D}$.

Note that the predicate of a possibly sortable atom can be equality, $Poss$ or any predicate appeared in $\mathcal{D}$.

Given any $\mathcal{D}$ in order-sorted SC, it is easy to see that every atom (term, respectively) in $TR(\mathcal{D})$ that can be considered as well-sorted wrt $\mathcal{D}$ is always a possibly sortable atom (term, respectively); while a possibly sortable atom (term, respectively) is not necessarily well-sorted wrt $\mathcal{D}$. We provide some simple examples of the terms and atoms defined in Def. 22.

EXAMPLE 23 We continue with Example 18. The query
$$\exists x. \exists y. InCity(x, twinCity(y), do(load(B_1, T_1), S_0))$$
in $TR(\mathcal{D})$ is ill-sorted wrt $\mathcal{D}$, but is possibly sortable wrt $\mathcal{D}$. The query
$$\exists x. InCity(x, twinCity(B_1), do(load(B_1, T_1), S_0))$$
in $TR(\mathcal{D})$ is not possibly sortable wrt $\mathcal{D}$, because $twinCity(B_1)$ is not a possibly sortable term wrt $TR(\mathcal{D})$.

Now, we have the following property.

PROPERTY 24 *Assume that* $W = F(\vec{t}, do([\alpha_1, \cdots, \alpha_n], S_0))$ *is an atomic fluent in* $\mathcal{L}_{sc}$ *that is not possibly sortable with respect to* $\mathcal{D}$. *Then, it takes at most linear time (with respect to the length of the whole formula) to terminate reasoning* $TR(\mathcal{D}) \models^{ms} W$ *by checking whether* W *is possibly sortable and computing the corresponding* $os(W)$ *(which is* $false$*). However, in the worst-case scenario, it takes exponential time (with respect to the length of the whole formula) to determine* $TR(\mathcal{D}) \models^{ms} W$ *by using the usual regression in Reiter's SC.*

Note that the worst-case scenario mentioned in Property 24 often happens when a BAT is not context-free. That is, it is common that the usual regression operator leads to a regressed query whose length is exponential in the length of the original formula. Furthermore, even the corresponding $os(W)$ of any query W is not $false$, according to the previous discussion, we sometimes still can gain further computational advantages during computing $\mathcal{R}^{os}[os(W)]$ when reasoning by order-sorted regression in $\mathcal{L}^{OS}$ instead of reasoning by regression in $\mathcal{L}_{sc}$.

7 Conclusions

We propose a logical theory for reasoning about actions wrt a taxonomy of objects based on OSL. We also define a regression-based reasoning mechanism that takes advantages of sort theories, and discuss the computational advantages. It is well-known that *PDDL* supports typed (sorted) variables and many implemented planners can take advantage of types. [Classen, Eyerich, Lakemeyer, and Nebel 2007] propose formal semantics for the typed ADL subset of PDDL using ES, a dialect of SC, where types are represented using unary predicates. Our work also contributes towards a formal logical foundation of PDDL, but in a different way using order-sorted logic. A possible future work can be extending our logic to hybrid order-sorted logic [Cohn 1989; Bierle, Hedtstück, Pletat, Schmitt, and Siekmann 1992; Weidenbach 1996]. Another possibility is to consider efficient reasoning in our framework by identifying specialized classes of queries or decidable fragments [Abadi, Rabinovich, and Sagiv 2007].

References

Abadi, A., A. M. Rabinovich, and M. Sagiv [2007]. Decidable fragments of many-sorted logic. In *LPAR*, Volume 4790 of *Lecture Notes in Computer Science*, pp. 17–31. Springer.

Bacchus, F., J. Y. Halpern, and H. J. Levesque [1999]. Reasoning about noisy sensors and effectors in the situation calculus. *Artif. Intell. 111*(1-2), 171–208.

Bierle, C., U. Hedtstück, U. Pletat, P. H. Schmitt, and J. Siekmann [1992]. An order-sorted logic for knowledge representation systems. *Artif. Intell. 55*(2-3), 149–191.

Brachman, R. J. and H. J. Levesque [1982]. Competence in knowledge representation. In *AAAI*, pp. 189–192.

Brachman, R. J., H. J. Levesque, and R. Fikes [1983]. Krypton: Integrating terminology and assertion. In *AAAI*, pp. 31–35.

Classen, J., P. Eyerich, G. Lakemeyer, and B. Nebel [2007]. Towards an integration of Golog and planning. In *20th International Joint Conference on Artificial Intelligence (IJCAI-07)*. AAAI Press.

Cohn, A. G. [1987]. A more expressive formulation of many sorted logic. *J. Autom. Reason. 3*(2), 113–200.

Cohn, A. G. [1989]. Taxonomic reasoning with many sorted logics. *Artificial Intelligence Review 3*(2-3), 89–128.

Goguen, J. A. and J. Meseguer [1987]. Remarks on remarks on many-sorted equational logic. *SIGPLAN Notices 22*(4), 41–48.

Hayes, P. J. [1971]. A logic of actions. *Machine Intelligence 6*, 495–520.

Herbrand, J. [1971]. *Logical Writings*. Cambridge: Harvard University Press. Warren D. Goldfarb (ed.).

Jouannaud, J.-P. and C. Kirchner [1991]. Solving equations in abstract algebras: A rule-based survey of unification. In *Computational Logic - Essays in Honor of Alan Robinson*, pp. 257–321. MIT Press.

Lakemeyer, G. and H. J. Levesque [2011]. A semantic characterization of a useful fragment of the situation calculus with knowledge. *Artif. Intell. 175*(1), 142–164.

Levesque, H. and G. Lakemeyer [2008]. Chapter 23 Cognitive Robotics. In V. L. Frank van Harmelen and B. Porter (Eds.), *Handbook of Knowledge Representation*, Volume 3 of *Foundations of Artificial Intelligence*, pp. 869 – 886. Elsevier.

Levesque, H., R. Reiter, Y. Lespérance, F. Lin, and R. Scherl [1997]. GOLOG: A logic programming language for dynamic domains. *Journal of Logic Programming 31*, 59–84.

Levesque, H. J. [1986]. Making believers out of computers. *Artif. Intell. 30*(1), 81–108.

Levesque, H. J. [1994]. Knowledge, action, and ability in the situation calculus. In R. Fagin (Ed.), *TARK*, pp. 1–4. Morgan Kaufmann.

McSkimin, J. R. [1976]. *The use of semantic information in deductive question-answering systems. PhD Thesis*. Ph.D. thesis, University of Maryland at College Park, College Park, MD, USA.

McSkimin, J. R. and J. Minker [1977]. The use of a semantic network in a deductive question-answering system. In *IJCAI'77: Proceedings of the 5th international joint conference on Artificial intelligence*, San Francisco, CA, USA, pp. 50–58. Morgan Kaufmann Publishers Inc.

Oberschelp, A. [1962]. Untersuchungen zur mehrsortigen quantorenlogik (in German). *Mathematische Annalen 145*, 297–333.

Oberschelp, A. [1990]. Order sorted predicate logic. In *Sorts and Types in Artificial Intelligence*, Volume 418 of *Lecture Notes in Computer Science*, pp. 8–17. Springer.

Reiter, R. [1977a]. An approach to deductive question-answering. BBN Technical Report 3649 (Accession Number : ADA046550), Bolt Beranek and Newman, Inc.

Reiter, R. [1977b]. An approach to deductive question-answering systems. *SIGART Bull.* (61), 41–43.

Reiter, R. [2001]. *Knowledge in Action: Logical Foundations for Describing and Implementing Dynamical Systems*. MIT Press.

Scherl, R. B. and H. J. Levesque [2003]. Knowledge, action, and the frame problem. *Artif. Intell. 144*(1-2), 1–39.

Schmidt, A. [1938]. Über deduktive theorien mit mehreren soften von grunddingen. *Mathematische Annalen 115*, 485–506.

Schmidt, A. [1951]. Die Zulässigkeit der Behandlung mehrsortiger Theorien mittels der üblichen einsortigen prädikatenlogik. *Mathematische Annalen 123*, 187–200.

Schmidt-Schauβ, M. [1989]. *Computational aspects of an order-sorted logic with term declarations*. New York: Springer-Verlag.

Walther, C. [1987]. *A many-sorted calculus based on resolution and paramodulation*. San Francisco: Morgan Kaufmann.

Wang, H. [1952]. Logic of many sorted theories. *Symbolic Logic 17*(2), 105–116.

Weidenbach, C. [1996]. Unification in sort theories and its applications. *Annals of Math. and AI 18*(2/4), 261–293.

18

That's All I know: A Logical Characterization of Iterated Admissibility

JOSEPH Y. HALPERN AND RAFAEL PASS

ABSTRACT. Brandenburger, Friedenberg, and Keisler provide an epistemic character-ization of iterated admissibility (i.e., iterated deletion of weakly dominated strategies) where uncertainty is represented using LPSs (lexicographic probability sequences). Their characterization holds in a rich structure called a *complete* structure, where all types are possible. Here, a logical characterization of iterated admissibility is given that involves only standard probability and holds in all structures, not just complete structures. Roughly speaking, our characterization shows that iterated admissibility captures the intuition that "all the agent knows" is that agents satisfy the appropriate rationality assumptions.

1 Introduction

One of Hector Levesque's main early breakthroughs was the introduction of the notion of "all I know" [Levesque 1990]. Hector used this notion as a way to capture nonmonotonic reasoning. We give here a somewhat surprising different application of it: as a way to capture solution concepts in game theory.

Admissibility is an old criterion in decision making. A strategy for player i is admissible if it is a best response to some belief of player i that puts positive probability on all the strategy profiles for the other players. Part of the interest in admissibility comes from the observation (due to Pearce [1984]) that a strategy σ for player i is admissible iff it is not weakly dominated; that is, there is no strategy σ' for player i that gives i at least as high a payoff as σ no matter what strategy the other players are using, and sometimes gives i a higher payoff.

It seems natural to ignore strategies that are not admissible. But there is a concep-tual problem when it comes to dealing with *iterated* admissibility (i.e., iterated deletion of weakly dominated strategies). As Mas-Colell, Whinston, and Green [1995, p. 240] put in their textbook when discussing iterated deletion of weakly dominated strategies:

> [T]he argument for deletion of a weakly dominated strategy for player i is that he contemplates the possibility that every strategy combination of his rivals occurs with positive probability. However, this hypothesis clashes with the logic of iterated deletion, which assumes, precisely, that eliminated strategies are not expected to occur.

Brandenburger, Friedenberg, and Keisler [2008] (BFK from now on) resolve this para-dox in the context of iterated deletion of weakly dominated strategies by assuming that strategies are not really eliminated. Rather, they assumed that strategies that are weakly dominated occur with infinitesimal (but nonzero) probability. (Formally, this is captured by using an LPS—*lexicographically ordered probability sequence*.) They define a notion of belief (which they call *assumption*) appropriate for their setting, and show that strategies that survive k rounds of iterated deletion are ones that are played in states where there there is kth-order mutual belief in rationality; that is, everyone assumes that everyone assumes ...($k-1$ times) that everyone is rational. However, they prove only that their characterization of iterated admissibility holds in particularly rich structures called *complete* structures (defined formally in Section 5), where all types are possible.

Here, we provide an alternate logical characterization of iterated admissibility. The characterization has the advantage that it holds in all structures, not just complete structures, and assumes that agents represent their uncertainty using standard probability measures, rather than LPSs or nonstandard probability measures (as is done in a characterization of Rajan [1998]). Moreover, while complete structures must be uncountable, we show that our formula characterizing iterated admissibility is satisfiable in a structure with finitely many states.

Roughly speaking, instead of assuming only that agents know (or assume) that all other agents satisfy appropriate levels of rationality, we assume that "all the agents know" is that the other agents satisfy the appropriate rationality assumptions. We are using the phrase "all agent i knows" here in essentially the same sense that it is used by Levesque [1990] and Halpern and Lakemeyer [2001]. We formalize this notion by requiring that the agent ascribes positive probability to all formulas of some language $\mathcal{L}$ that are consistent with his rationality assumptions. (This admittedly fuzzy description is made precise in Section 4.) As we show, when the language $\mathcal{L}$ is sufficiently rich, our logical formula characterizes iterated admissibility. More precisely, let $\mathcal{L}$-RAT_i be true iff player i is playing a best response to his belief and he considers possible all formulas in $\mathcal{L}$. Define $\mathcal{L}$-RAT_i^{k+1} to be true iff $\mathcal{L}$-RAT_i holds, player i is playing a strategy constant with $\mathcal{L}$-RAT_i^k, and for all $j \neq i$, i knows that $\mathcal{L}$-RAT_j^k holds, and that is "all that agent i knows" about players $j \neq i$. That is, $\mathcal{L}$-RAT_i^{k+1} holds (i.e., player i is $(k+1)$-level rational with respect to $\mathcal{L}$) iff player i is playing a best response to his beliefs, using a strategy consistent with k-level rationality, and the only thing he knows about the other players is that they are k-level rational with respect to $\mathcal{L}$, so that he considers possible all formulas in $\mathcal{L}$ compatible with k-level rationality for the other players. As we show, for natural choices of languages $\mathcal{L}$, a strategy σ for player i survives k rounds of iterated deletion of weakly dominated strategies iff there is a structure and a state where σ is played by player i and the formula $\mathcal{L}$-RAT_i^k holds.

For this result to hold, $\mathcal{L}$ must be reasonably expressive; in particular, it must be possible to express in $\mathcal{L}$ a player's beliefs about the strategies that other players are playing. Interestingly, for less expressive languages $\mathcal{L}$, $\mathcal{L}$-RAT_i^k instead characterizes iterated deletion of *strongly* dominated strategies. For instance, if $\mathcal{L}$ is empty, then "all agent i knows is φ" is equivalent to "agent i knows φ"; it then easily follows (given standard assumptions about knowledge) that $\mathcal{L}$-RAT_i^{k+1} is equivalent to the statement that player i is rational in the standard sense (i.e., player i is making a best response to his beliefs, with no constraints on the beliefs) and knows that everyone is rational and knows that everyone knows that everyone knows ...($k-1$ times) that everyone is rational. Tan and Werlang [1988] and Brandenburger and Dekel [1987] show that this formula characterizes rationalizability, and hence, by results of Pearce [1984], that it also characterizes iterated deletion of strongly dominated strategies. Thus, essentially the same logical formula characterizes both iterated removal of strongly and weakly dominated strategies; in a sense, the only difference between these notions is the expressiveness of the language used by the players to reason about each other.

For all choices of k and all languages $\mathcal{L}$ that we consider, we can find a state in a countable structure where $\mathcal{L}\text{-}RAT_i^k$ holds for all agents. Indeed, for some choices of $\mathcal{L}$ that characterize iterated admissibility, we can take the structure to be finite. (Interestingly, depending on the choice of $\mathcal{L}$, we cannot necessarily find a state where $\mathcal{L}\text{-}RAT^k$ holds for all k. It is not the case that $\mathcal{L}\text{-}RAT^{k+1}$ implies $\mathcal{L}\text{-}RAT^k$.) There are computational and conceptual advantages to having these formulas satisfied in finite structures; these are much easier for an agent to work with. However, in earlier work on "all I know" (e.g., [Halpern and Lakemeyer 2001]) there has beeen a great deal of emphasis on working with what is called the *canonical* structure, which has a state corresponding to every satisfiable collection of formulas. Of course, this requirement means that the canonical has an uncountable number of states. The complete structures used by BFK are also quite rich. We can show that the canonical structure is complete in the sense of BFK. Moreover, under a technical assumption, every complete structure is essentially canonical (i.e., it has a state corresponding to every satisfiable collection of formulas). This sequence of results allows us to connect iterated admissibility, complete structures, canonical structures, and the notion of "all I know".

The rest of the paper is organized as follows. In Section 2 we introduce the formal model (which is essentially the standard Kripke model for probability, adapted to the game-theoretic setting), and provide some initial characterizations of rationalizability and iterated deletion of strongly dominated strategies. Section 3 and 4 contain our main characterization of iterated admissibility using "all I know". In Section 3 we define "all I know" using a simple language; Section 4 considers the effect of using more expressive languages. In Section 5 we compare our results to those of BFK. We conclude with some discussion of the results in Section 6.

2   Probability Structures, Rationalizability, and Admissibility

We consider normal-form games with n players. Given a (normal-form) n-player game Γ, let $\Sigma_i(\Gamma)$ denote the strategies of player i in Γ. We omit the parenthetical Γ when it is clear from context or irrelevant. Let $\vec{\Sigma} = \Sigma_1 \times \cdots \times \Sigma_n$.

Let $\mathcal{L}^1$ be the language where we start with *true* and the special primitive proposition RAT_i and close off under modal operators B_i and $\langle B_i \rangle$, for $i = 1, \ldots, n$, conjunction, and negation. We think of $B_i \varphi$ as saying that φ holds with probability 1, and $\langle B_i \rangle \varphi$ as saying that φ holds with positive probability. As we shall see, $\langle B_i \rangle$ is definable as $\neg B_i \neg$ if we make the appropriate measurability assumptions.

To reason about the game Γ, we consider a class of probability structures corresponding to Γ. A *probability structure M appropriate for* Γ is a tuple $(\Omega, \mathsf{s}, \mathcal{F}, \mathcal{PR}_1, \ldots, \mathcal{PR}_n)$, where Ω is a set of states; s associates with each state $\omega \in \Omega$ a pure strategy profile $\mathsf{s}(\omega)$ in the game Γ; $\mathcal{F}$ is a σ-algebra over Ω; and, for each player i, $\mathcal{PR}_i$ associates with each state ω a probability distribution $\mathcal{PR}_i(\omega)$ on $(\Omega, \mathcal{F})$. Intuitively, $\mathsf{s}(\omega)$ is the strategy profile used at state ω and $\mathcal{PR}_i(\omega)$ is player i's probability distribution at state ω. As is standard, we require that each player knows his strategy and his beliefs. Formally, we require that (1) for each strategy σ_i for player i, $[\![\sigma_i]\!]_M = \{\omega : \mathsf{s}_i(\omega) = \sigma_i\} \in \mathcal{F}$, where $\mathsf{s}_i(\omega)$ denotes player i's strategy in the strategy profile $\mathsf{s}(\omega)$; (2) $\mathcal{PR}_i(\omega)([\![\mathsf{s}_i(\omega)]\!]_M) = 1$; (3) for each probability measure π on $(\Omega, \mathcal{F})$ and player i, $[\![\pi, i]\!]_M = \{\omega : \mathcal{PR}_i(\omega) = \pi\} \in \mathcal{F}$; and (4) $\mathcal{PR}_i(\omega)([\![\mathcal{PR}_i(\omega), i]\!]_M) = 1$.

The semantics is given as follows:

- $(M, \omega) \models$ *true* (so *true* is vacuously true).

- $(M, \omega) \models RAT_i$ if $\mathbf{s}_i(\omega)$ is a best response, given player i's beliefs on the strategies of other players induced by $\mathcal{PR}_i(\omega)$. (Because we restrict to appropriate structures, a players expected utility at a state ω is well defined, so we can talk about best responses.)

- $(M, \omega) \models \neg\varphi$ if $(M, \omega) \not\models \varphi$.

- $(M, \omega) \models \varphi \wedge \varphi'$ iff $(M, \omega) \models \varphi$ and $(M, \omega) \models \varphi'$.

- $(M, \omega) \models B_i\varphi$ if there exists a set $F \in \mathcal{F}$ such that $F \subseteq [\![\varphi]\!]_M$ and $\mathcal{PR}_i(\omega)(F) = 1$, where $[\![\varphi]\!]_M = \{\omega : (M, \omega) \models \varphi\}$.

- $(M, \omega) \models \langle B_i \rangle \varphi$ if there exists a set $F \in \mathcal{F}$ such that $F \subseteq [\![\varphi]\!]_M$ and $\mathcal{PR}_i(\omega)(F) > 0$.

Given a language (set of formulas) $\mathcal{L}$, M is *$\mathcal{L}$-measurable* if M is appropriate (for some game Γ) and $[\![\varphi]\!]_M \in \mathcal{F}$ for all formulas $\varphi \in \mathcal{L}$. It is easy to check that in an $\mathcal{L}^1$-measurable structure, $\langle B_i \rangle \varphi$ is equivalent to $\neg B_i \neg \varphi$.

To put our results on iterated admissibility into context, we first consider rationalizability [Bernheim 1984; Pearce 1984]. Pearce [1984] gives two definitions of rationalizability, which give rise to different epistemic characterizations. We repeat the definitions here, using the notation of Osborne and Rubinstein [1994].

DEFINITION 1. A strategy σ for player i in game Γ is *rationalizable* if, for each player j, there is a set $\mathcal{Z}_j \subseteq \Sigma_j(\Gamma)$ and, for each strategy $\sigma' \in \mathcal{Z}_j$, a probability measure $\mu_{\sigma'}$ on $\Sigma_{-j}(\Gamma)$ whose support is a subset of $\mathcal{Z}_{-j}$ such that

- $\sigma \in \mathcal{Z}_i$; and

- for each player j and strategy $\sigma' \in \mathcal{Z}_j$, strategy σ' is a best response to (the beliefs) $\mu_{\sigma'}$.

□

The second definition characterizes rationalizability in terms of iterated deletion.

DEFINITION 2. A strategy σ for player i in game Γ is *rationalizable* if, for each player j, there exists a sequence $X_j^0, X_j^1, X_j^2, \ldots$ of sets of strategies for player j such that $X_j^0 = \Sigma_j$ and, for each strategy $\sigma' \in X_j^k$, $k \geq 1$, a probability measure $\mu_{\sigma',k}$ whose support is a subset of $\vec{X}_{-j}^{k-1}$ such that

- $\sigma \in \cap_{j=0}^{\infty} X_i$; and

- for each player j, each strategy $\sigma' \in X_j^k$ is a best response to the beliefs $\mu_{\sigma',k}$.

□

Intuitively, X_j^1 consists of strategies that are best responses to some belief of player j, and X_j^{h+1} consists of strategies in X_j^h that are best responses to some belief of player j

with support X^h_{-j}; that is, beliefs that assume that everyone else is best responding to some beliefs assuming that everyone else is responding to some beliefs assuming ... (h times).

PROPOSITION 3. [Pearce 1984] *A strategy is rationalizable iff it is rationalizable'.*

We now state the epistemic characterization of rationalizability due to Tan and Werlang [1988] and Brandenburger and Dekel [1987] in our language; it just says that a strategy is rationalizable iff it can be played in a state where rationality is common knowledge.

Let RAT be an abbreviation of $RAT_1 \wedge \ldots \wedge RAT_n$; let $E\varphi$ be an abbreviation of $B_1\varphi \wedge \ldots \wedge B_n\varphi$; and define $E^k\varphi$ for all k inductively by taking $E^0\varphi$ to be φ and $E^{k+1}\varphi$ to be $E(E^k\varphi)$. Common knowledge of φ holds iff $E^k\varphi$ holds for all $k \geq 0$.

THEOREM 4. *The following are equivalent:*

   (a) *σ is a rationalizable strategy for i in game Γ;*

   (b) *there exists an $\mathcal{L}_1$-measurable structure M that is appropriate for Γ and a state ω such that $\mathbf{s}_i(\omega) = \sigma$ and $(M, \omega) \models B_i E^k RAT$ for all $k \geq 0$;*

   (c) *there exists a structure M that is appropriate for Γ and a state ω such that $\mathbf{s}_i(\omega) = \sigma$ and $(M, \omega) \models B_i E^k RAT$ for all $k \geq 0$.*

Proof. The proof is similar in spirit to that of Tan and Werlang [1988]; we include it here for completeness. Suppose that σ is rationalizable. Choose $\mathcal{Z}_j \subseteq \Sigma_j(\Gamma)$ and measures $\mu_{\sigma'}$ for each strategy $\sigma' \in \mathcal{Z}_j$ guaranteed to exist by Definition 1. Define an appropriate structure $M = (\Omega, \mathbf{s}, \mathcal{F}, \mathcal{PR}_1, \ldots, \mathcal{PR}_n)$, where

- $\Omega = \mathcal{Z}_1 \times \cdots \times \mathcal{Z}_n$;

- $\mathbf{s}_i(\vec{\sigma}) = \sigma_i$;

- $\mathcal{F}$ consist of all subsets of Ω;

- $\mathcal{PR}_i(\vec{\sigma})(\vec{\sigma}')$ is 0 if $\sigma'_i \neq \sigma_i$ and is $\mu_{\sigma_i}(\sigma'_{-i})$ otherwise.

Since each player is best responding to his beliefs at every state, it is easy to see that $(M, \vec{\sigma}) \models RAT$ for all states $\vec{\sigma}$. It easily follows (formally, by induction on k) that $(M, \vec{\sigma}) \models E^k RAT$. Since $(M, \vec{\sigma}) \models E^{k+1} RAT$, it must be the case that $(M, \vec{\sigma}) \models B_i E^k RAT$. Clearly M is $\mathcal{L}_1$-measurable. This shows that (a) implies (b).

The fact that (b) implies (c) is immediate.

Finally, to see that (c) implies (a), suppose that M is a structure appropriate for Γ and ω is a state in M such that $\mathbf{s}_i(\omega) = \sigma$ and $(M, \omega) \models B_i E^k RAT$ for all $k \geq 0$. Let $X^0_j = \Sigma_j$ and let $X^{k+1}_j = \{\mathbf{s}_j(\omega') : (M, \omega') \models B_j E^k RAT\}$ for $k \geq 0$. We prove by induction that the set X^k_j satisfy the requirements of Definition 2. This is true by definition for $k = 0$, since $X^0_j = \Sigma_j$. If $\sigma' \in X^k_j$ for $k \geq 1$, choose some state ω' such that $(M, \omega') \models B_j E^{k-1} RAT$ and $\mathbf{s}_j(\omega') = \sigma'$. Let $\mu = \mathcal{PR}_j(\omega')$. Note that $B_j E^k \varphi \Rightarrow B_j B_j E^{k-1} \varphi$ is valid for all k. Thus, at all states ω'' in the support of μ, we have $\mathbf{s}_j(\omega'') \in X^{k-1}_j$, by the induction hypothesis. Let $\mu_{\sigma', k}$ to be the projection of μ onto Σ_{-j}. Since players know their own strategies and beliefs, and $(M, \omega') \models B_j RAT$, it easily follows that $(M, \omega') \models RAT_j$, so σ'_j must be a best response to the beliefs $\mu_{\sigma', k}$, as desired. Thus, by Definition 2, σ is rationalizable' and, by Proposition 3, σ is rationalizable. $\qquad\square$

We now consider iterated deletion of strongly dominated (resp., weakly dominated) strategies.

DEFINITION 5. Strategy σ for player is i *strongly dominated by* σ' *with respect to* $\Sigma'_{-i} \subseteq \Sigma_{-i}$ if $u_i(\sigma', \tau_{-i}) > u_i(\sigma, \tau_{-i})$ for all $\tau_{-i} \in \Sigma'_{-i}$. Strategy σ for player is i *weakly dominated by* σ' *with respect to* $\Sigma'_{-i} \subseteq \Sigma_{-i}$ if $u_i(\sigma', \tau_{-i}) \geq u_i(\sigma, \tau_{-i})$ for all $\tau_{-i} \in \Sigma'_{-i}$ and $u_i(\sigma', \tau'_{-i}) > u_i(\sigma, \tau'_{-i})$ for some $\tau'_{-i} \in \Sigma'_{-i}$.

Strategy σ for player i survives k rounds of iterated deletion of strongly dominated (resp., weakly dominated) strategies if, for each player j, there exists a sequence $X_j^0, \ldots, X_j^k$ of sets of strategies for player j such that $X_j^0 = \Sigma_j$ and, if $h < k$, then X_j^{h+1} consists of the strategies in X_j^h not strongly (resp., weakly) dominated by any strategy with respect to X_{-j}^h, and $\sigma \in X_i^k$. Strategy σ survives iterated deletion of strongly dominated (resp., weakly dominated) strategies if it survives k rounds of iterated deletion for all k. □

The following well-known result connects strong and weak dominance to best responses.

PROPOSITION 6. [Pearce 1984]

- *A strategy σ for player i is not strongly dominated by any strategy with respect to Σ'_{-i} iff there is a belief μ_σ of player i whose support is a subset of Σ'_{-i} such that σ is a best response with respect to μ_σ.*

- *A strategy σ for player i is not weakly dominated by any strategy with respect to Σ'_{-i} iff there is a belief μ_σ of player i whose support is all of Σ'_{-i} such that σ is a best response with respect to μ_σ.*

It immediately follows from Proposition 6 (and is well known) that a strategy is rationalizable iff it survives iterated deletion of strongly dominated strategies. Thus, the characterization of rationalizability in Theorem 4 is also a characterization of strategies that survive iterated deletion of strongly dominated strategies. We now give a slightly different characterization that allows us to relate iterated deletion of strongly and weakly dominated strategies. For each player i, define the formulas C_i^k inductively by taking C_i^0 to be *true* and C_i^{k+1} to be an abbreviation of $RAT_i \wedge B_i(\wedge_j C_j^k)$. That is, C_i^{k+1} holds (i.e., player i is $k + 1$-level rational) iff player i is playing a best response to his beliefs, and he knows that all players are k-level rational.

Two formulas φ and ψ are *logically equivalent* if $\varphi \Leftrightarrow \psi$ is valid (i.e., true in all states in all appropriate structures).

LEMMA 7. *The following formulas are logically equivalent for all $k \geq 0$.*

(a) C_i^{k+1};

(b) $B_i C^{k+1}$;

(c) $RAT_i \wedge B_i(\wedge_{j \neq i} C_j^k)$;

(d) $RAT_i \wedge B_i E^{k-1} RAT$ *(taking $E^{-1}\varphi$ to be the formula true).*

Moreover, $C_i^{k+1} \Rightarrow C_i^k$ is valid for all $k \geq 0$.

Proof. A straightforward induction on k. □

The following alternative characterization of iterated deletion of strongly dominated strategies can be proved in much as the same way as Theorem 4; we omit details here.

THEOREM 8. *The following are equivalent:*

(a) *the strategy σ for player i survives k rounds of iterated deletion of strongly dominated strategies in game Γ;*

(b) *there is an $\mathcal{L}^1$-measurable structure M^k appropriate for Γ and a state ω^k in M^k such that $\mathbf{s}_i(\omega^k) = \sigma$ and $(M^k, \omega^k) \models C_i^k$;*

(c) *there is a structure M^k appropriate for Γ and a state ω^k in M^k such that $\mathbf{s}_i(\omega^k) = \sigma$ and $(M^k, \omega^k) \models C_i^k$.*

The following corollary now follows from Theorem 8, Lemma 7, and the fact that the deletion procedure converges after a finite number of steps.

COROLLARY 9. *The following are equivalent:*

(a) *The strategy σ for player i survives iterated deletion of strongly dominated strategies in game Γ;*

(b) *there exists a measurable structure M that is appropriate for Γ and a state ω such that $\mathbf{s}_i(\omega) = \sigma$ and $(M, \omega) \models C_i^k$ for all $k \geq 0$;*

(c) *there exists a structure M that is appropriate for Γ and a state ω such that $\mathbf{s}_i(\omega) = \sigma$ and $(M, \omega) \models C_i^k$ for all $k \geq 0$.*

We next turn to characterizing iterated deletion of weakly dominated strategies.

3 Characterizing Iterated Admissibility

In the standard treatment (which is essentially the one considered in Section 2), player i is taken to be $(k + 1)$-level rational iff player i is rational (i.e., playing a best response to his beliefs), and knows that all other player are k-level rational.[1] But what else do players know?

We want to consider a situation where, intuitively, *all* an agent knows about the other agents is that they satisfy the appropriate rationality assumptions. More precisely, we modify the formula C_i^{k+1} to require that not only does player i know that the players are k-level rational, but this is the *only* thing that he knows about the other players. That is, we say that agent i is $(k + 1)$-level rational if player i is rational, he knows that the players are k-level rational, and this is all player i knows about the other players. We here use the phrase "all agent i knows" in essentially the same sense that it is used by Levesque [1990] and Halpern and Lakemeyer [2001], but formalize it a bit differently. Roughly speaking, we interpret "all agent i knows is φ" as meaning that agent i believes φ, and considers possible every *formula* about the other players that is consistent with φ. Thus, what "all I know" means is very sensitive to the choice of the language. To stress this point, we talk about "all I knows *with respect to language $\mathcal{L}$*".

To formalize this, consider the modal operator $\Diamond$ defined as follows:

---

[1] We should perhaps say "believes" here rather than "knows", since a player can be mistaken. We are deliberately blurring the subtle distinctions between "knowledge" and "belief" here.

- $(M, \omega) \models \Diamond \varphi$ iff there is some structure M' appropriate for Γ and state ω' such that $(M', \omega') \models \varphi$.

Intuitively, $\Diamond \varphi$ is true if there is some state and structure where φ is true; that is, if φ_i is satisfiable. Note that if $\Diamond \varphi$ is true at some state, then it is true at all states in all structures. Define $O_i^{\mathcal{L}} \varphi$ (read "all agent i knows with respect to the language $\mathcal{L}$") to be an abbreviation of

$$B_i \varphi \wedge (\wedge_{\psi \in \mathcal{L}} (\Diamond(\varphi \wedge \psi) \Rightarrow \langle B_i \rangle \psi)).$$

Since $O_i^{\emptyset}$ (where $\emptyset$ denotes the empty language) is equivalent to B_i (under the standard identification of the empty conjunction with the formula *true*), it follows that C_i^{k+1} is just $RAT_i \wedge O_i^{\emptyset}(\wedge_j C_j^k)$. We next consider a slightly richer language, whose formulas can talk about strategies (but not beliefs) of the players. (In Section 4, we consider languages in which we can talk about both the strategies and beliefs of the players.) Define the primitive proposition $play_i(\sigma)$ as follows:

- $(M, \omega) \models play_i(\sigma)$ iff $\omega \in [\![\sigma]\!]_M$.

Let $play(\vec{\sigma})$ be an abbreviation of $\wedge_{j=1}^n play_j(\sigma_j)$, and let $play_{-i}(\sigma_{-i})$ be an abbreviation of $\wedge_{j \neq i} play_j(\sigma_j)$. Intuitively, $(M, \omega) \models play(\vec{\sigma})$ iff $s(\omega) = \sigma$, and $(M, \omega) \models play_{-i}(\sigma_{-i})$ if, at ω, the players other than i are playing strategy profile σ_{-i}.

Let $\mathcal{L}^0(\Gamma)$ be the language whose only formulas are (Boolean combinations of) formulas of the form $play_i(\sigma)$, $i = 1, \ldots, n$, $\sigma \in \Sigma_i$. Let $\mathcal{L}_i^0(\Gamma)$ consist of just the formulas of the form $play_i(\sigma)$, and let $\mathcal{L}_{-i}^0(\Gamma) = \cup_{j \neq i} \mathcal{L}_j^0(\Gamma)$. Again, we omit the parenthetical Γ when it is clear from context or irrelevant.

We would now like to define D_i^{k+1} by replacing B_i by $O_i^{\mathcal{L}_{-i}^0}$ in the definition of C_i^{k+1}. That is, rather than just saying that agent i believes that all agents are rational up to level k, we want to say that this is *all* agent i believes. However, if we used this definition, the resulting formulas would in general be inconsistent! For example, D_i^3 would then imply D_i^2, which would require that i consider possible (i.e., assign positive probability to) all strategies for the other players consistent with 1-rationality, while D_i^3 would also require that player i believe that the other players are 2-rational (and thus player i should ascribe probability 0 to all strategies for the other players that are 1-rational but not 2-rational). Thus, this is not the definition we use.

The first step in getting an appropriate analogue to C_i^k is to remove the conjunct D_i^k from the scope of $O_i^{\mathcal{L}}$. As Lemma 7(c) shows, this change would have no impact on the definition of C_i^{k+1}. With this change, it is no longer the case that D_i^{k+1} implies D_i^k (which is a good thing, since the two formulas are still inconsistent). However, since we want D_i^{k+1} to be true at a state if the strategy used by player i at that state survives k rounds of iterated deletion of weakly dominated strategies, we need to add a conjunct to D_i^{k+1} that guarantees that the strategy used is k'-rational for $k' < k + 1$. Thus, for each player i, define the formulas D_i^k inductively by taking D_i^0 to be *true* and D_i^{k+1} to be an abbreviation of

$$RAT_i \wedge D_i'^k \wedge O_i^{\mathcal{L}_{-i}^0}(\wedge_{j \neq i} D_j^k),$$

where $D_i'^k$ (read "player i plays a strategy consistent with k-level rationality") is an abbreviation of $(\wedge_{\sigma \in \Sigma_i(\Gamma)} play_i(\sigma) \Rightarrow \Diamond(play_i(\sigma) \wedge D_i^k))$. That is, D_i^{k+1} holds (i.e., player i is $(k+1)$-level rational) iff player i is rational, plays a strategy that is consistent with k-level rationality, knows that other players are k-level rational, and that is all player i knows about the *strategies* of the other players.[2]

By expanding the modal operator O, it easily follows that D_i^{k+1} implies $RAT_i \wedge B_i(\wedge_{j \neq i} D_j^k)$; an easy induction on k then shows that D_i^{k+1} implies C_i^{k+1}. But D_i^{k+1} requires more; it requires that player i assigns positive probability to each strategy profile for the other players that is compatible with D_{-i}^k (i.e., with level-k rationality). As we now show, the formula D_i^k characterizes strategies that survive iterated deletion of weakly dominated strategies.

THEOREM 10. *The following are equivalent:*

(a) *the strategy σ for player i survives k rounds of iterated deletion of weakly dominated strategies;*

(b) *there is an $\mathcal{L}^2$-measurable structure M^k appropriate for Γ and a state ω^k in M^k such that $\mathbf{s}_i(\omega^k) = \sigma$ and $(M^k, \omega^k) \models D_i^k$;*

(c) *there is a structure M^k appropriate for Γ and a state ω^k in M^k such that $\mathbf{s}_i(\omega^k) = \sigma$ and $(M^k, \omega^k) \models D_i^k$.*

In addition, there is a finite structure $\overline{M}^k = (\Omega^k, \mathbf{s}, \mathcal{F}, \mathcal{PR}_1, \dots, \mathcal{PR}_n)$ such that $\Omega^k = \{(k', i, \vec{\sigma}) : k' \leq k, 1 \leq i \leq n, \vec{\sigma} \in X_1^{k'} \times \cdots \times X_n^{k'}\}$, $\mathbf{s}(k', i, \vec{\sigma}) = \vec{\sigma}$, $\mathcal{F} = 2^{\Omega^k}$, $X_j^{k'}$ consists of all strategies for player j that survive k' rounds of iterated deletion of weakly dominated strategies and, for all states $(k', i, \vec{\sigma}) \in \Omega^k$, $(\overline{M}^k, (k', i, \vec{\sigma})) \models \wedge_{j \neq i} D_j^{k'}$.

Proof. We proceed by induction on k, proving both the equivalence of (a), (b), and (c) and the existence of a structure $\overline{M}^k$ with the required properties.

The result clearly holds if $k = 0$. Suppose that the result holds for k; we show that it holds for $k + 1$. We first show that (c) implies (a). Suppose that $(M^{k+1}, \omega^{k+1}) \models D_i^{k+1}$ and $\mathbf{s}_i(\omega^{k+1}) = \sigma_i$. It follows that σ_i is a best response to the belief μ_{σ_i} on the strategies of other players induced by $\mathcal{PR}_i^{k+1}(\omega)$. Since $(M^{k+1}, \omega^{k+1}) \models B_i(\wedge_{j \neq i} D_j^k)$, it follows from the induction hypothesis that the support of μ_{σ_i} is contained in X_{-i}^k. Since $(M^{k+1}, \omega^{k+1}) \models \wedge_{\sigma_{-i} \in \Sigma_{-i}} (\Diamond(play_{-i}(\sigma_{-i}) \wedge (\wedge_{j \neq i} D_j^k)) \Rightarrow \langle B_j \rangle (play_{-i}(\sigma_{-i})))$, it follows from the induction hypothesis that the support of μ_{σ_i} is all of X_{-i}^k. Since $(M^{k+1}, \omega^{k+1}) \models D_i'^k$, it follows from the induction hypothesis that $\sigma_i \in X_i^k$. Thus, since $(M^{k+1}, \omega^{k+1}) \models RAT_i$, it follows by Proposition 6 that $\sigma_i \in X_i^{k+1}$.

We next construct the structure $\overline{M}^{k+1} = (\Omega^{k+1}, \mathbf{s}, \mathcal{F}, \mathcal{PR}_1, \dots, \mathcal{PR}_n)$. As required, we define $\Omega^{k+1} = \{(k', i, \vec{\sigma}) : k' \leq k+1, 1 \leq i \leq n, \vec{\sigma} \in X_1^{k'} \times \cdots \times X_n^{k'}\}$, $\mathbf{s}(k', i, \vec{\sigma}) = \vec{\sigma}$, $\mathcal{F} = 2^{\Omega^{k+1}}$. For a state ω of the form $(k', i, \vec{\sigma})$, since $\sigma_j \in X_j^{k'}$, by Proposition 6, there exists a distribution μ_{k', σ_j} whose support is all of $X_{-j}^{k'-1}$ such that σ_j is a best response to μ_{σ_j}. Extend μ_{k', σ_j} to a distribution μ'_{k', i, σ_j} on Ω^{k+1} as follows:

- if $i \neq j$, then $\mu'_{k', i, \sigma_j}(k'', i', \vec{\tau}) = \mu_{k', \sigma_j}(\vec{\tau}_{-j})$ if $i' = j, k'' = k' - 1$, and $\tau_j = \sigma_j$, and 0 otherwise;

- $\mu'_{k', j, \sigma_j}(k'', i', \vec{\tau}) = \mu_{k', \sigma_j}(\vec{\tau}_{-j})$ if $i' = j, k'' = k'$, and $\tau_j = \sigma_j$, and 0 otherwise.

---

[2] D_i^k is essentially the formula RAT_i^k from the introduction. However, since we consider a number of variants of RAT_i^k, we find it useful to distinguish them.

Let $\mathcal{PR}_j(k', i, \vec{\sigma}) = \mu_{k', i, \sigma_j}$. We leave it to the reader to check that this structure is appropriate. An easy induction on k' now shows that $(\overline{M}^{k+1}, (k', i, \vec{\sigma})) \models \wedge_{j \neq i} D_j^{k'}$ for $i = 1, \ldots, n$.

To see that (a) implies (b), suppose that $\sigma_j \in X_j^{k+1}$. Choose a state ω in $\overline{M}^{k+1}$ of the form $(k+1, i, \vec{\sigma})$, where $i \neq j$. As we just showed, $(\overline{M}^{k+1}, (k', i, \vec{\sigma})) \models D_j^{k'}$, and $\mathbf{s}_j(k', i, \vec{\sigma}) = \sigma_j$. Moreover, $\overline{M}^{k+1}$ is measurable (since $\mathcal{F}$ consists of all subsets of Ω^{k+1}). Clearly (b) implies (c). $\qquad \square$

Note that there is no analogue of Corollary 9 here. This is because there is no state where D_i^k holds for all $k \geq 0$; it cannot be the case that i places positive probability on all strategies (as required by D_1^k) and that i places positive probability only on strategies that survive one round of iterated deletion (as required by D_2^k), unless all strategies survive one round on iterated deletion. We can say something slightly weaker though. There is some k^* such that the process of iterated deletion converges after k^* steps; that is, $X_j^{k^*} = X_j^{k^*+1}$ for all j (and hence $X_j^{k^*} = X_j^{k'}$ for all $k' \geq k^*$). That means that there is a state where $D_i^{k'}$ holds for all $k' > k^*$. Thus, we can show that a strategy σ for player i survives iterated deletion of weakly dominated strategies iff there exists a k^* and a state ω such that $\mathbf{s}_i(\omega) = \sigma$ and $(M, \omega) \models D_i^{k'}$ for all $k' > k^*$. Since C_i^{k+1} implies C_i^k, an analogous result holds for iterated deletion of strongly dominated strategies, with $D_i^{k'}$ replaced by $C_i^{k'}$.

It is also worth stressing that, unlike the BFK construction, in a state where D^k holds, an agent does *not* consider all strategies possible, but only the ones consistent with the appropriate level of rationality. We could require the agent to consider all strategies possible by using LPSs or nonstandard probability. The only change that this would make to our characterization is that, if we are using nonstandard probability, we would interpret $B_i \varphi$ to mean that φ holds with probability infinitesimally close to 1, while $\langle B_i \rangle \varphi$ would mean that φ holds with probability whose standard part is positive (i.e., non-infinitesimal probability). We do not pursue this point further.

4 Richer Languages

The formula D_i^{k+1} was defined with respect to the language $\mathcal{L}_{-i}^O$, and thus required that player i assign positive probability to all and only strategies consistent with D_{-i}^k. But why focus just on strategies? We now consider richer languages that can talk about players' beliefs; this requires players to ascribe positive probability to all beliefs that the other agents could have as well as all the strategies they could be using (which are consistent with appropriate levels of rationality).

First, let $\mathcal{L}^2(\Gamma)$ be the extension of $\mathcal{L}^1$ that includes a primitive proposition $play_i(\sigma)$ for each player i and strategy $\sigma \in \Sigma_i$.

To relate our results to those of BFK, even the language $\mathcal{L}^2$ is too weak, since it does not allow an agent to express probabilistic beliefs. Let $\mathcal{L}^3(\Gamma)$ be the language that extends $\mathcal{L}^2(\Gamma)$ by allowing formulas of the form $pr_i(\varphi) \geq \alpha$ and $pr_i(\varphi) > \alpha$, where α is a rational number in $[0, 1]$; $pr_i(\varphi) \geq \alpha$ can be read as "the probability of φ according to i is at least α", and similarly for $pr_i(\varphi) > \alpha$. We allow nesting here, so that we can have a formula of the form $pr_j(play_i(\sigma) \wedge pr_k(play_i(\sigma'))) > 1/3) \geq 1/4$. As we would expect,

- $(M, \omega) \models pr_i(\varphi)$ iff $\mathcal{PR}_i(\omega)(\llbracket \varphi \rrbracket_M) \geq \alpha$.

The restriction to α being rational allows the language to be countable. However, as we now show, it is not too serious a restriction.

Let $\mathcal{L}^4(\Gamma)$ be the language that extends $\mathcal{L}^2(\Gamma)$ by closing off under countable conjunctions, so that if $\varphi_1, \varphi_2, \ldots$ are formulas, then so is $\wedge_{m=1}^{\infty} \varphi_m$, and formulas of the form $pr_i(\varphi) > \alpha$, where α is a real number in $[0, 1]$. (We can express $pr_i(\varphi) \geq \alpha$ as the countable conjunction $\wedge_{\beta < \alpha, \beta \in Q \cap [0,1]} pr_i(\varphi) > \beta$, where Q is the set of rational numbers, so there is no need to include formulas of the form $pr_i(\varphi) \geq \alpha$ explicitly in $\mathcal{L}^4(\Gamma)$.) We omit the parenthetical Γ in $\mathcal{L}^3(\Gamma)$ and $\mathcal{L}^4(\Gamma)$ when the game Γ is clear from context. A subset Φ of $\mathcal{L}^3$ is $\mathcal{L}^3$-*realizable* if there exists an appropriate structure M for Γ and state ω in M such that, for all formulas $\varphi \in \mathcal{L}^3$, $(M, \omega) \models \varphi$ iff $\varphi \in \Phi$.[3] We can similarly define what it means for a subset of $\mathcal{L}^4$ to be $\mathcal{L}^4$-realizable.

LEMMA 11. *Every $\mathcal{L}^3$-realizable set can be uniquely extended to an $\mathcal{L}^4$-realizable set.*

Proof. It is easy to see that every $\mathcal{L}^3$-realizable set can be extended to an $\mathcal{L}^4$-realizable set. For suppose that Φ is $\mathcal{L}^3$-realizable. Then there is some state ω and structure M such that, for every formula $\varphi \in \mathcal{L}^3$, we have that $(M, \omega) \models \varphi$ iff $\varphi \in \Phi$. Let Φ' consist of the $\mathcal{L}^4$ formulas true at ω. Then clearly Φ' is an $\mathcal{L}^4$-realizable set that extends Φ.

To show that the extension is unique, suppose that there are two $\mathcal{L}^4$-realizable sets, say Φ_1 and Φ_2, that extend Φ. We want to show that $\Phi_1 = \Phi_2$. To do this, we consider two language, $\mathcal{L}^5$ and $\mathcal{L}^6$, intermediate between $\mathcal{L}^3$ and $\mathcal{L}^4$.

Let $\mathcal{L}^5$ be the language that extends $\mathcal{L}^2$ by closing off under countable conjunctions and formulas of the form $pr_i(\varphi) > \alpha$, where α is a rational number in $[0, 1]$. Thus, in $\mathcal{L}^5$, we have countable conjunctions and disjunctions, but can talk explicitly only about rational probabilities. Nevertheless, it is easy to see that for every formula $\varphi \in \mathcal{L}^4$, there is an formula equivalent formula $\varphi' \in \mathcal{L}^5$, since if α is a real number, then $pr_i(\varphi) > \alpha$ is equivalent to $\vee_{\beta > \alpha, \beta \in [0,1] \cap Q} pr_i(\varphi) > \beta$ (an infinite disjunction $\vee_{i=1}^{\infty} \varphi_i$ can be viewed as an abbreviation of $\neg \wedge_{i=1}^{\infty} \neg \varphi_i$).

Next, let $\mathcal{L}^6$ be the result of closing off formulas in $\mathcal{L}^3$ under countable conjunction and disjunction. Thus, in $\mathcal{L}^6$, we can apply countable conjunction and disjunction only at the outermost level, not inside the scope of pr_i. We claim that for every formula $\varphi \in \mathcal{L}^5$, there is an equivalent formula in $\mathcal{L}^6$. More precisely, for every formula $\varphi \in \mathcal{L}^5$, there exist formulas $\varphi_{ij} \in \mathcal{L}^3$, $1 \leq i, j < \infty$ such that φ is equivalent to $\wedge_{m=1}^{\infty} \vee_{n=1}^{\infty} \varphi_{mn}$. We prove this by induction on the structure of φ. If φ is RAT_i, $play_i(\sigma)$, or *true*, then the statement is clearly true. The result is immediate from the induction hypothesis if φ is a countable conjunction. If φ has the form $\neg \varphi'$, we apply the induction hypothesis, and observe that $\neg(\wedge_{m=1}^{\infty} \vee_{n=1}^{\infty} \varphi_{mn})$ is equivalent to $\vee_{m=1}^{\infty} \wedge_{n=1}^{\infty} \neg \varphi_{mn}$. We can convert this to a conjunction of disjunctions by distributing the disjunctions over the conjunctions in the standard way (just as $(E_1 \cap E_2) \cup (E_3 \cap E_4)$ is equivalent to $(E_1 \cup E_3) \cap (E_1 \cup E_4) \cap (E_2 \cup E_3) \cap (E_2 \cup E_4)$). Finally, if φ has the form $pr_i(\varphi') > \alpha$, we apply the induction hypothesis, and observe that $pr_i(\wedge_{m=1}^{\infty} \vee_{n=1}^{\infty} \varphi_{mn}) > \alpha$ is equivalent to

$$\vee_{\alpha' > \alpha, \alpha' \in Q \cap [0,1]} \wedge_{M=1}^{\infty} \vee_{N=1}^{\infty} pr_i(\wedge_{m=1}^{M} \vee_{n=1}^{N} \varphi_{mn}) > \alpha'.$$

The desired result follows, since if two states agree on all formulas in $\mathcal{L}^3$, they must agree on all formulas in $\mathcal{L}^6$, and hence on all formulas in $\mathcal{L}^5$ and $\mathcal{L}^4$. $\square$

---

[3]For readers familiar with standard completeness proofs in modal logic, if we had axiomatized the logic we are implicitly using here, the $\mathcal{L}^3$-realizable sets would just be the maximal consistent sets in the logic.

The choice of language turns out to be significant for a number of our results; we return to this issue at various points below.

With this background, let $\mathcal{L}_i^3$ consist of all formulas in $\mathcal{L}^3$ of the form $pr_i(\varphi) \geq \alpha$ and $pr_i(\varphi) > \alpha$ (φ can mention pr_j, $j \neq i$; it is only the outermost modal operator that must be i). Intuitively, $\mathcal{L}_i^3$ consists of the formulas describing i's beliefs. Let $\mathcal{L}_{i+}^3$ consist of $\mathcal{L}_i^3$ together with formulas of the form *true*, RAT_i, and $play_i(\sigma)$, for $\sigma \in \Sigma_i$. Let $\mathcal{L}_{(-i)+}^3$ be an abbreviation of $\cup_{j \neq i} \mathcal{L}_{j+}^3$. We can similarly define $\mathcal{L}_i^4$ and $\mathcal{L}_{i+}^4$.

Note that if $\varphi \in \mathcal{L}_{(-i)+}^3$, then $O_i^{L_{(-i)+}^3} \varphi$ is an abbreviation of the formula

$$B_i\varphi \wedge (\wedge_{\psi \in \mathcal{L}_{(-i)+}^3} \Diamond(\varphi \wedge \psi) \Rightarrow \langle B_j \rangle \psi).$$

Thus, $O_i^{\mathcal{L}_{(-i)+}^3} \varphi$ holds if agent i believes φ but does not know anything beyond that; he ascribes positive probability to all formulas in $\mathcal{L}_{(-i)+}^3$ consistent with φ. This is very much in the spirit of the Halpern-Lakemeyer [2001] definition in the context of epistemic logic. Of course, we could go further and define a notion of "all i knows" for the language $\mathcal{L}^4$. Doing this would give a definition that is even closer to that of Halpern and Lakemeyer. Unfortunately, we cannot require than agent i ascribe positive probability to all the formulas in $\mathcal{L}_{(-i)+}^4$ consistent with φ; in general, there will be an uncountable number of distinct and mutually exclusive formulas consistent with φ, so they cannot all be assigned positive probability. This problem does not arise with $\mathcal{L}^3$, since it is a countable language. Halpern and Lakemeyer could allow an agent to consider an uncountable set of worlds possible, since they were not dealing with probabilistic systems. In the sequel, we thus focus on the language $\mathcal{L}^3$ and let O_i denote $O_i^{\mathcal{L}_{(-i)+}^3}$.[4]

Define the formulas F_i^k inductively by taking F_i^0 to be the formula *true*, and F_i^{k+1} to an abbreviation of

$$RAT_i \wedge F_i'^k \wedge O_i(\wedge_{j \neq i} F_j^k),$$

where $F_i'^k$ is an abbreviation of $(play_i(\sigma) \Rightarrow \Diamond(play_i(\sigma) \wedge F_i^k))$. Thus, F_i^{k+1} says that i is rational, plays a k-level rational strategy, knows that all the other players satisfy level-k rationality (i.e., F_j^k), and that is all that i knows. It is easy to see that F_i^{k+1} implies D_j^{k+1}. The difference is that instead of requiring just that j assign positive probability to all strategy profiles compatible with F_{-j}^k, it requires that j assign positive probability to all formulas in $\mathcal{L}_{(-i)+}^3$ compatible with F_{-j}^k.

The next result shows that F_i^k characterizes iterated admissibility, just as D_i^k does.

THEOREM 12. *The following are equivalent:*

(a) *the strategy σ for player i survives k rounds of iterated deletion of weakly dominated strategies;*

(b) *there exists an $\mathcal{L}^3$-measurable structure M^k appropriate for Γ and a state ω^k in M^k such that $s_i(\omega^k) = \sigma$ and $(M^k, \omega^k) \models F_i^k$;*

(c) *there is a structure M^k appropriate for Γ and a state ω^k in M^k such that $s_i(\omega^k) = \sigma$ and $(M^k, \omega^k) \models F_i^k$.*

---

[4]Note that the modal operator $\Diamond$ is not in the language $\mathcal{L}^3$ or $\mathcal{L}^4$. None of our results would be affected if we had considered a language that also included $\Diamond$; for ease of exposition, we have decided not to include $\Diamond$ here.

Proof. The proof is similar in spirit to the proof of Theorem 10. We again proceed by induction on k. The result clearly holds for $k = 0$. If $k = 1$, the proof that (c) implies (a) is essentially identical to that of Theorem 10; we do not repeat it here.

To prove that (a) implies (b), we need three lemmas. The first shows that a formula is always satisfied in a state that has probability 0; the second shows that that we can get a new structure with a world where agent i ascribes positive probability to each of a countable collection of satisfiable formulas in $\mathcal{L}^3_{-i}$; and the third shows that formulas in $\mathcal{L}^4_{i+}$ for different players i are independent (i.e., if $\varphi_i \in \mathcal{L}^4_{i+}$ is satisfiable, then so is $\varphi_1 \wedge \ldots \wedge \varphi_n$).

LEMMA 13. *If $\varphi \in \mathcal{L}^4$ is satisfiable in a measurable structure, then there exists an $\mathcal{L}^3$-measurable structure M and state ω such that $(M, \omega) \models \varphi$, $\{\omega\}$ is measurable, $PR_j(\omega)(\{\omega\}) = 0$ for $j = 1, \ldots, n$.*

Proof. Suppose that $(M', \omega') \models \varphi$, where $M' = (\Omega', \mathbf{s}', \mathcal{F}', PR'_1, \ldots, PR'_n)$. Let $\Omega = \Omega' \cup \{\omega\}$, where where ω is a fresh state; let $\mathcal{F}$ be the smallest σ-algebra that contains $\mathcal{F}$ and $\{\omega\}$; let $\mathbf{s}$ and PR_j agree with $\mathbf{s}'$ and PR'_j when restricted to states in Ω'; more precisely, if $\omega'' \in \Omega'$, then $PR_j(\omega'')(A) = PR'_j(\omega'')(A \cap \Omega')$ for $j = 1, \ldots, n$). Finally, define $\mathbf{s}_i(\omega) = \mathbf{s}_i(\omega')$, and take $PR_j(\omega)(A) = PR'_j(\omega')(A \cap \Omega')$ for $j = 1, \ldots, n$. Clearly $\{\omega\}$ is measurable, and $PR_j(\omega)(\{\omega\}) = 0$ for $j = 1, \ldots, n$. An easy induction on structure shows that for all formulas ψ, (a) $(M, \omega) \models \psi$ iff $(M, \omega') \models \psi$, and (b) for all states $\omega'' \in \Omega'$, we have that $(M, \omega'') \models \psi$ iff $(M', \omega'') \models \psi$. It follows that $(M, \omega) \models \varphi$, and that M is measurable. $\square$

LEMMA 14. *Suppose that $\vec{\sigma} \in \vec{\Sigma}$, Φ' is a countable collection of formulas in $\mathcal{L}^4_{-i}$, $\varphi \in \mathcal{L}^4_{-i}$, and Σ'_{-i} is a set of strategy profiles in Σ_{-i} such that (a) for each formula $\varphi' \in \Phi'$, there exists some profile $\sigma_{-i} \in \Sigma'_{-i}$ such that $\varphi \wedge \varphi' \wedge play_{-i}(\sigma_{-i})$ is satisfied in a measurable structure, and (b) for each profile $\sigma_{-i} \in \Sigma'_{-i}$, $play_{-i}(\sigma_{-i})$ is one of the formulas in Φ'. Then, if μ_i is a probability measure with support Σ'_{-i}, there exists an $\mathcal{L}^3$-measurable structure M and state ω such that $\mathbf{s}(\omega) = \vec{\sigma}$, $(M, \omega) \models pr_i(play_{-i}(\sigma_{-i})) \geq \alpha$ iff $\mu_i(\sigma_{-i}) \geq \alpha$ (that is, μ_i agrees with $PR_i(\omega)$ when marginalized to strategy profiles in Σ'_{-i}), and $(M, \omega) \models B_i \varphi \wedge \langle B_i \rangle \varphi'$ for all $\varphi' \in \Phi'$.*

Proof. Let Φ', Σ'_{-i}, and μ_i be as in the statement of the lemma. Suppose that $\Phi' = \{\varphi_1, \varphi_2, \ldots, \ldots\}$. By assumption, for each formula $\varphi_k \in \Phi'$, there exists some strategy profile $\sigma'_{-i} \in \Sigma'_{-i}$, an $\mathcal{L}^3$-measurable structure $M^k = (\Omega^k, \mathbf{s}^k, \mathcal{F}^k, PR^k_1, \ldots, PR^k_n)$, and a state $\omega^k \in \Omega^k$ such that $(M^k, \omega^k) \models \varphi \wedge \varphi_k \wedge play_{-i}(\sigma'_{-i})$, for $k = 1, 2, \ldots$. By Lemma 13, we can assume without loss of generality that $\{\omega^k\} \in \mathcal{F}^k$ and $PR^k_j(\omega^k)(\{\omega^k\}) = 0$. Define $M^\infty = (\Omega^\infty, \mathbf{s}^\infty, \mathcal{F}^\infty, PR^\infty_1, \ldots, PR^\infty_n)$ as follows:

- $\Omega^\infty = \cup_{k=0}^\infty \Omega^k \cup \{\omega\}$, where ω is a fresh state;

- $\mathcal{F}^\infty$ is the smallest σ-algebra that contains $\{\omega\} \cup \mathcal{F}_1 \cup \mathcal{F}_2 \cup \ldots$;

- $\mathbf{s}^\infty$ agrees with $\mathbf{s}^k$ when restricted to states in Ω^k, except that $\mathbf{s}^\infty_i(\omega^k) = \sigma_i$ and $\mathbf{s}^\infty(\omega) = \vec{\sigma}$;

- PR^∞_j agrees with PR^k_j when restricted to states in Ω^k; more precisely, if $\omega' \in \Omega^k$, then $PR^\infty_j(\omega')(A) = PR^k_j(\omega')(A \cap \Omega^k)$, except that $PR^\infty_i(\omega) = PR^\infty_i(\omega^1) = PR_i(\omega^2) = \cdots$ is defined to be a distribution with support $\{\omega^1, \omega^2, \ldots\}$ (so that all these states are given positive probability) such that $\mathcal{PR}_i^\infty(\omega)$ agrees μ when marginalized to profiles in Σ_{-i}, and $\mathcal{PR}_j^\infty(\omega)(\{\omega\}) = 1$ for $j \neq i$. It is easy to see that our assumptions guarantee that this can be done.

We can now prove by a straightforward induction on the structure of ψ that (a) for all formulas ψ, $k = 1, 2, 3, \ldots$, and states $\omega' \in \Omega^k - \{\omega^k\}$, we have that $(M^k, \omega') \models \psi$ iff $(M^\infty, \omega') \models \psi$; and (b) for all formulas $\psi \in \mathcal{L}_{(-i)+}^4$, $k = 1, 2, 3, \ldots$, and $(M^k, \omega^k) \models \psi$ iff $(M^\infty, \omega^k) \models \psi$. (Here it is important that $\mathcal{PR}_j^\infty(\omega^k)(\{\omega^k\}) = \mathcal{PR}_j^k(\omega)(\{\omega\}) = 0$ for $j \neq i$; this ensures that j's beliefs about i's strategies and beliefs is unaffected by the fact that $\mathbf{s}_i^k(\omega^k) \neq \mathbf{s}_i^\infty(\omega^k)$ and $\mathcal{PR}_i^k(\omega^k) \neq \mathcal{PR}_i^\infty(\omega^k)$.) It easily follows that $(M^\infty, \omega) \models B_i\varphi \wedge \langle B_i \rangle\varphi'$ for all $\varphi' \in \Phi'$. $\qquad\square$

LEMMA 15. *If $\varphi_i \in \mathcal{L}_{i+}^4$ is satisfiable in an $\mathcal{L}^3$-measurable structure for $i = 1, \ldots, n$, then $\varphi_1 \wedge \ldots \wedge \varphi_n$ is satisfiable in an $\mathcal{L}^3$-measurable structure.*

Proof. Suppose that $(M^i, \omega^i) \models \varphi_i$, where $M^i = (\Omega^i, \mathbf{s}^i, \mathcal{F}^i, \mathcal{PR}_1^i, \ldots, \mathcal{PR}_n^i)$ and $\varphi_i \in \mathcal{L}_{i+}^4$. By Lemma 13, we again assume without loss of generality that $\{\omega^i\} \in \mathcal{F}^i$ and $\mathcal{PR}_j(\omega^i)(\{\omega^i\}) = 0$. Let $M^* = (\Omega^*, \mathbf{s}^*, \mathcal{F}^*, \mathcal{PR}_1^*, \ldots, \mathcal{PR}_n^*)$, where

- $\Omega^* = \cup_{i=1}^n \Omega^i$;

- $\mathcal{F}^*$ is the smallest σ-algebra containing $\mathcal{F}^1 \cup \ldots \cup \mathcal{F}^n$;

- $\mathbf{s}^*$ agrees with $\mathbf{s}^j$ on states in Ω^j except that $\mathbf{s}_i^*(\omega^j) = \mathbf{s}_i^i(\omega^i)$ (so that $\mathbf{s}^*(\omega^1) = \cdots = \mathbf{s}^*(\omega^n)$);

- $\mathcal{PR}_i^*$ agrees with $\mathcal{PR}_i^j$ on states in Ω^j except that $\mathcal{PR}_i^*(\omega^j) = \mathcal{PR}_i^i(\omega^i)$ (so that $\mathcal{PR}_i^*(\omega^1) = \cdots = \mathcal{PR}_i^*(\omega^n) = \mathcal{PR}_i^i(\omega^i)$).

We can now prove by induction on the structure of ψ that (a) for all formulas $\psi \in \mathcal{L}_{i+}^4$, $i = 1, \ldots, n$, and states $\omega' \in \Omega^i$, we have that $(M^i, \omega') \models \psi$ iff $(M^*, \omega') \models \psi$; (b) for all formulas $\psi \in \mathcal{L}_{i+}^4$, $1 \leq i, j \leq n$, $(M^i, \omega^i) \models \psi$ iff $(M^*, \omega^j) \models \psi$ (again, here it is important that $\mathcal{PR}_i^*(\omega^j)(\{\omega^j\}) = 0$ for $j = 1, \ldots, n$). Note that part (b) implies that the states $\omega^1, \ldots, \omega^n$ satisfy the same formulas in M^*. It easily follows that $(M^*, \omega^i) \models \varphi_1 \wedge \ldots \wedge \varphi_n$ for $i = 1, \ldots, n$. $\qquad\square$

We can now prove the theorem. Again, let X_j^k be the strategies for player j that survive k rounds of iterated deletion of weakly dominated strategies. To see that (a) implies (b), suppose that $\sigma_i \in X_i^{k+1}$. By Proposition 6, there exists a distribution μ_i whose support is X_{-i}^k such that σ_i is a best response to μ_i. By the induction hypothesis, for each strategy profile $\tau_{-i} \in X_{-i}^k$, and all $j \neq i$, the formula $play_j(\tau_j) \wedge F_j^k$ is satisfied in a measurable structure. By Lemma 15, $play_{-j}(\tau_{-j}) \wedge (\wedge_{j \neq i} F_j^k)$ is satisfied in an $\mathcal{L}^3$-measurable structure. Taking φ to be $\wedge_{j \neq i} F_j^k$, by Lemma 14, there exists an $\mathcal{L}^3$-measurable structure M and state ω in M such that the marginal of $\mathcal{PR}_i(\omega)$ on X_{-i}^k is μ_i, $\mathbf{s}_i(\omega)$ is σ_i, and $(M, \omega) \models B_i(\wedge_{j \neq i} F_j^k) \wedge (\wedge_{\psi \in \mathcal{L}_{(-j)+}^3} \Diamond(\psi \wedge (\wedge_{j \neq i} F_j^k)) \Rightarrow \langle B_j \rangle\psi)$. It follows that $(M, \omega) \models RAT_i \wedge O_i(\wedge_{j \neq i} F_j^k)$. Moreover, since $\sigma_i \in X_i^{k+1}$, σ_i is also in X_i^k. By the induction hypothesis, $play(\sigma_i) \wedge F_i^k$ is satisfiable. Thus, $(M, \omega) \models F_i'^k$. Hence, $(M, \omega) \models F_i^{k+1}$, as desired.

It is immediate that (b) implies (c). $\qquad\square$

5 Complete and Canonical Structures

5.1 Canonical Structures

The intuition behind "all i knows is φ" goes back to Levesque [1990]. The idea is that all i knows is φ if (1) i knows φ (so that φ is true in all the worlds that i considers possible) and (2) i considers possible all worlds consistent with φ (so that φ is false in all worlds that i does not consider possible). In the single-agent case (which is what Levesque considered) it is relatively easy to make precise the set of worlds that i does not consider possible, since a world can be identified with a truth assignment. This is much more complicated in the multi-agent setting considered by Halpern and Lakemeyer [2001]. They made it precise by working in the canonical structure. The advantage of the canonical structure is that, in a sense it has all possible worlds, so it is clear what worlds an agent does not consider possible. Although our definition of "all i knows" is more language-dependent, our intuitions for the notion are still grounded in the canonical structure. Thus, in this section, we consider our definitions in the context of canonical structures. The reason we say "structures" here rather than "structure" is that the notion of canonical structure is also language dependent.

Define the *canonical structure* $M^c = (\Omega^c, \mathbf{s}^c, \mathcal{F}^c, \mathcal{PR}_1^c, \ldots, \mathcal{PR}_n^c)$ *for* $\mathcal{L}^4$ as follows:

- $\Omega^c = \{\omega_\Phi : \Phi \text{ is a realizable subset of } \mathcal{L}^4(\Gamma)\}$;

- $\mathbf{s}^c(\omega_\Phi) = \vec{\sigma}$ iff $play(\sigma) \in \Phi$;

- $\mathcal{F}^c = \{F_\varphi : \varphi \in \mathcal{L}^4\}$, where $F_\varphi = \{\omega_\Phi : \varphi \in \Phi\}$;

- $\mathrm{Pr}_i^c(\omega_\Phi)(F_\varphi) = \inf\{\alpha : pr_i(\varphi) > \alpha \in \Phi\}$.

LEMMA 16. *M^c is an appropriate $\mathcal{L}^3$-measurable structure for Γ.*

Proof. It is easy to see that $\mathcal{F}^c$ is a σ-algebra, since the complement of F_φ is $F_{\neg\varphi}$ and $\cap_{m=1}^\infty F_{\varphi_i} = F_{\wedge_{m=1}^\infty \varphi_m}$. Given a strategy σ for player i, $[\![\sigma]\!]_{M^c} = F_{play_i(\sigma)} \in \mathcal{F}$. Moreover, each realizable set Φ that includes $play_i(\sigma)$ must also include $pr_i(play_i(\sigma)) = 1$, so that $\mathcal{PR}_i(\omega_\Phi)(\mathbf{s}_i(\omega_\Phi)) = \mathcal{PR}_i(\omega_\Phi)(F_{play_i(\mathbf{s}_i(\omega_\Phi))}) = 1$. Similarly, if $\mathcal{PR}_i(\omega_\Phi) = \pi$, then $\{\omega \in \Omega^c : \mathcal{PR}_i(\omega) = \pi\} = \cap_{\varphi \in \mathcal{L}^3} \cap_{\{\alpha \in Q \cap [0,1] : \pi([\![\varphi]\!]_{M^c}) \geq \alpha\}} F_{\varphi \geq \alpha} \in \mathcal{F}^c$. Moreover, if $\alpha \in Q \cap [0,1]$, then $\pi([\![\varphi]\!]_{M^c}) \geq \alpha$ iff $pr_i(\varphi) \geq \alpha \in \Phi$. But if $pr_i(\varphi) \geq \alpha \in \Phi$, then $pr_i(pr_i(\varphi) \geq \alpha) = 1 \in \Phi$. It easily follows that $\mathcal{PR}_i(\omega_\Phi)(\{\omega : \mathcal{PR}_i(\omega) = \pi\}) = 1$. Finally, the definition of $\mathcal{F}^c$ guarantees that every set $[\![\varphi]\!]_{M^c}$ is measurable and that $\mathcal{PR}_i(\omega_\Phi)$ is indeed a probability distribution on $(\Omega^c, \mathcal{F}^c)$. $\square$

The following result is the analogue of the standard "truth lemma" in completeness proofs in modal logic.

PROPOSITION 17. *For $\psi \in \mathcal{L}^4$, $(M^c, \omega_\Phi) \models \psi$ iff $\psi \in \Phi$.*

Proof. A straightforward induction on the structure of ψ. $\square$

We have constructed a canonical structure for $\mathcal{L}^4$. It follows easily from Lemma 11 that the canonical structure for $\mathcal{L}^3$ (where the states are realizable $\mathcal{L}^3$ sets) is isomorphic to M^c. (In this case, the set $\mathcal{F}^c$ of measurable sets would be the smallest σ-algebra containing $[\![\varphi]\!]_M$ for $\varphi \in \mathcal{L}^3$.) Thus, the choice of $\mathcal{L}^3$ vs. $\mathcal{L}^4$ does not play an important role when constructing a canonical structure.

A strategy σ_i for player i survives iterated deletion of weakly dominated strategies iff the formula

$$undominated^k(\sigma_i) = play_i(\sigma_i) \wedge \vee_{k^*=}^{\infty} (\wedge_{k=k^*}^{\infty} F_i^k)$$

is satisfied at some state in the canonical structure. But there are other structures in which $undominated(\sigma_i)$ is satisfied. One way to get such a structure is by essentially "duplicating" states in the canonical structure. The canonical structure can be *embedded* in a structure M if, for all $\mathcal{L}^3$-realizable sets Φ, there is a state ω_Φ in M such that $(M, \omega_\Phi) \models \varphi$ iff $\varphi \in \Phi$. Clearly $undominated(\sigma_i)$ is satisfied in any structure in which the canonical structure can be embedded.

A structure in which the canonical structure can be embedded is in a sense larger than the canonical structure. But $undominated(\sigma_i)$ can be satisfied in structures smaller than the canonical structure. (Indeed, with some effort, we can show that it is satisfiable in a structure with countably many states.) There are two reasons for this. The first is that to satisfy $undominated(\sigma_i)$, there is no need to consider a structure with states where all the players are irrational. It suffices to restrict to states where at least one player is using a strategy that survives at least one round of iterated deletion. This is because players know their strategy; thus, in a state where a strategy σ_i for player i is admissible, player i must ascribe positive probability to all other strategies; however, in those states, player i still plays σ_i.

A perhaps more interesting reason that we do not need the canonical structure is our use of the language $\mathcal{L}_3$. The formulas F_i^k guarantee that player i ascribes positive probability to all formulas φ consistent with the appropriate level of rationality. Since a finite conjunction of formulas in $\mathcal{L}^3$ is also a formula in $\mathcal{L}^3$, player i will ascribe positive probability to all finite conjunctions of formulas consistent with rationality. But a state is characterized by a *countable* conjunction of formulas. Since $\mathcal{L}^3$ is not closed under countable conjunctions, a structure that satisfies $undominated(\sigma_i)$ may not have states corresponding to all $\mathcal{L}^3$-realizable sets of formulas. If we had used $\mathcal{L}^4$ instead of $\mathcal{L}^3$ in the definition of F_i^k (ignoring the issues raised earlier with using $\mathcal{L}^4$), then there would be a state corresponding to every $\mathcal{L}^4$-realizable (equivalently, $\mathcal{L}^3$-realizable) set of formulas. Alternatively, if we consider appropriate structures that are compact in a topology where all sets definable by formulas (i.e., sets of the form $[\![\varphi]\!]_M$, for $\varphi \in \mathcal{L}^3$) are closed (in which case they are also open, since $[\![\neg\varphi]\!]_M$ is the complement of $[\![\varphi]\!]_M$), then all states where at least one player is using a strategy that survives at least one round of iterated deletion will be in the structure.

Although, as this discussion makes clear, the formula F_i^k that characterizes iterated admissibility can be satisfied in structures quite different from the canonical structure, the canonical structure does seem to be the most appropriate setting for reasoning about statements involving "all agent i knows". Moreover, as we now show, canonical structures allow us to relate our approach to that of BFK.

5.2 Complete Structures

BFK worked with complete structures. We now want to show that M^c is complete, in the sense of BFK. To make this precise, we need to recall some notions from BFK (with some minor changes to be more consistent with our notation).

BFK considered what they called *interactive probability structures*. These can be viewed as a special case of probability structures. A *BFK-like structure* (for a game Γ) is a probability structure $M = (\Omega, \mathbf{s}, \mathcal{F}, \mathcal{PR}_1, \ldots, \mathcal{PR}_n)$ such that there exist spaces $T_1, \ldots, T_n$ (where T_i can be thought of as the *type space* for player i) such that Ω is isomorphic to

$\vec{\Sigma} \times \vec{T}$ via some isomorphism h, where $h(\omega) = (\vec{\sigma}^\omega, \vec{t}^\omega)$ and

- $\mathbf{s}(\omega) = \vec{\sigma}^\omega$;

- the support of $\mathcal{PR}_i(\omega)$ is contained in $\{\omega' : \mathbf{s}_i(\omega') = \sigma', t_i^{\omega'} = t_i^\omega\}$, so that $\mathcal{PR}_i(\omega)$ induces a probability on $\Sigma_{-i} \times T_{-i}$;

- $\mathcal{PR}_i(\omega)$ depends only on t_i^ω, in the sense that if $t_i^\omega = t_i^{\omega'}$, then $\mathcal{PR}_i(\omega)$ and $\mathcal{PR}_i(\omega')$ induce the same probability distribution on $\Sigma_{-i} \times T_{-i}$.

A BFK-like structure M whose state space is isomorphic to $\vec{\Sigma} \times \vec{T}$ is *complete* if, for every distribution μ_i over $\Sigma_{-i} \times T_{-i}$, (where the measurable sets are the ones induced by the isomorphism h and the measurable sets $\mathcal{F}$ on Ω), there is a state ω in M such that the probability distribution on $\Sigma_{-i} \times T_{-i}$ induced by $\mathcal{PR}_i(\omega)$ is μ_i.

PROPOSITION 18. M^c *is a complete BFK-like structure.*

Proof. A set $\Phi \subseteq \mathcal{L}_i^3$ is $\mathcal{L}_i^3$-*realizable* if there exists an appropriate structure M for Γ and state ω in M such that, for all formulas $\varphi \in \mathcal{L}_i^3$, $(M, \omega) \models \varphi$ iff $\varphi \in \Phi$. Take the type space T_i to consist of all $\mathcal{L}_i^3$-realizable sets of formulas. There is an isomorphism h between Ω^c and $\vec{\Sigma} \times \vec{T}$, such that $h(\omega) = (\vec{\sigma}^\omega, \vec{t}^\omega)$, $\vec{\sigma}^\omega = \mathbf{s}^c(\omega)$, and t_i^ω consists of all formulas of the form $pr_i(\varphi) \geq \alpha$ that are true at (M^c, ω). It follows easily from Lemma 13 that h is a surjection. We can identify Ω^c, the state space in the canonical structure, with $\vec{\Sigma} \times \vec{T}$.

To prove that M^c is complete, given a probability μ on $\Sigma_{-i} \times T_{-i}$, we must show that there is some state ω in M^c such that the probability induced by $\mathcal{PR}_i(\omega)$ on $\Sigma_{-i} \times T_{-i}$ is μ. Let $M^\mu = (\Omega^{\sigma,\mu}, \mathcal{F}^\mu, \mathbf{s}^\mu, \mathcal{PR}_1^\mu, \ldots, \mathcal{PR}_n^\mu)$, where

- $\Omega^\mu = \Omega^c \cup \Sigma \times \{\mu\} \times T_{-i}$;

- $\mathcal{F}^\mu$ is the smallest σ-algebra that contains $\mathcal{F}^c$ and all sets of the form $\vec{\sigma} \times \{\mu\} \times [\![\varphi]\!]'_{M^c}$, and $[\![\varphi]\!]'_{M^c}$ consists of the all type profiles t_{-i} such that, for some state ω in M^c, $(M^c, \varphi) \models \varphi$ and $t_{-i}^\omega = t_{-i}$;

- $\mathbf{s}^\mu(\omega) = \mathbf{s}^c(\omega)$ for $\omega \in \Omega^c$, and $\mathbf{s}^\mu(\vec{\sigma} \times \{\mu\} \times \vec{t}) = \vec{\sigma}$;

- $\mathcal{PR}_j^\mu(\omega) = \mathcal{PR}_j^c(\omega)$ for $\omega \in \Omega^c$, $j = 1, \ldots, n$; for $j \neq i$, $\mathcal{PR}_j^\mu(\vec{\sigma} \times \mu \times t_{-i}) = \mathcal{PR}_j(\omega)$, where $\mathbf{s}_j(\omega) = \sigma_j$ and $t_j^\omega = t_j$ (this is well defined, since if $\mathbf{s}_j(\omega') = \sigma_j$ and $t_j^{\omega'} = t_j$, then $\mathcal{PR}_j(\omega) = \mathcal{PR}_j(\omega')$); finally, $\mathcal{PR}_i^\mu(\vec{\sigma} \times \mu \times t_{-i})$ is a distribution whose support is contained in $\{\sigma_i\} \times \Sigma_{-i} \times \{\mu\} \times T_{-i}$, and $\mathcal{PR}_i^\mu(\vec{\sigma} \times \mu \times t_{-i})(\vec{\sigma} \times \mu \times [\![\varphi]\!]'_{M^c}) = \mu([\![\varphi]\!]'_{M^c})$.

Choose an arbitrary state $\omega \in \vec{\Sigma} \times \{\mu\} \times T_{-i}$. The construction of M^μ guarantees that for $\varphi \in \mathcal{L}_{(-i)+}^4$, $(M^\mu, \omega) \models pr_i(\varphi) > \alpha$ iff $\mu([\![\varphi]\!]'_{M^c}) > \alpha$. By the construction of M^c, there exists a state $\omega' \in \Omega^c$ such that $(M^c, \omega') \models \psi$ iff $(M^\mu, \omega) \models \psi$. Thus, the distribution on $\Sigma_{-i} \times T_{-i}$ induced by $\mathcal{PR}_i(\omega)$ is μ, as desired. This shows that M^c is complete. $\qquad\square$

We now would like to show that every $\mathcal{L}^3$-measurable complete BFK-like structure is the canonical model. This is not quite true because states can be duplicated in an interactive structure. This suggests that we should try to show that the canonical structure can be embedded in every measurable complete structure. We can essentially show this, except that we need to restrict to *strongly measurable* complete structures, where a structure is

303

strongly measurable if it is measurable and the only measurable sets are those defined by $\mathcal{L}_4$ formulas (or, equivalently, the set of measurable sets is the smallest set that contains the sets defined by $\mathcal{L}_3$ formulas). We explain where strong measurability is needed at the end of the proof of the following theorem.

THEOREM 19. *If M is a strongly measurable complete BFK-like structure, then the canonical structure can be embedded in M.*

Proof. Suppose that M is a strongly measurable complete BFK-like structure. We can assume without loss of generality that the state space of M has the form $\vec{\Sigma} \times \vec{T}$, so that a state ω in M has the form $(\vec{\sigma}^\omega, \vec{t}^\omega)$. To prove the result, we need the following lemmas.

LEMMA 20. *If M is BFK-like, the truth of a formula $\varphi \in \mathcal{L}_i^4$ at a state ω in M depends only on i's type; that is, if $t_i^\omega = t_i^{\omega'}$, then $(M, \omega) \models \varphi$ iff $(M, \omega') \models \varphi$. Similarly, the truth of a formula in $\mathcal{L}_{i+}$ in ω depends only on $\vec{\sigma}^\omega$ and t_i^ω, and the truth of a formula in $\mathcal{L}_{i+}^4$ in ω depends only on t_{-i}^ω.*

Proof. A straightforward induction on structure. □

Define a *basic formula* to be one of the form $\psi_1 \wedge \ldots \wedge \psi_n$, where $\psi_i \in \mathcal{L}_{i+}^3$ for $i = 1, \ldots, n$.

LEMMA 21. *Every formula in $\mathcal{L}^3$ is equivalent to a finite disjunction of basic formulas.*

Proof. A straightforward induction on structure. □

LEMMA 22. *Every formula in $\mathcal{L}_{i+}^3$ is equivalent to a disjunction of formulas of the form*

$$play_i(\sigma) \wedge (\neg)RAT_i \wedge (\neg)pr_i(\varphi_1) > \alpha_1 \wedge \ldots \wedge ((\neg)pr_i(\varphi_m) > \alpha_m)$$
$$\wedge ((\neg)pr_i(\psi_1) \geq \beta_1) \wedge \ldots \wedge ((\neg)pr_i(\psi_{m'}) \geq \beta_{m'}),$$

where $\varphi_1, \ldots, \varphi_m, \psi_1, \ldots, \psi_{m'} \in \mathcal{L}_{(-i)+}^3$ and the "$(\neg)$" indicates that the presence of negation is optional.

Proof. A straightforward induction on the structure of formulas, using the observation that $\neg play_i(\sigma)$ is equivalent to $\bigvee_{\{\sigma' \in \Sigma_i : \sigma' \neq \sigma\}} play_i(\sigma')$. □

LEMMA 23. *If $\varphi \in \mathcal{L}^3$ is satisfiable, then $[\![\varphi]\!]_M \neq \emptyset$.*

Proof. By Lemma 21, it suffices to prove the result for the case that φ is a basic formula. By Lemma 22, it suffices to assume that the "i-component" of the basic formula is a conjunction. We now prove the result by induction on the depth of nesting of the modal operator pr_i in φ. (Formally, define $D(\psi)$, the depth of nesting of pr_i in ψ, by induction on the structure of ψ. If ψ has the form $play_j(\sigma)$, RAT_j, or $true$, then $D(\psi) = 0$; $D(\neg\psi) = D(\psi)$; $D(\psi_1 \wedge \psi_2) = \max(D(\psi_1), D(\psi_2))$; and $D(pr_i(\psi) > \alpha) = D(pr_i(\psi) \geq \alpha) = 1 + D(\psi)$.) Because the state space Ω of M is essentially a product space, by Lemma 20, it suffices to prove the result for formulas in $\mathcal{L}_{(i)+}^3$. It is clear that φ possibly puts constraints on what strategy i is using, the probability of strategy profiles in Σ_{-i}, and the probability of formulas that appear in the scope of pr_i in φ. If $M' = (\Omega', s', \mathcal{F}', PR'_1, \ldots, PR'_n)$ is a structure and $\omega' \in \Omega'$, then $(M', \omega') \models \varphi$ iff $s'_i(\omega')$ and $PR'_i(\omega')$ satisfies these constraints. (We leave it to the reader to formalize this somewhat informal claim.) By the induction hypothesis, each formula in the scope of pr_i in φ that is assigned positive probability by $\mathcal{PR}_i(\omega')$ is satisfied in M. Since M is complete and measurable, there is a state ω in M such that $\mathbf{s}_i(\omega) = \mathbf{s}'_i(\omega')$ and $\mathcal{PR}_i(\omega)$ places the same constraints on formulas that appear in φ as $\mathcal{PR}_i$. We must have $(M, \omega) \models \varphi$. $\square$

Returning to the proof of the theorem, suppose that $M = (\Omega, \mathbf{s}, \mathcal{F}, \mathcal{PR}_1, \ldots, \mathcal{PR}_n)$. Given a state $\omega \in \Omega^c$, we claim that there must be a state ω' in M such that $\mathbf{s}(\omega') = \mathbf{s}^c(\omega)$ and, for all $i = 1, \ldots, n$, $\mathcal{PR}_i^c(\omega)(\llbracket \psi \rrbracket_{M^c}) = \mathcal{PR}_i(\omega')(\llbracket \psi \rrbracket_M)$. To show this, because Ω is a product space, and $\mathcal{PR}_i(\omega')$ depends only on $T_i(\omega')$, it suffices to show that, for each i, there exists a state ω_i in M such that, for each i, $\mathcal{PR}_i^c(\omega)(\llbracket \psi \rrbracket_{M^c}) = \mathcal{PR}_i(\omega_i)(\llbracket \psi \rrbracket_M)$. By Lemma 23, if $\llbracket \psi \rrbracket_{M^c} \neq \emptyset$, then $\llbracket \psi \rrbracket_M \neq \emptyset$. Thus, the existence of ω_i follows from the assumption that M is complete and strongly measurable.

Roughly speaking, to understand the need for strong measurability here, note that even without strong measurability, the argument above tells us that there exists an appropriate measure defined on sets of the form $\llbracket \varphi \rrbracket_M$ for φ in $\mathcal{L}^3_{(-i)+}$. We can easily extend μ to a measure μ' on sets of the form $\llbracket \varphi \rrbracket_M$ for φ in $\mathcal{L}^4_{(-i)+}$. However, if the set $\mathcal{F}$ of measurable sets in M is much richer than the sets definable by $\mathcal{L}^4$ formulas, it is not clear that we can extend μ' to a measure on all of $\mathcal{F}$. In general, a countably additive measure defined on a subalgebra of a set $\mathcal{F}$ of measurable sets cannot be extended to $\mathcal{F}$. For example, it is known that, under the continuum hypothesis, Lebesgue measure defined on the Borel sets cannot be extended to all subsets of $[0, 1]$ [Ulam 1930] (see [Keisler and Tarski 1964] for further discussion). Strong measurability allows us to avoid this problem. $\square$

6 Discussion

We have provided a logical formula that captures the intuition that "all a player knows" is that players satisfy appropriate rationality assumptions. Our formalization of "all player i knows is φ" is in terms of a language $\mathcal{L}$: roughly speaking, we require that i assigns positive probability to all formulas $\psi \in \mathcal{L}$ that are consistent with φ. We provided a formula that is intended to capture the intuition that all player i knows (in language $\mathcal{L}$) is that all the other players are k-level rational. We showed that when $\mathcal{L}$ expresses statements about the strategies played by the other players, our logical formula characterizes strategies that are iterated admissible (i.e., survive iterated deletion of weakly dominated strategies). On the other hand, when $\mathcal{L}$ is less expressive (e.g., empty) the same logical formula instead characterizes strategies surviving iterated deletion of strictly dominated strategies (and thus also characterizes rationalizable strategies). Thus, the expressiveness of the language used by the players to describe their beliefs about other players can be viewed as affecting how the game is played.

We are currently working on showing that the notion of "all I know" can also be used to capture Pearce's notion of *extensive form rationalizability*, a variant of iterated admissibility that seems appropriate for reasoning about extensive-form games. These results suggest that the notion of "all I know" is really at the results suggest that the notion of "all I know" is really at the heart of many of the intuitions underlying solution concepts.

We also plan to consider the effect of the players using other langauges to describe their beliefs. For example, we are interested in the solution concept that arises if the language includes RAT_i for each player i but does not include $play_i(\sigma)$, so that players can talk about the rationality of other players without talking about the strategies they use.

We would also like to consider other ways of incorporating restrictions on how players form beliefs about other players. For example, we could restrict players' beliefs to be consistent with a theory T; namely, we might require i to assign positive probability to all and only formulas ψ consistent with both rationality and T (or, essentially equivalently, to all and only formulas that can be satisfied in some class of Kripke structures). Such restrictions provide a straighforward way of capturing players' prior beliefs about other players. They may also be used to capture the way "boundedly rational" players reason about each other. But what are "natural" restrictions? And how do such restriction affect how the game will be played? We leave an exploration of these questions for future reseach.

Acknowledgements: This is a minor modification of a paper with the title "A Logical Characterization of Iterated Admissibility" that appeared in *Proceedings of Twelfth Conference on Theoretical Aspects of Rationality and Knowledge (TARK)*, 2009. The first author is supported in part by NSF grants ITR-0325453, IIS-0534064, and IIS-0812045, by AFOSR grants FA9550-08-1-0438 and FA9550-05-1-0055, and by ARO grant W911NF-09-1-0281. The second author is supported in part by NSF CAREER Award CCF-0746990, AFOSR Award FA9550-08-1-0197, BSF Grant 2006317 and I3P grant 2006CS-001-0000001-02. The second author wishes to thank Silvio Micali for helpful discussions and his excitment about the research direction.

References

Bernheim, B. D. [1984]. Rationalizable strategic behavior. *Econometrica 52*(4), 1007–1028.

Brandenburger, A. and E. Dekel [1987]. Rationalizability and correlated equilibria. *Econometrica 55*, 1391–1402.

Brandenburger, A., A. Friedenberg, and J. Keisler [2008]. Admissibility in games. *Econometrica 76*(2), 307–352.

Halpern, J. Y. and G. Lakemeyer [2001]. Multi-agent only knowing. *Journal of Logic and Computation 11*(1), 41–70.

Keisler, J. and A. Tarski [1964]. From accessible to inaccessible cardinals. *Fundamenta Mathematica 53*, 225–308.

Levesque, H. J. [1990]. All I know: a study in autoepistemic logic. *Artificial Intelligence 42*(3), 263–309.

Mas-Colell, A., M. Whinston, and J. Green [1995]. *Microeconomic Theory*. Oxford, U.K.: Oxford University Press.

Osborne, M. J. and A. Rubinstein [1994]. *A Course in Game Theory*. Cambridge, Mass.: MIT Press.

Pearce, D. G. [1984]. Rationalizable strategic behavior and the problem of perfection. *Econometrica 52*(4), 1029–1050.

Rajan, U. [1998]. Trembles in the Bayesian foundation of solution concepts. *Journal of Economic Theory 82*, 248–266.

Tan, T. and S. Werlang [1988]. The Bayesian foundation of solution concepts of games. *Journal of Economic Theory 45*(45), 370–391.

Ulam, S. [1930]. Zur masstheorie in der allgemeinen mengenlehre. *Fundamenta Mathematicae 16*, 140–150.

19

Eliminating Function Symbols
from a Nonmonotonic Causal Theory

VLADIMIR LIFSCHITZ AND FANGKAI YANG

ABSTRACT. Nonmonotonic causal logic is a knowledge representation language designed for describing domains that involve actions and change. Problems related to action domains described in this language can be often solved using computational methods of answer set programming. This idea has led researchers at the University of Potsdam to the development of COALA, a compiler from action languages to answer set programming. Our goal is to address a significant limitation of the theory that COALA is based on: it is not directly applicable to fluents with non-Boolean values, represented by function symbols, such as the location of an object. We show that a function symbol in a causal theory can be often eliminated in favor of a new predicate symbol in such a way that the resulting theory can be translated into an executable logic program.

1 Introduction

Nonmonotonic causal logic is a knowledge representation language designed for describing domains that involve actions and change. Its original version [McCain and Turner 1997] was propositional. The first-order formulation defined in [Lifschitz 1997] provides additional expressive power that is essential for describing the semantics of action descriptions with variables [Lifschitz and Ren 2007]. Nonmonotonic causal logic was used for defining the semantics of action description languages $\mathcal{C}$ [Giunchiglia and Lifschitz 1998], $\mathcal{C}+$ [Giunchiglia, Lee, Lifschitz, McCain, and Turner 2004], and MAD [Lifschitz and Ren 2006].

A causal theory consists of causal rules $F \Leftarrow G$, where F and G are first-order formulas. The rule reads "F is caused if G is true." For instance, the causal rule

$$loc(x, t+1) = y \Leftarrow move(x, y, t)$$

expresses that there is a cause for the object x to be located at y at time $t+1$ if x is moved to y at time t. (Executing the move action is the cause.) The distinction between being true and having a cause is used in [McCain and Turner 1997] as a basis for an elegant solution to the frame problem. For instance, the commonsense law of inertia for locations can be expressed by the causal law

$$loc(x, t+1) = y \Leftarrow loc(x, t) = y \land loc(x, t+1) = y$$

(if the location of x at time $t+1$ is the same as at time t then there is a cause for this; inertia is the cause).

Results of [McCain 1997; Lifschitz and Yang 2010; Ferraris, Lee, Lierler, Lifschitz, and Yang 2011] show that causal logic is closely related to logic programming under the stable model semantics [Gelfond and Lifschitz 1988; Gelfond and Lifschitz 1991; Ferraris, Lee, and Lifschitz 2007; Ferraris, Lee, and Lifschitz 2011]. For this reason, computational methods of answer set programming (ASP) [Marek and Truszczyński 1999; Niemelä 1999; Lifschitz 2002; Lifschitz 2008] can be used for answering questions about action domains described in causal logic and in related action languages[Doğandağ, Alpaslan, and Akman 2001; Doğandağ, Ferraris, and Lifschitz 2004]. This idea has led to the development of COALA, a compiler from action languages to ASP [Gebser, Grote, and Schaub 2010].

Our goal here is to address a significant limitation of the theory that COALA is based on: it is restricted to Boolean-valued fluents, represented in causal logic by predicate symbols. It is not directly applicable to fluents with non-Boolean values, represented by function symbols, such as the location of an object in the example above. There is a good reason for this limitation. Describing non-Boolean fluents by logic programs involves an additional difficulty in view of the fact that rules in a logic program containing function symbols cannot be used to characterize values of functions; they can only characterize extents of predicates.[1]

Our approach is to show how a function symbol in a causal theory can be eliminated in favor of a new predicate symbol. In classical logic, this process is well understood. For instance, addition in first-order arithmetic can be described using a ternary predicate symbol, instead of a binary function symbol: we can write $sum(x, y, z)$ instead of $x + y = z$. (This alternative leads to the use of cumbersome formulas, of course, when complex algebraic expressions are involved.) Extending this process to nonmonotonic causal logic is not straightforward, especially if we want to arrive eventually at an executable ASP program.

This paper touches on several topics within logic-based artificial intelligence: representing properties of actions, nonmonotonic reasoning, and declarative programming. We are very pleased to have it published in a collection honoring Hector Levesque, who has contributed so much to all these areas.

The paper is organized as follows. After reviewing the syntax and semantics of first-order causal logic, we will describe two procedures for eliminating function constants from a causal theory in favor of predicate constants, "general" and "definite." Then we will show how definite elimination can help us turn a causal theory into executable ASP code, and how it can be extended to rules that express the synonymity of function symbols.

2 Review of Causal Logic

According to [Lifschitz 1997], a first-order causal theory T is defined by

- a list $\mathbf{c}$ of distinct function and/or predicate constants,[2] called the *explainable symbols* of T, and

- a finite set of *causal rules* of the form $F \Leftarrow G$, where F and G are first-order formulas.

The semantics of causal theories is defined by a syntactic transformation that is somewhat similar to circumscription [McCarthy 1986]; its result is usually a second-order for-

---

[1] The language of [Lin and Wang 2008] is not an exception: it does not permit formulas like $loc(x, t + 1) = y$ in heads of rules.

[2] We view object constants as function constants of arity 0, so that they are allowed in $\mathbf{c}$. Similarly, propositional symbols are viewed as predicate constants of arity 0. Equality, on the other hand, may not be included in c.

mula. For each member c of $\mathbf{c}$, choose a new variable vc similar to c,[3] and let $v\mathbf{c}$ stand for the list of all these variables. By $T^\dagger(v\mathbf{c})$ we denote the conjunction of the formulas

$$\forall \mathbf{x}(G \to F_{v\mathbf{c}}^{\mathbf{c}}) \tag{1}$$

for all rules $F \Leftarrow G$ of T, where $\mathbf{x}$ is the list of all free variables of F, G. (The expression $F_{v\mathbf{c}}^{\mathbf{c}}$ denotes the result of substituting the variables $v\mathbf{c}$ for the corresponding constants $\mathbf{c}$ in F.) We view T as shorthand for the sentence

$$\forall v\mathbf{c}(T^\dagger(v\mathbf{c}) \leftrightarrow (v\mathbf{c} = \mathbf{c})). \tag{2}$$

(By $v\mathbf{c} = \mathbf{c}$ we denote the conjunction of the formulas $vc = c$ for all members c of the tuple $\mathbf{c}$.) Accordingly, by a model of the causal theory T we understand a model of (2) in the sense of classical logic. The models of T are characterized, informally speaking, by the fact that the interpretation of the explainable symbols $\mathbf{c}$ in the model is the only interpretation of these symbols that is "causally explained" by the rules of T.

EXAMPLE 1. Let T_1 be the causal theory

$$\bot \Leftarrow a = b,$$
$$c = a \Leftarrow c = a,$$
$$c = b \Leftarrow q,$$

where the object constant c is the only explainable symbol (so that the object constants a and b and the propositional symbol q are not explainable).[4] The first rule of T_1 says that a and b are different from each other. The second rule ("if $c = a$ then there is a cause for this") expresses, in the language of causal logic, that by default $c = a$. The last rule says that there is a cause for c to be equal to b if q is true. According to the semantics of causal logic, T_1 is shorthand for the sentence

$$\forall vc(((a = b \to \bot) \land (c = a \to vc = a) \land (q \to vc = b)) \leftrightarrow vc = c)$$

where vc is an object variable. (There are no second-order variables in this formula. More generally, (1) does not involve second-order variables whenever all explainable symbols are object constants.) This sentence is equivalent to the quantifier-free formula

$$a \neq b \land (q \to c = b) \land (\neg q \to c = a). \tag{3}$$

The second conjunctive term shows that if q holds then the value of c is different from its default value a.

Causal rules with the head $\bot$, such as the first rule of T_1, are called *constraints*. The causal theory obtained from a causal theory T by adding a constraint $\bot \Leftarrow F$ is equivalent to $T \land \widetilde{\forall} \neg F$. The effect of adding the rule $\neg F \Leftarrow \top$ to T is usually different, unless F does not contain explainable symbols. Since the first rule of T_1 does not contain explainable symbols, it can be equivalently replaced by the rule $a \neq b \Leftarrow \top$.

EXAMPLE 2. We would like to describe the effect of moving an object. For simplicity, we only consider the time instants 0, 1 and the execution of the move action at time 0.

---

[3]That is to say, if c is a function constant then vc should be a function variable of the same arity; if c is a predicate constant then vc should be a predicate variable of the same arity.

[4]By $\bot$ and $\top$ we denote the 0-place connectives *false* and *true*.

The formulation below includes the auxiliary symbol *none*, which is used as the value of $loc(x, t)$ when the arguments are "not of the right kind" (that is, when x is not a physical object or when t is not a time instant). The rules of the causal theory T_2 are

$$\bot \Leftarrow 0 = 1,$$
$$\bot \Leftarrow 0 = none,$$
$$\bot \Leftarrow 1 = none,$$
$$loc(x, 0) = y \Leftarrow loc(x, 0) = y \wedge obj(x) \wedge place(y),$$
$$loc(x, 1) = y \Leftarrow move(x, y) \wedge obj(x) \wedge place(y),$$
$$loc(x, 1) = y \Leftarrow loc(x, 0) = y \wedge loc(x, 1) = y \wedge obj(x) \wedge place(y),$$
$$loc(x, t) = none \Leftarrow \neg obj(x),$$
$$loc(x, t) = none \Leftarrow t \neq 0 \wedge t \neq 1$$

where the function constant *loc* is the only explainable symbol. The rule with $loc(x, 0)$ in the head says that initially an object x can be anywhere: whatever value $loc(x, 0)$ has, there is a cause for that. The next two rules describe the effect of moving objects and the inertia property of locations. Causal theory T_2 is shorthand for the formula

$$\forall vloc(T_2^\dagger(vloc) \leftrightarrow (vloc = loc)),$$

where *vloc* is a binary function variable.

3   Plain Causal Theories

Let f be a function constant. An atomic formula is f-*plain* if

- it does not contain f, or

- has the form $f(\mathbf{t}) = u$, where $\mathbf{t}$ is a tuple of terms not containing f, and u is a term not containing f.

A first-order formula, a causal rule, or a causal theory is f-*plain* if all its atomic subformulas are f-plain. For instance, the causal theory from Example 1 is c-plain, and the causal theory from Example 2 is *loc*-plain.

It is easy to transform any first-order formula into an equivalent f-plain formula. For instance, $p(f(f(x)))$ is equivalent to the f-plain formula

$$\exists yz(f(x) = y \wedge f(y) = z \wedge p(z)).$$

Any causal theory can be transformed into an equivalent f-plain causal theory by applying this transformation to the heads and bodies of all rules. In Sections 4–7 we will assume that the given causal theory is f-plain.

4   General Elimination

Let T be an f-plain causal theory, where f is an explainable function constant. The causal theory T' is obtained from T as follows:

(1) in the signature of T, replace f with a new explainable predicate constant p of arity $n + 1$, where n is the arity of f;

(2) in the rules of T, replace each subformula $f(\mathbf{t}) = u$ with $p(\mathbf{t}, u)$;

(3) add the rules

$$(\exists y)p(\mathbf{x}, y) \Leftarrow \top \qquad (4)$$

and

$$\neg p(\mathbf{x}, y) \vee \neg p(\mathbf{x}, z) \Leftarrow y \neq z, \qquad (5)$$

where $\mathbf{x}$ is a tuple of variables, and the variables $\mathbf{x}$, y, z are pairwise distinct.

Rule (5) expresses, in the language of causal logic, the uniqueness of y such that $p(\mathbf{x}, y)$.

THEOREM 1. *The sentence*

$$\forall \mathbf{x}y(p(\mathbf{x}, y) \leftrightarrow f(\mathbf{x}) = y) \qquad (6)$$

entails $T \leftrightarrow T'$.

EXAMPLE 1, CONTINUED. The result T_1' of applying this transformation to T_1 and to the object constant c as f is the causal theory

$$
\begin{aligned}
\bot &\Leftarrow a = b, \\
p(a) &\Leftarrow p(a), \\
p(b) &\Leftarrow q, \\
(\exists y)p(y) &\Leftarrow \top, \\
\neg p(y) \vee \neg p(z) &\Leftarrow y \neq z
\end{aligned}
$$

with the explainable symbol p. According to Theorem 1, the equivalence between T_1 and T_1' is entailed by the sentence

$$\forall y(p(y) \leftrightarrow c = y). \qquad (7)$$

EXAMPLE 2, CONTINUED. The result T_2' of applying this transformation to T_2 and to the function constant loc as f is the causal theory

$$
\begin{aligned}
\bot &\Leftarrow 0 = 1, \\
\bot &\Leftarrow 0 = none, \\
\bot &\Leftarrow 1 = none, \\
at(x, 0, y) &\Leftarrow at(x, 0, y) \wedge obj(x) \wedge place(y), \\
at(x, 1, y) &\Leftarrow move(x, y) \wedge obj(x) \wedge place(y), \\
at(x, 1, y) &\Leftarrow at(x, 0, y) \wedge at(x, 1, y) \wedge obj(x) \wedge place(y), \\
at(x, t, none) &\Leftarrow \neg obj(x), \\
at(x, t, none) &\Leftarrow t \neq 0 \wedge t \neq 1, \\
(\exists y)at(x, t, y) &\Leftarrow \top, \\
\neg at(x, t, y) \vee \neg at(x, t, z) &\Leftarrow y \neq z
\end{aligned}
$$

with the explainable symbol at. According to Theorem 1, the equivalence between T_2 and T_2' is entailed by the sentence

$$\forall xty(at(x, t, y) \leftrightarrow loc(x, t) = y) \qquad (8)$$

By repeated applications of this process we can eliminate all explainable function symbols, provided that T is f-plain for each explainable symbol f.

The following corollary shows that there is a simple 1–1 correspondence between models of T and models of T'. Recall that the signature of T' is obtained from the signature of T by replacing f with p. For any interpretation I of the signature of T, by I_p^f we denote the interpretation of the signature of T' obtained from I by replacing the function f^I with the set p^I that consists of the tuples

$$\langle \xi_1, \ldots, \xi_n, f^I(\xi_1, \ldots, \xi_n) \rangle$$

for all $\xi_1, \ldots, \xi_n$ from the universe of I.

COROLLARY 2. *(a) An interpretation I of the signature of T is a model of T iff I_p^f is a model of T'. (b) An interpretation J of the signature of T' is a model of T' iff $J = I_p^f$ for some model I of T.*

Part (a) follows from the fact that the "union" of I and I_p^f satisfies (6). To show that any model of T' can be represented in the form I_p^f for some interpretation I, note that T' contains rules (4), (5) and consequently entails

$$\forall \mathbf{x} (\exists! y) p(\mathbf{x}, y). \tag{9}$$

5 Definite Elimination

Unfortunately, the elimination process described in the previous section does not help us turn a causal theory with explainable function symbols into a logic program. The problem is that the translation from causal logic into logic programming described in [Ferraris, Lee, Lierler, Lifschitz, and Yang 2011, Section 5] does not apply to causal rules with an existential quantifier in the head, such as (4). We will now describe an alternative elimination process, which is limited to causal theories of a special form but does not add rules containing quantifiers.

Consider an f-plain causal theory T, where f is an explainable function constant f satisfying the following condition: the head of any rule of T either does not contain f or has the form $f(\mathbf{t}) = u$, where $\mathbf{t}$ is a tuple of terms not containing explainable symbols, and u is a term not containing explainable symbols. The causal theory T'' is obtained from T as follows:

(1) in the signature of T, replace f with a new explainable predicate constant p of arity $n + 1$, where n is the arity of f;

(2) in the rules of T, replace each subformula $f(\mathbf{t}) = u$ with $p(\mathbf{t}, u)$;

(3′) add the rule

$$\neg p(\mathbf{x}, y) \Leftarrow \neg p(\mathbf{x}, y), \tag{10}$$

where $\mathbf{x}$ is a tuple of variables, and the variables $\mathbf{x}, y$ are pairwise distinct.

Rule (10) expresses the "closed world assumption" for p: by default, $p(\mathbf{x}, y)$ is false.[5]

We call this process "definite elimination" because all new causal rules that it introduces are definite in the sense of [Lifschitz 1997, Section 5]. Definite elimination is applicable to the causal theories from Examples 1 and 2, but it cannot be applied, for instance, to a theory containing a rule with the head of the form $f(x) = y \lor f(x) = z$ or $\neg(f(x) = y)$.

---

[5]It is somewhat similar to the second rule of Example 1 (Section 2).

THEOREM 3. *The sentences (6) and*

$$\exists xy(x \neq y) \tag{11}$$

entail $T \leftrightarrow T''$.

Formula (11) expresses that the universe contains at least two elements. Without this assumption, the statement of Theorem 3 would be incorrect. Indeed, consider the causal theory T_3 with an explainable function symbol f that consists of the single rule $\top \Leftarrow \top$. It is easy to check that T_3 is equivalent to $\forall xy(x = y)$. On the other hand, T_3'' consists of the rules

$$\top \Leftarrow \top,$$
$$\neg p(\mathbf{x}, y) \Leftarrow \neg p(\mathbf{x}, y),$$

and is equivalent to $\forall \mathbf{x} y \neg p(\mathbf{x}, y)$. The interpretation with a singleton universe that makes p identically true satisfies (6) and is a model of T_3, but it is not a model of T_3''.

COROLLARY 4. *If T contains a constraint of the form $\bot \Leftarrow t_1 = t_2$, where t_1, t_2 don't contain explainable function symbols, then (6) entails $T \leftrightarrow T''$.*

Indeed, if T contains the constraint $\bot \Leftarrow a = b$ then T'' contains it also, so that (11) is entailed both by T and by T''.

EXAMPLE 1, CONTINUED. The result T_1'' of applying definite elimination to T_1 and to the object symbol c is the theory

$$\begin{aligned}
\bot &\Leftarrow a = b, \\
p(a) &\Leftarrow p(a), \\
p(b) &\Leftarrow q, \\
\neg p(y) &\Leftarrow \neg p(y)
\end{aligned} \tag{12}$$

with the explainable symbol p. By Corollary 4, the equivalence between T_1 and T_1'' is entailed by sentence (7). Using the completion theorem from [Lifschitz 1997, Section 5], it is easy to check that causal theory (12) is equivalent to the first-order sentence

$$a \neq b \wedge (q \leftrightarrow p(b)) \wedge \forall y(p(y) \leftrightarrow (y = a \vee y = b)).$$

Under assumption (9), which can be written in this case as

$$(\exists! y)p(y), \tag{13}$$

this sentence can be equivalently transformed into a formula conveying the same information as (3):

$$a \neq b \wedge (q \rightarrow p(b)) \wedge (\neg q \rightarrow p(a)). \tag{14}$$

EXAMPLE 2, CONTINUED. The result T_2'' of applying definite elimination to theory T_2 and to function symbol loc is the theory

$$\begin{aligned}
\bot &\Leftarrow 0 = 1, \\
\bot &\Leftarrow 0 = none, \\
\bot &\Leftarrow 1 = none, \\
at(x, 0, y) &\Leftarrow at(x, 0, y) \wedge obj(x) \wedge place(y), \\
at(x, 1, y) &\Leftarrow move(x, y) \wedge obj(x) \wedge place(y), \\
at(x, 1, y) &\Leftarrow at(x, 0, y) \wedge at(x, 1, y) \wedge obj(x) \wedge place(y), \\
at(x, t, none) &\Leftarrow \neg obj(x), \\
at(x, t, none) &\Leftarrow t \neq 0 \wedge t \neq 1, \\
\neg at(x, t, y) &\Leftarrow \neg at(x, t, y)
\end{aligned} \tag{15}$$

with the explainable symbol *at*. By Corollary 4, the equivalence between T_2 and T_2'' is entailed by sentence (8). Using the completion theorem from [Lifschitz 1997] we can show that under assumption (9), which can be written in this case as

$$\forall x t (\exists! y) at(x, t, y), \tag{16}$$

theory (15) can be equivalently transformed into the conjunction of the universal closures of the formulas

$$\neg(0 = 1), \ \neg(0 = none), \ \neg(1 = none),$$
$$\forall xy(at(x, 0, y) \land \neg obj(x) \to y = none),$$
$$\forall xy(at(x, 0, y) \land obj(x) \to place(y)),$$
$$\forall xy(at(x, 1, y) \leftrightarrow ((move(x, y) \land obj(x) \land place(y))$$
$$\lor (at(x, 0, y) \land obj(x) \land place(y) \land \neg \exists w(move(x, w) \land place(w)))$$
$$\lor (y = none \land \neg obj(x)))),$$
$$\forall xyt((t \neq 0 \land t \neq 1) \to (at(x, t, y) \leftrightarrow y = none)).$$

The equivalence with the left-hand side $at(x, 1, y)$ is similar to successor state axioms in the sense of [Reiter 1991].

By repeated applications of this process we can eliminate all explainable function symbols provided that

- T is f-plain for each explainable function symbol f, and

- the head of each rule of T containing an explainable function symbol f has the form $f(\mathbf{t}) = u$, where $\mathbf{t}$ is a tuple of terms not containing explainable symbols, and u is a term not containing explainable symbols.

We saw in Section 4 that the mapping $I \mapsto I_p^f$ is a 1–1 correspondence between the class of models of T and the class of models of T'. For definite elimination, this mapping establishes a 1–1 correspondence between the models of T with the universe of cardinality > 1 and the models of T'' with the universe of cardinality > 1 that satisfy additional condition (9); that condition, generally, is not entailed by T''. By $T^{\exists!}$ we denote the causal theory obtained from T'' by adding the constraint

$$\perp \Leftarrow (\exists! y) p(\mathbf{x}, y). \tag{17}$$

It is clear that $T^{\exists!}$ is equivalent to the conjunction of T'' with (9).

COROLLARY 5. *(a) An interpretation I of the signature of T with the universe of cardinality > 1 is a model of T iff I_p^f is a model of $T^{\exists!}$. (b) An interpretation J of the signature of $T^{\exists!}$ with the universe of cardinality > 1 is a model of $T^{\exists!}$ iff $J = I_p^f$ for some model I of T.*

For instance, the mapping $I \mapsto I_p^c$ establishes a 1–1 correspondence between the models of T_1 and the models of $T_1^{\exists!}$. Similarly, the mapping $I \mapsto I_{at}^{loc}$ establishes a 1–1 correspondence between the models of T_2 and the models of $T_2^{\exists!}$.

As discussed at the beginning of this section, the definite elimination process is limited to causal rules satisfying an additional syntactic restriction: if the head of a rule of T

contains f then it should be an atomic formula. Without this restriction, the assertion of Theorem 3 would be incorrect. Consider, for instance, the causal theory T_4 with the rules

$$\bot \Leftarrow a = b,$$
$$c = a \lor c = b \Leftarrow \top$$

and explainable c. It is easy to check that T_4 is inconsistent. On the other hand, T_4'' is

$$\bot \Leftarrow a = b,$$
$$p(a) \lor p(b) \Leftarrow \top,$$
$$\neg p(x) \Leftarrow \neg p(x).$$

This causal theory is equivalent to

$$a \neq b \land (\forall x(p(x) \leftrightarrow x = a) \lor \forall x(p(x) \leftrightarrow x = b)).$$

The interpretation with the universe $\{a, b\}$ that interprets c as a and p as $\{a\}$ satisfies (6), (11), and T_4'', but does not satisfy T_4.

Another restriction imposed at the beginning of this section is that in formulas $f(\mathbf{t}) = u$ in the heads of rules, $\mathbf{t}$ and u don't contain explainable symbols. Without this restriction, the assertion of Theorem 3 would be incorrect. Let T_5 be the causal theory obtained from T_4 by adding the rule

$$d = c \Leftarrow \top,$$

with both c and d explainable. It is easy to check that T_5 is inconsistent. Consider the result T_5'' of applying definite elimination to T_5 and d:

$$\bot \Leftarrow a = b,$$
$$c = a \lor c = b \Leftarrow \top,$$
$$p(c) \Leftarrow \top,$$
$$\neg p(x) \Leftarrow \neg p(x),$$

with p and c explainable. This causal theory is equivalent to

$$(a \neq b) \land (\forall x(p(x) \leftrightarrow x = a) \lor \forall x(p(x) \leftrightarrow x = b)) \land p(c).$$

The interpretation with the universe $\{a, b\}$ that interprets c as a, d as a, and p as $\{a\}$ satisfies (6), (11), and T_5'', but does not satisfy T_5.

6 Modified Definite Elimination

As discussed in Section 5, rule (10) expresses the closed world assumption for p. In "modified definite elimination," (10) is replaced by a definite counterpart of the uniqueness rule (5):

$$\neg p(\mathbf{x}, y) \Leftarrow p(\mathbf{x}, z) \land y \neq z$$

($\mathbf{x}$ is a tuple of variables, and the variables $\mathbf{x}$, y, z are pairwise distinct). We will denote the result of applying the modified definite elimination process to T by T'''. For instance, T_1''' is

$$\bot \Leftarrow a = b,$$
$$p(a) \Leftarrow p(a),$$
$$p(b) \Leftarrow q,$$
$$\neg p(y) \Leftarrow p(z) \land y \neq z.$$

Causal theories T'' and T''' are essentially equivalent to each other. To be precise, formula (6) entails $T'' \leftrightarrow T'''$. Indeed, $(T''')^\dagger$ can be obtained from $(T'')^\dagger$ by replacing

$$\forall \mathbf{x}y(\neg p(\mathbf{x}, y) \to \neg vp(\mathbf{x}, y)) \tag{18}$$

with

$$\forall \mathbf{x}yz(p(\mathbf{x}, z) \wedge y \neq z \to \neg vp(\mathbf{x}, y)). \tag{19}$$

Formula (19) can be rewritten as

$$\forall \mathbf{x}y(\exists z(p(\mathbf{x}, z) \wedge y \neq z) \to \neg vp(\mathbf{x}, y)).$$

In the presence of (6), the antecedent of this formula is equivalent to the antecedent of (18). Consequently $(T'')^\dagger \leftrightarrow (T''')^\dagger$, so that $T'' \leftrightarrow T'''$.

This fact implies that the assertions of Theorem 3 and Corollary 4 will remain valid if we replace T'' in their statements with T'''.

7 From Causal Logic to ASP

7.1 Turning Causal Theories into Logic Programs

The version of the stable model semantics that we will refer to in this section is defined in [Ferraris, Lee, and Lifschitz 2011]. That paper defines, for any first-order sentence F and any tuple $\mathbf{p}$ of predicate constants, which models of F are $\mathbf{p}$-*stable*. The predicate symbols from $\mathbf{p}$ are called intensional, and the other predicate symbols are extensional. (Intensional predicates are somewhat similar to minimized predicates in circumscription and to explainable symbols in causal logic.)

The translation defined in [Ferraris, Lee, Lierler, Lifschitz, and Yang 2011, Section 5] transforms a causal theory T satisfying some syntactic conditions into a first-order sentence F that has "the same meaning under the stable model semantics" as the theory T. (One of the conditions on T is that its explainable symbols are predicate constants, not function constants.) To be more precise, if the explainable symbols of T, along with the auxiliary predicate symbols introduced by the translation, are taken to be intensional then the stable models of F are identical to the models of T, provided that the interpretations of the auxiliary predicate symbols are "forgotten." In many cases, this translation can be applied to theories obtained by the definite elimination process described above.

EXAMPLE 1, CONTINUED. Consider the causal theory $T_1^{\exists!}$, which, as we have seen, is "isomorphic" to T_1. It consists of the rules

$$
\begin{aligned}
\bot &\Leftarrow a = b, \\
p(a) &\Leftarrow p(a), \\
p(b) &\Leftarrow q, \\
\neg p(y) &\Leftarrow \neg p(y), \\
\bot &\Leftarrow \neg(\exists! y)p(y).
\end{aligned}
$$

The result of applying the translation from [Ferraris, Lee, Lierler, Lifschitz, and Yang 2011,

Section 5.2] to this theory is the conjunction of the universal closures of the formulas

$$
\begin{aligned}
& a \neq b, \\
& \neg\neg p(a) \ \rightarrow\ p(a), \\
& \neg\neg q \ \rightarrow\ p(b), \\
& \neg\neg\neg p(x) \ \rightarrow\ \widehat{p}(x), \\
& \neg\neg(\exists! y)p(y), \\
& \neg(p(x) \wedge \widehat{p}(x)), \\
& \neg(\neg p(x) \wedge \neg\widehat{p}(x)),
\end{aligned}
\tag{20}
$$

where $\widehat{p}$ is an auxiliary predicate.[6] Theorem 5 from [Ferraris, Lee, and Lifschitz 2011] shows that these formulas can be rewritten as

$$
\begin{aligned}
& a \neq b, \\
& \neg\widehat{p}(a) \ \rightarrow\ p(a), \\
& q \ \rightarrow\ p(b), \\
& \neg p(x) \ \rightarrow\ \widehat{p}(x), \\
& \neg\neg(\exists! y)p(y), \\
& \neg(p(x) \wedge \widehat{p}(x)), \\
& \neg(\neg p(x) \wedge \neg\widehat{p}(x)),
\end{aligned}
\tag{21}
$$

without changing the class of stable models. Thus the stable models of (the conjunction of the universal closures of) formulas (21) turn into the models of $T_1^{\exists!}$ as soon as the interpretation of $\widehat{p}$ is dropped. It follows that the class of stable models of (21) is "isomorphic" to the class of models of T_1.

EXAMPLE 2, CONTINUED. The result of applying the translation from [Lifschitz and Yang 2010] to $T_2^{\exists!}$ becomes, and simplifications,

$$
\begin{aligned}
& 0 \neq 1,\ 0 \neq none,\ 1 \neq none, \\
& \neg\widehat{at}(x,0,y) \wedge obj(x) \wedge place(y) \ \rightarrow\ at(x,0,y), \\
& move(x,y) \wedge obj(x) \wedge place(y) \ \rightarrow\ at(x,1,y), \\
& \neg\widehat{at}(x,0,y) \wedge \neg\widehat{at}(x,1,y) \wedge obj(x) \wedge place(y) \ \rightarrow\ at(x,1,y), \\
& \neg at(x,t,y) \ \rightarrow\ \widehat{at}(x,t,y), \\
& \neg obj(x) \ \rightarrow\ at(x,t,none), \\
& t \neq 0 \wedge t \neq 1 \ \rightarrow\ at(x,t,none), \\
& \neg\neg(\forall xt\exists! y)at(x,t,y), \\
& \neg(at(x,t,y) \wedge \widehat{at}(x,t,y)), \\
& \neg(\neg at(x,t,y) \wedge \neg\widehat{at}(x,t,y)).
\end{aligned}
\tag{22}
$$

The class of stable models of (22) is "isomorphic" to the class of models of T_2.

7.2 Turning Causal Theories into Executable Code

In many cases, answer set solvers such as CLINGO[7] allow us to generate the Herbrand stable models of a given sentence. Consequently they can be sometimes used to generate models of causal theories.

---

[6]This predicate is analogous to the classical negation of p in the sense of [Gelfond and Lifschitz 1991]. The last of formulas (20) is a consistency and completeness condition.

[7]http://potassco/sourceforge.net/

EXAMPLE 1, CONTINUED. We would like to find all models of T_1 with the universe $\{a, b\}$ in which the constants a, b represent themselves. These models correspond to the Herbrand stable models of (21). We can find them by running CLINGO on the following input:

```
u(a;b).  #domain u(X).
{q}.
p(a) :- not -p(a).
p(b) :- q.
-p(X) :- not p(X).
:- not 1{p(Z):u(Z)}1.
:- not p(X), not -p(X).
```

The first line expresses that the universe u consists of a and b, and that X is a variable for arbitrary elements of u. The choice rule in the second line says that q can be assigned an arbitrary value. The other lines correspond to five of the formulas (21), with the classical negation -p representing $\widehat{p}$. There is no need to include a constraint corresponding to $a \neq b$, because the unique name assumption is true in all Herbrand models and thus is taken by CLINGO for granted. A constraint corresponding to $\neg(p(x) \wedge \widehat{p}(x))$ would be redundant as well, since $\widehat{p}$ is represented by classical negation.

Given this input, CLINGO generates two stable models: one containing q and p(b), the other containing p(a). Consequently T_1 has two models of the kind that we are interested in: in one of them q is true and the value of c is b; in the other, q is false and the value of c is a.

EXAMPLE 2, CONTINUED. Consider the dynamic domain consisting of two objects o_1, o_2 that can be located in any of two places l_1, l_2. What are the possible locations of the objects after moving o_1 to l_2, for each possible initial state? To answer this question, we will find the models of T_2 with the universe

$$\{o_1, o_2, l_1, l_2, 0, 1, none\}$$

such that

- each of the constants 0, 1, *none* represents itself,

- the extent of *obj* is $\{o_1, o_2\}$,

- the extent of *place* is $\{l_1, l_2\}$, and

- the extent of *move* is $\{\langle o_1, l_2 \rangle\}$.

To this end, we will find the stable models of (22) that satisfy all these conditions.

This computational task is equivalent to finding the Herbrand models of the sentence obtained by conjoining the universal closures of formulas (22) with the formulas

$$obj(o_1),\ obj(o_2),\ place(l_1),\ place(l_2),\ move(o_1, l_2)$$

(o_1, o_2, l_1, l_2 are new object constants), with *obj*, *place* and *move* included in the list of intensional predicates along with *at* and $\widehat{at}$.[8] To find these models, we run CLINGO on the following input:

---

[8]This claim can be justified using the splitting theorem. See [Ferraris, Lee, Lifschitz, and Palla 2009, Sections 6, 7].

```
u(o1;o2;l1;l2;0;1;none).
#domain u(X).  #domain u(T).  #domain u(Y).
at(X,0,Y) :- not -at(X,0,Y), obj(X), place(Y).
at(X,1,Y) :- move(X,Y), obj(X), place(Y).
at(X,1,Y) :- not -at(X,0,Y), not -at(X,1,Y), obj(X), place(Y).
-at(X,T,Y) :- not at(X,T,Y).
at(X,T,none) :- not obj(X).
at(X,T,none) :- T!=0, T!=1.
:- not 1{at(X,T,Z):u(Z)}1.
:- not at(X,T,Y), not -at(X,T,Y).
obj(o1;o2).  place(l1;l2).  move(o1,l2).
```

CLINGO generates 4 stable models, one for each possible combination of the locations of o1 and o2 at time 0. In every stable model, at time 1 object o1 is at l2, and object o2 is at the same place where it was at time 0.

8 Synonymity Rules

In this section we extend the definite elimination process (Section 5) to the case when several explainable function constants are eliminated in favor of predicate constants simultaneously, and the causal theory may contain rules of the form

$$f_1(\mathbf{t}^1) = f_2(\mathbf{t}^2) \Leftarrow G,$$

where f_1 and f_2 are two of the symbols that are eliminated. This rule expresses that there is a cause for $f_1(\mathbf{t}^1)$ to be "synonymous" to $f_2(\mathbf{t}^2)$ under condition G. Such "synonymity rules" play an important role in reasoning about actions [Erdoğan and Lifschitz 2006; Lifschitz and Ren 2006; Lee, Lierler, Lifschitz, and Yang 2010].

Consider a causal theory T and a tuple $\mathbf{f}$ of explainable function constants such that the bodies of the rules of T are f-plain for all members f of $\mathbf{f}$, and the head of any rule of T

- does not contain members of $\mathbf{f}$, or

- has the form $f(\mathbf{t}) = u$, where f is a member of $\mathbf{f}$, $\mathbf{t}$ is a tuple of terms not containing explainable symbols, and u is a term not containing explainable symbols, or

- has the form $f_1(\mathbf{t}^1) = f_2(\mathbf{t}^2)$, where f_1, f_2 are members of $\mathbf{f}$, and $\mathbf{t}^1$, $\mathbf{t}^2$ are tuples of terms not containing explainable symbols.

The causal theory T'' is obtained from T as follows:

(1) in the signature of T, replace each member f of $\mathbf{f}$ with a new explainable predicate constant p of arity $n + 1$, where n is the arity of f;

(2a) in the rules of T, replace each subformula $f(\mathbf{t}) = u$ such that f is a member of $\mathbf{f}$ and u doesn't contain members of $\mathbf{f}$, with $p(\mathbf{t}, u)$;

(2b) in the heads of rules of T, replace each formula $f_1(\mathbf{t}^1) = f_2(\mathbf{t}^2)$ such that f_1, f_2 are members of $\mathbf{f}$, with $p_1(\mathbf{t}^1, y) \leftrightarrow p_2(\mathbf{t}^2, y)$, where y is a new variable;

(3') add, for every new predicate p, the rule

$$\neg p(\mathbf{x}, y) \Leftarrow \neg p(\mathbf{x}, y),$$

where $\mathbf{x}$ is a tuple of variables, and the variables $\mathbf{x}$, y are pairwise distinct.

THEOREM 6. *Sentences (6) for all f from* **f** *and sentence (11) entail* $T \leftrightarrow T''$.

EXAMPLE 3. Consider the causal theory T_6

$$
\begin{aligned}
f(x) = y &\Leftarrow a(x, y), \\
g(x) = y &\Leftarrow b(x, y), \\
f(x) = g(x) &\Leftarrow c(x)
\end{aligned}
$$

with the explainable symbols f, g. Its translation T_6'' is

$$
\begin{aligned}
p(x, y) &\Leftarrow a(x, y), \\
q(x, y) &\Leftarrow b(x, y), \\
p(x, y) \leftrightarrow q(x, y) &\Leftarrow c(x), \\
\neg p(x, y) &\Leftarrow \neg p(x, y), \\
\neg q(x, y) &\Leftarrow \neg q(x, y)
\end{aligned}
\tag{23}
$$

with the explainable symbols p, q.

Theorem 6 turns into Theorem 3 in the special case when **f** is a single function symbol and T does not contain synonymity rules.

For the class of causal theories studied in Section 5, the mapping $I \mapsto I_p^f$ establishes a 1–1 correspondence between the non-singleton models of T and the non-singleton models of $T^{\exists!}$ (Corollary 5). This assertion can be generalized to theories with synonymity rules as follows.

Recall that the signature of T'' is obtained from the signature of T by replacing the members of the tuple **f** of function constants with new predicate constants; we will denote the list of these predicate constants by **p**. For any interpretation I of the signature of T, by I_p^f we denote the interpretation of the signature of T'' obtained from I by replacing, for each $f \in$ **f**, the function f^I with the set p^I that consists of the tuples

$$
\langle \xi_1, \ldots, \xi_n, f^I(\xi_1, \ldots, \xi_n) \rangle
$$

for all $\xi_1, \ldots, \xi_n$ from the universe of I.

By $T^{\exists!}$ we denote the causal theory obtained from T'' by adding constraints (17) for all members p of **p**. It is clear that $T^{\exists!}$ is equivalent to the conjunction of T'' with formulas (9) for all $p \in$ **p**.

COROLLARY 7. *(a) An interpretation I of the signature of T with the universe of cardinality > 1 is a model of T iff I_p^f is a model of $T^{\exists!}$. (b) An interpretation J of the signature of $T^{\exists!}$ with the universe of cardinality > 1 is a model of $T^{\exists!}$ iff $J = I_p^f$ for some model I of T.*

EXAMPLE 3, CONTINUED. $T_6^{\exists!}$ consists of the rules

$$
\begin{aligned}
p(x, y) &\Leftarrow a(x, y), \\
q(x, y) &\Leftarrow b(x, y), \\
p(x, y) \leftrightarrow q(x, y) &\Leftarrow c(x), \\
\neg p(x, y) &\Leftarrow \neg p(x, y), \\
\neg q(x, y) &\Leftarrow \neg q(x, y), \\
\bot &\Leftarrow \neg(\exists! y)p(x, y), \\
\bot &\Leftarrow \neg(\exists! y)q(x, y).
\end{aligned}
$$

The mapping $I \mapsto I_{pq}^{fg}$ establishes a 1–1 correspondence between the non-singleton models of T_6 and the non-singleton models of $T_6^{\exists!}$.

The (simplified) result of applying the transformation from [Ferraris, Lee, Lierler, Lifschitz, and Yang 2011, Section 5.3] to $T_6^{\exists!}$ is

$$
\begin{aligned}
a(x,y) &\to p(x,y), \\
b(x,y) &\to q(x,y), \\
c(x) \wedge p(x,y) &\to q(x,y), \\
c(x) \wedge q(x,y) &\to p(x,y), \\
c(x) \wedge \widehat{p}(x,y) &\to \widehat{q}(x,y), \\
c(x) \wedge \widehat{q}(x,y) &\to \widehat{p}(x,y), \\
\neg p(x,y) &\to \widehat{p}(x,y), \\
\neg q(x,y) &\to \widehat{q}(x,y), \\
&\neg\neg(\exists!y)p(x,y), \\
&\neg\neg(\exists!y)q(x,y), \\
&\neg(p(x) \wedge \widehat{p}(x)), \\
&\neg(\neg p(x) \wedge \neg\widehat{p}(x)), \\
&\neg(q(x) \wedge \widehat{q}(x)), \\
&\neg(\neg q(x) \wedge \neg\widehat{q}(x)).
\end{aligned}
$$

The non-singleton stable models of the conjunction of the universal closures of these formulas, with the intensional predicates p, q, $\widehat{p}$, $\widehat{q}$, turn into the non-singleton models of $T_6^{\exists!}$ as soon as the interpretations of the auxiliary predicates $\widehat{p}$ and $\widehat{q}$ are dropped.

9 Related Work

The problem addressed in this paper is similar to the problem of eliminating multi-valued propositional constants from a multi-valued causal theory [Giunchiglia, Lee, Lifschitz, McCain, and Turner 2004]. In this sense, our general elimination and modified definite elimination are similar to the elimination methods proposed in [Lee 2005, Section 6.4.2]. On the other hand, modified definite elimination does not introduce rules similar to constraint (6.26) from [Lee 2005], and our proofs (not included in this note) are entirely different: the semantics of multi-valued propositional constants is based on a fixpoint construction and does not refer to syntactic transformations.

Eliminating function constants in the framework of a different nonmonotonic formalism — a version of the stable model semantics — is discussed in [Lin and Wang 2008].

10 Conclusion

In this paper we investigated some of the cases when an explainable function symbol can be eliminated from a first-order causal theory in favor of a predicate symbol. This is a step towards the goal of creating a compiler from the modular action language MAD [Lifschitz and Ren 2006] into answer set programming. It will differ from the current version of COALA [Gebser, Grote, and Schaub 2010] in that it will be applicable to action descriptions that involve non-Boolean fluents and synonymity rules.

Acknowledgments: We are grateful to Joohyung Lee for useful comments. This research was supported by the National Science Foundation under grant IIS-0712113.

References

Doğandağ, S., F. N. Alpaslan, and V. Akman [2001]. Using stable model semantics (SMODELS) in the Causal Calculator (CCALC). In *Proceedings 10th Turkish Symposium on Artificial Intelligence and Neural Networks*, pp. 312–321.

Doğandağ, S., P. Ferraris, and V. Lifschitz [2004]. Almost definite causal theories. In *Proceedings of International Conference on Logic Programming and Nonmonotonic Reasoning (LPNMR)*, pp. 74–86.

Erdoğan, S. T. and V. Lifschitz [2006]. Actions as special cases. In *Proceedings of International Conference on Principles of Knowledge Representation and Reasoning (KR)*, pp. 377–387.

Ferraris, P., J. Lee, Y. Lierler, V. Lifschitz, and F. Yang [2011]. Representing first-order causal theories by logic programs. *Theory and Practice of Logic Programming*. To appear.

Ferraris, P., J. Lee, and V. Lifschitz [2007]. A new perspective on stable models. In *Proceedings of International Joint Conference on Artificial Intelligence (IJCAI)*, pp. 372–379.

Ferraris, P., J. Lee, and V. Lifschitz [2011]. Stable models and circumscription. *Artificial Intelligence 175*, 236–263.

Ferraris, P., J. Lee, V. Lifschitz, and R. Palla [2009]. Symmetric splitting in the general theory of stable models. In *Proceedings of International Joint Conference on Artificial Intelligence (IJCAI)*, pp. 797–803.

Gebser, M., T. Grote, and T. Schaub [2010]. Coala: a compiler from action languages to ASP. In *Proceedings of European Conference on Logics in Artificial Intelligence (JELIA)*, pp. 169–181.

Gelfond, M. and V. Lifschitz [1988]. The stable model semantics for logic programming. In R. Kowalski and K. Bowen (Eds.), *Proceedings of International Logic Programming Conference and Symposium*, pp. 1070–1080. MIT Press.

Gelfond, M. and V. Lifschitz [1991]. Classical negation in logic programs and disjunctive databases. *New Generation Computing 9*, 365–385.

Giunchiglia, E., J. Lee, V. Lifschitz, N. McCain, and H. Turner [2004]. Nonmonotonic causal theories. *Artificial Intelligence 153(1–2)*, 49–104.

Giunchiglia, E. and V. Lifschitz [1998]. An action language based on causal explanation: Preliminary report. In *Proceedings of National Conference on Artificial Intelligence (AAAI)*, pp. 623–630. AAAI Press.

Lee, J. [2005]. *Automated Reasoning about Actions*[9]. Ph.D. thesis, University of Texas at Austin.

Lee, J., Y. Lierler, V. Lifschitz, and F. Yang [2010]. Representing synonymity in causal logic and in logic programming[10]. In *Proceedings of International Workshop on Nonmonotonic Reasoning (NMR)*.

Lifschitz, V. [1997]. On the logic of causal explanation. *Artificial Intelligence 96*, 451–465.

---

[9]http://peace.eas.asu.edu/joolee/papers/dissertation.pdf
[10]http://userweb.cs.utexas.edu/users/vl/papers/syn.pdf

Lifschitz, V. [2002]. Answer set programming and plan generation. *Artificial Intelligence 138*, 39–54.

Lifschitz, V. [2008]. What is answer set programming? In *Proceedings of the AAAI Conference on Artificial Intelligence*, pp. 1594–1597. MIT Press.

Lifschitz, V. and W. Ren [2006]. A modular action description language. In *Proceedings of National Conference on Artificial Intelligence (AAAI)*, pp. 853–859.

Lifschitz, V. and W. Ren [2007]. The semantics of variables in action descriptions. In *Proceedings of National Conference on Artificial Intelligence (AAAI)*, pp. 1025–1030.

Lifschitz, V. and F. Yang [2010]. Translating first-order causal theories into answer set programming. In *Proceedings of the European Conference on Logics in Artificial Intelligence (JELIA)*, pp. 247–259.

Lin, F. and Y. Wang [2008]. Answer set programming with functions. In *Proceedings of International Conference on Principles of Knowledge Representation and Reasoning (KR)*, pp. 454–465.

Marek, V. and M. Truszczyński [1999]. Stable models and an alternative logic programming paradigm. In *The Logic Programming Paradigm: a 25-Year Perspective*, pp. 375–398. Springer Verlag.

McCain, N. [1997]. *Causality in Commonsense Reasoning about Actions*[11]. Ph.D. thesis, University of Texas at Austin.

McCain, N. and H. Turner [1997]. Causal theories of action and change. In *Proceedings of National Conference on Artificial Intelligence (AAAI)*, pp. 460–465.

McCarthy, J. [1986]. Applications of circumscription to formalizing common sense knowledge. *Artificial Intelligence 26*(3), 89–116.

Niemelä, I. [1999]. Logic programs with stable model semantics as a constraint programming paradigm. *Annals of Mathematics and Artificial Intelligence 25*, 241–273.

Reiter, R. [1991]. The frame problem in the situation calculus: a simple solution (sometimes) and a completeness result for goal regression. In V. Lifschitz (Ed.), *Artificial Intelligence and Mathematical Theory of Computation: Papers in Honor of John McCarthy*, pp. 359–380. Academic Press.

---

[11]ftp://ftp.cs.utexas.edu/pub/techreports/tr97-25.ps.gz

20

Some Techniques for Proving Strong Termination of Some Golog Programs

FANGZHEN LIN

ABSTRACT. A nondeterministic program terminates if there is a way to execute it to a termination. It strongly terminates if it terminates however it is executed. In an earlier work, we looked at how to check if a nondeterministic Golog while-loop terminates. Here we look at how to check such a programs strongly terminates. Specifically, like in our earlier work, we propose some methods for reducing the problem of checking if a program will strongly terminate in an often infinite set of initial states to that of verifying whether the program will strongly terminate in some small initial states.

1 introduction

Many Golog programs [Levesque, Reiter, Lespérance, Lin, and Scherl 1997] are highly nondeterministic. The following is a simple example in the blocks world that can be used to achieve the goal $clear(a)$ in any given initial state:

```
While -clear(a) do
  pi(x, if holding(x) then putdown(x)) |
  pi(x,y, if on(x,y)&clear(x) then unstack(x,y);
end-while
```

Here "|" is the nondeterministic choice operator, and pi(x, ..., z, P) nondeterministically chooses some values for $x, ..., z$ and perform P with these values.

One can distinguish two kinds of terminations for such a nondeterministic program. We can say that it terminates (*weakly*) if there are some ways to perform it to a termination. This can be expressed in the situation calculus using the Do predicate [Levesque, Reiter, Lespérance, Lin, and Scherl 1997] as follows: let P be the above program, P weakly terminates in S if:

$$\exists s.Do(P, S, s).$$

This is the notion of termination defined in [Levesque, Reiter, Lespérance, Lin, and Scherl 1997].

However, we can define another kind of termination that requires the program be terminated *however* it is executed: for a while loop like the above one, at every iteration, there may be more than one action that can be executed, so "however" here means that it does not matter which one is chosen, eventually the program will terminate. In the following, we shall call this kind of termination *strong* termination. It cannot be formalized with only the Do predicate because this predicate concerns only the starting and ending situations of a program, whereas to formalize this notion of strong termination, one needs to talk about what happens during the execution of the program.

In [Lin 2008], we obtained some results that can be used to prove the weak termination of an iterative program like the above one. The basic idea is that for certain action theories, to show a program weakly terminates in all possible initial situations, it suffices to verify that this is the case for a set of initial situations up to certain size. Here we attempt to do the same for strong termination.

2 Logical preliminaries

We assume a first-order language with some function symbols for denoting actions. We assume that an action theory is given as a transition system: the states are interpretations of the first-order language, and the transition function σ is such that for a state M, an action $A(x)$, and a tuple u of objects in the domain of M, if $A(u)$ is executable in M, then $\sigma(A(u), M)$ is the state resulted from performing $A(u)$ in M, otherwise $\sigma(A(u), M)$ is undefined.

3 Strong termination

Given an action theory, we consider Golog programs of the form:

```
while -G do
  pi(x, if C1 then A1(x')) |
  ... |
  pi(y, if Ck then Ak(y'))
end-while
```

where G is a sentence, $C_1, ..., C_k$ formulas, $A_1, ..., A_k$ basic actions in the given action theory, x the union of x' and the variables that are free in C_1, and similarly for y.

Given a state M, an *execution trace* of a program P of the above form in M is a sequence of states $M_0, M_1, ..., M_n$ such that $M_0 = M$, and if $M \models G$ then $n = 0$, otherwise for each $0 \le i \le n$:

- $M_i \models \neg G$ and

- there is a $1 \le j \le k$ such that for some tuple u in the domain of M_{i-1}, $M_i = \sigma(A_j(u), M_{i-1})$, and $M_{i-1} \models C_j(u)$. Notice that by our convention, $M_i = \sigma(A_j(u), M_{i-})$ implies that $A_j(u)$ is executable in M_{i-1}.

We say that P strongly terminates in M if there is a finite number N such that all execution traces of P in M are of length smaller than N, and that P strongly terminates in a set of states if it strongly terminates in every state in the set.

If we consider program P as a possible plan to achieve the goal G, then strong termination is not enough to guarantee that the plan would work correctly. For instance, according to our definition, P strongly terminates if none of the conditionals inside the loop are executable. Thus the loop with the empty body: `while -P do end-while` always strongly terminates. To have the loop work properly, we also need to ensure that if a state does not satisfies G, then at least one of the conditionals can be executed. This condition is typically easier to check.

4 Transition system graphs

Let P be a Golog program of the above form. Without loss of generality, assume that if $i \ne j$, then A_i and A_j are distinct (function symbols). Now consider the following action theory: the actions in this domain are $A'_1, ..., A'_k$, corresponding to $A_1, ..., A_k$ in the program. For each i, the arity of A'_i is (x, y), where x is the arity of A_i, and y the variables in C_i but not in x. Its precondition is

$$Poss(A'_i(x, y)) \equiv Poss(A_i(x)) \wedge C_i(y) \wedge \neg G,$$

where $Poss(A_i(x))$ is the precondition of $A_i(x)$. For each state M, if $M \models Poss(A'_i(u, v))$, then $\sigma(A'_i(u, v), M) = \sigma(A_i(u), M)$.

It is clear that in any state M, P strongly terminates under the original action theory iff the following program P' strongly terminates under the new action theory:

```
while true do
  pi(x, Al' (x)) |
  ... |
  pi(y, Ak' (y)
end-while
```

Assuming that M is finite (i.e. it has a finite domain), then P' strongly terminates iff the following transition system graph does not have a cycle: the nodes of the graph are states that can be reached from M by a sequence of executable actions under the new action theory, and there is an edge M_1 to M_2 if there is an action $A'_i(u)$ such that $M_2 = \sigma(A'_i(u), M_1)$ (this implies that $A'_i(u)$ is executable in M_1). In the following, we call this the *transition system graph* at M, and denote it by G_M.

Given a program of the form

```
while -G do
  pi(x, if Cl then Al(x')) |
  ... |
  pi(y, if Ck then Ak(y'))
end-while
```

our strategy for proving that it strongly terminates in a set $\mathcal{M}$ of finite states is then as follows:

1. Construct the new actions $A'_1, ..., A'_k$ as described above.

2. Find a small subset $\mathcal{M}_0$ of $\mathcal{M}$ such that if none of the graphs in $\{G_M \mid M \in \mathcal{M}_0\}$ have cycles, then none of the graphs in $\{G_M \mid M \in \mathcal{M}\}$ have cycles.

3. Verify that the graphs in $\{G_M \mid M \in \mathcal{M}\}$ have no cycles.

For the reduction step (step 2 above), our main technique is graph homomorphism.

A graph *homomorphism* from a graph $G_1 = (V_1, E_1)$ to $G_2 = (V_2, E_2)$ is a function h from V_1 to V_2 such that if $(v, v') \in E_1$, then $(h(v), h(v')) \in E_2$. If there is a homomorphism from G_1 to G_2, then we say that G_1 is homomorphic to G_2. Clearly, if G_1 is homomorphic to G_2, then G_2 has a cycle if G_1 has one.

PROPOSITION 1. *Let M be a state, and $\mathcal{M}$ a set of states. If either*

1. *G_M is homomorphic to G_K for some $K \in \mathcal{M}$; or*

2. *for all K such that $(M, K) \in G_M$, $K \in \mathcal{M}$*

then if none of the graphs in $\{G_K \mid K \in \mathcal{M}\}$ have cycles, then G_M has no cycles either.

5 Examples

Consider the blocks world example given earlier:

```
While -clear(a) do
  pi(x, if holding(x) then putdown(x)) |
  pi(x,y, if on(x,y)&clear(x) then unstack(x,y);
end-while
```

We show that this program strongly terminates in all legal finite states: those with finite numbers of blocks arranged into single column towers of blocks on a table.

Let $putdown'(x)$ be like $putdown(x)$ but with the following precondition axiom:

$$Poss(putdown'(x)) \equiv \neg clear(a) \wedge holding(x),$$

and $unstack'(x, y)$ be like $unstack(x, y)$ but with the following precondition axiom:

$$Poss(unstack'(x, y)) \equiv \neg clear(a) \wedge on(x, y) \wedge clear(x) \wedge handempty.$$

As we discussed in the last section, the program strongly terminates in a state iff the transition system graph of the state under the new actions has no cycles. In the following, we show that for any legal state M, its transition system graph G_M under the new actions does not have any cycles.

Suppose M has n blocks in its domain D. We show that it can be reduced to a set S of states with fewer number of blocks. There are several cases:

1. $M \models clear(a)$. In this case let $S = \emptyset$.

2. Not the above case but there is a block $u \in D$ such that $M \models clear(u) \wedge ontable(u)$. Let M' be the restriction of M on $D \setminus \{u\}$. Then G_M is homomorphic to $G_{M'}$ under the following h: for any vertex K in G_M, let $h(K)$ be the restriction of K on $D \setminus \{u\}$. First of all, $h(K)$ is a vertex in $G_{M'}$: the same sequence of actions that gets to K from M will get $h(K)$ from M'. This is a homomorphic function because for any $K \in G_M$, if an action is executable in K, then this action cannot mention u, and the same action is executable in $h(K)$. Thus in this case, we can let $S = \{M'\}$.

3. Not the above two cases, but $holding(u)$ is true for some block u. Let $M' = \sigma(putdown(u), M)$, and M'' the restriction of M' on $D \setminus \{u\}$, and $S = \{M''\}$.

4. None of the above case, which means that $handempty$ is true and for some block $u \neq a$ and v, $on(u, v)$ and $clear(u)$ is true. In this case, let

$$M' = \sigma(putdown(u), \sigma(unstack(u, v), M)),$$

and M'' the restriction of M' on $D \setminus \{u\}$, and $S = \{M''\}$.

So in all cases, we have a S which, if not empty, consists of states that have fewer blocks in their domains than M has, and according to Proposition 1, has the property that if none of the transition system graphs in them have cycles, then G_M has no cycles. Thus by induction, G_M has no cycles if for every state with exactly one block, its transition system graph has no cycles. The latter is true and can be verified easily.

This example is a bit artificial as it is not really a good program for achieving $clear(a)$. For a more realistic example, consider the logistics domain where the goal is to move packages around. Assume that we have the following actions:

- $unload(x, y, z)$ - unload package x from y at location z, provided x is in y, and y is at z.

- $load(x, y, z)$ - the reverse of $unload(x, y, z)$, provided that both x and y are at z.

- $move(x, y, z)$ - move x from y to z, provided that when x is a truck, y and z are in the same city, and when x is an airplane, y and z are airports.

We use the following predicates: $incity(x, y)$ - location x is in city y; $airport(x)$ - x is an airport; $truck(x)$ - x is a truck; $airplane(x)$ - x is an airplane; $at(x, y)$ - x is at location y; $in(x, y)$ - package x is inside vehicle y. In the following, we use $samecity(x, y)$ to stand for

$$\exists z. incity(x, z) \wedge incity(y, z),$$

and $samelocation(x, y)$ to stand for

$$\exists z. at(x, z) \wedge at(y, z).$$

Now consider moving a package to a target location. Assume that a is a constant standing for a package, b a location. So the goal is $at(a, b)$. The following program is nondeterministic, but in a harmless way: it terminates strongly, meaning that however it is executed, if the goal can be achieved, it can be achieved by executing the program. It basically does case analysis: if a is in a truck at b, then just unload it, else if it is in a truck at another location in the same city, then move the truck to location b, etc.

```
while -at(a,b) do
  pi(x,y, if C1(x,y) then unload(a,x,y)) |
  pi(x,y, if C2(x,y) then load(a,x,y)) |
  pi(x,y,z, if C3(x,y,z) then move(x,y,z))
```

where $C_1(x, y)$ is

$$in(a, x) \wedge y = b \vee$$
$$samecity(y, b) \wedge \neg airport(b) \wedge airplane(x) \vee$$
$$\neg samecity(y, b) \wedge truck(x) \wedge airport(y),$$

$C_2(x, y)$ is the conjunction of $\neg(\exists x, y) C_1(x, y)$ and the following formula

$$(samecity(y, b) \vee \neg airport(y)) \wedge truck(x) \vee$$
$$airport(y) \wedge airplane(x) \wedge (\neg samecity(y, b) \vee airport(b)),$$

and $C_3(x, y, z)$ is the conjunctions of $\neg \exists x, y(C_1(x, y) \vee C_2(x, y))$ and the following formula

$$in(a, x) \wedge y = b \vee$$
$$in(a, x) \wedge truck(x) \wedge \neg samecity(z, b) \wedge airport(z) \wedge (\exists u) airplane(u) \wedge at(u, z) \vee$$
$$in(a, x) \wedge truck(x) \wedge \neg samecity(z, b) \wedge airport(z) \wedge$$
$$\qquad \neg(\exists u, v) airplane(u) \wedge at(u, v) \wedge samecity(z, v) \vee$$
$$in(a, x) \wedge airplane(x) \wedge \neg airport(b) \wedge samecity(y, b) \vee$$
$$\neg(\exists u) in(a, u) \wedge at(a, z) \wedge truck(x) \wedge \neg(\exists u)(truck(u) \wedge samelocation(u, a))$$
$$\neg(\exists u) in(a, u) \wedge at(a, z) \wedge airplane(x) \wedge \neg(\exists u)(airplane(u) \wedge samelocation(u, a)).$$

To show that this program strongly terminates in all legal states, we show that any such state can be reduced to one with just two cities and a few locations. Here a state is legal if it is physically possible: no object is at two different locations, an airplane can only be at an airport, a package cannot be both in a vehicle and at a location, etc., and obeys type information: $airport(x)$ if true if x is assigned an airport in the domain (we assume that the domain of a state contains sort information), etc.

Let D be the multi-sort domain with one package, two cities, three airports, three locations, one airplane, and two trucks.

Given a state M', we show that M' is homomorphic to a state M with domain D. Let D' be the domain of M'. Consider any mapping h_1 from D' to D that satisfies the following conditions:

- h_1 preserves sort information: if x is of sort α in D', then $h_1(x)$ is also of sort α in D.

- for any location l such that $M' \models at(a, l) \vee [in(a, v) \wedge at(v, l)]$ for some v, $h_1(l) = h_1(b)$ iff $l = b$.

- Suppose in M', a is in city c_1 and b in city c_2, then

    - $h_1(c_1) = h_2(c_2)$ iff $c_1 = c_2$, and

    - for any airport ap_1 in c_1 and any airport ap_2 in c_2, if $c_1 \neq c_2$, then $h_1(ap_1) \neq h_1(ap_2)$.

    - for any truck t_1 in c_1 and any truck t_2 in c_2, if $c_1 \neq c_2$, then $h_1(t_1) \neq h_1(t_2)$.

Given such a mapping h_1, we can define a mapping h from the states that have D' as their domain to states that have D as their domain: let I be a state whose domain is D', then $h(I)$ is a state with domain D and satisfies the following conditions:

- If $I \models at(x, y)$, then $h(I) \models at(h_1(x), h_1(y))$.

- If $I \models incity(x, y)$, then $h(I) \models incity(h_1(x), h_1(y))$.

- If $I \models in(x, y)$, then $h(I) \models in(h_1(x), h_1(y))$.

- $I \models airport(x)$ iff $h(I) \models airport(h_1(x))$.

- $I \models airplane(x)$ iff $h(I) \models airplane(h_1(x))$.

- $I \models truck(x)$ iff $h(I) \models truck(h_1(x))$.

We can show that h is a homomorphic function from G_M to $G_{h(M)}$: the program uses at most one airplane and at most one truck each in a's and b's cities, respectively, thus all the others are 'irrelevant' as reflected in the mapping h_1. This reduces the strong termination problem in all finite legal states to that of in all states that have D as their domain. There are only a small number of legal states with domain D, and it can be verified that the program strongly terminate in these states.

6 Conclusions

We consider the problem of how to prove that a given Golog program in the form of a nondeterministic while-loop strongly terminates in all possible initial states. As in our earlier work [Lin 2004; Lin 2007; Lin and Chen 2007; Lin 2008; Tang and Lin 2009] and also [Hu and Levesque 2010], we aim to reduce this general theorem proving problem to that of model checking in some small domains. For this, we first show that whether a program strongly terminates under a given action theory is equivalent to checking whether a transition system graph with respect to a revised action theory has cycles. To check whether a graph has cycles, we use the technique of graph homomorphism and that of reducing a graph to a set of subgraphs reachable from a vertex in the graph. We illustrate our methods by two examples, one from the blocks world domain and the other from the logistics domain.

Acknowledgments: This work was supported in part by Hong Kong RGC GRF 616208.

References

Hu, Y. and H. J. Levesque [2010]. A correctness result for reasoning about one-dimensional planning problems. In *Proceedings of the Twelfth International Conference on Principles of Knowledge Representation and Reasoning (KR2010)*.

Levesque, H., R. Reiter, Y. Lespérance, F. Lin, and R. Scherl [1997]. GOLOG: A logic programming language for dynamic domains. *Journal of Logic Programming, Special issue on Reasoning about Action and Change 31*, 59–84.

Lin, F. [2004]. Discovering state invariants. In *Proceedings of the Nineth International Conference on Principles of Knowledge Representation and Reasoning (KR2004)*, pp. 536–544.

Lin, F. [2007]. Finitely-verifiable classese of sentences. In *Proceedings of the 2007 AAAI Spring Symposium on Logical Formalization of Commonsense Reasoning.* http://www.ucl.ac.uk/commonsense07/.

Lin, F. [2008]. Proving goal achievability. In *Proceedings of the Eleventh International Conference on Principles of Knowledge Representation and Reasoning (KR2008)*, pp. 621–628.

Lin, F. and Y. Chen [2007]. Discovering classes of strongly equivalent logic programs. *Journal of Artificial Intelligence Research 28*, 431–451.

Tang, P. and F. Lin [2009]. Discovering theorems in game theory: Two-person games with unique pure nash equilibrium payoffs. In C. Boutilier (Ed.), *IJCAI*, pp. 312–317.

21

On First-order Definability and Computability of Progression for Local-effect Actions and Beyond

YONGMEI LIU AND GERHARD LAKEMEYER

ABSTRACT. In a seminal paper, Lin and Reiter introduced the notion of progression for basic action theories in the situation calculus. Unfortunately, progression is not first-order definable in general. Recently, Vassos, Lakemeyer, and Levesque showed that in case actions have only local effects, progression is first-order representable. However, they could show computability of the first-order representation only for a restricted class. Also, their proofs were quite involved. In this paper, we present a result stronger than theirs that for local-effect actions, progression is always first-order definable and computable. We give a very simple proof for this via the concept of forgetting. We also show first-order definability and computability results for a class of knowledge bases and actions with non-local effects. Moreover, for a certain class of local-effect actions and knowledge bases for representing disjunctive information, we show that progression is not only first-order definable but also efficiently computable.

Prologue

It is a great pleasure to contribute a paper to this book in honour of Hector Levesque's 60th birthday. Hector was our PhD advisor at the University of Toronto. Both of us worked on limited reasoning with Hector during our PhD years. With Hector, we started to explore the idea of solving the projection problem via progression and limited reasoning [Liu and Levesque 2005b]. In that paper, we proposed a weak variant of progression, and showed that for local-effect actions, weak progression of a certain form of incomplete first-order knowledge bases can be efficiently computed. This paper[1] continues the line of research. A feature of the paper is that we reduce progression to the well-explored concept of forgetting so that we obtain very simple proofs for the results in the paper. We are happy to present this paper as a birthday gift to Hector, because it well represents Hector's influence on our research.

1 Introduction

A fundamental problem in reasoning about action is *projection*, which is concerned with determining whether or not a formula holds after a number of actions have occurred, given a description of the preconditions and effects of the actions and what the world is like initially. Projection plays an important role in planning and in action languages such as Golog [Levesque, Reiter, Lespérance, Lin, and Scherl 1997] or $\mathcal{A}$ [Gelfond and Lifschitz 1993].

Two powerful methods to solve the projection problem are *regression* and *progression*. Roughly, regression reduces a query about the future to a query about the initial knowledge

---

[1] An earlier version of this paper appeared as [Liu and Lakemeyer 2009].

base (KB). Progression, on the other hand, changes the initial KB according to the effects of each action and then checks whether the formula holds in the resulting KB. One advantage of progression compared to regression is that after a KB has been progressed, many queries about the resulting state can be processed without any extra overhead. Moreover, when the action sequence becomes very long, as in the case of a robot operating for an extended period of time, regression simply becomes unmanageable. However, projection via progression has three main computational requirements which are not easy to satisfy: the new KB must be efficiently computed, its size should be at most linear in the size of the initial KB (to allow for iterated progression), and it must be possible to answer the query efficiently from the new KB.

As Lin and Reiter [1997] showed in the framework of Reiter's version of the situation calculus [Reiter 2001], progression is second order in general. And even if it is first-order (FO) definable, the size of the progressed KB may be unmanageable and even infinite. Recently, Vassos and Levesque [2008] also showed that the second-order nature of progression is in general inescapable, as a restriction to FO theories (even infinite ones) is strictly weaker in the sense that inferences about the future may be lost compared to the second-order version.

Nevertheless, for restricted action theories, progression can be FO definable and very effective. The classical example is STRIPS, where the initial KB is a set of literals and progression can be described via the usual *add* and *delete* lists. Since STRIPS is quite limited in expressiveness, it seems worthwhile to investigate more powerful action descriptions which still lend themselves to FO definable progression. Lin and Reiter [1997] already identified two such cases, and recently Vassos *et al.* [2008] were able to show that for so-called local-effect actions, which only change the truth values of fluent atoms with arguments mentioned by the actions, progression is always FO definable. However, they showed the computability of the FO representation only for a special case.

In this paper, we substantially improve and extend the results of Vassos *et al.*. By appealing to the notion of forgetting [Lin and Reiter 1994], we show that progression for arbitrary local-effect actions is always FO definable and computable. We extend this result to certain actions with non-local effects like the briefcase domain, where moving a briefcase implicitly moves all the objects contained in it. For the special case of so-called proper+ KBs [Lakemeyer and Levesque 2002] and a restricted class of local-effect actions we show that progression is not only first-order definable but also efficiently computable.

The rest of the paper is organized as follows. In the next section we introduce background material, including the notion of forgetting, Reiter's basic action theories, and progression. In Section 3 we present our result concerning local-effect actions. Section 4 deals with non-local effects and Section 5 with proper+ KBs. Then we conclude.

2 Preliminaries

We start with a first-order language $\mathcal{L}$ with equality. The set of formulas of $\mathcal{L}$ is the least set which contains the atomic formulas, and if ϕ and ψ are in the set and x is a variable, then $\neg\phi$, $(\phi \wedge \psi)$ and $\forall x\phi$ are in the set. The connectives $\vee$, $\supset$, $\equiv$, and $\exists$ are understood as the usual abbreviations. To improve readability, sometimes we put parentheses around quantifiers. We use the "dot" notation to indicate that the quantifier preceding the dot has maximum scope, e.g., $\forall x.P(x) \supset Q(x)$ stands for $\forall x[P(x) \supset Q(x)]$. We often omit leading universal quantifiers in writing sentences. We use $\phi \Leftrightarrow \psi$ to mean that ϕ and ψ are logically equivalent. Let ϕ be a formula, and let μ and μ' be two expressions. We denote by $\phi(\mu/\mu')$ the result of replacing every occurrence of μ in ϕ with μ'.

2.1 Forgetting

Lin and Reiter [1994] defined the concept of forgetting a ground atom or predicate in a theory. Intuitively, the resulting theory should be weaker than the original one, but entail the same set of sentences that are "irrelevant" to the ground atom or predicate.

DEFINITION 1. Let μ be either a ground atom $P(\vec{t})$ or a predicate symbol P. Let M_1 and M_2 be two structures. We write $M_1 \sim_\mu M_2$ if M_1 and M_2 agree on everything except possibly on the interpretation of μ.

DEFINITION 2. Let T be a theory, and μ a ground atom or predicate. A theory T' is a result of forgetting μ in T, denoted by $\mathrm{forget}(T, \mu) \Leftrightarrow T'$, if for any structure M, $M \models T'$ iff there is a model M' of T such that $M \sim_\mu M'$.

Clearly, if both T' and T'' are results of forgetting μ in T, then they are logically equivalent. Similarly, we can define the concept of forgetting a set of atoms or predicates. In this paper, we are only concerned with finite theories. So henceforth we only deal with forgetting for sentences.

Lin and Reiter [1994] showed that for any sentence ϕ and atom p, forgetting p in ϕ is FO definable and can be obtained from ϕ and p by simple syntactic manipulations. Here we reformulate their result in the context of forgetting a finite number of atoms. We first introduce some notation.

Let Γ be a finite set of ground atoms. We call a truth assignment θ of atoms from Γ a Γ-model. Clearly, a Γ-model θ can be represented by a conjunction of literals. We use $\mathcal{M}(\Gamma)$ to denote the set of all Γ-models. Let ϕ be a formula, and θ a Γ-model. We use $\phi[\theta]$ to denote the result of replacing every atom $P(\vec{t})$ in ϕ by the following formula:

$$(\vec{t} = \vec{t}_1 \wedge v_1) \vee \ldots \vee (\vec{t} = \vec{t}_m \wedge v_m) \vee (\vec{t} \neq \vec{t}_1 \wedge \ldots \wedge \vec{t} \neq \vec{t}_m \wedge P(\vec{t})),$$

where $P(\vec{t}_1), \ldots, P(\vec{t}_m)$ are all the P-atoms in Γ, and for each $j = 1, \ldots, m$, v_j is the truth value θ assigns to $P(\vec{t}_j)$.

PROPOSITION 3. Let ϕ be a formula, $M \sim_\Gamma M'$, $\theta \in \mathcal{M}(\Gamma)$, and $M \models \theta$. Then for any variable assignment σ, $M, \sigma \models \phi$ iff $M', \sigma \models \phi[\theta]$.

THEOREM 4. $\mathrm{forget}(\phi, \Gamma) \Leftrightarrow \bigvee_{\theta \in \mathcal{M}(\Gamma)} \phi[\theta]$.

Proof. Let M be a structure. We show that $M \models \bigvee_{\theta \in \mathcal{M}(\Gamma)} \phi[\theta]$ iff there is a model M' of ϕ s.t. $M \sim_\Gamma M'$. Suppose the latter. Let $\theta \in \mathcal{M}(\Gamma)$ s.t. $M' \models \theta$. By Proposition 3, $M \models \phi[\theta]$. Now suppose $M \models \phi[\theta]$, where $\theta \in \mathcal{M}(\Gamma)$. Let M' be the structure s.t. $M \sim_\Gamma M'$ and $M' \models \theta$. By Proposition 3, $M' \models \phi$. $\square$

COROLLARY 5. $\mathrm{forget}(\phi, \Gamma) \Leftrightarrow \bigvee_{\theta \in \mathcal{M}(\Gamma) \text{ and } \phi \wedge \theta \text{ is consistent}} \phi[\theta]$.

Proof. By Proposition 3, $\phi \wedge \theta$ entails $\phi[\theta]$. Thus if $\phi \wedge \theta$ is inconsistent, so is $\phi[\theta]$. $\square$

EXAMPLE 6. Let $\phi = \forall x.clear(x)$, and $\Gamma = \{clear(A), clear(B)\}$. Then
$\phi[clear(A) \wedge clear(B)] = \forall x. x = A \wedge true \vee x = B \wedge true \vee x \neq A \wedge x \neq B \wedge clear(x)$,
which is equivalent to $\forall x. x = A \vee x = B \vee x \neq A \wedge x \neq B \wedge clear(x)$. Similarly,
$\phi[clear(A) \wedge \neg clear(B)] \Leftrightarrow \forall x. x = A \vee x \neq A \wedge x \neq B \wedge clear(x)$,
$\phi[\neg clear(A) \wedge clear(B)] \Leftrightarrow \forall x. x = B \vee x \neq A \wedge x \neq B \wedge clear(x)$, and

335

$\phi[\neg clear(A) \wedge \neg clear(B)] \Leftrightarrow \forall x.x \neq A \wedge x \neq B \wedge clear(x)$.

Thus $forget(\phi, \Gamma) \Leftrightarrow \forall x.x = A \vee x = B \vee x \neq A \wedge x \neq B \wedge clear(x)$, which is equivalent to $\forall x.x \neq A \wedge x \neq B \supset clear(x)$.

We now assume $\mathcal{L}^2$, the second-order extension of $\mathcal{L}$.

THEOREM 7. $forget(\phi, P) \Leftrightarrow \exists R.\phi(P/R)$, here R is a second-order predicate variable.

EXAMPLE 8. Let $\phi_1 = \forall x.clear(x) \vee \exists y.on(y, x)$. Then $forget(\phi_1, clear) \Leftrightarrow \exists R \forall x. R(x) \vee \exists y.on(y, x)$, which is equivalent to $true$. Let $\phi_2 = \exists x.clear(x) \wedge \exists y.on(x, y)$. Then $forget(\phi_2, clear) \Leftrightarrow \exists R \exists x.R(x) \wedge \exists y.on(x, y)$, which is equivalent to $\exists x \exists y.on(x, y)$.

In general, forgetting a predicate is not FO definable. Naturally, by Theorem 7, second-order quantifier elimination techniques can be used for obtaining FO definability results for forgetting a predicate. In fact, Doherty *et al.* [2001] used such techniques for computing strongest necessary and weakest sufficient conditions of FO formulas, which are concepts closely related to forgetting. As surveyed in [Nonnengart, Ohlbach, and Szalas 1999], the following is a classical result on second-order quantifier elimination due to Ackermann (1935). A formula ϕ is positive (resp. negative) wrt a predicate P if $\neg P$ (resp. P) does not occur in the negation normal form of ϕ.

THEOREM 9. *Let P be a predicate variable, and let ϕ and $\psi(P)$ be FO formulas such that $\psi(P)$ is positive wrt P and ϕ contains no occurrence of P at all. Then*

$$\exists P.\forall \vec{x}(\neg P(\vec{x}) \vee \phi(\vec{x})) \wedge \psi(P)$$

is equivalent to $\psi(P(\vec{x}) \leftarrow \phi(\vec{x}))$, denoting the result of replacing each occurrence of $P(\vec{t})$ in ψ with $\phi(\vec{t})$, and similarly if the sign of P is switched and ψ is negative wrt P.

2.2 Basic action theories

The language $\mathcal{L}_{sc}$ of the situation calculus [Reiter 2001] is a many-sorted first-order language suitable for describing dynamic worlds. There are three disjoint sorts: *action* for actions, *situation* for situations, and *object* for everything else. $\mathcal{L}_{sc}$ has the following components: a constant S_0 denoting the initial situation; a binary function $do(a, s)$ denoting the successor situation to s resulting from performing action a; a binary predicate $Poss(a, s)$ meaning that action a is possible in situation s; action functions, *e.g.*, $move(x, y)$; a finite number of relational fluents, *i.e.*, predicates taking a situation term as their last argument, *e.g.*, $ontable(x, s)$; and a finite number of situation-independent predicates and functions. For simplicity of presentation, we do not consider functional fluents in this paper.

Often, we need to restrict our attention to formulas that refer to a particular situation. For this purpose, we say that a formula ϕ is uniform in a situation term τ, if ϕ does not mention any other situation terms except τ, does not quantify over situation variables, and does not mention $Poss$.

A particular domain of application will be specified by a basic action theory of the following form:

$$\mathcal{D} = \Sigma \cup \mathcal{D}_{ap} \cup \mathcal{D}_{ss} \cup \mathcal{D}_{una} \cup \mathcal{D}_{S_0}, \text{ where}$$

1. Σ is the set of the foundational axioms for situations.

2. $\mathcal{D}_{ap}$ is a set of action precondition axioms.

3. $\mathcal{D}_{ss}$ is a set of successor state axioms (SSAs), one for each fluent, of the form

$$F(\vec{x}, do(a, s)) \equiv \gamma_F^+(\vec{x}, a, s) \vee (F(\vec{x}, s) \wedge \neg \gamma_F^-(\vec{x}, a, s)),$$

where $\gamma_F^+(\vec{x}, a, s)$ and $\gamma_F^-(\vec{x}, a, s)$ are uniform in s.

4. $\mathcal{D}_{una}$ is the set of unique names axioms for actions: $A(\vec{x}) \neq A'(\vec{y})$, and $A(\vec{x}) = A(\vec{y}) \supset \vec{x} = \vec{y}$, where A and A' are distinct action functions.

5. $\mathcal{D}_{S_0}$, usually called the initial database, is a finite set of sentences uniform in S_0. We call $\mathcal{D}_{S_0}$ the initial KB.

2.3 Progression

Lin and Reiter [1997] formalized the notion of progression. Let $\mathcal{D}$ be a basic action theory, and α a ground action. We denote by S_α the situation term $do(\alpha, S_0)$.

DEFINITION 10. Let M and M' be structures with the same domains for sorts *action* and *object*. We write $M \sim_{S_\alpha} M'$ if the following two conditions hold: (1) M and M' interpret all situation-independent predicate and function symbols identically. (2) M and M' agree on all fluents at S_α: For every relational fluent F, and every variable assignment σ, $M, \sigma \models F(\vec{x}, S_\alpha)$ iff $M', \sigma \models F(\vec{x}, S_\alpha)$.

We denote by $\mathcal{L}_{sc}^2$ the second-order extension of $\mathcal{L}_{sc}$. The notion of uniform formulas carries over to $\mathcal{L}_{sc}^2$.

DEFINITION 11. Let $\mathcal{D}_{S_\alpha}$ be a set of sentences in $\mathcal{L}_{sc}^2$ uniform in S_α. $\mathcal{D}_{S_\alpha}$ is a progression of the initial KB $\mathcal{D}_{S_0}$ wrt α if for any structure M, M is a model of $\mathcal{D}_{S_\alpha}$ iff there is a model M' of $\mathcal{D}$ such that $M \sim_{S_\alpha} M'$.

Lin and Reiter [1997] proved that progression is always second-order definable. They used an old version of SSAs in their formulation of the result; here we reformulate their result using the current form of SSAs.

We let $\mathcal{D}_{ss}[\alpha, S_0]$ denote the instantiation of $\mathcal{D}_{ss}$ wrt α and S_0, i.e., the set of sentences $F(\vec{x}, do(\alpha, S_0)) \equiv \Phi_F(\vec{x}, \alpha, S_0)$. Let $F_1, \ldots, F_n$ be all the fluents. We introduce n new predicate symbols $P_1, \ldots, P_n$. We use $\phi \uparrow S_0$ to denote the result of replacing every occurrence of $F_i(\vec{t}, S_0)$ in ϕ by $P_i(\vec{t})$. We call P_i the lifting predicate for F_i. When Σ is a finite set of formulas, we denote by $\wedge\Sigma$ the conjunction of its elements.

THEOREM 12. *The following is a progression of $\mathcal{D}_{S_0}$ wrt α:*

$$\exists \vec{R}.\{\bigwedge(\mathcal{D}_{una} \cup \mathcal{D}_{S_0} \cup \mathcal{D}_{ss}[\alpha, S_0]) \uparrow S_0\}(\vec{P}/\vec{R}),$$

where $R_1, \ldots, R_n$ are second-order predicate variables.

Therefore, by Theorem 7, if ϕ is uniform in S_α and ϕ is a result of forgetting the lifting predicates $\vec{P}$ in $\bigwedge(\mathcal{D}_{una} \cup \mathcal{D}_{S_0} \cup \mathcal{D}_{ss}[\alpha, S_0]) \uparrow S_0$, then it is a progression of $\mathcal{D}_{S_0}$ wrt α.

3 Progression for local-effect actions

In this section, we show that for local-effect actions, progression is always FO-definable and computable.

We first show an intuitive result concerning forgetting a predicate: if a sentence ϕ entails that the truth values of two predicates P and Q are different at only a finite number of certain instances, then forgetting the predicate Q in ϕ can be obtained from forgetting the Q atoms of these instances in ϕ and then replacing Q by P in the result.

Let $\vec{x}$ be a variable vector, and let $\Delta = \{\vec{t}_1, \ldots, \vec{t}_m\}$ be a set of vectors of ground terms, where all the vectors have the same length. We use $\vec{x} \in \Delta$ to denote the formula

$\vec{x} = \vec{t}_1 \vee \ldots \vee \vec{x} = \vec{t}_m$. Let P and Q be two predicates. We let $Q(\Delta)$ denote the set $\{Q(\vec{t}) \mid \vec{t} \in \Delta\}$, and we let $P \approx_\Delta Q$ denote the sentence $\forall \vec{x}.\vec{x} \notin \Delta \supset P(\vec{x}) \equiv Q(\vec{x})$.

PROPOSITION 13. *Let ϕ be a formula, $M \models P \approx_\Delta Q$, and $\theta \in \mathcal{M}(Q(\Delta))$. Then for any variable assignment σ, $M, \sigma \models \phi[\theta](Q/P)$ iff $M, \sigma \models \phi[\theta]$.*

THEOREM 14. *Let P and Q be two predicates, and Δ a finite set of vectors of ground terms. If $forget(\phi, Q(\Delta)) \Leftrightarrow \psi$, then $forget(\phi \wedge (P \approx_\Delta Q), Q) \Leftrightarrow \psi(Q/P)$.*

Proof. Let $\Gamma = Q(\Delta)$. By Theorem 4, $forget(\phi, \Gamma) \Leftrightarrow \bigvee_{\theta \in \mathcal{M}(\Gamma)} \phi[\theta]$. Let M be a structure. We show that $M \models \bigvee_{\theta \in \mathcal{M}(\Gamma)} \phi[\theta](Q/P)$ iff there is a model M' of $\phi \wedge (P \approx_\Delta Q)$ s.t. $M \sim_Q M'$. Suppose the latter. Let $\theta \in \mathcal{M}(\Gamma)$ s.t. $M' \models \theta$. Since $M' \models \theta$ and $M' \models \phi$, by Proposition 3, $M' \models \phi[\theta]$. Since $M' \models P \approx_\Delta Q$, by Proposition 13, $M' \models \phi[\theta](Q/P)$. Since $M \sim_Q M'$, $M \models \phi[\theta](Q/P)$.

Now suppose $M \models \phi[\theta](Q/P)$, where $\theta \in \mathcal{M}(\Gamma)$. Let M' be the structure s.t. $M \sim_Q M'$, $M' \models P \approx_\Delta Q$, and $M' \models \theta$. Since $M \sim_Q M'$ and $M \models \phi[\theta](Q/P)$, $M' \models \phi[\theta](Q/P)$. Since $M' \models P \approx_\Delta Q$, by Proposition 13, $M' \models \phi[\theta]$. Since $M' \models \theta$, by Proposition 3, $M' \models \phi$. $\qquad \square$

Actions in many dynamic domains have only local effects in the sense that if an action $A(\vec{c})$ changes the truth value of an atom $F(\vec{d}, s)$, then $\vec{d}$ is contained in $\vec{c}$. This contrasts with actions having non-local effects such as moving a briefcase, which will also move all the objects inside the briefcase without having mentioned them.

DEFINITION 15. An SSA is local-effect if both $\gamma_F^+(\vec{x}, a, s)$ and $\gamma_F^-(\vec{x}, a, s)$ are disjunctions of formulas of the form $\exists \vec{z}[a = A(\vec{u}) \wedge \phi(\vec{u}, s)]$, where A is an action function, $\vec{u}$ contains $\vec{x}$, $\vec{z}$ is the remaining variables of $\vec{u}$, and ϕ is called a context formula. An action theory is local-effect if each SSA is local-effect.

EXAMPLE 16. Consider a simple blocks world. We use a single action, $move(x, y, z)$, moving a block x from block y to block z. We use two fluents: $clear(x, s)$, block x has no blocks on top of it; $on(x, y, s)$, block x is on block y; $eh(x, s)$, the height of block x is even. Clearly, the following SSAs are local-effect:

$$clear(x, do(a, s)) \equiv (\exists y, z)a = move(y, x, z) \vee$$
$$\quad clear(x, s) \wedge \neg(\exists y, z)a = move(y, z, x),$$
$$on(x, y, do(a, s)) \equiv (\exists z)a = move(x, z, y) \vee$$
$$\quad on(x, y, s) \wedge \neg(\exists z)a = move(x, y, z),$$
$$eh(x, do(a, s)) \equiv (\exists y, z)[a = move(x, y, z) \wedge \neg eh(z, s)] \vee$$
$$\quad eh(x, s) \wedge \neg(\exists y, z)[a = move(x, y, z) \wedge eh(z, s)].$$

By using the unique names axioms, the instantiation of a local-effect SSA on a ground action can be simplified. Suppose the SSA for F is local-effect. Let $\alpha = A(\vec{t})$ be a ground action. Then each of $\gamma_F^+(\vec{x}, \alpha, s)$ and $\gamma_F^-(\vec{x}, \alpha, s)$ is equivalent to a formula of the following form:

$$\vec{x} = \vec{t}_1 \wedge \psi_1(s) \vee \ldots \vee \vec{x} = \vec{t}_n \wedge \psi_n(s),$$

where $\vec{t}_i$ is a vector of ground terms contained in $\vec{t}$, and $\psi_i(s)$ is a formula whose only free variable is s. Without loss of generality, we assume that for a local-effect SSA, $\gamma_F^+(\vec{x}, \alpha, s)$ and $\gamma_F^-(\vec{x}, \alpha, s)$ have the above simplified form. In the case of our blocks world, we have:

$$clear(x, do(move(c_1, c_2, c_3), s)) \equiv x = c_2 \vee clear(x, s) \wedge \neg(x = c_3),$$
$$on(x, y, do(move(c_1, c_2, c_3), s)) \equiv x = c_1 \wedge y = c_3 \vee on(x, y, s) \wedge \neg(x = c_1 \wedge y = c_2),$$

$eh(x, do(move(c_1, c_2, c_3), s)) \equiv x = c_1 \wedge \neg eh(c_3, s) \vee eh(x, s) \wedge \neg(x = c_1 \wedge eh(c_3, s))$.

Following Vassos *et al.* [2008], we define the concepts of argument set and characteristic set:

DEFINITION 17. Let $\mathcal{D}$ be local-effect, and α a ground action. The argument set of fluent F wrt α is the following set:

$$\Delta_F = \{\vec{t} \mid \vec{x} = \vec{t} \text{ appears in } \gamma_F^+(\vec{x}, \alpha, s) \text{ or } \gamma_F^-(\vec{x}, \alpha, s)\}.$$

The characteristic set of α is the following set of atoms:

$$\Omega(s) = \{F(\vec{t}, s) \mid F \text{ is a fluent and } \vec{t} \in \Delta_F\}.$$

We let $\mathcal{D}_{ss}[\Omega]$ denote the instantiation of $\mathcal{D}_{ss}$ wrt Ω, i.e., the set of sentences $F(\vec{t}, S_\alpha) \equiv \Phi_F(\vec{t}, \alpha, S_0)$, where $F(\vec{t}, s) \in \Omega$. We let $\mathcal{D}_{ss}[\overline{\Omega}]$ denote the set of sentences $\vec{x} \notin \Delta_F \supset F(\vec{x}, S_\alpha) \equiv F(\vec{x}, S_0)$. Then we have

PROPOSITION 18. *Let $\mathcal{D}$ be local-effect, and α a ground action. Then*
$\mathcal{D}_{una} \models \mathcal{D}_{ss}[\alpha, S_0] \equiv \mathcal{D}_{ss}[\Omega] \cup \mathcal{D}_{ss}[\overline{\Omega}]$.

THEOREM 19. *Let $\mathcal{D}$ be local-effect, and α a ground action. Let $\Omega(s)$ be the characteristic set of α. Then the following is a progression of $\mathcal{D}_{S_0}$ wrt α:*

$$\bigwedge \mathcal{D}_{una} \wedge \bigvee_{\theta \in \mathcal{M}(\Omega(S_0))} (\mathcal{D}_{S_0} \cup \mathcal{D}_{ss}[\Omega])[\theta](S_0/S_\alpha).$$

Proof. By Proposition 18, Theorems 4, 12 and 14. $\square$

The size of the progression is $O(2^m n)$, where m is the size of the characteristic set, and n is the size of the action theory. When we do iterative progression wrt a sequence δ of actions, the size of the resulting KB is $O(2^{lm} n)$, where l is the length of δ, and m is the maximum size of the characteristic sets.

COROLLARY 20. *Progression for local-effect actions is always FO definable and computable.*

We remark that this result is strictly more general than the one obtained by Vassos *et al.* [2008], who only showed that progression for local-effect actions is FO definable in a non-constructive way, i.e., they left open the question whether the FO representation is computable or even finite. Moreover, while their proof is quite involved, having to appeal to Compactness of FO logic, ours is actually fairly simple.

Example 16 continued. Let $\alpha = move(A, B, C)$. Then $\Omega = \{clear(B, s), clear(C, s),$ $on(A, B, s), on(A, C, s), eh(A, s)\}$. $\mathcal{D}_{ss}[\Omega]$ is simplified to the following set:
$\{clear(B, S_\alpha), \neg clear(C, S_\alpha), \neg on(A,B,S_\alpha), on(A,C,S_\alpha), eh(A,S_\alpha) \equiv \neg eh(C, S_0)\}$.

Let $\mathcal{D}_{S_0}$ be the following set of sentences: $A \neq B, A \neq C, B \neq C, clear(A, S_0),$ $on(A, B, S_0), clear(C, S_0), clear(x, S_0) \supset eh(x, S_0), on(x, y, S_0) \supset \neg clear(y, S_0)$. Then $\mathcal{D}_{S_0}$ entails $\vartheta : \neg clear(B, S_0) \wedge clear(C, S_0) \wedge on(A, B, S_0) \wedge \neg on(A, C, S_0)$. Thus there are only two $\theta \in \mathcal{M}(\Omega(S_0))$ which are consistent with $\mathcal{D}_{S_0}$: $\theta_1 = \vartheta \wedge eh(A, S_0)$ and $\theta_2 = \vartheta \wedge \neg eh(A, S_0)$.

For example, let $\phi = clear(x, S_0) \supset eh(x, S_0)$. Then $\phi[\theta_1] \Leftrightarrow clear(x, S_0)[\vartheta] \supset x = A \vee x \neq A \wedge eh(x, S_0)$, and $\phi[\theta_2] \Leftrightarrow clear(x, S_0)[\vartheta] \supset x \neq A \wedge eh(x, S_0)$.

Thus $\phi[\theta_1] \vee \phi[\theta_2]$ is equivalent to $x = C \vee x \neq B \wedge x \neq C \wedge clear(x, S_0) \supset x = A \vee x \neq A \wedge eh(x, S_0)$.

By Theorem 19 and Corollary 5, the following is a progression of $\mathcal{D}_{S_0}$ wrt α:
$move(x_1, y_1, z_1) = move(x_2, y_2, z_2) \supset x_1 = x_2 \wedge y_1 = y_2 \wedge z_1 = z_2$,
$A \neq B, A \neq C, B \neq C, clear(A, S_\alpha)$,
$x = C \vee x \neq B \wedge x \neq C \wedge clear(x, S_\alpha) \supset x = A \vee x \neq A \wedge eh(x, S_\alpha)$,
$x = A \wedge y = B \vee (x \neq A \vee y \neq B) \wedge (x \neq A \vee y \neq C) \wedge on(x, y, S_\alpha) \supset$
$$\neg(y = C \vee y \neq B \wedge y \neq C \wedge clear(y, S_\alpha)),$$
$clear(B, S_\alpha), \neg clear(C, S_\alpha), \neg on(A, B, S_\alpha), on(A, C, S_\alpha), eh(A, S_\alpha) \equiv \neg eh(C, S_\alpha)$.

4 Progression for normal actions

In the last section, we showed that for local-effect actions, progression is FO definable and computable. An interesting observation about non-local-effect actions is that their effects often do not depend on the fluents on which they have non-local effects, that is, they normally have local effects on the fluents that appear in every γ_F^+ and γ_F^-. For example, moving a briefcase will move all the objects in it as well without affecting the fluent *in*. We will call such an action a normal action. In this section, we show that for a normal action α, if the initial KB has the property that for each fluent F on which α has non-local effects, the only appearance of F is in the form of $\phi(\vec{x}) \supset F(\vec{x}, S_0)$ or $\phi(\vec{x}) \supset \neg F(\vec{x}, S_0)$, then progression is FO definable and computable.

Our result is inspired by a result by Lin and Reiter [1997] that for context-free action theories, that is, action theories where every predicate appearing in every γ_F^+ and γ_F^- is situation-independent, if the initial KB has the property that for each fluent F, the only appearance of F is in the form of $\phi(\vec{x}) \supset F(\vec{x}, S_0)$ or $\phi(\vec{x}) \supset \neg F(\vec{x}, S_0)$, then progression is FO definable and computable. Incidentally, their result can be considered as an application of a simple case of the classical result by Ackermann (see Theorem 9). To prove our result, we combine the application of this simple case and the proof idea behind our result for local-effect actions.

We first present this simple case of Ackermann's result:

DEFINITION 21. We say that a finite theory T is semi-definitional wrt a predicate P if the only appearance of P in T is in the form of $P(\vec{x}) \supset \phi(\vec{x})$, where we call $\phi(\vec{x})$ a necessary condition of P, or $\phi(\vec{x}) \supset P(\vec{x})$, where we call $\phi(\vec{x})$ a sufficient condition of P. We use WSC_P (meaning weakest sufficient condition) to denote the disjunction of $\phi(\vec{x})$ such that $\phi(\vec{x}) \supset P(\vec{x})$ is in T, and we use SNC_P (meaning strongest necessary condition) to denote the conjunction of $\phi(\vec{x})$ such that $P(\vec{x}) \supset \phi(\vec{x})$ is in T.

THEOREM 22. *Let T be finite and semi-definitional wrt P. Let T' be the set of sentences in T that contains no occurrence of P. Then*
$forget(T, P) \Leftrightarrow T' \wedge \forall \vec{x}.\text{WSC}_P(\vec{x}) \supset \text{SNC}_P(\vec{x})$.

Proof. Clearly, $\exists R.T(P/R) \models T' \wedge \forall \vec{x}.\text{WSC}_P(\vec{x}) \supset \text{SNC}_P(\vec{x})$. To prove the opposite entailment, simply use the definition $\forall \vec{x}.P(\vec{x}) \equiv \text{WSC}_P(\vec{x})$. □

The following proposition shows that the SSA for a fluent F is semi-definitional wrt the predicate $F(\vec{x}, S_0)$ provided that F does not appear in γ_F^+ or γ_F^-.

PROPOSITION 23. *The sentence $F(\vec{x}, S_\alpha) \equiv \gamma_F^+(\vec{x}, \alpha, S_0) \vee F(\vec{x}, S_0) \wedge \neg\gamma_F^-(\vec{x}, \alpha, S_0)$ is equivalent to the following sentences: $\neg\gamma_F^+ \wedge F(\vec{x}, S_\alpha) \supset F(\vec{x}, S_0)$, $F(\vec{x}, S_0) \supset \gamma_F^- \vee F(\vec{x}, S_\alpha)$, $\gamma_F^+ \supset F(\vec{x}, S_\alpha)$, and $\neg\gamma_F^+ \wedge \gamma_F^- \supset \neg F(\vec{x}, S_\alpha)$.*

We now formalize our constraints on the actions and the initial KBs.

DEFINITION 24. We say that a ground action α has local effects on a fluent F, if by using $\mathcal{D}_{una}$, each of $\gamma_F^+(\vec{x}, \alpha, s)$ and $\gamma_F^-(\vec{x}, \alpha, s)$ can be simplified to a disjunction of formulas of the form $\vec{x} = \vec{t} \wedge \psi(s)$, where $\vec{t}$ is a vector of ground terms, and $\psi(s)$ is a formula whose only free variable is s. We denote by $LE(\alpha)$ the set of all fluents on which α has local effects.

DEFINITION 25. We say that α is normal if for each fluent F, all the fluents that appear in γ_F^+ and γ_F^- are in $LE(\alpha)$.

Clearly, both context-free and local-effect actions are normal actions.

DEFINITION 26. We say that $\mathcal{D}_{S_0}$ is normal wrt α if for each fluent $F \notin LE(\alpha)$, $\mathcal{D}_{S_0}$ is semi-definitional wrt F.

Thus any fluent $F \in LE(\alpha)$ can appear in $\mathcal{D}_{S_0}$ in an arbitrary way. We now have the main result of this section:

THEOREM 27. *Let $\mathcal{D}_{S_0}$ be normal wrt a normal action α. Then progression of $\mathcal{D}_{S_0}$ wrt α is FO definable and computable.*

Proof. By Theorem 12, we need to forget the lifting predicates in $\bigwedge(\mathcal{D}_{una} \cup \mathcal{D}_{S_0} \cup \mathcal{D}_{ss}[\alpha, S_0]) \uparrow S_0$. Since α is normal, for each fluent F, all the fluents that appear in γ_F^+ or γ_F^- are in $LE(\alpha)$. By Proposition 23, for each $F \notin LE(\alpha)$, $\mathcal{D}_{ss}[\alpha, S_0]$ is semi-definitional wrt $F(\vec{x}, S_0)$. Since $\mathcal{D}_{S_0}$ is normal wrt α, for each fluent $F \notin LE(\alpha)$, $\mathcal{D}_{S_0}$ is semi-definitional wrt F. By applying Theorem 22, we forget the lifting predicate for $F \notin LE(\alpha)$. Now by applying Theorem 19, we forget the lifting predicate for $F \in LE(\alpha)$. $\square$

EXAMPLE 28. The following is $\mathcal{D}_{ss}$ for the briefcase domain:
$$at(x, l, do(a, s)) \equiv (\exists b)[a = move(b, l) \wedge (x = b \vee in(x, b, s))] \vee$$
$$at(x, l, s) \wedge \neg(\exists b, m)[a = move(b, m) \wedge (x = b \vee in(x, b, s))],$$
$$in(x, b, do(a, s)) \equiv a = putin(x, b) \vee in(x, b, s) \wedge \neg a = getout(x, b).$$
For a ground action $\alpha = move(c_1, c_2)$, by using $\mathcal{D}_{una}$, $\mathcal{D}_{ss}[\alpha, S_0]$ can be simplified as follows:
$$at(x, l, do(\alpha, S_0)) \equiv l = c_2 \wedge (x = c_1 \vee in(x, c_1, S_0)) \vee$$
$$at(x, l, S_0) \wedge \neg(x = c_1 \vee in(x, c_1, S_0)),$$
$$in(x, b, do(\alpha, S_0)) \equiv in(x, b, S_0).$$
Clearly, α has local effects on in, and it is a normal action.

Now let $\mathcal{D}_{S_0}$ be as follows:
$$\exists x \forall y \neg in(x, y, S_0), \neg in(A_1, B_1, S_0), in(A_2, B_2, S_0) \vee in(A_2, B_3, S_0),$$
$$at(b, l, S_0) \wedge in(x, b, S_0) \supset at(x, l, S_0), at(x, l', S_0) \wedge l \neq l' \supset \neg at(x, l, S_0),$$
$$b = B_1 \wedge l = L_1 \vee b = B_2 \wedge l = L_2 \supset at(x, l, S_0),$$
$$b = B_3 \wedge (l = L_1 \vee l = L_2) \supset \neg at(x, l, S_0).$$
Then $\mathcal{D}_{S_0}$ is normal wrt $\alpha = move(B_1, L_2)$. To progress it wrt α, we first apply Theorem 22 to forget the lifting predicate for $at(x, l, s)$, and obtain a set Σ of sentences as follows:

1. If $\phi \in \mathcal{D}_{S_0}$ does not mention fluent at, then $\phi \in \Sigma$.

2. Add to Σ the following sentences:
   $$l = L_2 \wedge (x = B_1 \vee in(x, B_1, S_0)) \supset at(x, l, S_\alpha),$$
   $$l \neq L_2 \wedge (x = B_1 \vee in(x, B_1, S_0)) \supset \neg at(x, l, S_\alpha).$$

341

3. If $\phi \supset at(x, l, S_0)$ is in $\mathcal{D}_{S_0}$, then add to Σ the sentence

$$\phi \wedge \neg(x = B_1 \vee in(x, B_1, S_0)) \supset at(x, l, S_\alpha).$$

4. If $\phi \supset \neg at(x, l, S_0)$ is in $\mathcal{D}_{S_0}$, add to Σ the sentence

$$\phi \wedge (l \neq L_2 \vee x \neq B_1 \wedge \neg in(x, B_1, S_0)) \supset \neg at(x, l, S_\alpha).$$

Now since we have $in(x, b, S_\alpha) \equiv in(x, b, S_0)$, we simply replace each occurrence of S_0 in Σ with S_α; the result together with $\mathcal{D}_{una}$ is a progression of $\mathcal{D}_{S_0}$ wrt α.

5 Progression of proper$^+$ KBs

In Sections 3 and 4, we showed that for local-effect and normal actions, progression is FO definable and computable. However, the progression may not be efficiently computable. In this section, we show that for local-effect and normal actions, progression is not only FO definable but also efficiently computable under the two constraints that the initial KB is in the form of the so-called proper$^+$ KBs, which represent first-order disjunctive information, and the successor state axioms are essentially quantifier-free.

Proper$^+$ KBs were proposed by Lakemeyer and Levesque [2002] as a generalization of proper KBs, which were proposed by Levesque [1998] as an extension of databases. Intuitively, a proper$^+$ KB is equivalent to a (possibly infinite) set of ground clauses. A tractable limited reasoning service has been developed for proper$^+$ KBs [Liu, Lakemeyer, and Levesque 2004; Liu and Levesque 2005a]. What is particularly interesting about our results here is that progression of proper$^+$ KBs is definable as proper$^+$ KBs, so that we can make use of the available tractable reasoning service.

To formally define proper$^+$ KBs, we use a FO language $\mathcal{L}_c$ with equality, a countably infinite set of constants, which are intended to be unique names, and no other function symbols. We let e range over ewffs, *i.e.*, quantifier-free formulas whose only predicate is equality. We denote by $\mathcal{E}$ the axioms of equality and the set of formulas $\{(c \neq c') \mid c \text{ and } c' \text{ are distinct constants}\}$. We let $\forall \phi$ denote the universal closure of ϕ.

DEFINITION 29. Let e be an ewff and d a clause. Then a formula of the form $\forall(e \supset d)$ is called a $\forall$-clause. A KB is called *proper$^+$* if it is a finite non-empty set of $\forall$-clauses.

EXAMPLE 30. Consider our blocks world. The following is a initial KB $\mathcal{D}_{S_0}$ which is proper$^+$:

$on(x, y, S_0) \supset \neg clear(y, S_0),$
$on(x, y, S_0) \wedge eh(y, S_0) \supset \neg eh(x, S_0),$
$x = A \vee x = C \supset clear(x, S_0),$
$x = D \vee x = E \vee x = F \supset \neg eh(x, S_0),$
$x = A \wedge y = B \vee x = B \wedge y = F \supset on(x, y, S_0),$
$on(C, D, S_0) \vee on(C, E, S_0).$

We begin with forgetting in proper$^+$ KBs. We first introduce some definitions and propositions.

DEFINITION 31. Let ϕ be a sentence, and p a ground atom. We say that p is irrelevant to ϕ if $forget(\phi, p) \Leftrightarrow \phi$.

PROPOSITION 32. *Let p be a ground atom. Let ϕ_1, ϕ_2, ϕ_3 be sentences such that p is irrelevant to them. Then $forget((\phi_1 \supset p) \wedge (p \supset \phi_2) \wedge \phi_3, p) \Leftrightarrow (\phi_1 \supset \phi_2) \wedge \phi_3$.*

PROPOSITION 33. *Let $\phi = \forall(e \supset d)$ be a $\forall$-clause, and $P(\vec{c})$ a ground atom. Suppose that for any $P(\vec{t})$ appearing in d, $e \wedge \vec{t} = \vec{c}$ is unsatisfiable. Then $P(\vec{c})$ is irrelevant to ϕ.*

DEFINITION 34. Let Σ be a proper$^+$ KB, and $P(\vec{c})$ a ground atom. We say that Σ is in normal form wrt $P(\vec{c})$, if for any $\forall(e \supset d) \in \Sigma$, and for any $P(\vec{t})$ appearing in d, either $\vec{t}$ is $\vec{c}$ or $e \wedge \vec{t} = \vec{c}$ is unsatisfiable.

PROPOSITION 35. *Let $P(\vec{c})$ be a ground atom. Then every proper$^+$ KB can be converted into an equivalent one which is in normal form wrt $P(\vec{c})$. This can be done in $O(n + 2^w m)$ time, where n is the size of Σ, m is the size of sentences in Σ where P appears, and w is the maximum number of appearances of P in a sentence of Σ.*

Proof. Let $\phi = \forall(e \supset d)$ be a $\forall$-clause. Let $P(\vec{t_1}), \ldots, P(\vec{t_k})$ be all the appearances of P in ϕ, and let $\Theta = \{\bigwedge_{i=1}^{k} \vec{t_i} \circ_i \vec{c} \mid \circ_i \in \{=, \neq\}\}$. Let $\theta \in \Theta$. We let $d[\theta]$ denote d with each $P(\vec{t_i})$, $1 \le i \le k$, replaced by $P(\vec{c})$ if θ contains $\vec{t_i} = \vec{c}$. We use $\phi[\theta]$ to denote $\forall(e \wedge \theta \supset d[\theta])$. Obviously, ϕ is equivalent to the theory $\{\phi[\theta] \mid \theta \in \Theta\}$, which we denote by $\text{NF}(\phi, P(\vec{c}))$. For a proper$^+$ KB Σ, we convert it into the union of $\text{NF}(\phi, P(\vec{c}))$ where $\phi \in \Sigma$. $\square$

In the above proof, we can remove those generated $\forall$-clauses $\forall(e \supset d)$ where d contains complementary literals or e is unsatisfiable wrt $\mathcal{E}$. For example, the ewff $x = y \wedge x = A \wedge y = B$ is unsatisfiable wrt $\mathcal{E}$. Also, a $\forall$-clause $\forall(e \wedge t = c \supset d)$ can be simplified to $\forall(e \supset d)(t/c)$. Finally, an ewff can be simplified by use of $\mathcal{E}$.

DEFINITION 36. Let $\phi_1 = \forall(e_1 \supset d_1 \vee P(\vec{t}))$ and $\phi_2 = \forall(e_2 \supset d_2 \vee \neg P(\vec{t}))$ be two $\forall$-clauses, where $\vec{t}$ is a vector of constants or a vector of distinct variables. Without loss of generality, we assume that ϕ_1 and ϕ_2 do not share variables other than those contained in $\vec{t}$. We call the $\forall$-clause $\forall(e_1 \wedge e_2 \supset d_1 \vee d_2)$ the $\forall$-resolvent of the two input clauses wrt $P(\vec{t})$.

THEOREM 37. *Let Σ be a proper$^+$ KB, and $P(\vec{c})$ a ground atom. Then the result of forgetting $P(\vec{c})$ in Σ is definable as a proper$^+$ KB and can be computed in $O(n + 4^w m^2)$ time, where n, w, and m are as above.*

Proof. We first convert Σ into normal form wrt $P(\vec{c})$. Then we compute all $\forall$-resolvents wrt $P(\vec{c})$ and remove all clauses with $P(\vec{c})$. This results in a proper$^+$ KB, which, by Propositions 32 and 33, is a result of forgetting $P(\vec{c})$ in Σ. $\square$

THEOREM 38. *Let Σ be a proper$^+$ KB semi-definitional wrt predicate P. Then the result of forgetting P in Σ is definable as a proper$^+$ KB and can be computed in $O(n + m^2)$ time, where n is the size of Σ, and m is the size of sentences in Σ where P appears.*

Proof. We compute all $\forall$-resolvents wrt $P(\vec{x})$ and remove all clauses containing $P(\vec{x})$. This results in a proper$^+$ KB, which, by Theorem 22, is a result of forgetting P in Σ. $\square$

In the above theorems (Theorems 37 and 38), it is reasonable to assume that $w = O(1)$ and $m^2 = O(n)$. Under this assumption, both forgetting can be computed in $O(n)$ time.

Based on the above theorems, we have the following results concerning progression of proper$^+$ KBs. We first introduce a constraint on successor state axioms.

DEFINITION 39. An SSA is essentially quantifier-free if for each ground action α, by using $\mathcal{D}_{una}$, each of $\gamma_F^+(\vec{x}, \alpha, s)$ and $\gamma_F^-(\vec{x}, \alpha, s)$ can be simplified to a quantifier-free formula.

For example, the SSAs for our blocks world and briefcase examples are essentially quantifier-free. For a local-effect SSA, if each context-formula is quantifier-free, then it is essentially quantifier-free. In general, if both $\gamma_F^+(\vec{x}, a, s)$ and $\gamma_F^-(\vec{x}, a, s)$ are disjunctions of formulas of the form $\exists \vec{z}[a = A(\vec{u}) \wedge \phi(\vec{x}, \vec{z}, s)]$, where $\vec{u}$ contains $\vec{z}$, and ϕ is quantifier-free, then the SSA is essentially quantifier-free.

PROPOSITION 40. *Suppose $\mathcal{D}_{ss}$ is essentially quantifier-free. Then $\mathcal{D}_{ss}[\alpha, S_0]$ is definable as a proper$^+$ KB.*

THEOREM 41. *Suppose that $\mathcal{D}$ is local-effect, $\mathcal{D}_{ss}$ is essentially quantifier-free, and $\mathcal{D}_{S_0}$ is proper$^+$. Then progression of $\mathcal{D}_{S_0}$ wrt any ground action α is definable as a proper$^+$ KB and can be efficiently computed.*

THEOREM 42. *Suppose that $\mathcal{D}_{ss}$ is essentially quantifier-free, α is a normal action, and $\mathcal{D}_{S_0}$ is a proper$^+$ KB which is normal wrt α. Then progression of $\mathcal{D}_{S_0}$ wrt α is definable as a proper$^+$ KB and can be efficiently computed.*

Example 30 continued. We now progress $\mathcal{D}_{S_0}$ wrt $\alpha = move(A, B, C)$. For simplicity, we remove the second sentence from $\mathcal{D}_{S_0}$. The characteristic set of α is $\Omega = \{clear(B, s), clear(C, s), on(A, B, s), on(A, C, s), eh(A, s)\}$. $\mathcal{D}_{ss}[\Omega]$ is simplified to the following proper$^+$ KB: $\{clear(B, S_\alpha), \neg clear(C, S_\alpha), \neg on(A, B, S_\alpha), on(A, C, S_\alpha), eh(A, S_\alpha) \vee eh(C, S_0), \neg eh(A, S_\alpha) \vee \neg eh(C, S_0)\}$.

We convert $\mathcal{D}_{S_0}$ into normal form wrt $\Omega(S_0)$, and obtain after simplification:
$y \neq B \wedge y \neq C \supset (on(x, y, S_0) \supset \neg clear(y, S_0))$,
$x \neq A \supset (on(x, B, S_0) \supset \neg clear(B, S_0))$,
$x \neq A \supset (on(x, C, S_0) \supset \neg clear(C, S_0))$,
$on(A, B, S_0) \supset \neg clear(B, S_0)$,
$on(A, C, S_0) \supset \neg clear(C, S_0)$,
$x = D \vee x = E \vee x = F \supset \neg eh(x, S_0)$,
$clear(A, S_0), clear(C, S_0), on(A, B, S_0), on(B, F, S_0)$,
$on(C, D, S_0) \vee on(C, E, S_0)$.

We now do resolution on $\mathcal{D}_{S_0} \cup \mathcal{D}_{ss}[\Omega]$ wrt atoms in $\Omega(S_0)$, delete all clauses with some atom from $\Omega(S_0)$, and obtain the following set Σ:
$clear(B, S_\alpha), \neg clear(C, S_\alpha), \neg on(A, B, S_\alpha), on(A, C, S_\alpha)$,
$eh(A, S_\alpha) \vee eh(C, S_0), \neg eh(A, S_\alpha) \vee \neg eh(C, S_0)$,
$y \neq B \wedge y \neq C \wedge (x \neq A \vee y \neq B \wedge y \neq C) \supset (on(x, y, S_0) \supset \neg clear(y, S_0))$,
$x \neq A \supset \neg on(x, C, S_0), x = D \vee x = E \vee x = F \supset \neg eh(x, S_0)$,
$clear(A, S_0), on(B, F, S_0), on(C, D, S_0) \vee on(C, E, S_0)$.

We replace every occurrence of S_0 in Σ with S_α; the result together with $\mathcal{D}_{una}$ is a progression of $\mathcal{D}_{S_0}$ wrt α.

6 Conclusions

In this paper, we have presented the following results. First, we showed that for local-effect actions, progression is FO definable and computable. This result is stronger than the one obtained by Vassos *et al.* [2008], and our proof is a very simple one via the concept of forgetting. Next, we went beyond local-effect actions, and showed that for normal actions, *i.e.*, actions whose effects do not depend on the fluents on which the actions have non-local effects, if the initial KB is semi-definitional wrt these fluents, progression is FO definable and computable. Third, we showed that for local-effect actions whose successor state axioms are essentially quantifier-free, progression of proper$^+$ KBs is definable as proper$^+$

KBs and can be efficiently computed. Thus we can utilize the available tractable limited reasoning service for proper$^+$ KBs. As an extension of our first result, we have shown that for finite-effect actions, which change the truth values of fluents at only a finite number of instances, progression is FO definable. We have also shown that in the presence of functional fluents, our first and second results still hold. For lack of space, these results will be presented in a longer version of the paper. For the future, we would like to implement a Golog interpreter based on progression of proper$^+$ KBs, which we expect will lead to a more efficient version of Golog compared to the current implementation based on regression.

Acknowledgments

This work was supported in part by the European Union Erasmus Mundus programme.

References

Doherty, P., W. Lukaszewicz, and A. Szalas [2001]. Computing strongest necessary and weakest sufficient conditions of first-order formulas. In *Proc. IJCAI-01*.

Gelfond, M. and V. Lifschitz [1993]. Representing actions and change by logic programs. *Journal of Logic Programming 17*(2–4), 301–323.

Lakemeyer, G. and H. J. Levesque [2002]. Evaluation-based reasoning with disjunctive information in first-order knowledge bases. In *Proc. KR*.

Levesque, H. J. [1998]. A completeness result for reasoning with incomplete first-order knowledge bases. In *Proc. KR-98*.

Levesque, H. J., R. Reiter, Y. Lespérance, F. Lin, and R. Scherl [1997]. Golog: A logic programming language for dynamic domains. *J. of Logic Programming 31*, 59–84.

Lin, F. and R. Reiter [1994]. Forget it! In *Working Notes of AAAI Fall Symposium on Relevance*.

Lin, F. and R. Reiter [1997]. How to progress a database. *Artificial Intelligence 92*(1–2), 131–167.

Liu, Y. and G. Lakemeyer [2009]. On first-order definability and computability of progression for local-effect actions and beyond. In *Proc. IJCAI-09*.

Liu, Y., G. Lakemeyer, and H. J. Levesque [2004]. A logic of limited belief for reasoning with disjunctive information. In *Proc. KR-04*.

Liu, Y. and H. J. Levesque [2005a]. Tractable reasoning in first-order knowledge bases with disjunctive information. In *Proc. AAAI-05*.

Liu, Y. and H. J. Levesque [2005b]. Tractable reasoning with incomplete first-order knowledge in dynamic systems with context-dependent actions. In *Proc. IJCAI-05*.

Nonnengart, A., H. J. Ohlbach, and A. Szalas [1999]. Elimination of predicate quantifiers. In *Logic, Language and Reasoning, Part I*, pp. 159–181. Kluwer Academic Press.

Reiter, R. [2001]. *Knowledge in Action: Logical Foundations for Specifying and Implementing Dynamical Systems*. MIT Press.

Vassos, S., G. Lakemeyer, and H. J. Levesque [2008]. First-order strong progression for local-effect basic action theories. In *Proc. KR-08*.

Vassos, S. and H. J. Levesque [2008]. On the progression of situation calculus basic action theories: Resolving a 10-year-old conjecture. In *Proc. AAAI-08*.

22

Knowledge Representation, Search Problems and Model Expansion

DAVID G. MITCHELL AND EUGENIA TERNOVSKA

ABSTRACT. Arguments for logic-based knowledge representation often emphasize the primacy of entailment in reasoning, and traditional logic-based formulations of AI tasks were frequently in terms of entailment. More recently, practical progress in satisfiability-based methods has encouraged formulation of problems as model finding. Here, we argue for the formalization of search problems, which abound in AI as well as other areas, as a particular form of model finding called *model expansion*. An important conceptual part of this proposal is the formalization of the problem instance as a *structure*, rather than as a formula. Adopting this view leads naturally to taking descriptive complexity theory as the starting point for developing a theory of languages for representing search problems. We explain the formalization of search as model expansion, and the reasons we consider it an appropriate basis for such a theory. We emphasize the role of model expansion for first order logic, with extensions, in specifying NP search problems, and describe the formalization of arithmetic in this context.

This paper is dedicated to Hector J. Levesque.

1 Introduction

Search Problems as Model Finding

A classical argument in favour of explicit logic-based knowledge representation invokes the primacy of entailment in reasoning, and the fact that deduction can be automated because of the correspondence between the semantic notion of entailment and the syntactic notion of proof. As important as this is, not all reasoning involves deduction, and indeed it seems more natural to view search problems, in which we are asked to exhibit an object satisfying certain properties, as model finding rather than theorem proving. In this paper, we expand on a proposal first made by the authors in [Mitchell and Ternovska 2005], toward a logic-based theory of languages for representing search problems.

Traditional AI practice frequently cast problem solving as entailment, or theorem proving, as in the early work of Green [1969]. There, a user axiomatizes the claim that a solution exists, and a theorem prover is used to show the claim is true. As in much Prolog-based problem solving, a side effect of proving the theorem is construction of a term which describes a solution. Selman, Levesque and Mitchell [1992], in the paper which introduced the GSAT algorithm for SAT, argued that many AI problems traditionally cast as theorem proving should be re-formulated as model finding, concluding that in certain cases "model finding has a clear advantage over theorem proving and may lead us to AI methods that scale up more gracefully in practice". This has turned out to be true, and perhaps more generally than suggested in that paper.

Satisfiability-based techniques have since become widely successful and influential, providing computational engines for a variety of reasoning tasks. However, they did not initially offer much help with *representation*. For example, normal practice in SAT-based solving involves explicitly designing and implementing a reduction from the problem at hand to SAT, with each CNF formula describing the solutions for a particular instance of the problem. In 1999, three papers introduced logic programming-based languages for representing *problems*, with a distinct separation of problem description and individual instances. Cadoli *et al* [1999] proposed an extension to Datalog programs, called NP-Spec. Niemela [1999] and Marek and Truszczynski [1999] independently proposed representing search problems as normal logic programs under the stable model semantics, the approach now known as answer set programming (ASP). As described in [Marek and Truszczynski 1999], a specification for search problem Π consists of a normal logic program P_{Π}, and an instance I is represented by a set of atoms $edb_{\Pi}(I)$. A solution to I is obtained by finding a stable model of the program $P_{\Pi} \cup edb_{\Pi}(I)$. Thus [Marek and Truszczynski 1999] point out the importance of separation of specification from data (instance), but both are formulas (logic programs), and the distinction is in a sense one of convention.

In [Mitchell and Ternovska 2005], the authors proposed the logical task of *model expansion* as a formalization of search problems upon which to develop a theory of languages for specifying search problems (and systems to support these languages in practice). An important conceptual element of this proposal is the formal distinction of instances and specifications in which *instances are taken to be structures*. In the theory of computing, a search problem is a binary relation on strings, defining the solutions for each instance. The strings are assumed to be encodings of objects which, viewed abstractly, and in the language of logic, are structures. Taking instances to be structures allowed us to depart from the use of Herbrand models, which are at the heart of the logic programming-based methods mentioned above. From a knowledge representation point of view, taking instances as structures amounts to committing to what some part of the world is like, whereas the primary value of using formulas is to allow multiple models, that is, to give approximate descriptions of the world.

The abstraction of search problems as model expansion brings us directly to descriptive complexity theory, the study of expressiveness of languages in terms of the computational complexity of classes of structures definable in a language [Immerman 1999] (see also [Grädel 2007]). Knowing the expressiveness of a language, in this sense, is essential to good implementation and to appropriate selection of a language for a given task. Brachman and Levesque, in their paper "Expressiveness and tractability in knowledge representation and reasoning" [1987], demonstrated the inherent tradeoff between complexity and expressiveness in KR languages. While such a tradeoff was well established in theoretical work, the paper served as a wake-up call to AI researchers. Later, Levesque and Lakemeyer [2001] emphasized the *interplay between representation and reasoning* and the importance of the "study of how knowledge can at the same time be represented as comprehensively as possible *and* be reasoned with as effectively as possible." Taking descriptive complexity theory as central in the study of languages provides a basis for developing languages with controlled expressiveness, to support effective representation and reasoning.

The Theorems of Cook and Fagin

Cook's theorem [Cook 1971], which states that every problem in the complexity class NP can be reduced in polynomial time to propositional satisfiability (SAT), suggested the following scheme for solving NP search problems (also known as problems in FNP):

1. For each problem Π, implement a polytime reduction f to SAT;
2. Given an instance of Π, use a SAT solver to find a satisfying assignment for $f(\Pi)$, and thus a solution for Π, if there is one.

For some time this scheme was widely considered an idea of purely theoretical interest. However, progress in building and applying SAT solvers and related technology, such as solvers for ASP and satisfiability modulo theories (SMT, see e.g. Barret et al [2008]), has been so significant that researchers have begun to refer to NP as "the new tractable".

Fagin's theorem [Fagin 1974] says that the problems which can be axiomatized in the existential fragment of second order logic ($\exists$SO) are exactly those in NP. This result shows that the first step above, designing and implementing a reduction, can be replaced with declarative modelling. That is, rather than implement a reduction from Π to SAT, we may axiomatize Π with an $\exists$SO formula ϕ_Π. An instance of Π is a structure $\mathcal{A}$, and deciding if $\mathcal{A}$ has a solution is then equated with checking if $\mathcal{A} \models \phi_\Pi$. Since this is in NP, it can be done by reduction to SAT. In particular, we can find solutions to $\mathcal{A}$ by grounding:

1. Produce a ground formula $Gnd(\phi_\Pi, \mathcal{A})$ representing the solutions to $\mathcal{A}$.
2. Use a SAT solver to find a satisfying assignment for $Gnd(\phi_\Pi, \mathcal{A})$, if there is one.

It is easy to verify that there are grounding algorithms which, for every fixed formula ϕ, produce the grounding $Gnd(\phi, \mathcal{A})$ in time polynomial in the size of $\mathcal{A}$. An implementation of such an algorithm provides a universal automated reduction to SAT for problems in NP. That is, for each fixed ϕ_Π, the unary function $Gnd_\phi(\mathcal{A}) = Gnd(\phi, \mathcal{A})$ is essentially a polytime reduction from Π to SAT.

The theorems by Cook and Fagin suggest a conceptually elegant general method for solving NP search and decision problems, a fact which has certainly been observed by others. Our proposal amounts to taking this method seriously as the starting point for a general theory underlying new and existing languages and systems for representing and solving search problems. There are many foundational and practical issues between the idea and its realization, and also many issues in development and use of actual systems. In this paper, we describe early steps in developing the theory.

In the Section 2, we discuss the formalization of search as model expansion, and in particular NP-search as model expansion for first order logic. In Sections 3 and 4, we discuss several extensions of FO, which bring us closer to practical languages for representing NP search problems. Section 5 discusses a number of implemented systems for representing and solving search problems, and Section 6 briefly concludes the paper.

2   Search Problems as Model Expansion

We begin this section with a review of some necessary definitions. A vocabulary is a set τ of relation and function symbols, each with an associated arity. Constant symbols are zero-ary function symbols. A structure $\mathcal{A}$ for vocabulary τ (or, τ-structure) is a tuple containing a universe or domain A, and a relation (function) for each relation (function) symbol of τ. If R is a relation symbol of vocabulary τ, the relation corresponding to R in a τ-structure $\mathcal{A}$ is denoted $R^{\mathcal{A}}$. For example, we write

$$\mathcal{A} = (A;\ R_1^{\mathcal{A}}, \dots R_n^{\mathcal{A}},\ c_1^{\mathcal{A}}, \dots c_k^{\mathcal{A}},\ f_1^{\mathcal{A}}, \dots f_m^{\mathcal{A}}),$$

where the R_i are relation symbols, f_i function symbols, and constant symbols c_i are 0-ary function symbols. For a formula ϕ, we write $vocab(\phi)$ for the collection of exactly those function and relation symbols which occur in ϕ.

Let σ and τ be vocabularies, with $\sigma \subset \tau$, and let $\mathcal{A}$ be a σ-structure. A τ-structure $\mathcal{B}$ is an *expansion* of $\mathcal{A}$ to τ if $A = B$ (i.e., their domains are the same), and for every relation symbol R and in σ, $R^{\mathcal{A}} = R^{\mathcal{B}}$ and for every function symbol f in σ, $f^{\mathcal{A}} = f^{\mathcal{B}}$.

Definition: Model Expansion (MX)

For any logic $\mathcal{L}$, (e.g., FO, SO, FO(ID), FO(LFP), an ASP language, etc.,) the $\mathcal{L}$ model expansion problem, $\mathcal{L}$-MX, is:

> Instance: A pair $\langle \phi, \mathcal{A} \rangle$, where
> 1) ϕ is an $\mathcal{L}$ formula, and
> 2) $\mathcal{A}$ is a finite σ-structure, for vocabulary $\sigma \subseteq vocab(\phi)$.

> Problem: Find a structure $\mathcal{B}$ such that
> 1) $\mathcal{B}$ is an expansion of $\mathcal{A}$ to $vocab(\phi)$, and
> 2) $\mathcal{B} \models \phi$.

The vocabulary $\varepsilon := vocab(\phi) \setminus \sigma$, of symbols not interpreted by the structure, is called the *expansion vocabulary*. Symbols of the expansion vocabulary are second-order free variables.

MX, Fixed Formula Setting

We consider two natural settings for MX, the *combined setting*, in which we take both the formula and the structure as input, and the *fixed formula setting*, in which we consider a fixed formula and take structures as inputs. We will briefly discuss the combined setting below, but for most of what follows we focus on the fixed formula setting. In this setting, we consider a fixed formula vocabulary σ, and take only a σ-structure as input. The setting corresponds to the notion of *data complexity* as defined in [Vardi 1982].

The fixed formula setting provides a natural formalization of search problems. Here, ϕ is a problem description or specification, and a problem instance is a finite σ-structure $\mathcal{A}$. The formula specifies the relationship between an instance and its solutions. Finding a solution amounts to finding a structure $\mathcal{B}$ which is an expansion of $\mathcal{A}$ to the vocabulary of ϕ, and satisfies ϕ. A solution for instance $\mathcal{A}$ is given by the interpretations of the expansion vocabulary in structure $\mathcal{B}$.

EXAMPLE 1 (K-Colouring as FO MX, formula fixed). The problem is: Given a graph and a set K of colours, find a proper K-colouring of the graph. We will have two sorts, vertices V and colours K. The instance vocabulary is $\sigma := \{E\}$. The expansion vocabulary $\varepsilon := \{colour\}$, where $colour$ is a function from verticies to colours. The formula is:

$$\phi := \forall x \forall y \, (E(x, y) \supset (colour(x) \neq colour(y))),$$

where $colour$ is a free second-order variable. An instance is a structure $\mathcal{A} = (V \cup K; E^{\mathcal{A}})$, and a solution is an interpretation for $colour$ such that

$$\underbrace{(V \cup K; \overbrace{E^{\mathcal{A}}}^{\mathcal{A}}, colour^{\mathcal{B}})}_{\mathcal{B}} \models \phi.$$

Assume a standard encoding of languages as classes of structures as done in descriptive complexity (see, e.g., [Libkin 2004]). FO model expansion can define exactly the same classes of structures that ∃SO can, which immediately implies the following theorem.

THEOREM 2 ([Fagin 1974]). *The first-order model expansion problem in the data complexity setting captures* NP.

The property means that FO model expansion is in NP, and moreover that *every (search) problem in NP* can be represented as FO MX, with a fixed formula. Thus, FO MX is a *universal framework* for representing NP-search problems. Combined with an automated reduction from FO model expansion to SAT, which may be provided by a grounding algorithm, we have a *universal method* for modelling and solving NP-search problems.

While FO MX captures NP, for practical purposes it must be extended with features to make representation of complex domains convenient. In [Mitchell and Ternovska 2005], we proposed using FO augmented with non-monotonic inductive definitions, as described in Section 3. In other work, we extend FO with arithmetic, including aggregate operators, as described in Section 4.

It can be seen that model expansion is the task underlying a wide range of practical tools for solving search problems. For example, we have shown that MX underlies the high-level specification language ESSENCE (see [Mitchell and Ternovska 2008]), although the logic required to have the same expressive power as unrestricted ESSENCE is much more expressive than FO.

REMARK 3. Since FO MX can specify the same problems that ∃SO can, one may question the need for the term MX. We use it because we are interested in MX for a variety of logics other than FO, and because for each logic $\mathcal{L}$ there are several tasks of interest.

The Importance of Capturing Complexity Classes for Declarative Programming

We claimed, in [Mitchell and Ternovska 2005], that an important property for a specification language is capturing a complexity class. To illustrate, if a language $\mathcal{L}$ captures NP, we know that:

(a) $\mathcal{L}$ *can express every problem in NP* – which gives the user an assurance of universality of the language for the given complexity class,

(b) *no more than NP can be expressed* – thus solving can be achieved by universal polytime transformation to an NP-complete problem, for example by polytime grounding to SAT.

The capturing property is of clear practical importance. This does not mean that specification languages should necessarily have restricted power, but that we should understand which fragments of an expressive language have restricted power, and understand the possible consequences of using more expressive fragments. While many users would not be familiar with complexity theory and its practical implications, they could still be provided with fragments of specification languages corresponding to different complexity classes, and advised to use a "safe" fragment when possible.

Generalizing to Other Complexity Classes

Results analogous to Fagin's theorem followed for other complexity classes, most notably for P [Immerman 1982; Vardi 1982; Livchak 1983]. As with Fagin's theorem and NP, these results show that logics can be seen as modelling languages for problems in corresponding complexity classes, and provide the basis for universal solving methods for problems in these classes. The following properties for MX in the data complexity setting are immediate from the definition of MX and results in [Grädel 1992] and [Stockmeyer 1977].

- On ordered structures, FO universal Horn MX captures P,
- On ordered structures, FO universal Krom MX captures NL,
- Π_{k-1}^1 MX captures the Σ_k^P level of the Polynomial Hierarchy.

Notice the requirement of ordered structures above. Ordered structures are those with a total ordering on domain elements. For precise details, see [Ebbinghaus and Flum 1995] or [Libkin 2004]. Intuitively, the order is necessary to mimic the computation of a Turing machine on a tape. While for NP the order can be represented by an existentially quantified second-order variable, it is conjectured that there is no logic capturing the class P on arbitrary structures. The conjecture is due to Gurevich, and the positive resolution of this conjecture (i.e., a proof that no such logic exists) would imply P$\neq$NP. For further discussion and references see [Immerman 1999; Libkin 2004; Grädel 2007].

MX, Combined Setting

In this setting, both formula ϕ and the structure $\mathcal{A}$ are part of the instance. The setting corresponds to the notion of *combined complexity* [Vardi 1982]. Here, we may still have a formula which is fixed for all instances, but an instance will consist of a formula together with a finite structure (which may be just the universe). The expansions must satisfy the conjunction of the two formulas. We have the following.

THEOREM 4. *First-order model expansion, in the combined setting, is* NEXPTIME-*complete.*

This property is equivalent to a result in [Vardi 1982], and can be shown by an easy reduction from Bernays-Schönfinkel satisfiability or combined complexity of $\exists$SO over finite structures, as described in [Mitchell and Ternovska 2005].

Combined setting is very useful for knowledge representation when one needs to represent a part of an instance by a formula. This is especially convenient when non-determinism is involved. In [Mitchell, Ternovska, Hach, and Mohebali 2006], we give an example of a planning problem represented as MX in the combined setting.

Model Expansion in the Context of Related Tasks

Model expansion is closely related to the more studied tasks of model checking (MC) and finite satisfiability. For a given logic $\mathcal{L}$,

1. *Model Checking* (MC): given $(\mathcal{A}, \phi)$, where ϕ is a sentence in $\mathcal{L}$ and $\mathcal{A}$ is a finite structure for $vocab(\phi)$, does $\mathcal{A} \models \phi$?

2. *Model Expansion* (MX): given $(\mathcal{A}, \phi)$, where ϕ is a sentence in $\mathcal{L}$, $\mathcal{A}$ is a finite σ-structure and $\sigma \subseteq vocab(\phi)$, is there $\mathcal{B}$ for $vocab(\phi)$ which expands $\mathcal{A}$ and $\mathcal{B} \models \phi$?

3. *Finite Satisfiability*: given a sentence ϕ in $\mathcal{L}$, is there a finite $\mathcal{A}$ for $vocab(\phi)$ such that $\mathcal{A} \models \phi$?

In model checking, the entire structure is given and we ask if the structure satisfies the formula; in model expansion part of a structure is given and we ask for an expansion satisfying the formula; in finite satisfiability we ask is any finite structure satisfies the formula. Thus, for any logic $\mathcal{L}$, the complexity of model expansion — in both fixed formul and combined cases — lies between that of the other two tasks:

$$MC(\mathcal{L}) \leq MX(\mathcal{L}) \leq \text{Satisfiability}(\mathcal{L}).$$

In the case of FO, we have the following. In the *fixed formula* (data complexity) setting FO MC is in AC_0: capturing AC_0 requires extending the logic, for example to $FO(<,BIT)$ or $FO(+,\times)$ (see [Immerman 1999]). FO MX captures NP. In the *combined* setting (combined complexity), FO MC is PSPACE-complete [Stockmeyer 1974], and FO MX is NEXPTIME-complete. FO satisfiability in the finite is undecidable [Trahtenbrot 1950]. Model expansion avoids undecidability of FO by specifying the finite universe as part of the instance. We discuss the complexity of MC, MX and satisfiability for some fragments and extensions of FO in [Kolokolova, Liu, Mitchell, and Ternovska 2010]. In particular, we consider guarded fragments and logics with fixpoints and inductive definitions, and study both data and combined complexity.

3 Practical Knowledge Representation Features

The formulation of search problems as MX for classical logic has some elegance, and results showing capturing of complexity classes show that simple languages are sufficient for specifying large classes of problems. However, languages for practical use require a number of extensions for convenience of users. In this section, we discuss the extension of FO MX with sorts, order relations, and inductive definitions. None of these extensions affects the expressive power, or the universal polytime reduction to SAT, but they are all important for practical purposes. In the following section, we discuss the extension with arithmetic.

Ordered Structures and Multi-Sorted FO

An ordered structure is one that contains an order relation $\leq$ on its domain. Having an ordered domain can be very convenient in axiomatizing problems. While in many examples order is not needed, we find that having this relation as built-in saves a lot of effort in axiomatizing problems. In addition, it helps significantly in symmetry breaking thus indirectly reducing solving time. The order relation is also necessary for certain theoretical properties, such as defining a logic that captures the complexity class P. Since assuming ordered structures seems to cause no complications, we may always do so when convenient.

In multi-sorted logic, we assume the domain of a structure to be partitioned into a set of sorts, and that each variable ranges over the entities of one particular sort. Axiomatizations in multi-sorted logic tend to be more understandable than those in a single-sorted logic. For example, we use two sorts in Example 1 (K-Colouring), one for vertices and one for colours. Having multiple sorts also helps with efficient grounding. For theoretical considerations, the assumption about having multi-sorted logic can always be removed by the standard re-writing of formulas, using unary sort predicates, as explained in most textbooks on logic. We assume that equality is always present in the language, as is common in knowledge representation.

Inductive Definitions

Induction is very important in practice because many applications involve inductive properties, such as reachability and temporal properties of dynamic systems. However, such properties are not easily expressed as FO MX, and the methods for expressing them tend to produce specifications which are not handled well by current solvers. (Although reachability is not FO definable, it can be defined in $\exists$SO, so in theory we do not need to add an explicit mechanism for induction, but for practical purposes we do.) Therefore, in [Mitchell and Ternovska 2005], we proposed using the extension of classical logic with inductive definitions described below.

In our view, it is more appropriate to extend classical logic with a mechanism for (non-monotone) induction than to base an entire language or system on a non-monotonic logic. The reason is that the classical semantics is much simpler to work with, both for language users and system designers. In particular, classical logic is modular, that is, combining FO specifications is easy, because no special conditions are required to ensure the semantic content of a formula remains unchanged in combination with others. Since the large majority of constraints in most applications are classical - that is, do not involve minimization or closure - we may use the more complex semantics of induction *when and to the extent we need it*. With a purely non-monotonic logic, most of the time we must simulate the classical semantics using the more complex non-classical semantics.

Theoretical work can also benefit from use of classical logic. In our experience it is easier to verify that a classical theoretical result continue to hold when adding inductive definitions to classical logic than to prove the result from scratch for a non-classical semantics. For example, it is very easy to show that the main complexity and capturing properties for FO MX, which are mentioned in Section 2, hold also when inductive definitions are added to FO MX [Mitchell and Ternovska 2005; Kolokolova, Liu, Mitchell, and Ternovska 2010].

ID-logic is a logic which augments classical logic with a non-classical construction to represent induction, developed in [Denecker 2000; Denecker and Ternovska 2004b; Denecker and Ternovska 2008]. The construct allows for natural modelling of monotone and non-monotone inductive definitions, including iterated induction and induction over a well-founded order. Such constructions occur frequently in mathematics and a variety of applications, as well as in common-sense reasoning. The logic is suitable to model non-monotonicity, causality and temporal reasoning [Denecker and Ternovska 2004a; Denecker and Ternovska 2007], and as such is a good knowledge representation language. FO(ID) is the fragment of ID-logic in which the classical part is FO. We proposed this as the logic of choice for modelling NP-search problems as model expansion in [Mitchell and Ternovska 2005]. In the remainder of this section, we briefly describe the syntax of FO(ID). Details can be found in the cited papers.

Syntax of FO(ID)

The formulas of FO(ID) are constructed from atoms $P(\bar{t})$, and definitions Δ, closed under the standard use of $\wedge, \vee, \neg, \forall, \exists$. A definition Δ is a set of rules, which we illustrate by example.

EXAMPLE 5 (Transitive Closure of an Edge Relation).

$$\left\{ \begin{array}{l} \forall x \forall y \; (TC(x,y) \leftarrow E(x,y)), \\ \forall x \forall y \; (TC(x,y) \leftarrow \exists z \; (TC(x,z) \wedge E(z,y)) \end{array} \right\}$$

The syntax is quite general: the body of the rules may be unrestricted FO formulas. Disjunctions of definitions, multiple definitions for the same predicate, etc., are allowed. Predicates appearing in the heads of the rules are called defined, and those which are not defined are called open.

Semantics of ID-logic

The $\leftarrow$ operator in rules of definitions is the *definitional implication*, with semantics provided by the well-founded semantics from logic programming [Van Gelder, Ross, and Schlipf 1991]. Satisfaction is defined by the following induction:

$$
\begin{array}{lll}
A \models P(\bar{t}) & \text{if} & \bar{t}^A \in P^A \\
A \models \phi \circ \psi \ (\text{or} \ \circ \psi) & \text{if} & \circ \in \{\wedge, \vee, \neg, \forall, \exists\}, \text{ by the standard induction} \\
A \models \Delta & \text{if} & A \text{ is the 2-valued well-founded model of } \Delta.
\end{array}
$$

A definition can be used in multiple ways. For example, the definition of Example 5 can be used to compute the transitive closure of an edge relation given in the instance, or to construct an edge relation who's transitive closure has been given in the instance. In general, both instance and expansion predicates can appear as open predicates of definitions used in FO(ID) MX axiomatizations.

4 Adding Arithmetic to FO MX

Most practical constraint modelling languages have arithmetic, and it is important to analyze the expressive power of such languages with respect to the main computational task solved, which is model expansion. In [Ternovska and Mitchell 2009], we began formalizing the principles of using built-in arithmetic and other built-in operations in specification languages for search problems. The main challenge is that arithmetical structures are infinite, and adding arithmetic operations increases the expressive power tremendously. However, expressive power needs to be controlled for efficient solving, and a mathematical framework was needed. We formulated the notion of an MX *embedded in an infinite background structure*, similar to the notion of embedded structures in database research. The background structure can in principle be any infinite structure with operations built-in to practical languages.

Embedded MX

Embedded finite model theory (see [Libkin 2004]), the study of finite structures whose domain is drawn from some infinite structure, was motivated by the need to study databases that contain numbers and numerical constraints. Rather than think of a database as a finite structure, we take it to be a set of finite relations over an infinite domain.

DEFINITION 6. A structure $\mathcal{A}$ is *embedded* in an infinite *background* structure $\mathcal{M} = (U; \bar{M})$ if it is a structure $\mathcal{A} = (U; \bar{R})$ with a finite set $\bar{R}$ of finite relations and functions (incl. 0-ary functions) taking non-zero values on a finite set of elements, $\bar{M} \cap \bar{R} = \emptyset$. The set of elements of U that occur in some relation of $\mathcal{A}$ is the active domain of $\mathcal{A}$, denoted $adom_{\mathcal{A}}$.

In database research, embedded structures are used with logics for expressing queries. We use them in logics for MX specifications. The vocabularies for such logics consist of (1) the vocabulary of $\mathcal{A}$, which is our instance vocabulary σ; (2) the vocabulary ν of an infinite secondary structure $\mathcal{M} = (U; \bar{M})$, such as the arithmetical structure defined below; and (3) an expansion vocabulary ε. A formula ϕ over $\sigma \cup \nu \cup \varepsilon$ constitutes an MX specification. The task of model expansion remains the same: expand an embedded σ-structure to satisfy formula ϕ.

The framework was used to analyze the expressive power of the constraint modelling language Essence [Mitchell and Ternovska 2008]. For arithmetical structures, namely structures containing at least $(\mathbb{N}; 0, 1, \chi, <, +, ., min, max, \Sigma, \Pi)$, we have shown a capturing NP property. The theorem works for structures where the values of numbers in the input are limited in comparison with the domain size n by $2^{poly(n)}$.

Inexpressibility Results

While most problems in NP fall within the value restriction, some cannot be formalized naturally. One such problem is integer factorization. The value of the input is unlimited with respect to the domain size (which equals to one). While this is a restriction in the theorem, is something analogous present in practical system languages? Tasharrofi and Ternovska analyzed the input languages of several systems in [Tasharrofi and Ternovska 2010a] such as input languages of some ASP systems [Syrjänen 2000; Gebser, Kaminski, Kaufmann, Ostrowski, Schaub, and Thiele 2008], as well as the input language of the IDP system [Wittocx and Marien 2008]. They showed that integer factorization and other similar problems involving numbers are not expressible naturally in these languages, under reasonable complexity-theoretic assumptions. While the expressivity limitation can, in principle, be overcome with compact domain representations, or if all potentially useful numbers are given in the input, this is not a practical solution. Allowing compact domain representations leads to exponential time grounding, and giving all numbers as input is impractical especially for factorization, where very large numbers are needed e.g. for cryptographic applications. That paper also demonstrates that the system languages we analyzed capture NP with built-in arithmetic (also under small-cost restriction) when certain operations are present in the language. In [Tasharrofi and Ternovska 2010b], the small cost requirement in the capturing NP result of [Ternovska and Mitchell 2009] was eliminated by using a new logic, PBINT, which overcomes the expressiveness limitations of the existing languages, and can be viewed as an idealized constraint modelling language. The capturing NP with arithmetic theorem is proved for arbitrary embedded input structures, however a slightly different set of built-in operations is needed.

The Importance of Guarded Fragments

An important step towards formalization of MX in the presence of infinite background structures is the use of guarded logics. We use an adaptation of the guarded fragment GF_k of FO [Gottlob, Leone, and Scarcello 2001]. In formulas of GF_k, a conjunction of up to k atoms acts as a *guard* for each quantified variable. For example, in the following formula the universal quantifiers are guarded: $\forall \bar{x} \, (G_1 \wedge \ldots \wedge G_m \supset \phi)$.

We need to limit both the range of quantified variables and the range of expansion predicates. In addition to "lower guards" above, we introduced, for each expansion predicate $E \in \varepsilon$, an *upper guard* axiom of the form: $\forall \bar{x}(E(\bar{x}) \supset G_1(\bar{x}_1) \wedge \cdots \wedge G_k(\bar{x}_k))$, where the union of free variables in the G_i is precisely $\bar{x}$. For expansion functions, the upper guard axiom is on the graph of the function, i.e., $\forall \bar{x} \forall y \, (f(\bar{x}) = y \supset \phi)$, where ϕ is a conjunction of atoms, and every variable in $\bar{x}, y$ occurs free in ϕ.

In the simplest variant of the logic, we require all upper and lower guards to be from the instance vocabulary, so ranges of variables and expansion predicates are explicitly limited to $adom_\mathcal{A}$. We then relax this restriction, adding a mechanism for "user-defined" guard relations that may contain elements not in $adom_\mathcal{A}$. Details can be found in [Ternovska and Mitchell 2009].

Introducing guards is a very important step. Without them, embedded model expansion task is undecidable. Guards are needed to limit the expressive power of logics used for MX axiomatizations, and to limit the range of the quantifiers.

In order to illustrate the simplicity of knowledge representation in the presence of guards, we give examples of embedded MX specifications for search versions of two common optimization problems. The background structure is the arithmetical structure, a structure $\mathcal{N}$ containing at least $(\mathbb{N}; 0, 1, \chi, <, +, ., min, max, \Sigma, \Pi)$ (see [Ternovska and Mitchell 2009] for details).

EXAMPLE 7. KNAPSACK: Instance vocabulary $\sigma := \{O, w, v, b, k\}$, where: O is the set of objects, w is the weight function, v is the value function, b is the weight bound, and k is the value bound. Expansion vocabulary $\varepsilon := \{O'\}$, where O' is the set of selected objects. Upper guard axiom: $\forall x (O'(x) \supset O(x))$.   Axioms:

$$\Sigma_x(w(x) \,:\, O(x) \wedge O'(x)) \leq b$$
$$k \leq \Sigma_x(v(x) \,:\, O(x) \wedge O'(x)),$$

where $t_1 \leq t_2$ is an abbreviation for $t_1 < t_2 \vee t_1 = t_2$. Here, Σ_x is like a quantifier, binding x, and $O(x)$ is the lower guard for $O(x) \wedge O'(x)$.

REMARK 8. Upper and lower guards provide a logical formalization of the type systems of some existing constraint modelling languages [Mitchell and Ternovska 2008]. For example, the guard O in Example 7, corresponds to introducing a type "objects", and the upper guard axiom corresponds to declaring O' to be of this type. Lower guards correspond to types of variables.

EXAMPLE 9. MACHINE SCHEDULING PROBLEM [Hooker 2000]: The task is to assign jobs to machines so that constraints on release and due date are satisfied and a cost bound is achieved. The input structure lists jobs, machines, possible start times, the release date and due date for each job, the cost and duration for running each job on each machine, and the cost bound. The instance vocabulary, σ, consists of: $Job(j)$ – the set of jobs to be scheduled; $Machine(m)$ – the set of machines to perform jobs; $Time(t)$ – all possible starting times; $ReleaseDate(j)$ – each job has a release date; $DueDate(j)$ – each job has a due date; $Cost(j, m)$ – cost of performing job j on machine m; $Duration(j, m)$ – duration of executing j on m; and k – the cost bound. The active domain consists of all time points, costs, due and release dates, durations, jobs and machines. Expansion vocabulary consists of two functions: $Assignment(j)$ maps jobs to machines; $StartTime(j)$ maps jobs to start times.

Upper guard axioms:
$\forall j \forall m \, (Assignment(j) = m \supset Machine(m) \wedge Job(j))$
$\forall j \forall t \, (StartTime(j) = t \supset Time(t) \wedge Job(j))$

Axioms:
$\Sigma_j(Cost(j, Assignment(j)) \,:\, Job(j)) \leq k$
$\forall j (Job(j) \supset StartTime(j) \geq ReleaseDate(j))$
$\forall j (Job(j) \supset StartTime(j) + Duration(j) \leq DueDate(j)) \, \forall t \, (Time(t) \supset$
$(\forall m \, (Machine(m) \supset max_j(count_j(\psi(j, m, t))) = 1)).$

In the last axiom, which specifies that at most one job is on a machine at a time, $count_j(\psi(j, m, t))$ is an abbreviation for $\Sigma_j(\chi[\psi(j, m, t)])$, where χ is the characteristic function, and ψ defines the set of jobs being executed on machine m at time t, that is,

$\psi(j, m, t)$ is:

$$Job(j) \wedge Assignment(j) = m \wedge Time(StartTime(j))$$
$$\wedge\, StartTime(j) \leq t$$
$$\wedge\, t < StartTime(j) + Duration(j, Assignment(j)).$$

An optimization version (outside the scope of this paper) would include the objective function: *minimizing:* $\Sigma_j(Cost(j, Assignment(j)) : Job(j))$.

5 Systems

The literature on languages and systems for representing and solving NP search and decision problems is very large. Here, we limit our attention to a number of proposed languages for representing search problems. We would argue that MX is the underlying task being specified in the most typical use of all of these languages. That is, in normal use, the fragment of the language used for describing problem instances (often called "data") describes a finite structure, and the fragment of the language used for defining problems defines a class expansions of these structures.

REMARK 10. An important detail to recognize, when comparing languages of implemented systems with theoretical work, is that even systems intended exclusively for NP-search problems typically can express problems in the much larger complexity class EXP-TIME (and sometimes even more), because they have compact domain descriptions. That is, for user convenience, they permit describing a domain of size n by giving its size or an interval in $\mathbb{Z}$, rather than by enumeration. Thus, we do not examine the expressiveness here.

Explicitly FO-MX Systems

At least three systems have been built for languages explicitly formulated as model expansion for extensions of multi-sorted FO extended with arithmetic and inductive definitions. These are MXG [Mitchell, Ternovska, Hach, and Mohebali 2006], the IDP System [Wittocx and Marien 2008] and Enfragmo[Aavani, Tasharrofi, Unel, Ternovska, and Mitchell 2010]. All three of these systems involve grounding to SAT or extensions of SAT, although they use different grounding techniques.

Constraint Modelling Languages

In the literature on constraints (constraint satisfaction and constraint programming), a number of "constraint modelling languages" have been proposed over the last decade. Examples of these are Zinc[Marriott, Nethercote, Rafeh, Stuckey, de la Banda, and Wallace 2008], ESSENCE[Frisch, Harvey, Jefferson, Hernández, and Miguel 2008] and ESRA [Flener, Pearson, and Ågren 2003]. None of these are formulated in terms of logic, and they are all able to express more than NP. We showed in [Mitchell and Ternovska 2008] that a natural fragment of ESSENCE is essentially a syntactic variant of FO, defining FO MX problems, and believe that the same can be done for Zinc and ESRA. We also believe a similar treatment of the full languages is possible.

ASP and other Declarative Programming Languages

Under the rubric of "declarative programming", we find NP-Spec [Cadoli, Palopoli, Schaerf, and Vasile 1999], and the languages of ASP grounders such as Lparse [Syrjänen 2000], Gringo [Gebser, Kaminski, Kaufmann, Ostrowski, Schaub, and Thiele 2008] and DLV

[Leone, Pfeifer, Faber, Eiter, Gottlob, Perri, and Scarcello 2006], and the language ASPPS of East and Truszczynski [2006].

Answer set programming [Marek and Truszczynski 1999; Niemela 1999] is based on the language of logic programming, with the stable model semantics [Gelfond and Lifschitz 1991]. The languages of the grounders are extended significantly, with arithmetic and "weight constraints". The built-in recursion in ASP is often convenient, particularly for axiomatizing problems involving sequences of events, such as in verification and planning problems. A problem, in our view, is that the entire formula (logic program) is involved in the recursion. Many properties which can be easily and naturally expressed classically require a less natural expression within this recursion.

The modelling language NP-SPEC [Cadoli, Palopoli, Schaerf, and Vasile 1999] is stratified Datalog, with several extensions, including a kind of second order variable the brings the expressive power up to NP. An instance is a finite set of ground atoms, and models are Herbrand models. One implementation translated problems to Prolog, while another used reduction to SAT, but only for a restricted fragment of the language [Cadoli and Schaerf 2005].

East and Truszczynski's ASPPS [East and Truszczynski 2006] is the first system we know of based on the idea of extending classical logic with a recursion mechanism. The modelling language consists of a restricted form of first order formulas, written in an ASP-like syntax, augmented by arithmetic, weight constraints, and Horn rules with minimum-model semantics. The language could be viewed as a restricted form of FO(ID) with arithmetic and weight constraints. Semantically, ASPPS is based on Herbrand models. An implementation supports grounding to SAT, and also to propositional CNF extended with cardinality constraints.

6   Conclusion

We have outlined the formalization of search problems as the logical task of model expansion, and some of the reasons we believe this formalization and descriptive complexity theory are an appropriate basis for a theory of languages for representing search problems. The goal of this work is to develop a rigorous foundation for such languages, including practical languages and the systems that use them. To the best of our knowledge, we were the first to propose this program of research. We also described some initial steps toward building such a theory, namely the formalization and analysis of languages with arithmetic. Work has also been done, by our research group and others, on various aspects of system implementation and application, but we have not discussed this work here.

Acknowledgments: The authors are grateful to the Natural Sciences and Engineering Research Council of Canada (NSERC), and the Isaac Newton Institute for Mathematical Sciences, and to many colleagues for useful discussions.

References

Aavani, A., S. Tasharrofi, G. Unel, E. Ternovska, and D. Mitchell [2010]. Speed-up techniques for negation in grounding. In *Proc., 16th Int'l Conf. on Logic for Programming, Artificial Intelligence and Reasoning (LPAR-16)*, LNCS.

Barrett, C., R. Sebastiani, S. A. Seshia, and C. Tinelli [2008]. Satisfiability modulo theories. In A. Biere, M. Heule, H. van Maaren, and T. Walsh (Eds.), *Handbook of Satisfiability*, Chapter 12, pp. 737–797. IOS Press.

Cadoli, M., L. Palopoli, A. Schaerf, and D. Vasile [1999]. NP-SPEC: An executable specification language for solving all problems in NP. In *Proc., First Int'l Workshop on Practical Aspects of Declarative Languages (PADL)*, pp. 16–30.

Cadoli, M. and A. Schaerf [2005]. Compiling problem specifications into SAT. *Artificial Intelligence 162*, 89–120.

Cook, S. [1971]. The complexity of theorem proving procedures. In *Proc. 3rd Ann. ACM Symp. on Theory of Computing (STOC)*, New York, pp. 151–158. Association for Computing Machinery.

Denecker, M. [2000]. Extending classical logic with inductive definitions. In *Proc., First Int'l Conference on Computational Logic (CL-2000)*, pp. 703–717. Springer.

Denecker, M. and E. Ternovska [2004a]. Inductive situation calculus. In *Proc., Ninth Int'l Conf. on Principles of Knowledge Representation and Reasoning (KR-04)*, pp. 545–553. AAAI Press.

Denecker, M. and E. Ternovska [2004b]. A logic of non-monotone inductive definitions and its modularity properties. In *Proc., 7th Int'l Conf., on Logic Programming and Nonmonotonic Reasoning (LPNMR-04)*, pp. 47–60. Springer. LNCS 2923.

Denecker, M. and E. Ternovska [2007]. Inductive situation calculus. *Artificial Intelligence 171*(5-6), 332–360.

Denecker, M. and E. Ternovska [2008]. A logic of non-monotone inductive definitions. *ACM transactions on computational logic (TOCL) 9*(2), 1–51.

East, D. and M. Truszczynski [2006]. Predicate-calculus based logics for modeling and solving search problems. *ACM Transactions on Computational Logic (TOCL) 7*(1), 38–83.

Ebbinghaus, H.-D. and J. Flum [1995]. *Finite model theory*. Springer Verlag.

Fagin, R. [1974]. Generalized first-order spectra and polynomial-time recognizable sets. In *Complexity of Comput.*, pp. 43–73.

Flener, P., J. Pearson, and M. Ågren [2003]. Introducing ESRA, a relational language for modelling combinatorial problems. In *Proc., 9th Int'l Conf. on Principles and Practice of Constraint Programming (CP-03)*, pp. 971. Springer. LNCS 2833.

Frisch, A. M., W. Harvey, C. Jefferson, B. M. Hernández, and I. Miguel [2008]. ESSENCE: A constraint language for specifying combinatorial problems. *Constraints 13*(3), 268–306.

Gebser, M., R. Kaminski, B. Kaufmann, M. Ostrowski, T. Schaub, and S. Thiele [2008, November]. *A User's Guide to gringo, clasp, clingo, and iclingo*. University of Potsdam. (At: http://potassco.sourceforge.net/).

Gelfond, M. and V. Lifschitz [1991]. Classical negation in logic programs and disjunctive databases. *New Generation Computing 9*, 365–385.

Gottlob, G., N. Leone, and F. Scarcello [2001]. Robbers, marshals, and guards: game theoretic and logical characterizations of hypertree width. In *Proc., Twentieth ACM SIGMOD-SIGACT-SIGART symposium on Principles of database systems (PODS-01)*, pp. 195–206. ACM.

Grädel, E. [1992]. Capturing Complexity Classes by Fragments of Second Order Logic. *Theor. Comp. Sc. 101*, 35–57.

Grädel, E. [2007]. *Finite Model Theory and Descriptive Complexity*, pp. 125–230. Springer.

Green, C. [1969]. Theorem proving by resolution as a basis for question-answering systems. In B. Meltzer and D. Michie (Eds.), *Machine Intelligence 4*, pp. 183–205. Edinburgh University Press.

Hooker, J. [2000]. *Logic-based methods for optimization: combining optimization and constraint satisfaction*, Chapter 19, pp. 389–422. Wiley and Sons.

Immerman, N. [1982]. Relational queries computable in polytime. In *14th ACM Symp. on Theory of Computing (STOC)*, pp. 147–152. Springer Verlag.

Immerman, N. [1999]. *Descriptive complexity*. Springer.

Kolokolova, A., Y. Liu, D. Mitchell, and E. Ternovska [2010]. On the complexity of model expansion. In *Proc., 17th Int'l Conf. on Logic for Programming, Artificial Intelligence and Reasoning (LPAR-17)*, pp. 447–458. Springer. LNCS 6397.

Leone, N., G. Pfeifer, W. Faber, T. Eiter, G. Gottlob, S. Perri, and F. Scarcello [2006]. The DLV system for knowledge representation and reasoning. *ACM Transactions on Computational Logic (TOCL) 7*(3), 499–562.

Levesque, H. J. and R. J. Brachman [1987]. Expressiveness and tractability in knowledge representation and reasoning. *Computational Intelligence 3*, 78–93.

Levesque, H. J. and G. Lakemeyer [2001]. *The Logic of Knowledge Bases*. The MIT Press.

Libkin, L. [2004]. *Elements of Finite Model Theory*. Springer.

Livchak, A. [1983]. The relational model for process control. (in russian). *Automated Documentation and Mathematical Linguistics* (4), 27–29.

Marek, V. W. and M. Truszczynski [1999]. *Stable logic programming - an alternative logic programming paradigm*, pp. 375–398. Springer-Verlag. In: The Logic Programming Paradigm: A 25-Year Perspective, K.R. Apt, V.W. Marek, M. Truszczynski, D.S. Warren, Eds.

Marriott, K., N. Nethercote, R. Rafeh, P. J. Stuckey, M. G. de la Banda, and M. Wallace [2008]. The design of the zinc modelling language. *Constraints 13*(3), 229–267.

Mitchell, D. and E. Ternovska [2005]. A framework for representing and solving NP search problems. In *Proc. of the 20th National Conf. on Artif. Intell. (AAAI)*, pp. 430–435.

Mitchell, D., E. Ternovska, F. Hach, and R. Mohebali [2006]. Model expansion as a framework for modelling and solving search problems. Technical Report TR 2006-24, Simon Fraser University, School of Computing Science.

Mitchell, D. G. and E. Ternovska [2008]. Expressiveness and abstraction in ESSENCE. *Constraints 13(2)*, 343–384.

Niemela, I. [1999]. Logic programs with stable model semantics as a constraint programming paradigm. *Annals of Mathematics and Artificial Intelligence 25*(3,4), 241–273.

Selman, B., H. Levesque, and D. Mitchell [1992]. A new method for solving hard satisfiability problems. In *Proc., Tenth National Conference on Artificial Intelligence (AAAI-92)*, San Jose, CA, pp. 440–446. AAAI Press/MIT Press.

Stockmeyer, L. [1974]. *The Complexity of Decision Problems in Automata Theory.* Ph.D. thesis, MIT.

Stockmeyer, L. J. [1977]. The polynomial-time hierarchy. *Theoretical Computer Science 3*, 1–22.

Syrjänen, T. [2000]. *Lparse 1.0 User's Manual.* Helsinki University of Technology. http://www.tcs.hut.fi/Software/smodels/lparse.ps.gz.

Tasharrofi, S. and E. Ternovska [2010a, October). Built-in arithmetic in knowledge representation languages. In *NonMon at 30 (Thirty Years of Nonmonotonic Reasoning).*

Tasharrofi, S. and E. Ternovska [2010b, October). PBINT, a logic for modelling search problems involving arithmetic. In *Proc. of the 17th Conference on Logic for Programming, Artificial Intelligence and Reasoning (LPAR'17)*, Yogyakarta, Indonesia.

Ternovska, E. and D. G. Mitchell [2009]. Declarative programming of search problems with built-in arithmetic. In *Proc. of 21st International Joint Conference on Artificial Intelligence (IJCAI-09)*, pp. 942–947.

Trahtenbrot, B. [1950]. The impossibility of an algorithm for the decision problem for finite domains. *Doklady Academii Nauk SSSR 70*, 569–572.

Van Gelder, A., K. Ross, and J. Schlipf [1991]. The well-founded semantics for general logic programs. *Journal of Assoc. Comput. Mach. 38*(3), 620–650.

Vardi, M. Y. [1982]. The complexity of relational query languages (extended abstract). In *14th ACM Symp.on Theory of Computing, Springer Verlag*, pp. 137–146.

Wittocx, J. and M. Marien [2008, June). *The IDP System.* Katholieke Universiteit Leuven. (Manual at www.cs.kuleuven.be/˜dtai/krr/software/idpmanual.pdf).

23

A Decidable First-Order Logic for Knowledge Representation

PETER F. PATEL-SCHNEIDER

ABSTRACT. Decidable first-order logics with reasonable model-theoretic semantics have several benefits for knowledge representation (KR). These logics have the expressive power of standard first-order logic (FOL) along with an inference algorithm that will always terminate, both important considerations for KR. KR systems that include a faithful implementation of one of these logics can also use its model-theoretic semantics to provide meaning for the data they store. One such logic, a variant of a simple type of first-order relevance logic, is developed and its properties given. This logic, although extremely weak, does capture a well-motivated set of inferences that can be entrusted to a KR system.

Note: *To this day there continue to be attempts to to use logics similar to the logic described in this paper, even though such logics are so weak and inference engines have improved so dramatically since this paper was written.*

Knowledge representation has always played a major role in AI research. However, the exact role of knowledge representation (KR) systems in AI programs is not well-defined. In some knowledge-based systems the KR subsystem is used simply as a package for manipulating data structures with no commitment to what these data structures mean. In other knowledge-based systems the KR subsystem is not distinguished from the system as a whole and has responsibility for many tasks, such as planning and resource management.

Following the lead of Levesque [1984a], my view of KR is that a KR system will always be just a part of some larger knowledge-based system. Its job is to represent and manage the knowledge required by this larger system. This view makes a KR system much more than simply a collection of routines to manipulate uninterpreted data. Instead, there must be some sense in which the KR system is actually representing knowledge.

Since any KR system is part of a computer program and is thus a formal object, it can only store and manipulate expressions taken from some formal language. For the system to really represent knowledge there must be some rigorous external method of providing meaning for these expressions.

Tarskian or model theoretic semantics provide one way of doing precisely this.[1] They take expressions in a formal language and determine their truth with respect to some model. The knowledge that the KR system represents is then some function of this semantics and

---

[1] The importance of a formal account of the meaning of a KR system is in accordance with the views of McDermott who argues that any notation used in an AI system must have a denotation and that reasoning in the system must be sound (and preferably complete) with respect to that denotation [McDermott 1978]. It also emphasizes the role of semantics as espoused by Hayes [1977] and by Moore, who states "Whatever else a formalism may be, at least some of its expressions must have *referential semantics* if the formalism is really to be a representation of *knowledge*" [Moore 1982, p. 428]. Along these lines Frisch argues that any AI program should have a complete model-theoretic description [Frisch 1985].

the expressions that the KR system stores in its knowledge base (KB). The simplest such function is that each stored expression is taken to be "true" (although other methods of determining the knowledge in a KB are possible).

A KR system therefore has to manage a KB in some representation language, having both a formal syntax and a model-theoretic semantics. Part of the job of the KR system is to determine just what knowledge it is currently holding. This will involve determining the semantic consequences of the KB and, for most interesting semantic theories, will require inference of some sort.

The processing performed by the KR system to determine what knowledge it is holding must be both sound and complete with respect to the semantics. (Otherwise, the KR system is not implementing the representation language but is instead implementing some modified language with a different semantics.) Every answer given by the system, including answering "unknown" or even refusing to answer, must be sanctioned by the semantics of the representation language. Questions answered with "unknown" or not answered at all must be exactly those questions whose answers are not determined by the semantics of the representation language. In particular, the behavior of the KR system in answering questions cannot depend on considerations such as length of derivations or elapsed time for, if there is a dependence on these sorts of conditions, then specifying exactly what the KR system is doing is extremely hard without appealing directly to the algorithms in the system.

The central issue in this formulation of knowledge representation is then the selection or creation of a suitable representation language. This language should possess certain properties. First, it should be expressively powerful. Second, its semantics should reflect our intuitive notions about the language's constructs. Third, it should have a decidable and reasonably fast procedure for deciding whether statements follow from a KB. Therefore this language will be expressively powerful, in the sense of having a rich language, perhaps the same as the language of standard first-order logic (FOL), but deductively weak, in the sense of sanctioning a different set of inferences than standard logics do.

Obviously a representation language in which it is impossible to formulate many problem definitions and questions is inadequate. However, frame-based KR systems, as popular as they are, suffer from just this inability. The kind of language typically used in such systems cannot adequately express most disjunctions or similar weak statements, which are required in many contexts [Brachman and Levesque 1982; Moore 1982]. This shortcoming has often been side-stepped by adding extra-linguistic hooks to frame-based systems, for example, user-definable procedures that perform some part of the inference process. Such actions destroy the semantics of the representation language and systems that resort to such subterfuges do not satisfy the criteria for KR systems given here.

It is generally accepted in KR that the expressive power of at least FOL is needed in a general-purpose representation language. Moore [1982, p. 430] convincingly argues that "a general-purpose representation formalism [needs (at least)] all the features of first-order classical logic with equality". FOL with equality is itself a good candidate for the representation language because it has a semantics that corresponds well with our intuitive ideas about the world. Unfortunately, this logic has a severe problem—determining whether one formula follows from another is, in general, undecidable. It is this undecidability that makes a KR system built on FOL unsuitable for AI systems that depend on receiving answers.

This problem with FOL has led to many attempts to create KR systems based on FOL that always produce answers. Most of these systems retain the syntax of FOL while modifying its inferences in some way. The crudest of them simply take a theorem prover for FOL and place some *ad hoc* restrictions on it. In the simplest case, this consists of terminating the search for a proof after a pre-set amount of time or a certain number of proof steps, cutting choices out of the proof search space, or performing some other extra-linguistic modification of the theorem prover to guarantee that it will always terminate. Such modifications produce systems that, although they have the syntax of FOL, cannot be given an adequate semantics and have no means of completely characterizing the meaning of answers except by referring to the actions of the modified theorem prover. This results in systems that are not based on logic at all and thus do not really represent knowledge.

A more principled approach is to devise a semantics for first-order formulae that can model something like these limitations on inferences. For example, Konolige's system [Konolige 1985] includes a set of logically sound inference rules in the model structure. Such systems have been characterized in [Levesque 1984b] as *syntactic* variants because their semantic structures have to include syntactic entities. Aside from the problem of mixing syntax and semantics, these systems suffer from having to make far too many choices in their semantics. As shown in [Levesque 1984b] this makes these systems too fine-grained, allowing them to make distinctions between any two set of syntactic entities. For example, it would be possible to have the equivalence of $\alpha \vee \beta$ and $\beta \vee \alpha$ depend on the exact syntactic form of α or β. Of course the rules of these systems can be tuned to avoid these problems but there is nothing in their formalisms to prevent these strange dependencies. Because the semantics of such systems directly models the inference process, every part of this process shows up in the semantics. This makes for a very complicated semantics which only reflects the inference process and does not indicate what a formula means.

What is needed is a semantics that does not involve the inference process but is instead based on the usual model-theoretic ideas of interpretations and truth and falsity. Then the semantics will dictate the correct inferences as opposed to the other way around. However, if this semantics is to have any utility for KR, it must correspond with some part of our intuitions about how the world works while also having decidable inference. The semantics might not be as intuitive as the semantics of FOL, but this trade off is made to gain decidability and, hopefully, tractability of inference.

This semantics should also be deductively weaker than the semantics of FOL so that all reasoning in the logic is sound with respect to FOL. A KR system based on such a semantics would then perform a subset of the inferences of FOL.[2] The goal is to weaken just enough to achieve a logic with a decidable inference process that can be easily implemented. Although the valid inferences of this logic might not be all the inferences that one might like to perform, it would be a well-motivated set that can be entrusted to a KR system.

1 Relevance Logic

One candidate for this purpose is the logic of tautological entailments [Anderson and Belnap 1975, pp. 150–162], a simple and extremely weak type of relevance logic.[3] Proposi-

---

[2] It is possible to carry this weakening process too far, as pointed out by Pat Hayes. Logics where the interpretation of the logical symbols are not fixed can be constructed and if this is done in a particular way the logic will have no valid inferences. Such a logic is not the goal of this effort.

[3] Relevance logic has been used by Shapiro and several of his students [Shapiro and Wand 1976] as part of a semantic network KR system and a system for reasoning about belief [Martins and Shapiro 1988]. In fact, one of

tional tautological entailment has been used by Levesque [1984b] to model explicit propositional belief. Levesque also suggested that it could be used for KR.

The syntax of the logic of propositional tautological entailment is the same as that of standard propositional logic (PL), but without an implication operator. Sentences are built up from propositional letters in the usual way with the logical connectives ∧, ∨, and ¬. Much more interesting than the syntax of the logic is its semantics. This semantics [Dunn 1976] is based on four-valued *setups* (as opposed to the two-valued *assignments* of PL).

DEFINITION 1. A *setup*, s, consists of a mapping from propositional letters into the set of subsets of $\{t, f\}$. For any propositional letter, p, s supports the truth of p, $s \models_t p$, if $t \in s(p)$ and supports the falsity of p, $s \models_f p$, if $f \in s(p)$.

A setup then supports the truth of some letters and the falsity of some others. However, for some letters it may support neither their truth nor their falsity and for others it may support both their truth and their falsity. An assignment, on the other hand, supports the truth or the falsity of each letter, but never both.

In some ways these setups model the state of affairs that may be encountered by KR systems better than two-valued assignments. Such systems may have insufficient information to determine whether some proposition is true or false. They may also have been (inadvertently) told that some other proposition is both true and false [Belnap 1975, p. 30–36] and this local contradiction does not pollute the entire KB.

The support relationships are then extended to arbitrary sentences using the following rules:

1. $s \models_t \neg\alpha$ iff $s \models_f \alpha$
   $s \models_f \neg\alpha$ iff $s \models_t \alpha$

2. $s \models_t \alpha \vee \beta$ iff $s \models_t \alpha$ or $s \models_t \beta$
   $s \models_f \alpha \vee \beta$ iff $s \models_f \alpha$ and $s \models_f \beta$

3. $s \models_t \alpha \wedge \beta$ iff $s \models_t \alpha$ and $s \models_t \beta$
   $s \models_f \alpha \wedge \beta$ iff $s \models_f \alpha$ or $s \models_f \beta$

This is equivalent to the standard way of defining the truth or falsity of a sentence, aside from the differences between setups and assignments. A key difference between this semantics and that used in PL is that there are no sentences which are true in all setups nor are there any which are false in all setups. For example, $p \vee \neg p$ is neither true nor false in a setup that supports neither p's truth nor p's falsity.

From this basis, the idea of entailment (a relevance logic analogue of implication) can be defined.

DEFINITION 2. If α and β are propositional sentences, then α entails β (written $\alpha \rightarrow \beta$) iff for all setups, s, if $s \models_t \alpha$ then $s \models_t \beta$ and if $s \models_f \beta$ then $s \models_f \alpha$.

It turns out (as is well known) that if $s \models_t \alpha$ implies $s \models_t \beta$ for all setups s, then $s \models_f \beta$ implies $s \models_f \alpha$ for all setups s, and vice versa. Therefore, at least in propositional tautological entailment, it is sufficient to consider only the truth-preserving part of entailments.

---

the main proponents of relevance logic suggested that it would be a useful logic for KR [Belnap 1975] [Belnap 1977].

Now, if $\alpha \to \beta$ then β follows from α in PL, because the set of two-valued assignments is included in the set of setups. However, entailment is a much weaker notion than material implication. For example, $a \wedge \neg a \not\to b$ and $a \not\to b \vee \neg b$ show that a contradiction does not entail everything nor does every sentence entail a tautology.[4] One advantage of this is that an algorithm for computing entailment does not have to be able to determine if a sentence is a tautology. On the other hand, many beneficial inferences are not valid entailments. For example, $a \wedge (\neg a \vee b) \not\to b$, showing that *modus ponens* is not a valid rule for entailment.

There is a simple algorithm for determining if one sentence entails another in this logic.[5] To determine if $\alpha \to \beta$ first put α and β into conjunctive normal form (CNF). Then $\alpha_1 \wedge \ldots \wedge \alpha_n \to \beta_1 \wedge \ldots \wedge \beta_m$ (where each α_j and β_i are disjunctions of primitive propositions) iff for each β_i there is an α_j such that $\alpha_j \subseteq \beta_i$ (treating α_j and β_i as sets of primitive propositions). This algorithm is a bit disappointing at first glance because putting α and β into CNF can cause an exponential increase in their size and the main rationale for using tautological entailment was to have a faster algorithm.

Unfortunately, there is no way to have a fast algorithm for propositional tautological entailment between arbitrary formulae (unless P=NP) because it is possible to simulate PL in propositional tautological entailment. This simulation consists of adding pieces to both the left and right side of an entailment to ensure that every propositional letter of interest is assigned exactly one of true or false. The size of these pieces is proportional to the number of propositional letters in the formulae and, because propositional material implication is co-NP complete, so is propositional tautological entailment. More formally,

THEOREM 3. (Propositional Entailment co-NP Completeness Theorem)
The problem of determining whether $\alpha \to \beta$, in propositional tautological entailment, is co-NP complete, for α and β arbitrary propositional sentences.[6]

This does not affect the complexity of computing entailment between formulae in CNF because converting the sentences used in the proof to CNF can increase their size exponentially.

The possible exponential size increase in converting formulae into CNF is not a severe problem in KR. KBs are almost always a conjunction of many facts, each of which is much smaller than the KB as a whole. This means that placing KBs in CNF usually does not increase their size drastically. Then, if α and β are in CNF, computing whether $\alpha \to \beta$ takes time proportional to the product of their sizes [Levesque 1984b]. On the other hand the problem of determining if α implies β in PL is co-NP complete even if α and β are in CNF.

Propositional tautological entailment thus sanctions a subset of the inferences of PL that is easy to compute, for sentences in CNF and thus for most KBs, and, moreover,

---

[4]The superiority of relevance logic entailment over material implication for modeling the intuitive notion of implication is argued extensively by Anderson and Belnap [1975] (especially pp. 1–23).

[5]Another way of describing entailment here is via the following sound and complete axiomatization:

1. if $\alpha \to \beta$ and $\beta \to \gamma$ then $\alpha \to \gamma$

2. $\alpha \to \alpha_{CNF}$
   $\alpha_{CNF} \to \alpha$,   where α_{CNF} is α in conjunctive normal form (CNF)

3. $\alpha \to \alpha \vee \beta$
   $\alpha \wedge \beta \to \alpha$

This characterization is not emphasized here because, for the purposes of KR, an algorithmic characterization is much more important than an axiomatic one.

[6]The proofs of the theorems in the paper can be found in [Patel-Schneider 1990].

corresponds to a semantics that is plausible for KR. This well-motivated set of inferences can be entrusted to a simple KR system which will always produce answers quickly.

However, propositional logic is not expressive enough for a KR system. As stated before, FOL is generally considered to have the minimum expressive power for KR. This leads immediately to the idea of using first-order tautological entailment instead of propositional tautological entailment as a logic for KR.

First-order tautological entailment is to FOL as propositional tautological entailment is to PL. The syntax of first-order tautological entailment is the same as that of FOL, again without an implication sign. Formulae contain terms built in the usual way from variables and function letters (including constant letters), atomic formulae which are applications of predicate letters to terms, and conjunction, disjunction, negation, and universal and existential quantification. The semantics are defined in the obvious way, as a cross between the semantics of propositional tautological entailment and FOL. Thus the main difference between first-order tautological entailment and FOL is the difference between the two-valued models of FOL and the, otherwise similar, four-valued situations of first-order tautological entailment. Entailment in first-order tautological entailment is defined in a way similar to entailment in the logic of propositional tautological entailment.

First-order tautological entailment captures a similar set of inferences to those of the propositional version, where reasoning by contradiction and *modus ponens* are not valid. Based on the success with propositional tautological entailment, one would hope that the first-order version is decidable. Such is not the case, however. The same sort of mapping that worked in the propositional case to simulate PL can be performed to simulate FOL as shown by the following theorem:

THEOREM 4. (First-Order Entailment Undecidability Theorem)
The problem of determining whether $\alpha \rightarrow \beta$, in first-order tautological entailment, is undecidable, for α and β arbitrary first-order sentences.

This unfortunate finding destroys most of the utility of this logic for KR.[7] It means that there can be no algorithm for computing entailment in this logic that is guaranteed to terminate.

2 A Variant of First-Order Relevance Logic

First-order tautological entailment, even though it is undecidable, is a very interesting starting place for further investigation. It forms a subset of FOL, based on a semantics that is reasonable for KR, that is deductively much weaker than FOL. What is needed then is a yet-weaker variant of first-order tautological entailment which is decidable.

One computational problem with first-order tautological entailment, and also with FOL, is that quantification is equivalent to infinite conjunction or disjunction. It is this equivalence that makes quantification too powerful. For example, $(Pa \wedge Qa) \vee (Pb \wedge Qb)$ entails $\exists x\, Px \wedge Qx$ in first-order tautological entailment because whenever $(Pa \wedge Qa) \vee (Pb \wedge Qb)$ is true either $(Pa \wedge Qa)$ or $(Pb \wedge Qb)$ is true and so $Px \wedge Qx$ is true for some x (not necessarily the same individual all the time). This is a reasonable argument but it prevents the creation of an entailment algorithm for regular first-order tautological entailment similar to the one for propositional tautological entailment.

---

[7]While first-order tautological entailment is undecidable in the worst case, its average case behavior should be much better than that of material implication.

To invalidate this sort of entailment a semantics can be created where "$\exists x \; Px$ is true" would be read "there exists a particular individual for which P is true" and, similarly, "$\forall x \; Px$ is false" would be read "there exists a particular individual for which P is false". Under this reading the above entailment would not hold because there is no particular individual for which $Px \wedge Qx$ is true. (Note that $\exists x \; Px$ still would entail $\exists y \; Py$ because the individual would be the one from $\exists x \; Px$.) This reading makes quantification different from infinite conjunction or disjunction.

To formalize this change requires a significant modification of first-order relevance logic semantics. Instead of talking about situations (the first-order tautological entailment analogue of possible worlds), sets of situations are needed. For $\exists x \; Px$ to be true in a set of situations there must be some domain object, a single object common across all the situations, such that P is true for that object in each of the situations.

2.1 Syntax and Semantics

The syntax of this variant is the same as that of first-order tautological entailment, and thus almost the same as FOL. More formally,

DEFINITION 5. A *well-formed formula* is constructed as follows. The primitive elements are *variables*, written x_j, *function letters*, written f_j^n for $n \geq 0$, and *predicate letters*, written A_j^n for $n \geq 0$. A *term* is either a variable or a *function application*, written $f_j^n(t_1, \ldots, t_n)$ where f_j^n is a function letter and the t_i are terms. An *atomic formula* is a *predicate application*, written $A_j^n(t_1, \ldots, t_n)$ where A_j^n is a predicate letter and the t_i are terms. A well-formed formula is then either an atomic formula or $(\neg \alpha)$, $(\alpha \vee \beta)$, $(\alpha \wedge \beta)$, $(\forall x \alpha)$, or $(\exists x \alpha)$ where α and β are well-formed formulae and x is a variable. A *sentence* is then a well-formed formula with no variables occurring free, where free and bound occurances of variables are defined in the usual manner.

The semantics of this variant starts out the same as the semantics of regular first-order tautological entailment, which is very much like the semantics of FOL except for the provisions for the four relevance logic truth values. Of course, this change is a very significant one, weakening the logic severely, much as in the propositional case. The semantics starts with first-order situations, the analogue of first-order models.[8]

DEFINITION 6. A *situation* consists of a non-empty set D, the *domain* of the situation, and three mappings, h, t, and f. h maps each function letter, f_j^n, into a function from D^n to D, and t and f map each predicate letter, A_j^n, into an n–ary relation on D.

The mapping h is the usual mapping of function letters (and constant letters) into functions over the domain (and constants in the domain). The mappings t and f determine which atomic formulae are true and which are false and correspond to the interpretation function of FOL. Two mappings are needed because an atomic formula can be assigned true, false, neither, or both, as in propositional relevance logic. This change from two to four truth values means that there are many more situations than models, and models correspond to those situations with no "missing information" and no contradictions.

The semantics proceeds with the definition of variable maps and support relationships. Variable maps determine, in the usual way, which elements of the domain variables map into.

---

[8]Although much of this development could be by appealing to the analogous semantic constructs of FOL, there are some differences and it is best to be very careful here so that difficulties do not arise later.

DEFINITION 7. A *variable map* is a mapping from variables into some non-empty set. If ν is a variable map into D, x is a variable, and d is an element of D, then ν_d^x is a variable map into D with

$$\nu_d^x(y) = d, \quad \text{if } y = x,$$

$$\nu_d^x(y) = \nu(y), \quad \text{otherwise.}$$

Variable maps can then be extended, again in the usual way, into mappings for arbitrary terms.

DEFINITION 8. Given a situation, s, and a variable map, ν, into the domain of s, a mapping, ν_s^*, from terms into the domain of s can be defined as follows:

$$\nu_s^*(x) = \nu(x), \quad \text{if } x \text{ is a variable,}$$

$$\nu_s^*(f_j^n(t_1, \ldots, t_n)) = (h(f_j^n))(\nu_s^*(t_1), \ldots, \nu_s^*(t_n)), \quad \text{otherwise.}$$

Atomic formulae are given meaning in the obvious way, through support relationships. Because formulae can be neither true nor false, and can also be both true and false, there are two support relationships, $s, \nu \models_t \alpha$ which reads as "s supports the truth of α under ν", and $s, \nu \models_f \alpha$ which reads as "s supports the falsity of α under ν",

DEFINITION 9. The support relationships of first-order relevance logic between situations, variable maps, and atomic formulae are defined as follows:

$$s, \nu \models_t A_j^n(t_1, \ldots, t_n) \quad \text{iff} \quad \langle \nu_s^*(t_1), \ldots, \nu_s^*(t_n) \rangle \in t_s(A_j^n)$$

$$s, \nu \models_f A_j^n(t_1, \ldots, t_n) \quad \text{iff} \quad \langle \nu_s^*(t_1), \ldots, \nu_s^*(t_n) \rangle \in f_s(A_j^n)$$

where s is a situation, ν is a variable map into the domain of s, A_j^n is a predicate letter, and t_i is a term.

At this point the semantics of the two logics diverges. The semantics of regular first-order tautological entailment is included here for reference purposes and also for the proof of Theorem 4. This semantics proceeds by extending the support relationships to arbitrary first-order formulae using the following rules:

1. $s, \nu \models_t \neg\alpha \quad \text{iff} \quad s, \nu \models_f \alpha$
   $s, \nu \models_f \neg\alpha \quad \text{iff} \quad s, \nu \models_t \alpha$

2. $s, \nu \models_t \alpha \vee \beta \quad \text{iff} \quad s, \nu \models_t \alpha \text{ or } s, \nu \models_t \beta$
   $s, \nu \models_f \alpha \vee \beta \quad \text{iff} \quad s, \nu \models_f \alpha \text{ and } s, \nu \models_f \beta$

3. $s, \nu \models_t \alpha \wedge \beta \quad \text{iff} \quad s, \nu \models_t \alpha \text{ and } s, \nu \models_t \beta$
   $s, \nu \models_f \alpha \wedge \beta \quad \text{iff} \quad s, \nu \models_f \alpha \text{ or } s, \nu \models_f \beta$

4. $s, \nu \models_t \forall x\alpha \quad \text{iff} \quad \text{for all } d \in D \quad s, \nu_d^x \models_t \alpha$
   $s, \nu \models_f \forall x\alpha \quad \text{iff} \quad \text{for some } d \in D \quad s, \nu_d^x \models_f \alpha$

5. $s, \nu \models_t \exists x\alpha \quad \text{iff} \quad \text{for some } d \in D \quad s, \nu_d^x \models_t \alpha$
   $s, \nu \models_f \exists x\alpha \quad \text{iff} \quad \text{for all } d \in D \quad s, \nu_d^x \models_f \alpha$

It is easy to see that this semantics for regular first-order tautological entailment is very similar to standard Tarskian semantics except for the change from truth values of true and false to truth values of subsets of the set containing true and false. As argued on page 4, this change has some benefits with respect to KR. As in the propositional case, there are no tautologies (formulae which are true in all situations) in this semantics.

The entailment relationship in regular first-order tautological entailment is then defined similarly to entailment in propositional tautological entailment:

DEFINITION 10. If α and β are first-order formulae, α entails β ($\alpha \rightarrow \beta$), in regular first-order tautological entailment, iff for all situations, s, and all variable maps, ν, if $s, \nu \models_t \alpha$ then $s, \nu \models_t \beta$ and if $s, \nu \models_f \beta$ then $s, \nu \models_f \alpha$.

The semantics of the variant of first-order tautological entailment is more complicated. First, *compatible* sets of situations are defined.

DEFINITION 11. A compatible set of situations is a set of situations with the same domain and the same mapping of function letters to functions.

In other words, the situations in a compatible set of situations differ only on their assignments of truth and falsity to atomic formulae. In this logic the meaning of formulae will be determined with respect to these compatible sets of situations, and not simply with respect to single situations, as in regular first-order tautological entailment.

Given S, a compatible set of situations each with domain D, and ν, a variable map into D, the two support relations for this logic, $S, \nu \models_t \alpha$ and $S, \nu \models_f \alpha$, are defined as follows:

1. $S, \nu \models_t a$ iff   for all $s \in S$ $s, \nu \models_t a$,   for a atomic
   $S, \nu \models_f a$ iff   for all $s \in S$ $s, \nu \models_f a$,   for a atomic

2. $S, \nu \models_f \neg\alpha$ iff $S, \nu \models_t \alpha$
   $S, \nu \models_t \neg\alpha$ iff $S, \nu \models_f \alpha$

3. $S, \nu \models_t \alpha \wedge \beta$ iff $S, \nu \models_t \alpha$ and $S, \nu \models_t \beta$
   $S, \nu \models_f \alpha \wedge \beta$ iff $\exists S_1 S_2$ $S_1 \cup S_2 = S$ $S_1, \nu \models_f \alpha$ and $S_2, \nu \models_f \beta$

4. $S, \nu \models_t \alpha \vee \beta$ iff $\exists S_1 S_2$ $S_1 \cup S_2 = S$ $S_1, \nu \models_t \alpha$ and $S_2, \nu \models_t \beta$
   $S, \nu \models_f \alpha \vee \beta$ iff $S, \nu \models_f \alpha$ and $S, \nu \models_f \beta$

5. $S, \nu \models_t \forall x \alpha$ iff   for all $d \in D$ $S, \nu_d^x \models_t \alpha$
   $S, \nu \models_f \forall x \alpha$ iff   for some $d \in D$ $S, \nu_d^x \models_f \alpha$

6. $S, \nu \models_t \exists x \alpha$ iff   for some $d \in D$ $S, \nu_d^x \models_t \alpha$
   $S, \nu \models_f \exists x \alpha$ iff   for all $d \in D$ $S, \nu_d^x \models_f \alpha$

Under this semantics $\exists x \alpha$ is true in S under variable map ν if there is some domain element, common across all the situations in S, which, when taken as the mapping of x, makes α true in each situation. The same formula is false if all domain elements, when taken as the mapping of x, make α false in each situation. This means that it is fairly easy for an existential to be true (although more difficult than in regular first-order tautological entailment), but quite hard for it to be false.

This definition of the support relations is slightly strange in its definition of $S, \nu \models_t \alpha \vee \beta$ and $S, \nu \models_f \alpha \wedge \beta$. However, it is necessary to carry compatible sets of situations throughout in order to ensure that quantification is not equivalent to infinite conjunction or disjunction.

For quantifier-free formulae, this semantics is equivalent to the semantics for regular first-order tautological entailment, as shown by the following theorem:

THEOREM 12. Quantifier-Free Equivalence Theorem

If α is a quantifier-free formula, S is a compatible set of situations, and ν is a variable map into the common domain of elements of S then $S, \nu \models_t \alpha$ iff $\forall s \in S \; s, \nu \models_t \alpha$. Also $S, \nu \models_f \alpha$ iff $\forall s \in S \; s, \nu \models_f \alpha$.

PROOF. By induction on the size of α.

If α is atomic then the result is trivial.

If α is of the form $\neg\beta$ then $S, \nu \models_t \alpha$ iff $S, \nu \models_f \beta$ iff $\forall s \in S \; s, \nu \models_f \beta$ (from the inductive hypothesis) iff $\forall s \in S \; s, \nu \models_t \alpha$. Similarly for f.

If α is of the form $\beta_1 \wedge \beta_2$ then $S, \nu \models_t \alpha$ iff $S, \nu \models_t \beta_1$ and $S, \nu \models_t \beta_2$ iff $\forall s \in S \; s, \nu \models_t \beta_1$ and $\forall s \in S \; s, \nu \models_t \beta_2$ (from the inductive hypothesis) iff $\forall s \in S \; s, \nu \models_t \alpha$. Also $S, \nu \models_f \alpha$ iff $\exists S_1 S_2 \; S_1 \cup S_2 = S \; S_1, \nu \models_f \beta_1$ and $S_2, \nu \models_f \beta_2$ iff $\exists S_1 S_2 \; S_1 \cup S_2 = S \; \forall s \in S_1 \; s, \nu \models_f \beta_1$ and $\forall s \in S_2 \; s, \nu \models_f \beta_2$ (from the inductive hypothesis) iff $\forall s \in S \; s, \nu \models_f \alpha$. Similarly for α of the form $\beta_1 \vee \beta_2$. ∎

Therefore this semantics is only unusual in its treatment of quantifiers, as it was designed to be.

One side-effect of this change in the meaning of quantification is that quantifiers cannot be moved around and combined in all the ways they can in FOL while maintaining equivalent formulae. For example, $\exists x \, Px \vee \exists y \, Qy$ is not equivalent to $\exists z \, Pz \vee Qz$ but is only equivalent to $\exists x \exists y \, Px \vee Qy$. To illustrate this, consider the following compatible set of situations (being rather sloppy in notation) $S = \{s_1, s_2\}$ with domain $D = \{d_1, d_2\}$ where $P(d_1)$ is true in s_1 and $Q(d_2)$ is true in s_2. Then $\exists x \, Px \vee \exists y \, Qy$ is true in S, as is $\exists x \exists y \, Px \vee Qy$, but $\exists z \, Pz \vee Qz$ is not true in S.

Further, there are formulae which cannot be converted into an equivalent prenex form by simply moving quantifiers.[9] For example, $\forall y \exists x \, Pyx \vee \forall y' \exists x' \, Qy'x'$ cannot be treated in this fashion, since any prenex form created by just moving quantifiers would have one existential in the scope of both universals, as in $\forall y \exists x \forall y' \exists x' \, Pyx \vee Qy'x'$, which will be true in more compatible sets of situations than the original formula. This is demonstrated by $S = \{s_1, s_2, s_3\}$ and $D = \{d_1, d_2\}$ where $P(d_1, d_1)$ and $P(d_2, d_1)$ are true in s_1, $P(d_1, d_1)$, $Q(d_1, d_2)$, and $Q(d_2, d_2)$ are true in s_2, and $P(d_2, d_1)$, $Q(d_1, d_1)$, and $Q(d_2, d_1)$ are true in s_3. This gives $\forall y \exists x \forall y' \exists x' \, Pyx \vee Qy'x'$ true in S but $\forall y \exists x \, Pyx \vee \forall y' \exists x' \, Qy'x'$ not true.

Now that the support relations have been defined, the notion of entailment can be defined in this logic. As it turns out, there are three different versions of entailment possible. Recall that if β follows from α, then β must be true whenever α is and, conversely, α must be false whenever β is. These two conditions are equivalent by definition in standard two-valued logic where being false is equivalent to not being true and also happen to be equivalent in propositional tautological entailment and in regular first-order tautological entailment. However, the conditions are not equivalent in this variant, giving rise to three versions of

---

[9] I believe that there are formulae which have no equivalent prenex form at all but I have not proved this conjecture.

entailment, *t-entailment* (written $\rightarrow_t$), which carries the first condition, *f-entailment* (written $\rightarrow_f$), which carries the second condition, and *tf-entailment* (written $\rightarrow$), which carries both conditions. More formally,

DEFINITION 13. For α and β first-order formulae, the three variants of entailment are defined as follows:

1. $\alpha \rightarrow_t \beta$ iff   for all compatible sets of situations, S, and variable maps, ν, if $S, \nu \models_t \alpha$ then $S, \nu \models_t \beta$

2. $\alpha \rightarrow_f \beta$ iff   for all compatible sets of situations, S, and variable maps, ν, if $S, \nu \models_f \beta$ then $S, \nu \models_f \alpha$

3. $\alpha \rightarrow \beta$ iff $\alpha \rightarrow_t \beta$ and $\alpha \rightarrow_f \beta$

So what are these three versions of entailment like? First of all, they are all very weak— weaker than entailment in regular first-order tautological entailment and much weaker than material implication. Second, when determining whether one formula entails another, the order of conjuncts and disjuncts does not matter. Third, as in other relevance logics, *modus ponens* is not a valid rule. This, in large part, is what makes the logic very weak. Fourth, formulae can be weakened by removing elements of conjunctions and adding elements to disjunctions. In fact, for quantifier-free formulae each version of entailment is the same and they are all equivalent to regular first-order tautological entailment and to propositional tautological entailment. (This is an easy consequence of Theorem 12.) Finally, the entailments for quantifiers can be summed up as follows:

$$
\begin{array}{llll}
\forall x\, Px & \rightarrow & Pa & \qquad Pa & \rightarrow & \exists x\, Px \\
\forall x\, Px & \rightarrow_t & Pa \wedge Pb & \qquad Pa \vee Pb & \not\rightarrow_t & \exists x\, Px \\
\forall x\, Px & \not\rightarrow_f & Pa \wedge Pb & \qquad Pa \vee Pb & \rightarrow_f & \exists x\, Px \\
\forall x\, Px & \not\rightarrow & Pa \wedge Pb & \qquad Pa \vee Pb & \not\rightarrow & \exists x\, Px
\end{array}
$$

This shows that all versions of entailment have a universal entailing a single instantiation of itself and a single instantiation entailing an existential. Also, a universal *t*-entails the conjunction of any number of instantiations whereas a disjunction of instantiations does not *t*-entail an existential. For *f*-entailment the opposite is true, and so, of course, neither way works for *tf*-entailment.

Of these three versions of entailment, one should be chosen as the most suitable for KR. The version called *t*-entailment seems to be the best for this purpose. It is preferable to *tf*-entailment, because it is stronger. It and *f*-entailment are duals but the entailments retained by *t*-entailment (such as $\forall x\, Px \rightarrow_t Pa \wedge Pb$) are more natural for KR than those retained by *f*-entailment (such as $Pa \vee Pb \rightarrow_f \exists x\, Px$). This is especially true when taking the view that to demonstrate the existence of an object, a particular object must be produced (not just presenting a set of objects and saying that the desired object is among them). Further, this last type of entailment is a sort of reasoning by cases, a type of reasoning generally not valid in relevance logic. These considerations indicate that *t*-entailment is the best of the three variants for KR so it will be emphasized from now on.

2.2 Entailment Algorithm

This logic would not be very interesting if it did not have a decidable algorithm for determining entailment. The algorithm developed here unfortunately works only on formulae in prenex form, although an algorithm for arbitrary formulae is under development.

The first step in the development of this algorithm is an equivalence theorem stating that prenex form formulae can be converted into conjunctive normal form.

THEOREM 14. (Prenex to Conjunctive Normal Form Theorem)
If α is a formula in prenex form and α_{CNF} is α in prenex conjunctive normal form, then $\alpha \to \alpha_{CNF}$ and $\alpha_{CNF} \to \alpha$.

The next step is a Skolemization theorem for entailment which states that for all three versions of entailment $\alpha \to \beta$ for α and β in prenex form iff $\alpha_{S\exists} \to \beta_{S\forall}$, where $\alpha_{S\exists}$ is α with all existentially quantified variables Skolemized and $\beta_{S\forall}$ is β with all universals Skolemized. Then the following theorem characterizes t-entailment between formulae in normal form:

THEOREM 15. *If α and β are formulae in prenex conjunctive Skolem form ($\alpha = \forall \vec{z} \bigwedge \alpha_j$ and $\beta = \exists \vec{x} \bigwedge \beta_i$) then $\alpha \to_t \beta$ iff there exists θ, a substitution for $\vec{x}$, such that for each β_i there exists some α_j and ψ, a substitution for $\vec{z}$, such that $\alpha_j \psi \subseteq \beta_i \theta$ (treating $\alpha_j \psi$ and $\beta_i \theta$ as sets of literals).*

An algorithm for computing entailment between two formulae in this form can now be derived. Throughout this algorithm free variables in α and β will be treated as constants.

Consider α and β as above. For each α_j and β_i, compute a set of most general substitutions Θ_{ij} such that for $\theta \in \Theta_{ij}$, $\alpha_j \theta \subseteq \beta_i \theta$. For each element of Θ_{ij}, define a new substitution the same as that element except that occurrences of elements of $\vec{z}$ are systematically replaced by variables occurring nowhere else. Let Ψ_{ij} be the set of these substitutions and let $\Psi_i = \bigcup_j \Psi_{ij}$.

Now each element of Ψ_i is a way of making β_i match some α_j. (It is easy to see that Ψ_i is non-empty just in case there exists θ, a substitution for $\vec{x}$, and ψ, a substitution for $\vec{z}$, such that $\alpha_j \psi \subseteq \beta_i \theta$ for some α_j.) What is needed is some way of producing compatible matchings (compatible in that each variable in $\vec{x}$ has the same mapping) for each β_i. Such mappings well exist if and only if there is some substitution ϕ which is the most general unifier of some $\psi_i \in \Psi_i$, for each β_i. (Again it is easy to see that such a ϕ exists if and only if there exists θ, a substitution of ground terms for $\vec{x}$, such that for each β_i there exists some α_j and ψ, a substitution of ground terms for $\vec{z}$, such that $\alpha_j \psi \subseteq \beta_i \theta$.)

This algorithm does demonstrate that t-entailment in this logic is decidable, at least for formulae in prenex form. Although its worst case behavior is exponential in the size of α and β it will always terminate and will be quite fast under normal conditions, where clauses are not long and there are many different predicates. This characterization also gives a good indication of how t-entailment works for formulae in prenex form showing that the five informal comments above are the basis for a complete description of t-entailment. Note, however, that it is the semantics that *defines* t-entailment, not this algorithm.

An interesting side-effect of this algorithm is that it automatically performs answer extraction. The substitution ϕ gives a mapping for each of the existential variables in β, thus providing a set of answers. If more than one answer or the complete set of answers is required different ϕ can be searched for.

3 Conclusion

So what has been gained from this new logic? In essence, the logic provides a decidable subset of material implication that does not contain *any* of the harder types of inference. It does not have the rule of *modus ponens* and no chaining can be done. It also does not do reasoning by contradiction or reasoning by cases. The benefit of the logic is that it has a simple algorithm, derived directly from the semantics, that implements a form of entailment in the logic. This algorithm has a disappointing worst case running time, being doubly exponential in the size of the formulae, but it performs no undirected searching and can be easily implemented. Therefore, it can be placed inside a knowledge-based system without worrying about how to guide it or worrying about it failing to terminate. The algorithm performs *retrieval* of first-order sentences in prenex form from a knowledge base specified as a first-order sentence in prenex form, much as database systems perform retrieval of facts from a database. Many inferences are not provided but this is the price that must be paid to guarantee decidability.

This decidability of entailment is not the only feature of the logic that makes it suitable for use in a KR system, although it is the most important one. The syntax of the logic is the same as that of FOL so, in some sense, it is possible to state in the new logic anything that can be stated in FOL. Also, because a subset of material implication is provided, knowledge-based systems can treat sentences as sentences of FOL provided that they realize that the inferences provided are not complete. Further, since the semantics of the logic is closely related to regular first-order tautological entailment semantics, the logic does not suffer from some of the problems and paradoxes of material implication. Finally, the semantics is firmly based on the usual model theoretic ideas of truth and falsity, at least the relevance logic versions thereof, and serves to dictate the allowable inferences and not *vice versa*. This allows a simple characterization of entailment in terms of the truth of formulae. Together these points make the logic more suitable for KR than systems built by *ad hoc* modifications to a theorem prover or systems based on a particular set of inferences (and otherwise semantically unmotivated).

However, the logic presented above is simply the start of work on a decidable first-order logic-based KR system. Although this logic, especially *t*-entailment, has many of the right properties required for such a system there may be stronger logics that retain decidability. For example, there are stronger versions of relevance logic that might be used as the basis of such a logic. The search for such a logic is, of course, aided by this demonstration of one logic meeting the requirements.

Further, some KR researchers believe that FOL is inadequate for KR and have turned to modal or higher-order logics to represent causality, time, *et cetera*. The main stumbling block with using modal or higher-order logics in KR is that proof procedures for them are even worse than those for FOL. A decidable version of these logics would enable at least the simple inferences in such logics to be reliably made.

One way to strengthen the logic developed here without destroying decidability is by adding to it a terminological reasoner like the terminological component of KRYPTON [Brachman, Gilbert, and Levesque 1985]. To do this correctly would require building a full semantics for the combination of terminological reasoning and *t*-entailment, perhaps similar to the semantics by Brachman *et al* [1985]. One problem with such an approach is that the type of reasoning done by KRYPTON is based on theory resolution [Stickel 1985], which provides its theoretical foundation. Theory resolution only addresses the problem of adding this sort of reasoning onto a resolution-based theorem prover, so a separate account of terminological reasoning in the new logic will have to be done. In fact, this account will be harder than theory resolution because theory resolution does not add any power to a system whereas the idea here will be to increase power via terminological reasoning.

Another wrinkle in developing terminological reasoners for this new logic is determining which language to use. As shown by Brachman and Levesque [1984], terminological reasoning can be computationally intractable for even very simple terminological languages under standard semantics. If the KR system as a whole is to perform adequately, the terminological component must be tractable. One possibility for producing tractable terminological reasoners is to perform the same sort of reduction on the terminological component as the logic devised here does on the assertional component. This would produce a semantics of frame subsumption that only catches the easy subsumptions and leaves the hard and time-consuming ones out.

Any KR system based on this new logic is going to be quite weak even with terminological reasoning added. One way of making such a system stronger is to make entailment be only the first method for answering questions. If entailment fails to produce an answer then a stronger method could be used, perhaps some domain specific method. The advantage of this over having some arbitrary primary method is that entailment is a semantically motivated and well-defined notion so that it is possible to understand what its failure to produce an answer means.

An interesting idea for the fall-back method is to use a stronger logical system. The stronger system would take longer to produce answers and might take a very long time indeed but would be employed only in those cases where a stronger answer is really required and cannot be otherwise determined. A first idea for such a fall-back is, of course, FOL, but other logics, such as regular first-order tautological entailment, could be used. A big problem for a system with several levels of reasoners like this is how to retain the work done on the earlier, failed, steps when retreating to a fall-back method so that the resources expended there are not lost. However, such a system would be a considerable improvement over just using FOL.

In summary, the decidable first-order logic presented here forms an important first step toward building decidable, semantically-motivated KR systems.

Acknowledgments This is a minor revision of a paper that appeared in the *Journal of Automated Reasoning* **6**, 1990, pages 361–388.

References

American Association for Artificial Intelligence [1982, August). *Proceedings of the Second National Conference on Artificial Intelligence*, Pittsburgh, Pennsylvania. American Association for Artificial Intelligence.

American Association for Artificial Intelligence [1984, August). *Proceedings of the Fourth National Conference on Artificial Intelligence*, Austin, Texas. American Association for Artificial Intelligence.

Anderson, A. R. and N. D. Belnap, Jr. [1975]. *Entailment: The Logic of Relevance and Necessity*, Volume I. Princeton, New Jersey: Princeton University Press.

Belnap, Jr., N. D. [1975]. How a computer should think. In *Contemporary Aspects of Philosophy: Proceedings of the Oxford International Symposium*, pp. 30–56.

Belnap, Jr., N. D. [1977]. A useful four-valued logic. In G. Epstein and J. M. Dunn (Eds.), *Modern Uses of Multiple-Valued Logic*, pp. 8–37. Reidel.

Brachman, R. J., V. P. Gilbert, and H. J. Levesque [1985, August]. An essential hybrid reasoning system: Knowledge and symbol level accounts of KRYPTON. See International Joint Committee on Artificial Intelligence [1985], pp. 532–539.

Brachman, R. J. and H. J. Levesque [1982, August). Competence in knowledge representation. See American Association for Artificial Intelligence [1982], pp. 189–192.

Brachman, R. J. and H. J. Levesque [1984, August). The tractability of subsumption in frame-based description languages. See American Association for Artificial Intelligence [1984], pp. 34–37.

Dunn, J. M. [1976]. Intuitive semantics for first-degree entailments and 'coupled trees'. *Philosophical Studies 29*, 149–168.

Frisch, A. M. [1985, August). Using model theory to specify AI programs. See International Joint Committee on Artificial Intelligence [1985], pp. 148–154.

Hayes, P. J. [1977, August). In defence of logic. In *Proceedings of the Fifth International Joint Conference on Artificial Intelligence*, Cambridge, Massachusetts, pp. 559–565. International Joint Committee on Artificial Intelligence.

International Joint Committee on Artificial Intelligence [1985, August). *Proceedings of the Ninth International Joint Conference on Artificial Intelligence*, Los Angeles, California. International Joint Committee on Artificial Intelligence.

Konolige, K. [1985]. Belief and incompleteness. In J. R. Hobbs and R. C. Moore (Eds.), *Formal Theories of the Commonsense World*. Norwood, N. J.: Ablex.

Levesque, H. J. [1984a, May). A fundamental tradeoff in knowledge representation and reasoning. In *Proceedings of the Fifth Biennial Conference of the Canadian Society for Computational Studies of Intelligence*, London, Ontario, pp. 141–152. Canadian Society for Computational Studies of Intelligence. A revised version available as FLAIR Technical Report Number 35, Fairchild Laboratory for Artificial Intelligence Research, August 1984.

Levesque, H. J. [1984b, August). A logic of implicit and explicit belief. See American Association for Artificial Intelligence [1984], pp. 198–202.

Martins, J. P. and S. C. Shapiro [1988, May). A model for belief revision. *Artificial Intelligence 35*(1), 25–79.

McDermott, D. V. [1978, July–September). Tarskian semantics, or no notation without denotation! *Cognitive Science 2*(3), 277–282.

Moore, R. C. [1982, August). The role of logic in knowledge representation and commonsense reasoning. See American Association for Artificial Intelligence [1982], pp. 428–433.

Patel-Schneider, P. F. [1990]. A decidable first-order logic for knowledge representation. *Journal of Automated Reasoning 6*, 361–388.

Shapiro, S. C. and M. Wand [1976, November). The relevance of relevance. Technical Report 46, Computer Science Department, Indiana University.

Stickel, M. E. [1985]. Automated deduction by theory resolution. *Journal of Automated Reasoning 1*, 455–458. 333–355.

24

Knowledge, Action, and Cartesian Situations in the Situation Calculus

Ronald P. A. Petrick

ABSTRACT. We formalize the notion of a Cartesian situation in the situation calculus, a property that imposes strong structural conditions on the configuration of a set of possible worlds. Focusing on action theories that use the standard Scherl and Levesque account of knowledge and action, we show how Cartesian situations give rise to a set of decomposition properties for simplifying epistemic formulae (in particular, certain disjunctive and existentially quantified formulae) into equivalent components that only mention fluent literals. Moreover, we describe certain expressive classes of action theories that preserve the Cartesian property through action. This work offers the possibility of identifying action theories that can be compiled into alternative accounts of knowledge with similar representational restrictions, but without possible worlds.

1 Introduction and motivation

Agents that act in dynamic worlds often do so with incomplete information about their environments. Agents that also sense the world have the ability to gather new information and extend their knowledge about the world. As a result, logical accounts that model the behaviour of agents in such settings, for instance for the purpose of planning or high-level control, therefore require the ability to reason effectively about knowledge and action.

The problem of reasoning about knowledge and action has been widely studied and is relatively well understood. In the situation calculus [McCarthy and Hayes 1969], the standard approach to this problem due to Moore [1985] views knowledge in terms of an accessibility relation over a set of possible worlds [Hintikka 1962]. Scherl and Levesque [1994, 2003] adapt this approach in Reiter's version of the situation calculus [Reiter 2001] and provide a solution to the frame problem [McCarthy and Hayes 1969] for knowledge change.

While the Scherl and Levesque approach is representationally rich, it is computationally problematic: reasoning whether n atomic formulae are known potentially involves reasoning about 2^n possible worlds. As a result, alternative approaches have explored representations of knowledge and action that do not require an accessibility relation and possible worlds (e.g., [Funge 1998], [Demolombe and Pozos Parra 2000], [Soutchanski 2001], [Son and Baral 2001], [Petrick and Levesque 2002], [Liu and Levesque 2005], [Vassos and

The main results of this paper were established while the author was a PhD student at the University of Toronto, under the supervision of Hector Levesque. (Earlier versions of this work appear in [Petrick 2006; Petrick 2008].) It is therefore not surprising that this work intersects with many of the themes at the core of Hector's own research—knowledge, action, and the situation calculus—and is also related to the research of many of Hector's other former students and colleagues, some of whom are cited in this paper. Working with Hector, one cannot help but be inspired by his excitement for research: a quality he successfully instilled in this former student. It is a great privilege to contribute this paper to the *Festschrift* for Hector Levesque—a well-deserved honour for an extraordinary researcher.

Levesque 2007]). To achieve more tractable reasoning, many of these accounts sacrifice expressiveness by restricting the types of knowledge that can be represented. A common approach is to limit disjunctive knowledge and model an agent's knowledge state by sets of fluents that are known to be true or known to be false.

How do the limitations of these alternate representations relate to the standard Scherl and Levesque account of knowledge? Can we restrict Scherl and Levesque action theories to bring about similar representations? What must we give up to do so? McCarthy [1979a, 1979b] points out that a Cartesian product structure on a space of possible worlds introduces an independence property that removes dependencies between individual domain fluents. In this paper we formalize the notion of a *Cartesian situation*, a property inspired by the ordinary Cartesian product over sets, that imposes strong structural conditions on the configuration of a set of possible worlds. We show that Cartesian situations give rise to a collection of "decomposition" properties that let us simplify epistemic formulae—in particular, certain disjunctive and existentially quantified formulae—into equivalent components that only mention fluent literals. We also describe certain expressive classes of action theories that preserve the Cartesian property through action.

Treating knowledge in terms of fluent literals closely corresponds to the restrictions made by some of the alternative accounts of knowledge in the literature. While the formal properties we establish will help us better understand the nature of Cartesian situations, this work also offers the possibility of identifying standard action theories that can be compiled into these alternative representations supporting more tractable reasoning.

This paper is organized as follows. We begin by describing the situation calculus (Section 2) and the Scherl and Levesque model of knowledge and action (Section 3). We then define Cartesian situations (Section 4) and establish a set of decomposition properties for particular types of knowledge formulae (Section 5). Next, we identify certain classes of action theories that preserve the Cartesian property through action (Section 6), and give examples of such theories (Section 7). We close with a discussion of the importance of this work, and its connection to related approaches (Section 8).

2 Situation calculus

The situation calculus (as described in [Reiter 2001]) is a first-order, many-sorted language (with some second-order features) specifically designed for modelling dynamically changing worlds. All changes in the world result from the application of named *actions*, which are denoted by function symbols and may be parameterized. A first-order term called a *situation* represents a sequence of actions or a *possible world history*. An *object* is a first-order term used to denote anything that isn't an action or situation. A special constant S_0 indicates the *initial situation* in which no actions have been performed. A distinguished binary function $do(a, s)$ denotes the successor situation resulting from performing action a in situation s. The binary predicate $Poss(a, s)$ is used to indicate that it is possible to perform action a is situation s.

Relations whose truth values can change from situation to situation are referred to as *relational fluents*. Likewise, functions whose mappings can change from situation to situation are referred to as *functional fluents*. A fluent (relational or functional) is denoted by including a situation argument as its last parameter, indicating the fluent's value at that situation. Formulae without situation terms are referred to as *situation independent* formulae and are unchanged by action.

Domain theories are formally defined as follows.

DEFINITION 1. A *basic action theory (BAT)* consists of the following axioms:

1. For each action A, an *action precondition axiom*:

$$Poss(A(\vec{x}), s) \equiv \Pi_A(\vec{x}, s),$$

where Π_A is a first-order formula.[1]

2. For each relational fluent F, a *successor state axiom*:

$$F(\vec{x}, do(a, s)) \equiv \gamma_F^+(\vec{x}, a, s) \vee F(\vec{x}, s) \wedge \neg\gamma_F^-(\vec{x}, a, s),$$

where $\gamma_F^\pm$ are first-order formulae (see below).

3. For each functional fluent f, a *successor state axiom*:

$$f(\vec{x}, do(a, s)) = y \equiv \gamma_f(\vec{x}, y, a, s) \vee f(\vec{x}, s) = y \wedge \neg(\exists y')\gamma_f(\vec{x}, y', a, s),$$

where γ_f is a first-order formulae (see below).

4. A set of *initial situation axioms*, first-order sentences that syntactically only mention the situation term S_0 or no situation term at all.

5. *Unique names axioms* for actions and a set of domain independent *foundational axioms* (see [Reiter 2001] for more details).

In Definition 1, (1) characterizes the conditions required for an action A to be applied in s. We will typically ignore *Poss* in this paper and assume that actions are always executable. (2) and (3) provide a solution to the *frame problem*: the value of a fluent at $do(a, s)$ is completely determined by a formula that only mentions situation s. For a relational fluent F, γ_F^+ (similarly, γ_F^-) describes all the ways of making F true (similarly, false) in the situation $do(a, s)$. For a functional fluent f, γ_f characterizes how the mapping of f changes in the situation $do(a, s)$. (4) specifies the initial values of fluents.

3 A K fluent in the situation calculus

The situation calculus formalism described above does not distinguish between what is *true* in a situation and what is *known* in a situation. Scherl and Levesque [1994, 2003] formalize the idea of *knowledge* in the situation calculus by adapting a possible worlds model of knowledge as was done by Moore [1985]. A special binary relation over situations is introduced, $K(s', s)$, read informally as "s' is accessible from s."[2] The K fluent is treated like any other fluent, with the last argument being the official situation argument.

Informally, $K(s', s)$ holds when as far as an agent in situation s knows, it could be in situation s'. Using K we can define the expression **Knows**(ϕ, s) as the abbreviation:

$$\textbf{Knows}(\phi, s) \stackrel{\text{def}}{=} (\forall s').K(s', s) \supset \phi[s'].$$

Here, **Knows**(ϕ, s) means "ϕ is known to be true in situation s." ϕ can be any first-order formula and may contain the special situation term *now* meaning "at this moment." The

---

[1] Axioms with "free" variables can be thought of as being universally quantified from outside the axiom.

[2] We may also say that s' is "K-related" to s.

notation $\phi[s]$ indicates that all occurrences of *now* in ϕ should be replaced with s when **Knows**(ϕ, s) is expanded. Like other fluents, the K fluent also requires a successor state axiom that defines how it changes due to action.

DEFINITION 2. A *successor state axiom for K* has the form:

$$K(s'', do(a, s)) \equiv (\exists s').s'' = do(a, s') \wedge K(s', s) \wedge \bigwedge_{i=1}^{m} \varphi_i(a, s, s'),$$

where each $\varphi_i(a, s, s')$ is defined to have the form:

$$(\forall \vec{x}_i)\big(a = \alpha_i(\vec{u}_i) \supset \bigwedge_{j=1}^{l} (\psi_{ij}(\vec{v}_{ij}, s) \supset (\forall y).\theta_{ij}(\vec{w}_{ij}, y, s) \equiv \theta_{ij}(\vec{w}_{ij}, y, s'))\big).$$

Here, $\theta_{ij}(\vec{w}_{ij}, y, s)$ has the form $F(\vec{w}_{ij}, s)$ for a relational fluent symbol F, or $f(\vec{w}_{ij}, s) = y$ for a functional fluent f.

Each α_i is an action term denoting a *knowledge producing* action. The vectors of variables $\vec{u}_i$, $\vec{v}_{ij}$, and $\vec{w}_{ij}$ must all be (possibly empty) subsets of the variables of $\vec{x}_i$. ψ_{i1} is a situation-independent formula, and each ψ_{ij} $(1 < j \leq l)$ is a situation-independent formula or a formula of the form:

$$\psi_{ij}(\vec{v}_{ij}, s) \stackrel{\text{def}}{=} \psi_{ik}(\vec{v}_{ik}, s) \wedge \big([\neg]\theta_{ik}(\vec{w}_{ik}, y/t, s)\big),$$

where $1 \leq k < j$, and y is replaced by a ground, situation-independent term t in θ_{ik}.

We will assume that each action is either a knowledge-producing action or an ordinary *physical* action, but not both. Updates to K depend on the type of action that is applied. When a physical action A is applied, the K axiom reduces to the form $K(s'', do(A, s)) \equiv (\exists s').s'' = do(A, s'') \wedge K(s', s)$, meaning the size of the set of K-accessible situations is unchanged. Physical actions can cause changes to the values of regular fluents, however, thereby changing the agent's knowledge.

The α_i denote knowledge producing or *sensing* actions that only change the agent's knowledge. (We will typically encode knowledge-producing actions so they leave the values of ordinary fluents unchanged.) Each φ_i ranges over a universally-quantified formula that encodes the effects of a knowledge-producing action α_i, as a conjunction of j "guarded" expressions. Each ψ_{ij} is a *guard condition* for an effect that (conditionally) senses the value of a simple fluent literal expression, θ_{ij}. Guard conditions are expressed in terms of situation s and may not mention other situations. Each fluent in a guard formula must itself be sensed as an effect of the same action, and the guard conditions of that effect must be included as part of any guard formula that references that fluent. Thus, guard conditions that mention fluents are tightly related to other sensing effects of the same action (see Example 3 below). Applying α_i at s updates K for each guard condition ψ_{ij} that evaluates to "true" so that α_i observes θ_{ij}'s value and the agent comes to know whether θ_{ij} holds or not at $do(\alpha_i, s)$.[3]

Our K axiom differs from the standard K axiom in [Scherl and Levesque 1994] which does not include guard conditions. However, the Scherl and Levesque axiom allows more general expressions to be sensed, since each θ_{ij} can be an arbitrary first-order formula. Our axiom is similar to that of [Petrick and Levesque 2002], which lets us model interesting actions with conditional or universal sensory effects.

---

[3]We note that $(\forall y).f(\vec{x}, s) = y \equiv f(\vec{x}, s') = y$ is logically equivalent to $f(\vec{x}, s) = f(\vec{x}, s')$ and so a θ_{ij} expression of the form $f(\vec{x}, s) = y$ models an effect that senses the referent of f.

EXAMPLE 3. Consider the K successor state axiom:

$$K(s'', do(a, s)) \equiv (\exists s').s'' = do(a, s') \wedge K(s', s) \wedge$$
$$(\forall x)(a = sense_1(x) \supset (F(x, s) \equiv F(x, s'))) \wedge$$
$$(\forall x)(a = sense_2 \supset (F(x, s) \equiv F(x, s'))) \wedge$$
$$(\forall x)(a = sense_3(x) \supset$$
$$(f(x, s) = f(x, s')) \wedge$$
$$(f(x, s) = c \supset (F(x, s) \equiv F(x, s')))).$$

This axiom encodes the effects of three knowledge-producing actions. $sense_1$ uncondition-ally senses the truth value of a fluent F for the specified x. $sense_2$ is a more complex action with a *universal* sensory effect: the universal quantification of x results in the unconditional sensing of F for every value of x. Finally, $sense_3$ has a compound effect: it unconditionally senses the value of $f(x)$ and also *conditionally* senses $F(x)$, provided $f(x, s) = c$ is true. Here, $f(x, s) = c$ acts as the guard to the second effect and must evaluate as true before $F(x)$ is sensed. (Definition 2 requires that $f(x)$ be sensed as part of $sense_3$ before it can be used as a guard condition of the same action.) Actions like $sense_3$ allow the properties of a restricted set of objects to be sensed, contingent on the truth of another property.

The addition of K also requires new foundational axioms (see [Scherl and Levesque 1994]) and axioms that describe the possible world alternatives for K in the initial situation. To this end, $Init(s)$ is defined as the abbreviation:

$$Init(s) \overset{\text{def}}{=} \neg(\exists a, s')s = do(a, s').$$

$Init(s)$ indicates that "s is an initial situation." Ordinary initial situation axioms are also permitted to specify the values of fluents at any initial situation, not just S_0. Together these axioms describe what is known and not known initially.

Finally, by restricting the accessibility of K, we can model the properties of particular modal logics of knowledge (see, e.g., [Fagin, Halpern, Moses, and Vardi 1995]). The *initial accessibility properties* of K are defined by a subset of four axioms describing whether K is (i) reflexive: $(\forall s).Init(s) \supset K(s, s)$, (ii) transitive: $(\forall s_1, s_2, s_3).Init(s_1) \wedge Init(s_2) \wedge Init(s_3) \supset (K(s_2, s_1) \wedge K(s_3, s_2) \supset K(s_3, s_1))$, (iii) symmetric: $(\forall s_1, s_2).Init(s_1) \wedge Init(s_2) \supset (K(s_1, s_2) \supset K(s_2, s_1))$, or (iv) Euclidean: $(\forall s_1, s_2, s_3).Init(s_1) \wedge Init(s_2) \wedge Init(s_3) \supset (K(s_2, s_1) \wedge K(s_3, s_1) \supset K(s_2, s_3))$. Scherl and Levesque [1994] show that it is sufficient to limit K's accessibility on initial situations alone, to ensure the same proper-ties are preserved through action. We establish a similar result for our modified K axiom.

THEOREM 4. *If the initial accessibility properties defined for K include a subset of the reflexive, transitive, symmetric, and Euclidean properties, then these same properties hold for all situations, not just initial situations.*

Proof. (Sketch) To extend the initial accessibility properties of K from initial situations to all situations, we use the form of the K axiom to argue that a set of K-accessible situ-ations only changes in a non-increasing manner. Thus, when $K(do(a, s'), do(a, s))$ holds it must be the case that $K(s', s)$ also holds. The proof considers the conditions that could hold on successor and predecessor situations, as encoded in K, to show each accessibility relationship is preserved when an action is applied. □

We will refer to the above action theories with K as *KBATs*, and focus on the form of certain axioms like the successor state axioms, while assuming all other axioms have been appropriately defined. KBATs are important for modelling action theories with knowledge, and give rise to a number of basic properties for manipulating **Knows** expressions.

THEOREM 5. *Let Σ be a KBAT. If ϕ and ψ are first-order sentences where s appears free in ϕ and ψ, and $\phi(\vec{x}, s)$ is a first-order formula where $\vec{x}$ and s are the only free variables of ϕ, then the following properties hold:*

1. $\Sigma \models (\forall s).\mathbf{Knows}(\phi(now), s) \wedge \mathbf{Knows}(\psi(now), s) \equiv \mathbf{Knows}(\phi(now) \wedge \psi(now), s)$,

2. $\Sigma \models (\forall s).\mathbf{Knows}(\phi(now), s) \vee \mathbf{Knows}(\psi(now), s) \supset \mathbf{Knows}(\phi(now) \vee \psi(now), s)$,

3. $\Sigma \models (\forall s).(\forall \vec{x})\mathbf{Knows}(\phi(\vec{x}, now), s) \equiv \mathbf{Knows}((\forall \vec{x})\phi(\vec{x}, now), s)$,

4. $\Sigma \models (\forall s).(\exists \vec{x})\mathbf{Knows}(\phi(\vec{x}, now), s) \supset \mathbf{Knows}((\exists \vec{x})\phi(\vec{x}, now), s)$.

Proof. The proof details are mechanical and follow from the definition of **Knows** and ordinary first-order logic. $\square$

While the properties in Theorem 5 are not exhaustive, they are nevertheless important since they provide us with a useful set of tools for reasoning about **Knows**. For instance, properties like (1) and (3) provide us with a way of "deconstructing" certain **Knows** formulae into equivalent forms that remove conjunction and universal quantification from the scope of the **Knows** operator. We note that, in general, we cannot perform similar transformations for disjunction and existential quantification, as shown by properties (2) and (4). In the remainder of the paper we will investigate KBATs that give rise to Cartesian situations, and describe a set of additional properties that result from such theories.

4 Cartesian situations

KBATs may include both relational and functional fluents, however, we will assume that such theories only contain a finite number of fluent symbols. We will also consider simple formulae that characterize primitive assertions about relations and functions. These formulae will act as the building blocks for constructing more complex expressions.

DEFINITION 6. Let Σ be a KBAT. A *primitive fluent literal (pfl)* is a formula of the form: (i) $F(t_1, t_2, \ldots, t_n, s)$, (ii) $f(t_1, t_2, \ldots, t_n, s) = t_{n+1}$, (iii) $\neg F(t_1, t_2, \ldots, t_n, s)$, or (iv) $f(t_1, t_2, \ldots, t_n, s) \neq t_{n+1}$, where F is an ordinary relational fluent symbol, f is a functional fluent symbol, the t_is are situation-free terms, and s is a situation term.

A pfl makes an assertion about the value of a fluent at a situation: a relational fluent is true or false, or a functional fluent is equal or unequal to some mapping. As notation, we will usually denote a pfl as $\theta(\vec{x}, s)$, regardless of the underlying fluent or sign (sometimes without the situation term). For a pfl that mentions a functional fluent, we will include the function mapping (i.e., t_{n+1}) as the final argument of $\vec{x}$. We will also use $\bar{\theta}$ to denote the complement of a pfl. It follows from Definition 6 that the complement of a pfl is also a pfl.

Typically, we will consider *collections* of pfls, $\theta_1, \theta_2, \ldots, \theta_n$ ($n \geq 2$), with the property that all arguments (and function mappings) are distinct variables. This will allow us to quantify individually over particular terms, and characterize those instances of a collection that we do not want certain properties to apply to. To do this we will use guard formulae that constrain the arguments of a given collection of pfls.

DEFINITION 7. Let $\theta_1, \theta_2, \ldots, \theta_n$ be a collection of pfls, where the arguments (and function mappings) of the pfls are all distinct variables. Let $\vec{x}_i = \langle x_{i1}, x_{i2}, \ldots, x_{i|\vec{x}_i|} \rangle$ be the non-situation arguments of θ_i, for each $i = 1, 2, \ldots, n$. The formula $DNCI(\theta_1, \theta_2, \ldots, \theta_n)$ is defined as follows:

$$DNCI(\theta_1, \theta_2, \ldots, \theta_n) \overset{\text{def}}{=} \bigwedge_{\substack{i=1,\ldots,n \\ j=i+1,\ldots,n}} \Gamma(\theta_i, \theta_j),$$

where $\Gamma(\theta_i, \theta_j)$ is defined by the rules:

1. If the same relational fluent symbol appears in θ_i and θ_j (regardless of sign) then
$$\Gamma(\theta_i, \theta_j) \overset{\text{def}}{=} \begin{cases} x_{i1} \neq x_{j1} \vee \ldots \vee x_{i|\vec{x}_i|} \neq x_{j|\vec{x}_j|}, & \text{if } |\vec{x}_i| > 0 \\ \bot, & \text{if } |\vec{x}_i| = 0. \end{cases}$$

2. If the same functional fluent symbol appears in θ_i and θ_j (regardless of sign) then
$$\Gamma(\theta_i, \theta_j) \overset{\text{def}}{=} \begin{cases} x_{i1} \neq x_{j1} \vee \ldots \vee x_{i|\vec{x}_i|-1} \neq x_{j|\vec{x}_j|-1}, & \text{if } |\vec{x}_i| > 1 \\ \bot, & \text{if } |\vec{x}_i| = 1. \end{cases}$$

3. If the fluent symbols that appear in θ_i and θ_j differ then $\Gamma(\theta_i, \theta_j) \overset{\text{def}}{=} \top$.

$DNCI(\theta_1, \theta_2, \ldots, \theta_n)$ is a situation-independent formula that will only evaluate as "true" for distinct and non-complementary instances of $\theta_1, \theta_2, \ldots, \theta_n$. The precise definition of $DNCI(\theta_1, \theta_2, \ldots, \theta_n)$ depends on whether or not a fluent symbol appears more than once in a collection of pfls. For instance, $DNCI(\theta_1, \theta_2, \ldots, \theta_n)$ is defined to be $\top$ if no fluent symbol appears multiple times in $\theta_1, \theta_2, \ldots, \theta_n$; $DNCI(\theta_1, \theta_2, \ldots, \theta_n)$ is $\bot$ if a fluent with only a situation argument appears more than once in $\theta_1, \theta_2, \ldots, \theta_n$. Otherwise, $DNCI(\theta_1, \theta_2, \ldots, \theta_n)$ evaluates as "false" for those instantiations of the arguments of $\theta_1, \theta_2, \ldots, \theta_n$ that lead to multiple instances of the same pfl, or an instance of a pfl and its complement, becoming true. For instance, consider the scenario in the following example.

EXAMPLE 8. If $F(x_1, s)$ and $\neg F(x_2, s)$ are pfls then $DNCI(F(x_1, s), \neg F(x_2, s)) \overset{\text{def}}{=} (x_1 \neq x_2)$ by Definition 7. If we consider a guarded formula:

$$(\forall x_1, x_2, s).DNCI(F(x_1, s), \neg F(x_2, s)) \supset \Phi_F(x_1, x_2, s),$$

where $\Phi_F(x_1, x_2, s)$ has the form $F(x_1, s) \wedge \neg F(x_2, s)$, then $DNCI(F(x_1, s), \neg F(x_2, s))$ guards against the inconsistent instances of $\Phi_F(x_1, x_2, s)$ when $x_1 = x_2$.

For pfls that mention functional fluents, the (in)equality mapping is not considered. Thus, we can use $DNCI$ to guard against instances of pfls that assert multiple mappings for the same function. We will also use Definition 7 to define the formulae arising from a collection of functional fluent terms (which strictly speaking aren't pfls) and will denote such formulae as $DNCI(f_1(\vec{x}_1), f_2(\vec{x}_2), \ldots, f_n(\vec{x}_n))$. In collections of "propositional" pfls—pfls that only have situation arguments—$DNCI(\theta_1, \theta_2, \ldots, \theta_n)$ will always be $\top$ or $\bot$.

We now consider what we mean by a *Cartesian situation*, a structural property that imposes certain conditions on the way a set of K-accessible situations is configured. Our notion of a Cartesian situation is inspired by a Cartesian product over sets, e.g., $\{a, b\} \times \{c, d\} = \{\langle a, c \rangle, \langle b, c \rangle, \langle b, d \rangle, \langle a, d \rangle\}$. In the mathematical sense, a Cartesian product defines a set that contains an ordered vector for every possible combination of elements taken from the original sets. Applying this idea to situations, we would like to ensure that certain configurations of fluent values exist across a set of K-related situations. In particular, if we consider the possible values that individual fluents have at the situations accessible from some situation, we would like to guarantee that there must also exist accessible situations for every possible combination of these fluents' values.

EXAMPLE 9. Let Σ be a KBAT with two relational fluents, F and G, and let t be a ground situation term such that

$$\Sigma \models (\exists s_1, s_2, s_3, s_4).K(s_1, t) \wedge K(s_2, t) \wedge K(s_3, t) \wedge K(s_4, t) \wedge$$
$$F(s_1) \wedge G(s_2) \wedge \neg F(s_3) \wedge \neg G(s_4).$$

In this case, we have situations K-related to t where F and G each individually take on positive and negative values. If t is a Cartesian situation then we would like to ensure that there exist situations K-related to t for each combination of the possible values of F and G. Thus, we would expect that

$$\Sigma \models (\exists s_1, s_2, s_3, s_4).K(s_1, t) \wedge K(s_2, t) \wedge K(s_3, t) \wedge K(s_4, t) \wedge$$
$$F(s_1) \wedge G(s_1) \wedge \neg F(s_2) \wedge G(s_2) \wedge F(s_3) \wedge \neg G(s_3) \wedge \neg F(s_4) \wedge \neg G(s_4).$$

We note that the necessity of certain fluent configurations is based solely on the values of individual fluents at other accessible situations: if $\neg G$ didn't hold at any situation accessible from t then we wouldn't require a situation where F and $\neg G$, or $\neg F$ and $\neg G$, held.

Another observation is that the elements of a mathematical Cartesian product set are all interrelated. In particular, each element has the property that it only differs from some other element by one component. If we extend this idea to situations, then the Cartesian property must ensure that individual situations are configured to be identical to other situations, except for certain "minimal" differences. It is this characterization that we will use as the motivation for our formal definition of a Cartesian situation.

DEFINITION 10. Let Σ be a KBAT with a finite number of fluent symbols. Let $F_1, F_2, \ldots, F_n$ be the relational fluent symbols of Σ, and let $f_1, f_2, \ldots, f_m$ be the functional fluent symbols of Σ. The abbreviation $Cart(s)$ is defined as the expression:

$$[\bigwedge_{i=1}^{n} (\forall s', s'', \vec{x}_i).K(s', s) \wedge K(s'', s) \wedge F_i(\vec{x}_i, s') \not\equiv F_i(\vec{x}_i, s'') \supset$$
$$(\exists s^*).K(s^*, s) \wedge F_i(\vec{x}_i, s^*) \equiv F_i(\vec{x}_i, s'') \wedge$$
$$(\forall \vec{x}_j)(\vec{x}_j \neq \vec{x}_i \supset F_i(\vec{x}_j, s^*) \equiv F_i(\vec{x}_j, s')) \wedge$$
$$\bigwedge_{\substack{k=1\\k\neq i}}^{n} (\forall \vec{x}_k)F_k(\vec{x}_k, s^*) \equiv F_k(\vec{x}_k, s') \wedge \bigwedge_{k=1}^{m} (\forall \vec{x}_k)f_k(\vec{x}_k, s^*) = f_k(\vec{x}_k, s')] \wedge$$
$$[\bigwedge_{i=1}^{m} (\forall s', s'', \vec{x}_i).K(s', s) \wedge K(s'', s) \wedge f_i(\vec{x}_i, s') \neq f_i(\vec{x}_i, s'') \supset$$
$$(\exists s^*).K(s^*, s) \wedge f_i(\vec{x}_i, s^*) = f_i(\vec{x}_i, s'') \wedge$$
$$(\forall \vec{x}_j)(\vec{x}_j \neq \vec{x}_i \supset f_i(\vec{x}_j, s^*) = f_i(\vec{x}_j, s')) \wedge$$
$$\bigwedge_{\substack{k=1\\k\neq i}}^{m} (\forall \vec{x}_k)f_k(\vec{x}_k, s^*) = f_k(\vec{x}_k, s') \wedge \bigwedge_{k=1}^{n} (\forall \vec{x}_k)F_k(\vec{x}_k, s^*) \equiv F_k(\vec{x}_k, s')].$$

Intuitively, $Cart(s)$ means that "situation s is a *Cartesian situation*." Our definition of $Cart(s)$ is based on identifying pairs of situations that differ on the value of some instantiated fluent. Such a pair implies the existence of an accessible situation that is a cross between the two situations: the "differing" fluent value from one situation is true at this new situation, while the values of all other fluents are otherwise identical to those of the second situation. Although we only consider theories with a finite number of fluent symbols, we do not restrict how these symbols may be instantiated.

Using Definition 10, we can also demonstrate that our notion of a Cartesian situation satisfies our earlier intuition concerning the existence of certain configurations of fluent values. We revisit the scenario from Example 9.

EXAMPLE 11. Consider the KBAT Σ in Example 9 where F and G are relational fluents, t is a ground situation term, and

$$\Sigma \models (\exists s_1, s_2, s_3, s_4).K(s_1, t) \wedge K(s_2, t) \wedge K(s_3, t) \wedge K(s_4, t) \wedge$$
$$F(s_1) \wedge G(s_2) \wedge \neg F(s_3) \wedge \neg G(s_4).$$

Say $\Sigma \models Cart(t) \wedge (\exists s_1).K(s_1, t) \wedge F(s_1) \wedge \neg G(s_1)$ (we will have similar conclusions if $G(s_1)$ holds). Since $\Sigma \models (\exists s_2).K(s_2, t) \wedge G(s_2)$, it then follows from Definition 10 that $\Sigma \models (\exists s_2').K(s_2', t) \wedge F(s_2') \wedge G(s_2')$. Also, since $\Sigma \models (\exists s_3).K(s_3, t) \wedge \neg F(s_3)$ we have that $\Sigma \models (\exists s_3').K(s_3', t) \wedge \neg F(s_3') \wedge G(s_3')$. Finally, since $\Sigma \models (\exists s_1).K(s_1, t) \wedge \neg G(s_1)$ it also follows that $\Sigma \models (\exists s_4').K(s_4', t) \wedge \neg F(s_4') \wedge \neg G(s_4')$. Thus, there exist situations K-related to t for each combination of the possible values of F and G. Say instead $\Sigma \models (\exists s_1, s_2).K(s_1, t) \wedge K(s_2, t) \wedge F(s_1) \wedge \neg G(s_1) \wedge G(s_2)$ but $\Sigma \not\models (\exists s').K(s', t) \wedge F(s') \wedge G(s')$. Thus, there is no K-related situation where both F and G hold, and it follows from Definition 10 that $\Sigma \not\models Cart(t)$.

The accessibility of K also plays a role in determining if a situation is Cartesian or not. Recall that K may be initially constrained by a subset of four accessibility properties, and that Theorem 4 guarantees that any subset of these properties also holds for all situations. If certain combinations of these properties are known to hold, then it follows that some sets of situations must necessarily be "mutually" Cartesian.

THEOREM 12. *Let Σ be a KBAT. If the initial accessibility properties of K satisfy the conditions that K is (i) symmetric and Euclidean, (ii) reflexive and Euclidean, (iii) symmetric and transitive, or (iv) transitive and Euclidean, then*
$$\Sigma \models (\forall s_1, s_2).\big(Cart(s_1) \wedge K(s_2, s_1)\big) \supset Cart(s_2).$$

Proof. (Sketch) Theorem 4 ensures that each set of accessibility properties holds for all situations. We then show that given a Cartesian situation s, s and all situations K-related to s share the same set of accessible situations. Thus, K acts as an equivalence relation over this set. (We can also verify that no single accessibility property is alone sufficient to establish a similar result. See [Petrick 2006] for details.) □

5 Properties of Cartesian situations

The structural restrictions of Cartesian situations give rise to a set of properties that can be used to decompose complex **Knows** formulae into simpler but equivalent components, letting us reason about a formula's constituent parts. Although the standard properties of **Knows** already let us do this to some extent (e.g., see Theorem 5), Cartesian properties are much richer, but come with certain restrictions.

We begin by formalizing our earlier intuition about Cartesian situations, that certain configurations of fluent values must occur across a set of K-related situations.

THEOREM 13. *Let Σ be a KBAT. If $\theta_1, \theta_2, \ldots, \theta_n$ ($n \geq 2$) is a collection of pfls, where the arguments of each θ_i are all variables, denoted by $\vec{x}_i$, and each variable of $\vec{x}_1, \vec{x}_2, \ldots, \vec{x}_n$ appears no more than once, then*

$$\Sigma \models (\forall s, \vec{x}_1, \vec{x}_2, \ldots, \vec{x}_n).Cart(s) \wedge DNCI(\theta_1, \theta_2, \ldots, \theta_n) \supset$$
$$(\bigwedge_{i=1}^{n}(\exists s_i)(K(s_i, s) \wedge \theta_i[s_i])) \supset (\exists s^*)(K(s^*, s) \wedge \bigwedge_{i=1}^{n} \theta_i[s^*]).$$

Proof. (Sketch) The proof is by induction on n. In the induction step we consider the case where k of the θ_is hold at a single K-related situation, and θ_{k+1} holds at some K-related situation. We then consider the form of θ_{k+1} and argue that for each of the four possible types of pfls, Definition 10 ensures the existence of a K-related situation at which all $k+1$ of the θ_is hold. $\square$

Theorem 13 illustrates a strong structural property of Cartesian situations, ensuring the existence of particular combinations of fluent values, as in Example 9 and Example 11. (This idea is analogous to the existence of particular vectors in standard Cartesian product sets.) The structure of a Cartesian situation also lets us establish a new set of properties for **Knows**, for decomposing many common types of epistemic formulae into simpler components that mention pfls.

THEOREM 14. *Let Σ be a KBAT. If $\theta_1, \theta_2, \ldots, \theta_n$ ($n \geq 2$) is a collection of pfls, where the arguments of each θ_i are all variables, denoted by $\vec{x}_i$, and each variable of $\vec{x}_1, \vec{x}_2, \ldots, \vec{x}_n$ appears no more than once; and $f_1(\vec{x}_1), f_2(\vec{x}_2), \ldots, f_n(\vec{x}_n)$ ($n \geq 2$) are terms, where each f_i is a functional fluent symbol, each $\vec{x}_i$ is a vector of variables with no variable appearing more than once in $\vec{x}_1, \vec{x}_2, \ldots, \vec{x}_n$, and y is a variable that doesn't appear in $\vec{x}_1, \vec{x}_2, \ldots, \vec{x}_n$; then the following properties hold:*

1. $\Sigma \models (\forall s, \vec{x}_1, \vec{x}_2, \ldots, \vec{x}_n).Cart(s) \wedge DNCI(\theta_1, \theta_2, \ldots, \theta_n) \supset$
   $\left[\mathbf{Knows}(\bigvee_{i=1}^{n} \theta_i(now), s) \equiv \bigvee_{i=1}^{n} \mathbf{Knows}(\theta_i(now), s) \right]$,

2. $\Sigma \models (\forall s, \vec{x}_1, \vec{x}_2).Cart(s) \wedge DNCI(\theta_1, \theta_2) \supset$
   $\left[\mathbf{Knows}(\theta_1(now) \supset \theta_2(now), s) \equiv (\mathbf{Knows}(\bar{\theta}_1(now), s) \vee \mathbf{Knows}(\theta_2(now), s)) \right]$,

3. $\Sigma \models (\forall s, \vec{x}_1, \vec{x}_2, \ldots, \vec{x}_n).Cart(s) \wedge DNCI(\theta_1, \theta_2, \ldots, \theta_n) \supset$
   $\left[\mathbf{Knows}(\bigwedge_{i=1}^{n-1} \theta_i(now) \equiv \theta_{i+1}(now), s) \equiv \right.$
   $\left. (\bigwedge_{i=1}^{n} \mathbf{Knows}(\theta_i(now), s) \vee \bigwedge_{i=1}^{n} \mathbf{Knows}(\bar{\theta}_i(now), s)) \right]$,

4. $\Sigma \models (\forall s, \vec{x}_1, \vec{x}_2, \ldots, \vec{x}_n).Cart(s) \wedge DNCI(f_1(\vec{x}_1), f_2(\vec{x}_2), \ldots, f_n(\vec{x}_n)) \supset$
   $\left[\mathbf{Knows}(\bigwedge_{i=1}^{n-1} f_i(\vec{x}_i, now) = f_{i+1}(\vec{x}_{i+1}, now), s) \equiv \right.$
   $\left. (\exists y).\bigwedge_{i=1}^{n} \mathbf{Knows}(f_i(\vec{x}_i, now) = y, s) \right]$,

5. $\Sigma \models (\forall s, \vec{x}_1, \vec{x}_2, \ldots, \vec{x}_n).Cart(s) \wedge DNCI(f_1(\vec{x}_1), f_2(\vec{x}_2), \ldots, f_n(\vec{x}_n)) \supset$
   $\left[\mathbf{Knows}(\bigvee_{i=1}^{n-1} f_i(\vec{x}_i, now) \neq f_{i+1}(\vec{x}_{i+1}, now), s) \equiv \right.$
   $\left. (\forall y)\bigvee_{i=1}^{n} \mathbf{Knows}(f_i(\vec{x}_i, now) \neq y, s) \right]$.

Proof. (Sketch) In (1), the $\Leftarrow$ direction follows as a consequence of the standard definition of **Knows**. The $\Rightarrow$ direction uses Theorem 13 and the intermediate result:

$$\Sigma \models (\forall s).\left[\bigwedge_{i=1}^{n} (\exists s')(K(s', s) \wedge \phi_i[s']) \supset (\exists s^*)(K(s^*, s) \wedge \bigwedge_{i=1}^{n} \phi_i[s^*])\right] \equiv$$
$$\left[\mathbf{Knows}(\bigvee_{i=1}^{n} \neg\phi_i(now), s) \supset \bigvee_{i=1}^{n} \mathbf{Knows}(\neg\phi_i(now), s)\right]$$

that holds for arbitrary first-order sentences ϕ_i, observing that the negation of a pfl is also a pfl and that a *DNCI* formula does not change for a collection of pfls if the underlying signs of the pfls change. In (2), the proof is a straightforward application of (1) and the standard properties of **Knows**. In (3), the proof uses (2) and is otherwise a straightforward application of the standard properties of **Knows**. In (4), the $\Leftarrow$ direction follows as a consequence of the ordinary definition and properties of **Knows**. The $\Rightarrow$ direction is established from first principles by arguing from Definition 10. In (5), the result follows from (1) and by arguing that particular *DNCI* formulae do not change if we change the underlying functional pfls from equality to inequality mappings. $\square$

Theorem 14 illustrates that when situations are Cartesian, certain types of knowledge (i.e., certain configurations of K-accessible situations) are prohibited from occurring, giving rise to a set of knowledge decomposition properties. The list of properties in Theorem 14 is not exhaustive, however, it does provide us with a set of tools for simplifying many types of epistemic formulae involving **Knows**.

For instance, (1) establishes a fundamental result for Cartesian situations: it allows us to decompose knowledge of a disjunctive formula into an equivalent form where we can reason about knowledge of the individual disjuncts. This property comes at a representational cost, namely that we require definite knowledge of some of the component parts that make up the disjunction. (1) does not prohibit disjunction entirely, but instead has the effect of removing inter-dependencies among the primitive components (i.e., fluents). Thus, certain "general" disjunctions cannot be represented in theories with Cartesian restrictions. In practical terms, this means that such theories will be inherently less expressive than theories without such restrictions. For example, we can no longer explicitly represent knowledge of $F \vee G$ without definite knowledge of F or definite knowledge of G.

(2) illustrates that knowledge of a logical implication reduces to a disjunction of **Knows** expressions. Applying this result to logical equivalence produces the disjunctive normal form in (3). In particular, if a set of pfls are known to be equivalent then either each pfl must be known to be true, or the complement of each pfl must be known to be true. Thus, we "know whether" each formula is true or not and coming to know the truth of one formula means we also come to know the truth of the others.

(4) and (5) apply to knowledge of functional equality and inequality, respectively. For (4) we note that the formula $\mathbf{Knows}(\bigwedge_{i=1}^{n-1} f_i(now) = f_{i+1}(now), s)$ is logically equivalent to $\mathbf{Knows}((\exists y).\bigwedge_{i=1}^{n} f_i(now) = y, s)$, and so the consequent of (4) lets us remove an existential quantifier from the context of the **Knows** operator. This relationship indicates a situation where *de dicto* and *de re* knowledge are equivalent. (This equivalence doesn't hold in general, nor is it desirable). (5) establishes the dual of (4) and ensures that when functions are unequal they share no common mappings across the space of K-related situations (the set of possible mappings across all accessible situations is disjoint).[4]

---

[4]E.g., for two functions f_1 and f_2, (5) is equivalent to the world-level property $(\forall s', s'').K(s', s) \wedge K(s'', s) \supset f_1(s') \neq f_2(s'')$, when f_1 and f_2 are known to be unequal at a Cartesian situation s.

As we have seen, Cartesian situations provide us with new properties that augment those in Theorem 5 for simplifying **Knows** formulae. In practice, however, we may first have to detect and reason away certain "hidden" logical constraints (such as tautologies), before a formula can be expressed in a form that permits further simplifications to be made. In general this can be a difficult task, however, we consider how this can be done for one particular class of epistemic formulae.

THEOREM 15. *Let Σ be a KBAT. Let ϕ be a ground, quantifier-free formula without equality that doesn't mention functional fluents or K, and only mentions the situation term s, where s is free. If $\Pi(\phi) = \{\pi_1, \pi_2, \ldots, \pi_n\}$ is the set of prime implicates of ϕ, where each π_i has the form $(\alpha_{i1} \vee \alpha_{i2} \vee \ldots \vee \alpha_{i|\pi_i|})$, then*

$$\Sigma \models (\forall s).Cart(s) \supset \big[\mathbf{Knows}(\phi(now), s) \equiv \bigwedge_{i=1}^{n} \big(\bigvee_{j=1}^{|\pi_i|} \mathbf{Knows}(\alpha_{ij}(now), s)\big)\big].$$

Proof. (Sketch) We note that $\Sigma \models (\forall s).\phi[s] \equiv \bigwedge_{i=1}^{n} \pi_i[s]$ follows by expressing the prime implicates π_i of ϕ in Blake Canonical Form. The result then follows from the standard properties of **Knows** and Theorem 14(1). $\square$

Theorem 15 identifies a particular class of epistemic formulae that can be decomposed as completely as possible, into **Knows** components that only mention fluent literals. This is done by generating a set of prime implicates. We do not focus here on methods for finding sets of prime implicates, which can be a computationally expensive process.

6   Progressing Cartesian situations through action

In general, Cartesian situations are "brittle" and even simple actions can often break the Cartesian structure when applied. However, we can also identify expressive classes of KBATs that preserve the Cartesian property through action. We will investigate some of these action theories by considering knowledge producing and physical actions separately.

For knowledge producing actions we need only consider the form of the K successor state axiom, since such actions only change K and not ordinary fluents. It turns out that the form of the K axiom we give in Definition 2 ensures that if a situation is Cartesian then applying a sensing action preserves the Cartesian property. We have the following result.

THEOREM 16. *Let Σ be a KBAT. If α is a ground action term denoting a knowledge-producing action then $\Sigma \models (\forall s).Cart(s) \supset Cart(do(\alpha, s))$.*

Proof. (Sketch) We establish this result by reasoning about Definition 10 and the form of the K axiom. Intuitively, we consider two situations s_1 and s_2 that are K-related to a situation $do(\alpha, s)$, where s is Cartesian and α is a sensing action. For relational fluents, if an instance of a fluent holds at s_1 and the negation of the same instance holds at s_2, then we use the property that sensing actions leave fluent values unchanged to conclude that the same instances hold at the predecessor situations of s_1 and s_2 respectively, and these predecessors are necessarily K-related to s. Using the Cartesian definition we establish the existence of a "cross-product" situation that is K-related to s and then argue from the form of the K axiom (and that it only senses particular instances of fluent literals) that its successor situation must be K-related to $do(\alpha, s)$. Functional fluents are similar. $\square$

This result is important since it means that sensing actions as we've defined them will not break the Cartesian property. This is a direct consequence of the form of the K axiom, which models actions that sense instances of fluent literals. If we consider more expressive knowledge-producing actions than those permitted by Definition 2, then it becomes easy to construct theories that do not preserve the Cartesian property through action. For instance, consider the more general K successor state axiom in the next example.

EXAMPLE 17. Let Σ be a theory defined by the axioms:

$$F(do(a, s)) \equiv F(s),$$
$$G(do(a, s)) \equiv G(s),$$
$$K(s'', do(a, s)) \equiv (\exists s').s'' = do(a, s') \wedge K(s', s) \wedge$$
$$\quad (a = sense \supset ((F(s') \wedge G(s')) \equiv (F(s) \wedge G(s)))),$$
$$(\exists s_1, s_2, s_3, s_4).K(s_1, S_0) \wedge K(s_2, S_0) \wedge K(s_3, S_0) \wedge K(s_4, S_0) \wedge$$
$$\quad F(s_1) \wedge G(s_1) \wedge F(s_2) \wedge \neg G(s_2) \wedge \neg F(s_3) \wedge G(s_3) \wedge \neg F(s_4) \wedge \neg G(s_4),$$
$$F(S_0), \quad \neg G(S_0).$$

Here *sense* is a knowledge-producing action that senses the truth of a simple conjunctive formula, $F \wedge G$. Initially nothing is known about F or G and $\Sigma \models Cart(S_0)$. Applying *sense* produces situations K-related to $do(sense, S_0)$ for every configuration of F and G, except where both F and G hold. Thus, $\Sigma \not\models Cart(do(sense, S_0))$. (We note that **Knows**$(\neg F(now) \vee \neg G(now), do(sense, S_0))$ holds but **Knows**$(\neg F(now), do(sense, S_0))$ and **Knows**$(\neg G(now), do(sense, S_0))$ do not hold.)

For ordinary physical actions, it is also easy to construct KBATs whose actions do not preserve the Cartesian property. For instance, consider the following example.

EXAMPLE 18. Let Σ be a KBAT defined by the axioms:

$$F(do(a, s)) \equiv (a = change \wedge G(s)) \vee F(s),$$
$$G(do(a, s)) \equiv G(s),$$
$$K(s'', do(a, s)) \equiv (\exists s').s'' = do(a, s') \wedge K(s', s),$$
$$(\exists s_1, s_2, s_3, s_4).K(s_1, S_0) \wedge K(s_2, S_0) \wedge K(s_3, S_0) \wedge K(s_4, S_0) \wedge$$
$$\quad F(s_1) \wedge G(s_1) \wedge F(s_2) \wedge \neg G(s_2) \wedge \neg F(s_3) \wedge G(s_3) \wedge \neg F(s_4) \wedge \neg G(s_4),$$
$$F(S_0), \quad G(S_0).$$

F and G are initially unknown and $\Sigma \models Cart(S_0)$. Applying *change* produces situations K-related to $do(change, S_0)$ for every configuration of F and G, except where both $\neg F$ and G hold. Thus, $\Sigma \not\models Cart(do(change, S_0))$. (We note that **Knows**$(F(now) \vee \neg G(now), do(change, S_0))$ holds but neither **Knows**$(F(now), do(change, S_0))$ nor **Knows**$(\neg G(now), do(change, S_0))$ hold.)

Since physical actions affect the ordinary fluents of a KBAT, we will consider certain classes of KBATs by restricting the form of the ordinary successor state axioms.

6.1 Context-free theories

The first class of KBATs we consider are formed by restricting the $\gamma_F^{\pm}$ and γ_f components of the successor state axioms in Definition 1 to be situation independent formulae (i.e., no fluents). Such axioms are referred to as *context-free successor state axioms* [Lin and Reiter 1997]. Quantification is still permitted, provided the quantifiers only range over the situation-independent formulae. If every successor state axiom is context free then a KBAT is said to be a *context-free theory*. We have the following result for context-free KBATs.

THEOREM 19. *If Σ is a context-free theory then it follows that $\Sigma \models (\forall s)(Init(s) \supset Cart(s)) \supset (\forall s)Cart(s)$.*

Proof. (Sketch) The proof is by induction over situations. The approach is similar to the proof of Theorem 16 and we need only consider the case of physical actions (Theorem 16 supplies the result for sensing actions). Given a pair of differing fluent instances at situations K-related to $do(\alpha, s)$, we can reason that α could not have changed these instances, since such changes would apply across the set of K-related situations due to the situation-independent components of the ordinary successor state axioms. We can then apply the Cartesian definition at s and again argue from the form of the ordinary successor state axioms that an appropriate cross-product successor situation must exist. □

Since situation-independent formulae are evaluated without reference to individual situations, updates to fluents resulting from context-free successor state axioms are made uniformly across a set of K-related situations. Thus, if all initial situations are Cartesian then all successor situations will remain Cartesian. Although context-free theories limit fluent references, these types of axioms are important and arise quite often in practice. For instance, STRIPS-style planning actions [Fikes and Nilsson 1971]—actions with unconditional effects—are a subset of this class of actions.

6.2 Fluent dependent theories

We next consider a class of KBATs that allows actions to have particular conditional effects that depend on the value of some fluent. These theories are inspired by successor state axioms of the form:

$$F(do(a, s)) \equiv (a = A \wedge G(s)) \vee F(s) \wedge \neg(a = A \wedge \neg G(s))$$
$$G(do(a, s)) \equiv a = A \vee G(s).$$

In this case, when A is applied the truth of F in a successor situation depends on the current truth value of G; A also causes G to become true in a successor situation. If we instead had the following successor state axiom for G:

$$G(do(a, s)) \equiv (a = A \wedge F(s)) \vee G(s) \wedge \neg(a = A \wedge \neg F(s))$$

then the value of the fluent F at a successor situation would depend on G, and the value of G would depend on F. If we look at the changes across a set of K-related situations we see that they amount to a permutation or "re-mapping" of the fluent values across the set. New values, when introduced, are made in a uniform manner to all situations.

In general, we consider theories that allow particular "chains" (similar to the first example) and "cycles" (similar to the second example) of dependencies among fluents.

DEFINITION 20. A *fluent dependent theory* restricts the form of the ordinary successor state axioms as follows.

1. For each physical action β, let $\vec{y}_\beta = \langle y_1, y_2, \ldots, y_{|\beta|} \rangle$ be an associated vector of variables that will always appear as the non-situation arguments of β.

2. A successor state axiom for each ordinary relational fluent F has the general form in Definition 1, where $\gamma_F^+(\vec{x}, a, s)$ and $\gamma_F^-(\vec{x}, a, s)$ each have the form:

$$\phi(\vec{x}, a) \vee \bigvee_{i=1}^{n} \psi_i(\vec{x}, a, s).$$

ϕ is any situation independent formula whose free variables are among those in $\vec{x}$ and a. Each ψ_i has the form:

$$\psi_i(\vec{x}, a, s) \stackrel{\text{def}}{=} (\exists \vec{w}).a = \beta(\vec{y}_\beta) \left\{ \wedge L(\vec{z}, s) \right\},$$

where β denotes a physical action with associated parameters $\vec{y}_\beta$; the variables of $\vec{x}$ and $\vec{z}$ must be variables of $\vec{y}_\beta$; the variables of $\vec{w}$ are the variables of $\vec{y}_\beta$ not mentioned in $\vec{x}$; and (optionally) L is a relational fluent literal. Also,

(a) For each ψ_i in γ_F^+ (similarly, γ_F^-) that mentions a fluent literal L, some ψ_j in γ_F^- (similarly, γ_F^+) must be syntactically identical to ψ_i with the exception that L is replaced with the complement of L.

(b) Each action name β can only be mentioned at most once in γ_F^+ or γ_F^-.

(c) If ψ_i mentions an action β and a fluent symbol G with arguments $\vec{z}$, then the successor state axiom for G must have the general syntactic form in Definition 1, and the action β must be mentioned in γ_G^+ or γ_G^- (subject to the above requirements of a successor state axiom).

(We have a similar definition for functional fluents.)

The strong syntactic restrictions in Definition 20 have important consequences for Cartesian situations. Intuitively, they ensure that actions produce "symmetric" behaviour across sets of K-related situations. As a result, each action "shuffles" certain components (i.e., fluent values) at each situation, while retaining the overall Cartesian structure.

THEOREM 21. *If Σ is a fluent dependent theory then it follows that $\Sigma \models (\forall s)(Init(s) \supset Cart(s)) \supset (\forall s)Cart(s)$.*

Proof. (Sketch) The proof is by induction over situations. In the induction step we need only consider the case when a physical action is applied. The tricky part is establishing the existence of the required "cross-product" situation in the case when the fluent in question is part of a dependency chain. We must reason about the existence of a predecessor situation (using the induction assumption) that when progressed through the dependency update, results in the desired configuration of fluents. □

6.3 Propositional theories

We close this section by considering *propositional KBATs (pKBATs)*, action theories without functional fluents where each ordinary relational fluent has the property that its only argument is the situation argument. Such theories are restrictive but important. For instance, theories with finite domains of discourse can be converted to this form. Many planning algorithms also use techniques that first "propositionalize" planning domains. By focusing on pKBATs, our definitions of pfls, *DNCI*, and Cartesian situations can be simplified due to the restrictions on functional fluents and fluent arguments. As a result, the requirements for Cartesian situations are also much simpler.

THEOREM 22. *Let Σ be a pKBAT and let t be a ground situation term. $\Sigma \models Cart(t)$ iff for every collection of pfls, $\theta_1, \theta_2, \ldots, \theta_n$, where no fluent appears more than once,*

$$\Sigma \models \mathbf{Knows}(\bigvee_{i=1}^{n} \theta_i(now), t) \supset \bigvee_{i=1}^{n} \mathbf{Knows}(\theta_i(now), t).$$

Proof. (Sketch) Given the restrictions on sets of propositional pfls without repeated fluent symbols, each *DNCI* formula will be $\top$. The $\Rightarrow$ direction can then be established by Theorem 14(1). In the $\Leftarrow$ direction we make use of the intermediate result from the proof of Theorem 14(1) and a second intermediate result that establishes the converse of Theorem 13. This second result follows from Definition 10 when subjected to the restrictions of proposition KBATs (i.e., no functional fluents and no non-situation arguments). $\qquad\square$

Theorem 22 provides an alternate definition for Cartesian situations, expressed at the "knowledge level" rather than the level of K-related situations. Thus, to determine if a situation is Cartesian it suffices to ensure that particular **Knows** formulae can be decomposed into their component parts. We also establish a related result for certain action sequences.

THEOREM 23. *Let Σ be a pKBAT. Let t be a ground situation term and let $\vec{A}$ be any finite sequence (possibly empty) of ground action terms denoting physical actions. $\Sigma \models Cart(do(\vec{A}, t))$ iff for every collection of pfls, $\theta_1, \theta_2, \ldots, \theta_n$, where no fluent symbol appears more than once,*

$$\Sigma \models \mathbf{Knows}(\bigvee_{i=1}^{n} \theta_i(do(\vec{A}, now)), t) \supset \bigvee_{i=1}^{n} \mathbf{Knows}(\theta_i(do(\vec{A}, now)), t).$$

Proof. (Sketch) The proof follows from Theorem 22 and the intermediate result that $\Sigma \models (\forall s).\mathbf{Knows}(\phi(do(\vec{A}, now)), s) \equiv \mathbf{Knows}(\phi(now), do(\vec{A}, s))$, where ϕ is a first-order sentence and $\vec{A}$ is a sequence of physical actions. (In temporal logic this is similar to $K \circ^n \phi \equiv \circ^n K\phi$.) The intermediate result itself holds by induction on the length of $\vec{A}$, and by reasoning about the definition of **Knows** and the observation that the K axiom does not change the size of a set of K-related situations when a physical action is applied. (If $K(s', s)$ holds then $K(do(\alpha, s'), do(\alpha, s))$ holds when α is a physical action.) $\qquad\square$

Theorem 23 differs from Theorem 22 by placing the knowledge requirements for Cartesian situations in terms of a predecessor situation, rather than the Cartesian situation itself. Thus, provided no action sequence gives rise to particular epistemic disjunctions that can't be decomposed, the resulting situation will be Cartesian. What Theorem 23 does *not* provide is an efficient method for detecting all the conditions that must hold to ensure a situation is Cartesian. As a result, we may potentially have to consider disjunctions that arise from any sequence of physical actions.

7 Examples

Although the KBATs in the previous section limit the form of their successor state axioms, they still let us model some interesting actions and domains. We now consider some examples of such theories. In each case, we only highlight the successor state axioms and simply assume that other axioms are appropriately defined. (E.g., initial situation axioms must be properly defined to ensure that $Cart(S_0)$ holds.)

EXAMPLE 24. In this example we consider a Linux operating system domain with three ordinary fluents and three actions. The fluent $indir(f, d, s)$ indicates that "file f is in directory d in situation s." The fluent $readable(f, s)$ denotes that "file f is readable in situation s." A functional fluent $size(f, s) = y$ indicates that "file f has size y in situation s." The action $chmod\text{+}r(f)$ is a physical action that makes file f readable. Similarly, $chmod\text{-}r(f)$ makes f unreadable. The action $ls(d)$ is a knowledge-producing action that provides information about the files in directory d. The successor state axioms are given as follows:

$$indir(f, d, do(a, s)) \equiv indir(f, d, s),$$
$$size(f, do(a, s)) = y \equiv size(f, s) = y,$$
$$readable(f, do(a, s)) \equiv a = chmod\text{+}r(f) \vee readable(f, s) \wedge \neg(a = chmod\text{-}r(f)),$$

$$K(s'', do(a, s)) \equiv (\exists s').s'' = do(a, s') \wedge K(s', s) \wedge$$
$$(\forall d).(a = ls(d) \supset$$
$$(\forall f)(indir(f, d, s) \equiv indir(f, d, s')) \wedge$$
$$(\forall f)(indir(f, d, s) \supset (size(f, s) = size(f, s'))) \wedge$$
$$(\forall f)(indir(f, d, s) \supset (readable(f, s) \equiv readable(f, s')))).$$

The ordinary successor state axioms meet the requirements of a context-free theory. In this case, $indir$ and $size$ are unchanged by action. The successor state axiom for $readable$ encodes the (unconditional) effects of the $chmod\text{+}r$ and $chmod\text{-}r$ actions on $readable$: $chmod\text{+}r(f)$ makes f readable, while $chmod\text{-}r(f)$ makes f unreadable. The K axiom has the form required by Definition 2 and describes two types of sensory effects for ls. First, it encodes a universal effect: ls senses the files f that are in directory d (i.e., all f that satisfy $indir(f, d, s)$). Second, it encodes a conditional sensing effect: besides sensing the contents of directory d, ls also senses the size and readability of the files in d (i.e., the truth of $readable(f, s)$ and the denotation of $size(f, s)$ for all f such that $indir(f, d, s)$ is true).

EXAMPLE 25. In this example we consider two switches, each of which can be turned on or off. We include two fluents, on_1 and on_2, that represent the state of each switch. We also have two actions, $flip_1$ and $flip_2$, each of which toggles the state of a particular switch. The effects of these actions are encoded in the pair of successor state axioms:

$$on_1(do(a, s)) \equiv (a = flip_1 \wedge \neg on_1(s)) \vee on_1(s) \wedge \neg(a = flip_1 \wedge on_1(s)),$$
$$on_2(do(a, s)) \equiv (a = flip_2 \wedge \neg on_2(s)) \vee on_2(s) \wedge \neg(a = flip_2 \wedge on_2(s)),$$

as part of a fluent dependent theory. Each axiom encodes a simple one-fluent "cyclic" dependency: flipping a switch that is off turns it on; flipping it again turns it off.

The axioms also meet the requirements of a propositional KBAT, which lets us use results like Theorem 22 to simplify the definition of a Cartesian situation. For instance, determining if S_0 is Cartesian reduces to verifying that

$$\mathbf{Knows}(\theta_1(now) \vee \theta_2(now), S_0) \supset (\mathbf{Knows}(\theta_1(now), S_0) \vee \mathbf{Knows}(\theta_2(now), S_0))$$

holds for each valid pair of pfls, θ_1 and θ_2.

EXAMPLE 26. In this final example, we consider the representation of a simple data buffer that can hold three units of data. Data is written to the buffer by the action $write(d)$.

Three functional fluents, B_1, B_2, and B_3, represent the data cells of the buffer. We have the following successor state axioms:

$$B_1(do(a,s)) = y \equiv (\exists d)(a = write(d) \land y = d) \lor$$
$$B_1(s) = y \land \neg(\exists y')((\exists d).a = write(d) \land y' = d),$$
$$B_2(do(a,s)) = y \equiv ((\exists d).a = write(d) \land B_1(s) = y) \lor$$
$$B_2(s) = y \land \neg(\exists y')((\exists d).a = write(d) \land B_1(s) = y'),$$
$$B_3(do(a,s)) = y \equiv ((\exists d).a = write(d) \land B_2(s) = y) \lor$$
$$B_3(s) = y \land \neg(\exists y')((\exists d).a = write(d) \land B_2(s) = y').$$

Here, new data is inserted at one end of the buffer (B_1) and existing data in the buffer is shifted when a new data item is written. (I.e., B_2 takes B_1's value and B_3 takes B_2's value.) Subsequent *write* actions cause further shifts, with the existing data being overwritten. These axioms satisfy the requirements of a fluent dependent theory.

8 Discussion and conclusions

Cartesian situations are important since they give rise to action theories that let us decompose epistemic formulae into more primitive expressions of knowledge based on fluent literals. These properties come at a representational cost, since they limit the expressiveness of certain types of knowledge (most notably disjunctive knowledge). This trade-off also offers some computational advantages, however. For instance, by focusing on theories that reduce the general problem of reasoning about knowledge to the problem of reasoning about knowledge of fluent literals, we can look to alternative representations that limit the types of knowledge they can represent in similar ways (and potentially avoid the computational drawbacks of possible worlds reasoning). This work offers the possibility of identifying candidate theories that can be compiled into such alternative representations. As a result, this work is also closely related to a number of approaches that model knowledge and action without possible worlds and an accessibility relation.

For instance, Petrick and Levesque [2002] transform standard Scherl and Levesque theories into an equivalent form based on *knowledge fluents* [Demolombe and Pozos Parra 2000]—relational fluents that explicitly represent the agent's knowledge of ordinary fluents that are known to be true and known to be false. In this case, determining whether n atomic formulae are known reduces from reasoning about 2^n possible worlds (in the worst case) to reasoning about the truth of $3n$ fluents. Furthermore, knowledge change is treated as ordinary fluent change without the need for a K fluent. This approach relies on theories that limit disjunctive knowledge in a similar way to Cartesian situations. Soutchanski [2001] explores a similar approach, but focuses on less expressive theories than we consider here.

Liu and Levesque [2005] investigate the projection problem for a generalized database called a *proper KB*. Proper KBs do not permit general disjunctions and, thus, appear to be closely related to Cartesian-restricted theories. Liu and Levesque use a simpler notion of sensing than the model presented here, however, and work with relational fluents. Moreover, they primarily focus on actions with local effects, while we permit actions with more general effects. More work is needed to explore the connection between these approaches.

Another related approach is that of Son and Baral [2001], who describe the notion of an *a-state* or "approximate state" that represents an agent's knowledge by sets of fluents known to be true and known to be false. They also define a *0-approximation* that provides a simple method for progressing an a-state based on fluents that *must* become known, and those that *may* be true or false after an action is applied. Son and Baral provide a translation to situation calculus theories, however, the language they consider is propositional.

The focus of the work in this paper is somewhat different, with an emphasis on formally characterizing the conditions that underlie certain knowledge-restricted theories. The decomposition properties of Cartesian situations also help us better understand what we give up representationally in these accounts. Furthermore, this work offers the possibility of identifying action theories that can be compiled into some of the alternative accounts of knowledge and action mentioned above—and extending these approaches to new classes of action theories that haven't been previously considered.

For instance, the compilation techniques of [Petrick and Levesque 2002] extend to the action theories in this paper (see [Petrick 2006] for details). As an example, the axioms

$$Kreadable(f, do(a, s)) \equiv$$
$$a = chmod+r(f) \vee ((\exists d).a = ls(d) \wedge indir(f, d, s) \wedge readable(f, s)) \vee$$
$$Kreadable(f, s) \wedge a \neq chmod\text{-}r(f),$$
$$K\neg readable(f, do(a, s)) \equiv$$
$$a = chmod\text{-}r(f) \vee ((\exists d).a = ls(d) \wedge indir(f, d, s) \wedge \neg readable(f, s)) \vee$$
$$K\neg readable(f, s) \wedge a \neq chmod+r(f),$$

result from a (partial) translation of Example 24 into equivalent successor state axioms, where *Kreadable* and *K¬readable* are knowledge fluents that directly represent when an agent knows *readable* to be true, and knows *readable* to be false, without using a K fluent.

Finally, this work has important consequences for *planning*, where it can be used to help explain translation methods like those of [Palacios and Geffner 2007]. It also provides a practical tool for transforming the standard domain representations used by planners that plan with incomplete information and sensing (e.g., Contingent-FF [Hoffmann and Brafman 2005]), to those of knowledge-based planners like PKS [Petrick and Bacchus 2002].

References

Demolombe, R. and M. P. Pozos Parra [2000]. A simple and tractable extension of situation calculus to epistemic logic. In *Proceedings of the Twelfth International Symposium on Methodologies for Intelligent Systems (ISMIS-2000)*, pp. 515–524.

Fagin, R., J. Y. Halpern, Y. Moses, and M. Y. Vardi [1995]. *Reasoning About Knowledge*. Cambridge, MA: MIT Press.

Fikes, R. E. and N. J. Nilsson [1971]. STRIPS: A new approach to the application of theorem proving to problem solving. *Artificial Intelligence 2*, 189–208.

Funge, J. [1998]. Interval-valued epistemic fluents. In *AAAI Fall Symposium on Cognitive Robotics*, Orlando, FL, pp. 23–25.

Hintikka, J. [1962]. *Knowledge and Belief*. Ithaca, NY, USA: Cornell University Press.

Hoffmann, J. and R. Brafman [2005]. Contingent planning via heuristic forward search with implicit belief states. In *Proceedings of the International Conference on Automated Planning and Scheduling (ICAPS-2005)*, pp. 71–80.

Lin, F. and R. Reiter [1997]. How to progress a database. *Artificial Intelligence 92*(1–2), 131–167.

Liu, Y. and H. J. Levesque [2005]. Tractable reasoning with incomplete first-order knowledge in dynamic systems with context-dependent actions. In *Proc. of the International Joint Conference on Artificial Intelligence (IJCAI-05)*, pp. 522–527.

McCarthy, J. [1979a]. Ascribing mental qualities to machines. In M. Ringle (Ed.), *Philosophical Perspectives in Artificial Intelligence*. Harvester Press. Reprinted in (1990) *Formalizing Common Sense*, Ablex, Norwood, NJ.

McCarthy, J. [1979b]. First order theories of individual concepts and propositions. *Machine Intelligence 9*, 129–148.

McCarthy, J. and P. J. Hayes [1969]. Some philosophical problems from the standpoint of artificial intelligence. *Machine Intelligence 4*, 463–502.

Moore, R. C. [1985]. A formal theory of knowledge and action. In J. R. Hobbs and R. C. Moore (Eds.), *Formal Theories of the Commonsense World*, pp. 319–358. Norwood, NJ: Ablex Publishing.

Palacios, H. and H. Geffner [2007]. From conformant into classical planning: Efficient translations that may be complete too. In *Proceedings of the International Conference on Automated Planning and Scheduling (ICAPS-07)*, pp. 264–271.

Petrick, R. P. A. [2006]. *A Knowledge-level approach for effective acting, sensing, and planning*. Ph.D. thesis, Department of Computer Science, University of Toronto, Toronto, Ontario, Canada.

Petrick, R. P. A. [2008]. Cartesian situations and knowledge decomposition in the situation calculus. In *Proceedings of the International Conference on Principles of Knowledge Representation and Reasoning (KR-2008)*, pp. 629–639.

Petrick, R. P. A. and F. Bacchus [2002]. A knowledge-based approach to planning with incomplete information and sensing. In *Proceedings of the International Conference on Artificial Intelligence Planning and Scheduling (AIPS-2002)*, pp. 212–221.

Petrick, R. P. A. and H. Levesque [2002]. Knowledge equivalence in combined action theories. In *Proceedings of the International Conference on Principles of Knowledge Representation and Reasoning (KR-2002)*, pp. 303–314.

Reiter, R. [2001]. *Knowledge In Action: Logical Foundations for Specifying and Implementing Dynamical Systems*. MIT Press.

Scherl, R. B. and H. J. Levesque [1994]. The frame problem and knowledge-producing actions. Technical report, University of Toronto.

Scherl, R. B. and H. J. Levesque [2003]. Knowledge, action, and the frame problem. *Artificial Intelligence 144*(1–2), 1–39.

Son, T. C. and C. Baral [2001]. Formalizing sensing actions – a transition function based approach. *Artificial Intelligence 125*(1–2), 19–91.

Soutchanski, M. [2001]. A correspondence between two different solutions to the projection task with sensing. In *Fifth Symposium on Logical Formalizations of Commonsense Reasoning (Commonsense 2001)*.

Vassos, S. and H. Levesque [2007]. Progression of situation calculus action theories with incomplete information. In *Proceedings of the International Joint Conference on Artificial Intelligence (IJCAI-07)*, pp. 2029–2034.

25

Inference about Actions: Levesque's view on Action-Ability and Dirichlet Processes

FIORA PIRRI AND MATIA PIZZOLI

ABSTRACT. In this paper we have considered Levesque's approach to modeling the belief that a cognitive robot has about observed actions. Modeling observations of actions is a basic step for any further inference about actions, when the inference concerns what a robot can do or how it can perform a certain task. This requires the robot to be able, for example, to correctly discern two observed actions, even when the two motion sequences are not identical. To treat sequences of point actions, we have interpreted actions as non parametric densities, defined by a countable mixture of normal densities, which can thus approximate any function. We have faced the empirical Bayes problem of estimating their parameters, using Dirichlet processes. Finally, on the basis of the axiomatization provided by Levesque et al. in [Shapiro, Pagnucco, Lespérance, and Levesque 2000], we have shown how beliefs can be used to compare two actions, as defined via countable mixtures.

1 Introduction

One of the main contributions of Levesque to cognitive robotics has been his work on the concept of computability of actions, as building blocks of the robot knowledge about what it can do. The introduction of knowledge and beliefs in the theory of actions, likewise the paradigm of robot programming, had a crucial motivation and has still an important impact on the development of cognitive robotics. Often Levesque has highlighted that the solution to the frame problem, while solving a logical problem, via successor state axioms in the Situation Calculus [McCarthy 1963; Reiter 2001], introduced false statements, not in the logical sense, but in the sense of action meaning and description, hence on the effective path towards execution. This is due to the fact that a successor state axiom, as an explicit definition of the causation properties of an action, cannot capture the effective requirements for execution, in terms of the specification of all the nuts-and-bolts needed for specifying actions and their dynamic properties. Much of Levesque's work , since the papers at AAAI-93 [Scherl and Levesque 1993] and at TARK-94 [Levesque 1994], has been directed to introducing a subjective view in action theory, from planning to the frame problem solution, so as to polish its rigid non-truthfulness.

Levesque's extension of first order logic with modalities, applied to the successor state axioms, and in general in a theory of actions represent the opinion of the robot on what is effectively computable from its knowledge state. In this sense the work of Levesque in the theory of actions can be compared to the Bayesian view of beliefs, and how it contrasts classical statistics. In fact, the subjective view, with all the knowledge and beliefs and modal devices, exhibits the distinction between states (as states of knowledge) and first order definitions of the consequences of actions, in so bringing great flexibility for managing preconditions for action execution. On the other hand, the pure first order statements (those without modalities) draw no bound between the states of knowledge and the truth, with a sort of tacit consensus that statements, free floating in a theory without knowledge, are the robot current knowledge.

These restrictions, in fact, affect planning too, and on this side Levesque foresees action execution as the core of the agent acting in the real world. His contribution has been to put forward the opinion of the subject, namely the computation machine, about its ability to compute what to do ([Lespérance, Levesque, Lin, and Scherl 2000; Lin and Levesque 1998]) providing a description of a state of knowledge. This state can be complete or incomplete, but from the point of view of the computation machine it can evolve smoothly and, in the limit, it indicates the machine rationale about what can be achieved and inferred. Beyond all the contributions on modal and epistemic logic, up to the recent [Lakemeyer and Levesque 2011], on epistemic Situation Calculus, the contribution of Levesque, since [Levesque 1987] has been directed towards this deep philosophical view of computation. The paradigm of action execution is specified in a double way: on one side by several fluxes of information, the combination of which cannot be described by assertions on the universal state of the world: there are actions irrelevant for the world, which however induce the placid evolution of the knowledge state. On the other side, the paradigm is specified by the computational abilities of the machine, this has been fully achieved with robot programming, Golog [Levesque, Reiter, Lesperance, Lin, and Scherl 1997] and all its extensions.

In general in theories of actions (see the 1998 special issue of [ETAI 1998]), actions are represented as simple terms of a formal language; yet, these simple terms carry a huge amount of requirements, addressing causation, executability, controllability, impact and consequentiality. The frame and qualification problems [Reiter 1991; McCarthy and Hayes 1969] account for some of these requirements. In [Reiter 1996], Reiter has generalized the concept of action to that of a process, specified by a *start* and an *end* action, both showing the connection with programming and execution, and accounting for time and concurrency (see also [Pirri and Reiter 2000]). These concepts have been further developed into a new generation of dynamic programming constructs, going beyond the era of the logic of programs (e.g. [Harel, Meyer, and Pratt 1977]), still rooting the new concepts in that paradigm. In particular, these new concepts of executability, streaming from the theory of actions, found a thorough impact in Golog [Levesque, Reiter, Lesperance, Lin, and Scherl 1997], and its extensions. These new programming languages (see also [Thielscher 2005; Gelfond and Lifschitz 1998; Alechina, Dastani, Logan, and Meyer 2007], and many others) have conveyed the significance of representing actions at the core of computation: starting with primitive actions, more complex objects can be manipulated using suitable programming constructs, which become even more involved terms in ConGolog [DeGiacomo, Lespérance, and Levesque 2000]. Actions, as terms of the language, grow into the basic constituents of a new effective paradigm of computability, via executability, and achievability, see [Lin and Levesque 1998; Levesque 1998], in so becoming the bricks of robot programming. On the other hand, but contemporary to this endeavor, a wealth of literature, pushed also by Levesque's work, has faced all the implications of theory of actions concerning intention, causality, deliberation, and the social relevance of action execution, such as awareness, and free will (see, for example, [Cohen and Levesque 1990; Ascombe 2000; Bratman 2006; Lifschitz 2006; McCarthy 2007; McCarthy 1968]), in so enriching with social, psychological and philosophical meaning the nature of this new computation paradigm.

All these contributions are invaluable, for having introduced, via the action theory relevant concepts, a paradigm of computation relating programs with the real world, and creating the basis of cognitive robotic programming.

Still, strictly related to the ground concept of executability, actions can be defined as a class of learnable functions. This concept has not yet been specified in terms of action computation, while apparently it is strictly connected with the notions of *ability* and *knowing how*.

A further insight in the concept of ability, as the cornerstone for understanding the executability of a sequence - or a set of sequences - of actions, is that of learning an action structure [Lakemeyer and Levesque 1998].

Continuing the issues already posed in [Pirri 2011], in this paper we address the problem of action learning from perception and providing a suitable representation to be uploaded in an action theory. On the other hand, to honor Levesque's view of beliefs, as the effective measures of a state of knowledge, from perception to the construction of its representation, we follow a non parametric Bayesian approach to the identification of actions.

In the next section we shall introduce our approach following the early works of [Ferguson 1973; Blackwell and MacQueen 1973; Antoniak 1974; Ferguson 1983; Kuo 1986] in which we use a non parametric analysis of the empirical Bayes problem of making inference about the parameters of an unknown distribution. Then, in the third section, we connect the construction with the belief representation in Levesque's work and relate that to probabilistic automata. Finally we review some approach to action recognition.

2 Estimating action structure

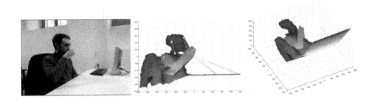

Figure 25.1. Images of a subject drinking water. Hints on the features gathered for coding observations, using the Kinect for 3D dense extraction. The way observations are coded is not relevant (in this paper), while it is relevant the detail and amount of observations.

An action can be described by an arbitrary density $f(x)$. We do not make any particular assumption or requirement on the structure of the observations. In other words, we can simply assume that any video sequence (see for example[Pirri 2011; Fanello, Gori, and Pirri 2010]) can be suitably transformed such that it can be represented by a collection of events $X_1, \ldots, X_n$. Given observations $x_1, \ldots, x_n$, where each x_i, the observed value of X_i is a sample from a known kernel $h(x_i|\mu_i, \sigma_i)$, an action can be represented, for example, on the real line, as:

$$f(x) = \sum_{i=1}^{\infty} p_i h(x \,|\theta_i) \tag{25.1}$$

Here $h(x|\theta_i)$ is the density of $\mathcal{N}(x|\theta_i)$, the normal distribution. The function $f(x)$ is an infinite mixture with a countable number of parameters $par = (p_1, p_2, \ldots, \theta_1, \theta_2, \ldots)$. Each observation x_i can be seen as the observation of a point action and it is associated with a parameter θ_i. Given the θ_i, the x_i are independent. The parameters can be seen as the states of the system while observing, which could be also defined as functions of other parameters (or states), that is, we might have $\theta_i = g(\lambda_i \gamma + \beta_i)$. This means, for example, that all the requirements concerning causality, temporal relations etc, might be hidden in the parameters (or states), from which the observations are dependent. Despite the number of parameters is, in principle, infinite, we can actually make only a finite amount of observations. Indeed, the estimation of the parameters $\theta_1, \theta_2, \ldots$ will let us see that many of them are identical, therefore an action can be characterized by the number of observations.

Consider, for example, the cognitive robot observing the action *cooking* (someone cooking pasta). Suppose, that it is a beginner chef who knows that each single movement has certain properties (comes from a known kernel), but does not know the way the whole cooking is done nor it knows how each primitive action is achieved. This state of knowledge is captured by the existence of an unknown distribution about the single gestures needed for cooking pasta, according to some special recipe. On the other hand, suppose our robot is a good chef (at least for a pasta) then to it a glance to the action sequence would be sufficient to understand what is going on. As it trusts its observations, few gestures, hence few observations will be needed to define $f(x)$, namely the action of cooking. Therefore its state of knowledge shall differ from the beginner, for whom every single gesture will be important. This tradeoff between prior information and states is captured by the unknown distribution that is put as a prior on the states affecting the robot beliefs about the observations. In Levesque's view there are two kind of actions: 1) actions affecting the knowledge state of the robot; 2) actions that the robot does to affect the world. We add: actions of both types can be learned, thus there is a preliminary stage in which the action, as a learnable function, affects the robot knowledge state.

To assess this view, we follow the idea of Ferguson [Ferguson 1973; Ferguson 1983] that the unknown distribution, say G, is itself a parameter taking values in a class of distributions on a measurable space. Moreover, from the Bayesian point of view, G is itself a random variable. The unknown distribution G is chosen by a Dirichlet process with two parameters G_0 and M which, in turn, may or may not have a further hierarchical structure. G_0 is the prior guess of G, or the *belief* of the *observer* about its knowledge of the unknown measure G, while M represents the strength of the belief. In modal logic or in a first order logic with knowledge and belief ([Scherl and Levesque 1993]) this would amount to express properties of knowledge and beliefs, such as S4 or KD5 etc. (see for example [Goldblatt 2003]).

In fact, we can note that the countable number of parameters $(\theta_1, \theta_2, \ldots)$ makes the need to establish the number of states, or regimes, irrelevant. This fact is rather important, because the way the robot observes the world is made dependent on its beliefs towards G and its prior guess at G_0.

Indeed, given a probability space and the set of distributions over it, a Dirichlet process is a way of defining prior distributions on this set. In some sense these concepts are close to what Levesque developed with a multi tree semantics for situations for the knowledge underlying the operators added to the successor state axioms [Scherl and Levesque 1993; Levesque, Pirri, and Reiter 1998], to interpret an agent abilities.

The Dirichlet processes, as class of priors are defined by Ferguson [Ferguson 1973] in several ways, one of which is as follows. Let α be a finite measure on a measurable space $(\mathbb{R}, \mathcal{A})$ then $D(\alpha)$ is a Dirichlet process with parameter α if, for any partition $A_1, \ldots, A_n$ of $\mathbb{R}$, $(G(A_1), \ldots, G(A_n))$ has a Dirichlet distribution with parameters $(\alpha(A_1), \ldots, \alpha(A_n))$.

Another definition has been provided by Sethuraman and Tiwari [Sethuraman and Tiwari 1981] as follows: let $q_1, q_2, \ldots$, be i.i.d. with a beta distribution, $Beta(M, l)$, $M > 0$; let $\theta_1, \theta_2, \ldots$ be i.i.d. G_0, and let the $\{\theta_i\}$ and $\{q_i\}$ be independent. Define

$$p_1 = (1 - q_1)$$
$$p_j = (\prod_{i=1}^{j-1} q_i)(1 - q_j), \quad j > 1 \tag{25.2}$$

Otherwise called the *stick breaking process*. Then,

$$G(\theta) = \sum p_j \delta_{\theta_j}(\theta) \tag{25.3}$$

is a Dirichlet process $D(\alpha)$ with parameters $\alpha = MG_0$, with G_0 the prior guess of G. Here $\delta_{x_j}(x)$ represents the distribution giving mass one to the point $x_j = x$.

This representation is equivalent to those given by Ferguson in [Ferguson 1973] but simpler, see also Blackwell and McQueen [Blackwell and MacQueen 1973].

According to the above representation, $f(x)$ can be expressed as the convolution of the known kernel $h(x_i|\theta_i)$ and the unknown G, namely $f(x) = \int h(x|\theta)dG(\theta)$. The priors distribution of the parameters can be defined as follows (see [Ferguson 1983]):

1. $p_1, p_2, \ldots$ and $\theta_1, \theta_2, \ldots$ are independent.

2. $p_1, p_2, \ldots$ are defined as in (eq. 25.2), where the q_i have common distribution the Cdf Beta, with parameters M, and 1, hence the pdf is:

$$b_{q_i}(x|M, 1) = M(1 - x)^{M-1}, \quad 0 < x < 1 \tag{25.4}$$

The p_is, tell the contribution of the i-th normal pdf $h_i(x|\theta_i)$ to $f(x)$. In fact, for an M small there will be most of the p_i close to zero and the first, according to the stick breaking process, will have higher values. On the other hand an M with a large value will flat the p_i to a small contribution each.

3. Let $\theta_i = (\mu_i, \sigma_i^2)$, with $((\mu_1, \sigma_1), (\mu_2, \sigma_2), \ldots)$ i.i.d. with common distribution G_0. Choosing a conjugate prior for h, G_0 is the inverse normal gamma distribution, with parameters $\alpha, \beta, \tau, \lambda$:

$$\frac{\sqrt{\tau}}{\sigma \sqrt{2\pi}} \frac{\beta^\alpha}{\Gamma(\alpha)} \left(\frac{1}{\sigma^2}\right)^{\alpha+1} \exp\left(-\frac{2\beta + \tau(\mu - \lambda)^2}{2\sigma^2}\right) \tag{25.5}$$

The following are known: $E(\mu_i) = \mu$, $Var(\mu_i) = E(\sigma_i^2)/\tau$, $E(\sigma_i^2) = \beta/(\alpha - 1)$. As argued above, we want to estimate the θ_i given the known distribution for the $h(x|\theta_i)$, the observations, $G_0(\theta)$, and the unknown distribution G, chosen from a Dirichlet process with parameter α. The Bayes estimator of θ_i is

$$\hat{\theta}_i = E(\theta_i|X_1, \ldots, X_n) \tag{25.6}$$

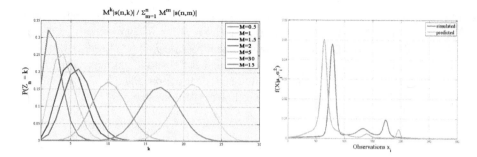

Figure 25.2. Left image: M varying. Right image: in red the simulated mixture, from which 30 observations are sampled, in green the mixtures obtained from the estimated parameters, the mixing values are obtained as in (eq. 25.2), here $M = 3$ and the simulated mixtures is generated with 3 components.

Note that G is the prior of the θ_i, and $G \sim D(\alpha)$ with $\alpha = MG_0$. Ferguson [Ferguson 1973] has shown that, if $G \sim D(\alpha)$ and $\theta_1, \ldots, \theta_n$ are samples from G, then the Bayes estimate $\hat{G}_n = (M/(M + n))G_0 + (1 - M/(M + n))(1/n) \sum \delta_{\theta_i}(\theta_i)$. According to Antoniak [Antoniak 1974], the posterior distribution of G, given $X_1, \ldots, X_n$ is a mixture of Dirichlet processes with parameters $\alpha + \sum \delta_{\theta_j}(\theta)$ and mixing distribution $H(\theta_1, \ldots, \theta_n | X_1, \ldots, X_n)$:
$G | X_1, \ldots, X_n \sim \int D(\alpha + \sum \delta_{\theta_j}(\theta)) dH(\theta_1, \ldots, \theta_n | X_1, \ldots, X_n)$. The expectation of $D(\alpha + \sum \delta_{\theta_j}(\theta))$ is $\hat{G}_n$ and the marginal $H(\theta_1, \ldots, \theta_n)$ of the distribution of the θ_i, gives:

$$dH(\theta_1, \ldots, \theta_n) = \left[\prod_{i=1}^{n} (MG_0 + \sum \delta_{\theta_j}(\theta_i))(d\theta_i) \right] / M^{(n)} \tag{25.7}$$

Here $M^{(n)} = M(M + 1) \cdots (M + n - 1)$. But $H(\theta_1, \ldots, \theta_n | X_1, \ldots, X_n)$ is the posterior distribution of $(\theta_1, \ldots, \theta_n)$, given $(X_1, \ldots, X_n)$. Lo, in [Lo 1978] has found a formula for this posterior distribution, from which:

$$dH(\theta_1, \ldots, \theta_n | X_1, \ldots, X_n) = \frac{\prod_{i=1}^{n} f(X_i | \theta_i) dH(\theta_1, \ldots, \theta_n)}{\int \cdots \int \prod_{i=1}^{n} f(X_i | \theta_i) dH(\theta_1, \ldots, \theta_n)} \tag{25.8}$$

Using (25.7) and (25.8), and considering for each θ_i the pdf h, the Bayes estimator for θ_i is:

$$\begin{aligned} \hat{\theta}_i &= E(\theta_i | X_1, \ldots, X_n) = \int \cdots \int \theta_i dH(\theta_1, \ldots, \theta_n | X_1, \ldots, X_n) \\ &= \frac{\int \cdots \int \left(\theta_i \prod_{i=1}^{n} h(x_i | \theta_i) \right) \prod_{i=1}^{n} (MG_0 + \sum \delta_{\theta_j}(\theta_i))(d\theta_i)}{\int \cdots \int \prod_{i=1}^{n} h(x_i | \theta_i) \prod_{i=1}^{n} (MG_0 + \sum \delta_{\theta_j}(\theta_i))(d\theta_i)} \end{aligned} \tag{25.9}$$

The numerator in equation (25.7), can be expanded yielding $n!$ terms, many of which are equal.

Some considerations on the repetition of identical parameters are in order and concern the magnitude of M. According to Ferguson [Ferguson 1973] and Antoniak findings [Antoniak 1974], the repetition of the θ_i is due to a *virtual memory* [Antoniak 1974] of the process, namely the characterization of the posterior parameter $\alpha + \sum_{\theta_j}(\theta)$, which is what

404

Blackwell and McQueen highlighted in [Blackwell and MacQueen 1973] with the Polya urn scheme. In particular, Antoniak shows that this depends only on the dimension of M:

$$P(\theta_1 = \theta_2 = \cdots = \theta_n) = 1^{(n-1)}/\{(M+1)^{(n-1)}\}$$
where $x^{(n)} = x(x+1)\cdots(x+n-1)$
and, similarly:
$$P(\theta_k \notin \{\theta_1, \ldots, \theta_{k-1}\}) = M/(M+k-1)$$

(25.10)

Therefore the rate at which new distinct values appear depends only on the magnitude of M and not on the shape of $\alpha = MG_0$. Letting Z_n be the number of different θ_i in the first n observations, Antoniak [Antoniak 1974] obtains:

$$P(Z_n = k) = (|s(n,k)|M^k)/M^{(n)})$$

(25.11)

Here $|s(n,k)|$ is the absolute value of the Stirling number of the first kind, and $P(Z_n = k)$ is the probability that, given n values, k of them are different.

To get rid of the Stirling numbers we take the asymptotic behavior according to [Abramowitz and Stegun 1964], pg. 824

$$|s(n,k)| \sim (n-1)!(\gamma + \log n)^{(k-1)}/(k-1)! \text{ for } k = o(\log(n))$$

(25.12)

Here $\gamma \sim 0.577215665$ is the Euler constant. Also, as $\sum_{i=1}^n |s(n,k)|M^k = \Gamma(M+n)/\Gamma(M)$, for $k = o(\log(n))$:

$$P(Z_n = k) \sim \left((n-1)!(\gamma + \log n)^{(k-1)}/(k-1)!\right) M^k \Gamma(M)/\Gamma(M+n)$$

(25.13)

However as Antoniak shows, $E(Z_n) \sim M(\log((M+n)/M))$, therefore new parameters will come out in the limit, although the more and more rarely.

Thus with this knowledge we can consider the computation of the $\hat{\theta}_k$ following the ideas of Kuo [Kuo 1986].

Given the observations $x_1, \ldots x_n$ these are partitioned in equivalence classes, so that if x_i and x_j are in the same set in a partition Q_h, then the parameters θ_i and θ_j are the same, hence $\delta_{\theta_j}(\theta_i)$ shall sum over all the θ_i of the set. The partitions are generated using (25.10).

More specifically, given the observations $x_1, \ldots, x_n$, for $i = 1, \ldots, n$ the x_{i+1} initiates a new set with probability $M/(M+i)$ otherwise x_{i+1} is placed in an already existing set with probability $w/(M+i)$, where w is the number of elements in the set. At every cycle u, of the Monte Carlo technique, a partition is created, the observations are randomly permuted and, among the most numerous sets, the choice of an already existing sets favors the one that minimizes the square difference, that is, $\sum_{x \in K}(x - x_{i+1})^2$. In fact, given the above limit (eq. 25.13), if M is fixated, then for each partition Q the number of the sets in Q should be such that $P(|Q| = k) = P(Z_n = k)$.

For each partition, together with each set K_j, $j = 1, \ldots m$, of observations, also the partition of indices $\{1, \ldots, n\}$ is recorded, because they play the role of the δ_{θ_j}.

Once the partition is computed then the following integrals are evaluated, exploiting the conjugacy of G_0. Let K be a set with index k:

$$W(u,k) = \int \theta \prod_{x_i \in K} h(x_i|\theta)G_0(d\theta) = \frac{2^{\alpha-1}\beta^\alpha \left(\frac{1}{Z}\right)^{\frac{|K|+1}{2}+\alpha} \Gamma\left(\frac{|K|+1}{2}+\alpha\right)}{\pi^{\frac{|K|}{2}}\Gamma(\frac{1}{2})\Gamma(\alpha)}$$

where $|K|$ is the cardinality of K, and
$$Z = 2\beta + \lambda^2 - \tau + \mu^2(|K| + \tau) + \sum_{i \in K} x_i - 2\mu(\lambda\tau + \sum_{i \in K} x_i^2)$$

(25.14)

$$\lambda = \frac{1}{|K|}\sum_{i \in K} x_i, \mu \sim N(\lambda, \sigma^2/\tau), \alpha, \beta, \tau > 0$$

1000 iterations IS MC							
	M=1, #S		M=2, #S		M=5, #S		M=30, #S
$P(Z_{30} = 1)$	0.0333	-	0.0022	-	3.59e-006	-	~ 0	-
$P(Z_{30} = 3)$	0.2347	2	0.0606	3	6.33e-004	3	~ 0	[5, 11]
$P(Z_{30} = 5)$	0.1861	2	0.1921	3	0.0125	[3, 9]	~ 0	[6, 12]
$P(Z_{30} = 8)$	0.0139	[3, 4]	0.1144	3	0.1167	[3, 9]	1.54e-007	[7, 14]
$P(Z_{30} = 21)$	~ 0	[3, 8]	~ 0	[4, 8]	9.63e-006	[4, 9]	0.1657	[7, 15]

Table 25.1. The table illustrates $P(Z_{30} = k)$, the probability of the number of different parameters given 30 observations, according to the value of M. It also shows the number of sets in each of the partitions generated in 1000 iterations of the importance sampling algorithm. Observations are obtained by sampling from mixtures of normal distributions with size k, with $k = 1, 3, 5, 8, 21$.

$$Y(u, Q) = \int \prod_{x_i \in K} h(x_i|\theta)G_0(d\theta) = \frac{2^{\alpha - \frac{1}{2}}\beta^\alpha \left(\frac{1}{Z}\right)^{\frac{|K|+2}{2} + \alpha} \Gamma\left(\frac{|K|+2}{2} + \alpha\right)}{\pi^{\frac{|K|}{2}}\Gamma(\frac{1}{2})\Gamma(\alpha)}$$ (25.15)

where $|K|$, Z, λ and μ are defined as above

Then at any cycle u, given a partition Q, the following are computed:

$$U(u, k, Q) = W(u, k) \prod_{\substack{k' \in Q \\ k' \neq k}} Y(u, Q)$$
$$V(u, Q) = \prod_{k \in Q} Y(u, Q)$$ (25.16)

Finally

$$\theta_k = \frac{\sum_u^N U(u, k, Q)/N}{\sum_u^n V(u, Q)/N}$$ (25.17)

In Figure 25.2 the behavior of $P(Z_n = k)$, on n observations, making k varying between 1 and n and M varying from 0.5 to 30, is illustrated. We can see that with $M = 30$, although the probability is less peaked, k has a non zero probability to be > 25.

These considerations show that M depends only on the number of observations and on the belief, by the computation machine, on the knowledge needed to represent them. In other words, M is the detail level at which the action function should be defined in order for the agent to be able to know what the action is representing. In fact, we can see that, by construction, the number of non empty sets in each partition is related to $P(Z_n = k)$, namely on the size of M, hence the mean of the occupied sets of the partition, generated at each iteration, gives the value k from which M can be effectively evaluated.

Some results are shown in Figure 25.2 and in Table 25.1. The results shown are obtained as follows. We generate a mixture with varying numbers of components. From this we sample H values. We use these H values as observations for the Bayesian estimator of the θs. Further quantitative details shall not be given here.

3 To belief control and probabilistic state machines

From the previous section we have seen that the flexibility of Dirichlet processes allowed us to define an action as a function with n states (or parameters), where n depends from the

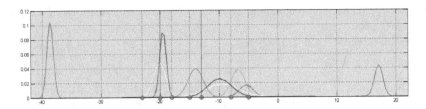

Figure 25.3. The Figure illustrates observed point actions on the real line and $f(x)$. Here we have highlighted the components of $f(x)$ with different colors.

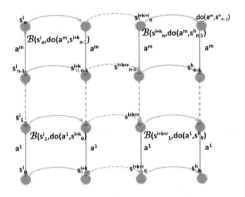

Figure 25.4. The structure of the belief fluent, and all its related situations according to Levesque in [Shapiro, Pagnucco, Lespérance, and Levesque 2000].

observations. For each observation x_i, the parameters θ_i can be computed, via the Bayes estimator. Although several values will be the same, since $P(Z_n = k)$ tends to 0 even for M very large, however there will always be new parameters as the number of observations tend to infinity, therefore the more the observations the more well-founded will be the defined action.

Now, if the parameters are the states, a knowledge state at the representation level can be defined only when the action is learned and thus is made distinguishable from any other action. Assume that actions a and a' are given, and the problem is to establish if they are the same or not. If a and a', namely the observations constituting them concern the same action, what kind of variability in the parameters we expect? An interesting inference would be to determine if the same set of parameters can be adapted to different sequences and what happens when the order of presentation of an observation is changed.

In principle we could say that a and a' are the same action if they have the same defining function, hence the same parameters. Let φ be a predicate defining the condition at which two actions can be considered the same, the belief of φ, in Levesque axiomatization [Shapiro, Pagnucco, Lespérance, and Levesque 2000], is:

$$Bel(\varphi, s) \stackrel{def}{=} \forall s'\, [B(s', s) \wedge (\forall s''.B(s'', s) \supset pl(s') \leq pl(s''))] \supset \varphi[s'] \qquad (25.18)$$

Now, we have different choices for the definition of φ. In comparing two sequences we can either compare each point action, or we can compare the plausibility of the state with respect to the action.

More specifically, consider the belief axiom:

$$B(s'', do(a, s)) \equiv \exists s'(B(s', s) \wedge s'' = do(a, s') \wedge S f(a, s) = S f(a, s')) \qquad (25.19)$$

and let us consider the point action observations x_i as actions, so that x_i is x_a. Now, two point actions can be the same yet the two actions they are part of could be different, or they could be different yet the actions be the same. For example in the cooking pasta action the action does not change if the salt is put in the water before or after throwing the pasta in it.

However, as we can see from Figure 3, if we had a sequence a_1, a_2, a_3 and a sequence a_3, a_1, a_2 then we would obtain $s'' = do(a_3, s')$ and $a_2 \neq a_3$ hence, although the action might be the same, because of the rearrangement of the sequence, the robot would not believe in the similarity between the two actions. On the other hand, since the observation of the point action x_a is a sensing action, we could specify when two sensing actions, with different arguments, are equal, such as, for example:

$$sense(x_a) = sense(x_b) \equiv \exists \theta. d_M(x_a, x_b | \theta) < \epsilon \wedge f(x_a | \theta) f(x_b | \theta) > 0. \qquad (25.20)$$

Here $d_M(x_a, x_b | \theta)$ is the Mahalanobis distance given that $\theta = (\mu, \sigma^2)$ and:

$$d_M(x_a, x_b | \theta) = \left[\left(\frac{(x_a - \mu)^2}{\sigma^2} \right) + \left(\frac{(x_b - \mu)^2}{\sigma^2} \right) \right]^{1/2} \qquad (25.21)$$

In this case the observations can be exchanged freely and what counts is the evaluation of their distance with respect to a common mode, of one of the components of the countable mixture. In fact, if x_a is a point action of action a and if x_b is a point action of action b then, if a and b come from the same distribution, and $f_a = f_b$ models them, then $f_a(x_a) f_b(x_b)$, shall not be zero and their distance from the mode will be sufficiently small. Thus the situations related by beliefs are all those situations in which the sequence of observations can be similarly evaluated by the mixture, at each point action. We need now to give a suitable definition of plausibility of a situation. This can be done using the Kullback-Leibler divergence measure, defined as $KL(f_a \| f_b) = \int f_a(x) \log \left(\frac{f_a(x)}{f_b(x)} \right) dx$, where $KL(f_a \| f_b) = 0$ iff $f_a = f_b$. This measure will tell us if the two probability measures (or the two random variables) are similar, however since this measure is not a distance being not symmetric, that is $KL(f_a \| f_b) \neq KL(f_b \| f_a)$, and since the belief fluent is symmetric, we shall take, the symmetric version, which is the Jensen-Shannon divergence, namely:

$$KL_S (f_a \| f_b) = \frac{1}{2} KL(f_a \| f_{a,b}) + \frac{1}{2} KL(f_b \| f_{a,b}) \qquad (25.22)$$

Here $f_{a,b} = \frac{1}{2}(f_a + f_b)$. We can now define first the plausibility term and then the fluent φ. Then:

$$pl(f_a, f_b, s) = p \quad \equiv \quad \begin{aligned} & KL_S (f_a(x_i) \| f_b(x_i)) \leq \epsilon \wedge p = 1 \vee \\ & KL_S (f_a(x_i) \| f_b(x_i)) > \epsilon \wedge p = 0 \end{aligned} \qquad (25.23)$$

We should note, here, that in (eq. 25.23) x_i can be any, since there is no need to compare the pdfs exactly at the observations as in (eq. 25.20). Here the belief is completely independent of the sequence of observations as it is comparing the two functions. Therefore the successor state axiom for plausibility, namely:

$$pl(f_a, f_b, do(a, s)) = pl(f_a, f_b, s) \qquad (25.24)$$

is satisfied. In fact, if two functions are similar, under the Jensen-Shannon divergence, they will remain similar, no matter the observations made, specifically concerning the functions. We are, now, free to define the fluent $\varphi(s)$. We want to express a condition of similarity between the actions using both the conditions, namely: the condition of equality between sensing actions, to ensure that the actions observed are step by step similar, and the condition induced by the plausibility, that concerns the actions as general functions (for which we have indeed chosen a non parametric model). Therefore, φ shall be defined as follows:

$$
\begin{aligned}
\varphi(a, b, s) \;\stackrel{def}{=}\; & a = [x_a^1, \ldots x_a^n] \wedge b = [x_b^1, \ldots, x_b^n] \wedge \\
& \exists s' \exists s''. s = (do(sense(x_a^n), \ldots, do(sense(x_a^1), \ldots), s'))) \wedge \\
& s'' = (do(sense(x_b^n), \ldots, do(sense(x_b^1), \ldots), s_b))) \wedge \\
& sense(x_a^1) = sense(x_b^1) \wedge \ldots \wedge sense(x_a^n) = sense(x_b^n) \wedge \\
& B(s', s_b)
\end{aligned}
\tag{25.25}
$$

Here $B(s', s_b)$ is the belief fluent stating that the situations at which the actions initiated where belief related. This fact is necessary to cope with the lattice shown in Figure 25.3 graphically representing axiom (25.19). We can see that while the plausibility can be verified in all situations, φ requires the belief to evaluate the point actions. Suppose that $pl(s)$ is verified in all s because the two functions are similar, but φ is not. This is possible as the definition is not parametric and, although at the modes the two functions might be similar, inducing a good Jensen-Shannon measure, the point actions observed might be quite distant. For example, the action of cooking while the ingredients are collected, and the action of cooking while the sauce is prepared. Then it is fully rationale that the robot does not believe that the two actions are the same. On the other hand two short sequences might be similar, yet their Jensen-Shannon divergence might be high. This is the advantage of having two constraints in the belief macro, that fully capture the flexibility of a non parametric definition of an action.

The last option is to build from the function $f(x)$ a probabilistic state machine PFA and make the fluent φ take as argument two state machines and verify the relation between them. A probabilistic automaton is defined by $\langle \Sigma, W, \delta, \iota \rangle$ where Σ is the alphabet, W is the set of states and $\delta : w \times \sigma \times W \mapsto [0, 1]$. We can see that, to cope with the definition, we need both a transition distribution and a function mapping the observations to the alphabet. This last problem requires an observation $x_1, \ldots, x_n$ to be denoted first as an n-uple, to keep the ordering, and eventually to a string.

To represent a distribution on states, we shall associate a state w_i to each p_i, whose meaning is the contribution of state w_i to the function $f(x)$. We define the transition function $\delta(w_i, a, w_j) : W \mapsto [0, 1]$, and an initial probability distribution on the states $\iota(w) \mapsto [0, 1]$:

$$
\begin{aligned}
& \sum_{w \in W} (\iota(w)) = 1, \quad \forall w \in W \\
& \delta(w, [\,], w') = \begin{cases} 1 & \text{if } w = w' \\ 0 & \text{otherwise} \end{cases} \\
& \delta(w, [a, b], w') = \sum_{w'' \in W} \delta(w, a, w'') \delta(w'', b, w')
\end{aligned}
\tag{25.26}
$$

The distribution over the automaton is, thus:

$$
P_A(a) = \sum_{w, w' \in W} \iota(q) \delta(q, a, q') \prod_{x < a} f(x_i)
\tag{25.27}
$$

Now, to build a belief concerning two actions a and b, according to Levesque's model, and in particular the axioms given in [Shapiro, Pagnucco, Lespérance, and Levesque 2000], we shall add the above conditions, specified in (eq. 25.26) as axioms, by suitably extending the signature of the language used, so for example we need to extend $\iota(w)$ to $\iota(w,s)$. Furthermore we need to assess the following:

$$\forall s.Init(s) \supset \exists w.\iota(w,s) \leq 1 \tag{25.28}$$

And for the plausibility functional fluent we shall define, similarly as in (eq. 25.23):

$$pl(a,b,s) = p \quad \equiv KL_s(P_A(a)\|P_B(b)) \leq \epsilon \wedge p = 1 \vee KL_s(P_A(a)\|P_B(b)) > \epsilon \wedge p = 0. \tag{25.}$$

Now, for a φ that can state the property that two actions are similar, and that can be believed or not by the robot, we shall take into account also the transitions, and define it as follows:

$$
\begin{aligned}
\varphi(a,b,s) \stackrel{def}{=} \; & a = (x_a^1 \ldots x_a^n) \wedge b = (x_b^1, \ldots, x_b^n) \wedge \\
& \exists s' \exists s''.s = (do(sense(x_a^n)), \ldots, do(sense(x_a^1, \ldots)), s'))) \wedge \\
& s'' = (do(sense(x_b^n)), \ldots, do(sense(x_b^1, \ldots)), s_b))) \wedge \\
& sense(x_a^1) = sense(x_b^1) \wedge \ldots \wedge sense(x_a^n) = sense(x_b^n) \wedge \\
& \exists w_A \exists w_B.\|\iota(w_A, s) - \iota(w_B, s)\| \leq \epsilon \wedge \\
& \exists w'_A \exists w'_B.\|\delta(w_A, a, w'_A) - \delta(w_B, b, w'_B)\| \leq \epsilon
\end{aligned}
\tag{25.30}
$$

In this way we shall compare the whole transition via φ and the evaluation of the probabilities via the plausibility function. We can see that if the two automata have the same model then the cognitive robot shall believe that the two actions are the same. Indeed, if the observed action is believed to be the same as one already estimated, then the cognitive robot state of knowledge is that the observed action is known.

4 Background on inference about perceived actions

Inference about perceived actions brings together several research issues, mainly related to cognitive vision and cognitive robotics, but it also involves other areas of computer vision, image processing and robotics. The most well known paradigm concerns "action recognition". Nevertheless, a common understanding of the concept of action as acquired by observations has not been yet emerged. A clear provision of the difference between actions and gestures, or on the specification of a recognizable action, in terms of representation, model of computation and class of learnable functions, is still lacking. For reviews and discussions on action recognition, see [Moeslund, Hilton, and Krüger 2006; Poppe 2010; Krger, Kragic, and Geib 2007]. From our point of view, we are interested in the problem of obtaining a model for an action, so as to represent it and being able to both reproduce it, by executing it, and to interpret and understand other agents activities and their effects, to trigger a loop of actions as actions effects and causes. Under these desiderata the problem can be decomposed into three main subproblems.

The first subproblem is related to modeling the dynamic system specifying the action. This implies extraction, from a video, of the features necessary to identify the states of an action, in terms of velocity, trajectories, time specification, and body parts involved. Of course each of these aspects can be faced and solved in several different ways, for example in 2D or 3D, hence using or not view points, using different methodologies for tracking and extracting optical flow, or for capturing the limb dynamics of motion. Already for the generality of this subproblem there are two different main views, namely those based on modeling the human body, such as [Davis and Sharma 2004], and those preferring a direct pattern recognition method based on 2D data features. A pioneering work of this approach is [Efros, Berg, Mori, and Malik 2003].

The second subproblem is related to the definition and representation of the primitives, as multivariate stochastic variables, and the space they reside in [Bobick 1997]. This is, in general, related to the problem of the representation model. Action denotations can be specified as continuous functions or as discrete terms, compositionally, for example defining primitives, or not. This largely depends on the descriptors used, such as, for example, based on motion energy [Bobick and Davis 2001], on silhouette ([Li, Zhang, and Liu 2008] or on local features tracking [Laptev, Caputo, Schüldt, and Lindeberg 2007].

The last subropoblem is classification, here three different approaches can be distinguished: direct classification (see e.g. [Bobick and Davis 2001]), based on distance methods, and explicitly avoiding models of actions; discriminative classification (see e.g. [Jhuang, Serre, Wolf, and Poggio 2007]), modeling the unobserved states via the observed features but without introducing action models; finally, generative models, pioneered by [Newtson, Gretchen, and Joyce 1977], use observations to directly model the action states. Generative models are mainly exploited by the learning by imitation community, see [Schaal, Ijspeert, and Billard 2003]. The most used are the Hidden Markov Models (HMM). After the seminal work by Yamato et al. [Yamato, Ohya, and Ishii 1992] HMMs have been very successful in the field of gesture recognition. Feng and Perona [Feng and Perona 2002], without explicitly resorting to time, model an HMM with key-poses corresponding to states. Ahmad and Lee [Ahmad and Lee 2008] take into account multiple viewpoints and use a multi-dimensional HMM to deal with different observations. Also Ikizler and Forsyth [İkizler and Forsyth 2008] construct HMMs for legs and arms individually, where limbs 3D trajectories are taken to be the observations. Similar approaches are based on stochastic regular grammars and probabilistic finite automata, see for example [Cho, Cho, and Um 2004] and [Bobick and Ivanov 1998].

In both [Pirri 2011] and [Fanello, Gori, and Pirri 2010] we have illustrated the whole path from early attention to to the model of action, via a PFA in [Fanello, Gori, and Pirri 2010] and up to a theory of actions, passing through HMMs and a probabilistic logical model in [Pirri 2011]. Indeed, in [Fanello, Gori, and Pirri 2010] we considered attention to and focus towards the human motion, as distinct from non-human motion, either natural or mechanical. This aspect, in particular, resorts to the theory of motion coherence and structured motion ([Adelson and Bergen 1985],[Wildes and Bergen 2000]), for which oriented filters have been proven to be appropriate [Adelson and Movshon 1983]. We showed how a bank of 3D Gabor filters can be tuned to respond selectively to some specific human motion. Thus, focus on regions of the scene interested by human motion provides the robot with a natural segmentation of where to look at for learning behaviors. In particular, attention to distinct human motion seems to be explicitly dependent on scale, frequency, direction, but not on shape. This fact suggested us to define descriptors based only on these features, from which we extracted principally the directions of the arms and hands movements (see also [Braddick, OBrien, Wattam-Bell, Atkinson, and Turner 2000]).

The very simple structure of the descriptors enabled a straightforward classification that includes all direction dependent primitives, such as *up*, *tilt*, *release*, *grasp* and so on. The basic classification can been extended to any legal sequence of actions for which a deterministic accepting automata exists. In this sense, according to the classification on human motion analysis, as provided by Moeslund et al. [Moeslund, Hilton, and Krüger 2006], our approach falls into the category of *action primitives and grammars*, as no explicit reference to human model is used in the behavior modeling. Indeed, the early attentional phase makes implicit reference to the human kinematics, or better to the smoothness and coherence of human motion for which receptive fields are tuned to (analogously to face recognition), in so showing how human models can be subsumed by the motion model. On the other hand, although no mention is made to body parts, the selection of just arms and hand behaviors introduces an apparent implicit human model. Still, the receptive fields mechanism of selection could be extended to distinguish a gait from a grasping, as well.

The purpose of our work encompasses the classification and regression problem (see [Poppe 2010]). The purpose made explicit in [Fanello, Gori, and Pirri 2010; Pirri 2011] has been to enable robot action learning, by learning the primitives and their structured progression. This can be considered as a form of imitation learning (we refer the reader to the review of [Krger, Kragic, and Geib 2007]) although an important generalization inference has been done in the construction of the accepting automaton on one side and on a theory of action, via a probabilistic model, on the other.

Our work, in both [Pirri 2011] and [Fanello, Gori, and Pirri 2010] has been tested and several examples are given. Here we slightly modify the approach and, assuming that primitives of actions, in the form of univariate distributions are given, we exploit a Bayesian approach to estimate the number of components of a HMM and then show how to map this into a probabilistic finite state machine, which is the model of the specific action.

5 Conclusions

In this paper we have considered Levesque's approach to modeling the belief that a cognitive robot has about observed actions. Modeling observations of actions is a basic step for any further inference about actions, when the inference concerns what a robot can do or how it can perform a certain task. This requires the robot to be able, for example, to distinguish two observed actions, even when the two motion sequences are not identical. To treat sequences of point actions, we have interpreted actions as non parametric densities, defined by a countable mixture of normal densities, which can thus approximate any function. We use the method introduced by Ferguson [Ferguson 1973] to define the empirical Bayes problem of estimating their parameters. Finally, on the basis of the axiomatization provided by Levesque et al. in [Shapiro, Pagnucco, Lespérance, and Levesque 2000], we have shown how beliefs can be used to compare two actions, as defined via countable mixtures.

6 Acknowledgments

This paper is dedicated with fond friendship to Hector Levesque, to his deep insight into the role of knowledge and beliefs in logics, to his extraordinary contribution to the AI community and for having given birth, having nurtured and made grow up into a strong community, all of us: the cognitive roboticists.

References

Abramowitz, M. and I. A. Stegun (1964). *Handbook of Mathematical Functions with Formulas, Graphs, and Mathematical Tables*. New York: Dover Publications.

Adelson, E. H. and J. R. Bergen (1985). Spatiotemporal energy models for the perception of motion. *J. of the Optical Society of America A 2*(2), 284–299.

Adelson, E. H. and J. A. Movshon (1983). The perception of coherent motion in two-dimensional patterns. In *ACM Workshop on Motion: Representation and Perception*, pp. 11–16.

Ahmad, M. and S. Lee (2008). Human action recognition using shape and clg-motion flow from multi-view image sequences. *Pattern Recogn. 41*(7), 2237–2252.

Alechina, N., M. Dastani, B. Logan, and J.-J. C. Meyer (2007). A logic of agent programs. In *AAAI*, pp. 795–800.

Antoniak, C. (1974). Mixtures of dirichlet processes with applications to bayesian non-parametric problems. *Ann. Statist. 2*, 337–351.

Ascombe, E. (2000). *Intention*. Harvard University Press.

Blackwell, D. and J. MacQueen (1973). Ferguson distributions via polya urn schemes. *Ann. Statist. 1*, 353–355.

Bobick, A. and Y. Ivanov (1998, June). Action recognition using probabilistic parsing. In *CVPR, 1998*, pp. 196 –202.

Bobick, A. F. (1997). Movement, activity, and action: the role of knowledge in the perception of motion. *Philosophical Transactions of the Royal Society of London 352*, 12571265.

Bobick, A. F. and J. W. Davis (2001). The recognition of human movement using temporal templates. *IEEE Trans. on Pattern Analysis and Machine Intelligence 23(3)*, 257–267.

Braddick, O., J. OBrien, J. Wattam-Bell, J. Atkinson, and R. Turner (2000). Form and motion coherence activate independent, but not dorsal/ventral segregated, networks in the human brain. *Current Biology 10*, 731–734.

Bratman, M. (2006). *Structures of Agency*. Oxford University Press.

Cho, K., H. Cho, and K. Um (2004). Human action recognition by inference of stochastic regular grammars. In *Structural, Syntactic, and Statistical Pattern Recognition*, Volume 3138, pp. 388–396.

Cohen, P. R. and H. J. Levesque (1990). Intention is choice with commitment. *Artif. Intell. 42*(2-3), 213–261.

Davis, J. W. and V. Sharma (2004). Robust background-subtraction for person detection in thermal imagery. In *CVPRW '04*, pp. 128.

DeGiacomo, G., Y. Lespérance, and H. J. Levesque (2000). Congolog, a concurrent programming language based on the situation calculus. *Artif. Intell. 121*(1-2), 109–169.

Efros, A., A. Berg, G. Mori, and J. Malik (2003). Recognizing action at a distance. *Internatinal Conference on Computer Vision II*, 726–733.

ETAI (1998). *Electron. Trans. Artif. Intell. 2*, 193–210.

Fanello, S. R. F., I. Gori, and F. Pirri (2010). Arm-hand behaviours modelling: From attention to imitation. In *ISVC (2)*, pp. 616–627.

Feng, X. and P. Perona (2002). Human action recognition by sequence of movelet codewords. *3D Data Processing Visualization and Transmission, International Symposium on 0*, 717.

Ferguson, T. (1973). A bayesian analysis of some nonparametric problems. *Ann. Statist. 1*, 209–230.

Ferguson, T. (1983). Bayesian density estimation by mixtures of normal distributions. *Recent advances in Statist. 1*, 287–302.

Gelfond, M. and V. Lifschitz (1998). Action languages. *Electron. Trans. Artif. Intell. 2*, 193–210.

Goldblatt, R. (2003). Mathematical modal logic: A view of its evolution. *Journal of Applied Logic 1*(5-6), 309–392.

Harel, D., A. R. Meyer, and V. R. Pratt (1977). Computability and completeness in logics of programs (preliminary report). In *STOC*, pp. 261–268.

İkizler, N. and D. A. Forsyth (2008). Searching for complex human activities with no visual examples. *Int. J. Comput. Vision 80*(3), 337–357.

Jhuang, H., T. Serre, L. Wolf, and T. Poggio (2007). A biologically inspired system for action recognition. pp. 1–8.

Krger, V., D. Kragic, and C. Geib (2007). The meaning of action a review on action recognition and mapping. *Advanced Robotics 21*, 1473–1501.

Kuo, L. (1986). Computations of mixtures of dirichlet processes. *SIAM J. Sci. Stat. Comput. 7*, 60–71.

Lakemeyer, G. and H. J. Levesque (1998). Aol: A logic of acting, sensing, knowing, and only knowing. In *KR*, pp. 316–329.

Lakemeyer, G. and H. J. Levesque (2011). A semantic characterization of a useful fragment of the situation calculus with knowledge. *Artif. Intell. 175*(1), 142–164.

Laptev, I., B. Caputo, C. Schüldt, and T. Lindeberg (2007). Local velocity-adapted motion events for spatio-temporal recognition. *Comput. Vis. Image Underst. 108*(3), 207–229.

Lespérance, Y., H. J. Levesque, F. Lin, and R. B. Scherl (2000). Ability and knowing how in the situation calculus. *Studia Logica 66*(1), 165–186.

Levesque, H. J. (1987). All i know: An abridged report. In *AAAI*, pp. 426–431.

Levesque, H. J. (1994). Knowledge, action, and ability in the situation calculus. In *TARK*, pp. 1–4.

Levesque, H. J. (1998). What robots can do. In *KR*, pp. 651.

Levesque, H. J., F. Pirri, and R. Reiter (1998). Foundations for the situation calculus. *Electron. Trans. Artif. Intell. 2*, 159–178.

Levesque, H. J., R. Reiter, Y. Lesperance, F. Lin, and R. B. Scherl (1997). GOLOG: A logic programming langauge for dynamic domains. *Journal of Logic Programming 31*, 59–84.

Li, W., Z. Zhang, and Z. Liu (2008). Expandable data-driven graphical modeling of human actions based on salient postures. *Circuits and Systems for Video Technology, IEEE Transactions on 18*, 1499–1510.

Lifschitz, V. (2006). Actions, causation and logic programming. In *ILP*, pp. 1.

Lin, F. and H. J. Levesque (1998). What robots can do: Robot programs and effective achievability. *Artif. Intell.* *101*(1-2), 201–226.

Lo, A. Y. (1978). On a class of bayesian nonparametric estimates: I. density estimate. *Ann. Statist.* *12*, 351–357.

McCarthy, J. (1963). Situations, actions and causal laws. Technical report, Stanford Art Intelligence Project: Memo 2.

McCarthy, J. (1968). Programs with common sense. In *Semantic Information Processing*, pp. 403–418. MIT Press.

McCarthy, J. (2007). Elephant 2000: a programming language based on speech acts. In *OOPSLA Companion*, pp. 723–724.

McCarthy, J. and P. Hayes (1969). Some philosophical problems from the standpoint of artificial intelligence. *Machine Intelligence 4*, 463–502.

Moeslund, T. B., A. Hilton, and V. Krüger (2006). A survey of advances in vision-based human motion capture and analysis. *Computer Vision and Image Understanding 104*(2-3), 90–126.

Newtson, D., A. E. Gretchen, and B. Joyce (1977). The objective basis of behavior units. *Journal of Personality and Social Psychology 35*(12), 847 – 862.

Pirri, F. (2011). The well-designed logical robot: Learning and experience from observations to the situation calculus. *Artificial Intelligence 175*(1), 378–415.

Pirri, F. and R. Reiter (2000). *Logic-Based Artificial Intelligence*, Chapter Planning with natural actions in the Situation Calculus. Kluwer.

Poppe, R. (2010). A survey on vision-based human action recognition. *Image and Vision Computing 28*, 976–990.

Reiter, R. (1991). *Artificial Intelligence and Mathematical Theory of Computation: Papers in Honor of John McCarthy*, Chapter The frame problem in the situation calculus: a simple solution (sometimes) and a completeness result for goal regression. Academic Press.

Reiter, R. (1996). Natural actions, concurrency and continuous time in the situation calculus. In *KR*, pp. 2–13.

Reiter, R. (2001). *Knowledge in action : logical foundations for specifying and implementing dynamical systems*. MIT Press.

Schaal, S., A. Ijspeert, and A. Billard (2003). Computational approaches to motor learning by imitation. *Philosophical transaction of the Royal Society of London 358*, 537–547.

Scherl, R. B. and H. J. Levesque (1993). The frame problem and knowledge-producing actions. In *AAAI*, pp. 689–695.

Sethuraman, J. and R. Tiwari (1981). Convergence of dirichlet measures and the interpretation of their parameters. Technical Report M583, The Florida State University, Department of Satistics.

Shapiro, S., M. Pagnucco, Y. Lespérance, and H. J. Levesque (2000). Iterated belief change in the situation calculus. In *KR*, pp. 527–538.

Thielscher, M. (2005). FLUX: A logic programming method for reasoning agents. *Theory and Practice of Logic Programming 5*(4–5), 533–565.

Wildes, R. P. and J. R. Bergen (2000). Qualitative spatiotemporal analysis using an oriented energy representation. In *ECCV '00*, pp. 768–784.

Yamato, J., J. Ohya, and K. Ishii (1992). Recognizing human action in time-sequential images using hidden markov model. In *Proceedings CVPR '92.*, pp. 379–385.

26

Approximate Linear Programming for First-Order MDPs

SCOTT SANNER AND CRAIG BOUTILIER

ABSTRACT.   We introduce a new approximate solution technique for first-order Markov decision processes (FOMDPs). Representing the value function linearly w.r.t. a set of first-order basis functions, we compute suitable weights by casting the corresponding optimization as a *first-order linear program* and show how off-the-shelf theorem prover and LP software can be effectively used. This technique allows one to solve FOMDPs independent of a specific domain instantiation; furthermore, it allows one to determine bounds on approximation error that apply *equally to all domain instantiations*. We apply this solution technique to the task of elevator scheduling with a rich feature space and multi-criteria additive reward, and demonstrate that it outperforms a number of intuitive, heuristically-guided policies.

Preface (by Craig Boutilier)

My career as a researcher and academic was launched in a fairly direct fashion by Hector, though this all occurred before he know who I was, and certainly without his knowledge. During my undergraduate days at Halifax, a friend, Tim Lownie, was completing his Master's thesis under Hector's supervision at the University of Toronto. Tim would occasionally send research papers at the leading edge of knowledge representation (KR) our way. While "Reflection and Semantics in LISP" (Brian Smith), "An Overview of the KL-ONE Knowledge Representation System" (Ron Brachman and Jim Schmolze), among other topics, all seemed like heady stuff to a naive undergraduate, one paper that real made an impression on me was Hector's paper "Foundations of a Functional Approach to Knowledge Representation." This paper, more than anything, made the whole knowledge representation enterprise seem to me to be critical to the prospects of eventually creating intelligent systems. From there, it was a relatively short path to graduate studies in knowledge representation at Toronto. Though I eventually ended up working on nonmonotonic reasoning and belief revision with Ray Reiter, Hector not only served on my Ph.D. committee, but his foundational work on "only knowing" played a prominent role in the account of belief revision I described in my thesis.

After my Ph.D. studies, I gradually drifted out of abstract KR and reasoning as a research area to focus more on decision making under uncertainty. But KR remains in my blood, and even in these areas, an emphasis on *natural representations* of knowledge, that support (theoretically or practically) *tractable reasoning and decision making*, lies at the heart of most of my research in these areas, whether they are new representations for probabilistic knowledge (e.g., refinements of Bayesian networks), stochastic systems, or utility and preference functions. Hector's foundational research has of course influenced entire communities in this way, not just me.

The following contribution is a exemplar of this approach. This article, co-authored by my-then Ph.D. student Scott Sanner, originally appeared in UAI-05.[1] It describes a method for solving Markov decision processes that adapts state of the art approximate linear programming techniques to exploit the considerable structure that can be made evident through a first-order logical representation of the (stochastic) transition system and the reward function. This work draws not only on the situation calculus representation of dynamic systems that formed the basis the "Toronto approach" to cognitive robotics (as exemplified by Hector's work with Ray Reiter, Gerhard Lakemeyer, Yves Lesperance, and others); it also exploits a specific situation calculus representation of stochastic actions that was heavily influenced by the Bacchus, Halpern and Levesque [1995] article "Reasoning about Noisy Sensors in the Situation Calculus." Scott continued to develop additional, very interesting algorithms of this type. And first-order and relational Markov decision processes have been a quite a hot topic in recent years, in no small part due to Hector's (perhaps indirect) influence.

1 Introduction

Markov decision processes (MDPs) have become the *de facto* standard model for decision-theoretic planning problems. While classic dynamic programming algorithms for MDPs [Puterman 1994] require explicit state and action enumeration, recent techniques for exploiting propositional structure in factored MDPs [Boutilier, Dean, and Hanks 1999] avoid explicit state and action enumeration, thus making it possible to optimally solve extremely complex problems containing 10^8 states [Hoey, St-Aubin, Hu, and Boutilier 1999]. Approximation techniques for factored MDPs using basis functions have also been successful, yielding approximate solutions to problems with up to 10^{40} states [Guestrin, Koller, Parr, and Venktaraman 2003; Schuurmans and Patrascu 2001].

While such techniques for factored MDPs have proven effective, they cannot generally exploit first-order structure. Yet many realistic planning domains are best represented in first-order terms, exploiting the existence of domain objects, relations over these objects, and the ability to express objectives and action effects using quantification. As a result, a new class of algorithms have been introduced to explicitly handle MDPs with relational (RMDP) and first-order (FOMDP) structure.[2] *Symbolic dynamic programming (SDP)* [Boutilier, Reiter, and Price 2001], *first-order value iteration (FOVIA)* [Hölldobler and Skvortsova 2004], and the *relational Bellman algorithm (ReBel)* [Kersting, Otterlo, and Raedt 2004] are model-based algorithms for solving FOMDPs and RMDPs, using appropriate generalizations of value iteration [Puterman 1994]. *Approximate policy iteration* [Fern, Yoon, and Givan 2003] induces rule-based policies from sampled experience in small-domain instantiations of RMDPs and generalizes these policies to larger domains. In a similar vein, *inductive policy sselection using first-order regression* [Gretton and Thiebaux 2004] uses regression to provide the hypothesis space over which a policy is induced. *Approximate linear programming (for RMDPs)* [Guestrin, Koller, Gearhart, and Kanodia 2003] is an approximation technique using linear program optimization to find a best-fit value function over a number of sampled RMDP domain instantiations.[3]

---

[1] Scott Sanner and Craig Boutilier. Approximate Linear Programming for First-order MDPs. In the *Proceedings of the Twenty-first Conference on Uncertainty in Artificial Intelligence (UAI-05)*, pp.509–517, Edinburgh (2005).

[2] We use the term *relational MDP* to refer models that allow implicit existential quantification, and *first-order MDP* for those with explicit existential and universal quantification.

[3] Because the language in this work allows count aggregators in the transition and reward functions, it is more

In this paper, we provide a novel approach to approximation of FOMDP value functions by representing them as a linear combination of first-order basis functions and using a first-order generalization of approximate linear programming techniques for propositional MDPs [Guestrin, Koller, Gearhart, and Kanodia 2003; Schuurmans and Patrascu 2001] to solve for the required weights. While many FOMDP solution techniques need to instantiate a ground FOMDP w.r.t. a set of domain objects to obtain a propositional version of the FOMDP (usually performing this grounding operation multiple times for many different domain sizes), our approach exploits the power of full first-order reasoning taken by SDP, thus avoiding the exponential blowup that can result from domain instantiation. Since the use of first-order reasoning techniques allows our solution to abstract over all possible ground domain instantiations, the bounds that we derive for approximation error apply *equally to all domain instantiations* (i.e., the bounds are independent of domain size and and thus do not rely on a distribution over domain sizes).

2 Markov Decision Processes

Formally, a finite state and action MDP [Puterman 1994] is a tuple $\langle S, A, T, R \rangle$ where: S is a finite state space; A is a finite set of actions; $T : S \times A \times S \to [0, 1]$ is a transition function, with $T(s, a, \cdot)$ a distribution over S for all $s \in S, a \in A$; and $R : S \times A \to \mathbb{R}$ is a bounded reward function. Our goal is to find a policy π that maximizes the value function, defined using the infinite horizon, discounted reward criterion: $V_\pi(s) = E_\pi[\sum_{t=0}^\infty \gamma^t \cdot r^t | S_0 = s]$, where r^t is a reward obtained at time t, $0 \leq \gamma < 1$ is a discount factor, and S_0 is the initial state.

A stationary policy takes the form $\pi : S \to A$, with $\pi(s)$ denoting the action to be executed in state s. Policy π is optimal if $V_\pi(s) \geq V_{\pi'}(s)$ for all $s \in S$ and all policies π'. The optimal value function V^* is the value of any optimal policy and satisfies the following:

$$V^*(s) = R(s, a) + \max_a \gamma \sum_{t \in S} T(s, a, t) \cdot V^*(t) \tag{1}$$

In the approximate linear programming solution to MDPs [Guestrin, Koller, Parr, and Venktaraman 2003; Schuurmans and Patrascu 2001] on which our work is based, we represent the value function as the weighted sum of k basis functions $B_i(s)$:

$$V(s) = \sum_{i=1}^k w_i B_i(s) \tag{2}$$

Our goal is to find a good setting of weights that approximates the optimal value function. One way of doing this is to cast the optimization problem as a linear program that directly solves for the weights of an L_1-minimizing approximation of the optimal value function:

---

expressive than typical RMDPs.

Variables: $w_1, \ldots, w_k$

Minimize: $\displaystyle\sum_{s \in S} \sum_{i=1}^{k} w_i B_i(s)$

Subject to: $\displaystyle 0 \geq \left(R(s,a) + \gamma \sum_{t \in S} T(s,a,t) \cdot \sum_{i=1}^{k} w_i B_i(t) \right) -$

$$\sum_{i=1}^{k} w_i B_i(s) \; ; \; \forall a \in A, s \in S \qquad (3)$$

While the size of the objective and the number of constraints in this LP is proportional to the number of states (and therefore exponential), recent techniques use compact, factored basis functions and exploit the reward and transition structure of factored MDPs [Guestrin, Koller, Parr, and Venktaraman 2003; Schuurmans and Patrascu 2001], making it possible to avoid generating an exponential number of constraints (and rendering the entire LP compact).

3 First-Order Representation of MDPs

3.1 The Situation Calculus

The situation calculus [McCarthy 1963] is a first-order language for axiomatizing dynamic worlds. Its basic ingredients consist of *actions*, *situations*, and *fluents*. Actions are first-order terms involving action function symbols. For example, the action of moving an elevator e up one floor might be denoted by the action term $up(e)$. A situation is a first-order term denoting the occurrence of a sequence of actions. These are represented using a binary function symbol do: $do(\alpha, s)$ denotes the situation resulting from doing action α in situation s. In an elevator domain, the situation term

$$do(up(E), do(open(E), do(down(E), S_0)))$$

denotes the situation resulting from executing sequence *[down(E),open(E),up(E)]* in S_0. Relations whose truth values vary from state to state are called fluents, and are denoted by predicate symbols whose last argument is a situation term. For example, $EAt(e, 10, s)$ is a relational fluent meaning that elevator e is at floor 10 in situation s.[4]

A domain theory is axiomatized in the situation calculus with four classes of axioms [Reiter 2001]. For this paper, the most important of these axioms are the *successor state axioms (SSAs)*. There is one such axiom for each fluent $F(\vec{x}, s)$, with syntactic form $F(\vec{x}, do(a, s)) \equiv \Phi_F(\vec{x}, a, s)$ where $\Phi_F(\vec{x}, a, s)$ is a formula with free variables among $a, s, \vec{x}$. These characterize the truth values of the fluent F in the next situation $do(a, s)$ in terms of the current situation s, and embody a solution to the frame problem for deterministic actions [Reiter 2001].

---

[4]In contrast to states, situations reflect the entire history of action occurrences. As we will see, the specification of dynamics in the situation calculus will allow us to recover state properties from situation terms, and indeed, the form we adopt for successor state axioms renders the process Markovian.

Following is a a simple example of an SSA for *EAt* (assuming 1 is the ground floor):

$$EAt(e, 1, do(a, s)) \equiv$$
$$EAt(e, 2, s) \wedge a = down(e) \vee EAt(e, 1, s) \wedge a \neq up(e)$$

This states that an elevator e is at the first floor after doing action a in situation s if e was at the second floor in s and the action $down(e)$, or if e was already at the first floor in s and the action was not to move up.

The *regression* of a formula ψ through an action a is a formula ψ' that holds prior to a being performed iff ψ holds after a. Successor state axioms support regression in a natural way. Suppose that fluent F's SSA is $F(\vec{x}, do(a, s)) \equiv \Phi_F(\vec{x}, a, s)$. We inductively define the regression of a formula whose situation arguments all have the form $do(a, s)$ as follows:

$$
\begin{aligned}
Regr(F(\vec{x}, do(a, s))) &= \Phi_F(\vec{x}, a, s) \\
Regr(\neg\psi) &= \neg Regr(\psi) \\
Regr(\psi_1 \wedge \psi_2) &= Regr(\psi_1) \wedge Regr(\psi_2) \\
Regr((\exists x)\psi) &= (\exists x)Regr(\psi)
\end{aligned}
$$

For example, regressing $\neg EAt(e, 1, do(a, s))$ exploits the the SSA for *EAt* to obtain the following description of the conditions on the "pre-action" situation s under which this formula is true:

$$Regr(\exists e \ \neg EAt(e, 1, do(a, s))) =$$
$$\exists e \ \neg[EAt(e, 2, s) \wedge a = down(e) \vee EAt(e, 1, s) \wedge a \neq up(e)]$$

3.2 Stochastic Actions and the Situation Calculus

The classical situation calculus relies on the deterministic form of actions to allow compact and natural representation of dynamics using SSAs as described in the previous section. To generalize this to stochastic actions, as required by FOMDPs, we decompose stochastic actions (those in the control of some agent) into a *collection* of deterministic actions, each corresponding to a possible outcome of the stochastic action. We then specify a distribution according to which "nature" may choose a deterministic action from this set whenever that stochastic action is executed. As a consequence we need formulate SSAs using only these deterministic *nature's choices* [Bacchus, Halpern, and Levesque 1995; Boutilier, Reiter, and Price 2001], obviating the need for a special treatment of stochastic actions in SSAs.

We illustrate this approach with a simple example in an elevator domain consisting of elevators, floors, and people. At each time step, an elevator can move up, down, or can open its doors at the current floor. Suppose the *open* action is stochastic with two possible outcomes, success and failure (we suppose *up* and *down* action are deterministic and always succeed). We decompose *open* into two deterministic choices, *openS* and *openF*, corresponding to success and failure. The probability of each choice is then specified, for example:

$$EAt(e, 1, s) \supset prob(openS(e), open(e), s) = 0.85$$
$$EAt(e, 1, s) \supset prob(openF(e), open(e), s) = 0.15$$
$$EAt(e, f, s) \wedge f \neq 1 \supset prob(openS(e), open(e), s) = 0.9$$
$$EAt(e, f, s) \wedge f \neq 1 \supset prob(openF(e), open(e), s) = 0.1$$
$$prob(upS(e), up(e), s) = 1.0$$
$$prob(downS(e), down(e), s) = 1.0$$

Here we see that the probability of the door failing to open is slightly higher on the first floor (0.15) than on other floors (0.10); this is simply to illustrate the "style" of the representation. Next, we define the SSAs in terms of these deterministic choices. However, in order to formalize the Elevator domain independent of any fixed set of floors, we must first introduce an axiomatization of above and below functions ($fa(x)$ and $fb(x)$ respectively) as well as transitive closure predicates $above(x, y)$ and $below(x, y)$ over these functions:[5]

$$\forall x, y (\exists z \, below(x, z) \wedge below(z, y)) \supset below(x, y)$$
$$\forall x, y (\exists z \, above(x, z) \wedge above(z, y)) \supset above(x, y)$$
$$\forall x, y \, above(x, y) \supset \neg below(x, y)$$
$$\forall x, y \, above(x, y) \supset below(y, x)$$
$$\forall x, y \, below(x, y) \supset \neg above(x, y)$$
$$\forall x, y \, below(x, y) \supset above(y, x)$$
$$\forall x, y \, above(x, y) \supset x \neq y; \quad \forall x, y \, below(x, y) \supset x \neq y$$
$$\forall x, y \, x = y \supset \neg above(x, y); \quad \forall x, y \, x = y \supset \neg below(x, y)$$
$$\forall x \, above(fa(x), x); \quad \forall x \, below(fb(x), x)$$
$$\forall x \, x = fa(fb(x)); \quad \forall x \, x = fb(fa(x))$$

Given these foundational axioms, we can define the SSAs in a straightforward way.[6] For the fluent $EAt(e, f, s)$ (meaning elevator e is at floor f), we have the following SSA:

$$EAt(e, f, do(a, s)) \equiv$$
$$EAt(e, fb(f), s) \wedge a = upS(e) \vee$$
$$EAt(e, fa(f), s) \wedge a = downS(e) \vee$$
$$EAt(e, f, s) \wedge a \neq upS(e) \wedge a \neq downS(e)$$

Intuitively, e is at floor f after doing a only if e (successfully) moved up or down from an suitable floor above or below, or if e was already at f and did not (successfully) move

---

[5] While transitive closure is technically not first-order definable, the axiomatization of $above(x, y)$ and $below(x, y)$ given here suffices to yield a contradiction when an assertion conflicts with an inference derivable from these axioms. Inconsistency detection is all that is needed by our algorithms.

[6] SSAs can often be compiled directly from "effect" axioms that directly specify the effects of actions [Reiter 2001].

up or down. The SSA $PAt(p, f, s)$ (person p is at floor f) is:

$$PAt(p, f, do(a, s)) \equiv$$
$$[\exists e \ EAt(e, f, s) \land OnE(p, e, s) \land$$
$$Dst(p, f) \land a = openS(e))] \lor [PAt(p, f, s) \land$$
$$\neg(\exists e \ EAt(e, f, s) \land \neg Dst(p, f) \land a = openS(e))]$$

Intuitively, person p is at floor f in the situation resulting from doing a iff p has destination f and was on e, which had (stopped and successfully) opened its doors at f; or p was at f already and it's not the case that e opened its doors at f and p was not at their destination. Finally, $OnE(p, e, s)$ (person p is on elevator e) has the following SSA (with obvious meaning):

$$OnE(p, e, do(a, s))$$
$$[\exists f \ EAt(e, f, s) \land PAt(p, f, s) \land$$
$$\neg Dst(p, f) \land a = openS(e)] \lor [OnE(p, e, s) \land$$
$$\neg(\exists f \ EAt(e, f, s) \land Dst(p, f) \land a = openS(e))]$$

3.3 Case Representation

When working with first-order specifications of rewards, probabilities, and values, one must be able to assign values to different portions of state space. For this purpose we use the notation $t = case[\phi_1, t_1; \cdots ; \phi_n, t_n]$ as an abbreviation for the logical formula:

$$\bigvee_{i \leq n} \{\phi_i \land t = t_i\}.$$

Here the ϕ_i are *state formulae* (whose situation term does not use do) and the t_i are terms. We sometimes write this $case[\phi_i, t_i]$. Often the t_i will be constants and the ϕ_i will partition state space. For example, we may represent our reward function as:

$$case[\forall p, f \ PAt(p, f, s) \supset Dst(p, f) \ , \ 10;$$
$$\exists p, f \ PAt(p, f, s) \land \neg Dst(p, f) \ , \ 0]$$

This specifies that we receive a reward of 10 if all people are at their destinations and 0 if not. We introduce the following operators on case statements:

$$case[\phi_i, t_i : i \leq n] \otimes case[\phi_j, v_j : j \leq m] =$$
$$case[\phi_i \land \psi_j, t_i \cdot v_j : i \leq n, j \leq m]$$
$$case[\phi_i, t_i : i \leq n] \oplus case[\phi_j, v_j : j \leq m] =$$
$$case[\phi_i \land \psi_j, t_i + v_j : i \leq n, j \leq m]$$
$$case[\phi_i, t_i : i \leq n] \ominus case[\phi_j, v_j : j \leq m] =$$
$$case[\phi_i \land \psi_j, t_i - v_j : i \leq n, j \leq m]$$

Intuitively, to perform an operation on two case statements, we simply take the cross-product of their partitions and perform the corresponding operation on the resulting paired

Scott Sanner and Craig Boutilier

partitions. Letting each ϕ_i and ψ_j denote generic first-order formulae, we can perform the cross-sum $\oplus$ of two case statements in the following manner:

$$case[\phi_1, 10; \ \phi_2, 20] \oplus case[\psi_1, 1; \ \psi_2, 2] \ =$$
$$case[\phi_1 \wedge \psi_1, 11; \ \phi_1 \wedge \psi_2, 12; \ \phi_2 \wedge \psi_1, 21; \ \phi_2 \wedge \psi_2, \ 22]$$

Note that the conjunction of two first-order formalae in a partition may be inconsistent, implying that that partition formula can never be satisfied and thus discarded.

We can also represent choice probabilities using case notation. Letting $A(\vec{x})$ be a stochastic action with choices $n_1(\vec{x}), \cdots, n_k(\vec{x})$, we define:

$$pCase(n_j(\vec{x}), A(\vec{x}), s) = case[\phi_1^j(\vec{x}, s), p_1^j; \cdots ; \phi_n^j(\vec{x}, s), p_n^j],$$

where the ϕ_i^j partition state space and p_i^j is the probability of choice $n_i(\vec{x})$ being executed under condition $\phi_i^j(\vec{x}, s)$ when the agent executes stochastic action $A(\vec{x})$.

We can likewise use case statements (or sums over case statements) to represent the reward and value functions. We use the notation $rCase(s)$ and $vCase(s)$ to respectively denote these case representations.

3.4 Symbolic Dynamic Programming

Suppose we are given a value function in the form $vCase(s)$. Backing up this value function through an action $A(\vec{x})$ yields a case statement containing the logical description of states that would give rise to $vCase(s)$ after doing action $A(\vec{x})$, as well as the values thus obtained. This is analogous to classical goal regression, the key difference being that action $A(\vec{x})$ is stochastic. In MDP terms, the result of such a backup is a case statement representing a Q-function.

There are in fact two types of backups that we can perform. The first, $B^{A(\vec{x})}$, regresses a value function through an action and produces a case statement with *free variables* for the action parameters. The second, B^A, existentially quantifies over the free variables $\vec{x}$ in $B^{A(\vec{x})}$ and takes the max over the resulting partitions. The application of this operator results in a case description of the regressed value function indicating the best value that could be achieved by *any* instantiation of $A(\vec{x})$ in the pre-action state.

To define the $B^{A(\vec{x})}$ operator, we first define the *first-order decision theoretic regression (FODTR)* operator [Boutilier, Reiter, and Price 2001]:

$$FODTR(vCase(s), A(\vec{x})) \ =$$
$$\gamma \left[\oplus_j \{pCase(n_j(\vec{x}), s) \otimes Regr(vCase(do(n_j(\vec{x}), s))) \} \right]$$

(Unlike the original definition of FODTR [Boutilier, Reiter, and Price 2001], we do not include the sum of $rCase(s)$ in the FODTR definition since it is useful to keep it separate in this paper.) Thus, we can express the $B^{A(\vec{x})}$ operator as follows:

$$B^{A(\vec{x})}(vCase(s)) = rCase(s) \oplus FODTR(vCase(s), A(\vec{x})) \tag{4}$$

To obtain the result of the B^A operator on this same value function, we need only existentially quantify the free variables $\vec{x}$ in $B^{A(\vec{x})}$, which we denote as $\exists \vec{x} \ B^{A(\vec{x})}$. Since the case notation is disjunctively defined, we can distribute the existential quantification into each individual case partition. However, even if the partitions were mutually exclusive prior to the backup operation, the existential quantification can result in overlap between the partitions. Thus, we modify the existential quantification operator to ensure that the highest possible value is achieved in every state by sorting each partition $\langle\phi(\vec{x}), t_i\rangle$ by the t_i values (such that $i < j \supset t_i > t_j$) and ensuring that a partition can be satisfied only when no higher value partition can be satisfied:

$$\exists\vec{x}\ case[\phi_1(\vec{x}), t_1; \cdots ; \phi_n(\vec{x}), t_n] \equiv$$
$$case[\exists\vec{x}\ \phi_i(\vec{x}) \wedge \bigwedge_{j \leq i} \neg\exists\vec{x}\ \phi_j(\vec{x}), v_i : i \leq n]$$

Assuming that the $rCase(s)$ statement does not depend on the parameters of the action (this can be easily relaxed, if needed) and exploiting the fact $\oplus$ is conjunctively defined, we arrive at the definition of the B^A operator:

$$B^A(vCase(s)) = rCase(s) \oplus \exists\vec{x}\ FODTR(vCase(s), A(\vec{x}))$$

4 Approximate Linear Programming for FOMDPs

We now generalize the approximate linear programming techniques from propositional factored MDPs to FOMDPs.

4.1 Value Function Representation

We represent the value function as a weighted sum of k *first-order basis functions*, denoted $bCase_i(s)$, each of which is intended to contain *small* partitions of formulae that provide a first-order abstraction of the state space:

$$vCase(s) = \oplus_{i=1}^k\ w_i \cdot bCase_i(s) \tag{5}$$

As in the propositional case [Guestrin, Koller, Parr, and Venktaraman 2003; Schuurmans and Patrascu 2001], the fact that basis functions are represented compactly using logical formulae to specify partitions of state space means that the value function can be represented very concisely using Eq. 5. Our approach is distinguished by the use of first-order formulae to represent the partition specifications.

We can easily apply our previously defined backup operators $B^{A(\vec{x})}$ and B^A to this representation and obtain some simplification as a result of the structure in Eq. 5. For $B^{A(\vec{x})}$, we substitute value function expression in Eq. 5 into the definition $B^{A(\vec{x})}$ (Eq. 4). Exploiting the fact that $\oplus$ is conjunctively defined and that $Regr(\psi_1 \wedge \psi_2) = Regr(\psi_1) \wedge Regr(\psi_2)$, we push the regression down to the individual basis functions and reorder the summations. As a consequence, we find that the backup of a linear combination of basis functions is simply the linear combination of the FODTR of each basis function:

$$B^{A(\vec{x})}(\oplus_i w_i bCase_i(s)) =$$
$$rCase(s) \oplus (\oplus_i w_i FODTR(bCase_i(s), A(\vec{x}))) \tag{6}$$

By representing the value function as a linear combination of basis functions, we can often achieve a reasonable approximation of the exact value function by exploiting additive structure inherent in many real-world problems (e.g. additive reward functions or problems with independent subgoals). Unlike exact solution methods where value functions can grow exponentially in size during the solution process and must be logically simplified [Boutilier, Reiter, and Price 2001], here we necessarily achieve a compact form of the value function that requires no simplification, just the discovery of good weights.

To apply the B^A operator for a linear combination of basis functions, we let F_A denote the set of indices for basis functions affected by action $A(\vec{x})$ (thus having free variables $\vec{x}$), and let N_A denote the set of indices for basis functions not affected by the action.[7] Then, we again exploit the fact that $\oplus$ is defined by conjunction and push the $\exists \vec{x}$ through to the the sum over case statements with free variables $\vec{x}$. This yields the following definition of B^A:

$$B^A(\oplus_i w_i bCase_i(s)) = rCase(s) \oplus (\oplus_{i \in N_A} w_i bCase_i(s))$$
$$\oplus \exists \vec{x} \, (\oplus_{i \in F_A} w_i FODTR(bCase_i(s), A(\vec{x})))$$

It is important to note that the backup operator results in the "cross sum" $\oplus$ of a collection of case statements, performing the explicit sum only over those case statements affected by the action being backed up. This results in a representation of the backed-up value function that often retains most of its original additive structure.

As a concrete example of the application of backup operators $B^{A(x)}$ and B^A, suppose we have a reward function that assigns 10 for being in a state where all people are at their intended destination, and 0 otherwise:[8]

$$rCase(s) = case[\, \forall p, f \, PAt(p, f, s) \supset Dst(p, f), 10 \; ; \; \neg \text{''}, 0 \,]$$

Suppose our value function is represented by the following weighted sum of two basis functions (the first basis function indicates whether some person is at a floor that is not their destination, the second whether there is a person on an elevator stopped at a floor that is their destination):

$$vCase(s) =$$
$$w_1 \cdot case[\, \exists p, f \, PAt(p, f, s) \wedge \neg Dst(p, f), 1 \; ; \; \neg \text{''}, 0 \,] \oplus$$
$$w_2 \cdot case[\, \exists p, f, e \, Dst(p, f) \wedge OnE(p, f, s) \wedge EAt(e, f, s), 1 \; ;$$
$$\neg \text{''}, 0 \,]$$

Because $down(x)$ is a deterministic action, FODTR is simply the direct regression of each case statement through nature's choice $downS(x)$. Thus, we obtain the following expression for $B^{down(x)}$:

$$B^{down(x)}(vCase(s)) =$$
$$case[\, \forall p, f \, PAt(p, f, s) \supset Dst(p, f), 10 \; ; \; \neg \text{''}, 0 \,]$$
$$\oplus \gamma \, w_1 \cdot case[\, \exists p, f \, PAt(p, f, s) \wedge \neg Dst(p, f), 1 \; ; \; \neg \text{''}, 0 \,]$$
$$\oplus \gamma \, w_2 \cdot case[\, \exists p, f, e \, Dst(p, f) \wedge OnE(p, f, s) \wedge$$
$$((EAt(e, f, s) \wedge e \neq x) \vee (EAt(e, fa(f), s) \wedge e = x)), 1 \; ;$$
$$\neg \text{''}, 0 \,]$$

---

[7] Since an action typically affects only a few fluents and basis functions contain relatively few fluents, only a few of the basis functions will be affected by an action. This fact holds for many domains, including the elevator domain in this paper.

[8] We replace the partition formula with $\neg$ '' when the formula is just the negation of its predecessor in the case statement.

Applying B^{down} yields:

$$
B^{down}(vCase(s)) =
$$
$$
\exists x\ B^{down(x)}(vCase(s)) =
$$
$$
case[\ \forall p, f\ PAt(p, f, s) \supset Dst(p, f), 10\ ;\ \neg\text{''}, 0\ ]
$$
$$
\oplus \gamma\ w_1 \cdot case[\ \exists p, f\ PAt(p, f, s) \wedge \neg Dst(p, f), 1\ ;\ \neg\text{''}, 0\ ]
$$
$$
\oplus \gamma\ w_2 \cdot case[\ \exists x, p, f, e\ Dst(p, f) \wedge OnE(p, f, s) \wedge
$$
$$
((EAt(e, f, s) \wedge e \neq x) \vee (EAt(e, fa(f), s) \wedge e = x)), 1\ ;
$$
$$
\neg\text{''} \wedge \exists x\ \forall p, f, e\ \neg Dst(p, f) \vee \neg OnE(p, f, s) \vee
$$
$$
((\neg EAt(e, f, s) \vee e = x) \wedge
$$
$$
(\neg EAt(e, fa(f), s) \vee e \neq x)), 0\ ]
$$

Note that in the computation of B^{down} above, we push the existential into the only case statement containing free variable x. We then make the second partition in this case statement mutually exclusive of the first partition to ensure that the highest value gets assigned to every state. This results in a logical description of states that describes the maximum value that can be obtained by applying *any* instantiation of action $down(e)$ and subsequently obtaining value dictated by $vCase(s)$. Thus, as noted by Boutilier, Price and Reiter [2001], the symbolic dynamic programming approach provides *both state and action space abstraction*.

4.2 First-order Linear Program Formulation

We now generalize the approximate linear programming approach for propositional factored MDPs (i.e., Eq. 3) to first-order MDPs. If we simply substitute appropriate notation, we arrive at the following formulation of the first-order approximate linear programming approach (FOALP):

$$
\text{Variables: } w_i\ ;\ \forall i \leq k
$$
$$
\text{Minimize: } \sum_s \sum_{i=1}^{k} w_i \cdot bCase_i(s)
$$
$$
\text{Subject to: } 0 \geq B^A(\oplus_{i=1}^k w_i \cdot bCase_i(s)) \ominus
$$
$$
(\oplus_{i=1}^k w_i \cdot bCase_i(s))\ ;\ \forall\ A, s \qquad (7)
$$

Unfortunately, while the objective and constraints in the approximate LP for a propositional factored MDP range over a finite number of states s, this direct generalization to the FOALP approach for FOMDPs requires dealing with infinitely (or indefinitely) many situations s.

Since we are summing over infinitely many situations in the FOALP objective, it is ill-defined. Thus, we redefine the FOALP objective in a manner which preserves the intention of the original approximate linear programming solution for MDPs. In the propositional LP (Eq. 3), the objective weights each state equally and minimizes the sum of the values of all states. Consequently, rather than count ground states in our FOALP objective—of which there would be an infinite number per partition—we suppose that each basis function partition is chosen because it represented a potentially useful partitioning of state space,

Input: A set c of *case* statements being summed and an ordering $R_1 \ldots R_n$ of *all* relations in c.

Output: The maximum value achievable and the consistent partition which achieves that value.

1. Convert the first-order formulae in each case partition to a set of CNF clauses.

2. For each relation $R \in R_1 \ldots R_n$ (in order):

   (a) Remove all *case* statements in c containing R and store their explicit "cross-sum" $\oplus$ in tmp.

   (b) Do the following for each partition in tmp:

       • Resolve all clauses on relation R.

       • Remove all clauses containing R (sound because we are doing ordered resolution).

       • If a resolvent of $\emptyset$ exists in this partition, remove this partition from tmp and continue.

       • Remove dominated partitions whose clauses are a superset of another partition with greater value.

   (c) Insert tmp back into c.

3. Find the maximum partition in each of the case statements in c and return it along with its corresponding value.

Figure 1. The first-order factored max (FOMAX) algorithm for finding the max consistent partition in a first-order cost network.

and thus weight each case *partition* equally. Consequently, we write the above FOALP objective as the following:

$$\sum_{s}\sum_{i=1}^{k} w_i \cdot bCase_i(s) \;=\; \sum_{i=1}^{k} w_i \sum_{s} bCase_i(s) \tag{8}$$

$$\sim \sum_{i=1}^{k} w_i \sum_{\langle \phi_j, t_j \rangle \in bCase_i} \frac{t_j}{|bCase_i|}$$

With the issue of the infinite objective resolved, this leaves us with one final problem—the infinite number of constraints (i.e., one for every situation s). Fortunately, we can work around this since case statements are finite. Since the value t_i for each case partition $\langle \phi_i(s), t_i \rangle$ is constant over all situations satisfying the $\phi_i(s)$, we can explicitly sum over the $case_i(s)$ statements in each constraint to yield a single case statement. For this "flattened" case statement, we can easily verify that the constraint holds in each of the finite number of resulting first-order partitions of state space.

Furthermore, we can avoid generating the full set of case partitions in the explicit sum by using constraint generation [Schuurmans and Patrascu 2001]. But before we formally define the algorithm for solving first-order LPs via constraint generation, we first outline an important subroutine.

First-order Cost Network Maximization

One of the most important computations in the approximate FOMDP solution algorithms is efficiently finding the consistent set of case partitions which maximizes a first-order cost network of the following form:

$$\max_{s} \left(\oplus_j \, case_j(s) \right) \tag{9}$$

Recall that this is precisely the form of the constraints (the backed up linear value function) in Equation 7. The FOMAX algorithm given in Figure 1 efficiently carries out this computation. It is similar to variable elimination [Dechter 1999], except that we use first-order ordered resolution in place of propositional ordered resolution.[9]

First-Order Constraint Generation

In the first-order constraint generation (FOCG) algorithm given in Figure 2, we initialize our LP with an initial setting of weights, but no constraints. Then we alternate between generating constraints based on maximal constraint violations at the current solution and re-solving the LP with these additional constraints. This process repeats until no constraints are violated and we have found the optimal solution. In practice, this approach typically generates *far* fewer constraints than the full set. And using FOCG, we can now efficiently solve the FOALP given in Equation 7.

To make the FOMAX and FOCG algorithms more concrete, we provide a simple example of finding the most violated constraint in Figure 3.

4.3 Obtaining Error Bounds

Following Schuurmans and Patrascu [2001], we can also compute an upper bound on the error of the approximated value function:

$$\max_{s} \, vCase^*(s) \ominus vCase_{\pi_{greedy}(\oplus_i \, w_i \cdot bCase_i)}(s) \le$$

$$\frac{\gamma}{1-\gamma} \min_{A} \max_{s} \left[\left(\oplus_i \, w_i \cdot bCase_i(s) \right) \right.$$

$$\left. \ominus B^A \left(\oplus_i \, w_i \cdot bCase_i(s) \right) \right] \tag{10}$$

The inner $\max_s$ can be efficiently computed by the FOMAX algorithm (Figure 1) leaving only a few simple arithmetic operations to carry out. This yields an efficient means of computing a strict upper bound on the error of executing a greedy policy according to an approximated value function that applies *equally to all domain instantiations*.

5 Experimental Results

We have applied the first-order approximate linear programming algorithm to the FOMDP domain of elevator scheduling. We refer to Section 3 for a rough description of the FOMDP

---

[9]We note that the ordered resolution strategy is not refutation-complete as it may loop infinitely at an intermediate relation elimination step before finding a latter relation with which to resolve a contradiction. To prevent infinite looping, we simply bound the number of inferences performed at each relation elimination step. While a missed refutation in the FOMAX algorithm may result in the generation of unneccessary constraints during the FOCG algorithm, and this in turn may overconstrain the set of feasible solutions, this has not led to infeasibility problems in practice.

Input: An LP objective obj, a convergence tolerance tol, and a set of first-order constraints, each encoded as $0 \geq \max_s \oplus_i case_i(s)$ where there are w_i variables embedded in the values of the case partitions.

Output: Weights for the optimal LP solution (within tolerance tol).

1. Initialize the weights: $w_i = 0$; $\forall i \leq k$
2. Initialize the LP constraint set: $C = \emptyset$
3. Initialize $C_{new} = \emptyset$
4. For each constraint inequality:

   (a) Calculate $\varphi = \arg\max_s \oplus_i case_i(s)$ using FOMAX.

   (b) If $eval(\varphi) \geq tol$, let c encode the linear constraint for $0 \geq \oplus_i case_i(\varphi)$.

   (c) $C_{new} = C_{new} \cup \{c\}$

5. If $C_{new} = \emptyset$, terminate with w_i as the solution to this LP.
6. $C = C \cup C_{new}$
7. Solve the LP with objective obj and updated constraints C.
8. Return to step 3.

Figure 2. The first-order constraint generation (FOCG) algorithm for solving a first-order linear program.

specification of certain aspects of this domain. To this we have added a rich set of features used in elevator planning domains [Koehler and Schuster 2000]: $VIP(p)$ indicating that person p is a VIP, $Group(p, g)$ indicating that a person p is in group g (persons in certain differing groups should not share the same elevator), and $Attended(p)$ to indicate that person p requires someone to accompany them in the elevator. We have also added action costs of 1 for *open*, *up*, *down* and a simple *noop* action with cost 0.[10] We require that all policies observe typical elevator scheduling constraints [Koehler and Schuster 2000] by specifying appropriate action preconditions; for example, the elevator cannot reverse direction if it has a passenger whose destination is in the current direction of travel.

Given this domain specification, we have defined a multicriteria reward function $rCase(s)$ with positive additive rewards for satisfying each of the following conditions:

$$+2 : \forall p, f \ PAt(p, f, s) \supset Dst(p, f)$$
$$+2 : \forall p, f \ Dst(p, f) \wedge \neg PAt(p, f, s) \supset \exists e \ On(p, e, s)$$
$$+2 : \forall p, f \ VIP(p) \wedge PAt(p, f, s) \supset Dst(p, f)$$
$$+2 : \forall p, f \ VIP(p) \wedge Dst(p, f) \wedge \neg PAt(p, f, s)$$
$$\supset \exists e \ OnE(p, e, s)$$
$$+4 : \forall p, e \ OnE(p, e, s) \wedge Attended(p) \supset \exists p_2 OnE(p_2, e, s)$$
$$+8 : \forall p_1, p_2, g_1, g_2, e \ OnE(p_1, e, s) \wedge OnE(p_2, e, s)$$
$$\wedge p_1 \neq p_2 \wedge Group(p_1, g_1) \wedge Group(p_2, g_2) \supset g_1 = g_2$$

---

[10] Since these costs are constant for each action, the backup operators can be easily modified to handle action costs by subtracting the constant cost from *one* of the cases being backed up.

Suppose we are given the following constraint specification (using a tabular case notation) for a first-order linear program:

$$0 \geq \max_{s} \left(\begin{array}{|c|} \hline \forall p, f \, Dst(p, f) \supset PAt(p, f, s):10 \\ \hline \neg" \qquad\qquad\qquad : 0 \\ \hline \end{array} \oplus \begin{array}{|c|} \hline \exists p, f \, Dst(p, f) \wedge \neg PAt(p, f, s): \quad w_1 \\ \hline \neg" \qquad\qquad\qquad\qquad :-w_1 \\ \hline \end{array} \oplus \begin{array}{|c|} \hline \exists p, e \, OnE(p, e, s):w_2 \\ \hline \neg" \qquad\qquad\quad : 0 \\ \hline \end{array} \right) $$

Now, suppose that we are at step 4 of the FOCG algorithm and the last LP solution yielded weights $w_1 = 2$ and $w_2 = 1$. We can efficiently compute the most violated constraint (if one exists) by evaluating the weights in the constraint and applying the FOMAX algorithm. We begin with step 1 of FOMAX where we convert all first-order formulae to CNF:

$$0 \geq \max_{s} \left(\begin{array}{|c|} \hline \{\neg Dst(p, f) \vee PAt(p, f, s)\} \quad :10 \\ \hline \{Dst(c_1, c_2), \neg PAt(c_1, c_2, s)\}: 0 \\ \hline \end{array} \oplus \begin{array}{|c|} \hline \{Dst(c_3, c_4), \neg PAt(c_3, c_4, s)\}: \quad 2 \\ \hline \{\neg Dst(p, f) \vee PAt(p, f, s)\} \quad :-2 \\ \hline \end{array} \oplus \begin{array}{|c|} \hline \{OnE(c_5, c_6, s)\}:1 \\ \hline \{\neg OnE(p, e, s)\} :0 \\ \hline \end{array} \right) $$

Given relation elimination order PAt, Dst, OnE, we now proceed to step 2 of FOMAX. We begin by eliminating the PAt relation: we take the cross-sum $\oplus$ of case statements containing PAt, resolve all clauses in each partition, and remove all clauses containing PAt (indicated by struck-out text):

$$0 \geq \max_{s} \left(\begin{array}{|c|} \hline \{\neg Dst(p,f) \vee PAt(p,f,s), \neg Dst(p,f) \vee PAt(p,f,s), \emptyset\ \} : 12 \\ \hline \{\neg Dst(p,f) \vee PAt(p,f,s)\} \qquad\qquad\qquad : 8 \\ \hline \{Dst(c_1, c_2), Dst(c_3, c_4), \neg PAt(c_1, c_2, s), \neg PAt(c_3, c_4, s)\ \} : \quad 2 \\ \hline \{\neg Dst(p,f) \vee PAt(p,f,s), \neg Dst(p,f) \vee PAt(p,f,s), \emptyset\ \} :-2 \\ \hline \end{array} \oplus \begin{array}{|c|} \hline \{OnE(c_5, c_6, s)\} : 1 \\ \hline \{\neg OnE(p, e, s)\} : 0 \\ \hline \end{array} \right) $$

Because the partitions valued 12 and -2 contain the empty clause (i.e. they are inconsistent), we can remove them. And because the partition of value 8 dominates the partition of value 2 (i.e. $2 < 8$ and the clauses of the value 2 partition are a superset of the clauses of the value 8 partition), we can remove it as well. This yields the following simplified result:

$$0 \geq \max_{s} \left(\begin{array}{|c|} \hline \{\ \} : 8 \\ \hline \end{array} \oplus \begin{array}{|c|} \hline \{OnE(c_5, c_6, s)\} : 1 \\ \hline \{\neg OnE(p, e, s)\} : 0 \\ \hline \end{array} \right) $$

From here it is obvious that the Dst elimination step will have no effect and the OnE elimination step will yield a maximal consistent partition with value 9. Since this is a positive value and thus a violation of the original constraint, we can generate the new linear constraint $0 \geq 10 + -w_1 + w_2$ based on the original constituent partitions that led to this maximal constraint violation.

Figure 3. An example of using FOMAX to find the maximally violated constraint during an iteration of the FOCG algorithm.

Policy	F5	F10	F15	ME
{ }, {A}	116 ± 28	106 ± 27	105 ± 28	N/A
{V}, {V,A}	115 ± 30	108 ± 30	107 ± 28	N/A
{G}, {A,G}	**125 ± 24**	**119 ± 21**	114 ± 20	N/A
{V,G}, {V,A,G}	119 ± 30	114 ± 24	**115 ± 23**	N/A
Myopic 1	118 ± 10	119 ± 9	120 ± 13	N/A
Myopic 2	**123 ± 12**	**122 ± 5**	120 ± 12	N/A
FOALP {1,2}	133 ± 31	114 ± 32	112 ± 23	177
FOALP {3,4}	148 ± 26	129 ± 23	117 ± 23	159
FOALP {5}	147 ± 26	126 ± 17	120 ± 17	146
FOALP {6}	**154 ± 25**	**130 ± 19**	**125 ± 19**	92

Table 1. The performance of various policies in elevator domains with 5, 10, and 15 floors (denoted F5, F10, and F15, respectively), with passenger arrivals distributed according to $N(0.1, 0.35)$ per time-step, and discount rate $\gamma = 0.9$. Heuristic policies, shown at the top, are designated by the sets of constraints used: (V) prioritize VIP, (A) no attended conflict, and (G) no group conflict. For example, the empty set policy {} uses no constraints and picks up everyone at each floor it passes. The one- and two-step lookahead myopic policies are labeled *Myopic 1* and *2*. FOALP policies are labeled using the number of basis functions used, counting from the top of the above list of reward criteria (e.g., *FOALP {3,4}* represents two FOALP-derived policies using the first three and first four basis functions, respectively). Policies are grouped when they turn out to be identical (e.g., FOALP-3 and FOALP-4 are identical, as are heuristic policies {V} and {V, A}). Performance is measured using the the sum of discounted rewards obtained by time step 50, averaged over 100 random simulations. The best results for each group of policies and number of floors are in boldface. FOALP error bounds are computed according to Equation 10 and are shown in the max error (ME) column.

In our FOALP solution, we have used a set of basis functions that represent each of the additive reward criteria above, in addition to one constant basis function. We experimented with a variety of more complex compound basis functions in addition to the ones above but found that they received negligible weight and did not affect the resulting policy. Thus, it seems that a linear combination of these six basis functions may approximate a reasonable amount of structure in the optimal value function.

We evaluated two simple myopic policies that take the best action from the current state based on one- and two-step lookahead. We also evaluated several intuitive, heuristic policies that seem to offer reasonable performance in the elevator domain. The basic heuristic policy always picks up passengers at each floor. The other policies are specified by adding the following constraints to this basic policy in various combinations: always prioritize picking up VIPs over regular passengers (V); never allow an attended person to be unaccompanied in the elevator (A); and never allow a group conflict in the elevator (G).

We used the Vampire [Riazanov and Voronkov 2002] theorem prover and the CPLEX LP solver in our FOALP implementation.[11] The performance of various policies, measured

---

[11] We note that while the full FOALP specification with six basis functions contained over 250,000 constraints, only 40 of these constraints needed to be generated to determine the optimal solution. Nonetheless, a considerable time is spent in theorem proving deriving and ensuring the consistency of constraints. We note that running times ranged from 5 minutes for one basis function to 120 minutes for six basis functions on a 2Ghz Pentium-IV

using the average total discounted reward obtained over 100 simulated trials of 50 stages each, are shown in Table 1. While the results have relatively high standard deviation due to the stochasticity of the domain, the difference in performance between FOALP with six basis functions and the best of the heuristic policies is statistically significant at the 93% confidence level for each domain size. Thus, we can conclude that the FOALP approach with six basis functions outperforms the heuristic policies with some confidence. This appears to be due to the fact that the additive value function allows FOALP to tradeoff various domain features while the hard constraints must be applied in all situations (leading to suboptimal performance in many cases) and the myopic policies cannot see far enough ahead to make good choices for the long-term. We note that while the upper bounds on error are not tight, there is nevertheless a strong correlation between these values and actual policy performance.

6 Related Work

While our work incorporates elements of symbolic dynamic programming (SDP) [Boutilier, Reiter, and Price 2001], it uses this in an efficient approximation framework that represents the value function compactly and avoids the need for logical simplification. This is in contrast to an exact value iteration framework that proves intractable in many cases for SDP, FOVIA [Hölldobler and Skvortsova 2004], and ReBel [Kersting, Otterlo, and Raedt 2004] due to the combinatorial explosion of the value function representation and the need to perform complex logical simplifications. FOALP uses a model-based approach to avoid domain instantiation unlike the non-SDP approaches [Fern, Yoon, and Givan 2003; Guestrin, Koller, Gearhart, and Kanodia 2003] which need to instantiate domains. In addition, FOALP does not use inductive methods [Fern, Yoon, and Givan 2003; Gretton and Thiebaux 2004] that generally require substantial simulation to generalize from experience. The work by Guestrin *et al.* [Guestrin, Koller, Gearhart, and Kanodia 2003] is the only other RMDP approximation work that can provide bounds on policy quality, but these are PAC-bounds under the assumption that the probability of domains falls off exponentially with their size. In contrast, the policy bounds obtained by FOALP apply equally to all domains.

7 Concluding Remarks

We have introduced a novel approximate linear programming technique for FOMDPs that represents an approximately optimal value function as a linear combination of first-order basis functions. This technique allows us to efficiently solve for an approximation of the value function using off-the-shelf theorem prover and LP software. It offers us the advantage that we can approximately solve FOMDPs independent of a domain instantiation and it additionally allows us to determine upper bounds on approximation error that apply equally to all possible domain instantiations. We have applied our algorithm to the task of elevator scheduling and have found that it is not only efficient, but that its policies outperform a number of intuitive, heuristically-guided policies for this domain.

An interesting question for future work is whether the uniform weighting of partitions in the FOALP objective is the best approach. It would be informative to explore alternative FOALP objective specifications and to evaluate their impact on value function quality over a variety of domains. In addition, it is an interesting question as to whether dynamic

---

machine.

reweighting schemes could improve solution quality by identifying state partitions with large error and adjusting their weights so they receive more emphasis on the next LP iteration. Altogether, such improvements to FOALP could bolster an already promising approach for efficiently and compactly approximating FOMDP value functions.

References

Bacchus, F., J. Y. Halpern, and H. J. Levesque [1995]. Reasoning about noisy sensors in the situation calculus. In *Proceedings of the Fourteenth International Joint Conference on Artificial Intelligence (IJCAI-95)*, Montreal, pp. 1933–1940.

Boutilier, C., T. Dean, and S. Hanks [1999]. Decision-theoretic planning: Structural assumptions and computational leverage. *Journal of Artificial Intelligence Research* 11, 1–94.

Boutilier, C., R. Reiter, and B. Price [2001]. Symbolic dynamic programming for first-order MDPs. In *Proceedings of the Seventeenth International Joint Conference on Artificial Intelligence (IJCAI-01)*, Seattle, pp. 690–697.

Dechter, R. [1999]. Bucket elimination: A unifying framework for reasoning. In *Artificial Intelligence*, Volume 113, pp. 41–85.

Fern, A., S. Yoon, and R. Givan [2003]. Approximate policy iteration with a policy language bias. In *Advances in Neural Information Processing Systems 16 (NIPS-2003)*, Vancouver.

Gretton, C. and S. Thiebaux [2004]. Exploiting first-order regression in inductive policy selection. In *Proceedings of the Twentieth Conference on Uncertainty in Artificial Intelligence (UAI-04)*, Banff, AB, pp. 217–225.

Guestrin, C., D. Koller, C. Gearhart, and N. Kanodia [2003]. Generalizing plans to new environments in relational MDPs. In *Proceedings of the Eighteenth International Joint Conference on Artificial Intelligence (IJCAI-03)*, Acapulco, pp. 1003–1010.

Guestrin, C., D. Koller, R. Parr, and S. Venktaraman [2003]. Efficient solution methods for factored MDPs. *Journal of Artificial Intelligence Research* 19(1), 399–468.

Hoey, J., R. St-Aubin, A. Hu, and C. Boutilier [1999]. SPUDD: Stochastic planning using decision diagrams. In *Proceedings of the Fifteenth Conference on Uncertainty in Artificial Intelligence (UAI-99)*, Stockholm, pp. 279–288.

Hölldobler, S. and O. Skvortsova [2004]. On interacting defaults. In *Proceedings of the AAAI Workshop on Learning and Planning in Markov Processes—Advances and Challenges*, San Jose, CA, pp. 31–36.

Kersting, K., M. V. Otterlo, and L. D. Raedt [2004]. Bellman goes relational. In *Proceedings of the Twenty-first International Conference on Machine Learning (ICML-04)*, Banff, AB, pp. 59–66.

Koehler, J. and K. Schuster [2000]. Elevator control as a planning problem. In *Proceedings of the Fifth International Conference on AI Planning Systems*, Breckenridge, CO, pp. 331–338.

McCarthy, J. [1963]. Situations, actions and causal laws. Technical report, Stanford University. Reprinted in Semantic Information Processing (M. Minsky ed.), MIT Press, Cambridge, Mass., 1968, pages 410-417.

Puterman, M. L. [1994]. *Markov Decision Processes: Discrete Stochastic Dynamic Programming*. New York: Wiley.

Reiter, R. [2001]. *Knowledge in Action: Logical Foundations for Describing and Implementing Dynamical Systems*. Cambridge, MA: MIT Press.

Riazanov, A. and A. Voronkov [2002]. The design and implementation of Vampire. *AI Communcations* 15(2), 91–110.

Schuurmans, D. and R. Patrascu [2001]. Direct value approximation for factored MDPs. In *Advances in Neural Information Processing Systems 14 (NIPS-2001)*, Vancouver, pp. 1579–1586.

27

Golog Speaks the BDI Language

Sebastian Sardina and Yves Lespérance

ABSTRACT. In this paper, we relate two of the most well developed approaches to agent-oriented programming, namely, BDI (Belief-Desire-Intention) style programming and "Golog-like" high-level programming. In particular, we show how "Golog-like" programming languages can be used to develop BDI-style agent systems. The contribution of this paper is twofold. First, it demonstrates how *practical* agent systems can be developed using high-level languages like Golog or IndiGolog. Second, it provides BDI languages a clear classical-logic-based semantics and a powerful logical foundation for incorporating new reasoning capabilities not present in typical BDI systems.

Preamble

We dedicate this paper to Hector Levesque. Both of us were fortunate enough to have had Hector as our Ph.D. supervisor. The approach and technical foundations used in the paper draw a lot from Hector's work. Indeed, the paper straddles two areas of Artificial Intelligence (AI) where Hector has contributed enormously throughout his career, namely, agent theory and agent-oriented programing. Hector has had a great influence on the field of AI. He has contributed to numerous areas, including reasoning about action, agent theory and models of communication, knowledge based systems, description logic, and non-monotonic reasoning. One striking aspect of Hector is his passion and enthusiasm for research in AI. One cannot sit through one of his lectures without acquiring some of this enthusiasm. A research meeting with him is an experience that is difficult to forget: his quick grasp of technical details, clarity and precision in seeing through problems, and kindness and patience with others are remarkable. His vision, support, and advice has played a key role in our careers and those of numerous others. His dedication, integrity, and generosity, have been inspiring. Thanks Hector, for all you have given us.

1 Introduction

BDI (Belief-Desire-Intention) agent programming languages and platforms (e.g., PRS [Georgeff and Ingrand 1989], AgentSpeak and Jason [Rao 1996; Bordini, Hübner, and Wooldridge 2007], Jack [Busetta, Rönnquist, Hodgson, and Lucas 1999], and JADEX [Pokahr, Braubach, and Lamersdorf 2003]) and the situation calculus-based Golog high-level programming language and its successors (e.g., ConGolog [De Giacomo, Lespérance, and Levesque 2000], IndiGolog [Sardina, De Giacomo, Lespérance, and Levesque 2004; De Giacomo, Lespérance, Levesque, and Sardina 2009], and FLUX [Thielscher 2005]) are two of the most well developed approaches within the agent-oriented programming paradigm. In this paper, we analyze the relationship between these two families of languages and show that BDI agent programming languages are closely related to IndiGolog, a situation calculus based programming language supporting online execution of programs in dynamic environments, sensing actions to acquire information from the environment, and exogenous events.

BDI agent programming languages were conceived as a simplified and operationalized version of the BDI (Belief, Desire, Intention) model of agency, which is rooted in philosophical work such as Bratman's [Bratman 1987] theory of practical reasoning and Dennett's theory of intentional systems [Dennett 1987]. Practical work in the area has sought to develop programming languages that incorporate a simplified BDI semantics basis that has a computational interpretation. An important feature of BDI-style programming languages and platforms is their interleaved account of sensing, deliberation, and execution [Pollack 1992; Rao and Georgeff 1992]. In BDI systems, *abstract plans* written by programmers are combined and executed in real-time. By executing as they reason, BDI agents reduce the likelihood that decisions will be made on the basis of outdated beliefs and remain responsive to the environment by making adjustments in the steps chosen as they proceed. Because of this, BDI agent programming languages are well suited to implementing systems that need to operate in "soft" real-time scenarios [Ljungberg and Lucas 1992; Benfield, Hendrickson, and Galanti 2006; Belecheanu, Munroe, Luck, Payne, Miller, McBurney, and Pechoucek 2006]. Unlike in classical planning-based architectures, *execution* happens at each step. The assumption is that the careful crafting of plans' preconditions to ensure the selection of appropriate plans at execution time, together with a built-in mechanism for retrying alternative options, will usually ensure that a successful execution is found, even in the context of a changing environment.

In contrast to this, high-level programming languages in the Golog line aim for a middle ground between classical planning and normal programming. The idea is that the programmer may write a *sketchy* non-deterministic program involving domain specific actions and test conditions and that the interpreter will reason about these and search for a valid execution. The semantics of these languages is defined on top of the *situation calculus*, a popular predicate logic framework for reasoning about action [McCarthy and Hayes 1969; Reiter 2001]. The interpreter for the language uses an action theory representing the agent's beliefs about the state of the environment and the preconditions and effects of the actions to find a provably correct execution of the program. By controlling the amount of nondeterminism in the program, the high-level program execution task can be made as hard as classical planning or as easy as deterministic program execution. In IndiGolog, this framework is generalized to allow the programmer to control planning/lookahead and support on-line execution and sensing the environment.

In this paper, we show how a BDI agent can be built within the IndiGolog situation calculus-based programming framework. More concretely, we describe how to translate an agent programmed in a typical BDI programming language into a high-level IndiGolog program with an associated situation calculus action theory, such that *(i)* their ultimate behavior coincide; and *(ii)* the original structure of the propositional attitudes (beliefs, intentions, goals, etc.) of the BDI agent and the model of execution are preserved in the IndiGolog translation. We first do this (Section 3) for what we call the *core* engine of BDI systems, namely, the reactive context-sensitive expansion of events/goals. After this, in Section 4, we show how to accommodate more sophisticated BDI reasoning mechanisms such as goal failure recovery. Before presenting these results, in Section 2, we give a brief overview of BDI programming languages and Golog-related programming languages. The paper ends with a short discussion in Section 5, where we mention some potential advan-

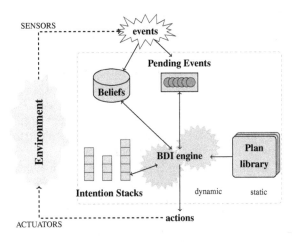

Figure 1. A typical BDI-style architecture.

tages of programming BDI agents in the situation calculus, in particular, various reasoning about action techniques that IndiGolog BDI agents could incorporate.

2 Preliminaries

2.1 BDI Programming

BDI agent systems were developed as a way of enabling *abstract plans* written by programmers to be combined and used in real-time, in a way that is both flexible and robust. A BDI system responds to *events*, the inputs to the system, by selecting a plan from the *plan library*, and placing it into the *intention base*, thus committing to the plan for responding to the event/goal in question. The execution of this plan-strategy may, in turn, post new subgoal events to be achieved. The plan library stands for a collection of pre-defined *hierarchical plans* indexed by goals (i.e., events) and representing the standard operations in the domain. Figure 1 depicts a typical BDI-style architecture.

There are a number of agent programming languages and development platforms in the BDI tradition, such as PRS [Georgeff and Ingrand 1989], AgentSpeak and Jason [Rao 1996; Bordini, Hübner, and Wooldridge 2007], Jack [Busetta, Rönnquist, Hodgson, and Lucas 1999], JADEX [Pokahr, Braubach, and Lamersdorf 2003], and 3APL/2APL [Hindriks, de Boer, van der Hoek, and Meyer 1999; Dastani 2008]. Our discussion is based on the CAN family of BDI languages [Winikoff, Padgham, Harland, and Thangarajah 2002; Sardina and Padgham 2007] (Conceptual Agent Notation), which are AgentSpeak-like languages with a semantics capturing the common essence of typical BDI systems.

A BDI agent configuration (or simply a BDI agent) Υ is a tuple $\langle \Pi, \mathcal{B}, \mathcal{A}, \Gamma \rangle$, where $\mathcal{B}$ stands for the agent's current beliefs about the world, generally a set of atoms, Π is the (static) plan-library, $\mathcal{A}$ is the sequence of actions executed so far, and Γ is the multi-set of intentions the agent is currently pursuing. The *plan library* contains plan rules of the form $e : \psi \leftarrow P$, where e is an event/goal that triggers the plan, ψ is the context for which the plan may be applied (i.e., the precondition of the rule), and P is the body of the plan rule— *P is a reasonable strategy in order to resolve the event/goal e when condition ψ is believed to hold*. The *plan-body P* is a program built from primitive actions A that the agent can execute directly (e.g., $drive(loc_1, loc_3)$), operations to add $+b$ and delete $-b$ beliefs, tests for conditions $?\phi$, and (internal) subgoaling event posting $!e$ (e.g., $!\,Travel(mel, yyz)$). Complex plan bodies are built with the usual sequence $;$ and concurrency $\|$ constructs. There are also a number of auxiliary constructs internally used when assigning semantics to programs: the empty (terminating) program nil; the construct $P_1 \rhd P_2$, which tries to execute P_1, falling back to P_2 if P_1 is not executable; and $(\!|\psi_1 : P_1, \ldots, \psi_n : P_n|\!)$, which encodes a set of alternative guarded plans. Lastly, the *intention base* Γ contains the current, partially instantiated, plan-body programs that the agent has already committed to for handling some events—since Γ is a multi-set it may contain a program more than once.

As with most BDI agent programming languages, the Plotkin-style operational semantics of CAN closely follows Rao and Georgeff's abstract interpreter for intelligent rational agents [Rao and Georgeff 1992]: *(i)* incorporate any pending external events; *(ii)* select an intention and execute a step; and *(iii)* update the set of goals and intentions. A *transition relation* $C \longrightarrow C'$, on so-called *configurations* is defined by a set of *derivation rules* and specifies that executing configuration C a *single step* yields configuration C'. A *derivation rule* consists of a, possibly empty, set of premises, typically involving the existence of transitions together with some auxiliary conditions, and a single transition conclusion derivable from these premises. Two transition relations are used to define the semantics of the CAN language. The first transition relation $\longrightarrow$ defines what it means to execute a single intention and is defined in terms of *intention-level* configurations of the form $\langle \Pi, \mathcal{B}, \mathcal{A}, P \rangle$ consisting of the agent's plan-library Π and belief base $\mathcal{B}$, the actions $\mathcal{A}$ executed so far, and the program P being executed. The second transition relation $\Longrightarrow$ is defined in terms of the first and characterizes what it means to execute a whole agent.

So, the following are some of the intention-level derivation rules for the language:[1]

$$\frac{\Delta = \{\psi : P \mid e : \psi \leftarrow P \in \Pi\}}{\langle \Pi, \mathcal{B}, \mathcal{A}, !e \rangle \longrightarrow \langle \Pi, \mathcal{B}, \mathcal{A}, (\!|\Delta|\!) \rangle} \; Ev \qquad \frac{\langle \Pi, \mathcal{B}, \mathcal{A}, P_1 \rangle \longrightarrow \langle \Pi, \mathcal{B}', \mathcal{A}', P_1' \rangle}{\langle \Pi, \mathcal{B}, \mathcal{A}, P_1 \rhd P_2 \rangle \longrightarrow \langle \Pi, \mathcal{B}', \mathcal{A}', P_1' \rhd P_2 \rangle} \; \rhd$$

$$\frac{\mathcal{B} \models \phi\theta}{\langle \mathcal{B}, \mathcal{A}, ?\phi \rangle \longrightarrow \langle \mathcal{B}, \mathcal{A}, nil \rangle} \; ? \qquad \frac{\psi : P \in \Delta \quad \mathcal{B} \models \psi\theta}{\langle \Pi, \mathcal{B}, \mathcal{A}, (\!|\Delta|\!) \rangle \longrightarrow \langle \Pi, \mathcal{B}, \mathcal{A}, P\theta \rhd (\!|\Delta \setminus \{\psi : P\}|\!) \rangle} \; Sel$$

Derivation rule Ev captures the first stage in the plan selection process for a (pending) event/goal e, in which the agent collects, from the plan library, the set $(\!|\Delta|\!)$ of the so-called "*relevant*" (guarded) plans that may be used to resolve the pending event. Such set is later used by rules Sel and $\rhd$ to commit to and execute, respectively, an *applicable* strategy/plan P (one whose condition ψ is believed true). Notice in rule Sel how the remaining non-selected plans are kept as backup plans as the second program in the $\rhd$ construct. Finally, rule $?$ accounts for transitions over a basic test program.

On top of these intention-level derivation rules, the set of agent-level derivation rules are defined. Basically, an agent transition involves either assimilating external events from the environment or executing an active intention. The set of external events $\mathcal{E}$ stands for those external events that were "sensed" by the agent, and it may include external achievement events of the form $!e$ as well as belief update events of the form $+b$ and $-b$. Also, in the rules below, the following auxiliary function is used to represent the set of achievement events caused by belief changes: $\Omega(\mathcal{B}, \mathcal{B}') = \{!b^- \mid \mathcal{B} \models b, \; \mathcal{B}' \not\models b\} \cup \{!b^+ \mid \mathcal{B} \not\models$

---

[1] Configurations must also include a variable substitution θ for keeping track of all bindings done so far during the execution of a plan-body. For legibility, we keep substitutions implicit in places where they need to be carried across multiple rules (e.g., in rule $?$).

$b, \mathcal{B}' \models b\}$.

$$\frac{\mathcal{E} \text{ is a set of external events} \quad \mathcal{B}' = (\mathcal{B} \setminus \{b \mid -b \in \mathcal{E}\}) \cup \{b \mid +b \in \mathcal{E}\}}{\langle \Pi, \mathcal{B}, \mathcal{A}, \Gamma \rangle \Longrightarrow \langle \Pi, \mathcal{B}', \mathcal{A}, \Gamma \uplus \{!e \mid !e \in \mathcal{E}\} \uplus \Omega(\mathcal{B}, \mathcal{B}')\rangle} \; A_{ext}$$

$$\frac{P \in \Gamma \quad \langle \Pi, \mathcal{B}, \mathcal{A}, P \rangle \longrightarrow \langle \Pi, \mathcal{B}', \mathcal{A}', P' \rangle}{\langle \Pi, \mathcal{B}, \mathcal{A}, \Gamma \rangle \Longrightarrow \langle \Pi, \mathcal{B}', \mathcal{A}', (\Gamma \setminus \{P\}) \uplus \{P'\} \uplus \Omega(\mathcal{B}, \mathcal{B}')\rangle} \; A_{exec}$$

Rule A_{exec} assimilates a set of external events, including achievements events of the form $!e$, as well as belief update events of the form $+b$ and $-b$, after which both belief and intention bases of the agent may be updated. Note that, by means of auxiliary function Ω, a new (achievement) event of the form $!b^+$ or $!b^-$ is posted for each belief b that changes due to an external belief update; such an event may in turn trigger some new behavior.

Rule A_{exec} states that the agent may evolve one step if an active intention P can be advanced one step with remaining intention P' being left to execute. In such a case, the intention base is updated by replacing P with P' and including the belief update events produced by potential changes in the belief base. Observe that the intention base is a *multiset*, which means that it may contain several occurrences of the same intention.

Relative to the above derivation rules, one can formally define the meaning of an agent as its possible execution traces. (See [Winikoff, Padgham, Harland, and Thangarajah 2002; Sardina and Padgham 2007] for the complete semantics.)

DEFINITION 1 (BDI Execution). A *BDI execution* E of an agent $\Upsilon_0 = \langle \Pi, \mathcal{B}_0, \mathcal{A}_0, \Gamma_0 \rangle$ is a, possibly infinite, sequence of agent configurations $\Upsilon_0 \cdot \Upsilon_1 \cdot \ldots \cdot \Upsilon_n \cdot \ldots$ such that $\Upsilon_i \Longrightarrow \Upsilon_{i+1}$, for all $i \geq 0$. $\qquad \square$

2.2 High-Level Programming in Golog

The *situation calculus* [McCarthy and Hayes 1969; Reiter 2001] is a logical language specifically designed for representing dynamically changing worlds in which all changes are the result of named *actions* The constant S_0 is used to denote the initial situation where no actions have yet been performed. Sequences of actions are built using the function symbol *do*: $do(a, s)$ denotes the successor situation resulting from performing action a in situation s. Relations whose truth values vary from situation to situation are called *fluents*, and are denoted by predicate symbols taking a situation term as their last argument, (e.g., $Holding(x, s)$). A special predicate $Poss(a, s)$ is used to state that action a is executable in s.

Within this language, we can formulate action theories describing how the world changes as the result of the available actions. For example, a *basic action theory* [Reiter 2001] includes domain-independent foundational axioms to describe the structure of situations, one successor state axiom per fluent (capturing the effects and non-effects of actions), one precondition axiom per action (specifying when the action is executable), and initial state axioms describing what is true initially (i.e., what is true in the initial situation S_0).

On top of situation calculus action theories, logic-based programming languages can be defined, which, in addition to the primitive actions, allow the definition of complex actions. Golog [Levesque, Reiter, Lespérance, Lin, and Scherl 1997], the first situation calculus-based agent programming language, provides all the usual control structures (e.g., sequence, iteration, conditional, etc.) plus some *nondeterministic constructs*. These non-deterministic constructs allow the loose specification of programs by leaving "*gaps*" that ought to be resolved by the reasoner/planner or executor. ConGolog extends Golog to support concurrency [De Giacomo, Lespérance, and Levesque 2000]. To provide an intuitive overview of the language, consider the following nondeterministic program for an agent that goes to work in the morning:[2]

> **proc** $goToWork$
> $ringAlarm; (hitSnooze; ringAlarm)^*; turnOffAlarm;$
> $(\pi food)[Edible(food)?; eat(food)];$
> $(haveShower \parallel brushTeeth);$
> $(driveToUniversity \mid trainToUniversity);$
> $(Time < 11 : 00)?$
> **endProc**

While this high-level program provides a general strategy for getting up and going to work, it is underspecified, and many details, such as what to eat and how to travel to work, are left open. Program $\delta_1 \mid \delta_2$ nondeterministically chooses between programs δ_1 and δ_2, $\pi x. \delta(x)$ executes program $\delta(x)$ for *some* legal binding for variable x, and δ^* performs δ zero or more times. Concurrency is supported by the following three constructs: $(\delta_1 \parallel \delta_2)$ expresses the concurrent execution (interpreted as interleaving) of programs δ_1 and δ_2; $\delta_1 \rangle\rangle \delta_2$ expresses the concurrent execution of δ_1 and δ_2 with δ_1 having higher priority; and $\delta^\parallel$ executes δ zero or more times concurrently. Note that a concurrent process may become (temporarily) blocked when it reaches a test/wait action ϕ? whose condition ϕ is false, or a primitive action whose preconditions are false. Test/wait actions can also be used to control which nondeterministic branches can be executed, e.g., $[(\phi?; \delta_1) \mid (\neg\phi?; \delta_2)]$, and to constrain the value of a nondeterministically bound variable, e.g., $\pi x.[\phi(x)?; \delta(x)]$. Finally, the language also accommodates the standard if-then-elses, while loops, and recursive procedures.

Finding a legal execution of a high-level program is at the core of the whole approach. Originally, Golog and ConGolog programs were intended to be executed *offline*, that is, a complete execution was obtained before committing even to the first action. However, IndiGolog [Sardina, De Giacomo, Lespérance, and Levesque 2004; De Giacomo, Lespérance, Levesque, and Sardina 2009], the latest language within the Golog family, provides a formal logic-based account of interleaved planning, sensing, and action by executing programs *online* and using a specialized new construct $\Sigma(\delta)$, the *search operator*, to perform local offline planning when required.

Roughly speaking, an *online* execution of a program finds a next possible action, executes it in the real world, then obtains sensing results and observed exogenous actions, and repeats the cycle until the program's execution is completed.

Formally, an online execution is a sequence of so-called online configuration of the form (δ, σ), where δ is a high-level program and σ is a history (see [De Giacomo, Lespérance, Levesque, and Sardina 2009] for its formal definition). A history contains the sequence of actions executed so far as well as the sensing information obtained. Online executions are characterized in terms of the following two predicates [De Giacomo, Lespérance, and Levesque 2000]: $Final(\delta, s)$ holds if program δ may legally terminate in situation s; and $Trans(\delta, s, \delta', s')$ holds if a single step of program δ in situation s may lead to situation s' with δ' remaining to be executed. In the next section, we will generalize the notion of online execution to suit our purposes.

---

[2]We thank Ryan Kelly and Adrian Pearce for allowing us to re-use their example.

3 BDI Programming in IndiGolog

Programming a BDI agent in the situation calculus amounts to developing a special basic action theory and a special IndiGolog high-level agent program to be executed with it. From now on, let $\Upsilon = \langle \Pi, \mathcal{B}, \mathcal{A}, \Gamma \rangle$ be the BDI agent to program in IndiGolog.

3.1 The BDI Basic Action Theory

We start by showing how to obtain an action theory $\mathcal{D}^{\Upsilon}$ for our agent Υ. We assume that Υ is stated over a first-order language $\mathcal{L}_{BDI}$ containing finitely many belief and event atomic relations, namely, $b_1(\vec{x_1}), \ldots, b_n(\vec{x_n})$ and $e_1(\vec{x_1}), \ldots, e_m(\vec{x_n})$.

Let us then define what the fluents and actions available in the situation calculus language $\mathcal{L}_{sitCalc}$ are. First, for every belief atomic predicate $b(\vec{x})$ in $\mathcal{L}_{BDI}$, the language $\mathcal{L}_{sitCalc}$ includes a relational fluent $b(\vec{x}, s)$ together with two primitive actions $add_b(\vec{x})$ and $del_b(\vec{x})$ which are meant to change the fluent's truth value. Second, for each achievement event type $e(\vec{x})$ in the domain, there is a corresponding action term $ach_e(\vec{x})$ in $\mathcal{L}_{sitCalc}$. Finally, for every action atom $A(\vec{x})$ in $\mathcal{L}_{BDI}$, there is a corresponding action function $A(\vec{x})$ in $\mathcal{L}_{sitCalc}$.

In addition, the language $\mathcal{L}_{sitCalc}$ shall include one auxiliary distinguished fluent and two actions to model external event handling. Fluent $PendingEv(s)$ stands for the multi-set of events that are "pending" and need to be handled, either belief update or achievement events. This fluent is affected by two actions. Whereas action $post(e)$ indicates the external posting of event e; action $serve(e)$ indicates that (pending) event e has been selected and is being handled. In both actions, argument e is of sort action.

Let us now construct the basic action theory $\mathcal{D}^{\Upsilon}$ corresponding to a BDI agent $\Upsilon = \langle \Pi, \mathcal{B}, \mathcal{A}, \Gamma \rangle$, as follows:

1. The initial description in $\mathcal{D}^{\Upsilon}$ is defined in the following way:

$$\mathcal{D}^{\Upsilon}_{S_0} = \bigcup_{i=1}^{n} \{\forall \vec{x}. b_i(\vec{x}, S_0) \equiv \vec{x} = t_i^{\vec{1}} \vee \ldots \vee \vec{x} = t_i^{\vec{k_i}}\} \cup$$
$$\{\forall a. Exog(a) \equiv (\exists a')a = post(a')\},$$

where for every $i \in \{1, \ldots, n\}$, $\mathcal{B} \models b_i(\vec{x}) \equiv [\vec{x} = t_i^{\vec{1}} \vee \ldots \vee \vec{x} = t_i^{\vec{k_i}}]$, for some $k_i \geq 0$—$b_i(t_i^{\vec{1}}), \ldots, b_i(t_i^{\vec{k_i}})$ are all the true belief atoms in $\mathcal{B}$ with respect to belief relation b_i (each $t_i^{\vec{j}}$ is a vector of ground terms).

2. The following precondition axioms, for every fluent $b(\vec{x})$ and action type $A(\vec{x})$:

$$Poss(serve(a), s) \equiv (a \in PendingEv(s)) \qquad Poss(A(\vec{x}), s) \equiv \texttt{True}$$
$$Poss(add_b(\vec{x}), s) \equiv Poss(del_b(\vec{x}), s) \equiv \texttt{True} \qquad Poss(post(a), s) \equiv \texttt{True}$$

3. For every domain fluent $b(\vec{x}, s)$, $\mathcal{D}^{\Upsilon}$ includes the following successor state axiom:

$$b(\vec{x}, do(a, s)) \equiv$$
$$a = add_b(\vec{x}) \vee a = post(add_b(\vec{x})) \vee b(\vec{x}, s) \wedge (a \neq del_b(\vec{x}) \wedge a \neq post(del_b(\vec{x})).$$

That is, the truth-value of fluent b is affected only by actions add_b and del_b, either internally executed or externally sensed from the environment.

More importantly, action theory $\mathcal{D}^{\Upsilon}$ includes a successor state axiom for fluent $PendingEv(do(a, s))$ specifying how the multi-set of pending events changes:

$$PendingEv(do(a, s)) = v \equiv [\gamma(a, v, s) \lor PendingEv(s) = v \land \neg \exists v'.\gamma(a, v', s)];$$

where:

$$\gamma(a, v, s) \stackrel{\text{def}}{=} \left(\bigvee_{i=1}^{n} [\gamma_i^+(a, v, s) \lor \gamma_i^-(a, v, s)] \lor \bigvee_{i=1}^{m} [\gamma_i^e(a, v, s)] \lor \exists a'.a = serve(a') \land v = PendingEv(s) \setminus \{a'\} \right);$$

$$\gamma_i^+(a, v, s) \stackrel{\text{def}}{=}$$
$$\exists \vec{x}.\, a \in \{add_{b_i}(\vec{x}), post(add_{b_i}(\vec{x}))\} \land \neg b_i(\vec{x}) \land v = PendingEv(s) \uplus \{add_{b_i}(\vec{x})\};$$

$$\gamma_i^-(a, v, s) \stackrel{\text{def}}{=}$$
$$\exists \vec{x}.\, a \in \{del_{b_i}(\vec{x}), post(del_{b_i}(\vec{x}))\} \land b_i(\vec{x}) \land v = PendingEv(s) \uplus \{del_{b_i}(\vec{x})\};$$

$$\gamma_i^e(a, v, s) \stackrel{\text{def}}{=} \exists \vec{x}.\, a = post(ach_{e_i}(\vec{x})) \land v = PendingEv(s) \uplus \{ach_{e_i}(\vec{x})\}.$$

That is, an actual change in the belief of an atom, either due to the execution of some intention or an external event, automatically produces a corresponding pending belief update event. Moreover, an external achievement event $ach_e(\vec{x})$ becomes pending when sensed. On the other hand, an event e ceases to be pending when action $serve(e)$ is executed.

4. Theory $\mathcal{D}^{\Upsilon}$ includes unique name axioms for all actions in $\mathcal{L}_{sitCalc}$, as well as the standard domain-independent foundational axioms for the situation calculus ([Reiter 2001]).

This concludes the construction of the BDI basic action theory $\mathcal{D}^{\Upsilon}$.

3.2 The BDI Agent Program

Let us now construct the IndiGolog BDI agent program δ^{Υ} that is meant to execute relative to the BDI action theory $\mathcal{D}^{\Upsilon}$. We start by showing how to inductively transform a BDI plan-body program P into an IndiGolog program δ_P, namely (remember that plan-bodies programs are used to build BDI plans in the plan library):

$$\delta_P = \begin{cases} P & \text{if } P = A \mid nil \\ \phi? & \text{if } P = ?\phi \\ add_b(\vec{t}) & \text{if } P = +b(\vec{t}) \\ del_b(\vec{t}) & \text{if } P = -b(\vec{t}) \\ handle(ach_e(\vec{t})) & \text{if } P = !e(\vec{t}) \\ (\delta_{P_1}; \delta_{P_2}) & \text{if } P = (P_1; P_2) \\ \delta_{P_1} & \text{if } P = P_1 \triangleright P_2 \\ achieve_e(\vec{x}\theta) & \text{if } P = (\!|\Delta|\!) \text{ and } \Delta \subseteq \{\psi\theta : P\theta \mid e : \psi \leftarrow P \in \Pi\}, \text{ for} \\ & \quad \text{some } \theta \text{ (partially) binding the parameters of event } e(\vec{x}). \end{cases}$$

Notice that achievement events $!e$ occurring in a plan are handled via simple plan invocation, by invoking procedure $handle$. Also, for now, the translation just ignores the second program in $P_1 \triangleright P_2$, as the version of CAN we consider in this section does not try P_2 when P_1 happens to fail. We shall revisit this later in Section 4. Finally, when the BDI program is a set of (relevant) alternatives of the form $(\!|\Delta|\!)$ for some event e, we map it to a procedure call that will basically amount to a non-deterministic choice among all such alternatives (see below).

Next, we describe how to transform the BDI plans in the agent's plan library. To that end, suppose that $e(\vec{x})$ is an event in the BDI language $\mathcal{L}_{BDI}$ such with the following $n \geq 0$ plans in Π ($\vec{v}_{\vec{t}}$ denotes all the distinct free variables in the terms $\vec{t}$):

$$e(\vec{t_i}) : \psi_i(\vec{v}_{\vec{t_i}}, \vec{y_i}) \leftarrow P_i(\vec{v}_{\vec{t_i}}, \vec{y_i}, \vec{z_i}), \text{ where } i \in \{1, \ldots, n\}.$$

Then, we build the following high-level Golog procedure with n non-deterministic choices (i.e., as many as plan-rules for the event):

> **proc** $achieve_e(\vec{x})$
> $\quad |_{i \in \{1, \ldots, n\}} \ [(\pi \vec{v}_{\vec{t_i}}, \vec{y_i}, \vec{z_i}).(\vec{x} = \vec{t_i} \wedge \psi_i(\vec{v}_{\vec{t_i}}, \vec{y_i}))?; \delta_{P_i}(\vec{v}_{\vec{t_i}}, \vec{y_i}, \vec{z_i})]$
> **endProc**

Roughly speaking, executing $achieve_e(\vec{x})$ involves nondeterministically choosing among the n available options in the plan library for event e. See that the first test statement in each option amounts to checking the relevance and applicability of the option. Thus, the execution of $achieve_e(\vec{x})$ is bound to *block* if no option is relevant or applicable. In particular, the procedure will *always* block if the agent Υ has no plan to handle the event in question—that is, if $n = 0$, the corresponding Golog procedure is simply $?(\texttt{False})$.

Let Δ_Π denote the set of Golog procedures as above, one per event in the BDI language, together with the following procedure:

> **proc** $handle(a)$
> $\quad |_{i=1}^{n} \ [(\exists \vec{x_i}.a = add_{b_i}(\vec{x_i}))?; \ achieve_{b_i^+}(\vec{x_i})] \ |$
> $\quad |_{i=1}^{n} \ [(\exists \vec{x_i}.a = del_{b_i}(\vec{x_i}))?; \ achieve_{b_i^-}(\vec{x_i})] \ |$
> $\quad |_{i=1}^{m} \ [(\exists \vec{x_i}.a = ach_{e_i}(\vec{x_i}))?; \ achieve_{e_i}(\vec{x_i})]$
> **endProc**

That is, when a is a legal event (belief update or achievement goal), procedure $handle(a)$ calls the appropriate procedure that is meant to *resolve* the event. Observe that this program contains two nondeterministic programs per belief atom in the domain (one to handle its addition and one to handle its deletion from the belief base), plus one nondeterministic program per achievement event in the domain.

Finally, we define the top-level IndiGolog BDI agent program as follows:

$$\delta^\Upsilon \stackrel{\text{def}}{=} \Delta_\Pi; [\delta_{env} \| (\delta_\Gamma \| \delta_{BDI})]; (\neg \exists e \ PendingEv(e))?, \tag{1}$$

where (assuming that $\Gamma = \{P_1, \ldots, P_n\}$):

$$\delta_\Gamma \stackrel{\text{def}}{=} \delta_{P_1} \| \cdots \| \delta_{P_n}; \quad \delta_{env} \stackrel{\text{def}}{=} (\pi a. Exog(a)?; a)^*; \quad \delta_{BDI} \stackrel{\text{def}}{=} [\pi a. serve(a); handle(a)]^\|.$$

The set of programs Δ_Π provides the environment encoding the BDI plan library. Program δ_Γ accounts for all current intentions in Υ; if $\Gamma = \emptyset$, then $\delta_\Gamma = nil$. In turn, program δ_{env} models the external environment, which can perform zero, one, or more actions of the form $post(a)$, representing an external achievement event goal ($a = ach_e(\vec{t})$) or a belief update event ($a = add_b(\vec{t})$ or $a = del_b(\vec{t})$).

The most interesting part of δ^{Υ} is indeed the ConGolog program δ_{BDI}, which implements (part of) the BDI execution cycle. More concretely, δ_{BDI} is responsible for selecting an external event and *spawning* a new "intention" concurrent thread for handling it. To that end, δ_{BDI} picks an event a (e.g., $add_{At}(23, 32)$ or $achieve_{moveTo}(0, 0)$) to be served and executes action $serve(a)$. Observe that an event can be served only if it is currently pending (see action precondition for action $serve(a)$ in Subsection 3.1). After the action $serve(a)$ has been successfully executed, the selected event a is actually handled, by calling procedure $handle(a)$ defined in Δ_{Π}. More importantly, this is done in a "new" concurrent thread, so that program δ_{BDI} is still able to serve and handle other pending events. The use of concurrent iteration to spawn a new intention from the "main BDI thread" is inspired from the server example application in [De Giacomo, Lespérance, and Levesque 2000].

Note that Δ_{Π} and δ_{Γ} are domain dependent, i.e., they are built relative to a particular BDI agent Υ, whereas programs δ_{BDI} and δ_{env} are independent of the BDI agent being encoded. Observe also that the whole high-level program δ^{Υ} may terminate only when no more events are pending.

From now on, let $\mathcal{G}^{\Upsilon} = \langle \mathcal{D}^{\Upsilon}, \delta^{\Upsilon}, \mathcal{A} \rangle$ denote the IndiGolog agent for BDI agent Υ.

3.3 LC-Online Executions

Once we have a BDI IndiGolog program $\mathcal{G}^{\Upsilon}$ on hand, we should be able to execute it and obtain the same behavior and outputs as with the original BDI agent. Unfortunately, we cannot execute $\mathcal{G}^{\Upsilon}$ online, as defined in [De Giacomo, Lespérance, Levesque, and Sardina 2009], as such executions may commit too early to free variables in a program—online executions are sequences of *ground* online configurations. For example, program $\pi x. turnOffAlarm; Edible(x)?; eat(x)$ may do a transition by executing the primitive action *turnOffAlarm* and instantiating x to *Clock*, yielding $Edible(Clock)?; eat(Clock)$ as the remaining program, which is bound to fail as the object picked is not edible. In fact, as no constraints are imposed on x in the first transition, any binding for it would be legal.

What we need, instead, is an account of execution that commits to free variables only when necessary. To that end, we generalize the online execution notion from [De Giacomo, Lespérance, Levesque, and Sardina 2009; Sardina, De Giacomo, Lespérance, and Levesque 2004] to what we call *least-committed* online executions. Below, we use $end[\sigma]$ to denote the situation term corresponding to the history σ; and $Axioms[\mathcal{D}, \sigma]$ to denote the complete set of axioms in the IndiGolog theory, which includes the action theory $\mathcal{D}$ for the domain, the sensing results gathered so far in history σ, and all axioms for *Trans* and *Final*. So, we first define two meta-theoretic versions of relations *Trans* and *Final* as follows:

$$mTrans(\delta(\vec{x}, \vec{y}), \sigma, \delta'(\vec{x}, \vec{z}), \sigma') \stackrel{\text{def}}{=}$$
$$Axioms[\mathcal{D}, \sigma] \models \exists \vec{y} \forall \vec{x}, \vec{z}. Trans(\delta(\vec{x}, \vec{y}), end[\sigma], \delta'(\vec{x}, \vec{z}), end[\sigma']);$$

$$mFinal(\delta(\vec{x}, \vec{y}), \sigma) \stackrel{\text{def}}{=} Axioms[\mathcal{D}, \sigma] \models \exists \vec{x}. Final(\delta(\vec{x}), end[\sigma]).$$

Here, $\delta(\vec{x})$ means that the vector of variables $\vec{x}$ contains *all* the free variables mentioned in the program; different variables vectors are assumed disjoint. Thus, $\vec{x}$ are the free variables in δ that are still free in δ'; $\vec{y}$ are the free variables in δ that have been instantiated and are not present in δ'; and $\vec{z}$ are the new free variables in δ' that did not appear in δ.

We can then define least-committed executions as follows.

DEFINITION 2 (LC-Online Execution). A *least-committed online* (lc-online) execution of an IndiGolog program δ starting from a history σ is a, possibly infinite, sequence of configurations $(\delta_0 = \delta, \sigma_0 = \sigma), (\delta_1, \sigma_1), \ldots$ such that for every $i \geq 0$:

1. $mTrans(\delta_i, \sigma_i, \delta_{i+1}, \sigma_{i+1})$ holds; and

2. for all δ' such that $mTrans(\delta_i, \sigma_i, \delta', \sigma_{i+1})$ and $\delta_{i+1} = \delta'\theta$ for some substitution θ, there exists θ' such that $\delta' = \delta_{i+1}\theta'$.

A finite lc-online execution $(\delta_0, \sigma_0), \ldots, (\delta_n, \sigma_n)$ is *terminating* iff $mFinal(\delta_n, \sigma_n)$ or for all δ', σ' $mTrans(\delta_n, \sigma_n, \delta', \sigma')$ does not hold. □

We notice that, as expected, it can be shown that an lc-online execution stands for all its ground online instances as defined in [De Giacomo, Lespérance, Levesque, and Sardina 2009]. However, by executing programs in a least committed way, we avoid premature binding of variables and eliminate some executions where the program is bound to fail.

3.4 BDI/IndiGolog Bisimulation

We are now ready to provide the main results of the paper. Namely, we show that given any BDI execution of an agent, there exists a matching execution of the corresponding IndiGolog agent, and vice-versa. In addition, the correspondence in the internal structure of the agents is always maintained throughout the executions.

We start by characterizing when a BDI agent and an IndiGolog agent configuration "match." To that end, we shall use relation $\Upsilon \approx \mathcal{G}$, which, intuitively, holds if a BDI agent Υ and an IndiGolog agent $\mathcal{G}$ represent the same (BDI) agent system. Formally, relation $\langle \Pi, \mathcal{B}, \mathcal{A}, \Gamma \rangle \approx \langle \mathcal{D}, \delta, \sigma \rangle$ holds iff

1. $\delta = \Delta_\Pi; [\delta_{env} \| (\delta_{\Gamma'} \| \delta_{BDI})]; ?(\neg \exists e\ PendingEv(e))$, for some $\Gamma' \subseteq \Gamma$ such that $\Gamma = \Gamma' \uplus \{a \mid Axioms[\mathcal{D}, \sigma] \models a \in PendingEv(end[\sigma])\}$;

2. $\mathcal{A}$ and σ contain the same sequence of *domain* actions;

3. for every ground belief atom b: $\mathcal{B} \models b$ iff $Axioms[\mathcal{D}, \sigma] \models b[end[\sigma]]$;

4. $\mathcal{D} = \mathcal{D}^{\Upsilon'}$, for some $\Upsilon' = \langle \Pi, \mathcal{B}', \mathcal{A}, \Gamma \rangle$.

The first condition states that the IndiGolog program is of the form shown in equation (1) above (see Section 3.2), but where some active intentions may still be "pending." In other words, some intentions in Γ that have not yet started execution may not show up yet as concurrent processes in δ, but they are implicitly represented as "pending" in fluent $PendingEv(s)$. The second requirement states that both agents have performed the same sequence of domain primitive actions, that is, actions other than the internal ones $serve(a)$, $post(a)$, $add_b(\vec{x})$, and $del_b(\vec{x})$. The third condition requires both agents to coincide on what they *currently* believe. Observe that the *initial* beliefs of the IndiGolog agent do not necessarily need to coincide with those of the BDI agent, as long as the *current* beliefs do (i.e., the beliefs that hold after history σ); in fact the BDI agent configuration does not specify what it believed initially, while the IndiGolog agent's action theory does. Lastly, the IndiGolog agent executes relative to a basic action theory whose dynamics are as described in Section 3.1.

First of all, it is possible to show that the encoding of initial BDI agents, that is agents that have not yet performed any action, into IndiGolog agents described above is in the $\approx$ relation with the original BDI agent.

THEOREM 3. *Let Υ be an initial BDI agent (that is, $\mathcal{A} = \epsilon$). Then, $\Upsilon \approx \langle \mathcal{D}^\Upsilon, \delta^\Upsilon, \mathcal{A} \rangle$.*

The importance of a BDI agent and an IndiGolog agent being in the $\approx$ relation is that their respective transitions can then always be simulated by the other type of agent To demonstrate that, we first show that any BDI transition can be replicated by the corresponding IndiGolog agent. Observe that IndiGolog may need several transitions to replicate the BDI transition when it comes to assimilating external events; whereas BDI agents incorporate sets of external events in a single transition, the IndiGolog agent incorporates one event per transition. Also, IndiGolog agents ought to execute the special action $serve(a)$ to start handling external achievement events.

THEOREM 4. *Let Υ be a BDI agent and $\langle \mathcal{D}, \delta, \sigma \rangle$ an IndiGolog agent where $\Upsilon \approx \langle \mathcal{D}, \delta, \sigma \rangle$. If $\Upsilon \Longrightarrow \Upsilon'$, then there exists a program δ' and a history σ' such that $mTrans^*(\delta, \sigma, \delta', \sigma')$ holds relative to action theory $\mathcal{D}$, and $\Upsilon' \approx \langle \mathcal{D}, \delta', \sigma' \rangle$.*

Furthermore, in the other direction, any step in a BDI IndiGolog execution can always be "mimicked" by the corresponding BDI agent.

THEOREM 5. *Let Υ and $\langle \mathcal{D}, \delta, \sigma \rangle$ be a BDI and an IndiGolog agents, respectively, such that $\Upsilon \approx \langle \mathcal{D}, \delta, \sigma \rangle$. Suppose that $mTrans(\delta, \sigma, \delta', \sigma')$ holds relative to action theory $\mathcal{D}$, for some IndiGolog program δ' and history σ'. Then, either $\Upsilon \approx \langle \mathcal{D}, \delta', \sigma' \rangle$ or there exists a BDI agent Υ' such that $\Upsilon \Longrightarrow \Upsilon'$ and $\Upsilon' \approx \langle \mathcal{D}, \delta', \sigma' \rangle$.*

So, when the IndiGolog agent performs a transition it remains "equivalent" to the BDI agent or to some evolution of the BDI agent. The former case applies only when the transition in question involved the execution of a $serve(a)$ action to translate a pending event into a concurrent process.

Putting both theorems together, our encoding allows IndiGolog to bisimulate BDI agents.

4 BDI Failure Handling

Since BDI systems are meant to operate in dynamic settings, plans that were supposed to work may fail due to changes in the environment. Indeed, a plan may fail because a test condition $?\phi$ is not believed true, an action cannot be executed, or a sub-goal event does not have any applicable plans. The BDI language we have discussed so far has no strategy towards failed plans or intentions, once an intention cannot evolve, it simply remains in the intention base *blocked*. In this section, we discuss how BDI programming languages typically address plan/intention failure and show how the above IndiGolog encoding can be extended accordingly. In particular, we show how agents can *abandon* failed intentions and *recover* from problematic plans by trying alternative options.

Before getting into technical details, we shall first introduce a new construct into the IndiGolog language. In "Golog-like" languages, a program that is *blocked* may not be dropped for the sake of another program. To overcome this, we introduce the construct $\delta_1 \triangleright \delta_2$ with the intending meaning that δ_1 should be executed, falling back to δ_2 if δ_1 becomes blocked:[3]

$$Trans(\delta_1 \triangleright \delta_2, s, \delta', s') \equiv (\exists \gamma.\, Trans(\delta_1, s, \gamma, s') \wedge \delta' = \gamma \triangleright \delta_2) \vee$$
$$\neg \exists \gamma, s''.\, Trans(\delta_1, s, \gamma, s'') \wedge Trans(\delta_2, s, \delta', s');$$

$$Final(\delta_1 \triangleright \delta_2, s, \delta', s') \equiv Final(\delta_1, s) \vee \neg \exists \gamma, s''.\, Trans(\delta_1, s, \gamma, s'') \wedge Final(\delta_2, s).$$

---

[3]One could easily extend these definitions to only allow dropping a blocked δ_1 under given conditions; this could be used to implement "time outs" or allow blocking for synchronization.

4.1 Dropping Impossible Intentions

It is generally accepted that intentions that cannot execute further may simply be *dropped* by the agent — rational agents should not pursue intentions/goals that are deemed impossible [Rao and Georgeff 1991; Cohen and Levesque 1990]. This is indeed the behavior of AgentSpeak agents.[4]

The BDI language of Section 2.1 can be easily extended to provide such an intention-dropping facility, by just adding the following agent-level operational rule:

$$\frac{P \in \Gamma \quad \langle \Pi, \mathcal{B}, \mathcal{A}, P \rangle \not\longmapsto}{\langle \Pi, \mathcal{B}, \mathcal{A}, \Gamma \rangle \Longrightarrow \langle \Pi, \mathcal{B}, \mathcal{A}, \Gamma \setminus \{P\} \rangle} A_{clean}$$

That is, an agent may choose to just drop an intention from its intention base if it cannot execute further in the current mental state. To mimic this behavior in our BDI IndiGolog formalization, we slightly modify the domain-independent program δ_{BDI} as follows:

$$\delta_{BDI} \stackrel{\text{def}}{=} [\pi a. \, serve(a); (handle(a) \rhd (\texttt{True})?)]^{\|}.$$

Here, a pending event is handled within the scope of a $\rhd$, which basically allows the intention thread to simply terminate if it becomes blocked. Notice that, as with BDI languages, for procedure $handle(a)$ to be blocked, every sub-goal event triggered by the handling of a (represented in the IndiGolog program as simple procedure calls) ought to be blocked. Observe also that in this approach, only the main program corresponding to a top-level event may be dropped, not lower-level instrumental subgoals.

4.2 BDI Goal Failure Recovery

Merely dropping a whole intention when it becomes *blocked* provides a rather weak level of commitment to goals. The failure of a plan should not be equated to the failure of its parent goal, as there could be alternative ways to achieve the latter. For example, suppose an agent has the goal to quench her thirst, and in the service of this goal, she adopts the subgoal of buying a can of soda [Sardina and Padgham 2007]. However, upon arrival at the store, she realizes that all the cans of soda are sold out. Fortunately though, the shop has bottles of water. In this situation, it is irrational for the agent to drop the whole goal of quenching her thirst just because soda is not available. Yet this is what an AgentSpeak agent would do. Similarly, we do not expect the agent to fanatically insist on her subgoal and just wait indefinitely for soda to be delivered. What we expect is the agent to merely drop her commitment to buy soda and adopt the alternative subgoal of buying a bottle of water, thereby achieving the *main* goal.

As a matter of fact, one of the typical features of implemented BDI languages is that of plan-goal failure recovery: if a plan happens to fail for a goal, usually due to unexpected changes in the environment, another plan is tried to achieve the goal. If no alternative strategy is available, then the goal is deemed failed and failure is propagated to higher-level motivating goals, and so on. This mechanism thus provides a stronger level of *commitment to goals*, by decoupling plan failure from goal failure.

To accommodate this approach to failure recovery, we further extend the BDI language of Section 2.1, by providing the following additional derivation rule for "try" construct $\rhd$:

$$\frac{\langle \Pi, \mathcal{B}, \mathcal{A}, P_1 \rangle \not\longmapsto \quad \langle \Pi, \mathcal{B}', \mathcal{A}', P_2' \rangle \longrightarrow \langle \Pi, \mathcal{B}', \mathcal{A}', P_2' \rangle}{\langle \Pi, \mathcal{B}, \mathcal{A}, P_1 \rhd P_2 \rangle \longrightarrow \langle \Pi, \mathcal{B}', \mathcal{A}', P_2' \rangle} \rhd_f$$

---

[4]There has been work on more sophisticated treatments of plan failure in extensions of AgentSpeak; see for instance [Bordini, Hübner, and Wooldridge 2007].

That is, if the current strategy P_1 is *blocked* but the alternative backup program P_2 is able to evolve, then it is legal to drop P_1 and switch to P_2. Observe that due to derivation rules *Ev* and *Sel*, $P_2 = (\!|\Delta|\!)$ will encode the set of *relevant* plans that have not yet been tried for the event being addressed. From now on, let the CAN language refer to our extended BDI language, with both new derivation rules A_{clean} and $\rhd_f$ for failure included.

Hence, due to the interaction between derivation rules *Ev*, *Sel* and $\rhd_f$, a CAN BDI agent executes a program $P_1 \rhd (\!|\Delta|\!)$ in order to resolve an goal event $!e$. When the current strategy P_1 being pursued is not able to make a step, the agent may check the set of alternatives $(\!|\Delta|\!)$ in the hope of finding another option P_2 for addressing e. If one is found, the agent may opt to abandon its strategy P_1 and continue with P_2. (Details can be found in [Winikoff, Padgham, Harland, and Thangarajah 2002; Sardina and Padgham 2007].)

Let us now describe how to replicate this failure recovery behavior within our IndiGolog framework of Section 3. For simplicity, we shall assume that, as with actions, only *ground* posting of subgoal events are allowed in the BDI language. This means that all variables $\vec{x}$ in an event posting $!e(\vec{x})$ are considered *inputs* to the event. If an event is meant to return data, it must do so by using of the belief base. To support failure recovery, we slightly modify how plans in the plan library Π are converted into ConGolog procedures. Specifically, for each event $e(\vec{x})$, we define the following procedure (and make procedure $achieve_e(\vec{x})$ simply call $achieve'_e(\vec{x}, [1, \ldots, 1])$):

> **proc** $achieve'_e(\vec{x}, \vec{w})$ // $\vec{w}$ is an n-long boolean vector
> $|_{i \in \{1, \ldots, n\}} \; [(\pi \vec{v_{t_i}}, \vec{y_i}, \vec{z_i}).(\vec{x} = \vec{t} \land \psi_i(\vec{v_{\vec{t}}}, \vec{y}) \land w = 1)?; \delta_{P_i}(\vec{v_{t_i}}, \vec{y_i}, \vec{z_i}) \rhd \Phi_i(\vec{x}, \vec{w})]$
> **endProc**

> where $\Phi_i(\vec{x}, \vec{w}) \stackrel{\text{def}}{=} achieve'_e(\vec{x}, [w_1, \ldots, w_{i-1}, 0, w_{i+1}, \ldots, w_n])$.

The boolean vector $\vec{w}$ has one component per plan rule in the library for the event in question; its i-th component w_i states whether the i-th plan in Π is *available* for selection. Condition $(\vec{x} = \vec{t} \land \psi_i(\vec{v_{\vec{t}}}, \vec{y}) \land w = 1)$ checks whether event $e(\vec{x})$ is relevant, applicable, and available. Program Φ_i determines the *recovery strategy*, in this case, recursively calling the procedure to achieve the event, but removing the current plan from consideration (by setting its component in $\vec{w}$ to 0). Due to the semantics of $\rhd$, recovery would only be triggered if procedure $achieve'_e(\vec{x}, \vec{w})$ may execute one step, which implies that there is indeed an available plan that is relevant and applicable for the event.

It turns out that these are the only modifications to the encoding of Section 3 required to mimic the behavior of CAN agents with failure handling in the IndiGolog high-level language.

THEOREM 6. *Theorems 4 and 5 hold for CAN agents that drop impossible intentions and perform failure recovery under the extended translations into IndiGolog agents.*[5]

More interestingly, the proposed translation can be adapted to accommodate several alternative accounts of execution and failure recovery. For example, goal failure recovery can be disallowed for an event by just taking $\Phi_i(\vec{x}, \vec{w}) \stackrel{\text{def}}{=} ?(\texttt{False})$ above. Similarly, a framework under which *any* plan may be (re)tried for achieving a goal event, regardless of previous (failed) executions, is obtained by taking $\Phi_i(\vec{x}, \vec{w}) \stackrel{\text{def}}{=} achieve_e(\vec{x})$. In this case, the event is "fully" re-posted within the intention.

---

[5]Note that for this to go through, we have to extend the translation δ_P (Section 3.2) of programs of the form $(\!|\Delta|\!)$ to set the bit vector w in $achieve_e(\vec{x}.w)$ properly (i.e., to 1 iff the alternative is in the $(\!|\Delta|\!)$).

The key point here is that, due to the fact that the BDI execution and recovery model is represented in our BDI IndiGolog at the *object* level, one can go even further and design more sophisticated accounts of execution and failure recovery for BDI agents. It is straightforward, for instance, to model the kind of goal failure recovery originally proposed for AgentSpeak, in which the system would automatically post a distinguished *failure goal* (denoted $!-g$); the programmer may then choose to provide handling plan could, for example, carry out some clean-up tasks and even re-post the failed event [Rao 1996; Bordini, Hübner, and Wooldridge 2007]. This type of behavior can be easily achieved by taking $\Phi_i(\vec{x}, \vec{w}) \stackrel{\text{def}}{=} ach_{fail\_e}(\vec{x}); ?(\texttt{False})$, and allowing the programmer to provide plan rules in the library for handling the special event $fail\_e(\vec{x})$. Notice that the event is *posted* so it would eventually create a new intention all-together; the current plan would then immediately be blocked/failed.

5 Discussion

In this paper, we have shown how one can effectively program BDI-style agent systems in the situation calculus-based IndiGolog high-level programming language. The benefits of this are many. First, we gain a better understanding of the common features of BDI agent programming languages and "Golog-like" high-level programming languages, as well as of what is specific to each type of language, and what is required to reproduce BDI languages in the latter. We also get a new classical-logic situation calculus-based semantics for BDI agent programming languages. This opens many avenues for enhancing the BDI programming paradigm with reasoning capabilities, for instance, model-based belief update capabilities, lookahead planning capabilities, plan/goal achievement monitoring capabilities, etc. In fact, the situation calculus and basic action theories provide a rich and and well-studied logical framework for specifying the belief update and planning part of agent programming languages.

It might be said that our bisimulation results are not very surprising, as both BDI languages and IndiGolog are universal programming languages. However, the simplicity and modularity of our encoding shows that the relationship between BDI languages and IndiGolog is actually fairly close. Representing a BDI-style plan library (including the association between a plan and the event/goal it addresses) in IndiGolog is quite straightforward: each plan becomes an alternative in an achievement procedure associated with the event/goal; event-directed plan invocation can be done by calling the event's achievement procedure. The key feature that is missing, BDI-style event-directed plan triggering, can in fact be added by incorporating into the IndiGolog program an "event server" that calls the event's achievement procedure; such a server can easily be programmed using IndiGolog's concurrent iteration construct. We also show how the simple BDI language approach to belief update can be modeled in the situation calculus, using action theories with a simple type of successor state axiom and add_{b_i} and del_{b_i} actions. This then gives us a nice framework for defining more sophisticated belief update approaches. As well, we have shown that failure recovery mechanisms can be added to IndiGolog by introducing constructs such as the "try" construct $\triangleright$, very much as has been done in some BDI-languages.

One could also argue that the approach we follow decreases the "separation of concerns" in that both the agent program and the generic BDI execution engine are encoded into a single IndiGolog high-level program, meaning that the agent program and the BDI interpreter are no longer separated. To avoid this, one could develop an alternative account where one gives a situation calculus semantics for the BDI execution cycle at the meta-level, by re-defining what a BDI-IndiGolog agent is and what counts as an "online" execution for such agents. In this paper, however, we intended to keep the IndiGolog framework as intact as possible. Also, by encoding the BDI engine in a logical/object language (as an IndiGolog program), one can formally express (and prove) properties of programs and of the programming language in the (situation calculus) object language. Nonetheless, the alternative approach is of interest and we are in fact working on it.

There has only been limited work on relating "Golog-like" and BDI programming languages. Hindriks *et al.* [Hindriks, Lespérance, and Levesque 2000] show that ConGolog can be bisimulated by the agent language 3APL under some conditions, which include the agent having complete knowledge. In [Hindriks, de Boer, van der Hoek, and Meyer 1998], it is also shown that AgentSpeak can be encoded into 3APL. Our results, thus, are complementary, in showing the inverse relationship.

Also related is the work of Gabaldon [Gabaldon 2002] on encoding Hierarchical Task Network (HTN) libraries in ConGolog. There are similarities between our work and his in the way procedural knowledge is encoded in ConGolog. This is is not surprising, as HTN planning systems and BDI agents have many similarities [Dix, Muñoz-Avila, Nau, and Zhang 2003]. But note that in HTNs, and hence in Gabaldon's translation, the objective is *planning* and not reactive execution. We on the other hand, focus on capturing the typical execution regime of BDI agent systems, rather than on performing lookahead planning to synthesize a solution. As a result, we address issues such as external events and plan failure that do not arise in HTN planning.

Finally, we note that in this work we have mostly focused on the core features of BDI systems, and have not dealt with more advanced features present in some more recent versions of BDI systems. For instance, the full CAN [Winikoff, Padgham, Harland, and Thangarajah 2002; Sardina and Padgham 2007] language as well as 2APL [Dastani 2008] and Jason [Bordini, Hübner, and Wooldridge 2007; Hübner, Bordini, and Wooldridge 2006] provide, in some way or other, support for so-called *declarative goals*, goals which go beyond "events" by decoupling goal success/failure from plan success/failure. However, the way such advanced features have been added to different BDI languages is not always uniform. More work is needed to handle those advanced features in an encoding of BDI programming into IndiGolog.

Acknowledgments

An earlier version of this paper originally appeared in [Sardina and Lespérance 2010] and is reprinted with kind permission of Springer Science+Business Media. This work was supported by Agent Oriented Software and the Australian Research Council (grant LP0882234) and the National Science and Engineering Research Council of Canada.

References

Belecheanu, R. A., S. Munroe, M. Luck, T. Payne, T. Miller, P. McBurney, and M. Pechoucek [2006, Hakodate, Japan). Commercial applications of agents: Lessons, experiences and challenges. In *Proceedings of Autonomous Agents and Multi-Agent Systems (AAMAS)*, pp. 1549–1555. ACM Press.

Benfield, S. S., J. Hendrickson, and D. Galanti [2006]. Making a strong business case for multiagent technology. In *Proceedings of Autonomous Agents and Multi-Agent Systems (AAMAS)*, New York, NY, USA, pp. 10–15. ACM Press.

Bordini, R. H., J. F. Hübner, and M. Wooldridge [2007]. *Programming Multi-agent*

Systems in AgentSpeak Using Jason. Wiley Series in Agent Technology. Wiley.

Bratman, M. E. [1987]. *Intentions, Plans, and Practical Reason.* Harvard University Press.

Busetta, P., R. Rönnquist, A. Hodgson, and A. Lucas [1999, January). JACK intelligent agents: Components for intelligent agents in Java. *AgentLink Newsletter 2*, 2–5. Agent Oriented Software Pty. Ltd.

Cohen, P. R. and H. J. Levesque [1990]. Intention is choice with commitment. *Artificial Intelligence Journal 42*, 213–261.

Dastani, M. [2008, June). 2APL: a practical agent programming language. *Autonomous Agents and Multi-Agent Systems 16*(3), 214–248.

De Giacomo, G., Y. Lespérance, and H. J. Levesque [2000]. ConGolog, a concurrent programming language based on the situation calculus. *Artificial Intelligence Journal 121*(1–2), 109–169.

De Giacomo, G., Y. Lespérance, H. J. Levesque, and S. Sardina [2009]. IndiGolog: A high-level programming language for embedded reasoning agents. In R. H. Bordini, M. Dastani, J. Dix, and A. E. Fallah-Seghrouchni (Eds.), *Multi-Agent Programming: Languages, Platforms and Applications*, Chapter 2, pp. 31–72. Springer.

Dennett, D. [1987]. *The Intentional Stance.* The MIT Press.

Dix, J., H. Muñoz-Avila, D. S. Nau, and L. Zhang [2003]. IMPACTing SHOP: Putting an AI planner into a multi-agent environment. *Annals of Mathematics and Artificial Intelligence 37*(4), 381–407.

Gabaldon, A. [2002, April). Programming hierarchical task networks in the situation calculus. In *Proc. of AIPS'02 Workshop on On-line Planning and Scheduling*, Toulouse, France.

Georgeff, M. P. and F. F. Ingrand [1989]. Decision making in an embedded reasoning system. In *Proceedings of the International Joint Conference on Artificial Intelligence (IJCAI)*, Detroit, USA, pp. 972–978.

Hindriks, K. V., F. S. de Boer, W. van der Hoek, and J.-J. Meyer [1998]. A formal semantics for an abstract agent programming language. In *Proceedings of the International Workshop on Agent Theories, Architectures, and Languages (ATAL)*, pp. 215–229.

Hindriks, K. V., F. S. de Boer, W. van der Hoek, and J.-J. Meyer [1999]. Agent programming in 3APL. *Autonomous Agents and Multi-Agent Systems 2*, 357–401.

Hindriks, K. V., Y. Lespérance, and H. J. Levesque [2000]. An embedding of Con-Golog in 3APL. In *Proceedings of the European Conference in Artificial Intelligence (ECAI)*, pp. 558–562.

Hübner, J. F., R. H. Bordini, and M. Wooldridge [2006]. Programming declarative goals using plan patterns. In *Proceedings of the International Workshop on Declarative Agent Languages and Technologies (DALT)*, Volume 4327 of *Lecture Notes in Computer Science (LNCS)*, pp. 123–140.

Levesque, H. J., R. Reiter, Y. Lespérance, F. Lin, and R. B. Scherl [1997]. GOLOG: A logic programming language for dynamic domains. *Journal of Logic Programming 31*, 59–84.

Ljungberg, M. and A. Lucas [1992]. The OASIS air-traffic management system. In *Proceedings of the Pacific Rim International Conference on Artificial Intelligence (PRICAI)*, Seoul, Korea.

McCarthy, J. and P. J. Hayes [1969]. Some philosophical problems from the standpoint of artificial intelligence. *Machine Intelligence 4*, 463–502.

Pokahr, A., L. Braubach, and W. Lamersdorf [2003, 9]. JADEX: Implementing a BDI-infrastructure for JADE agents. *EXP - in search of innovation (Special Issue on JADE) 3*(3), 76–85.

Pollack, M. E. [1992]. The uses of plans. *Artificial Intelligence Journal 57*(1), 43–68.

Rao, A. S. [1996]. Agentspeak(L): BDI agents speak out in a logical computable language. In W. V. Velde and J. W. Perram (Eds.), *Proceedings of the Seventh European Workshop on Modelling Autonomous Agents in a Multi-Agent World. (Agents Breaking Away)*, Volume 1038 of *Lecture Notes in Computer Science (LNCS)*, pp. 42–55. Springer-Verlag.

Rao, A. S. and M. P. Georgeff [1991]. Modeling rational agents within a BDI-architecture. In *Proceedings of Principles of Knowledge Representation and Reasoning (KR)*, pp. 473–484.

Rao, A. S. and M. P. Georgeff [1992]. An abstract architecture for rational agents. In *Proceedings of Principles of Knowledge Representation and Reasoning (KR)*, San Mateo, CA, pp. 438–449.

Reiter, R. [2001]. *Knowledge in Action. Logical Foundations for Specifying and Implementing Dynamical Systems*. The MIT Press.

Sardina, S., G. De Giacomo, Y. Lespérance, and H. J. Levesque [2004, August]. On the semantics of deliberation in IndiGolog – From theory to implementation. *Annals of Mathematics and Artificial Intelligence 41*(2–4), 259–299.

Sardina, S. and Y. Lespérance [2010]. GologSpeak: Golog speaks the BDI language. In L. Braubach, J.-P. Briot, and J. Thangarajah (Eds.), *Proceedings of the Programming Multiagent Systems Languages, Frameworks, Techniques and Tools workshop (PROMAS)*, Volume 5919 of *Lecture Notes in Computer Science (LNCS)*, Budapest, Hungary, pp. 82–89. Springer.

Sardina, S. and L. Padgham [2007, May]. Goals in the context of BDI plan failure and planning. In E. H. Durfee, M. Yokoo, M. N. Huhns, and O. Shehory (Eds.), *Proceedings of Autonomous Agents and Multi-Agent Systems (AAMAS)*, Hawaii, USA, pp. 16–23. ACM Press.

Thielscher, M. [2005]. FLUX: A logic programming method for reasoning agents. *Theory and Practice of Logic Programming 5*(4–5), 533–565. Special Issue of Theory and Practice of Logic Programming on Constraint Handling Rules.

Winikoff, M., L. Padgham, J. Harland, and J. Thangarajah [2002, April]. Declarative & procedural goals in intelligent agent systems. In *Proceedings of Principles of Knowledge Representation and Reasoning (KR)*, Toulouse, France, pp. 470–481.

28

Belief Change with Noisy Sensing and Introspection

STEVEN SHAPIRO

ABSTRACT. We model belief change due to noisy sensing, and belief introspection in the framework of the situation calculus. We give some properties of our axiomatization and show that it does not suffer from the problems with combining sensing, introspection, and plausibility update described in Shapiro *et al.* [2000].

1 Introduction

In this paper, we generalize the framework of Shapiro *et al.* [2000], where belief change due to sensing was combined with belief introspection in the situation calculus. In that framework, sensing was assumed to be infallible and the plausibilities of alternate situations (i.e., possible worlds) were fixed in the initial state, never to be updated. Here, we relax both assumptions. That is, we model noisy sensors whose readings may be inaccurate, and may return different values in subsequent readings. We also allow the plausibilities of situations to change over time, bringing the framework more in line with traditional models of belief change. We give some properties of our axiomatization and show that it does not suffer from the problems with combining sensing, introspection, and plausibility update described by Shapiro *et al.* In the next section, we present the situation calculus and Shapiro *et al.*'s framework for modelling belief change in it, and we briefly review belief revision and update. In Section 3, we present the formal details of our framework for belief change. In Section 4, we discuss some properties of our framework, and in Section 5, we conclude and discuss future work.

2 Situation Calculus

The basis of our framework for belief change is an action theory [Reiter 2001] based on the situation calculus [McCarthy and Hayes 1969], and extended to include a knowledge operator [Scherl and Levesque 1993]. The situation calculus is a predicate calculus language for representing dynamic domains. A situation represents a snapshot of the domain. There is a set of initial situations corresponding to the ways the agent[1] believes the domain might be initially. The actual initial state of the domain is represented by the distinguished initial situation constant, S_0. The term $do(a, s)$ denotes the unique situation that results from the agent performing action a in situation s. Thus, the situations can be structured

---

[1] Here we assume that there is a single agent, however it would not be difficult to generalize the framework to handle multiple agents.

into a set of trees, where the root of each tree is an initial situation and the arcs are actions. The initial situations are defined as those situations that do not have a predecessor $Init(s) \stackrel{\text{def}}{=} \neg\exists a, s'.s = do(a, s')$.

Predicates and functions whose value may change from situation to situation (and whose last argument is a situation) are called *fluents*. For instance, we use the fluent $INR_1(s)$ to represent that the agent is in room R_1 in situation s. The effects of actions on fluents are defined using *successor state axioms* [Reiter 2001], which provide a succinct representation for both effect axioms and frame axioms [McCarthy and Hayes 1969]. For example, assume that there are only two rooms, R_1 and R_2, and that the action LEAVE takes the agent from the current room to the other room. Then, the successor state axiom for INR_1 would be as follows:[2]

$$INR_1(do(a, s)) \equiv$$
$$((\neg INR_1(s) \wedge a = \text{LEAVE}) \vee (INR_1(s) \wedge a \neq \text{LEAVE})).$$

This axiom asserts that the agent will be in R_1 after doing some action iff either the agent is in R_2 ($\neg INR_1(s)$) and leaves it or the agent is currently in R_1 and the action is anything other than leaving it.

Moore [1985] defined a possible-worlds semantics for a modal logic of knowledge in the situation calculus by treating situations as possible worlds. Scherl and Levesque [1993] adapted the semantics to the action theories of Reiter [2001]. The idea was to have an accessibility relation on situations, $K(s', s)$, which holds if in situation s, the situation s' is considered possible by the agent. Note, the order of the arguments is reversed from the usual convention in modal logic.

Levesque [1996] introduced a predicate, $SF(a, s)$, to describe the result of performing the binary-valued sensing action a. $SF(a, s)$ holds iff the sensor associated with a returns the sensing value 1 in situation s. Each sensing action senses some property of the domain. The property sensed by an action is associated with the action using a *guarded sensed fluent axiom* [De Giacomo and Levesque 1999]. For example, suppose that there are lights in R_1 and R_2 and that $LIGHT_1(s)$ ($LIGHT_2(s)$, resp.) holds if the light in R_1 (R_2, resp.) is on. Then, the axioms:

$$INR_1(s) \supset (SF(\text{SENSELIGHT}, s) \equiv LIGHT_1(s)), \text{ and}$$
$$\neg INR_1(s) \supset (SF(\text{SENSELIGHT}, s) \equiv LIGHT_2(s)).$$

can be used to specify that the SENSELIGHT action senses whether the light in the room where the agent is currently located is on.

Shapiro et al. [2000] adapted Spohn's ordinal conditional functions [Spohn 1988; Darwiche and Pearl 1997] to the situation calculus by introducing plausibilities over situations using a function $pl(s)$ which returns a natural number representing plausibility of situation s. The lower the number, the more plausible the situation is considered. The plausibilities were fixed in the initial situation and were not allowed to change, i.e., they used this successor state axiom for pl: $pl(do(a, s)) = pl(s)$. They adopted Scherl and Levesque's [2003] successor state axiom for the accessibility relation (which they renamed B):[3]

$$B_{SPLL}(s'', do(a, s)) \equiv$$
$$\exists s'.B_{SPLL}(s', s) \wedge s'' = do(a, s') \wedge (SF(a, s') \equiv SF(a, s)).$$

---

[2]We adopt the convention that unbound variables are universally quantified in the widest scope.

[3]For simplicity, we assume here that all actions are always executable and omit the action precondition axioms and references to a *Poss* predicate that are normally included in situation calculus action theories.

In other words, the situations s'' that are accessible from $do(a, s)$ are the ones that result from doing action a in a situation s' that is accessible from s, such that the sensor associated with action a has the same value in s' as it does in s. In other words, after doing a, the agent's beliefs will be expanded to include what the value of the sensor associated with a is in s. If a is a sensing action, the agent's beliefs will also include the property associated with a in the guarded sensed fluent axiom for a. If a is a physical action, then the agent's beliefs will also include the effects of a as specified by the successor state axioms.

Shapiro et al. defined the beliefs of the agent to be the formulae true in the most plausible accessible situations:[4]

$$\mathbf{Bel}_{SPLL}(\phi, s) \stackrel{\text{def}}{=} \forall s'[B_{SPLL}(s', s) \wedge (\forall s''.B_{SPLL}(s'', s) \supset pl(s') \leq pl(s''))] \supset \phi[s'].$$

Shapiro et al. thus modelled belief change with infallible sensors. If the agent senses a property ϕ, and ϕ actually holds, then all the situations that satisfy $\neg\phi$ become inaccessible. For example, if the agent believes $\neg\phi$ and senses ϕ, then all the most plausible, accessible situations will become inaccessible. A new set of accessible situations will become most plausible, all of which satisfy ϕ, yielding belief in ϕ. However, Shapiro et al. did not allow for the possibility of the agent subsequently sensing $\neg\phi$.

There are various ways of axiomatizing dynamic applications in the situation calculus. Here we adopt a simple form of the guarded action theories described by De Giacomo and Levesque [1999] consisting of: (1) successor state axioms for each fluent, and guarded sensed fluent axioms for each action, as discussed above; (2) unique names axioms for the actions, and domain-independent foundational axioms (we adopt the ones given by Levesque et al. [1998] which accommodate multiple initial situations, but we do not describe them further here); and (3) initial state axioms, which describe the initial state of the domain and the initial beliefs of the agent. In what follows, we will use Σ to refer to a guarded action theory of this form.

Before formally defining a belief operator in this language, we briefly review the notion of belief change. Belief change, simply put, aims to study the manner in which an agent's doxastic (belief) state should change when the agent is confronted by new information. In the literature, there is often a clear distinction between two forms of belief change: *revision* and *update*. Both forms can be characterized by an axiomatic approach (in terms of rationality postulates) or through various constructions (e.g., epistemic entrenchment, possible worlds, etc.). The AGM theory [Gärdenfors 1988] is the prototypical example of belief revision while the KM framework [Katsuno and Mendelzon 1991] is often identified with belief update.

Intuitively speaking, belief revision is appropriate for modeling static environments about which the agent has only partial and possibly incorrect information. New information is used to fill in gaps and correct errors, but the environment itself does not undergo change. Belief update, on the other hand, is intended for situations in which the environment itself is changing due to the performing of actions.

---

[4]We use ϕ to denote a formula that may contain a distinguished situation constant, *Now*, as a placeholder for a situation, e.g., INR$_1$(*Now*). $\phi[s]$ denotes the formula that results from substituting s for *Now* in ϕ. Where the intended meaning is clear, we omit the placeholder.

3 Belief Change with Noisy Sensors

Shapiro *et al.* [2000] modelled belief change due to sensing but it was assumed that the sensors were always accurate. This is quite a strong assumption which we will relax here. If sensing is exact, then the sensors will never be contradicted and so belief revision is limited to revising the agent's *initial* beliefs. But once an initial belief is corrected, it will never change again. In this context, it seems reasonable to have a fixed plausibility relation. However, if the sensors can return different results over time, this approach will not work because after sensing two contradicting values for the same formula, the agent will have contradictory beliefs.

To model noisy sensing, we add another distinguished predicate $SR(a, s)$, which is similar to SF described previously. The idea is that while $SF(a, s)$ describes the property of the world *ideally* sensed by action a, the actual values returned by the sensor may not correspond exactly to the property described by SF. So, we will use $SR(a, s)$ to describe the value *actually* returned by the sensor associated with action a. Another way of describing SR is that it is the result of adding noise to the sensor described by SF. How to specify SR is still an unresolved issue. We want SR to be related to SF but perhaps only related by a stochastic relation. This problem is reserved for future work.

As with Shapiro *et al.*, we assume that the agent knows the history of actions it has taken. By that we mean that the agent only considers a situation possible if it agrees with the history of actions in the actual situation.[5] We further assume that the agent has privileged access to its sensors. That is, after the agent reads its sensor, it knows the value of the sensor and it remembers the sequence of sensor readings it has made to date. The agent only considers possible those situations that agree with the actual situation on the history of sensor readings.

To model plausibilities that can change, we dispense with the *pl* predicate used by Shapiro *et al.*, and instead add a plausibility argument to the accessibility relation. So, $B(s', n, s)$ will denote that s' is considered a possible situation by the agent with plausibility n in situation s.[6] As before, the lower the plausibility level the more plausible the agent considers the situation to be. The beliefs of the agent are determined by the situations with plausibility 0:

$$\mathbf{Bel}(\phi, s) \stackrel{\text{def}}{=} \forall s'.B(s', 0, s) \supset \phi[s'].$$

As previously mentioned, we have two further distinguished predicates: $SF(a, s)$ and $SR(a, s)$, both of which take an action and a situation as arguments. The former holds if the property ideally sensed by sensing action a holds in situation s, and the latter holds if the sensor associated with a actually returns the value 1 in s. We adopt the convention that if A is a non-sensing action, then $\forall s.SF(A, s) \land SR(A, s)$ holds.

The dynamics of the agent's beliefs are formalized by the following successor state axiom:

AXIOM 1.

$$B(s'', n'', do(a, s)) \equiv$$
$$\exists s', n'.B(s', n', s) \land s'' = do(a, s') \land$$
$$(SR(a, s') \equiv SR(a, s)) \land Update(n'', n', a, s', s),$$

---

[5] A treatment of exogenous actions that are hidden from the agent was given by Shapiro and Pagnucco [2004].

[6] Note that this representation was considered and rejected by Shapiro *et al.* due to a difficulty discussed below.

where $Update(n'', n', a, s', s)$ (defined below) holds if n'' is the updated plausibility of situation s' after the occurrence of action a, when the plausibility of s' with respect to s is n'.[7] This axiom states that s'' will be accessible from $do(a, s)$ with plausibility n'', iff there exist s' and n' such that s' was accessible from s with plausibility n', s' and s agree on the actual value of the sensor associated with a, and n'' is the result of updating the plausibility of s' with respect to s due to a. Note that situations that disagree with s on the value of the sensor associated with a, (and those whose last action is not a) are discarded altogether from the accessibility relation. This assumes that the agent will never come to believe that it was mistaken about its sensor readings (or about the history of action occurrences).

We update the plausibilities as follows. We say a situation's (say, s) sensor reading is *correct* with respect to the sensor associated with a, if its SF and SR values agree, i.e., $SR(a, s) = SF(a, s)$. Those situations whose sensor readings are correct will have their plausibility levels decreased (i.e., they will become *more* plausible) and the others will have their plausibility levels increased. For concreteness, we will use Darwiche and Pearl's [1997] update function, but others are possible.

$$Correct(a, s) \stackrel{\text{def}}{=} SR(a, s) \equiv SF(a, s)$$

$$Good(s', n', a, s) \stackrel{\text{def}}{=} B(s', n', s) \wedge (SR(a, s') \equiv SR(a, s)) \wedge \\ Correct(a, s')$$

$$Min(n, a, s) \stackrel{\text{def}}{=} (\exists s^* Good(s^*, n, a, s)) \wedge \\ \forall s', n'.Good(s', n', a, s) \supset n' \geq n$$

$$Update(n'', n', a, s', s) \stackrel{\text{def}}{=} \\ (Correct(a, s') \supset \exists n^*.Min(n^*, a, s) \wedge n'' = n' - n^*) \wedge \\ (\neg Correct(a, s') \supset n'' = n' + 1)$$

In other words, the situations whose sensor readings are incorrect have their plausibilities increased by 1. The situations whose readings are correct are updated by subtracting the *Min* value, which is the lowest plausibility among the accessible situations that agree with the actual situation on the correct sensor reading. The result is that the agent believes that its sensor reading is correct, since this will hold in all the 0-plausibility situations.

Following Shapiro *et al.*, we assume that $B(S', N, S)$ only holds, if S and S' have the same histories. This means that the agent knows what actions have occurred. To enforce this, we use the following axiom which says that the situations accessible from an initial situation are also initial.

AXIOM 2.
$$Init(s) \wedge B(s', n, s) \supset Init(s').$$

We formalize B as a relation so that we can exclude certain situations from being accessible altogether. We can think of these situations as completely implausible. However, for the situations that are assigned some plausibility, we want their plausibility to be unique:

---

[7]*Update* could be a function, however we found it more convenient to formulate it as a relation.

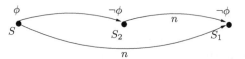

Figure 1. Introspection, exact sensing, and updating plausibilities clash

AXIOM 3.

$$Init(s) \land B(s', n, s) \land B(s', n', s) \supset n = n'.$$

To ensure that the agent has positive and negative introspection, we need to impose a constraint on the situations accessible from initial situations. As is well known (see, e.g., Fagin *et al.* [1995]), to get positive and negative introspection in contexts without plausibilities, it suffices for the accessibility relation to be transitive and Euclidean. Our constraint is a generalization of the combination of transitivity and Euclideanness that takes plausibilities into account. To get positive and negative introspection, we only need the accessibility relation to be transitive and Euclidean over situations with plausibility 0. However, since we are dealing with a dynamic framework, situations with higher plausibility levels could later have plausibility 0, therefore we enforce these constraints over all plausibility levels.

AXIOM 4.

$$Init(s) \supset (B(s', n, s) \supset \forall s'', n''.B(s'', n'', s') \equiv B(s'', n'', s)).$$

In other words, for initial s, if s' is accessible from s with some plausibility, then s and s' have the same accessible situations with the same plausibilities.

Shapiro *et al.* described a conflict in their framework between preserving this constraint and updating plausibilities, which is illustrated in Fig. 1. In this example, S_2 is accessible from S with some unspecified plausibility, and S_1 is accessible from both S and S_2 with plausibility n. Note that this example satisfies the constraint described in Axiom 4 (if we assume the situations are all initial). Now, recall that for Shapiro *et al.* sensing was assumed to be accurate. Therefore, if the agent senses ϕ, the plausibility level of S_1 with respect to S should increase because they disagree on the value of ϕ, whereas, the plausibility level of S_1 with respect to S_2 should decrease because they agree on the value of ϕ. Therefore, the generalization of the constraint described in Axiom 4 that omits the condition that s be initial will not be satisfied after sensing ϕ. This means that the agent may loose introspection.

The problem here is that in S, a sensor that senses ϕ would say that ϕ holds, while in S_2, the sensor would say that ϕ does not hold. So, loosely speaking, the agent in S_1 is being told to revise its beliefs with ϕ by S and with $\neg \phi$ by S_2. In our framework, this problem is avoided because all the situations that disagree with S on the value of the sensor will be dropped from the accessibility relation. In effect, the beliefs of the agent in all the surviving accessible situations are revised by the same formula. In other words, to avoid this problem, we (as well as Shapiro *et al.*) had to model agents that have privileged access to their sensors, i.e., they *know* the results of sensing. In the next section, we provide a theorem which says that we have indeed avoided this problem.

Note that this problem only arises when beliefs are changed due to sensing (and the agent is introspective). When an agent senses ϕ, it is told *whether* ϕ holds. In the traditional belief change setting, the agent is informed *that* ϕ holds. The crucial difference is that in the former case, the content of the belief-producing action depends on the actual situation, but not in the latter. As we stated earlier, the problem illustrated in Fig. 1 is that in S, the sensor says that ϕ holds, while in S_2, the sensor says that ϕ does not hold. If we were to model informing instead of sensing, e.g., using the action INFORM(ϕ), the value of ϕ in S and S_2 would be irrelevant. In both situations, the agents beliefs would be revised with ϕ. While there have been previous approaches to belief revision with unreliable observations, e.g., [Aucher 2005; Bacchus, Halpern, and Levesque 1999; Boutilier, Friedman, and Halpern 1998; Laverny and Lang 2004] almost all of them use informing as the belief-producing action rather than sensing. Bacchus *et al.* [1999] model sensing (also in the framework of the situation calculus) as the nondeterministic choice of inform actions, one for each possible value returned by the sensor, but they do not address introspection.

One issue that remains to be resolved is how to ensure that there is always at least one accessible situation. Since we are modelling noisy sensing, the agent's sensors could say that ϕ holds and later say that ϕ does not hold. How do we then prevent the agent's beliefs from lapsing into inconsistency? We need to ensure that regardless of the history of sensing results, for each action a, there is always an accessible situation (but not necessarily a most plausible one) that agrees with actual situation on the value returned by the sensor associated with a, and that value is correct, i.e.:

$$\forall a, s \exists s', n.B(s', n, s) \wedge (SR(a, s') \equiv SR(a, s)) \wedge Correct(a, s').$$

We believe we can achieve this using an axiom similar to the one given by Lakemeyer and Levesque [1998], and we will investigate this in future work.

4 Properties

In this section, we give some properties of our axiomatization of belief change and show that it does not suffer from the problem discussed by Shapiro *et al.* Let Σ be the foundational axioms together with the axioms of the previous section. First, we can show that the constraints imposed on the initial state given in Axioms 3 and 4 are preserved over all sequences of actions.

THEOREM 5.

$$\Sigma \models \forall n, n', s', s.B(s', n, s) \wedge B(s', n', s) \supset n = n',$$
$$\Sigma \models \forall n, n'', s, s', s''.B(s', n, s) \supset \forall s'', n''.B(s'', n'', s') \equiv B(s'', n'', s).$$

The latter property ensures that the agent always has full introspection, and shows that we do not suffer from the problem of combining sensing, introspection, and updating plausibilities discussed by Shapiro *et al.*

COROLLARY 6.

$$\Sigma \models \forall s.\mathbf{Bel}(\phi, s) \supset \mathbf{Bel}(\mathbf{Bel}(\phi), s),$$
$$\Sigma \models \forall s.\neg\mathbf{Bel}(\phi, s) \supset \mathbf{Bel}(\neg\mathbf{Bel}(\phi), s).$$

Shapiro *et al.* discussed a possible solution to their problem with updating plausibilities by setting the plausibility levels from all accessible situations to be the same as they are in the actual situation. However, they showed that this solution was unsatisfactory by giving an example using this scheme that entailed a counterintuitive property, namely, that the agent believes ϕ, but thinks that after sensing ϕ, it will believe $\neg\phi$. We can show that it is not possible to construct such an example in our framework that is reasonable. In particular, we show that in any such example, the agent believes that either its sensor is incorrect or that its beliefs will be inconsistent after sensing ϕ. The second alternative is clearly not reasonable. We would not want to model an agent that believes ϕ but also believes that after sensing ϕ its beliefs will become inconsistent. The first alternative does not make sense either because we are modelling agents that revise their beliefs according to what their sensors tell them. If the agent were to believe that its sensor is not correct, then it would not make sense to revise its beliefs according to what the sensor said. So, while the agent might be aware that its sensors are not always correct, we want to avoid situations where the agent actually believes that its sensor will return the wrong value. Accordingly, the next theorem says that if the agent believes ϕ and it thinks that it will believe $\neg\phi$ after sensing ϕ, and a is a sensing action for ϕ that does not change the value of ϕ if it holds initially, then the agent believes that either its sensor is incorrect or that its beliefs will be inconsistent after sensing ϕ.

THEOREM 7.

$$\Sigma \models \forall a, s.\mathbf{Bel}(\phi, s) \wedge \mathbf{Bel}(\mathbf{Bel}(\neg\phi, do(a, Now)), s) \wedge (\forall s'.SF(a, s') \equiv \phi(s')) \wedge$$
$$(\forall s'.\phi[s'] \supset \phi[do(a, s')]) \supset$$
$$\mathbf{Bel}([\neg Correct(a, Now) \vee Bel(FALSE, do(a, Now))], s).$$

Next, we show that the agent will revise its beliefs appropriately. If an action a (ideally) senses a property ϕ, and the sensor indicates that ϕ holds, then after sensing, the agent will believe that ϕ held before the sensing occurred. We first define what it means for ϕ to hold in the previous situation: $\mathbf{Prev}(\phi, s) \overset{\text{def}}{=} \exists a, s'.s = do(a, s') \wedge \phi[s']$.

THEOREM 8.

$$\Sigma \models \forall a, s.(\forall s'.SF(a, s') \equiv \phi[s']) \wedge SR(a, s) \supset \mathbf{Bel}(\mathbf{Prev}(\phi), do(a, s)).$$

If the agent also believes that a does change the value of ϕ, then the agent will believe ϕ after doing a.

Since the basis of our framework is a theory of action, belief updates are handled naturally as resulting from physical actions. We show that (as with Shapiro *et al.*'s framework) belief updates are handled appropriately. If a is a physical action (i.e., SF and SR are identically true) and situation s has at least one accessible situation with 0-plausibility, and the agent believes that a causes ϕ' to hold if ϕ holds initially, then the agent will believe that ϕ' holds after doing a in s, if it believes that ϕ holds in s.

THEOREM 9.

$$\Sigma \models \forall a, s.\exists s' B(s', 0, s) \wedge (\forall s'.SF(a, s') \wedge SR(a, s')) \wedge$$
$$\mathbf{Bel}([\phi(Now) \supset \phi'(do(a, Now))], s) \wedge \mathbf{Bel}(\phi, s) \supset$$
$$\mathbf{Bel}(\phi', do(a, s)).$$

Finally, we can show that Shapiro *et al.*'s framework is a special case of ours. If we assume that sensing is always accurate, and for every action a, every situation has an accessible situation that agrees with it on the value of the sensor associated with a, then Shapiro *et al.*'s axioms for B and pl combined and translated into our notation follow from our axioms.

THEOREM 10.

$$\Sigma \models (\forall a, s.Correct(a, s)) \wedge [\forall a, s \exists s'.B(s', 0, s) \wedge (SR(a, s') \equiv SR(a, s))] \supset$$
$$\forall a, s'', n'', s.B(s'', n'', do(a, s)) \equiv$$
$$\exists s'.B(s', n'', s) \wedge s'' = do(a, s') \wedge (SF(a, s') \equiv SF(a, s)).$$

5 Conclusions and Future Work

In this paper, we introduced a framework for modelling belief change as a result of noisy sensing in the situation calculus, where the agent has full introspection of its beliefs. Our framework allows the updating of plausibilities of situations, and we showed that we resolved the difficulty with combining all these elements discussed by Shapiro *et al.* As previously mentioned, there are some issues that are as yet unresolved. One is how to specify the SR predicate. Another is how to ensure that there are enough situations with the right properties to prevent the agent's beliefs from lapsing into inconsistency. We would also like to investigate the extent to which our framework satisfies the standard belief change postulates [Darwiche and Pearl 1997; Gärdenfors 1988; Katsuno and Mendelzon 1991]. Lastly, we would like to extend the framework to handle unreliable physical actions, i.e., noisy effects.

Acknowledgments: This paper follows up on work I did with Hector (who supervised both my M.Sc. and Ph.D.) and others, which in turn grew out of work Hector did on knowledge and sensing in the situation calculus. It has been an honour and a great pleasure to have had the opportunity to work closely with Hector, and to contribute a paper to this volume dedicated to him. Happy birthday, Hector!

A version of this paper originally appeared in [Shapiro 2005].

References

Aucher, G. [2005]. A combined system for update logic and belief revision. In M. W. Barley and N. Kasabov (Eds.), *Intelligent Agents and Multi-Agent Systems: 7th Pacific Rim International Workshop on Multi-Agents, PRIMA 2004, Auckland, New Zealand, August 8-13, 2004, Revised Selected Papers*, Volume 3371 of *LNAI*, pp. 1–17. Berlin: Springer-Verlag.

Bacchus, F., J. Y. Halpern, and H. J. Levesque [1999]. Reasoning about noisy sensors and effectors in the situation calculus. *Artificial Intelligence 111*(1–2), 171–208.

Boutilier, C., N. Friedman, and J. Y. Halpern [1998]. Belief revision with unreliable observations. In *Proceedings of the Fifteenth National Conference on Artificial Intelligence (AAAI-98)*, pp. 127–134.

Darwiche, A. and J. Pearl [1997]. On the logic of iterated belief revision. *Artificial Intelligence 89*(1–2), 1–29.

De Giacomo, G. and H. J. Levesque [1999]. Progression using regression and sensors. In *Proceedings of the Sixteenth International Joint Conference on Artificial Intelligence (IJCAI-99)*, pp. 160–165.

Fagin, R., J. Y. Halpern, Y. Moses, and M. Y. Vardi [1995]. *Reasoning about Knowledge.* MIT Press.

Gärdenfors, P. [1988]. *Knowledge in Flux.* The MIT Press: Cambridge, MA.

Katsuno, H. and A. O. Mendelzon [1991]. Propositional knowledge base revision and minimal change. *Artificial Intelligence 52*, 263–294.

Lakemeyer, G. and H. J. Levesque [1998]. AOL: a logic of acting, sensing, knowing, and only knowing. In *Principles of Knowledge Representation and Reasoning: Proceedings of the Sixth International Conference (KR-98)*, pp. 316–327.

Laverny, N. and J. Lang [2004]. From knowledge-based programs to graded belief-based programs — Part I: On-line reasoning. In *Proceedings of the 16th European Conference on Artificial Intelligence (ECAI-04)*, Amsterdam, pp. 368–372. IOS Press.

Levesque, H. J. [1996, August]. What is planning in the presence of sensing? In *Proceedings of the Thirteenth National Conference on Artificial Intelligence (AAAI-96)*, Portland, OR, pp. 1139–1146.

Levesque, H. J., F. Pirri, and R. Reiter [1998]. Foundations for a calculus of situations. *Electronic Transactions of AI (ETAI) 2*(3–4), 159–178.

McCarthy, J. and P. J. Hayes [1969]. Some philosophical problems from the standpoint of artificial intelligence. In B. Meltzer and D. Michie (Eds.), *Machine Intelligence 4*. Edinburgh University Press.

Moore, R. C. [1985]. A formal theory of knowledge and action. In J. R. Hobbs and R. C. Moore (Eds.), *Formal Theories of the Common Sense World*, pp. 319–358. Norwood, NJ: Ablex Publishing.

Reiter, R. [2001]. *Knowledge in Action: Logical Foundations for Specifying and Implementing Dynamical Systems.* Cambridge, MA: MIT Press.

Scherl, R. B. and H. J. Levesque [1993, July]. The frame problem and knowledge-producing actions. In *Proceedings of the Eleventh National Conference on Artificial Intelligence*, Washington, DC, pp. 689–695. AAAI Press/The MIT Press.

Scherl, R. B. and H. J. Levesque [2003]. Knowledge, action, and the frame problem. *Artificial Intelligence 144*(1–2), 1–39.

Shapiro, S. [2005]. Belief change with noisy sensing and introspection. In L. Morgenstern and M. Pagnucco (Eds.), *Working Notes of the IJCAI-05 on Nonmonotonic Reasoning, Action, and Change (NRAC 2005)*, pp. 84–89.

Shapiro, S. and M. Pagnucco [2004]. Iterated belief change and exogenous actions in the situation calculus. In R. López de Mántaras and L. Saitta (Eds.), *Proceedings of the 16th European Conference on Artificial Intelligence (ECAI-04)*, Amsterdam, pp. 878–882. IOS Press.

Shapiro, S., M. Pagnucco, Y. Lespérance, and H. J. Levesque [2000]. Iterated belief change in the situation calculus. In A. G. Cohn, F. Giunchiglia, and B. Selman (Eds.), *Principles of Knowledge Representation and Reasoning: Proceedings of the Seventh International Conference (KR2000)*, San Francisco, CA, pp. 527–538. Morgan Kaufmann.

Spohn, W. [1988]. Ordinal conditional functions: A dynamic theory of epistemic states. In W. Harper and B. Skyrms (Eds.), *Causation in Decision, Belief Change, and Statistics*, pp. 105–134. Kluwer Academic Publishers.

29

A Database-type Approach for Progressing
Action Theories with Bounded Effects

STAVROS VASSOS AND SEBASTIAN SARDINA

ABSTRACT. In this paper we study the progression of situation calculus action theories that are able to handle a class of actions that, while extremely simple conceptually and common in many settings, cannot be handled by previous approaches. Specifically, based on the notion of *safe-range queries* from database theory and *just-in-time* action histories, we present a new type of action theories that ensures that actions have bounded effects over a restricted range of objects. Such theories may represent incomplete information and can be progressed by directly updating the knowledge base in an algorithmic manner.

We dedicate this paper to Hector Levesque, who as much as anyone, promoted the logic-based approach to problems and the fundamental insight that expressivity can be traded for efficiency. Both aspects are strongly reflected in this work. This paper is an extended version of a recent result that appeared in [Vassos, Sardina, and Levesque 2009]. As such, the work we present here has been greatly influenced by Hector, both in terms of the specific research results reported as well as in the fact that it involves the contribution of two generations of students under his supervision. Indeed, we feel privileged to have had Hector as our Ph.D. supervisor. His remarkable qualities, both intellectual and personal, have had major impact on our careers in a way that is impossible to put it in text. Just one thing we can say: Thanks Hector!

1 Introduction

One of the requirements for building agents with a pro-active behavior is the ability to reason about action and change. The ability to *predict* how the world will be after performing a sequence of actions is the basis for offline automated planning, scheduling, web-service composition, etc. In the situation calculus [McCarthy and Hayes 1969; Reiter 2001] such reasoning problems are examined in the context of the so-called basic action theories (BATs). These are logical theories that specify the preconditions and effects of actions, and an initial knowledge base (KB) that represents the initial state of the world before any action has occurred.

A basic action theory can be used to solve offline problems as well as to equip a situated agent with the ability to *keep track* of the current state of the world. As a theory is a static entity, in the sense that the axioms do not change over time, reasoning about the current state is typically carried over using techniques based on *regression* that transform queries about the future into queries about the initial state [Reiter 2001]. This is an effective choice for some applications, but a poor one for many settings where an agent may act autonomously for long periods of time. In those cases, it is mandatory that the theory be (periodically) updated so that the initial KB be replaced by a new one reflecting the changes due to the actions that have already occurred. This is identified as the problem of *progression* for basic action theories [Lin and Reiter 1997].

In general, a KB in a basic action theory is an unrestricted first-order logical theory that offers great flexibility and expressiveness. The price to pay though is high, as it is hard to find practical solutions for the related reasoning problems. As far as progression is concerned, it was shown by Lin and Reiter [1997] that the updated KB requires second-order logic in the general case. For this reason, several restrictions on the theories have been proposed so that the updated KB is first-order representable. In particular, two recent results show that progression is practical provided that actions are *local-effect* [Vassos, Lakemeyer, and Levesque 2008] or *normal* [Liu and Lakemeyer 2009].

The local-effect assumption essentially means that all the objects that may be affected by the action are directly specified by the arguments of the action. For example, an action that results in the robot moving from location l_1 to location l_2 needs to explicitly mention l_1 and l_2 in its arguments in order to be local-effect, e.g., $move(l_1, l_2)$. On the other hand, a so-called normal action is more general in that it may also affect objects that are not included as arguments, as long as these are specified by information that exists in the KB in a particular way. In this manner, the above moving action may be allowed also to affect the location of objects being held by the robot.

Nonetheless, it turns out that there are many simple actions that do not qualify as local-effect or normal. For example, action *moveFwd* (with no arguments) that causes a robot to move forward by a fixed length is neither local-effect nor normal. These actions arise naturally in many robotic domains and grid-style games, for instance, and cannot be handled by previous techniques for progression unless further restrictions are assumed (e.g., a finite domain). Furthermore, these actions cannot, in general, be reformulated into a well-behaved local-effect version that includes all the necessary ground arguments as the required information, for example, the robot's current location, may be unknown.

In order to handle such actions, we present what we call *range-restricted* basic action theories (RR-BATs). These theories are able to represent actions that may not be local-effect or normal but whose (range of) effects can be "bounded." For such theories, we describe a method for progression such that the new KB is first-order and finite, and we prove that the method is logically correct and effectively computed via database-like evaluation for a special case of interest. To the best of our knowledge, it is the first result that accounts for a possibly infinite domain, incomplete information, sensing, and a particular class of simple actions that goes beyond the local-effect assumption.

The rest of the paper is organized as follows. In Section 2, we review the situation calculus and the progression task for unrestricted basic action theories. Then, in Section 3, we specify the structure for the theories of action and queries to be considered, and define the notion of *possible answers* for queries. In Section 4, we present a general method for progression for these theories and examine the complexity of the approach in a special case. We conclude by discussing related and future work, and drawing final conclusions.

2 Formal preliminaries

The language $\mathcal{L}$ of the situation calculus as presented by Reiter [2001] is a three-sorted first-order logic language with equality and some limited second-order features. The three sorts are the following: *action*, *situation*, and a catch-all sort *object* for everything else depending on the domain of application.

Similar to a normal one-sorted first-order language, $\mathcal{L}$ includes function and predicate symbols. In this case since there are three sorts, each of the symbols has a type that specifies the sorts for the arguments it takes. The situation calculus includes symbols only of certain types each of which has a special role in the representation of the world and its dynamics.

An action term or simply an *action* represents an atomic action that may be performed in the world. For example consider the action $move(l_1, l_2)$ that may be used to represent that a robot moves from location l_1 to location l_2. A situation term or simply a *situation* represents a world history as a sequence of actions. The constant S_0 is used to denote the *initial situation* where no actions have occurred. Sequences of actions are built using the function symbol do, such that $do(\alpha, \sigma)$ represents the successor situation resulting from performing action α in situation σ.

A *relational fluent* is a predicate whose last argument is a situation, and thus whose truth value can change from situation to situation. For example, $RobotAt(l, \sigma)$ may be used to represent that the robot lies at location l in situation σ. In order to simplify the analysis we have restricted the language $\mathcal{L}$ so that there are no functional fluent symbols in $\mathcal{L}$, that is, functions whose last argument is a situation. This is not a restriction on the expressiveness of $\mathcal{L}$ as functional fluents can be represented by relational fluents with a few extra axioms.

Actions need not be executable in all situations, and the predicate atom $Poss(\alpha, \sigma)$ states that action α is executable in situation σ. For example, $Poss(move(l_1, l_2), \sigma)$ is intended to represent that the action $move(l_1, l_2)$ can be executed in situation σ. Actions may also have a sensing result: the special function $SR(\alpha, \sigma)$ denotes the sensing outcome of action α when executed in situation σ [Scherl and Levesque 2003].

In this paper, we shall restrict our attention to a language $\mathcal{L}$ with a finite number of relational fluent symbols that only take arguments of sort object (apart their last situation argument), an infinite number of constant symbols of sort object, and a finite number of function symbols of sort action that take arguments of sort object. We adopt the following notation with subscripts and superscripts: α and a for terms and variables of sort action; σ and s for terms and variables of sort situation; t and x, y, z, w for terms and variables of sort object. Also, we use A for action function symbols, F, G for fluent symbols, and b, c, d, e, o for constants of sort object. Finally, we will typically write $\phi(\vec{x})$ to state that the free variables of the formula are among $\vec{x}$.

The well-formed first-order formulas of $\mathcal{L}$ are defined inductively similarly to a normal one-sorted language but also respecting that each parameter has a unique sort. As far as the second-order formulas of $\mathcal{L}$ are concerned, only quantification over relations is allowed and the well-formed formulas are defined inductively similarly to a normal second-order language.

Often we will focus on sentences that refer to a particular situation. For this purpose, for any situation term σ, we define the set of *uniform formulas in* σ to be all those (first-order or second-order) formulas in $\mathcal{L}$ that do not mention any other situation terms except for σ, do not mention $Poss$, and where σ is not used by any quantifier [Lin and Reiter 1997].

2.1 Basic action theories

Within the language $\mathcal{L}$, one can formulate action theories that describe how the world changes as the result of the available actions. We focus on a variant of the *basic action theories (BATs)* [Reiter 2001] of the following form:[1]

---

[1] For legibility, we typically omit leading universal quantifiers. The difference with Reiter's BATs is the incorporation of $\mathcal{D}_{sr}$ for sensing and the set $\mathcal{E}$.

$$\mathcal{D} = \mathcal{D}_{ap} \cup \mathcal{D}_{ss} \cup \mathcal{D}_{una} \cup \mathcal{D}_{sr} \cup \mathcal{D}_0 \cup \Sigma \cup \mathcal{E},$$

where:

1. $\mathcal{D}_{ap}$ is the set of action precondition axioms (PAs), one per action symbol A, of the form $Poss(A(\vec{y}), s) \equiv \Pi_A(\vec{y}, s)$, where $\Pi_A(\vec{y}, s)$ is first-order and uniform in s. PAs characterize the conditions under which actions are physically possible.

2. $\mathcal{D}_{ss}$ is the set of successor state axioms (SSAs), one per fluent symbol F, of the form $F(\vec{x}, do(a, s)) \equiv \Phi_F(\vec{x}, a, s)$, where $\Phi_F(\vec{x}, a, s)$ is first-order and uniform in s. SSAs describe how fluents change between situations as the result of actions.

3. $\mathcal{D}_{sr}$ is the set of sensing-result axioms (SRAs), one for each action symbol A, of the form $SR(A(\vec{y}), s) = r \equiv \Theta_A(\vec{y}, r, s)$, where $\Theta_A(\vec{y}, r, s)$ is first-order and uniform in s. SRAs relate sensing outcomes with fluents, or more generally, with complex properties of the domain.

4. $\mathcal{D}_{una}$ is the set of unique-names axioms for actions.

5. $\mathcal{D}_0$, the *initial knowledge base (KB)*, is a set of first-order sentences uniform in S_0 describing the initial situation S_0.

6. Σ is the set of domain independent axioms of the situation calculus, formally defining the legal situations. A second-order induction axiom is included in Σ.

7. $\mathcal{E}$ is an infinite set of unique-names axioms for object constants.

Probably the most interesting component of a basic action theory is the set successor state axioms, which together encode the dynamics of the domain being represented. Technically, SSAs are meant to capture the effects and non-effects of actions. To achieve that in a parsimonious way, one typically follows Reiter [1991]'s well-known solution to the frame problem and writes successor state axioms of the following form:

$$F(\vec{x}, do(a, s)) \equiv \gamma_F^+(\vec{x}, a, s) \vee F(\vec{x}, s) \wedge \neg\gamma_F^-(\vec{x}, a, s),$$

where both $\gamma_F^+(\vec{x}, a, s)$ and $\gamma_F^+(\vec{x}, a, s)$ are first-order formulas uniform in s encoding the positive and negative effects, respectively, of action a on fluent F at situation s.

As a running example we consider the simple scenario of a robot that is capable of navigating and moving objects around in a grid-like world while some objects such as boxes may also contain other objects. The robot is equipped with a basic action theory $\mathcal{D}$ that captures the initial state and the dynamics of the example domain. $\mathcal{D}$ uses fluents $RobotAt(x, s)$ and $RobotDir(x, s)$ to represent information about the location of the robot and the direction it is facing, and $Connected(x_1, x_2, x_3, s)$ to represent that location x_3 is adjacent to location x_1 wrt direction x_2 in s.

$\mathcal{D}$ includes the following SSA to capture the way fluent $RobotAt(x, s)$ is affected by the actions of the robot:

$$RobotAt(x, do(a, s)) \equiv \gamma_{RobotAt}^+(x, a, s) \vee RobotAt(x, s) \wedge \neg\gamma_{RobotAt}^-(x, a, s), \quad (1)$$

where $\gamma_{RobotAt}^+(x, a, s)$ is the formula

$$\exists z_1, z_2.\, a = moveFwd \wedge RobotAt(z_1, s) \wedge RobotDir(z_2, s) \wedge Connected(z_1, z_2, x, s),$$

and $\gamma_{RobotAt}^-(x, a, s)$ is the formula $a = moveFwd$.

In words, the robot is at location x after executing action a iff a is the action of moving forward, the robot is currently at location z_1 and facing towards direction z_2, and x is the next adjacent location to z_1 towards direction z_2, or the robot was already in location x (i.e., $RobotAt(x, s)$ holds) and it has not performed a move action.

The current location of objects is modeled using fluent $At(x_1, x_2, s)$: object x_1 is at location x_2 at situation s, and fluent $In(x_1, x_2, s)$ represents that object x_2 is inside x_1. When the robot moves, all objects being carried by the robot move as well. This is accounted in the following SSA for fluent $At(x_1, x_2, s)$ that is included in $\mathcal{D}$:

$$At(x_1, x_2, do(a, s)) \equiv \gamma_{At}^+(x_1, x_2, a, s) \vee At(x_1, x_2, s) \wedge \neg\gamma_{At}^-(x_1, x_2, a, s), \qquad (2)$$

where $\gamma_{At}^+(x_1, x_2, a, s)$ is the formula

$$\exists z_1, z_2 \big(a = moveFwd \wedge RobotAt(z_1, s) \wedge RobotDir(z_2, s) \wedge$$
$$Connected(z_1, z_2, x_2, s) \wedge Holding(x_1, s)\big) \vee$$
$$\exists z_1, z_2, z_3 \big(a = moveFwd \wedge RobotAt(z_1, s) \wedge RobotDir(z_2, s) \wedge$$
$$Connected(z_1, z_2, x_2, s) \wedge Holding(z_3, s) \wedge In(z_3, x_1, s)\big),$$

and $\gamma_{At}^-(x_1, x_2, a, s)$ is the formula

$$a = moveFwd \wedge Holding(x_1, s) \vee$$
$$\exists z.\, a = moveFwd \wedge Holding(z, s) \wedge In(z, x_1, s).$$

Formula $\gamma_{At}^+(x_1, x_2, a, s)$ that describes the positive effects, states that object x_1 is at location x_2 if the robot has successfully moved to location x_2 and either the robot is holding x_1 or x_1 is inside an object that the robot is holding. Similarly, the negative effect formula $\gamma_{At}^-(x_1, x_2, a, s)$ in the SSA states that object x_1 is not (anymore) at location x_2 if the robot has moved while carrying x_1 or some other object that has x inside it.

Although we do not list them here, fluents $RobotDir(x, w)$, $Holding(x, s)$, $In(x_1, x_2, s)$, and $Connected(x_1, x_2, x_3, s)$ are also meant to have their corresponding SSAs in basic action theory $\mathcal{D}$ of our running example. For instance, the SSA for $RobotDir(x, s)$ states that the current direction the robot is facing is affected by the turning action that the robot can perform. Similarly, appropriate action precondition axioms are assumed.

Although the agent can pick up containers that have other objects inside, it may have at certain point incomplete information on what is inside the containers, that is, incomplete information about fluent $In(x_1, x_2, s)$. Nonetheless, we assume the agent can sense, using a special device, whether there is gold inside a container object. Technically, the agent can execute sensing action $senseGold(y)$ that determines whether there is gold inside container object y:

$$Poss(senseGold(y), s) \equiv Holding(y, s),$$

$$SR(senseGold(y), s) = r \equiv [(r = 1) \equiv In(y, item_2, s)].$$

In words, the agent can sense for gold in an object whenever the agent is holding the object, and the outcome of such action is 1 iff the container object in question contains gold pieces inside.

Finally, the initial KB $\mathcal{D}_0$, represents the initial knowledge the agent has about the world using sentences uniform in S_0, that is, sentences that only refer to the initial situation such as $RobotAt(loc_1, S_0)$ and $RobotDir(north, S_0)$. We postpone the specification of $\mathcal{D}_0$ for our example until Section 3.1 where we introduce a specific form of initial KBs.

2.2 Progression

The progression of a basic action theory is the problem of *updating* the initial KB so that it reflects the current state of the world after some actions have been performed instead of the initial state of the world. In other words, in order to do a one-step progression of a basic action theory $\mathcal{D}$ with respect to a ground action α, we are to replace $\mathcal{D}_0$ in $\mathcal{D}$ by a suitable set $\mathcal{D}_\alpha$ of sentences (i.e., a new KB), so that the original theory $\mathcal{D}$ and the theory $(\mathcal{D} - \mathcal{D}_0) \cup \mathcal{D}_\alpha$ are equivalent with respect to how they describe the situation $do(\alpha, S_0)$ as well as all situations in the future of $do(\alpha, S_0)$ [Lin and Reiter 1997].

In this paper, instead of the model-theoretic definition of progression of Lin and Reiter [1997], we follow the definition of the so-called *strong progression* of Vassos et al. [2008] which we extend slightly in order to account for sensing actions.

Let $\mathcal{D}$ be a basic action theory over relational fluents $F_1, \ldots, F_n$, and let $Q_1, \ldots, Q_n$ be second-order predicate variables. For any formula ϕ in $\mathcal{L}$, let $\phi\langle \vec{F} : \vec{Q} \rangle$ be the formula that results from replacing any fluent atom $F_i(t_1, \ldots, t_n, \sigma)$ in ϕ, where σ is a situation term, with atom $Q_i(t_1, \ldots, t_n)$.

DEFINITION 1. Let $\mathcal{D}$ be a basic action theory over fluents $\vec{F}$ such that $\mathcal{D}_0$ is finite, α a ground action term of the form $A(\vec{c})$, and d a sensing outcome result. Let $Prog(\mathcal{D}, \alpha, d)$ be the following second-order sentence uniform in $do(\alpha, S_0)$:

$$\exists \vec{Q}. \, \mathcal{D}_0\langle \vec{F} : \vec{Q} \rangle \wedge \Theta_A(\vec{c}, d, do(\alpha, S_0)) \wedge \bigwedge_{i=1}^{n} \forall \vec{x}. \, F_i(\vec{x}, do(\alpha, S_0)) \equiv \left(\Phi_{F_i}(\vec{x}, \alpha, S_0)\langle \vec{F} : \vec{Q} \rangle \right).$$

Then, we say that a set of formulas $\mathcal{D}_\alpha$ uniform in situation $do(\alpha, S_0)$ is a *strong progression* of $\mathcal{D}$ wrt pair of action-outcome (α, d) iff $\mathcal{D}_\alpha \cup \mathcal{D}_{una} \cup \mathcal{E}$ is logically equivalent[2] to $\{Prog(\mathcal{D}, \alpha, d)\} \cup \mathcal{D}_{una} \cup \mathcal{E}$. □

Informally, set $\mathcal{D}_\alpha$ represents the updated initial KB after action α has been executed with sensing outcome d. Although $Prog(\mathcal{D}, \alpha, d)$ is defined in second-order logic we are interested in cases where we can find a $\mathcal{D}_\alpha$ that is *first-order representable*. In the sequel we shall present some restrictions that are sufficient conditions for doing this as well as a method for computing a finite $\mathcal{D}_\alpha$ for a special case of interest.

3 Range-restricted basic action theories

In this section, we present a new type of basic action theories in which the initial KB is a *database of possible closures* and the axioms in $\mathcal{D}_{ap}$, $\mathcal{D}_{ss}$, and $\mathcal{D}_{sr}$ are *range-restricted*. Essentially, we will be specifying the effects of actions in a way that resembles to logic-programming. In order to make progression work in first-order we also require a semantic assumption that may not hold in all situations but may be enforced with an appropriate account of sensing. Under these three assumptions, we will show later that a finite first-order updated (initial) KB can be computed.

3.1 A database of possible closures

Intuitively, we treat each fluent as a *multi-valued function*, where the last argument of sort object is considered as the "*output*" and the rest of the arguments of sort object as the

---

[2]Whenever we say that two formulas are logically equivalent we assume that the logical symbol $=$ is always interpreted as the true identity.

"*input*" of the function.[3] This distinction is important as we require that $\mathcal{D}_0$ expresses incomplete information only about the output of fluents.

DEFINITION 2. Let V be a set of constants and τ a fluent atom of the form $F(\vec{c}, w, S_0)$, where $\vec{c}$ is a vector of constants and w a variable. We say that τ has the *ground input* $\vec{c}$ and the *output* w. The *atomic closure* χ of τ on V is the sentence:

$$\forall w. F(\vec{c}, w, S_0) \equiv \bigvee_{e \in V} (w = e).$$

The *closure* of the fluent vector $\vec{\tau} = \langle \tau_1, \ldots, \tau_n \rangle$ of distinct atoms on a ground input vector $\vec{V} = \langle V_1, \ldots, V_n \rangle$ of sets of constants is the conjunction $(\chi_1 \wedge \ldots \wedge \chi_n)$, where each χ_i is the atomic closure of τ_i on V_i. $\square$

A closure of $\vec{\tau}$ expresses complete information about the output of all input-grounded fluents in $\vec{\tau}$. For example, consider fluents $RobotAt(x, s)$ and $RobotDir(x, s)$ that represent information about the location of the robot and the direction it is facing. Let χ_1 be the the atomic closure of $RobotAt(w, S_0)$ on $\{loc_1\}$ and χ_2 the atomic closure of $RobotDir(w, S_0)$ on $\{north\}$, that is:

$$\forall w. RobotAt(w, S_0) \equiv (w = loc_1); \qquad (\chi_1)$$
$$\forall w. RobotDir(w, S_0) \equiv (w = north). \qquad (\chi_2)$$

Then χ_1 and χ_2 express complete information about the location of the robot and the direction it is facing.

Consider now fluent $In(x, y, s)$ which represents that object y is inside the object x at situation s, and the input-grounded fluent atom $In(box_1, w, S_0)$. Let χ_3 be the following sentence:

$$\forall w. In(box_1, w, S_0) \equiv (w = item_1 \vee w = item_2). \qquad (\chi_3)$$

Then, χ_3 is the atomic closure of $In(box_1, w, S_0)$ on $\{item_1, item_2\}$ which states that there are *exactly* two objects inside box_1, namely $item_1$ and $item_2$.

Finally, let χ_4 be the the atomic closure of $In(box_2, w, S_0)$ on $\{gold\}$:

$$\forall w. In(box_2, w, S_0) \equiv (w = gold). \qquad (\chi_4)$$

Then, $(\chi_3 \wedge \chi_4)$ is a (non-atomic) closure of the vector of input-grounded fluent atoms

$$\langle In(box_1, w, S_0), In(box_2, w, S_0) \rangle$$

that expresses complete information about the contents of box_1 and box_2.

We now show how we can combine these closure statements in order to express incomplete information.

DEFINITION 3. A *possible closures axiom (PCA)* for a vector of input-grounded fluents $\vec{\tau}$ is a disjunction of the form $(\chi_1 \vee \cdots \vee \chi_n)$, where each χ_i is a closure of $\vec{\tau}$ on a distinct vector $\vec{V_n}$ of constants. $\square$

A PCA for $\vec{\tau}$ expresses *disjunctive information* about the output of all fluents in $\vec{\tau}$, by stating how such outputs can be combined together (in n possible ways). For example

---

[3]This is similar to *modes* in logic programming [Apt and Pellegrini 1994] and it can be easily generalized to multiple outputs.

consider again the input-grounded fluent atom $In(box_1, w, S_0)$, and let χ_5 be the closure of $In(box_1, w, S_0)$ on $\{item_1, gold\}$:

$$\forall w. In(box_1, w, S_0) \equiv (w = item_1 \vee w = gold). \tag{χ_5}$$

Then, $(\chi_3 \vee \chi_5)$ is a PCA for $In(box_1, w, S_0)$ stating that there are exactly two objects inside box_1, one being $item_1$ and the other being *either $item_2$ or gold*. Similarly, let χ_6 be the closure of $In(box_2, w, S_0)$ on $\{item_2\}$:

$$\forall w. In(box_2, w, S_0) \equiv (w = item_2). \tag{χ_6}$$

Then $(\chi_3 \wedge \chi_4) \vee (\chi_5 \wedge \chi_6)$ is a PCA for

$$\left\langle In(box_1, w, S_0,), In(box_2, w, S_0) \right\rangle,$$

which states that box_1 and box_2 contain the three items $item_1, item_2, gold$, but only two possible combinations are allowed: either $gold$ is in box_2 and the other two items are in box_1, or $item_2$ is in box_2 and the other two items are in box_1.

Using a set of PCAs, each one referring to different input-grounded fluent atoms, we are now able to define the form of our initial knowledge bases.

DEFINITION 4. A *database of possible closures (DBPC)* is a set $\mathcal{D}_0 = \{\Xi^{\vec{\tau}_1}, \ldots, \Xi^{\vec{\tau}_\ell}\}$, where each $\Xi^{\vec{\tau}_i}$ is a PCA for $\vec{\tau}_i$ such that $\vec{\tau}_i \cap \vec{\tau}_j = \emptyset$, for all distinct $i, j \in \{1, \ldots, \ell\}$. For every i, each disjunct of $\Xi^{\vec{\tau}_i}$ is called a *possible closure* wrt $\mathcal{D}_0$. $\square$

So, for every fluent atom τ with a ground input (e.g., $In(box_1, w, S_0)$), either the output of τ is completely unknown in S_0 (i.e., τ is not mentioned in $\mathcal{D}_0$) or there is just one PCA $\Xi_{\vec{\tau}_i}$ (with $\tau \in \vec{\tau}_i$) that specifies its output value in several possible "worlds" (one per disjunct in the PCA).

For our running example, the initial knowledge base is as follows:

$$\mathcal{D}_0 = \{\chi_1, \chi_2, (\chi_3 \wedge \chi_4) \vee (\chi_5 \wedge \chi_6)\} \cup \{\chi_7, \ldots, \chi_{13}\},$$

where $\chi_1, \ldots, \chi_6$ are defined above, χ_7 is the closure of $Connected(loc_1, north, w, S_0)$ on $\{loc_2\}$, χ_8 is the closure of $Holding(w, S_0)$ on the set $\{box_2\}$, and $\chi_9, \ldots, \chi_{13}$ are the atomic closures of $At(box_1, w, S_0), At(item_1, w, S_0), At(box_2, w, S_0), At(item_2, w, S_0)$, and $At(gold, w, S_0)$ on set $\{loc_1\}$, respectively. Note that each of the ten sentences in $\mathcal{D}_0$ is a PCA, and each of $\chi_1, \chi_2, \chi_3 \wedge \chi_4, \chi_5 \wedge \chi_6, \chi_7, \ldots, \chi_{13}$ is a possible closure wrt $\mathcal{D}_0$.

We now turn our attention to the so-called *possible answers* to a formula wrt a KB.

DEFINITION 5. Let $\mathcal{D}_0$ be a DBPC and $\gamma(\vec{x})$ a first-order formula uniform in S_0. The *possible answers* to γ wrt $\mathcal{D}_0$, denoted $\mathsf{pans}(\gamma, \mathcal{D}_0)$, is the smallest set of pairs $(\vec{c}, \chi)$ such that:

- χ is a closure of some fluent vector $\vec{\tau}$ such that $\mathcal{E} \cup \{\chi\} \models \gamma(\vec{c})$; and

- χ is consistent with $\mathcal{D}_0$ and minimal in the sense that every atomic closure in χ is necessary. $\square$

Intuitively, $\mathsf{pans}(\gamma, \mathcal{D}_0)$ characterizes all the cases where $\gamma(\vec{x})$ is satisfied in a model of $\mathcal{D}_0 \cup \mathcal{E}$ for some instantiation of $\vec{x}$. In our example, $\mathsf{pans}(In(box_1, x, S_0), \mathcal{D}_0)$ is the set

$$\{(item_1, \chi_3), (item_2, \chi_3), (item_1, \chi_5), (gold, \chi_5)\}.$$

3.2 Formulas with finite possible answers

Observe that the possible answers to a formula may be *infinite*. For instance, let $\gamma_1(x)$ be $In(box_3, x, S_0)$ and $\mathcal{D}_0$ as before. Since nothing is said about $In(box_3, x, S_0)$ in $\mathcal{D}_0$, for every constant c in $\mathcal{L}$, $(c, \chi_c) \in \mathsf{pans}(\gamma_1, \mathcal{D}_0)$, where χ_c is the closure of $In(box_3, w, S_0)$ on $\{c\}$. Similarly, let $\gamma_2(x)$ be $\neg In(box_1, x, S_0)$. Then, $\mathsf{pans}(\gamma_2, \mathcal{D}_0)$ includes the infinite set $\{(c, \chi_3) \mid c \neq item_1, c \neq item_2\}$, since everything but $item_1$ or $item_2$ is not in box_1 when $\mathcal{E} \cup \{\chi_3\}$ is assumed.

Following this observation we distinguish two potential sources of an infinite set of possible answers to γ wrt to $\mathcal{D}_0$: first, when γ includes a fluent atom of the form $F(\vec{c}, w, S_0)$ that is not mentioned in $\mathcal{D}_0$, and second, when γ includes negative literals. Our objective is to identify a class of formulas γ for which $\mathsf{pans}(\gamma, \mathcal{D}_0)$ is finite. To that end we appeal to the notions of *just-in-time* and *range-restricted*.

DEFINITION 6. Let $\mathcal{D}_0$ be a DBPC and $\gamma(\vec{x})$ a first-order formula uniform in S_0. Then $\gamma(\vec{x})$ is *just-in-time (JIT)* wrt $\mathcal{D}_0$ iff for every pair $(\vec{c}, \chi)$ in $\mathsf{pans}(\gamma, \mathcal{D}_0)$, there exists a consistent set T of possible closures wrt $\mathcal{D}_0$ such that $T \cup \mathcal{E} \models \chi$. □

Definition 6 implies that each of the possible answers to γ wrt $\mathcal{D}_0$ consists of information that is listed *explicitly* in $\mathcal{D}_0$. For example, consider again $\mathsf{pans}(In(box_1, x, S_0), \mathcal{D}_0)$ and note that χ_3 and χ_5 are implied by $\chi_3 \wedge \chi_4$ and $\chi_5 \wedge \chi_6$, respectively, which are both possible closures wrt $\mathcal{D}_0$. This provides a way to avoid the cases that are similar to γ_1 (note that γ_1 is not JIT wrt $\mathcal{D}_0$). Nonetheless, this is not enough to avoid an infinite set of possible answers (note that γ_2 is JIT wrt $\mathcal{D}_0$). We shall also require formulas to be *range-restricted* in the following sense.

DEFINITION 7. The first-order formula γ is *safe-range* wrt a set of variables X according to the following rules:

1. let $\vec{t}$ be a vector of variables and constants, c a constant, and x a variable of sort object, then:

   - $x = c$ is safe-range wrt $\{x\}$;
   - $F(\vec{t}, c, S_0)$ is safe-rage wrt $\{\}$;
   - $F(\vec{t}, x, S_0)$ is safe-range wrt $\{x\}$ if x is not included in $\vec{t}$, and safe-range wrt $\{\}$ otherwise;

2. if ϕ is safe-range wrt X_ϕ, ψ is safe-range wrt X_ψ, then:

   - $\phi \vee \psi$ is safe-range wrt $X_\phi \cap X_\psi$;
   - $\phi \wedge \psi$ is safe-range wrt $X_\phi \cup X_\psi$;
   - $\neg \phi$ is safe-range wrt $\{\}$;
   - $\exists x \phi$ is safe-range wrt $X / \{x\}$ provided that $x \in X$;

3. no other formula is safe-range.

A formula is said to be *range-restricted* iff it is safe-range wrt the set of its free variables. □

For example, $In(x, y, S_0)$ is safe-range wrt $\{y\}$ but not range-restricted and not JIT wrt $\mathcal{D}_0$ of our example. On the other hand, the formulas $In(x, y, S_0) \wedge x = box_1$ and $In(box_1, y, S_0)$ are both range-restricted and JIT wrt $\mathcal{D}_0$. Note also that γ_2 is not range-restricted.

We now state the main result of this section.

THEOREM 8. *Let $\mathcal{D}_0$ be a DBPC and $\gamma(\vec{x}, \vec{y})$ a first-order formula uniform in S_0 that is JIT wrt $\mathcal{D}_0$ and safe-range wrt the variables in $\vec{x}$. Then for every constant vector $\vec{d}$ that has the same size as $\vec{y}$, $\mathsf{pans}(\gamma(\vec{x}, \vec{d}), \mathcal{D}_0)$ is a finite set $\{(\vec{e}_1, \chi_1), \ldots, (\vec{e}_n, \chi_n)\}$ such that the following holds:*

$$\mathcal{D}_0 \cup \mathcal{E} \models \forall \vec{x}.\gamma(\vec{x}, \vec{d}) \equiv \bigvee_{i=1}^{n} (\vec{x} = \vec{e}_i \wedge \chi_i).$$

The proof is done by induction on the construction of γ. Since γ is safe-range wrt the variables in $\vec{x}$ we only need to consider the cases of Definition 7. Even though the ideas are straightforward, the actual proof is tedious and can be found in the appendix.

In other words, the range-restricted and JIT assumptions on a formula guarantee *finitely many possible answers*, and imply a syntactic normal form for the formula as shown next.

DEFINITION 9. Let γ be a first-order range-restricted formula uniform in S_0 that is JIT wrt $\mathcal{D}_0$. Then, the *possible answers normal form (PANS-NF)* of γ wrt $\mathcal{D}_0$ is a disjunction of formulas of the form $\vec{x} = \vec{e} \wedge \chi$ as implied by Theorem 8. □

For example, consider again $\mathsf{pans}(In(box_1, x, S_0), \mathcal{D}_0)$. The possible answers normal form of $In(box_1, x, S_0)$ wrt $\mathcal{D}_0$ is the formula

$$(x = item_1 \wedge \chi_3) \vee (x = item_2 \wedge \chi_3) \vee (x = item_1 \wedge \chi_5) \vee (x = gold \wedge \chi_5).$$

3.3 Range-restricted action theories

We now have all the necessary machinery to define our type of basic action theories.

DEFINITION 10. A successor state axiom for F is *range-restricted* iff $\gamma_F^+(\vec{x}, a, s)$ and $\gamma_F^-(\vec{x}, a, s)$ are disjunctions of formulas of the form:

$$\exists \vec{z}(a = A(\vec{y}) \wedge \phi(\vec{x}, \vec{z}, s)),$$

where $\vec{y}$ may contain some variables from $\vec{x}$, $\vec{z}$ corresponds to the remaining variables of $\vec{y}$, and ϕ (called the *context formula*) is uniform in s and is such that $\phi(\vec{x}, \vec{z}, S_0)$ is safe-range wrt a set that includes the variables in $\vec{x}$ that are not in $\vec{y}$.

An SRA for A is *range-restricted* iff $\Theta_A(\vec{y}, r, S_0)$ is safe-range wrt any set of variables. A *range-restricted basic action theory (RR-BAT)* is one such that all axioms in $\mathcal{D}_{ss}, \mathcal{D}_{sr}$ are range-restricted and $\mathcal{D}_0$ is a DBPC. □

For example, observe that the SSAs for *RobotAt* and *At*, i.e., (1) and (2) follow the structure implied by Definition 10 as they are built on appropriate disjunctions. Observe also that the context formulas of the SSAs have been carefully crafted so that they also satisfy the required condition about being range-restricted. In particular note that the instantiated context formula of $\gamma_{RobotAt}^+$, that is,

$$RobotAt(z_1, S_0) \wedge RobotDir(z_2, S_0) \wedge Connected(z_1, z_2, x, S_0),$$

is safe-range wrt $\{x\}$, as required by Definition 10.

The important property of RR-BATs is that when a γ_F^{*}[4] formula in the successor state axioms is instantiated with a ground action $A(\vec{c})$, $\mathcal{D}_{una}$ can be used to eliminate the action term $a = A(\vec{c})$ so that γ_F^{*} becomes range-restricted (similarly for Θ_A in SRAs). So,

---

[4]We use γ^* to denote either γ^+ or γ^-.

whenever these formulas are also JIT wrt $\mathcal{D}_0$ then we can simplify them into PANS-NF by means of Theorem 8.

In the next section we show how to progress an RR-BAT whenever the JIT assumption holds and give an algorithm for a special case that the theory is built on conjunctive formulas.

4   Just-in-time progression

The RR-BATs are defined so that when a JIT assumption holds, there is a finite set of ground fluent atoms that may be affected by a ground action. The intuition is that in this case we can progress $\mathcal{D}_0$ by appealing to the progression technique in [Vassos, Lakemeyer, and Levesque 2008] that works when the set of fluents that may be affected is fixed by the arguments of the action.

4.1   The progression method for the general case

The next definition captures the condition under which our method for progression is logically correct.

DEFINITION 11. An RR-BAT $\mathcal{D}$ is *just-in-time (JIT)* wrt (α, d), where α is a ground action and d is a sensing result, iff for all fluent symbols F, $\gamma_F^+(\vec{x}, \alpha, S_0)$ and $\gamma_F^-(\vec{x}, \alpha, S_0)$ are JIT wrt $\mathcal{D}_0$, and $\Theta_A(\vec{c}, d, S_0)$ is JIT wrt $\mathcal{D}_0$, where α is $A(\vec{c})$. □

Essentially, this condition requires that for all input-grounded fluent atoms that are needed for the specification of the effects of α, there should be disjunctive or complete information about their output in the form of some PCA in $\mathcal{D}_0$.

We introduce the following notation.

DEFINITION 12. Let $\mathcal{D}$ be an RR-BAT that is JIT wrt (α, d). Without loss of generality we assume that $\Theta_A(\vec{e}, d, S_0)$ and all $\gamma_F^+(\vec{x}, \alpha, S_0)$, $\gamma_F^-(\vec{x}, \alpha, S_0)$ are in PANS-NF wrt $\mathcal{D}_0$. Then, the *context set* of (α, d) wrt $\mathcal{D}$, denoted as $\mathcal{J}$, is the set of all $F(\vec{c}, w, S_0)$ such that one of the following is true:

1. $\vec{x} = \langle \vec{c}, e \rangle \wedge \chi$ is a disjunct of some $\gamma_F^*(\vec{x}, \alpha, S_0)$;

2. $F(\vec{c}, w, S_0)$ appears in some $\gamma_F^*(\vec{x}, \alpha, S_0)$;

3. $F(\vec{c}, w, S_0)$ appears in $\Theta_A(\vec{c}, d, S_0)$, where $\alpha = A(\vec{c})$. □

Intuitively, the context set $\mathcal{J}$ specifies all those atomic closures that need to be updated after the action is performed (case 1) as well as those on which the change is conditioned on (case 2), and the atomic closures for which some condition is sensed to be true (case 3). The important property of $\mathcal{J}$, which follows from Theorem 8, is that it is a *finite* set.

In the simple case of our running example and the ground action *moveFwd*,[5] the context set is the following:

$$\{ RobotAt(w, S_0), RobotDir(w, S_0), Connected(loc_1, north, w, S_0),$$
$$Holding(w, S_0), At(box_2, w, S_0), At(item_2, w, S_0), At(gold, w, S_0) \}.$$

Note that $At(box_1, w, S_0)$ and $At(item_1, w, S_0)$ are not needed in the progression as this information is neither affected when *moveFwd* is applied in $\mathcal{D}_0$, nor needed to specify some other effect of the action.

---

[5]For actions with no sensing result axioms, such as *moveFwd*, we typically do not refer to the sensing result d.

We now define the $\mathcal{J}$-models which will provide a way of separating $\mathcal{D}_0$ into two parts: one that remains unaffected after the action is performed and one that needs to be updated. The part that remains unaffected corresponds to all those PCAs that do not mention any atom in $\mathcal{J}$. For the part that needs updating, we will construct a large PCA that lists all the possibilities for the closures of the atoms in $\mathcal{J}$ as follows.

DEFINITION 13. Let $\mathcal{J} = \{\tau_1, \ldots, \tau_n\}$ be the context set of (α, d) wrt an RR-BAT $\mathcal{D}$. A $\mathcal{J}$-model θ is a sentence of the form $(\chi_1 \wedge \ldots \wedge \chi_m)$ such that every χ_i is a possible closure wrt $\mathcal{D}_0$ that mentions at least one of the atoms in $\mathcal{J}$, all atoms in $\mathcal{J}$ are mentioned in θ, θ is consistent, and atoms in θ appear in lexicographic order. □

Note that since $\mathcal{D}_0$ is finite there are finitely many $\mathcal{J}$-models. In the simple case of our running example there are two $\mathcal{J}$-models, namely:

$$\theta_1 : \chi_1 \wedge \chi_2 \wedge \chi_3 \wedge \chi_4 \wedge \chi_7 \wedge \chi_8 \wedge \chi_{11} \wedge \chi_{12} \wedge \chi_{13};$$

$$\theta_2 : \chi_1 \wedge \chi_2 \wedge \chi_5 \wedge \chi_6 \wedge \chi_7 \wedge \chi_8 \wedge \chi_{11} \wedge \chi_{12} \wedge \chi_{13}.$$

Note that χ_9 and χ_{10} that represent the location of box_1 and $item_1$ are not included in any of the $\mathcal{J}$-models as the corresponding atoms, that is, $At(box_1, w, S_0)$ and $At(item_1, w, S_0)$, are not included in the context set $\mathcal{J}$.

The disjunction ϕ of all the $\mathcal{J}$-models is a PCA that corresponds to the "cross-product" of the PCAs in $\mathcal{D}_0$ that hold information about $\vec{\tau}$. Essentially, ϕ corresponds to the part of $\mathcal{D}_0$ that needs updating, and the intuition is that we can progress $\mathcal{D}_0$ by progressing each of the $\mathcal{J}$-models separately.

DEFINITION 14. Let $\mathcal{J}$ be the context set of (α, d) wrt an RR-BAT $\mathcal{D}$, and θ a $\mathcal{J}$-model that is the closure of $\{F_1(\vec{c}_1, w, S_0), \ldots, F_n(\vec{c}_n, w, S_0)\}$ on $\langle V_1, \ldots, V_n \rangle$. We define Γ_i^+ as the smallest set of constants e such that:

1. $\vec{x} = \langle \vec{c}_i, e \rangle \wedge \chi$ is a disjunct of $\gamma_{F_i}^+(\vec{x}, \alpha, S_0)$;

2. $\{\chi \wedge \theta\} \cup \mathcal{E}$ is consistent.

The set Γ_i^- is defined similarly based on $\gamma_{F_i}^-(\vec{x}, \alpha, S_0)$. The *progression* of θ wrt (α, d) is the closure on the updated vector $\langle V_1', \ldots, V_n' \rangle$, where V_i' is the set $(V_i - \Gamma_i^-) \cup \Gamma_i^+$. The $\mathcal{J}$-model θ is *filtered* iff $\{\Theta_A(\vec{c}, d, S_0) \wedge \theta\} \cup \mathcal{E}$ is inconsistent, where α is $A(\vec{c})$. □

Each $\mathcal{J}$-model θ is updated based on the possible answers of the instantiated formulas $\gamma_{F_i}^*$. For every disjunct of $\gamma_{F_i}^*$ expressed in PANS-NF, the atom $w = e$ is either removed or added to the closure of $F_i(\vec{c}_i, w, S_0)$ provided that the condition χ for the change is consistent with θ. Similarly, a $\mathcal{J}$-model may be filtered out if it is not consistent with the conditions that are implied by the sensing result d.

For example, consider θ_1 and the atom $At(item_2, w, S_0)$ from $\mathcal{J}$. The corresponding Γ^+ and Γ^- for this atom are both the empty set $\{\}$ as $item_2$ is inside box_1 in model θ_1, and therefore its location remains intact. On the other hand, if one considers model θ_2, the corresponding Γ^+ and Γ^- for $At(item_2, w, S_0)$ are $\{loc_2\}$ and $\{loc_1\}$, respectively. Putting it all together, the updated $\mathcal{J}$-models θ_1' and θ_2' are as follows:

$$\theta_1' : \chi_1' \wedge \chi_2 \wedge \chi_3 \wedge \chi_4 \wedge \chi_7 \wedge \chi_8 \wedge \chi_{11}' \wedge \chi_{12} \wedge \chi_{13}',$$

$$\theta_2' : \chi_1' \wedge \chi_2 \wedge \chi_5 \wedge \chi_6 \wedge \chi_7 \wedge \chi_8 \wedge \chi_{11}' \wedge \chi_{12}' \wedge \chi_{13},$$

where χ_1', χ_{11}', χ_{12}', and χ_{13}' are the atomic closures of $RobotAt(w, S_0)$, $At(box_2, w, S_0)$, $At(item_2, w, S_0)$, and $At(gold, w, S_0)$ on $\{loc_2\}$, respectively.

We now state the main result of this section that illustrates how the new knowledge base is constructed from $\mathcal{D}_0$.

THEOREM 15. *Let $\mathcal{D}$ be an RR-BAT that is consistent and JIT wrt the ground action α and sensing result d, $\mathcal{J}$ the context set of (α, d) wrt $\mathcal{D}$, $\{\theta_1, \ldots, \theta_n\}$ the set of all the $\mathcal{J}$-models that are not filtered, and $\{\phi_1, \ldots, \phi_m\}$ the set of all PCAs in $\mathcal{D}_0$ that do not have any atoms in common with any $\mathcal{J}$-model. Let $\mathcal{D}_\alpha$ be the following set:*

$$\{ \bigvee_{i=1}^{n} \theta_i', \phi_1, \ldots, \phi_m \},$$

where θ_i' is the progression of θ_i wrt (α, d). Then, the set $\mathcal{D}_\alpha(S_0 / do(\alpha, S_0))$ is a strong progression of $\mathcal{D}$ wrt (α, d), where $\mathcal{D}_\alpha(\sigma / \sigma')$ denotes the result of replacing every occurrence of σ in every sentence in $\mathcal{D}_\alpha$ by σ'.

In the absence of sensing, the proof idea is to show that assuming $\mathcal{E}$, $\mathcal{D}_0$ is equivalent to $\{\bigvee_1^n \theta_i, \phi_1, \ldots, \phi_m\}$, and progressing $\mathcal{D}_0$ reduces to updating each of the sub-models θ_i appropriately. For the case of sensing we only need to remove every θ_i that is not consistent with the instantiated SRA. As far our running example is concerned, observe that $\{\theta_1' \vee \theta_2', \chi_9, \chi_{10}\}$ is a strong progression of $\mathcal{D}$ wrt *moveFwd*.

We close by noting two important points. First, observe that the progression of $\mathcal{D}_0$ is again a DBPC. On the other hand, the fact that $\mathcal{D}$ is JIT wrt some action α may not be preserved in general after α or some other action is performed. This is because the JIT assumption essentially requires that there is disjunctive or complete information in the KB about the output w of all the fluent atoms of the form $F(\vec{c}, w, S_0)$ that are mentioned in the instantiated $\gamma^*(\vec{x}, \alpha, S_0)$ formulas, and these atoms may be different each time. Nonetheless, checking whether the JIT assumption holds may be performed by the same method that computes the possible answers of formulas as we will see in the next section. Moreover, the JIT assumption may be enforced by an appropriate account of sensing so that complete or disjunctive knowledge is acquired for these atoms. Please refer to Section 6 for a short discussion on this matter.

4.2 The case of conjunctive queries

Our method for progression is based on the ability to compute the possible answers to the γ_F^* and Θ_A formulas wrt $\mathcal{D}_0$, as we have assumed that they are in PANS-NF. In order to give some insight on the practicality of our method, we examine the case that these formulas are similar to the *conjunctive queries* of database theory [Abiteboul, Hull, and Vianu 1994].

DEFINITION 16. A *conjunctive query* γ is a formula uniform in S_0 of the form

$$\exists \vec{x}(\phi_1 \wedge \cdots \wedge \phi_n),$$

where ϕ_i is a fluent atom with free variables not necessarily in $\vec{x}$. □

Conjunctive queries offer an intuitive way of building safe-range formulas using the output of an atom with a ground input as an input to another atom. For instance, note that the context formula of $\gamma_{RobotAt}^+$ in (1) is a conjunctive query, and observe that the output of the fluent atoms $RobotAt(z_1, s)$ and $RobotDir(z_2, s)$ is used to specify the input of $Connected(z_1, z_2, x, s)$. Similarly, observe that the γ_{At}^+ and γ_{At}^- formulas in (2) are built on conjunctive queries of this sort.

Given a conjunctive query γ as input and a DBPC $\mathcal{D}_0$, PANS$(\gamma, \mathcal{D}_0)$ that is described in Algorithm 1 checks whether γ is range-restricted and JIT wrt $\mathcal{D}_0$, and if so, computes

---

Algorithm 1: PANS($\gamma, \mathcal{D}_0$)

---

1 **if** γ is the empty conjunction **then**
2 | **return** {};   // query successfully processed

3 $\Gamma := \{F(\vec{c}, t, S_0) \in \gamma \mid F(\vec{c}, w, S_0)$ is mentioned in $\mathcal{D}_0\}$;
4 **if** $\Gamma = \emptyset$ **then**
5 | **return** *failure*;   // no atom to continue

6 $X := \emptyset$;   // initialize the set of possible answers
7 $\tau :=$ the first atom $F(\vec{c}, t, S_0)$ in Γ;
8 $\gamma' := \gamma$ without the atom τ;
9 **for** *possible closures* χ *of* τ *on* V *wrt* $\mathcal{D}_0$, **do**
10 | **if** t *is a constant* **then**
11 | | **if** $t \in V$ **then**
12 | | | $Y := \mathsf{PANS}(\gamma', \mathcal{D}_0)$;   // recursive call, t constant
13 | | | **if** $Y = $ *failure* **then**
14 | | | | **return** *failure*
15 | | | $Y' := \{(\omega, \chi \wedge \chi') \mid (\omega, \chi') \in Y, \chi \wedge \chi' \not\models \bot\}$;
16 | | | $X := X \cup Y'$;   // merge the possible answers
17 | **else**
18 | | **for** *constants* $e \in V$ **do**
19 | | | $Y := \mathsf{PANS}(\gamma'|_e^t, \mathcal{D}_0)$;   // recursive call, t variable
20 | | | **if** $Y = $ *failure* **then**
21 | | | | **return** *failure*
22 | | | $Y' := \{(t = e \wedge \omega, \chi \wedge \chi') \mid (\omega, \chi') \in Y, \chi \wedge \chi' \not\models \bot, t = e \wedge \omega \not\models \bot\}$;
23 | | | $X := X \cup Y'$;   // merge the possible answers

24 **return** (the projection of X on the free variables of γ)

---

the set $\mathsf{pans}(\gamma, \mathcal{D}_0)$. The algorithm works by i) selecting an atom for which a finite set of possible answers is guaranteed (lines 3–8), ii) specifying the possible answers to this atom and recursively finding those to the query without the selected atom (lines 9–14, 17–21), and iii) merging the answers (lines 15–16, 22–23), until all atoms in γ have been selected.

The next theorem states the correctness of PANS.

THEOREM 17. *Let $\mathcal{D}_0$ be a finite DBPC and γ a first-order conjunctive query uniform in S_0. Then, PANS($\gamma, \mathcal{D}_0$) always terminates, and moreover if γ is range-restricted and JIT wrt $\mathcal{D}_0$, it returns the set $\mathsf{pans}(\gamma, \mathcal{D}_0)$.*

Algorithm PANS($\gamma, \mathcal{D}_0$) can easily be extended to handle equalities as well as negated atoms. The first case can be addressed via standard unification procedures. For negated atoms the idea is that a *ground* literal of the form $\neg F(\vec{c}, d, S_0)$ such that $F(\vec{c}, w, S_0)$ is mentioned in $\mathcal{D}_0$, may also be selected by the algorithm. Then, the algorithm works in the same way as for the ground positive literal except that it iterates over the possible closures of $F(\vec{c}, w, S_0)$ for which $F(\vec{c}, d, S_0)$ is *not* true. (Observe the similarities with *negation as failure* of logic programming [Apt and Pellegrini 1994].)

We now turn our attention to the complexity of PANS($\gamma, \mathcal{D}_0$) and progression. First note that when γ is a range-restricted conjunctive query that is also JIT wrt $\mathcal{D}_0$, the size of $\mathsf{pans}(\gamma, \mathcal{D}_0)$ is $O(N^n)$, where n is the size of γ and N is the size of $\mathcal{D}_0$. This is because the assumptions for γ and Theorem 8 ensure that in the worst case for each atom τ in γ, the set $\mathsf{pans}(\tau, \mathcal{D}_0)$ corresponds to the information in the entire $\mathcal{D}_0$.

THEOREM 18. *Let γ be a range-restricted conjunctive query and $\mathcal{D}_0$ a finite DBPC. Let n be the number of atoms in γ, k the number of distinct atoms with a ground input in $\mathcal{D}_0$, m the maximum number of disjunct in a PCA in $\mathcal{D}_0$, and l the maximum number of constants in an atomic closure in $\mathcal{D}_0$. Then $\mathsf{PANS}(\gamma, \mathcal{D}_0)$ runs in time $O((k \cdot l + n^2) \cdot (m \cdot l)^n)$.*

Even though the complexity is exponential on the size n of γ, our progression method evaluates the formulas γ_F^* and Θ_A from $\mathcal{D}_{ss}$ and $\mathcal{D}_{sr}$, whose sizes do not grow. In fact, it is safe to assume that the sizes of these formulas are bounded by a small integer N, in which case the complexity of $\mathsf{PANS}(\gamma, \mathcal{D}_0)$ is a polynomial with degree N.

Assuming then that all the relevant formulas for progressing $\mathcal{D}_0$ wrt α are in PANS-NF by the use of $\mathsf{PANS}(\gamma, \mathcal{D}_0)$, the complexity of our method is dominated by the process of specifying all the $\mathcal{J}$-models, where $\mathcal{J}$ is the context set of α. In the worst case, the size of $\mathcal{J}$ is k, which essentially means that with the execution of α all the atoms in $\mathcal{D}_0$ become *mutually constrained*, and thus need to appear in the same PCA.

Nonetheless, we expect the size of $\mathcal{J}$ to be manageable in practice, as the robotic or agent domains we have in mind are quite different from logical puzzles of this sort. In particular, we expect that the amount of disjunctive knowledge that is represented in $\mathcal{D}_0$ always comes in (many) small "packages." Under these plausible assumptions, it follows that m is always bounded by some small constant M, in which case, the running time of our progression method and the size of $\mathcal{D}_\alpha$ are both polynomial to the size of $\mathcal{D}_0$ with a degree that depends on the actual values of N, M.

These assumptions are relatively strong and certainly do not apply to all applications we have in mind, but we believe that they are safe for many "organic" scenarios where the theory $\mathcal{D}$ is used to represent the effects of a domain that is similar to the physical world. Furthermore, as far as the assumptions on the incomplete information are concerned, we believe that they can be enforced with an appropriate account of sensing.

5 Related work

The notion of progression for basic action theories was first introduced by Lin and Reiter [1997]. The version we use here is due to Vassos et al. [2008], which we extended slightly to account for sensing. As far as a first-order definable progression is concerned, Lin and Reiter [1997] were first to investigate restrictions on the successor state axioms as they introduced a *context-free* assumption for actions.

Liu and Levesque [2005] introduced the *local-effect* assumption for actions when they proposed a weaker version of progression that is logically incomplete, but remains practical. Vassos et al. [2008] showed that, under this assumption, a logically correct first-order progression can actually be computed by updating a finite $\mathcal{D}_0$. Our restriction of Definition 10 is similar, but goes beyond local-effect. The key difference is that we do not require the arguments of a fluent F be included in those of the action. This allows us to handle actions like *turn* and *moveFwd* (with no arguments) affecting fluents like $RobotDir(x, s)$ and $RobotAt(x, s)$. To stay practical, though, we restricted the structure of $\mathcal{D}_0$. It is important to point out that it is not always possible to revert to a *ground* action of the form $move(x, y, z)$ as, for instance, the starting x or destination y may be unknown to the agent.

In [Liu and Lakemeyer 2009], the authors first showed how the result of Vassos et al. [2008] on local-effect progression relates to the notion of forgetting, and examined the more expressive so-called *normal* actions. Both our work on actions with bounded effects and that of Liu and Lakemeyer allow the representation of actions that go beyond local-effect, but each approach proceeds in a different direction.

The intuition behind normal actions is that a ground action α is allowed to have non-local effects on a fluent F as long as α has local effects on all fluents appearing in the effect formulas γ_F^*. For instance, consider the ground action $move(box_1, loc_1, loc_2)$ for moving object box_1 together with every object inside it from location loc_1 to location loc_2. Such action does have non-local effects on fluent $At(x, y, s)$ because it affects the location of objects that are inside box_1 and these are not named as arguments in the action. Nonetheless, assuming that the only mentioned fluent in γ_{At}^* is $In(x, y, s)$, the action is considered normal, since it trivially (i.e., has no effect whatsoever) has local-effects on $In(x, y, s)$.

While similar in its effects, the action *moveFwd* of our example, on the other hand, is not normal. This is because the action has non-local effects on fluent $RobotAt(x, y, s)$, and $RobotAt(x, y, s)$ is in fact mentioned in $\gamma_{RobotAt}^+$. To see this, one has to look carefully at the technical details of Definition 4.4 in [Liu and Lakemeyer 2009]: a ground action has local-effects on fluent F if using $\mathcal{D}_{una}$ and the action it is possible to simplify γ_F^+ and γ_F^- into a disjunction that explicitly states all potentially affected ground atoms of F. Since the agent may not know a-priori the boundaries of the world it is situated in, the number of potential locations is not bounded, hence such a simplification is not possible by only using $\mathcal{D}_{una}$. Our approach is different in that we also use $\mathcal{D}_0$ in order to achieve a similar simplification. As a result, we also require additional constraints on the structure of $\mathcal{D}_0, \mathcal{D}_{ss}, \mathcal{D}_{sr}$ as well as the JIT assumption.

Another approach to first-order progression involves imposing restrictions on the *queries* that can be performed over a basic action theory. This was first investigated by Lin and Reiter [1997] who showed that for answering projection queries that are uniform in some situation σ, a strong progression is always equivalent to a first-order set of sentences. Shirazi and Amir [2005] proved a similar result in the context of *logical filtering*. Vassos and Levesque [2008] extended this result showing that un-nested quantification over situations may also be allowed.

Outside of the situation calculus, Thielscher [1999] defined a dual representation for basic action theories based on state update axioms that explicitly define the direct effects of each action, and investigated progression in this setting. Unlike our work, where the sentences in $\mathcal{D}_0$ are replaced by an updated version, there, the update relies on expressing the changes using constraints which may need to be conjoined to the original database.

Finally, we observe that the notion of possible closures is a generalization of the *possible values* of Vassos and Levesque [2007]. The notions of the safe-range and range-restricted queries come from the database theory where this form of "safe" queries has been extensively studied [Abiteboul, Hull, and Vianu 1994]. The notion of just-in-time formulas was introduced for a different setting in [De Giacomo, Levesque, and Sardina 2001] and, in our case, is also related to the *active domain* of a database [Abiteboul, Hull, and Vianu 1994].

6 Discussion

We have proposed a type of basic action theories, where the KB is a *database of possible closures* and actions have *bounded effects*. For such theories, we showed a logically correct method for computing a finite first-order progression by directly updating the KB. Moreover, we developed an algorithm for the main progression task, namely computing possible answers, for the special case of conjunctive queries. Our progression account is, as far as we are aware, the first one to go beyond local-effect, while being able to handle a particular class of simple (though common) actions, sensing, and incomplete information with possibly infinite domains.

In particular, our approach is strictly more general than local-effect actions, but as far as the work of [Liu and Lakemeyer 2009] on normal actions is concerned, neither approach is strictly more general than the other. First, as we already illustrated, our RR-BATs are able to handle some very simple cases, like the action *moveFwd*, that are not normal. In addition, only our approach accounts for sensing actions. On the other hand, our approach relies on having disjunctive knowledge about the relevant part of the KB that needs to be updated, which is not always needed for a normal action. For example, the action $move(x_1, x_2)$ that results in the agent moving from x_1 to x_2 and taking along the objects she is holding is normal, and the KB can be progressed even if there is no information in the KB about the objects that the agent is holding. With our approach, however, we can only progress provided there is disjunctive (or complete) knowledge about the objects she is holding. There is also a mismatch between the structure of the KBs that are used. In the general case of Liu and Lakemeyer [2009], our RR-BATs are clearly more restricted, but in the case where the authors considered an implementation based on the so-called proper$^+$ KBs, both approaches have benefits depending on context.

Also, the two approaches are conceptually quite different. The work on normal actions is based on the notion of forgetting, while ours is rooted on database theory. Whereas the KBs used for normal actions are more general than our DBPCs, the latter are closer to *incomplete* databases, e.g., the so-called c-tables [Imielinski and Lipski 1984]. In fact, our work is motivated by the fact that almost all practical real agent systems or automated planners make use of database-type approaches to model the agent's beliefs. As incomplete/inconsistent databases is a very active area of research, we expect to be able to leverage on such technology (e.g., [Fuxman, Fazli, and Miller 2005]).

As far as future work is concerned, we envision work on practical systems that do not rely on one method for progression but on a collection of techniques that apply under different conditions. For example, one might build a system that uses a DBPC as a KB which is progressed wrt both normal and range-restricted actions depending on the conditions that are satisfied. In the cases that progression cannot work, the system may attempt to impose the JIT condition in order to perform a range-restricted progression or alternatively fall back to using regression (that accounts for much wider classes) for any reasoning about the future until the necessary conditions for progression hold. As far as RR-BATs are concerned, our future work focuses on a richer notion of sensing. At the moment, sensing can be used effectively to reduce the uncertainty as long as some "bounds" on such uncertainty is already available. In our example from Section 3.1, the $senseGold(y)$ action (see end of Sections 2.1) can be used to determine whether there is gold inside box_1, as the agent already has some information of what could be inside such box. In that way, sensing actions provide a "filtering" mechanism for what is already stated in the KB. We believe it is not difficult to generalize such sensing account to allow sensing of atomic closures (Definition 2) or even possible closures (Definition 3). However, one has to be careful on how to merge such general sensing outcomes with the knowledge already encoded in the KB.

References

Abiteboul, S., R. Hull, and V. Vianu [1994]. *Foundations of Databases: Logical Level.* Addison Wesley.

Apt, K. and A. Pellegrini [1994]. On the occur-check free Prolog program. *ACM Transactions on Programming Languages and Systems (TOPLAS) 16*(3), 687–726.

De Giacomo, G., H. J. Levesque, and S. Sardina [2001]. Incremental execution of guarded theories. *Computational Logic 2*(4), 495–525.

Fuxman, A., E. Fazli, and R. J. Miller [2005]. ConQuer: Efficient management of inconsistent databases. In *Proceedings of the ACM International Conference on Management of Data (SIGMOD)*, pp. 155–166.

Imielinski, T. and W. Lipski [1984]. Incomplete information in relational databases. *Journal of the ACM 31*(4), 761–791.

Lin, F. and R. Reiter [1997]. How to progress a database. *Artificial Intelligence 92*(1-2), 131–167.

Liu, Y. and G. Lakemeyer [2009]. On first-order definability and computability of progression for local-effect actions and beyond. In *Proceedings of the International Joint Conference on Artificial Intelligence (IJCAI)*, pp. 860–866.

Liu, Y. and H. J. Levesque [2005]. Tractable reasoning in first-order knowledge bases with disjunctive information. In *Proceedings of the National Conference on Artificial Intelligence (AAAI)*, pp. 639–644.

McCarthy, J. and P. J. Hayes [1969]. Some philosophical problems from the standpoint of artificial intelligence. *Machine Intelligence 4*, 463–502.

Reiter, R. [1991]. The frame problem in the situation calculus: A simple solution (sometimes) and a completeness result for goal regression. In V. Lifschitz (Ed.), *Artificial Intelligence and Mathematical Theory of Computation: Papers in Honor of John McCarthy*, pp. 359–380. San Diego, CA: Academic Press.

Reiter, R. [2001]. *Knowledge in Action. Logical Foundations for Specifying and Implementing Dynamical Systems*. The MIT Press.

Scherl, R. and H. J. Levesque [2003]. Knowledge, action, and the frame problem. *Artificial Intelligence 144*(1–2), 1–39.

Shirazi, A. and E. Amir [2005]. First-order logical filtering. In *Proceedings of the International Joint Conference on Artificial Intelligence (IJCAI)*, pp. 589–595.

Thielscher, M. [1999]. From situation calculus to fluent calculus: State update axioms as a solution to the inferential frame problem. *Artificial Intelligence 111*(1-2), 277–299.

Vassos, S., G. Lakemeyer, and H. J. Levesque [2008]. First-order strong progression for local-effect basic action theories. In *Proceedings of Principles of Knowledge Representation and Reasoning (KR)*, pp. 662–272.

Vassos, S. and H. Levesque [2007]. Progression of situation calculus action theories with incomplete information. In *Proceedings of the International Joint Conference on Artificial Intelligence (IJCAI)*, pp. 2029–2034.

Vassos, S. and H. J. Levesque [2008]. On the progression of situation calculus basic action theories: Resolving a 10-year-old conjecture. In *Proceedings of the National Conference on Artificial Intelligence (AAAI)*, pp. 1004–1009.

Vassos, S., S. Sardina, and H. Levesque [2009]. Progressing basic action theories with non-local effect actions. In *Proceedings of the Ninth International Symposium on Logical Formalizations of Commonsense Reasoning (CS'09)*, pp. 135–140.

A Proof of Theorem 8

Proof. We prove the theorem by induction on the construction of the formulas γ. Since γ is safe-range wrt the variables in $\vec{x}$ we only need to consider the cases of Definition 7.

Base case. We only show the case that $\gamma(x, y)$ is $F(\vec{c}_1, y, \vec{c}_2, x)$, where x, y are distinct variables and $\vec{c}_1, \vec{c}_2$ are vectors of constants of sort object, and the other cases are similar. Let d be an arbitrary constant of the language. Then $\gamma(x, d)$ is the formula $F(\vec{c}_1, d, \vec{c}_2, x)$. We first show that there is a possible closures axiom ϕ in $\mathcal{D}_0$ that mentions $F(\vec{c}_1, d, \vec{c}_2, w)$. Let e be a constant of $\mathcal{L}$. We take two cases as follows.

- Case i): $F(\vec{c}_1, d, \vec{c}_2, e)$ is consistent with $\mathcal{D}_0 \cup \mathcal{E}$. Then, by the fact that $\gamma(x, y)$ is just-in-time wrt $\mathcal{D}_0$ it follows that there is a closure χ such that $\{\chi\} \cup \mathcal{E} \models \gamma(e, d)$, where χ is a conjunction of closures each of which is a possible closure wrt $\mathcal{D}_0$. It follows that there is a possible closures axiom ϕ in $\mathcal{D}_0$ that mentions $F(\vec{c}_1, d, \vec{c}_2, w)$.

- Case ii): $F(\vec{c}_1, d, \vec{c}_2, e)$ is not consistent with $\mathcal{D}_0 \cup \mathcal{E}$. It follows that there is a possible closures axiom ϕ in $\mathcal{D}_0$ that mentions $F(\vec{c}_1, d, \vec{c}_2, w)$, such that $F(\vec{c}_1, d, \vec{c}_2, e)$ is not true in any of the possible closures wrt ϕ.

It follows that there is a possible closures axiom ϕ in $\mathcal{D}_0$ that mentions $F(\vec{c}_1, d, \vec{c}_2, w)$. Without loss of generality we assume that ϕ is a possible closures axiom for $F(\vec{c}_1, d, \vec{c}_2, w)$. We will show how to rewrite ϕ in the form that the theorem requires. The axiom ϕ has the form

$$\bigvee_{i=1}^{n} \chi_i,$$

where each χ_i is an atomic closure of $F(\vec{c}_1, d, \vec{c}_2, w)$ on some set of constants $\{e_1, \ldots, e_m\}$, i.e, a sentence of the form

$$\forall w. F(\vec{c}_1, d, \vec{c}_2, w) \equiv w = e_1 \vee \ldots \vee w = e_m.$$

For each χ_i of this form let χ_i' be the formula

$$\bigvee_{j=1}^{m} (x = e_j \wedge \chi_i),$$

and let ϕ' be

$$\forall x. F(\vec{c}_1, d, \vec{c}_2, x) \equiv \bigvee_{i=1}^{n} \chi_i'.$$

Note that ϕ' has the form that the theorem requires. So, in order to prove the theorem it suffices to show that $\mathcal{D}_0 \cup \mathcal{E} \models \phi'$. Let M be an arbitrary model of $\mathcal{D}_0 \cup \mathcal{E}$. Since ϕ is a sentence in $\mathcal{D}_0$ it follows that $M \models \phi$. Observe that this there is exactly one k, $1 \leq k \leq n$, such that $M \models \chi_k$. Observe that if we simplify χ_k to *true* and all the other χ_i to *false* in ϕ' we obtain the sentence χ_k. Therefore, $M \models \phi'$ and since M was an arbitrary model of $\mathcal{D}_0 \cup \mathcal{E}$, it follows that $\mathcal{D}_0 \cup \mathcal{E} \models \phi'$. Also, by the definition of a possible answer and the structure of ϕ' it follows that the set $\mathsf{pans}(\gamma(x, d), \mathcal{D}_0)$ is the set that the theorem requires. Finally, since d was arbitrary it follows that this holds for every d, thus the theorem holds for the case of $F(\vec{c}_1, d, \vec{c}_2, w)$.

Inductive step. We only show the case when γ is a disjunction of formulas and the other cases are similar. We assume the theorem holds for $\phi_1(\vec{x}_1, \vec{z}, \vec{y}_1)$ that is safe-range wrt the variables in $\vec{x}_1$ and $\vec{z}$, and does not mention any free variable other than $\vec{x}_1, \vec{z}, \vec{y}_1$. Similarly we assume that the theorem holds for $\phi_2(\vec{x}_2, \vec{z}, \vec{y}_2)$ that is safe-range wrt the variables in $\vec{x}_2$ and $\vec{z}$, and does not mention any free variable other than $\vec{x}_2, \vec{z}, \vec{y}_2$, where $\vec{x}_1$ and $\vec{x}_2$ share no variables. Let $\gamma(\vec{x}_1, \vec{y}_1, \vec{x}_2, \vec{y}_2, \vec{z})$ be the formula

$$\phi_1(\vec{x}_1, \vec{z}, \vec{y}_1) \vee \phi_2(\vec{x}_2, \vec{z}, \vec{y}_2).$$

We will show that the theorem holds for γ. By the definition of the safe-range formulas it follows that γ is safe-range for the variables in $\vec{z}$. Let $\vec{d}_1$ be a vector of constants of the same size as $\vec{y}_1$, and $\vec{d}_2$ be a vector of constants of the same size as $\vec{y}_2$. By the induction hypothesis it follows that $\mathsf{pans}(\phi_1(\vec{x}_1, \vec{z}, \vec{d}_1), \mathcal{D}_0)$ is a finite set of the form

$$\{(\langle \vec{c}_{11}, \vec{e}_{11} \rangle, \chi_{11}), \ldots, (\langle \vec{c}_{1n}, \vec{e}_{1n} \rangle, \chi_{1n})\}$$

and the following holds:

$$\mathcal{D}_0 \cup \mathcal{E} \models \forall \vec{x}_1. \forall \vec{z}. \phi_1(\vec{x}_1, \vec{z}, \vec{d}_1) \equiv \bigvee_{i=1}^{n} (\vec{x}_1 = \vec{c}_{1i} \wedge \vec{z} = \vec{e}_{1i} \wedge \chi_{1i}).$$

Similarly, $\mathsf{pans}(\phi_2(\vec{x}_2, \vec{z}, \vec{d}_2), \mathcal{D}_0)$ is a finite set of the form

$$\{(\langle \vec{c}_{21}, \vec{e}_{21} \rangle, \chi_{21}), \ldots, (\langle \vec{c}_{2m}, \vec{e}_{2m} \rangle, \chi_{2m})\},$$

and the following holds:

$$\mathcal{D}_0 \cup \mathcal{E} \models \forall \vec{x}_2. \forall \vec{z}. \phi_2(\vec{x}_2, \vec{z}, \vec{d}_2) \equiv \bigvee_{i=1}^{m} (\vec{x}_2 = \vec{c}_{2i} \wedge \vec{z} = \vec{e}_{2i} \wedge \chi_{2i}).$$

Let $\vec{b}_1$ be a vector of constants of the same size as $\vec{x}_1$ and $\vec{b}_2$ a vector of constants of the same size as $\vec{x}_2$. It follows that

$$\mathcal{D}_0 \cup \mathcal{E} \models \forall \vec{z}. \phi_1(\vec{b}_1, \vec{z}, \vec{d}_2) \vee \phi_2(\vec{b}_2, \vec{z}, \vec{d}_2) \equiv$$
$$\bigvee_{i=1}^{n} (\vec{b}_1 = \vec{c}_{1i} \wedge \vec{z} = \vec{e}_{1i} \wedge \chi_{1i}) \vee \bigvee_{i=1}^{m} (\vec{b}_2 = \vec{c}_{2i} \wedge \vec{z} = \vec{e}_{2i} \wedge \chi_{2i}).$$

By the uniqueness of names for the constants of sort object it follows that each of the atoms of the form $\vec{b}_1 = \vec{c}_{1i}$ and $\vec{b}_2 = \vec{c}_{2i}$ can be simplified to either *true* or *false*. Therefore the previous sentence can be simplified to the form that the theorem requires and the theorem follows. $\qquad\square$